A GUIDE TO QUANTUM GROUPS

T0339935

A GUIDE TO
QUANTUM GROUPS

Vyjayanthi Chari
University of California, Riverside

Andrew Pressley
King's College, London

CAMBRIDGE
UNIVERSITY PRESS

CAMBRIDGE UNIVERSITY PRESS
Cambridge, New York, Melbourne, Madrid, Cape Town, Singapore,
São Paulo, Delhi, Dubai, Tokyo, Mexico City

Cambridge University Press
The Edinburgh Building, Cambridge CB2 8RU, UK

Published in the United States of America by
Cambridge University Press, New York

www.cambridge.org
Information on this title: www.cambridge.org/9780521558846

© Cambridge University Press 1994

This publication is in copyright. Subject to statutory exception
and to the provisions of relevant collective licensing agreements,
no reproduction of any part may take place without the written
permission of Cambridge University Press.

First published 1994
First paperback edition (with corrections) 1995
Reprinted 1998

A catalogue record for this publication is available from the British Library

ISBN 978-0-521-43305-1 Hardback
ISBN 978-0-521-55884-6 Paperback

Cambridge University Press has no responsibility for the persistence or
accuracy of URLs for external or third-party internet websites referred to in
this publication, and does not guarantee that any content on such websites is,
or will remain, accurate or appropriate. Information regarding prices, travel
timetables, and other factual information given in this work is correct at
the time of first printing but Cambridge University Press does not guarantee
the accuracy of such information thereafter.

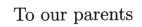
To our parents

Contents

7 Quantized function algebras   215

8 Structure of QUE algebras: the universal R-matrix 255

9 Specializations of QUE algebras 279

14 Canonical bases   475

15 Quantum group invariants of knots
and 3-manifolds   494

Introduction

Quantum groups first arose in the physics literature, particularly in the work of L. D. Faddeev and the Leningrad school, from the 'inverse scattering method', which had been developed to construct and solve 'integrable' quantum systems. They have excited great interest in the past few years because of their unexpected connections with such, at first sight, unrelated parts of mathematics as the construction of knot invariants and the representation theory of algebraic groups in characteristic p.

In their original form, quantum groups are associative algebras whose defining relations are expressed in terms of a matrix of constants (depending on the integrable system under consideration) called a quantum R-matrix. It was realized independently by V. G. Drinfel'd and M. Jimbo around 1985 that these algebras are Hopf algebras, which, in many cases, are deformations of 'universal enveloping algebras' of Lie algebras. A little later, Yu. I. Manin and S. L. Woronowicz independently constructed non-commutative deformations of the algebra of functions on the groups $SL_2(\mathbb{C})$ and SU_2, respectively, and showed that many of the classical results about algebraic and topological groups admit analogues in the non-commutative case.

Thus, although many of the fundamental papers on quantum groups are written in the language of integrable systems, their properties are accessible by more conventional mathematical techniques, such as the theory of topological and algebraic groups and Lie algebras. Our aim in this book is to present the theory of quantum groups from this latter point of view. In fact, we shall concentrate on the study of the 'Lie algebras' of quantum groups, which seems to be the approach which has proved most powerful, particularly in applications, but we shall also discuss, in rather less detail, their relation with 'non-commutative algebraic geometry and topology'.

We shall now describe what a quantum group is, beginning by trying to explain the motivation for the use of the adjective 'quantum'.

In classical mechanics, the phase space M of a dynamical system is a *Poisson manifold*. This means that the space $\mathcal{F}(M)$ of (differentiable) complex-valued functions on M is equipped with a Lie bracket $\{\ ,\ \} : \mathcal{F}(M) \times \mathcal{F}(M) \to \mathcal{F}(M)$ (satisfying certain additional conditions), called the Poisson bracket. The dynamical equations defining the time evolution of the system are equivalent to the equations

$$\frac{d}{dt} f(m(t)) = \{\mathcal{H}_{\text{cl}}, f\}(m(t))$$

for $f \in \mathcal{F}(M)$, where $\mathcal{H}_{\text{cl}}$ is a fixed function on M called the (classical) hamiltonian, and $m(t) \in M$ is the 'state' of the system at time t. For example, for a single particle moving along the real line, M is the cotangent bundle $T^*(\mathbb{R})$, and if q is the coordinate on $\mathbb{R}$ ('position') and p the coordinate in the fibre direction ('momentum'), the Poisson bracket is

$$\{f_1, f_2\} = \frac{\partial f_1}{\partial p}\frac{\partial f_2}{\partial q} - \frac{\partial f_2}{\partial p}\frac{\partial f_1}{\partial q}.$$

In particular, the Poisson bracket of the coordinate functions is

(1) $\{p, q\} = 1.$

In quantum mechanics, the space M is replaced by the set of rays in a complex Hilbert space V, and the space $\mathcal{F}(M)$ of functions on M by the algebra $\mathrm{Op}(V)$ of (not necessarily bounded) operators on V. The time evolution of an operator A is given by

$$\frac{dA}{dt} = [\mathcal{H}_{\mathrm{qu}}, A]$$

for some operator $\mathcal{H}_{\mathrm{qu}} \in \mathrm{Op}(V)$, called the (quantum) hamiltonian. For example, in the case of a single particle moving along the real line, V is the space $\mathrm{L}^2(\mathbb{R})$ of square-integrable functions of q, and the operators P and Q corresponding to the coordinate functions p and q are given by

$$P = -\sqrt{-1}h\frac{\partial}{\partial q}, \qquad Q = \text{multiplication by } q,$$

where h is $1/2\pi$ times Planck's constant. Note that

(2) $[P, Q] = -\sqrt{-1}h\,\mathrm{id}_V.$

The question is: how to pass from the classical to the quantum description of a system. This is the problem of *quantization*. Ideally, one would like a map $\mathcal{Q}$ which assigns to each function $f \in \mathcal{F}(M)$ an operator $\mathcal{Q}(f)$ on V. Moreover, since time evolution in the classical and quantum descriptions is given by taking the Poisson bracket and commutator with the hamiltonian, respectively, $\mathcal{Q}$ should satisfy the relation

$$\mathcal{Q}\{f_1, f_2\} = \frac{[\mathcal{Q}(f_1), \mathcal{Q}(f_2)]}{-\sqrt{-1}h}$$

(the normalization comes from (1) and (2)). Unfortunately, it is known that, even for the simplest case of a single particle moving along the real line, no such map $\mathcal{Q}$ exists.

There is, however, an alternative formulation of the quantization problem, introduced by J. E. Moyal in 1949. This begins by noting that the fundamental difference between the classical and quantum descriptions is that

$\mathcal{F}(M)$ is a commutative algebra, whereas $\mathrm{Op}(V)$ is non-commutative (when $\dim(V) > 1$). Moyal's idea is to try to reproduce the results of quantum mechanics by replacing the usual product on $\mathcal{F}(M)$ by a non-commutative product $*_h$, depending on a parameter h, such that $*_h$ becomes the usual product as $h \to 0$, just as 'quantum mechanics becomes classical mechanics as Planck's constant tends to zero', and such that

$$(3) \qquad \lim_{h \to 0} \frac{f_1 *_h f_2 - f_2 *_h f_1}{h} = \{f_1, f_2\}.$$

If we think of $\mathcal{F}(M)$ with the Moyal product $*_h$ as a non-commutative algebra of functions $\mathcal{F}_h(M)$, we find ourselves in the realm of non-commutative geometry in the sense of A. Connes. The philosophy here is that any 'space' is determined by the algebra of functions on it (with the usual product). For example, every affine algebraic variety over $\mathbb{C}$ is determined (up to isomorphism) by the commutative algebra of regular functions on it, whereas every compact topological space is determined by its commutative C*-algebra of complex-valued continuous functions. More precisely, the category of 'spaces' in these examples is dual to the category of the corresponding algebras. Thus, a non-commutative algebra should be viewed as the space of functions on a 'non-commutative space', and we can say that Moyal's construction gives a deformation of the classical phase space M to a family of non-commutative (or 'quantum') spaces M_h such that $\mathcal{F}_h(M)$ is the algebra of functions on M_h.

The category of quantum spaces, then, might be defined as the category dual to the category of associative, but not necessarily commutative, algebras. To define the notion of a quantum group, let us first return for a moment to the classical situation. If G is a group, the multiplication $\mu : G \times G \to G$ of G induces a homomorphism $\mu^* = \Delta : \mathcal{F}(G) \to \mathcal{F}(G \times G)$ of algebras of functions. Now, if we define the algebra $\mathcal{F}(G)$ and the tensor product appropriately, $\mathcal{F}(G \times G)$ will be isomorphic to $\mathcal{F}(G) \otimes \mathcal{F}(G)$ as an algebra. For example, if G is an affine algebraic group over $\mathbb{C}$, and $\mathcal{F}(G)$ is the algebra of regular functions on G, the ordinary algebraic tensor product will do. Thus, we have a comultiplication $\Delta : \mathcal{F}(G) \to \mathcal{F}(G) \otimes \mathcal{F}(G)$. (The reason for this terminology is that the multiplication on $\mathcal{F}(G)$ can be viewed as a map $\mathcal{F}(G) \otimes \mathcal{F}(G) \to \mathcal{F}(G)$.) Similarly, the inverse map $\iota : G \to G$ induces a map $\iota^* = S : \mathcal{F}(G) \to \mathcal{F}(G)$, called the antipode, and evaluation at the identity element of G is a homomorphism $\epsilon : \mathcal{F}(G) \to \mathbb{C}$, called the counit. The maps Δ, S and ϵ satisfy certain compatibility properties which reflect the defining properties of the inverse and the associativity of multiplication in G, and combine to give $\mathcal{F}(G)$ the structure of a *Hopf algebra*.

We might therefore *define the category of quantum groups to be the category dual to the category of (not necessarily commutative) Hopf algebras.* (We said 'might' here, and in our tentative definition of a quantum space, because, to ensure that the categories of quantum spaces and quantum groups have reasonable properties, it would be necessary to impose some restrictions on the class of algebras which are acceptable as 'quantized algebras of functions'. Manin suggests that one should work with 'Koszul algebras', but we shall not discuss this point here.) As is common practice in the literature, we shall often abuse terminology by referring to a Hopf algebra itself as a quantum group.

As the preceding discussion suggests, one way to try to construct non-classical examples of quantum groups is to look for deformations, in the category of Hopf algebras, of classical algebras of functions $\mathcal{F}(G)$. Just as the classical Poisson bracket can be recovered as the 'first order part' of Moyal's deformation (see (3)), so it turns out that the existence of a deformation $\mathcal{F}_h(G)$ of $\mathcal{F}(G)$ automatically endows the group G itself with extra structure, namely that of a *Poisson–Lie group*. This is a Poisson structure on G which is compatible with the group structure in a certain sense. Conversely, to construct deformations of $\mathcal{F}(G)$, it is natural to begin by describing the possible Poisson–Lie group structures on G and then to attempt to extend these 'first order deformations' to full deformations. This is the approach taken in this book. Poisson–Lie groups are also of interest in their own right, for they form the natural setting for the study of classical integrable systems with symmetry.

There is another Hopf algebra associated to any Lie group G, namely the universal enveloping algebra $U(\mathfrak{g})$ of its Lie algebra $\mathfrak{g}$. This is essentially the *dual* of $\mathcal{F}(G)$ in the category of Hopf algebras. In general, the vector space dual A^* of any finite-dimensional Hopf algebra A is also a Hopf algebra: the multiplication $A^* \otimes A^* \to A^*$ is dual to the comultiplication $\Delta : A \to A \otimes A$ of A, and the comultiplication of A^* is dual to the multiplication of A. Note that A^* is commutative if and only if A is cocommutative, i.e. if and only if $\Delta(A)$ is contained in the symmetric part of $A \otimes A$. If, as is usually the case in examples of interest, A is infinite dimensional, this duality often continues to hold provided the dual and tensor product are defined appropriately. To a deformation $\mathcal{F}_h(G)$ of $\mathcal{F}(G)$ through (not necessarily commutative) Hopf algebras therefore corresponds a deformation $U_h(\mathfrak{g})$ of $U(\mathfrak{g})$ through (not necessarily cocommutative) Hopf algebras.

In fact, only non-cocommutative deformations of $U(\mathfrak{g})$ are of interest, since any deformation of $U(\mathfrak{g})$ through cocommutative Hopf algebras is necessarily of the form $U(\mathfrak{g}_h)$ for some deformation $\mathfrak{g}_h$ of $\mathfrak{g}$ through Lie algebras. However, many interesting Lie algebras have no non-trivial deformations. This is the case, for example, if $\mathfrak{g}$ is a (finite-dimensional) complex semisimple Lie algebra, such as the Lie algebra $sl_2(\mathbb{C})$ of 2×2 complex matrices of trace zero. This follows from the fact that the condition of semisimplicity is open, so that any small deformation of $\mathfrak{g}$ will still be semisimple, whereas the semisimple Lie algebras are discretely parametrized (by their Dynkin diagrams, for example).

The first example of a non-cocommutative deformation of this type was discovered by P. P. Kulish and E. K. Sklyanin in 1981 in the case $\mathfrak{g} = sl_2(\mathbb{C})$ (although the importance of its Hopf structure was not realized until later). Note that $sl_2(\mathbb{C})$ has a basis

$$(4) \qquad \bar{X}^+ = \begin{pmatrix} 0 & 1 \\ 0 & 0 \end{pmatrix}, \qquad \bar{X}^- = \begin{pmatrix} 0 & 0 \\ 1 & 0 \end{pmatrix}, \qquad \bar{H} = \begin{pmatrix} 1 & 0 \\ 0 & -1 \end{pmatrix},$$

whose Lie brackets are given by

$$(5a) \qquad [\bar{X}^+, \bar{X}^-] = \bar{H}, \qquad [\bar{H}, \bar{X}^\pm] = \pm 2\bar{X}^\pm.$$

The comultiplication is given on these basis elements by

$$(5b) \qquad \Delta(\bar{H}) = \bar{H} \otimes 1 + 1 \otimes \bar{H}, \qquad \Delta(\bar{X}^\pm) = \bar{X}^\pm \otimes 1 + 1 \otimes \bar{X}^\pm,$$

an assignment which extends uniquely to an algebra homomorphism Δ : $U(sl_2(\mathbb{C})) \to U(sl_2(\mathbb{C})) \otimes U(sl_2(\mathbb{C}))$. The deformation $U_h(sl_2(\mathbb{C}))$ is generated by elements H, $X^\pm$, which satisfy the relations

$$(6a) \qquad X^+ X^- - X^- X^+ = \frac{e^{hH} - e^{-hH}}{e^h - e^{-h}}, \qquad HX^\pm - X^\pm H = \pm 2X^\pm.$$

It has a non-cocommutative comultiplication given on generators by

$$(6b) \qquad \begin{aligned} \Delta(H) &= H \otimes 1 + 1 \otimes H, \\ \Delta(X^+) &= X^+ \otimes e^{hH} + 1 \otimes X^+, \qquad \Delta(X^-) = X^- \otimes 1 + e^{-hH} \otimes X^-. \end{aligned}$$

Formally, at least, it is clear that (6a) and (6b) go over into (5a) and (5b) as $h \to 0$. The Hopf algebra defined in (6a,b) is called 'quantum $sl_2(\mathbb{C})$'. (See Chapter 6 for formulas for the antipode and counit of $U_h(sl_2(\mathbb{C}))$, and for a way to make sense of expressions such as e^{hH}.)

The Hopf algebra dual to $U_h(sl_2(\mathbb{C}))$, the 'algebra $\mathcal{F}_h(SL_2(\mathbb{C}))$ of functions on quantum $SL_2(\mathbb{C})$', was discovered by L. D. Faddeev and L. A. Takhtajan in 1985. It is the associative algebra generated by elements a, b, c, d with the following multiplicative relations:

$$(7) \qquad ab = e^{-h}ba, \qquad ac = e^{-h}ca, \qquad bd = e^{-h}db, \qquad cd = e^{-h}dc,$$

$$(8) \qquad bc = cb, \qquad ad - da + (e^h - e^{-h})bc = 0,$$

$$(9) \qquad ad - e^{-h}bc = 1,$$

and comultiplication

$$\Delta(a) = a \otimes a + b \otimes c, \qquad \Delta(b) = a \otimes b + b \otimes d,$$

$$\Delta(c) = c \otimes a + d \otimes c, \quad \Delta(d) = c \otimes b + d \otimes d.$$

Note that, when $h \to 0$, the relations (7), (8) and (9) just say that the matrix

$$\begin{pmatrix} a & b \\ c & d \end{pmatrix}$$

has commuting entries and determinant one. Thus, $\mathcal{F}_h(SL_2(\mathbb{C}))$ is a deformation of the algebra of functions on the group $SL_2(\mathbb{C})$ of 2×2 complex matrices of determinant one.

As we mentioned at the beginning of this introduction, the algebra structure of $\mathcal{F}_h(G)$ can be described by a matrix of constants, namely

$$(10) \qquad R = e^{-h/2} \begin{pmatrix} e^h & 0 & 0 & 0 \\ 0 & 1 & 0 & 0 \\ 0 & e^h - e^{-h} & 1 & 0 \\ 0 & 0 & 0 & e^h \end{pmatrix}.$$

In fact, if

$$T = \begin{pmatrix} a & b \\ c & d \end{pmatrix},$$

the relations (7) and (8) are equivalent to

$$(11) \qquad (T \otimes 1)(1 \otimes T)R = R(1 \otimes T)(T \otimes 1).$$

Note that $T \otimes 1$ and $1 \otimes T$ do not commute, since the entries of T do not commute (if $h \neq 0$); note also that R is most naturally viewed as an element of $\mathrm{End}(\mathbb{C}^2 \otimes \mathbb{C}^2)$. It is in the form (11) that quantum groups usually appear in the theory of integrable systems.

For the dual Hopf algebra $U_h(sl_2(\mathbb{C}))$, the quantum R-matrix expresses, as one would expect, the non-cocommutativity of the comultiplication. Namely, let $\Delta^{\mathrm{op}}(x)$ be the result of interchanging the order of the factors in $\Delta(x)$, for any $x \in U_h(sl_2(\mathbb{C}))$. It turns out that there is an invertible element $\mathcal{R} \in U_h(sl_2(\mathbb{C})) \otimes U_h(sl_2(\mathbb{C}))$, called the 'universal R-matrix', such that

$$\Delta^{\mathrm{op}}(x) = \mathcal{R}\Delta(x)\mathcal{R}^{-1}$$

for all $x \in U_h(sl_2(\mathbb{C}))$ (actually, $\mathcal{R}$ is a formal infinite sum of elements of the algebraic tensor product). The relation between $\mathcal{R}$ and R is very simple: the reader will easily verify that, if we replace $\bar{X}^\pm$ and $\tilde{H}$ by $X^\pm$ and H in (4), we obtain a matrix representation of $U_h(sl_2(\mathbb{C}))$; applying this representation to $\mathcal{R}$ gives the matrix R.

Quantum groups might have remained a curiosity to the mathematical community at large but for their surprising connections with other parts of mathematics, most notably the theory of invariants of links and 3-manifolds, and the representation theory of Lie algebras in characteristic p.

The former depends on the classical relation between braids and links. Recall that a *braid on m strands* is a collection of m non-intersecting strings in $\mathbb{R}^3$ joining m fixed points in a plane to m fixed points in another parallel plane. Joining corresponding points in the two planes in a standard way associates to any braid a link (called its 'closure'), i.e. a collection of non-intersecting circles in $\mathbb{R}^3$. Joining braids end to end makes the set of isotopy classes of braids into a group $\mathcal{B}_m$. The relation with quantum groups arises because there is a simple way to associate to any quantum R-matrix $R \in \text{End}(V \otimes V)$ a representation ρ_m of $\mathcal{B}_m$ on $V^{\otimes m}$ for all $m \geq 2$. This depends on the fact that R satisfies the *quantum Yang–Baxter equation*

$$R_{12}R_{13}R_{23} = R_{23}R_{13}R_{12};$$

here, R_{12} means $R \otimes \text{id} \in \text{End}(V^{\otimes 3})$, etc. To obtain an invariant of links, one needs a family of 'traces' $tr_m : \text{End}(V^{\otimes m}) \to \mathbb{C}$ such that $tr_m(\rho_m(b)) = tr_n(\rho_n(b'))$ whenever the closures of the braids $b \in \mathcal{B}_m$ and $b' \in \mathcal{B}_n$ are equivalent links. Thanks to a classical theorem of A. Markov, it is known precisely which pairs (b, b') have the latter property (and for this reason, the tr_m are usually called 'Markov traces'). Using the quantum R-matrix (10) and a suitable Markov trace, one obtains in this way the celebrated *Jones polynomial*. In fact, this is essentially Jones's original construction, except that he obtained his R-matrix by using a 'Hecke algebra' instead of a quantum group (but we shall see that Hecke algebras should probably be regarded as 'quantum' objects).

The application to 3-manifolds is based on the well-known fact that every compact, oriented, connected 3-manifold without boundary can be obtained, up to homeomorphism, by performing surgery on a link in the 3-dimensional sphere. One shows that a cleverly chosen combination of the quantum invariants of this link depends only on the 3-manifold, and not on the choice of the link along which surgery is performed.

The application of quantum groups to representations of Lie algebras in characteristic p is no less remarkable. It makes use of a certain 'standard' deformation $U_h(\mathfrak{g})$ of $U(\mathfrak{g})$, where $\mathfrak{g}$ is any finite-dimensional complex semisimple Lie algebra (and which reduces, when $\mathfrak{g} = sl_2(\mathbb{C})$, to the algebra found by Kulish and Sklyanin). To describe the relation with characteristic p, it is convenient to replace the deformation parameter h by $\epsilon = e^h$, and to write $U_\epsilon(\mathfrak{g})$ for $U_h(\mathfrak{g})$. It then turns out that the representation theory of $U_\epsilon(\mathfrak{g})$ depends crucially on whether ϵ is a root of unity or not. In the latter case, the theory is essentially the same as the representation theory of $\mathfrak{g}$ itself (over $\mathbb{C}$), but in the former it resembles the modular representation theory of $\mathfrak{g}$. This is more than an analogy: if ϵ is a primitive pth root of unity, where p is a prime, there is a ring homomorphism from $U_\epsilon(\mathfrak{g})$ to the enveloping algebra $U_{\mathbb{F}_p}(\mathfrak{g})$

of $\mathfrak{g}$ over the field $\mathbb{F}_p$ of p elements (this is obtained essentially by replacing ϵ by $1 \in \mathbb{F}_p$), and, under certain additional conditions, representations of $U_\epsilon(\mathfrak{g})$ can be 'specialized' to give representations of $U_{\mathbb{F}_p}(\mathfrak{g})$. Thus, both the modular representation theory of $\mathfrak{g}$ and the characteristic zero theory are reflected in the representation theory of the family of Hopf algebras $U_\epsilon(\mathfrak{g})$ (over $\mathbb{C}$). Using this relation, substantial progress has been made toward the solution of several long-standing conjectures in the modular theory.

We should also mention the roles played by quantum groups in physics, which go well beyond their origins in inverse scattering theory. Perhaps the most interesting of these is the relation between quantum groups and conformal and quantum field theories. The first evidence of this was the experimental observation that the 'fusion rules' of certain conformal field theories can be reproduced by considering the decomposition of tensor products of representations of the quantum groups $U_\epsilon(\mathfrak{g})$ when ϵ is a root of unity. Further evidence came from a remarkable theorem of T. Kohno and Drinfel'd on the relation between the Knizhnik–Zamolodchikov (KZ) equation and $U_h(\mathfrak{g})$. The KZ equation is a system of first order partial differential equations for a function of m complex variables with values in the tensor product $V^{\otimes m}$, where V is a representation of $\mathfrak{g}$. The KZ system has regular singularities along the hyperplanes $z_i = z_j$ in $\mathbb{C}^m$, for all $i \neq j$, and can be viewed as a connection on a bundle over the complement $\mathcal{D}_m$ of these hyperplanes, with fibre $V^{\otimes m}$. Moreover, this connection has a symmetry property which means that there is an induced connection on a bundle over the space $\mathcal{C}_m$ of orbits of the obvious action of the symmetric group Σ_m on $\mathcal{D}_m$. The crucial property of the KZ equation is that this connection is flat, which implies that the monodromy of its solutions defines a representation of the fundamental group of $\mathcal{C}_m$ on $V^{\otimes m}$. The latter group is exactly the braid group $\mathcal{B}_m$ that we discussed above, where we noted that a representation of $\mathcal{B}_m$ on $V^{\otimes m}$ could also be obtained by using a quantum R-matrix. According to Kohno and Drinfel'd, if we use the R-matrix given by the action of the universal R-matrix of $U_h(\mathfrak{g})$ on $V \otimes V$, where h is 'generic', these two representations of $\mathcal{B}_m$ coincide.

The importance of the KZ equation in conformal field theory is that it is satisfied by the 'm-point functions' of the theory. Thus, the Kohno–Drinfel'd theorem indicates a connection between $U_h(\mathfrak{g})$ and conformal field theory. This has recently been extended by D. Kazhdan and G. Lusztig, who have shown that the KZ equation is intimately related to the category of representations of the 'specialized' algebras $U_\epsilon(\mathfrak{g})$ when ϵ is a root of unity. This result provides an 'explanation' for the coincidence between the fusion rules arising in conformal field theory and those arising from quantum groups at roots of unity.

These and other examples to be discussed in this book show that the theory of quantum groups occupies an important place in the mainstream of mathematics and mathematical physics.

We now describe the contents of the book systematically. In Chapter 1, we give the definition and basic properties of Poisson–Lie groups, and of their infinitesimal counterparts, Lie bialgebras. We describe, among other examples, a standard family of Lie bialgebra structures on every (finite-dimensional) complex simple Lie algebra. These induce Poisson–Lie group structures on the associated compact Lie group K. Like all Poisson manifolds, K has a canonical decomposition into symplectic submanifolds, called its *symplectic leaves*, and it turns out that they have a beautiful description in terms of the so-called Bruhat decomposition of K.

In the last section of this chapter, we describe the formulation of quantization as a deformation of Poisson structure. We discuss only the simplest example of a single particle moving along the real line, since our aim is mainly to motivate the treatment of the deformation theory of Hopf algebras in Chapter 6.

Chapter 2 returns to the discussion of Lie bialgebras. We show that Lie bialgebra structures on a Lie algebra $\mathfrak{g}$ can be constructed from solutions of the 'classical Yang–Baxter equation' (CYBE), and that in some cases, for example when $\mathfrak{g}$ is complex semisimple, all Lie bialgebra structures arise in this way. We also discuss the relation between the CYBE and classical integrable systems.

Chapter 2 shows the importance of the problem of classifying the solutions of the CYBE. In Chapter 3, we give an essentially complete description of the solutions with values in $\mathfrak{g} \otimes \mathfrak{g}$, where $\mathfrak{g}$ is a complex simple Lie algebra. The discussion requires more familiarity with Lie theory than the other early chapters of the book, and, since most of the results will not be needed later, we recommend that this chapter be omitted on a first reading.

Chapter 4 begins with a summary of the general results about Hopf algebras we shall need. We have already mentioned that quantum groups are usually non-cocommutative as Hopf algebras. However, they (or their duals) are often 'quasitriangular': this means that the comultiplication of the Hopf algebra A is conjugate to the opposite comultiplication by an invertible element $\mathcal{R}$ of $A \otimes A$, called its universal R-matrix. The element $\mathcal{R}$ satisfies certain additional conditions, including the quantum Yang–Baxter equation (QYBE). We discuss the general properties of quasitriangular Hopf algebras and their relation with the QYBE.

One of the most important general properties of a Hopf algebra A is that there is a natural way to make the tensor product of two representations of A into another representation of A. This means that the category $\mathbf{rep}_A$ of representations of A is a *monoidal category*. If A has additional properties, these will be reflected in $\mathbf{rep}_A$ and vice versa. For example, if A is quasitriangular, $\mathbf{rep}_A$ is a quasitensor category, which means roughly that the tensor product operation is associative and commutative up to isomorphism. We discuss the basic properties of quasitensor categories in Chapter 5, and give a number of examples arising from algebra, topology and physics.

The idea that the properties of a Hopf algebra are encoded in, and might be recovered from, its representations is called *Tannaka–Krein duality*, and this will be a guiding principle throughout this book. We prove a Tannaka–Krein type theorem valid for a large class of Hopf algebras.

We meet our first examples of quantum groups in Chapters 6 and 7. We begin by discussing the general theory of deformations of Hopf algebras, concentrating on universal enveloping algebras and function algebras. We show that any deformation $U_h(\mathfrak{g})$ of $U(\mathfrak{g})$ gives rise to a Lie bialgebra structure on $\mathfrak{g}$, and that every finite-dimensional Lie bialgebra arises as the 'classical limit' of some deformation in this way. To make sense of formulas such as those in (6a,b), one interprets $U_h(\mathfrak{g})$ as an algebra over the ring $\mathbb{C}[[h]]$ of formal power series in an indeterminate h.

In Chapter 6, we 'derive' the deformation $U_h(sl_2(\mathbb{C}))$ described above and show that it can be extended to a deformation of the universal enveloping algebra of any symmetrizable Kac–Moody algebra, using the fact that such algebras are generated by certain $sl_2(\mathbb{C})$ subalgebras.

In Chapter 7, we construct deformations of the algebras of functions on the classical complex Lie groups using the quantum R-matrix method outlined above. We also discuss the duality relation between these quantized function algebras and the quantized universal enveloping algebras of Chapter 6. For a large class of quantum R-matrices, we construct an analogue of the de Rham complex, which can be viewed as a theory of 'differential calculus' on the associated quantum group. In the final section of this chapter, we discuss the relation between the quantum Yang–Baxter equation and certain models in statistical mechanics, where the QYBE plays the role of an integrability condition.

In Chapter 8, we obtain a formula for the universal R-matrix of the quantum groups $U_h(\mathfrak{g})$ defined in Chapter 6, assuming that $\mathfrak{g}$ is finite dimensional. We make use of an action of a (generalized) braid group on any quantum group of this type, which is analogous to the well-known action of (a finite covering of) the Weyl group of $\mathfrak{g}$ on $U(\mathfrak{g})$ in the classical situation.

Chapters 9, 10 and 11 are devoted to the structure and representation theory of the 'specialization' $U_\epsilon(\mathfrak{g})$, where $\epsilon \in \mathbb{C}^\times$. The definition of $U_\epsilon(\mathfrak{g})$ requires some effort, for $U_h(\mathfrak{g})$ is defined over the algebra of formal power series in h, so that the only specialization which appears to make sense is $h = 0$. In fact, as we show in Chapter 9, the specialization to $U_\epsilon(\mathfrak{g})$ can be carried out in two ways, called 'restricted' and 'non-restricted'. The two specializations coincide if ϵ is not a root of unity, but not otherwise. We also study the relation, hinted at above, between $U_\epsilon(\mathfrak{g})$ when ϵ is a root of unity and Lie algebras in characteristic p.

In Chapter 10, we give the classification of the finite-dimensional irreducible representations of $U_\epsilon(\mathfrak{g})$ when $\mathfrak{g}$ is a finite-dimensional complex simple Lie algebra and ϵ is not a root of unity. It turns out that, up to a rather trivial twisting by certain outer automorphisms, the parametrization, and even the characters, of such representations are exactly the same as in the classical case.

There are some differences when compared with the classical situation, however. We show, in particular, that in the quantum analogue of the classical Frobenius–Schur duality between representations of the (special or general) linear groups and symmetric groups, the role of the symmetric group is played by a *Hecke algebra*.

In Chapter 11, we study the representation theory at a root of unity. For the restricted specialization $U_\epsilon^{\mathrm{res}}(\mathfrak{g})$, the finite-dimensional irreducible representations are discretely parametrized as in the classical case (in characteristic zero), but their structure (and dimensions) are different in general. On the other hand, for the non-restricted specialization $U_\epsilon(\mathfrak{g})$, we can say, oversimplifying a little, that there is an irreducible representation of $U_\epsilon(\mathfrak{g})$ associated to each point of the Lie group G with Lie algebra $\mathfrak{g}$, and that representations which are associated to points of the same conjugacy class are equivalent up to twisting by an element of a certain infinite-dimensional group of automorphisms of $U_\epsilon(\mathfrak{g})$.

The tensor product of two irreducible representations of $U_\epsilon^{\mathrm{res}}(\mathfrak{g})$, when ϵ is a root of unity, is not in general completely reducible. However, essentially by discarding those indecomposable pieces of the tensor product which are not irreducible, we obtain a 'truncated' tensor product which has very interesting properties. We show that the resulting truncated representation ring is an example of a *fusion ring*, a new algebraic structure first encountered in conformal field theory. As we mentioned earlier in this introduction, this coincidence was one of the first indications of the connection between quantum groups and conformal and quantum field theories.

In Chapter 12, we return to the case of generic ϵ, but consider quantizations of some infinite-dimensional groups, namely quantum affine algebras and Yangians. It is important to classify the finite-dimensional irreducible representations of these quantum groups, for to such a representation one can associate a solution of the quantum Yang–Baxter equation with *spectral parameters*:

$$R_{12}(u_1-u_2)R_{13}(u_1-u_3)R_{23}(u_2-u_3) = R_{23}(u_2-u_3)R_{13}(u_1-u_3)R_{12}(u_1-u_2).$$

Here, $R(u)$ is a function of a complex variable u with values in $\mathrm{End}(V \otimes V)$; it is a rational or trigonometric function of u according as V is a representation of a Yangian or a quantum affine algebra, respectively.

We also describe some surprising connections between quantum affine algebras and Yangians, and the classical representation theory of p-adic groups.

In Chapter 13, we return to quantized algebras of functions. Classically, the algebra $\mathcal{F}(G)$ is commutative, so all of its irreducible representations are one-dimensional; in fact, there is a one-dimensional representation of $\mathcal{F}(G)$

associated to each point of G (given simply by evaluating a function at the point). However, $\mathcal{F}_h(G)$ is non-commutative in general, and the representation theory of $\mathcal{F}_h(G)$ replaces, to some extent, the study of G as a point set.

We study the case where G is a complex simple Lie group, and replace $\mathcal{F}_h(G)$ by a suitable specialization $\mathcal{F}_\epsilon(G)$. We classify the representations which are unitary with respect to a conjugate-linear involution on $\mathcal{F}_\epsilon(G)$ whose classical limit is the involution which defines the compact real form of G (for such representations to exist, it is necessary that $\epsilon \in \mathbb{R}$).

There is a natural C^*-norm on $\mathcal{F}_\epsilon(G)$, and the C^*-completion $C_\epsilon(G)$ of $\mathcal{F}_\epsilon(G)$ can be regarded as a quantization of the algebra of continuous functions on G. This makes contact with the theory of *compact matrix quantum groups* (or *matrix pseudogroups*) of Woronowicz. Compact quantum groups are closely related to 'Kac algebras', the theory of which is the culmination of a long effort to generalize Pontryagin duality to locally compact groups which are not necessarily abelian. Apart from a number of technical differences, perhaps the most important structural difference between the two theories is that in the theory of Kac algebras the existence of a Haar integral is postulated, whereas this is one of Woronowicz's main theorems about compact quantum groups. We give a simple definition of the Haar integral by making use of the representation theory of $U_\epsilon(\mathfrak{g})$.

In the final section of this chapter, we show that several q-special functions are related to the representations of compact or non-compact quantum groups. This was first noticed by L. L. Vaksman and Ya. S. Soibelman in 1986, who showed that the 'little q-Jacobi polynomials' are essentially the matrix elements of the irreducible unitary representations of $\mathcal{F}_\epsilon(SL_2(\mathbb{C}))$. Thus, almost 150 years after the birth of the theory of q-special functions, their representation-theoretic meaning was found.

In Chapter 14, we show how the results of Chapters 9 and 10 can be used to give a solution to a classical problem in the theory of complex semisimple Lie algebras, namely that of giving a basis of any finite-dimensional irreducible representation of such an algebra $\mathfrak{g}$ which is in some sense canonical. If $\mathfrak{g}$ is of type A, B, C or D, one solution to this problem is given by the Gel'fand–Tsetlin basis, although this cannot be extended to the algebras of exceptional type. We describe a solution to this problem, due independently to M. Kashiwara and Lusztig, for arbitrary $\mathfrak{g}$ in the quantum case, which, specializing to $h = 0$, also solves the classical problem. As a by-product, one obtains a graphical algorithm for decomposing the tensor product of two irreducible representations of $\mathfrak{g}$. It is perhaps worth emphasizing that this construction cannot be carried out purely in the classical setting: one must pass to the quantum case and then specialize.

Chapter 15 describes the application of quantum groups to the construction of invariants of links and 3-manifolds, outlined above. This is due to

N. Yu. Reshetikhin and V. G. Turaev, but strongly motivated by the work of V. F. R. Jones and E. Witten.

The final Chapter 16 describes a generalization, due to Drinfel'd, of the notion of a Hopf algebra, called a quasi-Hopf algebra. The category of quasi-Hopf algebras is in some ways simpler than the category of Hopf algebras because of the existence of a kind of 'gauge symmetry' on the former category, which often allows one to 'twist' a given quasi-Hopf algebra into a simpler one with the 'same' representation theory.

One of the main results of Chapter 16 is the theorem of Kohno and Drinfel'd mentioned earlier. This is proved by using the Knizhnik–Zamolodchikov equation to construct a certain quasi-Hopf algebra which is isomorphic to a twist of the QUE algebra $U_h(\mathfrak{g})$. We give an introduction to Kazhdan and Lusztig's generalization of this theorem, including the construction of their new tensor product of representations of affine Lie algebras.

We also describe briefly some recent work of Drinfel'd relating quasi-Hopf algebras to the absolute Galois group of $\mathbb{Q}$. This can be regarded as the fulfilment of part of A. Grothendieck's 'Esquisse d'un programme'.

The book concludes with an Appendix, which establishes the notation relating to complex simple Lie algebras (and, more generally, to Kac–Moody algebras), which will be used throughout the book.

As a distinct area of study, the theory of quantum groups is still less than 10 years old, and it is perhaps still possible to give an account of all its main lines of development from a single point of view. One of the peculiarities of our subject is that, even though most of the foundational results were stated by Drinfel'd in his 1986 talk at the International Congress of Mathematicians, for many of them there is no detailed exposition in print. On the other hand, the results obtained since 1986, including those due to Drinfel'd himself, are generally well-covered in journal articles. Although we certainly do not claim to have filled all the gaps in the existing literature, our aim has been to provide a full account of the foundations, while for the more advanced topics we have usually been content to state the most important results and give a sketch of the main ideas of the proof, together with detailed references to the literature. We have tried not to assume too much familiarity on the part of the reader with the classical theory of Lie groups and Lie algebras, at least in the first half of the book. We hope that, as a result, the book will serve as an introductory text for those readers meeting quantum groups for the first time, and will provide more experienced readers with a survey of all the major developments in the subject, together with a guide to where the details can be found. This approach has made it necessary to include a rather extensive bibliography, and, although we have tried to make this reasonably complete, some gaps are inevitable, and we apologise to those authors whose relevant work has inadvertently been omitted.

We have learned about quantum groups from many sources, but we would like to thank in particular V. G. Drinfel'd for two inspirational lectures at Luminy in 1988 which ignited our interest in the subject, N. Yu. Reshetikhin for drawing our attention to some of the relevant work in the physics literature, C. de Concini and V. G. Kac for explaining to us their work on quantum groups at roots of unity, and S. Donkin for help with the relation between roots of unity and modular Lie algebras.

1
Poisson–Lie groups
and Lie bialgebras

As we mentioned in the Introduction, one reason for the terminology used to describe 'quantum groups' is that their relation to ordinary Lie groups is analogous to that between quantum mechanics and classical mechanics. One of the principal messages of this book is that this is more than a formal analogy, and we begin our treatment of quantum groups by discussing the formulation of classical mechanics for a dynamical system whose phase space is a group.

The traditional geometric framework for classical dynamics is that of symplectic manifolds. Recall that a smooth manifold M is said to be *symplectic* if it is equipped with a closed 2-form ω which gives a non-degenerate bilinear form on the tangent space at each point of M. A hamiltonian system on M is then specified by giving a smooth real-valued function $\mathcal{H}$ on M, the *hamiltonian*. By using the 2-form ω, one can associate to the 1-form $d\mathcal{H}$ a vector field $X_{\mathcal{H}}$, whose flow describes the time evolution of the system.

It is clear that what is essential in this formulation is to have some way to associate to every 1-form on M a vector field on M. For this we need a bivector on M, i.e. a section of the second tensor power of the tangent bundle of M, for then we can contract the 1-form against the bivector to obtain a vector field. A *Poisson manifold* is a smooth manifold equipped with a bivector with certain properties analogous to those which assert that the symplectic form ω is skew and closed, but without any requirement of non-degeneracy.

Every symplectic manifold M has a natural Poisson structure, for the symplectic form ω allows one to identify 1-forms and vector fields, and hence 2-forms and skew bivectors. In fact, it is exactly the everywhere non-degenerate Poisson structures which arise in this way. The most general Poisson manifold, however, has a canonical decomposition into symplectic submanifolds (not all of the same dimension in general), called its *symplectic leaves*.

The Poisson manifolds which will be of most interest to us are the *Poisson–Lie groups*, introduced in Section 1.2. These are Lie groups G which have Poisson structures which are compatible with the group operations in a certain sense. We shall see that such Poisson structures never arise from symplectic structures.

The most important examples of Poisson–Lie groups, at least for the study of quantum groups, concern the case in which G is a compact group (or its complexification). We construct in Section 1.4 a canonical family of Poisson–Lie group structures on G whose symplectic leaves have a beautiful description in terms of the Bruhat decomposition of G. Moreover, there are induced Poisson structures on each coadjoint orbit $\mathcal{O}$ of G whose symplectic leaves are exactly the Schubert cells in $\mathcal{O}$.

The infinitesimal versions of Poisson–Lie groups are called *Lie bialgebras*. More precisely, if G is any Poisson–Lie group, the dual $\mathfrak{g}^*$ of its Lie algebra $\mathfrak{g}$ turns out to have a natural Lie algebra structure, which is compatible with that of $\mathfrak{g}$ itself, in a certain sense; together, these are said to form a Lie bialgebra structure on $\mathfrak{g}$. Conversely, if G is connected and simply-connected, every Lie bialgebra structure on $\mathfrak{g}$ arises from a unique Poisson–Lie group structure on G. The basic properties of Lie bialgebras are described in Sections 1.3 and 1.4, but we postpone the discussion of their classification until Chapters 2 and 3.

Section 1.4 describes an equivalent formulation of the notion of a Lie bialgebra in which $\mathfrak{g}$ and $\mathfrak{g}^*$ play symmetric roles; it allows us to give simple explicit descriptions of a number of important examples of Lie bialgebras. It also shows that $\mathfrak{g}^*$ and the direct sum (of vector spaces) $\mathfrak{g} \oplus \mathfrak{g}^*$ inherit natural Lie bialgebra structures from that on $\mathfrak{g}$ itself. It follows that the Poisson–Lie group G^* with Lie algebra $\mathfrak{g}^*$ acts on G. This so-called *dressing action*, studied in Section 1.5, is the basis for the application of Poisson–Lie groups in the theory of integrable systems, which will be described in Section 2.3. It also allows one to give a simple description of the symplectic leaves of a Poisson–Lie group.

Section 1.6 is an introduction to the interpretation of quantization of classical hamiltonian systems in terms of a deformation of the Poisson structure on the phase space of the system. The discussion provides some motivation for the adjective in the term 'quantum group'.

1.1 Poisson manifolds

In this section, we describe those aspects of the theory of Poisson manifolds that will be needed later.

A Definitions Let M be a smooth manifold of finite dimension m. We denote by $C^\infty(M)$ the algebra of smooth real-valued functions on M.

DEFINITION 1.1.1 *A Poisson structure on M is an $\mathbb{R}$-bilinear map*

$$\{\,,\,\}_M : C^\infty(M) \times C^\infty(M) \to C^\infty(M),$$

called the Poisson bracket, which satisfies the following conditions:

(1) $$\{f_1, f_2\} = -\{f_2, f_1\},$$
(2) $$\{f_1, \{f_2, f_3\}\} + \{f_2, \{f_3, f_1\}\} + \{f_3, \{f_1, f_2\}\} = 0,$$
(3) $$\{f_1 f_2, f_3\} = f_1\{f_2, f_3\} + \{f_1, f_3\}f_2,$$

for all f_1, f_2, $f_3 \in C^\infty(M)$. (We omit the subscript on the Poisson bracket whenever this will not lead to confusion.)

The skew-symmetry property (1) and the *Jacobi identity* (2) together mean that $\{\ ,\ \}$ is a Lie bracket on $C^\infty(M)$. Property (3), the *Leibniz identity*, means that, for all $f \in C^\infty(M)$, the map $g \to \{g, f\}$ is a derivation of $C^\infty(M)$; hence, there a vector field X_f on M such that $X_f(g) = \{g, f\}$ for all g. (A vector field which arises in this way is called a *hamiltonian vector field*.) In particular, $\{g, f\}$ depends only on the differential of g and, in view of the skew-symmetry of the Poisson bracket, on the differential of f. It follows that there is a well-defined map of bundles $B_M : T^*M \to TM$, where TM (resp. T^*M) denotes the tangent bundle (resp. the cotangent bundle) of M, such that $B_M(df) = X_f$ for all $f \in C^\infty(M)$. Equivalently, there is a skew-symmetric 2-tensor $w_M \in TM^{\otimes 2}$, called the *Poisson bivector*, such that $\{f, g\}_M = \langle (df \otimes dg), w_M \rangle$. (Here, and elsewhere in this book, $\langle\ ,\ \rangle$ denotes the natural pairing between a vector space, or bundle of vector spaces (in this case $TM^{\otimes 2}$), and its dual.) The Jacobi identity (2) for the Poisson bracket is equivalent to the vanishing of the so-called Schouten bracket $[w, w]$ (see Schouten (1940)).

It follows from these remarks that, in local coordinates $(x_1, \ldots, x_m)$ on M, the Poisson bracket can be written in the form

$$\{f, g\} = \sum_{i,j=1}^{m} w_{ij}(x) \frac{\partial f}{\partial x_i} \frac{\partial g}{\partial x_j},$$

where the locally defined functions w_{ij} are the coefficients of the Poisson bivector

$$w_x = \sum_{i,j=1}^{m} w_{ij}(x) \frac{\partial}{\partial x_i} \otimes \frac{\partial}{\partial x_j}.$$

Since w is skew,

$$w_{ij} = -w_{ji}$$

and the Jacobi identity is equivalent to

$$\sum_{r=1}^{m} \left(w_{ri} \frac{\partial w_{jk}}{\partial x_r} + w_{rj} \frac{\partial w_{ki}}{\partial x_r} + w_{rk} \frac{\partial w_{ij}}{\partial x_r} \right) = 0.$$

It was in this form that Poisson structures were first introduced by Lie (1888–93), although the special case in which $M = \mathbb{R}^{2n}$ with bracket

$$\{f,g\} = \sum_{i,j=1}^{n} \left(\frac{\partial f}{\partial x_i} \frac{\partial g}{\partial x_{i+n}} - \frac{\partial g}{\partial x_i} \frac{\partial f}{\partial x_{i+n}} \right)$$

was introduced by S. D. Poisson early in the 19th century.

B Functorial properties A smooth map $F : N \to M$ between Poisson manifolds is called a *Poisson map* if it preserves the Poisson brackets of M and N, i.e. if

$$(4) \qquad \{f_1, f_2\}_M \circ F = \{f_1 \circ F, f_2 \circ F\}_N$$

for all f_1, $f_2 \in C^\infty(M)$. Equivalently, F should take the Poisson bivector of N to that of M:

$$(5) \qquad (F'_x \otimes F'_x)(w_N)_x = (w_M)_{F(x)},$$

where F'_x is the tangent linear map of F at the point $x \in M$.

A submanifold S of dimension s of a Poisson manifold M has a Poisson structure such that the inclusion map $S \hookrightarrow M$ is a Poisson map if and only if, at each point $x \in S$, the Poisson bivector of M lies in the subspace $(T_x S)^{\otimes 2}$ of $(T_x M)^{\otimes 2}$, and in this case S is said to be a *Poisson submanifold* of M (the equivalence is easily seen by writing out the equation

$$\{f_1|_S, f_2|_S\}_S = \{f_1, f_2\}_M|_S$$

in coordinates $(q_1, \ldots, q_m)$ such that S is given locally by the equations $q_{s+1} = \cdots = q_m = 0$).

If M and N are Poisson manifolds, their product $M \times N$ is a Poisson manifold in a natural way: for f_1, $f_2 \in C^\infty(M \times N)$, $x \in M$, $y \in N$, the *product Poisson structure* is given by

$$(6) \quad \{f_1, f_2\}_{M \times N}(x, y) = \{f_1(\,, y), f_2(\,, y)\}_M(x) + \{f_1(x,\,), f_2(x,\,)\}_N(y).$$

Examples of Poisson submanifolds are given in 1.1.2 and 1.1.3. Product Poisson structures will be used in the next section.

C Symplectic leaves Every manifold admits the *trivial* Poisson structure $\{f, g\} \equiv 0$, for which the map $B = 0$. At the opposite extreme, M is said to be *symplectic* if B is an isomorphism of bundles, or, equivalently, if the Poisson bivector w is everywhere non-degenerate as a bilinear form on $T^* M$. In this case, $\omega = (B \otimes B)^{-1}(w)$ is a non-degenerate 2-form on M, and it is easy to check that the Jacobi identity for $\{\,,\,\}$ is equivalent to ω being closed.

Conversely, a non-degenerate 2-form ω on M defines a bundle isomorphism $B' : TM \to T^*M$ by $B'(X) = \omega(X, \ )$, and then $w = (B' \otimes B')^{-1}(\omega)$ is the Poisson bivector of a Poisson structure on M. In general, the *rank* of a Poisson structure at a point $x \in M$ is defined to be the rank of the linear map $B_x : T_x^*M \to T_xM$. Thus, a Poisson structure is symplectic precisely when its rank is everywhere equal to the dimension of M.

The first non-trivial result in this chapter asserts that every Poisson manifold is a disjoint union of symplectic manifolds in a canonical way.

DEFINITION–PROPOSITION 1.1.2 *Let M be a Poisson manifold. Define an equivalence relation $\sim$ on M as follows: $x \sim y$ if and only if there is a piecewise smooth curve in M joining x to y, each smooth segment of which is part of an integral curve of a hamiltonian vector field on M. Then, the equivalence classes of $\sim$ are (immersed) symplectic Poisson submanifolds of M, called its* symplectic leaves *(see Fig. 1).* ∎

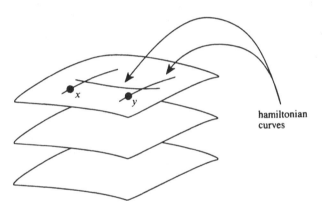

Fig.1 Symplectic leaves

Example 1.1.3 Let $\mathfrak{g}$ be any finite-dimensional (real) Lie algebra with Lie bracket $[\ ,\ ]$. Since $\mathfrak{g}^*$ is a vector space, the tangent space of $\mathfrak{g}^*$ at any point can be identified canonically with $\mathfrak{g}^*$ itself, so the differential df of any smooth function f on $\mathfrak{g}^*$ is a map $\mathfrak{g}^* \to (\mathfrak{g}^*)^* \cong \mathfrak{g}$; thus we can set

(7) $$\{f_1, f_2\}(\xi) = \langle [(df_1)_\xi, (df_2)_\xi], \xi \rangle$$

for all $\xi \in \mathfrak{g}^*$.

If $x_1, \ldots, x_m$ is a basis of $\mathfrak{g}$, we may regard $x_1, \ldots, x_m$ as (global) coordinates on $\mathfrak{g}^*$. It is easy to see that

$$\{f_1, f_2\}(\xi) = \sum_{i,j} w_{ij}(\xi) \frac{\partial f_1}{\partial x_i} \frac{\partial f_2}{\partial x_j},$$

where
$$w_{ij}(\xi) = \sum_k c_{ij}^k \langle \xi, x_k \rangle$$

and the c_{ij}^k are the structure constants of $\mathfrak{g}$:

$$[x_i, x_j] = \sum_k c_{ij}^k x_k.$$

Obviously, $w_{ij} = -w_{ji}$; on the other hand,

$$\sum_r \left(w_{ri} \frac{\partial w_{jk}}{\partial x_r} + w_{rj} \frac{\partial w_{ki}}{\partial x_r} + w_{rk} \frac{\partial w_{ij}}{\partial x_r} \right) = \sum_s \sum_r (c_{jk}^r c_{ri}^s + c_{ki}^r c_{rj}^s + c_{ij}^r c_{rk}^s) \langle \xi, x_s \rangle$$

is zero because the sum over r vanishes by the Jacobi identity in $\mathfrak{g}$. Thus, $\{\,,\,\}$ is a Poisson structure on $\mathfrak{g}^*$: it is called the *Lie–Poisson structure*.

To determine the symplectic leaves of $\mathfrak{g}^*$, let G be a connected Lie group with Lie algebra $\mathfrak{g}$. We denote by Ad (resp. Ad*) the adjoint (resp. coadjoint) representation of G on $\mathfrak{g}$ (resp. on $\mathfrak{g}^*$), and by ad (resp. ad*) the corresponding actions of $\mathfrak{g}$. With this notation, the integral curve η_i of the hamiltonian vector field X_{x_i}, associated to the function x_i on $\mathfrak{g}^*$, and passing through $\xi \in \mathfrak{g}^*$ at time $t = 0$, is given by $\eta_i(t) = \mathrm{Ad}^*(\exp(-tx_i))(\xi)$. Indeed, for any $x \in \mathfrak{g}$, we have

$$X_{x_i}(x)(\xi) = \langle [x, x_i], \xi \rangle = -\langle x, \mathrm{ad}_{x_i}^*(\xi) \rangle.$$

Thus, the integral curve of X_{x_i} passing through ξ at $t = 0$ is the solution of the differential equation

$$\frac{d\eta_i}{dt} = -\mathrm{ad}_{x_i}^*(\eta_i)$$

with initial condition $\eta_i(0) = \xi$, from which our assertion follows. Finally, since the image of the exponential map $\exp : \mathfrak{g} \to G$ generates G as a group, and since the hamiltonian vector field X_f associated to any smooth function $f \in C^\infty(\mathfrak{g}^*)$ is a linear combination of the X_{x_i},

$$X_f = \sum_{i=1}^m \frac{\partial f}{\partial x_i} X_{x_i},$$

it follows that *the symplectic leaves of $\mathfrak{g}^*$ are precisely the orbits of the coadjoint action of G*. In particular, we recover the well-known fact that the coadjoint orbits of any Lie group are symplectic manifolds. $\Diamond$

Remark A *complex Poisson manifold* is a Poisson manifold which has a complex structure J which preserves the Poisson bivector w of M:

$$(J \otimes J)(w) = w.$$

The symplectic leaves of a complex Poisson manifold M are complex submanifolds of M, and their complex and symplectic structures are compatible, i.e. if ω is the symplectic form, we have $\omega(JX, JY) = \omega(X, Y)$ for all tangent vectors X, Y. This means that each symplectic leaf is a pseudo-Kähler manifold. $\triangle$

1.2 Poisson–Lie groups

We now turn to the particular class of Poisson manifolds of interest to us.

A Definitions A Poisson–Lie group is a Lie group and a Poisson manifold, the two structures being compatible in the following sense.

DEFINITION 1.2.1 *(a) A Poisson–Lie group is a Lie group G which has a Poisson structure $\{\,,\,\}$ such that the multiplication map $\mu : G \times G \to G$ ($\mu(g_1, g_2) = g_1 g_2$) of G is a Poisson map (of course, $G \times G$ is given the product Poisson structure).*

(b) A homomorphism $\Phi : G \to H$ of Poisson–Lie groups is a homomorphism of Lie groups which is also a Poisson map.

Of course, the trivial Poisson structure on any Lie group is a Poisson–Lie group structure. For a non-trivial example, let $\mathfrak{g}$ be any Lie algebra and give its dual $\mathfrak{g}^*$ the Lie–Poisson structure described in 1.1.3. Then $\mathfrak{g}^*$, regarded as an abelian Lie group under addition, is a Poisson–Lie group. Moreover, if $\mathfrak{h}$ is another Lie algebra, the homomorphisms of Poisson–Lie groups $\mathfrak{h}^* \to \mathfrak{g}^*$ are exactly the duals of the Lie algebra homomorphisms $\mathfrak{g} \to \mathfrak{h}$. (Although these assertions can be checked directly, they follow immediately from 1.4.1 below.)

Warning If G is a Poisson–Lie group, the inversion map $\iota : G \to G$ ($\iota(g) = g^{-1}$) of G is *not* a Poisson map unless G has the trivial Poisson structure. In fact, it follows from 1.2.1 that it is an *anti*-Poisson map, i.e.

$$\{f_1 \circ \iota, f_2 \circ \iota\} = -\{f_1, f_2\} \circ \iota$$

for all f_1, $f_2 \in C^\infty(G)$.

Moreover, left translations $L_g : G \to G$ and right translations $R_g : G \to G$ by elements $g \in G$ are *not* Poisson maps in general. $\blacklozenge$

The condition that a Poisson structure on a Lie group G should be compatible with the group structure in the sense of 1.2.1 can be expressed simply in terms of the Poisson bivector w of G. In fact, recalling the definition of the product Poisson structure in (6), the condition is

$$\{f_1, f_2\}(gg') = \{f_1 \circ L_g, f_2 \circ L_g\}(g') + \{f_1 \circ R_{g'}, f_2 \circ R_{g'}\}(g)$$

for all f_1, $f_2 \in C^\infty(G)$, g, $g' \in G$. This is equivalent to

$$\langle w_{gg'}, (df_1)_{gg'} \otimes (df_2)_{gg'} \rangle = \langle w_{g'}, d(f_1 \circ L_g)_{g'} \otimes d(f_2 \circ L_g)_{g'} \rangle$$
$$+ \langle w_g, d(f_1 \circ R_{g'})_g \otimes d(f_2 \circ R_{g'})_g \rangle,$$

which in turn is equivalent to

$$(8) \qquad w_{gg'} = ((L_g)'_{g'} \otimes (L_g)'_{g'})(w_{g'}) + ((R_{g'})'_g \otimes (R_{g'})'_g)(w_g).$$

We have proved

PROPOSITION 1.2.2 *A Poisson structure on a Lie group G is a Poisson–Lie group structure if and only if, for all g, $g' \in G$, the value of its Poisson bivector at gg' is the sum of the left translate by g of its value at g' and the right translate by g' of its value at g.* ∎

Taking g and g' equal to the identity element e of G, we deduce

COROLLARY 1.2.3 *The rank of the Poisson structure of a Poisson–Lie group is zero at the identity element of the group. In particular, the Poisson structure of a Poisson–Lie group is never symplectic.* ∎

B Poisson homogeneous spaces A Lie subgroup H of a Poisson–Lie group G is said to be a *Poisson–Lie subgroup* of G if H is also a Poisson submanifold of G. In this case, it is clear that H is itself a Poisson–Lie group and that the inclusion $H \hookrightarrow G$ is a homomorphism of Poisson–Lie groups. The quotient G/H is then called a *Poisson homogeneous space*.

PROPOSITION 1.2.4 *Let G/H be a Poisson homogeneous space. Then G/H can be given a unique Poisson manifold structure such that the natural projection map $\pi : G \to G/H$ is a Poisson map.*

PROOF Define a bivector $w_{G/H}$ on G/H by

$$(9) \qquad (w_{G/H})_{gH} = (\pi'_g \otimes \pi'_g)(w_G)_g$$

where $g \in G$ and w_G is the Poisson bivector of G. To show that $w_{G/H}$ is well defined, note that

$$(10) \qquad (w_G)_{gh} = ((L_g)'_h \otimes (L_g)'_h)(w_G)_h + ((R_h)'_g \otimes (R_h)'_g)(w_G)_g$$

by 1.2.2, and that

$$\pi \circ L_g = L_g \circ \pi \qquad \text{and} \qquad \pi \circ R_h = \pi$$

for $g \in G$, $h \in H$ (we use L_g to denote both left translation by $g \in G$ on G and on G/H). Applying $\pi'_{gh} \otimes \pi'_{gh}$ to the second term on the right-hand side of (10) therefore gives $(w_G)_g$, while applying it to the first term gives

$$((L_g)'_{eH} \otimes (L_g)'_{eH})(\pi'_h \otimes \pi'_h)(w_G)_h,$$

which vanishes because $(w_G)_h \in T_h H \otimes T_h H$. This proves that

$$(w_{G/H})_{ghH} = (w_{G/H})_{gH}.$$

Define a bracket $\{\ ,\ \}_{G/H}$ on $C^\infty(G/H)$ by

$$\{f_1, f_2\}_{G/H}(gH) = \langle (w_{G/H})_{gH}, (df_1)_{gH} \otimes (df_2)_{gH} \rangle.$$

From the definition of $w_{G/H}$, one easily shows that

$$\{f_1, f_2\}_{G/H} \circ \pi = \{f_1 \circ \pi, f_2 \circ \pi\}_G.$$

From this, and the fact that π is a submersion, it follows immediately that $\{\ ,\ \}_{G/H}$ satisfies the Jacobi and Leibniz identities. ∎

The following example exhibits the Hopf fibration as a Poisson homogeneous space.

Example 1.2.5 Let $G = SU_2$, the group of 2×2 unitary matrices of determinant one, and let H be the diagonal subgroup of G. The homogeneous space G/H is diffeomorphic to the Riemann sphere $\mathbb{CP}^1$.

Introduce the elements

$$E = \begin{pmatrix} 0 & 1 \\ -1 & 0 \end{pmatrix}, \qquad F = \begin{pmatrix} 0 & \sqrt{-1} \\ \sqrt{-1} & 0 \end{pmatrix}$$

of the Lie algebra su_2, consisting of the 2×2 skew-hermitian matrices of trace zero. Define a bivector on SU_2 by

$$(w_G)_g = ((L_g)'_e \otimes (L_g)'_e - (R_g)'_e \otimes (R_g)'_e)(r),$$

where $r = E \otimes F - F \otimes E$. It follows from 2.1.2 that this is a Poisson bivector (this can also be verified directly). It is easy to check that $(\text{Ad}_g \otimes \text{Ad}_g)(r) = 0$ when $g \in H$, so that the bivector w_G vanishes on H. In particular, H is a Poisson–Lie subgroup of G.

Let us compute the induced Poisson structure on $\mathbb{CP}^1$. It follows from the definition of w_G that the value of $w_{G/H}$ at gH, for any $g \in SU_2$, is the left translate by L_g of the bivector

$$(\pi'_e \otimes \pi'_e)(r - (\text{Ad}_g^{-1} \otimes \text{Ad}_g^{-1})(r))$$

in $T_{eH}(G/H) \otimes T_{eH}(G/H)$. Choose a local coordinate system (x, y) such that any point near the 'north pole' $[1, 0] \in \mathbb{C}P^1$ has homogeneous coordinates $[1, x + \sqrt{-1}y]$. Then, for

$$g = \begin{pmatrix} a & b \\ -\bar{b} & \bar{a} \end{pmatrix},$$

we find that

$$(\pi'_e \otimes \pi'_e)(r - (\mathrm{Ad}_g^{-1} \otimes \mathrm{Ad}_g^{-1})(r)) = (1 - |a|^2 + |b|^2) \left(\frac{\partial}{\partial x} \otimes \frac{\partial}{\partial y} - \frac{\partial}{\partial y} \otimes \frac{\partial}{\partial x} \right).$$

Note that this vanishes precisely when $g \in H$. It follows that the induced Poisson structure on $\mathbb{C}P^1$ has rank zero at the north pole, but is non-degenerate elsewhere. $\Diamond$

We shall return to this example in Section 1.5.

1.3 Lie bialgebras

This section is devoted to the infinitesimal analogues of Poisson–Lie groups, called Lie bialgebras.

A The Lie bialgebra of a Poisson–Lie group It is natural to ask what structure on the Lie algebra $\mathfrak{g}$ of a Lie group G is associated to a Poisson–Lie group structure on G. In fact, if G is a Poisson–Lie group, there is a canonical Lie algebra structure on $\mathfrak{g}^*$: if ξ_1, $\xi_2 \in \mathfrak{g}^*$, choose f_1, $f_2 \in C^\infty(G)$ with $(df_i)_e = \xi_i$, and set

$$(11) \qquad [\xi_1, \xi_2]_{\mathfrak{g}^*} = (d\{f_1, f_2\})_e.$$

Assuming for a moment that $[\ ,\ ]_{\mathfrak{g}^*}$ is well defined, it is obviously skew-symmetric and satisfies the Jacobi identity because $\{\ ,\ \}$ has the same properties. That the right-hand side of (11) only depends on ξ_1, ξ_2 is a consequence of the following alternative description of $[\xi_1, \xi_2]_{\mathfrak{g}^*}$.

Let $w^R : G \to \mathfrak{g} \otimes \mathfrak{g}$ be the right translate of the Poisson bivector w of G to the identity, and let $\delta : \mathfrak{g} \to \mathfrak{g} \otimes \mathfrak{g}$ be the tangent linear map of w^R at e. Then, we have

$$(12) \qquad [\xi_1, \xi_2]_{\mathfrak{g}^*} = \delta^*(\xi_1 \otimes \xi_2).$$

In fact, for any $g \in G$,

$$\{f_1, f_2\}(g) = \langle w^R(g), ((R_g)'_e \otimes (R_g)'_e)^*((df_1)_g \otimes (df_2)_g) \rangle.$$

Differentiating this equation at $g = e$ in the direction $X \in \mathfrak{g}$, and using the fact that $w^R(e) = 0$ by 1.2.3, gives

$$\langle X, d\{f_1, f_2\}_e \rangle = \langle \delta(X), \xi_1 \otimes \xi_2 \rangle.$$

Furthermore, in terms of w^R, (8) becomes

$$w^R(gg') = (\mathrm{Ad}_g \otimes \mathrm{Ad}_g)(w^R(g')) + w^R(g).$$

This means that w^R is a 1-cocycle of G with values in $\mathfrak{g} \otimes \mathfrak{g}$ (on which G acts by the adjoint representation in each factor). Its derivative δ at e is therefore a 1-cocycle of $\mathfrak{g}$ with values in $\mathfrak{g} \otimes \mathfrak{g}$, i.e.

$$(13) \qquad \delta[X, Y] = (\mathrm{ad}_X \otimes 1 + 1 \otimes \mathrm{ad}_X)\delta(Y) - (\mathrm{ad}_Y \otimes 1 + 1 \otimes \mathrm{ad}_Y)\delta(X).$$

This motivates the first part of

DEFINITION 1.3.1 *Let $\mathfrak{g}$ be a Lie algebra.*

(a) A Lie bialgebra structure *on $\mathfrak{g}$ is a skew-symmetric linear map $\delta_\mathfrak{g} : \mathfrak{g} \to \mathfrak{g} \otimes \mathfrak{g}$, called the* cocommutator, *such that*

(a1) $\delta_\mathfrak{g}^* : \mathfrak{g}^* \otimes \mathfrak{g}^* \to \mathfrak{g}^*$ is a Lie bracket on $\mathfrak{g}^*$,

(a2) $\delta_\mathfrak{g}$ is a 1-cocycle of $\mathfrak{g}$ with values in $\mathfrak{g} \otimes \mathfrak{g}$.

(b) A homomorphism of Lie bialgebras *$\varphi : \mathfrak{g} \to \mathfrak{h}$ is a homomorphism of Lie algebras such that*

$$(14) \qquad (\varphi \otimes \varphi) \circ \delta_\mathfrak{g} = \delta_\mathfrak{h} \circ \varphi.$$

Remarks [1] In this definition, and in the rest of this chapter, Lie algebras may be defined over any field of characteristic zero, except where stated or implied otherwise (for example, for Lie algebras of Lie groups, which are usually regarded as real Lie algebras).

[2] A Lie bialgebra structure is determined by the Lie algebra structures on $\mathfrak{g}$ and $\mathfrak{g}^*$. Thus, we shall sometimes denote the Lie bialgebra by the pair $(\mathfrak{g}, \mathfrak{g}^*)$, and sometimes by the pair $(\mathfrak{g}, \delta_\mathfrak{g})$.

[3] If $(\mathfrak{g}, \delta_\mathfrak{g})$ is a Lie bialgebra, so is $(\mathfrak{g}, k\delta_\mathfrak{g})$ for any scalar k. In particular, taking $k = -1$ gives the *opposite* Lie bialgebra $\mathfrak{g}^{\mathrm{op}}$ of $\mathfrak{g}$.

[4] Condition (a1) makes sense even if $\mathfrak{g}$ is infinite dimensional, since $\mathfrak{g}^* \otimes \mathfrak{g}^* \subset (\mathfrak{g} \otimes \mathfrak{g})^*$.

[5] The kernel $\mathfrak{k}$ of a homomorphism of Lie bialgebras $\varphi : \mathfrak{g} \to \mathfrak{h}$ is a *Lie bialgebra ideal* of $\mathfrak{g}$, i.e. a Lie algebra ideal of $\mathfrak{g}$ such that $\delta_\mathfrak{g}(\mathfrak{k}) \subseteq \mathfrak{g} \otimes \mathfrak{k} + \mathfrak{k} \otimes \mathfrak{g}$. The quotient $\mathfrak{g}/\mathfrak{k}$ has a canonical Lie bialgebra structure such that the natural map $\mathfrak{g}/\mathfrak{k} \to \mathfrak{h}$ is an isomorphism of Lie bialgebras. The proof is obvious. $\triangle$

The motivation for part (b) of the definition derives from

THEOREM 1.3.2 *Let G be a Lie group with Lie algebra $\mathfrak{g}$. If G is a Poisson–Lie group, then $\mathfrak{g}$ has a natural Lie bialgebra structure, called the tangent Lie bialgebra of G. If $\Phi : G \to H$ is a homomorphism of Poisson–Lie groups, and $\mathfrak{h}$ is the tangent Lie bialgebra of H, the derivative of Φ at the identity element e of G is a homomorphism of Lie bialgebras $\mathfrak{g} \to \mathfrak{h}$.*

Conversely, if G is connected and simply-connected, every Lie bialgebra structure on $\mathfrak{g}$ is the tangent Lie bialgebra of a unique Poisson structure on G which makes G into a Poisson–Lie group. Moreover, if $\varphi : \mathfrak{g} \to \mathfrak{h}$ is a homomorphism of Lie bialgebras, and H is a Poisson–Lie group with tangent Lie bialgebra $\mathfrak{h}$, there is a unique homomorphism $\Phi : G \to H$ of Poisson–Lie groups with $\Phi'_e = \varphi$.

PROOF We have shown above that the Lie algebra of a Poisson–Lie group has a natural Lie bialgebra structure. The condition that a homomorphism $\Phi : G \to H$ is a Poisson map is

$$(15) \qquad (\Phi'_g \otimes \Phi'_g)(w_G)_g = (w_H)_{\Phi(g)}$$

for all $g \in G$. Using 1.2.3, we find that the derivatives of the left- and right-hand sides of this equation at $g = e$ are $(\varphi \otimes \varphi) \circ \delta_{\mathfrak{g}}$ and $\delta_{\mathfrak{h}} \circ \varphi$, respectively, whence φ satisfies (14).

For the converse, we use the fact that, if G is connected and simply-connected, every 1-cocycle of $\mathfrak{g}$ with values in $\mathfrak{g} \otimes \mathfrak{g}$ can be 'integrated' to a unique 1-cocycle of G with values in $\mathfrak{g} \otimes \mathfrak{g}$ (see Bourbaki (1972)). It follows immediately that, with these assumptions on G, every Lie bialgebra structure on $\mathfrak{g}$ comes from a unique Poisson–Lie group structure on G. Further, if $\varphi : \mathfrak{g} \to \mathfrak{h}$ is a homomorphism of Lie bialgebras, there is a unique homomorphism of Lie groups $\Phi : G \to H$ such that $\Phi'_e = \varphi$. To show that Φ is also a Poisson map, we must show that (15) holds. For this, it suffices to note that both sides define 1-cocycles of G with values in $\mathfrak{h} \otimes \mathfrak{h}$, and, by the previous argument, their associated Lie algebra cocycles are equal. ∎

Remark A *complex Poisson–Lie group* is a complex Lie group such that the tangent Lie bialgebra structure $\delta_{\mathfrak{g}} : \mathfrak{g} \to \mathfrak{g} \otimes \mathfrak{g}$ is complex linear. Theorem 1.3.2 holds as stated for complex Poisson–Lie groups and Lie bialgebras.

One must beware, however, that complex Poisson–Lie groups are *not* complex Poisson manifolds unless the Poisson bracket is identically zero! In fact, if J is the complex structure of G, differentiating the condition $(J \otimes J)(w_G)_g = (w_G)_g$ for a complex Poisson manifold at $g = e$, we see that $(J \otimes J) \circ \delta_{\mathfrak{g}} = \delta_{\mathfrak{g}}$. But, for a complex Poisson–Lie group, the left-hand side is equal to $\delta_{\mathfrak{g}} \circ J$. △

B Manin triples The definition we have given of a Lie bialgebra $(\mathfrak{g}, \mathfrak{g}^*)$ conceals the fact that $\mathfrak{g}$ and $\mathfrak{g}^*$ play a symmetric role. This is revealed by the following definition, which we shall see is equivalent to 1.3.1.

DEFINITION 1.3.3 *A Manin triple is a triple of Lie algebras* $(\mathfrak{p}, \mathfrak{p}_+, \mathfrak{p}_-)$ *together with a non-degenerate symmetric bilinear form* (,) *on* $\mathfrak{p}$ *invariant under the adjoint representation of* $\mathfrak{p}$, *such that*

(i) $\mathfrak{p}_+$ *and* $\mathfrak{p}_-$ *are Lie subalgebras of* $\mathfrak{p}$,

(ii) $\mathfrak{p} = \mathfrak{p}_+ \oplus \mathfrak{p}_-$ *as vector spaces,*

(iii) $\mathfrak{p}_+$ *and* $\mathfrak{p}_-$ *are isotropic for* (,).

PROPOSITION 1.3.4 *For any finite-dimensional Lie algebra* $\mathfrak{g}$, *there is a one-to-one correspondence between Lie bialgebra structures on* $\mathfrak{g}$ *and Manin triples* $(\mathfrak{p}, \mathfrak{p}_+, \mathfrak{p}_-)$ *such that* $\mathfrak{p}_+ = \mathfrak{g}$.

PROOF If $(\mathfrak{g}, \mathfrak{g}^*)$ is a Lie bialgebra, we set $\mathfrak{p}_+ = \mathfrak{g}$, $\mathfrak{p}_- = \mathfrak{g}^*$, and take the inner product on $\mathfrak{p} = \mathfrak{g} \oplus \mathfrak{g}^*$ determined by

$$(16) \qquad (X, \xi)_{\mathfrak{p}} = \langle X, \xi \rangle, \qquad (X, X)_{\mathfrak{p}} = (\xi, \xi)_{\mathfrak{p}} = 0$$

for $X \in \mathfrak{g}$, $\xi \in \mathfrak{g}^*$. For the converse, the inner product determines an isomorphism of vector spaces $\mathfrak{p}_- \cong \mathfrak{p}_+^*$, and hence a Lie algebra structure on $\mathfrak{p}_+^*$. Thus, both parts of the proposition follow from the following lemma. ∎

LEMMA 1.3.5 *Suppose that* $\mathfrak{g}$ *and its dual space* $\mathfrak{g}^*$ *are Lie algebras, and define an inner product on* $\mathfrak{g} \oplus \mathfrak{g}^*$ *by (16). There is a Lie algebra structure on* $\mathfrak{g} \oplus \mathfrak{g}^*$ *(direct sum of vector spaces) such that* $\mathfrak{g}$ *and* $\mathfrak{g}^*$ *are Lie subalgebras and which leaves the inner product invariant if and only if* $(\mathfrak{g}, \mathfrak{g}^*)$ *is a Lie bialgebra, and in this case the Lie algebra structure is unique.*

PROOF Choose a basis $\{X_r\}$ of $\mathfrak{g}$, and let $\{\xi^r\}$ be the dual basis of $\mathfrak{g}^*$. Suppose that

$$[X_r, X_s] = c_{rs}^t X_t, \quad [\xi^r, \xi^s] = \gamma_t^{rs} \xi^t.$$

(Throughout the proof, summation over repeated indices is understood.) To determine $[\xi^r, X_s]$, we use the fact that the bilinear form (,) on $\mathfrak{g} \oplus \mathfrak{g}^*$ is invariant under $\mathfrak{g}$ and $\mathfrak{g}^*$. This implies that

$$([\xi^r, X_s], \xi^t) = -(X_s, [\xi^r, \xi^t]),$$

so that the $\mathfrak{g}$-component of $[\xi^r, X_s]$ is $-\gamma_s^{rt} X_t$. Similarly, the $\mathfrak{g}^*$-component is $c_{st}^r \xi^t$, so that

$$[\xi^r, X_s] = c_{st}^r \xi^t - \gamma_s^{rt} X_t.$$

Note that the bracket on $\mathfrak{g} \oplus \mathfrak{g}^*$ is uniquely determined. We must prove that it satisfies the Jacobi identity if and only if the dual $\delta_{\mathfrak{g}}$ of the Lie bracket on $\mathfrak{g}^*$ is a 1-cocycle on $\mathfrak{g}$ with values in $\mathfrak{g} \otimes \mathfrak{g}$. Clearly, $\delta_{\mathfrak{g}}(X_r) = \gamma_r^{st} X_s \otimes X_t$. Hence, the cocyle condition

$$\delta_{\mathfrak{g}}([X_r, X_s]) = X_r . \delta_{\mathfrak{g}}(X_s) - X_s . \delta_{\mathfrak{g}}(X_r)$$

can be written

$$c_{rs}^t \gamma_t^{ab} X_a \otimes X_b = \gamma_s^{pq} X_r.(X_p \otimes X_q) - \gamma_r^{pq} X_s.(X_p \otimes X_q)$$
$$= \gamma_s^{pq}(c_{rp}^a X_a \otimes X_q + c_{rq}^b X_p \otimes X_b)$$
$$- \gamma_r^{pq}(c_{sp}^a X_a \otimes X_q + c_{sq}^b X_p \otimes X_b).$$

Equating coefficients of $X_a \otimes X_b$, the cocycle condition is

$$(17) \qquad c_{rs}^t \gamma_t^{ab} = c_{rp}^a \gamma_s^{pb} + c_{rq}^b \gamma_s^{aq} - c_{sp}^a \gamma_r^{pb} - c_{sq}^b \gamma_r^{aq}.$$

On the other hand, to verify the Jacobi identity for the bracket on $\mathfrak{g} \oplus \mathfrak{g}^*$, we must show that

$$[\xi^a, [X_r, X_s]] + [X_s, [\xi^a, X_r]] + [X_r, [X_s, \xi^a]] = 0,$$

together with a similar relation involving two ξ's and one X. Expanding the left-hand side, we get

$$c_{rs}^t(c_{tp}^a \xi^p - \gamma_t^{ap} X_p) - c_{rq}^a(c_{sp}^q \xi^p - \gamma_s^{qp} X_p) - \gamma_r^{aq} c_{sq}^p X_p$$
$$+ \gamma_r^{aq} c_{sq}^p X_p + c_{sq}^a(c_{rp}^q \xi^p - \gamma_r^{qp} X_p) + \gamma_s^{aq} c_{rq}^p X_p = 0.$$

Equating coefficients of ξ^p gives a relation equivalent to the Jacobi identity in $\mathfrak{g}$; equating coefficients of X_p gives a relation which is easily seen, on suitably re-naming the indices, to be equivalent to (17). Similarly, the other Jacobi relation in $\mathfrak{g} \oplus \mathfrak{g}^*$ is equivalent to (17) together with the Jacobi identity in $\mathfrak{g}^*$. ∎

Remark If $\dim(\mathfrak{g}) = \infty$, the formulations in 1.3.1 and 1.3.3 are not equivalent. If $(\mathfrak{p}, \mathfrak{p}_+, \mathfrak{p}_-)$ is a Manin triple with $\mathfrak{p}_+ = \mathfrak{g}$, the Lie bracket on $\mathfrak{p}_-$ induces a map $\mathfrak{p}_-^* \to (\mathfrak{p}_- \otimes \mathfrak{p}_-)^*$. This gives a Lie bialgebra structure on $\mathfrak{g}$ provided that it maps $\mathfrak{g} \subset \mathfrak{p}_-^*$ into $\mathfrak{g} \otimes \mathfrak{g} \subset \mathfrak{p}_-^* \otimes \mathfrak{p}_-^* \subset (\mathfrak{p}_- \otimes \mathfrak{p}_-)^*$. △

C Examples The definition in terms of Manin triples is very convenient for constructing examples of Lie bialgebras.

Example 1.3.6 Let G be a Lie group with the trivial Poisson structure. Its tangent Lie bialgebra $(\mathfrak{g}, \mathfrak{g}^*)$ is obtained by giving $\mathfrak{g}^*$ the zero Lie bracket.

Dually, let $\mathfrak{g}$ be a finite-dimensional abelian Lie algebra. Then, the Lie bialgebra structures on $\mathfrak{g}$ are in one-to-one correspondence with the Lie algebra structures on $\mathfrak{g}^*$. ◇

Example 1.3.7 Let $\mathfrak{g}$ be the two-dimensional non-abelian Lie algebra. Then, *every* skew-symmetric linear map $\delta : \mathfrak{g} \to \mathfrak{g} \otimes \mathfrak{g}$ is a Lie bialgebra structure on

$\mathfrak{g}$, but, up to isomorphism and scalar multiples, there are only two distinct non-trivial structures. Indeed, one can choose a basis $\{X, Y\}$ of $\mathfrak{g}$ such that $[X, Y] = X$; then $\delta(X) = \alpha X \wedge Y$, $\delta(Y) = \beta X \wedge Y$ for some scalars α, β ($X \wedge Y$ means $X \otimes Y - Y \otimes X$). If $\alpha \neq 0$, setting $X' = X$, $Y' = Y - \frac{\beta}{\alpha}X$ gives $\delta(X') = \alpha X' \wedge Y'$, $\delta(Y') = 0$ and $[X', Y'] = X'$. It is easy to check that in both cases we do indeed have a Lie bialgebra structure. Drinfel'd (1987) calls the second structure the 'non-degenerate' case, but this should not be confused with our use of this term in the next chapter. $\Diamond$

Example 1.3.8 Let $\mathfrak{g} = \mathfrak{g}(A)$ be the Kac–Moody algebra associated to a symmetrizable generalized Cartan matrix A. We use the notation in the Appendix. There is a unique Lie bialgebra structure on $\mathfrak{g}$, called the *standard structure*, with cocommutator

(18) $$\delta(H_i) = 0, \quad \delta(D_i) = 0, \quad \delta(X_i^{\pm}) = d_i\, X_i^{\pm} \wedge H_i.$$

Uniqueness is immediate from the cocycle property of δ and the fact that the H_i, D_i and $X_i^{\pm}$ generate $\mathfrak{g}$ as a Lie algebra. To prove the existence of this Lie bialgebra structure, we use Manin triples.

Let $\mathfrak{p} = \mathfrak{g} \oplus \mathfrak{g}$ (direct sum of Lie algebras), let $\mathfrak{p}_+$ be the diagonal subalgebra of $\mathfrak{p}$, let

$$\mathfrak{p}_- = \{(x, y) \in \mathfrak{b}_- \oplus \mathfrak{b}_+ \mid \mathfrak{h}\text{--component of } x + y \text{ is zero}\},$$

and define an inner product on $\mathfrak{p}$ in terms of the standard inner product $(\ ,\ )_{\mathfrak{g}}$ on $\mathfrak{g}$ by

$$((x, y), (x', y'))_{\mathfrak{p}} = (x, x')_{\mathfrak{g}} - (y, y')_{\mathfrak{g}}$$

for $x, x', y, y' \in \mathfrak{g}$. Using the facts that $\mathfrak{g} = \mathfrak{b}_+ + \mathfrak{b}_-$, $\mathfrak{h} = \mathfrak{b}_+ \cap \mathfrak{b}_-$ and $\mathfrak{n}_{\pm}$ are isotropic for $(\ ,\ )_{\mathfrak{g}}$, it is easy to check that $(\mathfrak{p}, \mathfrak{p}_+, \mathfrak{p}_-)$ is a Manin triple. Thus, if $\dim(\mathfrak{g}) < \infty$, it follows that $\mathfrak{g} \cong \mathfrak{p}_+$ has a Lie bialgebra structure. In fact, the following computation of the cocommutator δ shows, in view of the remark at the end of Subsection 1.3B, that this is true even if $\dim(\mathfrak{g}) = \infty$.

The first equation in (18) follows from the equation

$$\langle \delta(H_i, H_i), (X, Y) \otimes (X', Y') \rangle = -(H_i, [X, X']) + (H_i, [Y, Y']) = 0$$

for $X, X' \in \mathfrak{b}_+$, $Y, Y' \in \mathfrak{b}_-$, which in turn follows from the fact that $(\mathfrak{h}, \mathfrak{n}_{\pm}) = 0$. The proof of the second equation is similar. Finally, we must have $\delta(X_i^+) = X_i^+ \wedge H$ for some $H \in \mathfrak{h}$, since the only elements of $\mathfrak{g}$ which have non-zero inner product with X_i^+ are multiples of X_i^-, and since the equation $[X, X'] = X_i^+$ can be satisfied with $X, X' \in \mathfrak{b}^+$ only if $X, X' \in \mathbb{C}.X_i^+ + \mathfrak{h}$. To determine H, we compute both sides of the equation

$$(\delta(X_i^+, X_i^+), (0, X_i^-) \otimes (H_j, -H_j))_{\mathfrak{p} \otimes \mathfrak{p}} = ((X_i^+, X_i^+), [(0, X_i^-), (H_j, -H_j)])_{\mathfrak{p}}.$$

This gives
$$2d_i^{-1}(H, H_j) = d_i^{-1}a_{ji},$$
from which it follows that $H = \frac{1}{2}d_iH_i$. Thus, the standard structure can be obtained from the Manin triple $(\mathfrak{p}, \mathfrak{p}_+, \mathfrak{p}_-)$ by multiplying the inner product on $\mathfrak{p}$ by two.

As we shall see in the next chapter, these are not the only Lie bialgebra structures on $\mathfrak{g}$ (up to isomorphism and scalar multiples), even when $\mathfrak{g} = sl_2(\mathbb{C})$. They are characterized by the fact that the two-dimensional Lie subalgebras $\mathfrak{b}_i^\pm$ of $\mathfrak{g}$ spanned by $H_i, X_i^\pm$, equipped with the non-degenerate Lie bialgebra structure of 1.3.7 (suitably normalized), are sub-Lie bialgebras of $\mathfrak{g}$ (i.e. the inclusions $\mathfrak{b}_i^\pm \hookrightarrow \mathfrak{g}$ are homomorphisms of Lie bialgebras). $\diamondsuit$

Example 1.3.9 Let $\mathfrak{a}$ be a finite-dimensional complex simple Lie algebra, and set $\mathfrak{g} = \mathfrak{a}[u] = \mathfrak{a} \underset{\mathbb{C}}{\otimes} \mathbb{C}[u]$, where u is an indeterminate. The Lie bracket on $\mathfrak{g}$ is defined pointwise (viewing $\mathfrak{g}$ as the space of polynomial maps $\mathbb{C} \to \mathfrak{a}$). Fix an invariant inner product $(\ ,\ )_\mathfrak{a}$ on $\mathfrak{a}$, and let $t \in \mathfrak{a} \otimes \mathfrak{a}$ be the Casimir element, i.e. t corresponds to the identity element $\mathrm{id} \in \mathrm{End}(\mathfrak{a}, \mathfrak{a}) \cong \mathfrak{a}^* \otimes \mathfrak{a}$ under the isomorphism $\mathfrak{a}^* \cong \mathfrak{a}$ induced by $(\ ,\ )_\mathfrak{a}$. Define $\delta : \mathfrak{g} \to \mathfrak{g} \otimes \mathfrak{g}$ by

$$(19) \qquad \delta(f)(u,v) = (\mathrm{ad}_{f(u)} \otimes 1 + 1 \otimes \mathrm{ad}_{f(v)})\left(\frac{t}{u-v}\right).$$

Note that, although the right-hand side of (19) appears to be a rational function of u and v with a pole along $u = v$, it is actually a polynomial because
$$(\mathrm{ad}_X \otimes 1 + 1 \otimes \mathrm{ad}_X)(t) = 0$$
for all $X \in \mathfrak{a}$ (this is equivalent to the $\mathfrak{a}$-invariance of $(\ ,\ )_\mathfrak{a}$).

The corresponding Manin triple is $(\mathfrak{a}((u^{-1})), \mathfrak{a}[u], \mathfrak{a}[[u^{-1}]])$, i.e. $\mathfrak{p}_+ = \mathfrak{g}$, $\mathfrak{p}_-$ is the space of formal power series in u^{-1} with coefficients in $\mathfrak{a}$, and $\mathfrak{p} = \mathfrak{p}_+ \oplus \mathfrak{p}_-$. The inner product on $\mathfrak{p}$ is

$$(f,g)_\mathfrak{p} = -\mathrm{res}_0(f(u), g(u))_\mathfrak{a},$$

where 'res$_0$' means the u^{-1} coefficient.

This example is related to the quantum groups called Yangians, which we shall study in Chapter 12. $\diamondsuit$

Example 1.3.10 Let $\mathfrak{g}$ be a Lie algebra with non-trivial centre $\mathfrak{z}$, and assume that $[\mathfrak{g}, [\mathfrak{g}, \mathfrak{g}]] \neq 0$. Then, $\mathfrak{g}$ has a family of non-trivial Lie bialgebra structures given by

$$\delta_{Z,Y}(X) = Z \wedge [X,Y],$$

for any $Y \in \mathfrak{g}$, $Z \in \mathfrak{z}$. It is straightforward to check that this does define a Lie bialgebra structure on $\mathfrak{g}$ (this is most easily seen using 2.1.2).

For example, we can take $\mathfrak{g}$ to be the *Virasoro algebra* Vir, the central extension (with one-dimensional centre) of the Lie algebra $\text{Vect}(S^1)$ of smooth complex vector fields on the circle S^1 defined by the 2-cocycle

$$\omega\left(\xi(\theta)\frac{d}{d\theta}, \eta(\theta)\frac{d}{d\theta}\right) = \frac{\sqrt{-1}}{12}\int_0^{2\pi}(\xi'''(\theta) + \xi'(\theta))\eta(\theta)d\theta.$$

This means that $Vir = \text{Vect}(S^1) \oplus \mathbb{C}.c$ as a vector space, where the element c is central and

$$\left[\xi(\theta)\frac{d}{d\theta}, \eta(\theta)\frac{d}{d\theta}\right] = (\xi(\theta)\eta'(\theta) - \xi'(\theta)\eta(\theta))\frac{d}{d\theta} + \omega\left(\xi\frac{d}{d\theta}, \eta\frac{d}{d\theta}\right)c. \quad \diamond$$

Example 1.3.11 Let ψDO be the vector space of formal pseudo-differential operators of the form $P = \sum_{k=-\infty}^{\text{finite}} p_k(\theta)D^k$, where the p_k lie in the space $C^\infty(S^1)$ of smooth functions on the circle and $D = d/d\theta$. The residue of P is defined by $\text{res}(P) = p_{-1}$.

Leibniz's rule provides an associative algebra structure on ψDO: from $D \circ f = df/d\theta + f \circ D$ (where f and $df/d\theta \in C^\infty(S^1)$ are regarded as multiplication operators), we obtain $D^{-1} \circ f = f \circ D^{-1} - D^{-1} \circ \frac{df}{d\theta} \circ D^{-1}$, and hence by iteration

$$D^{-1} \circ f = \sum_{k=0}^{\infty}(-1)^k\frac{d^k f}{dx^k}D^{-k-1}.$$

We regard ψDO as a Lie algebra in the usual way: $[P, Q] = P \circ Q - Q \circ P$. It is clear that the subspaces DO of differential operators and INT of integral operators (i.e. those formal series involving no negative powers of D, or only negative powers of D, respectively) are subalgebras of ψDO.

There is a derivation $\mathcal{D}$ of ψDO such that $\mathcal{D}(D^k) = 0$ for all k and

$$\mathcal{D}(f) = \sum_{k=1}^{\infty}\frac{(-1)^{k+1}}{k}\frac{d^k f}{d\theta^k}D^{-k}.$$

Formally, the right-hand side of this equation is obtained by 'integrating $[D^{-1}, f]$ with respect to D', so $\mathcal{D}$ may be thought of as 'log(D)'.

The formula

$$\omega(P, Q) = \int_0^{2\pi}\text{res}\left(\mathcal{D}(P) \circ Q\right)d\theta$$

is a 2-cocycle on ψDO, and thus defines a central extension $\widetilde{\psi DO}$ of ψDO by adjoining a central element c. (This is a generalization of the central extension in the preceding example: it is easy to see that the restriction of ω to the subalgebra of ψDO consisting of the smooth vector fields, i.e. the elements of ψDO which are homogeneous of degree 1 in D, is a multiple of the cocycle considered in 1.3.10.) The derivation $\mathcal{D}$ extends to $\widetilde{\psi DO}$ by setting $\mathcal{D}(c) = 0$, and we let $\mathfrak{g}$ be the semidirect product $\widetilde{\psi DO} \rtimes \mathbb{C}.\mathcal{D}$.

Let $\mathfrak{g}_\pm$ be the subalgebras of $\mathfrak{g}$ defined by $\mathfrak{g}_+ = DO \oplus \mathbb{C}.c$, $\mathfrak{g}_- = INT \oplus \mathbb{C}.\mathcal{D}$. We then have $\mathfrak{g} = \mathfrak{g}_+ \oplus \mathfrak{g}_-$ as vector spaces. Further, the formula

$$(P_1 + \lambda_1 \mathcal{D} + \mu_1 c, \ P_2 + \lambda_2 \mathcal{D} + \mu_2 c) = 2 \int_0^{2\pi} \operatorname{res}(P_1 \circ P_2)\, d\theta + \lambda_1 \mu_2 + \lambda_2 \mu_1$$

defines a non-degenerate invariant symmetric bilinear form on $\mathfrak{g}$ for which $\mathfrak{g}_\pm$ are isotropic subspaces. Hence, $(\mathfrak{g}, \mathfrak{g}_+, \mathfrak{g}_-)$ is a Manin triple.

There is a group G_- which, formally at least, has $\mathfrak{g}_-$ as its Lie algebra. An element of G_- is a formal expression $P = (1 + \sum_{k=-\infty}^{-1} p_k(\theta) D^k) \circ D^\alpha$, where $\alpha \in \mathbb{R}$. The multiplication in G_- is defined by

$$D^\alpha \circ f = \sum_{k=0}^\infty \frac{\alpha(\alpha-1)\dots(\alpha-k+1)}{k!} \frac{d^k f}{d\theta^k} D^{\alpha-k}.$$

Taking $\alpha = 0$, we obtain a subgroup with Lie algebra INT, while the tangent vector at $D^0 = 1$ to the one-parameter group $\alpha \mapsto D^\alpha$ is formally given by $(d/d\alpha)|_{\alpha=0}(D^\alpha) = \log(D) = \mathcal{D}$.

Thus, 1.3.2 suggests that G_- should have a Poisson–Lie group structure whose tangent Lie bialgebra is $\mathfrak{g}_-$. Khesin & Zakharevich (1992) have shown that this is formally the case, and that the required Poisson structure on G_- is (a generalization of) the so-called 'second Adler–Gel'fand–Dickii' Poisson structure, which is of central importance in the theory of soliton equations. See, for example, Gel'fand & Dickii (1976) and Adler (1979). ◇

D Derivations An automorphism of a Lie bialgebra $(\mathfrak{g}, \mathfrak{g}^*)$ is a Lie algebra automorphism $\alpha : \mathfrak{g} \to \mathfrak{g}$ such that $\alpha^* : \mathfrak{g}^* \to \mathfrak{g}^*$ is also a Lie algebra automorphism. Thus, it is natural to make the following definition.

DEFINITION 1.3.12 *A* derivation *of a Lie bialgebra* $(\mathfrak{g}, \mathfrak{g}^*)$ *is a linear map* $D : \mathfrak{g} \to \mathfrak{g}$ *such that D is a derivation of* $\mathfrak{g}$ *as a Lie algebra, and D^* is a derivation of* $\mathfrak{g}^*$ *as a Lie algebra.*

If D is a derivation of a finite-dimensional Lie bialgebra $(\mathfrak{g}, \mathfrak{g}^*)$, then e^{tD} is a one-parameter group of automorphisms of $(\mathfrak{g}, \mathfrak{g}^*)$. This follows immediately from the corresponding statement for Lie algebras.

The purpose of this subsection is to observe that every Lie bialgebra has a canonical derivation (which may be zero).

PROPOSITION 1.3.13 *Let* $(\mathfrak{g}, \delta)$ *be a Lie bialgebra, and let $D : \mathfrak{g} \to \mathfrak{g}$ be the composite of the cocommutator* $\delta : \mathfrak{g} \to \mathfrak{g} \otimes \mathfrak{g}$ *and the Lie bracket* $[\ ,\ ] : \mathfrak{g} \otimes \mathfrak{g} \to \mathfrak{g}$. *Then, D is a derivation of* $(\mathfrak{g}, \delta)$.

PROOF Let x, $y \in \mathfrak{g}$ and write

$$\delta(x) = \sum_i x_i \otimes x'_i, \qquad \delta(y) = \sum_j y_j \otimes y'_j.$$

Then, using the cocyle property of δ, we have

$$
\begin{aligned}
D([x,y]) &= [\ ,\ ] \circ \delta([x,y]) \\
&= [\ ,\ ] \circ ([\delta(x), y \otimes 1 + 1 \otimes y] + [x \otimes 1 + 1 \otimes x, \delta(y)]) \\
&= \sum_i ([[x_i, y], x'_i] + [x_i, [x'_i, y]]) + \sum_j ([[x, y_j], y'_j] + [y_j, [x, y'_j]]) \\
&= \left[\sum_i [x_i, x'_i], y \right] + \left[x, \sum_j [y_j, y'_j] \right]
\end{aligned}
$$

by the Jacobi identity. The first term on the right-hand side of the last equation is $[D(x), y]$, the second $[x, D(y)]$.

This proves that D is a derivation of $\mathfrak{g}$. Applying the preceding argument to $\mathfrak{g}^*$ instead of $\mathfrak{g}$ proves that D^* is a derivation of $\mathfrak{g}^*$. ∎

Examples 1.3.14 For the trivial Lie bialgebra structure on a Lie algebra $\mathfrak{g}$, we have $D = 0$.

For the standard Lie bialgebra structure on a complex simple Lie algebra $\mathfrak{g}$, we find that $D = \mathrm{ad}_H$, where H is the element of the Cartan subalgebra $\mathfrak{h}$ of $\mathfrak{g}$ such that $\langle \alpha_i, H \rangle = -4d_i$ for all simple roots α_i ($-\frac{1}{2}H$ corresponds to the sum of the positive roots of $\mathfrak{g}$ under the isomorphism $\mathfrak{g} \cong \mathfrak{g}^*$ induced by the standard invariant bilinear form on $\mathfrak{g}$ – see the Appendix).

For the Lie bialgebra structure on the polynomial Lie bialgebra $\mathfrak{a}[u]$ defined in 1.3.9, $D = d/du$. ◊

1.4 Duals and doubles

A Duals of Lie bialgebras and Poisson–Lie groups The formulation in terms of Manin triples shows that the notion of a Lie bialgebra is self-dual. In fact, if $(\mathfrak{p}, \mathfrak{p}_+, \mathfrak{p}_-)$ is a Manin triple, it is obvious that $(\mathfrak{p}, \mathfrak{p}_-, \mathfrak{p}_+)$ is one too (with the same inner product). It follows that, if $\mathfrak{g}$ is a finite-dimensional Lie bialgebra, so is $\mathfrak{g}^*$, and that $\mathfrak{g}^{**} \cong \mathfrak{g}$ as Lie bialgebras. In view of 1.3.2, there is a contravariant functor $G \to G^*$ on the category of connected, simply-connected (finite-dimensional) Poisson–Lie groups, such that $G^{**} \cong G$, covering the functor $\mathfrak{g} \to \mathfrak{g}^*$ on the level of tangent Lie bialgebras. In general, a Poisson–Lie group G^* with tangent Lie bialgebra $\mathfrak{g}^*$ is called, following Drinfel'd, a *dual Poisson–Lie group* of G.

Here is a simple example of a dual of a Poisson–Lie group; we give some more interesting examples later in this section.

Example 1.4.1 From 1.3.6, it follows that a dual of the trivial Poisson structure on a Lie group G is given by the vector space $\mathfrak{g}^*$, regarded as an abelian Lie group under addition, with the Lie–Poisson structure of 1.1.3. ◊

B The classical double As well as showing that the notion of a Lie bialgebra is self-dual, the formulation in terms of Manin triples also shows that, if $(\mathfrak{g}, \mathfrak{g}^*)$ is a Lie bialgebra, there is a Lie algebra structure on $\mathfrak{g} \oplus \mathfrak{g}^*$ which leaves invariant the natural inner product. In fact, more is true.

PROPOSITION 1.4.2 *Let* $(\mathfrak{g}, \mathfrak{g}^*)$ *be a finite-dimensional Lie bialgebra. Then there is a canonical Lie bialgebra structure on* $\mathfrak{g} \oplus \mathfrak{g}^*$ *such that the inclusions*

$$\mathfrak{g} \hookrightarrow \mathfrak{g} \oplus \mathfrak{g}^* \hookleftarrow (\mathfrak{g}^*)^{\mathrm{op}}$$

into the two summands are homomorphisms of Lie bialgebras.

With this Lie bialgebra structure, $\mathfrak{g} \oplus \mathfrak{g}^*$ is called the *double* of $\mathfrak{g}$, and is denoted by $\mathcal{D}(\mathfrak{g})$. By 1.3.2, there is a corresponding notion of the double $\mathcal{D}(G)$ of a Poisson–Lie group G.

PROOF Let $r \in \mathfrak{g} \otimes \mathfrak{g}^* \subset \mathcal{D}(\mathfrak{g}) \otimes \mathcal{D}(\mathfrak{g})$ correspond to the identity map $\mathfrak{g} \to \mathfrak{g}$. The cocommutator of $\mathfrak{g} \oplus \mathfrak{g}^*$ is defined by

$$\delta_{\mathcal{D}}(u) = (\mathrm{ad}_u \otimes \mathrm{id} + \mathrm{id} \otimes \mathrm{ad}_u)(r)$$

for all $u \in \mathcal{D}(\mathfrak{g})$. We shall prove in 2.1.11 that this is a Lie bialgebra structure on $\mathcal{D}(\mathfrak{g})$. Let us verify that it has the stated properties.

With the notation used in the proof of 1.3.5, and using the summation convention, we have $r = X_t \otimes \xi^t$, so that

$$\begin{aligned}
\delta_{\mathcal{D}}(X_s) &= [X_s, X_t] \otimes \xi^t + X_t \otimes [X_s, \xi^t] \\
&= c_{st}^p X_p \otimes \xi^t + X_t \otimes (\gamma_s^{tp} X_p - c_{sp}^t \xi^p) \\
&= \gamma_s^{tp} X_t \otimes X_p,
\end{aligned}$$

since the remaining terms cancel after interchanging the summation indices p and t.

Comparing with the proof of 1.3.5, we see that $\delta_{\mathcal{D}}(X_s) = \delta_{\mathfrak{g}}(X_s)$, so that the inclusion $\mathfrak{g} \hookrightarrow \mathcal{D}(\mathfrak{g})$ is a homomorphism of Lie bialgebras. The proof for the inclusion $(\mathfrak{g}^*)^{\mathrm{op}} \hookrightarrow \mathcal{D}(\mathfrak{g})$ is similar. ∎

Example 1.4.3 The Lie bialgebra structure on a symmetrizable Kac–Moody algebra $\mathfrak{g}$ defined in 1.3.8 can almost be realized as a double, as follows. With the notation in 1.3.8, let $\mathfrak{q} = \mathfrak{g} \oplus \mathfrak{h}$ (direct sum of Lie algebras) and

$$\mathfrak{q}_\pm = \{(x, h) \in \mathfrak{q} \mid x \in \mathfrak{b}_\pm, \ \pm h = \mathfrak{h}\text{–component of } x\}.$$

Note that $\mathfrak{q}_\pm \cong \mathfrak{b}_\pm$. Define an inner product on $\mathfrak{q}$ by

$$((x,h),(x',h'))_\mathfrak{q} = (x,x')_\mathfrak{g} - (h,h')_\mathfrak{g}.$$

As in 1.3.8, it follows from the orthogonality properties of the inner product on $\mathfrak{g}$ that $(\mathfrak{q}, \mathfrak{q}_+, \mathfrak{q}_-)$ is a Manin triple. Thus, we have Lie bialgebra structures on $\mathfrak{q}_\pm$ and on $\mathcal{D}(\mathfrak{q}_+) \cong \mathfrak{g} \oplus \mathfrak{h}$. The cocommutator can be found by a calculation similar to that in 1.3.8: one finds that $\delta|_{\mathfrak{h} \oplus \mathfrak{h}} = 0$ and

$$\delta(X_i^+, 0) = \frac{1}{2} d_i(X_i^+, 0) \wedge (H_i, H_i), \quad \delta(X_i^-, 0) = \frac{1}{2} d_i(X_i^-, 0) \wedge (H_i, -H_i).$$

Thus, $0 \oplus \mathfrak{h}$ is a Lie bialgebra ideal of $\mathcal{D}(\mathfrak{q}_+)$, and since $\mathcal{D}(\mathfrak{q}_+)/(0 \oplus \mathfrak{h}) \cong \mathfrak{g}$ as Lie algebras, we have a Lie bialgebra structure on $\mathfrak{g}$ given by

$$\delta(X_i^\pm) = \frac{1}{2} d_i X_i^\pm \wedge H_i, \quad \delta|_\mathfrak{h} = 0.$$

Multiplying by two gives the standard structure. $\Diamond$

We shall now show that the complex simple Lie algebra $\mathfrak{g}$, regarded as a *real* Lie algebra, is exactly the double of a Lie bialgebra structure on its compact real form.

C Compact Poisson–Lie groups Let G be the complexification of a connected, simply-connected *compact* Lie group K. The Lie algebra $\mathfrak{g}$ of G is then a complex semisimple Lie algebra, and we shall assume that it is actually simple. Moreover, if $\mathfrak{t}$ is a maximal abelian subalgebra of the Lie algebra $\mathfrak{k}$ of K, then $\mathfrak{h} = \mathfrak{t} \oplus \sqrt{-1}\mathfrak{t}$ is a Cartan subalgebra of $\mathfrak{g}$, $\mathfrak{b} = \sqrt{-1}\mathfrak{t} \oplus \mathfrak{n}_+$ is a solvable (real) Lie subalgebra of $\mathfrak{g}$, and $\mathfrak{g} = \mathfrak{k} \oplus \mathfrak{b}$ as (real) vector spaces. In terms of the standard $\mathbb{C}$-basis $\{H_i\}_{i=1,\dots,n} \cup \{X_\alpha^\pm\}_{\alpha \in \Delta^+}$ of $\mathfrak{g}$ described in the Appendix, we can take $\mathfrak{k}$ to be the $\mathbb{R}$-linear span of the set

$$\{\sqrt{-1}H_i\}_{i=1,\dots,n} \cup \{E_\alpha\}_{\alpha \in \Delta^+} \cup \{F_\alpha\}_{\alpha \in \Delta^+},$$

where $E_\alpha = X_\alpha^+ - X_\alpha^-$, $F_\alpha = \sqrt{-1}(X_\alpha^+ + X_\alpha^-)$. Then, $\{\sqrt{-1}H_i\}_{i=1,\dots,n}$ is an $\mathbb{R}$-basis of $\mathfrak{t}$, and $\mathfrak{k}$ is generated by $\{\sqrt{-1}H_i, E_{\alpha_i}, F_{\alpha_i}\}_{i=1,\dots,n}$.

For example, if $K = SU_{n+1}$ then $\mathfrak{g} = sl_{n+1}(\mathbb{C})$, $\mathfrak{h}$ is the diagonal matrices in $sl_{n+1}(\mathbb{C})$, $\mathfrak{t}$ is the matrices in $\mathfrak{h}$ with purely imaginary diagonal entries, and $\mathfrak{n}_+$ is the strictly upper triangular matrices in $\mathfrak{g}$.

Returning to the general case, set $\mathfrak{p} = \mathfrak{g}$, regarded as a real Lie algebra, $\mathfrak{p}_+ = \mathfrak{k}$ and $\mathfrak{p}_- = \mathfrak{b}$. As inner product on $\mathfrak{p}$, we take

$$(x_1, x_2)_\mathfrak{p} = \Im(x_1, x_2),$$

the imaginary part of the standard inner product on $\mathfrak{g}$. Using the formulas for this inner product given in the Appendix, and the above bases, it is easy to check that

$$(\mathfrak{k}, \mathfrak{k}) \subset \mathbb{R}, \qquad (\mathfrak{n}_+, \mathfrak{n}_+) = 0, \qquad (\mathfrak{n}_+, \mathfrak{t}) = 0,$$

and it follows immediately that $(\mathfrak{p}, \mathfrak{p}_+, \mathfrak{p}_-)$ is a Manin triple. This defines Lie bialgebra structures on $\mathfrak{g}$, $\mathfrak{k}$ and $\mathfrak{b}$, and Poisson–Lie group structures on G, K and the connected subgroup B of G with Lie algebra $\mathfrak{b}$. Thus, G is a double of K and B is a dual of K.

The Lie bialgebra structure $\delta_{\mathbb{R}}$ on $\mathfrak{g}$ defined in this way can be computed explicitly by the method used to calculate the standard Lie bialgebra structure δ in 1.3.8. We find that $\delta_{\mathbb{R}} = \sqrt{-1}\delta$. The presence of the factor $\sqrt{-1}$ guarantees that this Lie bialgebra structure takes $\mathfrak{k}$ into $\mathfrak{k} \otimes \mathfrak{k}$. In fact,

$$\delta_{\mathbb{R}}(H_i) = 0, \quad \delta_{\mathbb{R}}(E_{\alpha_i}) = d_i E_{\alpha_i} \otimes \sqrt{-1}H_i, \quad \delta_{\mathbb{R}}(F_{\alpha_i}) = d_i F_{\alpha_i} \otimes \sqrt{-1}H_i.$$

Remark It is shown by Ginzburg and Weinstein (1992) that B is isomorphic as a Poisson manifold to the Lie–Poisson structure on $\mathfrak{k}^*$ defined in 1.1.3. The proof uses cohomological techniques.

Of course, B and $\mathfrak{k}^*$ are not isomorphic as Poisson–Lie groups, since $\mathfrak{k}^*$ is abelian and B is not. $\triangle$

We shall study compact Poisson–Lie groups further in Subsection 1.5C, where we shall describe their symplectic leaves.

1.5 Dressing actions and symplectic leaves

Having introduced several examples of Lie bialgebras in the preceding sections, we now study the geometrical properties of the associated Poisson–Lie groups. The main tool needed to describe their symplectic leaves is an action on the groups of their dual groups.

A Poisson actions We remarked in Subsection 1.2A that, if G is a Poisson–Lie group, left translations by elements of G do not usually preserve the Poisson structure of G. However, they do define Poisson actions of G on itself in the sense of the following definition.

DEFINITION 1.5.1 *Let G be a Poisson–Lie group and M a Poisson manifold. A left action $\lambda : G \times M \to M$ of G on M is called a Poisson action if λ is a Poisson map, $G \times M$ being given the product Poisson structure. Right Poisson actions are defined similarly.*

Thus, left (and right) translations in a Poisson–Lie group G are Poisson actions of G on itself. More generally, let H be a Poisson–Lie subgroup of G.

Then, by 1.2.4, the homogeneous space G/H has a unique Poisson structure such that the natural projection $\pi : G \to G/H$ is a Poisson map. The natural left action of G on G/H is a Poisson action. Indeed, this follows immediately from the following commutative diagram, in which the horizontal arrows are the natural actions of G on itself and on G/H, since all the arrows other than the map $G \times G/H \to G/H$ are already known to be Poisson maps:

$$
\begin{array}{ccc}
G \times G & \longrightarrow & G \\
{\scriptstyle \mathrm{id} \times \pi} \downarrow & & \downarrow {\scriptstyle \pi} \\
G \times G/H & \longrightarrow & G/H
\end{array} \quad .
$$

B Dressing transformations and symplectic leaves The Poisson actions which are of most interest to us in this section arise from the notion of the double of a Poisson–Lie group, introduced in Subsection 1.4B.

Let G be a connected Poisson–Lie group, let G^* be its connected, simply-connected dual Poisson–Lie group, and let $\mathcal{D}(G)$ be the connected, simply-connected double of G. (Without the assumption of simple-connectedness, the following constructions can be carried out locally in a neighbourhood of the identity of the groups concerned.) By 1.3.2 and 1.4.2, there are homomorphisms of Lie groups

$$G \hookrightarrow \mathcal{D}(G) \hookleftarrow G^*$$

lifting the inclusion maps

$$\mathfrak{g} \hookrightarrow \mathcal{D}(\mathfrak{g}) \hookleftarrow \mathfrak{g}^*$$

of Lie algebras. Thus, we can define a product map $G \times G^* \to \mathcal{D}(G)$, which is a diffeomorphism onto a neighbourhood of the identity in $\mathcal{D}(G)$. We shall *assume from now on that this map is a (global) diffeomorphism of $G \times G^*$ onto $\mathcal{D}(G)$*. This holds, for example, if G is compact and the image of G^* in $\mathcal{D}(G)$ is closed (see Lu & Weinstein (1990) for the simple proof). (Again, even if this hypothesis is not satisfied, the constructions we are going to make can still be carried out locally.) We shall identify $\mathcal{D}(G)$ with $G \times G^*$ as a manifold from now on, and denote by $\mathrm{pr}_G : \mathcal{D}(G) \to G$ and $\mathrm{pr}_{G^*} : \mathcal{D}(G) \to G^*$ the projections onto the two factors.

If $g \in G$ and $\gamma \in G^*$, the product $\gamma.g$ in $\mathcal{D}(G)$ can be factorized

$$\gamma.g = g^\gamma . \gamma^g$$

for some $g^\gamma \in G$, $\gamma^g \in G^*$. If $\gamma_1, \gamma_2 \in G^*$, we can compute $\gamma_1.\gamma_2.g \in \mathcal{D}(G)$ in two ways; this gives

$$g^{\gamma_1 . \gamma_2} . (\gamma_1 . \gamma_2)^g = (g^{\gamma_2})^{\gamma_1} . (\gamma_1)^{g^{\gamma_2}} . \gamma_2^g.$$

Projecting onto G shows that the map

$$\lambda : G^* \times G \to G$$

defined by

$$\lambda(g, \gamma) = g^\gamma$$

is a left action of G^* on G. (Similarly, we get a right action of G on G^*.)

DEFINITION 1.5.2 *The right action of G^* on G given by*

$$\rho(\gamma, g) = g^{(\gamma^{-1})},$$

for $g \in G$, $\gamma \in G^*$, is called the dressing action *of G^* on G.*

One of the most important geometric applications of dressing transformations is to the determination of the symplectic leaves in Poisson homogeneous spaces. We begin with the case of Poisson–Lie groups.

PROPOSITION 1.5.3 *The dressing action of G^* on G is a Poisson action whose orbits are exactly the symplectic leaves of G.* ∎

COROLLARY 1.5.4 *The symplectic leaf of G passing through $g \in G$ is the image under pr_G of the left coset $G^* \cdot g$ in $\mathcal{D}(G)$.*

PROOF This follows immediately from the proposition, together with the definition of the dressing action. ∎

The dressing action gives a simple characterization of Poisson–Lie subgroups of a given Poisson–Lie group.

PROPOSITION 1.5.5 *A closed Lie subgroup H of a Poisson–Lie group G is a Poisson–Lie subgroup of G if and only if H is invariant under the dressing action of G^* on G.* ∎

It follows that, if H is a Poisson–Lie subgroup of a Poisson–Lie group G, then the dressing action of the dual Poisson–Lie group G^* on G induces a right action of G^* on G/H. We have the following generalization of 1.5.3. See Semenov-Tian-Shansky (1985) for a proof.

THEOREM 1.5.6 *Let H be a Poisson–Lie subgroup of a Poisson–Lie group G. Then, the dressing action of the dual Poisson–Lie group G^* of G on G/H is a (right) Poisson action, and its orbits are exactly the symplectic leaves of G/H.* ∎

C Symplectic leaves in compact Poisson–Lie groups Our aim in this subsection is to apply the results on dressing transformations we have just described to determine the symplectic leaves of the compact Poisson–Lie groups K described in Subsection 1.4C, and of certain of their Poisson homogeneous spaces. We use the notation introduced in Subsection 1.4C.

Recall that the product map $K \times B \to G$ is a (global) diffeomorphism (see Helgason (1978)). Hence, by 1.5.4, the symplectic leaf of K passing through $k \in K$ is the image of the left coset Bk under the projection pr_K of $G = K \times B$ onto K. To describe this more explicitly, let $B_+ = TB = BT$ and recall the *Bruhat decomposition* of G into double cosets modulo B_+:

$$G = \coprod_{w \in W} B_+ w B_+.$$

(See Helgason (1978) again.) Here, w runs through the Weyl group $W = N(T)/T$ of K, where $N(T)$ is the normalizer of T in K (as usual, we do not distinguish between an element of $N(T)$ and its coset modulo T). Since W normalizes T, the Bruhat decomposition is equivalent to

$$G = \coprod_{\substack{w \in W \\ t \in T}} tBwB.$$

For each $w \in W$, let Σ_w be the symplectic leaf $\mathrm{pr}_K(Bw)$; it is clear that $\Sigma_w \subset BwB$. The most general B-orbit in G is of the form $tBwb$ for $t \in T$, $w \in W$ and $b \in B$, and its projection onto K is obviously $t.\Sigma_w$. Thus, we have:

PROPOSITION 1.5.7 *The symplectic leaves of K are parametrized by $W \times T$. The leaf Σ_w with parameter $(w, 1)$ is the projection of $Bw \subset G \cong K \times B$ onto K; that with parameter (w, t) is $t.\Sigma_w$. The symplectic leaf $t.\Sigma_w$ is contained in the piece $K_w = K \cap B_+ w B_+$ of the Bruhat decomposition of K.* ∎

We can give a still more explicit description of Σ_w, but we first discuss the simplest example $K = SU_2$. Since the Weyl group of SU_2 has two elements $\{\pm 1\}$, the symplectic leaves are parametrized by the disjoint union of two circles. Those associated to the identity element 1 are the single points

$$\begin{pmatrix} t & 0 \\ 0 & t^{-1} \end{pmatrix}$$

for $t \in S^1 = \{t \in \mathbb{C} \mid |t| = 1\}$; those associated to -1 are the two-dimensional discs

$$D_t = \left\{ \begin{pmatrix} z & -\overline{w} \\ w & \overline{z} \end{pmatrix} \;\middle|\; |z|^2 + |w|^2 = 1, \, \arg w = \arg t \right\}.$$

The zero-dimensional leaves form a common 'binding circle' for the 2-dimensional leaves ('pages'); see Fig. 2.

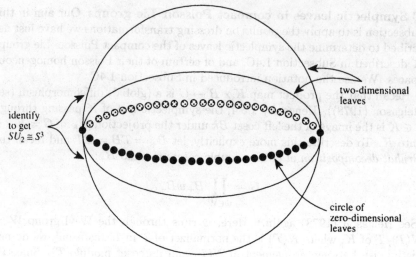

identify
to get
$SU_2 \cong S^3$

two-dimensional
leaves

circle of
zero-dimensional
leaves

Fig. 2 Symplectic leaves in SU_2

Returning to the general case, recall from the Appendix that, for each $i = 1, \ldots, n$, there is a subalgebra $\mathfrak{g}^{(i)} \hookrightarrow \mathfrak{g}$ isomorphic to $sl_2(\mathbb{C})$. This restricts to a homomorphism $su_2 \hookrightarrow \mathfrak{k}$, which lifts to a homomorphism $\varphi_i : SU_2 \to K$. As we shall see shortly, the φ_i are homomorphisms of Poisson–Lie groups (for φ_i, one should use the inner product on the Lie algebra su_2 induced from a fixed inner product on $\mathfrak{k}$). Let $\Sigma_i = \varphi_i(\Sigma_{-1})$. Then, we have

PROPOSITION 1.5.8 (a) $\Sigma_i = \Sigma_{s_i}$, where s_i is the simple reflection corresponding to the simple root α_i.
(b) If w is an arbitrary element of the Weyl group, and

$$w = s_{i_1} s_{i_2} \cdots s_{i_{\ell(w)}}$$

is an expression for w of minimal length as a product of simple reflections, then

$$\Sigma_w = \Sigma_{i_1} . \Sigma_{i_2} . \cdots . \Sigma_{i_{\ell(w)}}.$$

In particular, $\dim(\Sigma_w) = 2\ell(w)$. ∎

We now consider certain homogeneous spaces of K, specifically its coadjoint orbits. The most general coadjoint orbit is diffeomorphic to $G/P \cong K/K \cap P$, where P is some (closed) Lie subgroup of G containing B_+. We claim that $K \cap P$ is a Poisson–Lie subgroup of K. The assertion made before 1.5.8 is the special case where P is the subgroup with Lie algebra

$$\mathfrak{p}_i = \mathfrak{b}_+ \oplus \mathfrak{g}_{-\alpha_i},$$

where $\mathfrak{g}_{-\alpha_i}$ is the root space corresponding to the root $-\alpha_i$. To prove the claim, let $k \in K \cap P$ and $b \in B$, and write $bk = k'b'$ for some $k' \in K$, $b' \in B$.

Then,
$$\mathrm{pr}_K(bk) = k' = bk(b')^{-1} \in K \cap P.$$

Thus, $K \cap P$ is invariant under the dressing action of B on K and hence, by 1.5.5, is a Poisson–Lie subgroup of K.

PROPOSITION 1.5.9 *Every coadjoint orbit $\mathcal{O}$ of K has a canonical Poisson structure such that the natural map $K \to \mathcal{O}$ is a Poisson map. Moreover, if $\mathcal{O} \cong G/P$, where P is a Lie subgroup of G containing B_+, the decomposition of $\mathcal{O}$ into its symplectic leaves coincides with the Schubert cell decomposition of G/P:*
$$G/P = \bigcup_{w \in W/W_P} BwP/P,$$

where W_P is the Weyl group of P.

PROOF The first part follows immediately from 1.5.6. For the second part, we simply recall that the Bruhat decomposition of G induces the following decomposition of G/P:
$$G/P = \bigcup_{w \in W/W_P} B_+ wP/P = \bigcup_{w \in W/W_P} BTwP/P = \bigcup_{w \in W/W_P} BwP/P,$$

and that the pieces BwP/P of this decomposition are topological cells (see Helgason (1978)). ∎

Warning These Poisson structures on the coadjoint orbits are *not* symplectic, in general. In particular, they are not the same as the (symplectic) Poisson structures described in 1.1.3. Khoroshkin, Radul & Rubtsov (1993) show that the Poisson structures given by 1.1.3 and 1.5.9 are *compatible*, in the sense that any linear combination of them is also a Poisson structure, if and only if $\mathcal{O}$ is a hermitian symmetric space. ◆

To illustrate 1.5.9, we return to the example $K = SU_2$. Up to diffeomorphism, there are two types of coadjoint orbit, namely single points and 2-spheres. Let $\mathcal{O}$ be any 2-sphere orbit. The projection map $\pi : SU_2 \to \mathcal{O} \cong SU_2/T$ is the Hopf fibration, (cf. 1.2.5). There are two symplectic leaves in $\mathcal{O}$, a single point, say N (the 'north pole'), and the complementary 2-disc $\mathcal{O} \backslash \{N\}$. The map π sends each zero-dimensional leaf in SU_2 to N, and maps each two-dimensional leaf D_t diffeomorphically onto $\mathcal{O} \backslash \{N\}$, preserving the symplectic structures.

D The twisted case The Poisson–Lie group structure on a compact group K we have described is not the only one (up to a scalar multiple), except in the case $K = SU_2$. In fact, we have the following result, which is a consequence of the classification of the Lie bialgebra structures on complex simple Lie algebras described in Chapter 3.

PROPOSITION 1.5.10 *Let $\delta_{\mathbf{R}} : \mathfrak{k} \to \mathfrak{k} \otimes \mathfrak{k}$ be the tangent Lie bialgebra structure of the Poisson–Lie group structure on the compact group K defined above. Then, for any $a \in \mathbb{R}$ and any $u \in \wedge^2(\mathfrak{t})$, the formula*

$$\delta_{a,u}(x) = a\delta_{\mathbf{R}}(x) + (\mathrm{ad}_x \otimes 1 + 1 \otimes \mathrm{ad}_x)(u)$$

defines a Lie bialgebra structure on $\mathfrak{k}$. Conversely, every Lie bialgebra structure on $\mathfrak{k}$ is isomorphic to $\delta_{a,u}$ for some a, u. ∎

The results proved above dealt with the case $a \neq 0$, $u = 0$. We now summarize briefly the corresponding results in the remaining cases.

First, suppose that a and u are both non-zero. By regarding $u \in \mathfrak{t} \otimes \mathfrak{t}$ as a linear map $\mathfrak{t}^* \to \mathfrak{t}$ and using the standard inner product, we obtain an endomorphism $\tilde{u}$ of $\mathfrak{t}$. For any element w of the Weyl group W, let

$$\mathfrak{t}_w = (\tilde{u} - w^{-1}\tilde{u}w)(\mathfrak{t}),$$

and let $T_w = \exp(\mathfrak{t}_w)$ be the corresponding subgroup of T.

PROPOSITION 1.5.11 *Let K be a compact Lie group with the Poisson–Lie group structure defined by $\delta_{a,u}$, where a and u are non-zero. Then, if $t \in T$ and $w \in W$, the symplectic leaf passing through the point $t.w \in K$ is $T_w.t.\Sigma_w$, where Σ_w is as in 1.5.7. The leaf $T_w.t.\Sigma_w$ is contained in the piece K_w of the Bruhat decomposition.* ∎

Remark For generic u and $w \neq 1$, the symplectic leaf $T_w.t.\Sigma_w$ is equal to K_w if n is even, and dense in K_w if n is odd. In particular, there is always a symplectic leaf which is dense in K (namely, that associated to the longest element w_0 of the Weyl group W). △

Consider finally the case $a = 0$, $u \neq 0$. The symplectic leaves are now unrelated to the Bruhat decomposition. To describe them, let $\mathfrak{n}$ be the orthogonal complement of $\mathfrak{t}$ in $\mathfrak{k}$ with respect to the standard inner product. Then we have

$$\mathfrak{k}^* = \mathfrak{t}^* \oplus \mathfrak{n}^*,$$

with the understanding that linear functionals on $\mathfrak{t}$ or $\mathfrak{n}$ are extended to $\mathfrak{k}$ by making them vanish on the complementary subspace. If $\xi \in \mathfrak{k}^*$, we denote by $\langle \xi, u \rangle$ the element $(\xi \otimes \mathrm{id})(u) \in \mathfrak{t}$.

PROPOSITION 1.5.12 *Let K be a compact Lie group with the Poisson–Lie group structure defined by $\delta_{0,u}$, where $u \in \wedge^2(\mathfrak{t})$ is non-zero. Then, the symplectic leaf passing through a point $k \in K$ is*

$$\Sigma_k = \{\exp\left(\langle \xi, u \rangle\right) . k . \exp\left(-\langle \mathrm{Ad}^*_{k^{-1}}(\xi), u \rangle\right) \mid \xi \in \mathfrak{k}^*\}.$$

(Here, if $u = \sum_r t_r \otimes t'_r$, with t_r, $t'_r \in \mathfrak{t}$, then $\langle \xi, u \rangle = \sum_r \langle \xi, t_r \rangle t'_r$.) ∎

Remark The double of K in this case is the semidirect product $K \ltimes \mathfrak{t}^*$, where K acts on the additive group $\mathfrak{t}^*$ by the coadjoint representation Ad^*. △

1.6 Deformation of Poisson structures and quantization

In this section we give a brief introduction to the formulation of quantization as deformation of Poisson structure, due to Moyal (1949). Since our main aim is to motivate the study of deformation of Hopf algebras in Chapter 6, we shall not attempt to be completely precise.

A Deformations of Poisson algebras If A is an associative algebra, a *deformation* of A is a family of bilinear associative maps

$$*_h : A \times A \to A,$$

depending on a parameter h, such that $*_0$ is the given multiplication of A. Ideally, we would like $*_h$ to depend smoothly or analytically on h in some sense. However, most of the algebras A one is interested in deforming are infinite dimensional, in which case even the meaning of 'smooth' or 'analytic' is problematic. Thus, we shall ignore such difficulties by considering only *formal* deformations. This means that we treat h as an indeterminate and expand $*_h$ in a formal power series

$$a_1 *_h a_2 = \sum_{n=0}^{\infty} h^n \pi_n(a_1, a_2)$$

for certain bilinear maps $\pi_n : A \times A \to A$. The associativity of $*_h$ is equivalent to

$$\sum_{r+s=n} \pi_r(\pi_s(a_1, a_2), a_3) = \sum_{r+s=n} \pi_r(a_1, \pi_s(a_2, a_3))$$

for all a_1, a_2, $a_3 \in A$ and all $n \geq 0$.

Now suppose that A is a commutative *Poisson algebra*, i.e. A is commutative as an algebra, and, in addition to its associative algebra structure, A is equipped with a skew-symmetric bilinear map $\{\ ,\ \} : A \times A \to A$, the Poisson bracket, which satisfies the Jacobi identity

$$\{a_1, \{a_2, a_3\}\} + \{a_2, \{a_3, a_1\}\} + \{a_3, \{a_1, a_2\}\} = 0$$

and the Leibniz identity

$$\{a_1 a_2, a_3\} = a_1 \{a_2, a_3\} + \{a_1, a_3\} a_2$$

(cf. (1)–(3); of course, the algebra of C^∞ functions on any Poisson manifold is a Poisson algebra). Then, we say that $*_h$ is a *Poisson algebra deformation*, or *quantization*, of A if

$$\frac{a_1 *_h a_2 - a_2 *_h a_1}{h} \equiv \{a_1, a_2\} \pmod{h}.$$

This makes sense because

$$a_1 *_h a_2 \equiv a_1 a_2 \pmod{h}$$

so that

$$a_1 *_h a_2 - a_2 *_h a_1 \equiv 0 \pmod{h}$$

because A is commutative.

We shall have much more to say about Poisson algebra deformations in Chapter 6, but to justify this definition, and to give an important example, we turn now to the interpretation of quantization in terms of deformation theory.

B Weyl quantization The case of interest in the study of quantization is $A = C^\infty(M)$, where M is a Poisson manifold. In Chapter 6, we shall be interested in the case where M is a Poisson–Lie group, but for now we consider only the simplest case $M = \mathbb{R}^2$ with coordinates p, q and the (symplectic) Poisson structure

$$(20) \qquad \{f_1, f_2\}(p, q) = \frac{\partial f_1}{\partial p}\frac{\partial f_2}{\partial q} - \frac{\partial f_2}{\partial p}\frac{\partial f_1}{\partial q}.$$

We can interpret M as the phase space of a single particle moving along the real line, with position q and momentum p.

Quantum mechanics textbooks tell us that to quantize this system, one wants to associate to the coordinate functions p and q of $\mathbb{R}^2$ self-adjoint operators P and Q on some Hilbert space $\mathcal{H}$ such that the commutator $[P, Q] = PQ - QP$ is given by

$$(21) \qquad [P, Q] = -\sqrt{-1}h\,\mathrm{id}_{\mathcal{H}},$$

where h is $1/2\pi$ times Planck's constant. More generally, one would like to associate to any real-valued function f on $\mathbb{R}^2$ a self-adjoint operator F on $\mathcal{H}$ such that $[F_1, F_2]$ is, in some sense, approximately the same as the operator associated to the Poisson bracket $\{f_1, f_2\}$ (we have already mentioned in the Introduction that 'approximately' cannot be replaced by 'exactly' here).

Equation (21) means that P and Q determine a hermitian representation ϕ of the three-dimensional *Heisenberg Lie algebra* $\mathfrak{h}$, the vector space $\sqrt{-1}\mathbb{R} \oplus \sqrt{-1}\mathbb{R} \oplus \mathbb{R}$ with the Lie bracket

$$[(x_1, x_2, a), (y_1, y_2, b)] = (0, 0, x_1 y_2 - x_2 y_1),$$

namely

$$\phi(x_1, x_2, a) = x_1 P + x_2 Q - \sqrt{-1}ha\,\mathrm{id}_{\mathcal{H}}.$$

Under mild assumptions, ϕ can be 'exponentiated' to a unitary representation Φ of the *Heisenberg group* H, the space $\mathbb{R} \oplus \mathbb{R} \oplus \mathbb{R}$ with multiplication

$$\mu((x_1, x_2, a), (y_1, y_2, b)) = (x_1 + y_1, x_2 + y_2, a + b + x_1 y_2).$$

Note that H has one-dimensional centre, spanned by $c = (0, 0, 1)$.

The quantization procedure of Weyl (1927) is essentially to extend Φ to the group algebra of H. More precisely, for a function $f(p, q)$, Weyl takes the associated operator to be

$$\Phi(f) = \iint \hat{f}(\xi, \eta)\Phi(\exp[\sqrt{-1}(\xi P + \eta Q)])\ d\xi\ d\eta,$$

where

$$\hat{f}(\xi, \eta) = \left(\frac{1}{2\pi}\right)^2 \iint f(p, q)\exp[-\sqrt{-1}(\xi p + \eta q)]\ dp\ dq$$

is the Fourier transform of f. The operator $\Phi(f)$ is, at least formally, self-adjoint if f is real-valued, and is well-defined and of trace class if f is in the Schwartz space $\mathcal{S}(\mathbb{R}^2)$, i.e. if f is C^∞ and decreases, together with all its derivatives, faster than the reciprocal of any polynomial at infinity. Of course, one wants to define $\Phi(f)$ for f's more general than this (for example, for the coordinate functions p and q!). If f is C^∞ and grows, together with all its derivatives, at most polynomially at infinity, then $\Phi(f)$ will in general be a densely defined unbounded operator.

Every unitary representation of H on a Hilbert space $\mathcal{H}$ on which the central element c acts as $\exp(-\sqrt{-1}h)\mathrm{id}_{\mathcal{H}}$ is isomorphic to a direct sum of copies of the representation on the space $L^2(\mathbb{R})$ of square-integrable functions ψ of q given by

$$\Phi(x, 0, 0)(\psi)(q) = \psi(q + hx), \quad \Phi(0, y, 0)(\psi)(q) = \exp(\sqrt{-1}qy)\psi(q).$$

In this representation, to the coordinate functions p and q are associated the operators $-\sqrt{-1}h\partial/\partial q$ and multiplication by q, respectively. More generally, the operator associated to any polynomial f in p and q is the differential operator obtained by replacing each monomial $p^m q^n$ in f by the average, over all orderings of the $m + n$ terms, of the product of m differentiation operators $-\sqrt{-1}h\partial/\partial q$ and n operators of multiplication by q. The map Φ is one-to-one on the space of functions f which do not grow too fast at infinity.

C Quantization as deformation Moyal (1949) interpreted the Weyl quantization procedure as a deformation of the algebra $A = C^\infty(\mathbb{R}^2)$. In fact, he computed the product $*_h$ on A such that

$$\Phi(f_1 *_h f_2) = \Phi(f_1)\Phi(f_2)$$

as follows (h is now treated as an indeterminate). The right-hand side is given by

$$\iiiint \hat{f}_1(\xi_1, \eta_1) \hat{f}_2(\xi_2, \eta_2) \exp[\sqrt{-1}(\xi_1 P + \eta_1 Q)] \exp[\sqrt{-1}(\xi_2 P + \eta_2 Q)]$$
$$\times \, d\xi_1 d\eta_1 d\xi_2 d\eta_2.$$

The product of the two exponentials may be written as a single exponential by using the Baker–Campbell–Hausdorff formula for H. This is quite simple since $[\mathfrak{h}, [\mathfrak{h}, \mathfrak{h}]] = 0$; one finds that

$$\exp(X_1)\exp(X_2) = \exp(X_1 + X_2 + \frac{1}{2}[X_1, X_2])$$

for $X_1, X_2 \in \mathfrak{h}$. Writing $\xi = \xi_1 + \xi_2$, $\eta = \eta_1 + \eta_2$, the integral therefore becomes

$$(22) \quad \iiiint \hat{f}_1(\xi_1, \eta_1) \hat{f}_2(\xi_2, \eta_2) \exp[\sqrt{-1}h(\xi_1\eta_2 - \xi_2\eta_1)/2]$$
$$\times \, \exp[\sqrt{-1}(\xi P + \eta Q)] d\xi_1 d\eta_1 d\xi_2 d\eta_2.$$

If we now expand the exponential $\exp[\sqrt{-1}h(\xi_1\eta_2 - \xi_2\eta_1)/2]$ in powers of ξ_1, η_1, ξ_2 and η_2, we find that

$$\hat{f}_1(\xi_1, \eta_1) \hat{f}_2(\xi_2, \eta_2) \exp[\sqrt{-1}h(\xi_1\eta_2 - \xi_2\eta_1)/2] =$$
$$\sum_{r,s=0}^{\infty} \left(\frac{\sqrt{-1}h}{2} \right)^{r+s} \frac{(-1)^r}{r!s!} \hat{f}_1^{s,r}(\xi_1, \eta_1) \hat{f}_2^{r,s}(\xi - \xi_1, \eta - \eta_1),$$

where, for any $f \in C^\infty(\mathbb{R}^2)$, and any $r, s \geq 0$, we set

$$f^{r,s}(p, q) = \left(-\sqrt{-1}\frac{\partial}{\partial p} \right)^r \left(-\sqrt{-1}\frac{\partial}{\partial q} \right)^s f(p, q).$$

When we perform the integrals in (22) over ξ_1 and η_1, we therefore find the convolution of $\hat{f}_1^{s,r}$ and $\hat{f}_2^{r,s}$, i.e. the Fourier transform of the product $f_1^{s,r} f_2^{r,s}$. It follows that

$$(f_1 *_h f_2)(p, q) = \sum_{r,s=0}^{\infty} \left(\frac{\sqrt{-1}h}{2} \right)^{r+s} \frac{(-1)^r}{r!s!} \left(-\sqrt{-1}\frac{\partial}{\partial p} \right)^s \left(-\sqrt{-1}\frac{\partial}{\partial q} \right)^r f_1(p, q)$$
$$\times \left(-\sqrt{-1}\frac{\partial}{\partial p} \right)^r \left(-\sqrt{-1}\frac{\partial}{\partial q} \right)^s f_2(p, q).$$

This equation can be written more concisely by regarding the Poisson bracket (20) as a bidifferential operator Π acting on $C^\infty(\mathbb{R}^2) \times C^\infty(\mathbb{R}^2)$ by

$$\Pi(f_1, f_2) = \{f_1, f_2\}.$$

Then, writing $\Pi^0(f_1, f_2)$ for the ordinary product $f_1 f_2$, the preceding equation is equivalent to

$$f_1 *_h f_2 = \sum_{n=0}^{\infty} \frac{1}{n!} \left(\frac{-\sqrt{-1}h}{2} \right)^n \Pi^n(f_1, f_2),$$

i.e.

$$(23) \qquad f_1 *_h f_2 = \exp\left(\frac{-\sqrt{-1}h}{2} \Pi \right)(f_1, f_2).$$

It is clear that $*_h$ is a deformation of the associative algebra $C^\infty(\mathbb{R}^2)$ in the sense of the definition given at the beginning of this section, for

$$f_1 *_h f_2 \equiv f_1 f_2 \pmod{h}$$

and, since Φ is one-to-one, the associativity of $*_h$,

$$f_1 *_h (f_2 *_h f_3) = (f_1 *_h f_2) *_h f_3,$$

follows from the associativity of multiplication of the operators $\Phi(f_1)$, $\Phi(f_2)$ and $\Phi(f_3)$,

$$\Phi(f_1)(\Phi(f_2)\Phi(f_3)) = (\Phi(f_1)\Phi(f_2))\Phi(f_3).$$

The *Moyal bracket* $\{\ ,\ \}_h$ on $C^\infty(\mathbb{R}^2)$ is the Lie bracket defined by the multiplication $*_h$:

$$\{f_1, f_2\}_h = f_1 *_h f_2 - f_2 *_h f_1.$$

It is clear from the formula for $*_h$ given in (23) that

$$\frac{\{f_1, f_2\}_h}{h} \equiv \{f_1, f_2\} \pmod{h}.$$

Thus, *the product $*_h$ is a Poisson algebra deformation of $C^\infty(\mathbb{R}^2)$ with the Poisson bracket (20)*.

The discussion we have given can be generalized from $\mathbb{R}^2$ to arbitrary symplectic manifolds (see the notes below), but for our purposes we need not pursue this discussion any further (see, however, Subsection 6.2C).

Bibliographical notes

1.1 The general notion of a Poisson manifold was introduced by Lie (1888–93). In this work, Lie also defined the Lie–Poisson structure on the dual of a Lie algebra, and gave the proof of 1.1.2 for Poisson structures of constant rank; the general case is due to Kirillov (1976). For further information on Poisson manifolds in general, see Weinstein (1983). For the application of Poisson and symplectic manifolds in hamiltonian mechanics, see Abraham & Marsden (1978) and Arnold (1978).

1.2 The basic references for Poisson–Lie groups are Drinfel'd (1983a,b), in which they are called Hamilton–Lie groups, and Semenov-Tian-Shansky (1985). Further details may be found in Kosmann-Schwarzbach & Magri (1988) and Lu & Weinstein (1990). There are a number of generalizations of the notion of a Poisson–Lie group: for *quasi-Poisson-Lie groups*, see Kosmann-Schwarzbach (1991b), and, for *symplectic groupoids*, see Coste, Dazord & Weinstein (1987) and Weinstein & Xu (1992).

1.3 The basic references for this subsection are the same as those mentioned in the previous paragraph.

1.4 The notion of the double of a Lie bialgebra is a special case of that of a *matched pair* of Lie algebras, introduced by Takeuchi (1981). A pair of Lie algebras (possibly infinite dimensional) is matched if there is a weakly non-degenerate pairing between them, and if they act on each other in a 'compatible' fashion. If each Lie algebra is the dual of the other, and both actions are coadjoint actions, one recovers the situation discussed in this section. See also Majid (1988).

For further information on compact Poisson–Lie groups, see Lu & Weinstein (1990).

1.5 For the proofs of 1.5.3, 1.5.5 and 1.5.6, see Semenov-Tian-Shansky (1985) and Lu & Weinstein (1990), which also contain further information about Poisson actions and dressing transformations.

The description we have given of the symplectic leaves in a compact Lie group G is due to Vaksman & Soibelman (1988) in the case of SU_2, and Lu & Weinstein (1990), Levendorskii & Soibelman (1991a) and Soibelman (1991a) in the general case. For the results on the 'twisted' structures see Levendorskii (1990, 1992) and Levendorskii & Soibelman (1991a).

1.6 The transition from functions on phase space to operators on a Hilbert space is, of course, basic to quantum mechanics. The formulation we have given appears in Weyl (1927). Its interpretation in terms of a deformation of the algebra structure on the space of functions is due to Moyal (1949).

It is natural to ask whether the Poisson algebra deformation of $C^\infty(\mathbb{R}^2)$ described in Section 1.6 is the only one. The work of Vey (1975), Bayen *et al.* (1978), Lichnerowicz (1981, 1982), and others, has shown that the Moyal deformation is indeed the only one, at least if one imposes some reasonable conditions on the deformations one is willing to consider. In the other direction, de Wilde & Lecomte (1988) have extended the construction of the deformed product to arbitrary symplectic manifolds. For more recent work on Poisson deformations, see Rieffel (1989, 1990a,b), Huebschmann (1990) and Karasev & Maslov (1993).

2
Coboundary Poisson–Lie groups and the classical Yang–Baxter equation

As we mentioned in the Introduction, the existence of a quantization of a Lie group G implies the existence of a Lie bialgebra structure on its Lie algebra $\mathfrak{g}$. We recall from Chapter 1 that a Lie bialgebra structure on $\mathfrak{g}$ is a linear map $\delta : \mathfrak{g} \to \mathfrak{g} \otimes \mathfrak{g}$, which is, among other things, a 1-cocycle. Among the 1-cocycles of $\mathfrak{g}$ with values in $\mathfrak{g} \otimes \mathfrak{g}$ are the 1-coboundaries, for which $\delta(X) = [X\,,\,r]$ for some $r \in \mathfrak{g} \otimes \mathfrak{g}$. An element $r \in \mathfrak{g} \otimes \mathfrak{g}$ defines a Lie bialgebra structure in this way if and only if the symmetric part of r is a $\mathfrak{g}$-invariant element of $\mathfrak{g} \otimes \mathfrak{g}$, and the expression

$$[[r,r]] = [r_{12}, r_{13}] + [r_{12}, r_{23}] + [r_{13}, r_{23}]$$

is a $\mathfrak{g}$-invariant element of $\mathfrak{g} \otimes \mathfrak{g} \otimes \mathfrak{g}$. (The meaning of r_{12} etc. is fairly obvious, but will be explained in Section 2.1.) The latter condition is satisfied, in particular, if $[[r,r]] = 0$: this is the *classical Yang–Baxter equation* (CYBE). The CYBE first appeared explicitly in the literature on integrable hamiltonian systems, but it is a special case of the Schouten bracket in differential geometry, introduced in 1940.

These results are proved in Sections 2.1 and 2.2, in which we also re-examine the examples of Lie bialgebras given in Chapter 1 to see which of them are coboundary.

In the final Section 2.3, we show the importance of the CYBE in the study of classical hamiltonian systems, where it plays the role of an integrability condition. We illustrate the discussion with an example of Toda lattice type.

2.1 Coboundary Lie bialgebras

A Definitions We recall from 1.2.4 that a Lie bialgebra structure on a Lie algebra $\mathfrak{g}$ is a linear map $\delta : \mathfrak{g} \to \mathfrak{g} \otimes \mathfrak{g}$, its cocommutator, which satisfies the following conditions:

(a) δ is skew-symmetric,

(b) $\delta^* : \mathfrak{g}^* \otimes \mathfrak{g}^* \to \mathfrak{g}^*$ is a Lie bracket on $\mathfrak{g}^*$,

(c) δ is a 1-cocycle.

(As in the previous chapter, the Lie algebras we consider are assumed to be defined over any field of characteristic zero, except where stated or implied otherwise.)

From now on, we shall write the action of an element $X \in \mathfrak{g}$ on an element v of a representation V of $\mathfrak{g}$ as $X.v$, provided the action is understood. Thus, the cocycle condition (c) is

$$\delta([X, Y]) = X.\delta(Y) - Y.\delta(X)$$

for all X, $Y \in \mathfrak{g}$, with the adjoint action in each factor of the tensor product $\mathfrak{g} \otimes \mathfrak{g}$ understood.

Among the 1-cocycles of $\mathfrak{g}$ with values in $\mathfrak{g} \otimes \mathfrak{g}$ are the 1-coboundaries, for which

$$(1) \qquad\qquad\qquad \delta(X) = X.r$$

for some $r \in \mathfrak{g} \otimes \mathfrak{g}$ and all $X \in \mathfrak{g}$.

DEFINITION 2.1.1 *A* coboundary Lie bialgebra *is a Lie bialgebra $\mathfrak{g}$ whose cocommutator is a 1-coboundary.*

PROPOSITION 2.1.2 *Let $\mathfrak{g}$ be a Lie algebra and let $r \in \mathfrak{g} \otimes \mathfrak{g}$. The map δ defined by (1) is the cocommutator of a Lie bialgebra structure on $\mathfrak{g}$ if and only if the following conditions are satisfied:*

(i) $r_{12} + r_{21}$ is a $\mathfrak{g}$-invariant element of $\mathfrak{g} \otimes \mathfrak{g}$,

(ii) $[[r, r]] \equiv [r_{12}, r_{13}] + [r_{12}, r_{23}] + [r_{13}, r_{23}]$ is a $\mathfrak{g}$-invariant element of $\mathfrak{g} \otimes \mathfrak{g} \otimes \mathfrak{g}$. (The action of $\mathfrak{g}$ on $\mathfrak{g} \otimes \mathfrak{g}$ and on $\mathfrak{g} \otimes \mathfrak{g} \otimes \mathfrak{g}$ is by the adjoint representation in each factor.)

Before proving this result, let us explain the notation. Suppose that

$$(2) \qquad\qquad\qquad r = \sum_i a_i \otimes b_i$$

for some a_i, $b_i \in \mathfrak{g}$. Then in (i), r_{12} is simply r, while

$$r_{21} = \sigma(r) = \sum_i b_i \otimes a_i$$

is obtained by applying to r the flip σ of the two factors in the tensor product. In (ii), we have

$$[r_{12}, r_{13}] = \sum_{i,j} [a_i, a_j] \otimes b_i \otimes b_j,$$

$$[r_{12}, r_{23}] = \sum_{i,j} a_i \otimes [b_i, a_j] \otimes b_j,$$

$$[r_{13}, r_{23}] = \sum_{i,j} a_i \otimes a_j \otimes [b_i, b_j].$$

Remark It is obvious from (1) that, if $r \in \mathfrak{g} \otimes \mathfrak{g}$ defines a Lie bialgebra structure on $\mathfrak{g}$, and if $r' \in \mathfrak{g} \otimes \mathfrak{g}$ differs from r by a $\mathfrak{g}$-invariant element, then r' defines the *same* Lie bialgebra structure as r. By condition (i) in 2.1.2, it follows that every coboundary Lie bialgebra can be obtained from a *skew-symmetric* r. $\triangle$

We now turn to the proof of 2.1.2.

PROOF The proof that r satisfies (i) if and only if the map δ defined by (1) satisfies condition (a) above is completely straightforward. To prove that (ii) is equivalent to (b) (in the presence of (a)), we first introduce some notation. For any linear map $\delta : \mathfrak{g} \to \mathfrak{g} \otimes \mathfrak{g}$, and any $X \in \mathfrak{g}$, let

$$\mathrm{Jac}_\delta(X) = \sum_{\mathrm{c.p.}} (\delta \otimes \mathrm{id}) \delta(X),$$

where '$\sum_{\mathrm{c.p.}}$' means the sum of the term which follows the summation sign and the two similar terms obtained from it by cyclic permutation of the factors in the tensor product $\mathfrak{g} \otimes \mathfrak{g} \otimes \mathfrak{g}$. If we define a bracket on $\mathfrak{g}^*$ by

$$[\xi, \eta]_{\mathfrak{g}^*} = \delta^*(\xi \otimes \eta),$$

then, since

$$[[\xi, \eta]_{\mathfrak{g}^*}, \zeta]_{\mathfrak{g}^*} = \delta^*(\delta^* \otimes \mathrm{id})(\xi \otimes \eta \otimes \zeta),$$

it is clear that $[\ ,\ ]_{\mathfrak{g}^*}$ satisfies the Jacobi identity if and only if Jac_δ is the zero map. It therefore suffices to prove

LEMMA 2.1.3 *Let $\mathfrak{g}$ be a Lie algebra and let $r \in \mathfrak{g} \otimes \mathfrak{g}$ have $\mathfrak{g}$-invariant symmetric part. Define $\delta : \mathfrak{g} \to \mathfrak{g} \otimes \mathfrak{g}$ by (1). Then*

$$(3) \qquad\qquad \mathrm{Jac}_\delta(X) + X.[[r, r]] = 0$$

for all $X \in \mathfrak{g}$.

PROOF By direct computation, making repeated use of the invariance property of r and the Jacobi identity in $\mathfrak{g}$. If we write out the left-hand side of (3) explicitly, we find that $\mathrm{Jac}_\delta(X)$ is a sum of twelve terms and that $X.[[r, r]]$ is a sum of nine terms. After rearranging the terms suitably, we obtain, using the expression (2) for r and with a summation over repeated indices understood,

$$[[X.a_i], a_j] \otimes b_i \otimes b_j + [[X, a_i], b_j] \otimes b_i \otimes a_j + [X, [a_i, a_j]] \otimes b_i \otimes b_j$$

$$+[X, b_i] \otimes [a_i, a_j] \otimes b_j + [X, a_i] \otimes [b_i, a_j] \otimes b_j$$

$$+[X, b_i] \otimes a_j \otimes [a_i, b_j] + [X, a_i] \otimes a_j \otimes [b_i, b_j]$$

$$+[a_i, b_j] \otimes [X, b_i] \otimes a_j + [a_i, a_j] \otimes [X, b_i] \otimes b_j$$

$$+[a_i, a_j] \otimes b_j \otimes [X, b_i] + [a_i, a_j] \otimes b_i \otimes [X, b_j]$$
$$+a_j \otimes [[X, a_i], b_j] \otimes b_i + a_j \otimes [a_i, b_j] \otimes [X, b_i] + a_i \otimes [X, [b_i, a_j]] \otimes b_j$$
$$+a_i \otimes [b_i, a_j] \otimes [X, b_j] + a_i \otimes [X, a_j] \otimes [b_i, b_j] + a_i \otimes a_j \otimes [X, [b_i, b_j]]$$
$$+b_j \otimes b_i \otimes [[X, a_i], a_j] + b_j \otimes [X, b_i] \otimes [a_i, a_j]$$
$$+b_i \otimes [[X, a_i], a_j] \otimes b_j + b_i \otimes a_j \otimes [[X, a_i], b_j].$$

Interchanging the indices i and j in the first term and using the Jacobi identity in $\mathfrak{g}$, the sum of the first three terms becomes

$$-a_j \otimes b_i \otimes [[X, a_i], b_j] - b_j \otimes b_i \otimes [[X, a_i], a_j].$$

Next, the sum of the fourth and fifth terms is

$$-((1 \otimes \mathrm{ad}_{a_j})([X, b_i] \otimes a_i + [X, a_i] \otimes b_i)) \otimes b_j$$
$$= ((1 \otimes \mathrm{ad}_{a_j})(a_i \otimes [X, b_i] + b_i \otimes [X, a_i])) \otimes b_j$$
$$= a_i \otimes [a_j, [X, b_i]] \otimes b_j + b_i \otimes [a_j, [X, a_i]] \otimes b_j,$$

using the invariance property

$$X.(a_i \otimes b_i + b_i \otimes a_i) = 0.$$

Similarly, the sum of the sixth and seventh terms becomes

$$a_i \otimes a_j \otimes [b_j, [X, b_i]] + b_i \otimes a_j \otimes [b_j, [X, a_i]],$$

and that of the eighth and ninth terms

$$-b_j \otimes [X, b_i] \otimes [a_i, a_j] - a_j \otimes [X, b_i] \otimes [a_i, b_j].$$

Finally, the sum of the tenth and eleventh terms is obviously zero (interchange the indices i and j in one term).

Inserting these results, we find that the expression on the left-hand side of (3) can be written in the form $\sum_i (a_i \otimes u_i + b_i \otimes v_i)$, where u_i, $v_i \in \mathfrak{g}$. In fact,

$$u_i = [[X, a_j], b_i] \otimes b_j + [X, [b_i, a_j]] \otimes b_j + [a_j, [X, b_i]] \otimes b_j$$
$$+ [a_j, b_i] \otimes [X, b_j] + [b_i, a_j] \otimes [X, b_j]$$
$$+ [X, a_j] \otimes [b_i, b_j] - b_j \otimes [[X, a_j], b_i] - [X, b_j] \otimes [a_j, b_i]$$
$$+ a_j \otimes [X, [b_i, b_j]] + a_j \otimes [b_j, [X, b_i]].$$

On the right-hand side, the sum of the first three terms is zero by the Jacobi identity in $\mathfrak{g}$, and that of the next two terms is zero by the skew-symmetry of the Lie bracket in $\mathfrak{g}$. The sum of the next three terms is

$$(1 \otimes \mathrm{ad}_{b_i})([X, a_j] \otimes b_j + b_j \otimes [X, a_j] + [X, b_j] \otimes a_j)$$
$$= -(1 \otimes \mathrm{ad}_{b_i})(a_j \otimes [X, b_j]) = -a_j \otimes [b_i, [X, b_j]],$$

using the invariance property of r again. Hence,

$$u_i = a_j \otimes ([X, [b_i, b_j]] + [b_j, [X, b_i]] - [b_i, [X, b_j]]),$$

which vanishes by the Jacobi identity in $\mathfrak{g}$. A similar, but shorter, argument proves that $v_i = 0$. ∎

Remarks [1] We have already observed that, in considering coboundary Lie bialgebras, we may restrict attention to skew-symmetric elements r. It is not hard to show that in this case $[[r, r]]$ is skew too (i.e. $[[r, r]] \in \bigwedge^3 \mathfrak{g}$ if $r \in \bigwedge^2 \mathfrak{g}$).

[2] An argument similar to that used to prove the lemma shows that, *if r, $s \in \mathfrak{g} \otimes \mathfrak{g}$ are such that r is skew and s is $\mathfrak{g}$-invariant, then $[[r + s, r + s]] = [[r, r]] + [[s, s]]$.* This result will be used later.

[3] Two coboundary Lie bialgebras $\mathfrak{g}_1$ and $\mathfrak{g}_2$ defined by $r_1 \in \mathfrak{g}_1 \otimes \mathfrak{g}_1$ and $r_2 \in \mathfrak{g}_2 \otimes \mathfrak{g}_2$ are isomorphic if and only if there is an isomorphism of Lie algebras $\alpha : \mathfrak{g}_1 \to \mathfrak{g}_2$ such that $(\alpha \otimes \alpha) r_1 - r_2$ is $\mathfrak{g}_2$-invariant. △

B The classical Yang–Baxter equation Of course, the simplest way to satisfy condition (ii) in 2.1.2 is to assume that

$$[[r, r]] = 0.$$

This is the *classical Yang–Baxter equation* (CYBE). A solution of the CYBE is often called a *classical r-matrix*, the terminology deriving from the example $\mathfrak{g} = \mathrm{End}(V)$, where V is a vector space, in which case $r \in \mathrm{End}(V \otimes V)$ may be viewed as a matrix. A (coboundary) Lie bialgebra structure which arises from a solution of the CYBE is said to be *quasitriangular*; if it arises from a skew solution of the CYBE, it is said to be *triangular*. This terminology is borrowed from the 'quantized' situation, discussed in Chapter 4.

The following result will be needed later. Its proof, like that of 2.1.2, is elementary, and we leave it to the reader.

PROPOSITION 2.1.4 *Suppose that $(\mathfrak{g}, \delta)$ is a quasitriangular Lie bialgebra, where δ is given by (1) for a solution $r \in \mathfrak{g} \otimes \mathfrak{g}$ of the CYBE. Let $\rho : \mathfrak{g}^* \to \mathfrak{g}$ be the linear map associated to r by*

$$\langle \rho(\xi), \eta \rangle = \langle r, \xi \otimes \eta \rangle,$$

where $\xi, \eta \in \mathfrak{g}^*$. Then, ρ is a homomorphism of Lie algebras ($\mathfrak{g}^*$ has the Lie algebra structure defined by δ^*). ∎

More generally, suppose that $\mathfrak{g}$ is equipped with a non-degenerate symmetric bilinear form $(\,,\,)$, invariant under the adjoint action of $\mathfrak{g}$. Then, there is a canonical $\mathfrak{g}$-invariant element of $\bigwedge^3 \mathfrak{g}^*$, namely

$$(4) \qquad\qquad (X, Y, Z) \mapsto ([X, Y], Z).$$

If, in addition, $\dim(\mathfrak{g}) < \infty$, we can identify $\mathfrak{g}$ and $\mathfrak{g}^*$ as representations of $\mathfrak{g}$, so we obtain a canonical $\mathfrak{g}$-invariant element ω of $\bigwedge^3 \mathfrak{g}$. Then condition (ii) in 2.1.2 will be satisfied if

$$(5) \qquad\qquad [[r,r]] = -\omega.$$

This is the *modified classical Yang–Baxter equation* (MCYBE). It is clear that the CYBE is the limiting case of the MCYBE as the bilinear form (,) on $\mathfrak{g}$ 'tends to infinity'.

There is a close relationship between the solutions of the CYBE and those of the MCYBE. The bilinear form may be regarded as an element of $(\mathfrak{g} \otimes \mathfrak{g})^* \cong \mathfrak{g}^* \otimes \mathfrak{g}^*$, and hence, using the isomorphism $\mathfrak{g} \cong \mathfrak{g}^*$, as an element $t \in \mathfrak{g} \otimes \mathfrak{g}$ (t is called the *Casimir element*). Then t is $\mathfrak{g}$-invariant and symmetric, and a simple computation shows that

$$(6) \qquad\qquad [[t,t]] = \omega.$$

Remark [2] following the proof of 2.1.2 implies that *r is a skew-symmetric solution of the MCYBE if and only if $r + t$ is a solution of the CYBE*. (This is the reason for the choice of sign in (5).)

C Examples We shall now interpret the examples of Lie bialgebras given in Chapter 1 in the light of 2.1.1.

Example 2.1.5 If $\mathfrak{g}$ is abelian, only the trivial Lie bialgebra structure is coboundary. $\Diamond$

Example 2.1.6 We saw in 1.3.7 that the two-dimensional non-abelian Lie algebra $\mathfrak{g}$ has, up to isomorphism and scalar multiples, exactly two non-trivial Lie bialgebra structures. Choosing a basis $\{X, Y\}$ of $\mathfrak{g}$ such that $[X, Y] = X$, the structures are given by

$$\delta(X) = 0, \qquad \delta(Y) = X \wedge Y$$

and

$$\delta(X) = X \wedge Y, \qquad \delta(Y) = 0.$$

The first structure is coboundary (in fact, triangular): it arises from the solution $r = Y \wedge X$ of the CYBE. The second structure is not coboundary. $\Diamond$

Example 2.1.7 If $\mathfrak{g}$ is a finite-dimensional complex simple Lie algebra, every 1-cocycle of $\mathfrak{g}$ with values in any finite-dimensional representation of $\mathfrak{g}$ is a coboundary by Whitehead's lemma (see Jacobson (1962), page 77). In particular, every Lie bialgebra structure on $\mathfrak{g}$ is coboundary. In fact, since

$\mathfrak{g}$ has, up to a scalar multiple, exactly one non-degenerate invariant bilinear form, it follows from the remarks at the end of the preceding subsection that every Lie bialgebra structure on $\mathfrak{g}$ is quasitriangular.

For the standard Lie bialgebra structure defined in 1.3.8, the r-matrix can be defined as follows. Let t be the Casimir element of $\mathfrak{g} \otimes \mathfrak{g}$, and write $t = t_{+-} + t_0 + t_{-+}$, where $t_0 \in \mathfrak{h} \otimes \mathfrak{h}$, $t_{+-} \in \mathfrak{n}_+ \otimes \mathfrak{n}_-$ and $t_{-+} \in \mathfrak{n}_- \otimes \mathfrak{n}_+$. Then the standard structure is defined by

$$r = t_0 + 2t_{+-}.$$

It is quasitriangular but not triangular. We postpone the proof of these assertions until after 2.1.11.

The description of the Lie bialgebra structures given in 1.3.8 is valid for any symmetrizable Kac–Moody algebra $\mathfrak{g}$. However, the element r which should serve as the r-matrix is now given by an *infinite* sum of elements of $\mathfrak{g} \otimes \mathfrak{g}$. Drinfel'd describes $\mathfrak{g}$ as 'pseudotriangular' in this case. The example of (untwisted) affine Lie algebras is discussed in 2.1.10. $\Diamond$

Example 2.1.8 We mentioned in 1.3.8 that the standard structures on a finite-dimensional complex simple Lie algebra $\mathfrak{g}$ do not exhaust the (non-trivial) Lie bialgebra structures on $\mathfrak{g}$. To illustrate this, we shall describe *all* the Lie bialgebra structures in the simplest case $\mathfrak{g} = sl_2(\mathbb{C})$.

It suffices to find elements $r \in \bigwedge^2 sl_2(\mathbb{C})$ such that $[[r,r]] \in \bigwedge^3 sl_2(\mathbb{C})$ is $sl_2(\mathbb{C})$-invariant. But since $sl_2(\mathbb{C})$ is simple and $\dim(\bigwedge^3 sl_2(\mathbb{C})) = 1$, the whole of $\bigwedge^3 sl_2(\mathbb{C})$ is $sl_2(\mathbb{C})$-invariant. Thus, condition (ii) in 2.1.2 imposes no condition on r. Furthermore, $\bigwedge^2 sl_2(\mathbb{C})$ is isomorphic to the adjoint representation of $sl_2(\mathbb{C})$ (the isomorphism is given by the Lie bracket), so the $sl_2(\mathbb{C})$-invariant part of $\bigwedge^2 sl_2(\mathbb{C})$ is zero. Hence, the Lie bialgebra structures on $sl_2(\mathbb{C})$ are parametrized by the three-dimensional vector space $\bigwedge^2 sl_2(\mathbb{C})$. Moreover, it is not difficult to check that r satisfies the CYBE if and only if, when r is regarded as an element of $sl_2(\mathbb{C})$, we have $\det(r) = 0$. The solutions of the CYBE thus form a quadric cone in $\bigwedge^2 sl_2(\mathbb{C})$.

By Remark [3] following the proof of 2.1.2, the Lie bialgebra structures defined by r_1, $r_2 \in \bigwedge^2 sl_2(\mathbb{C})$ are isomorphic if and only if there exists an automorphism α of $sl_2(\mathbb{C})$ such that $r_2 - (\alpha \otimes \alpha)r_1$ is $sl_2(\mathbb{C})$-invariant. By the discussion in the preceding paragraph, this is possible only if $r_2 = (\alpha \otimes \alpha)r_1$. Since every automorphism of $sl_2(\mathbb{C})$ is given by the adjoint action of some element of the group $SL_2(\mathbb{C})$, we see that *the isomorphism classes of Lie bialgebra structures on $sl_2(\mathbb{C})$ are parametrized by the orbits of the adjoint action of $SL_2(\mathbb{C})$ on $sl_2(\mathbb{C})$*. These orbits are of three types:

(i) The one-parameter family of non-zero semisimple orbits, represented by

$$\begin{pmatrix} \lambda & 0 \\ 0 & -\lambda \end{pmatrix}$$

for $\lambda \in \mathbb{C}^\times$. This corresponds to

$$r = \lambda X^+ \wedge X^-$$

where

$$X^+ = \begin{pmatrix} 0 & 1 \\ 0 & 0 \end{pmatrix}, \quad X^- = \begin{pmatrix} 0 & 0 \\ 1 & 0 \end{pmatrix}.$$

These are multiples of the standard structure.

(ii) The single orbit consisting of the non-zero nilpotent elements, represented by X^+. This corresponds to

$$r = \frac{1}{2} H \wedge X^+,$$

where

$$H = \begin{pmatrix} 1 & 0 \\ 0 & -1 \end{pmatrix},$$

which is a skew solution of the CYBE, and defines a triangular Lie bialgebra structure on $sl_2(\mathbb{C})$.

(iii) The orbit consisting of the zero element, corresponding to the trivial structure $r = 0$. ◊

Example 2.1.9 Let $\mathfrak{g} = \mathfrak{a}[u]$, where $\mathfrak{a}$ is a finite-dimensional complex simple Lie algebra and u is an indeterminate. In 1.3.9 we defined a Lie bialgebra structure on $\mathfrak{g}$. Formally, this arises from the r-matrix

(7)
$$r = \frac{t}{u_1 - u_2},$$

where $t \in \mathfrak{a} \otimes \mathfrak{a}$ is the Casimir element. Here, we view $\mathfrak{g} \otimes \mathfrak{g} = \mathfrak{a}[u] \otimes \mathfrak{a}[u] \cong (\mathfrak{a} \otimes \mathfrak{a})[u_1, u_2]$ as the space of polynomials in u_1, u_2 with coefficients in $\mathfrak{a} \otimes \mathfrak{a}$. Of course, r is not a polynomial, i.e. r does not lie in the algebraic tensor product $\mathfrak{g} \otimes \mathfrak{g}$. Nevertheless, it gives rise, as we have seen in 1.3.9, to a genuine Lie bialgebra structure on $\mathfrak{g}$. In fact, this structure is pseudotriangular, for r satisfies the CYBE. Indeed, this is an easy consequence of the relations

$$[t_{12}, t_{13} + t_{23}] = [t_{23}, t_{12} + t_{13}] = 0,$$

which follow from the fact that t is invariant under the adjoint action of $\mathfrak{a}$.

The r-matrix in (7) is the simplest rational solution of the CYBE. We shall discuss the classification of such rational solutions in Chapter 3, and shall meet the r-matrix in (7) again in Chapters 12 and 16. ◊

Example 2.1.10 Suppose that $\mathfrak{g}$ is the untwisted affine Lie algebra associated to a complex simple Lie algebra $\mathfrak{a}$. We recall from the Appendix that

$$\mathfrak{g} = \mathfrak{a}\,[u, u^{-1}] \oplus \mathbb{C}.c \oplus \mathbb{C}.D$$

as a vector space, where c is central and

$$[D, f](u) = \frac{df}{du}$$

for $f \in \mathfrak{a}[u, u^{-1}]$. The Lie bracket on $\mathfrak{g}$ is given in terms of that of $\mathfrak{a}$ by

$$[f, g](u) = [f(u), g(u)] + \mathrm{res}_0\left(\frac{df}{du}, g(u)\right),$$

where $(\ ,\ )$ is an invariant inner product on $\mathfrak{a}$, and 'res$_0$' means the u^{-1} coefficient.

To describe the standard Lie bialgebra structure on $\mathfrak{g}$, one checks first that

$$(f + \lambda c + \mu D, g + \xi c + \eta D) = \mathrm{res}_0(f(u), g(u)) + \lambda\eta + \mu\xi$$

is a non-degenerate invariant inner product on $\mathfrak{g}$. Hence, in the notation of 2.1.7, we have

$$t_0^{\mathfrak{g}} = t_0^{\mathfrak{a}} + c \otimes D + D \otimes c,$$

and, since

$$\mathfrak{n}_\pm^{\mathfrak{g}} = \mathfrak{n}_\pm^{\mathfrak{a}} \oplus u^{\pm 1}\mathfrak{a}[u^{\pm 1}],$$

we have

$$t_{+-}^{\mathfrak{g}} = t_{+-}^{\mathfrak{a}} + \sum_\lambda \sum_{k>0} (X_\lambda \otimes X_\lambda)\left(\frac{u_1}{u_2}\right)^k,$$

where $\{X_\lambda\}$ is an orthonormal basis of $\mathfrak{a}$. Formally, this gives the standard r-matrix of $\mathfrak{g}$ in terms of that of $\mathfrak{a}$ by

$$r_{\mathfrak{g}} = r_{\mathfrak{a}} + \frac{1}{2}(c \otimes D + D \otimes c) + t^{\mathfrak{a}}\left(\frac{u_1}{u_2 - u_1}\right).$$

As in the previous example, we see a rational function of u_1 and u_2 with a pole along $u_1 = u_2$. $\diamond$

D The classical double The following result will enable us to justify 2.1.7 and to complete the proof of 1.4.2.

PROPOSITION 2.1.11 *Let $\mathfrak{g}$ be a finite-dimensional Lie bialgebra. Then, the double $\mathcal{D}(\mathfrak{g})$ of $\mathfrak{g}$ is a quasitriangular Lie bialgebra.*

PROOF The Lie bialgebra structure of $\mathcal{D}(\mathfrak{g})$ was defined in 1.4.2 (although we have still to prove that it *is* a Lie bialgebra structure). We continue to use the notation introduced in the proofs of 1.3.5 and 1.4.2.

Thus,

$$[[r,r]] = \sum_{s,t}\left([X_s, X_t]\otimes\xi^s\otimes\xi^t + X_s\otimes[\xi^s, X_t]\otimes\xi^t + X_s\otimes X_t\otimes[\xi^s,\xi^t]\right)$$

$$= \sum_{s,t,p}\left(c_{st}^p X_p\otimes\xi^s\otimes\xi^t - \gamma_t^{sp}X_s\otimes X_p\otimes\xi^t + c_{tp}^s X_s\otimes\xi^p\otimes\xi^t\right.$$

$$\left. + \gamma_p^{st}X_s\otimes X_t\otimes\xi^p\right).$$

After suitably re-naming the indices, and using the skew-symmetry property $c_{st}^p = -c_{ts}^p$, we find that $[[r,r]] = 0$.

The $\mathcal{D}(\mathfrak{g})$-invariance of the symmetric part of r is also easily checked; in fact, the symmetric part of r is half the Casimir element of $\mathcal{D}(\mathfrak{g})\otimes\mathcal{D}(\mathfrak{g})$. ∎

We can now prove the assertions made in 2.1.7, using the fact, proved in 1.4.3, that the standard Lie bialgebra structure on a complex simple Lie algebra $\mathfrak{g}$ can 'almost' be realized as the double $\mathcal{D}(\mathfrak{b}_+)$. With the notation in the Appendix, the r-matrix of $\mathcal{D}(\mathfrak{q}_+)$ is

$$r_\mathcal{D} = \frac{1}{2}\sum_{i=1}^{n} d_i(H_i, H_i)\otimes(H_i, -H_i) + \sum_{\alpha\in\Delta_+} d_\alpha(X_\alpha^+, 0)\otimes(X_\alpha^-, 0).$$

To get the r-matrix of $\mathfrak{g}$, we must project onto $\mathfrak{g}\otimes\mathfrak{g}$ and then multiply the result by two, since the Lie bialgebra structure on $\mathcal{D}(\mathfrak{q}_+)$ induces $\frac{1}{2}$ that on $\mathfrak{g}$. Thus,

$$r_\mathfrak{g} = \sum_{i=1}^{n} d_i H_i\otimes H_i + 2\sum_{\alpha\in\Delta_+} d_\alpha X_\alpha^+\otimes X_\alpha^-.$$

This is the r-matrix written down in 2.1.7. Note that it follows from 2.1.11 without further calculation that $r_\mathfrak{g}$ satisfies the CYBE.

2.2 Coboundary Poisson–Lie groups

In this section, we describe the special properties of those Poisson–Lie groups whose tangent Lie bialgebras are coboundary.

A The Sklyanin bracket If $\mathfrak{g}$ is a finite-dimensional coboundary Lie bialgebra, its cocommutator is given, for $X \in \mathfrak{g}$, by

$$\delta(X) = (\mathrm{ad}_X \otimes 1 + 1 \otimes \mathrm{ad}_X)(r),$$

where we may and shall assume, without loss of generality, that $r \in \mathfrak{g} \otimes \mathfrak{g}$ is *skew-symmetric* (see the remark following 2.1.2). If G is a Lie group with Lie algebra $\mathfrak{g}$, to describe the Poisson structure on G associated to δ we must 'integrate' δ to a 1-cocycle of G with values in $\mathfrak{g} \otimes \mathfrak{g}$. But this is easy: it is trivial to check that

$$w^R(g) = (\mathrm{Ad}_g \otimes \mathrm{Ad}_g)(r) - r$$

has the correct derivative at the identity element of G and the cocycle property

$$w^R(gg') = (\mathrm{Ad}_g \otimes \mathrm{Ad}_g)(w^R(g')) + w^R(g)$$

for g, $g' \in G$. We recall from Section 1.2 that the Poisson bivector $w_g \in T_g(G) \otimes T_g(G)$ is the right translate of $w^R(g)$ to g, and that the associated Poisson bracket is given, for f_1, $f_2 \in C^\infty(G)$, by

$$\{f_1, f_2\}(g) = \langle w_g, (df_1)_g \otimes (df_2)_g \rangle.$$

Hence, we have proved

PROPOSITION 2.2.1 *Let G be a finite-dimensional Lie group with Lie algebra $\mathfrak{g}$. Suppose that $r \in \bigwedge^2 \mathfrak{g}$ defines a Lie bialgebra structure on $\mathfrak{g}$ via (1). Then the corresponding Poisson–Lie group structure on G is given by*
(8)
$$\{f_1, f_2\}(g) = \langle r, (((L_g)'_e)^* \otimes ((L_g)'_e)^* - ((R_g)'_e)^* \otimes ((R_g)'_e)^*)((df_1)_g \otimes (df_2)_g) \rangle,$$

for f_1, $f_2 \in C^\infty(G)$. ∎

Remark A Poisson bracket of the form (8) associated to a skew-symmetric r-matrix is called a *Sklyanin bracket* on G. △

It will be useful to express (8) a little more explicitly. Let $\{X_s\}$ be a basis of $\mathfrak{g}$ and let $\{X_s^L\}$ (resp. $\{X_s^R\}$) be the corresponding left (resp. right) invariant vector fields on G. Then, if

$$r = \sum_{s,t} r^{st} X_s \otimes X_t,$$

we have

$$\{f_1, f_2\} = \sum_{s,t} r^{st}((X_s^L f_1)(X_t^L f_2) - (X_s^R f_1)(X_t^R f_2)).$$

Remark We can be even more explicit if $G = GL_n(\mathbb{R})$ (or, more generally, if G is a matrix group). Let t_{ij} be the matrix entries in $GL_n(\mathbb{R})$, regarded as functions on $GL_n(\mathbb{R})$. The Poisson bracket is clearly determined by the brackets of these matrix elements. We find that

$$X^L(t_{ij}) = (TX)_{ij}, \quad X^R(t_{ij}) = (XT)_{ij},$$

where $T = (t_{ij})$. It follows that

$$\{t_{ij}, t_{k\ell}\} = \sum_{a,b}(r^{aj\,b\ell}t_{ia}t_{kb} - r^{ia\,kb}t_{aj}t_{b\ell}).$$

One usually writes

$$\{t_{ij}, t_{k\ell}\} = \{T \otimes T\}_{ij\,k\ell},$$

so that the preceding equation can be written

$$\{T \otimes T\} = [T \otimes T, r].$$

This is the basic equation of *classical inverse scattering theory*. See Section 2.3. $\triangle$

An element $r \in \mathfrak{g} \otimes \mathfrak{g}$ gives rise to other Poisson structures on G, as the following proposition shows.

PROPOSITION 2.2.2 *Let G be a finite-dimensional Lie group with Lie algebra $\mathfrak{g}$. Let $r \in \mathfrak{g} \otimes \mathfrak{g}$ be skew-symmetric. Using the above notation, define, for f_1, $f_2 \in C^\infty(G)$,*

(9)
$$\{f_1, f_2\}^L = \sum_{s,t} r^{s\,t}(X_s^L f_1)(X_t^L f_2),$$

$$\{f_1, f_2\}^R = \sum_{s,t} r^{s\,t}(X_s^R f_1)(X_t^R f_2).$$

Then the following are equivalent:

(i) $\{\ ,\ \}^L$ is a Poisson structure on G,

(ii) $\{\ ,\ \}^R$ is a Poisson structure on G,

(iii) r is a solution of the CYBE.

PROOF This follows immediately from

LEMMA 2.2.3 *Let $\{V_\alpha\}$ be any finite set of (smooth) vector fields on a Lie group G, and define*

$$\{f_1, f_2\}^V = \sum_{\alpha,\beta} \rho^{\alpha\beta}(V_\alpha f_1)(V_\beta f_2)$$

for f_1, $f_2 \in \mathrm{C}^\infty(G)$, where the constants $\rho^{\alpha\beta}$ satisfy $\rho^{\alpha\beta} = -\rho^{\beta\alpha}$. Then,

$$\{f_1, \{f_2, f_3\}^V\}^V + \{f_3, \{f_1, f_2\}^V\}^V + \{f_2, \{f_3, f_1\}^V\}^V$$
$$= \sum_{\alpha,\beta,\gamma,\delta} \rho^{\alpha\beta}\rho^{\gamma\delta}(([V_\gamma, V_\alpha]f_1)(V_\delta f_2)(V_\beta f_3) + (V_\gamma f_1)([V_\delta, V_\alpha]f_2)(V_\beta f_3)$$
$$+ (V_\gamma f_1)(V_\alpha f_2)([V_\delta, V_\beta]f_3)).$$

PROOF Straightforward, making use of the skew-symmetry of the coefficients $\rho^{\alpha\beta}$. ∎

Remark The Poisson bracket $\{\ ,\ \}^L$ (resp. $\{\ ,\ \}^R$) on G arising from 2.2.2 is left (resp. right) invariant, in the sense that left (resp. right) translations are Poisson maps $G \to G$. However, neither of these Poisson brackets is a Poisson–Lie group structure on G (unless $r = 0$). △

B r-matrices and 2-cocycles The following corollary of 2.2.2 gives an interesting alternative geometric interpretation of the CYBE. It applies to elements $r \in \mathfrak{g} \otimes \mathfrak{g}$ which are *non-degenerate*, which means that the associated linear map $\mathfrak{g}^* \to \mathfrak{g}$ is a vector space isomorphism.

COROLLARY 2.2.4 *Let G be a finite-dimensional Lie group with Lie algebra $\mathfrak{g}$, and let $r \in \mathfrak{g} \otimes \mathfrak{g}$ be skew-symmetric and non-degenerate. Then, r is a solution of the CYBE if and only if the inverse of the isomorphism $\mathfrak{g}^* \to \mathfrak{g}$ induced by r, regarded as a bilinear form on $\mathfrak{g}$, is a 2-cocycle.*

We recall that a bilinear form ω on a Lie algebra $\mathfrak{g}$ is a *2-cocycle* if

$$\omega([X,Y],Z) + \omega([Y,Z],X) + \omega([Z,X],Y) = 0$$

for all $X, Y, Z \in \mathfrak{g}$.

PROOF By 2.2.2, r satisfies the CYBE if and only if (9) defines a left-invariant Poisson structure on G. It is clear that the Poisson bivector of (9) is everywhere non-degenerate if and only if the matrix (r^{st}) is invertible. As we mentioned in Section 1.1, a non-degenerate Poisson structure is the same thing as a symplectic structure. Finally, a left-invariant symplectic form on G is determined by its value at the identity element of G, which is a 2-cocycle on $\mathfrak{g}$. ∎

Of course, there exist non-zero elements $r \in \mathfrak{g} \otimes \mathfrak{g}$ which are both skew and non-degenerate only if $\dim(\mathfrak{g})$ is even. Apart from this, it might be thought that non-degeneracy is a serious restriction. If one is looking for solutions of the CYBE, however, this is not so. For example, we have

PROPOSITION 2.2.5 *If $\mathfrak{g}$ is a finite-dimensional complex semisimple Lie alge-*
bra, every skew-symmetric solution of the CYBE in $\mathfrak{g} \otimes \mathfrak{g}$ is degenerate.

PROOF Fix a non-degenerate invariant symmetric bilinear form $(\, , \,)$ on $\mathfrak{g}$.
Suppose for a contradiction that $r \in \mathfrak{g} \otimes \mathfrak{g}$ is non-degenerate, skew-symmetric
and satisfies the CYBE. Let ω be the 2-cocycle on $\mathfrak{g}$ associated to r by 2.2.4.
Since $(\, , \,)$ is non-degenerate, we can write $\omega(X, Y) = (\mathcal{L}(X), Y)$ for some
linear map $\mathcal{L} : \mathfrak{g} \to \mathfrak{g}$. One checks that the cocycle property of ω is equivalent
to $\mathcal{L}$ being a derivation of $\mathfrak{g}$. As $\mathfrak{g}$ is semisimple, $\mathcal{L}$ is an inner derivation, i.e.
$\mathcal{L}(X) = [L, X]$ for some $L \in \mathfrak{g}$ and all $X \in \mathfrak{g}$. But then L lies in the kernel of
ω, contradicting non-degeneracy. ∎

On the other hand, non-degeneracy can always be achieved by passing to
a subalgebra:

PROPOSITION 2.2.6 *Let $\mathfrak{g}$ be a finite-dimensional Lie algebra, and let $r \in \mathfrak{g} \otimes \mathfrak{g}$*
be a skew-symmetric solution of the CYBE. Then there is a Lie subalgebra
$\mathfrak{g}_0$ of $\mathfrak{g}$ such that r lies in $\mathfrak{g}_0 \otimes \mathfrak{g}_0$ and is non-degenerate when regarded as an
element of this subspace of $\mathfrak{g} \otimes \mathfrak{g}$.

PROOF If r is non-degenerate there is nothing to prove. So let $m \geq 1$ be
the dimension of the kernel of r, regarded as a skew-symmetric bilinear form
on $\mathfrak{g}^*$. Then it follows from elementary linear algebra that we may choose a
basis

$$\{X_1, \ldots, X_n, Y_1, \ldots, Y_n, Z_1, \ldots, Z_m\}$$

of $\mathfrak{g}$ such that

$$r = \sum_{i=1}^{n} X_i \wedge Y_i.$$

We must show that the linear subspace $\mathfrak{g}_0$ of $\mathfrak{g}$ spanned by the X_i and Y_i for
$1 \leq i \leq n$ is a Lie subalgebra of $\mathfrak{g}$.

Since r satisfies the CYBE, we have

$$\begin{aligned}
[[r, r]] = \sum_{i,j=1}^{n} &\big([X_i, X_j] \otimes Y_i \otimes Y_j - [X_i, Y_j] \otimes Y_i \otimes X_j - [Y_i, X_j] \otimes X_i \otimes Y_j \\
&+ [Y_i, Y_j] \otimes X_i \otimes X_j + X_i \otimes [Y_i, X_j] \otimes Y_j - X_i \otimes [Y_i, Y_j] \otimes X_j \\
&- Y_i \otimes [X_i, X_j] \otimes Y_j + Y_i \otimes [X_i, Y_j] \otimes X_j + X_i \otimes X_j \otimes [Y_i, Y_j] \\
&- X_i \otimes Y_j \otimes [Y_i, X_j] - Y_i \otimes X_j \otimes [X_j, Y_i] + Y_i \otimes Y_j \otimes [X_i, X_j]\big) = 0.
\end{aligned}$$

Suppose, for example, that $[X_i, X_j]$ in the first term in the sum involves some
Z_k. A moment's inspection shows that none of the other eleven terms involves
$Z_k \otimes Y_i \otimes Y_j$, which is a contradiction. ∎

We conclude this section with a very interesting application of the idea of
2.2.4 to the problem of constructing Lie bialgebra structures on the infinite-
dimensional Lie algebra of vector fields on the circle.

Example 2.2.7 Witten (1988a) attempts to construct Lie bialgebra structures on the Lie algebra $\mathrm{Vect}_{\mathbb{R}}(S^1)$ of smooth real-valued vector fields on the circle S^1 by inverting 2-cocycles ω on $\mathrm{Vect}_{\mathbb{R}}(S^1)$. Note that, for the proof of 2.2.4 to go through in this infinite-dimensional situation, ω must be non-degenerate in the strong sense, that the linear map $\mathrm{Vect}_{\mathbb{R}}(S^1) \to \mathrm{Vect}_{\mathbb{R}}(S^1)^*$ induced by ω must have a well-defined inverse.

The first observation is that *a 2-cocycle ω on $\mathrm{Vect}_{\mathbb{R}}(S^1)$ can be non-degenerate only if it is a coboundary.* The dual space $\mathrm{Vect}_{\mathbb{R}}(S^1)^*$ is a space of quadratic differentials on S^1 (pairing a vector field and a quadratic differential gives a 1-form. which can be integrated over S^1). Then, ω is a coboundary if there is a quadratic differential Φ such that

$$\omega(X,Y) = \langle \Phi, [X,Y] \rangle$$

for all $X,Y \in \mathrm{Vect}_{\mathbb{R}}(S^1)$. If we write $X = f(d/d\theta)$, $Y = g(d/d\theta)$, $\Phi = \varphi\, d\theta \otimes d\theta$. where f, g and φ are functions on S^1, we find, after an integration by parts,

$$(10) \qquad \omega(f,g) = \int_0^{2\pi} (2\varphi g' + \varphi' g) f \, d\theta.$$

The most general 2-cocycle is obtained by adding to (10) a term $c \int fg''' d\theta$ for some $c \in \mathbb{R}$ (see Kirillov (1976)). The kernel of the resulting cocycle is the space of solutions of the differential equation

$$(11) \qquad cg''' + 2\varphi g' + \varphi' g = 0$$

which have period 2π. One may argue heuristically, with Witten, that if $c \neq 0$ and $\varphi \equiv 0$, the only periodic solutions of (11) are $g = $ constant, so the kernel has dimension 1. But the dimension modulo 2 of the kernel of a skew form is a topological invariant, so if $c \neq 0$ the kernel has dimension 1 or 3 for all φ. For a rigorous argument, see Kirillov (1976).

We assume from now on that $c = 0$. We must also assume that φ has at least one zero, for otherwise $g = \varphi^{-1/2}$ is a solution of (11) and so ω is still degenerate. With this assumption, the map $\mathrm{Vect}_{\mathbb{R}}(S^1) \to \mathrm{Vect}_{\mathbb{R}}(S^1)^*$ induced by ω, which is given by

$$(12) \qquad \omega(g) = 2\varphi g' + \varphi' g,$$

is injective. To prove surjectivity, we must solve, for any given ψ, the differential equation

$$2\varphi g' + \varphi' g = \psi,$$

i.e.

$$(13) \qquad \frac{d}{d\theta}(g\sqrt{\varphi}) = \frac{\psi}{2\sqrt{\varphi}}.$$

Formally, the inverse of ω is therefore the map $r : \text{Vect}_{\mathbb{R}}(S^1)^* \to \text{Vect}_{\mathbb{R}}(S^1)$, given by

$$(14) \qquad r(\psi) = \frac{1}{2\sqrt{\varphi}} \int_0^{2\pi} \frac{\psi}{\sqrt{\varphi}} d\theta.$$

The preceding argument is only heuristic, partly because the existence of zeros of φ makes it unclear whether (14) makes sense, and partly because we have been too vague about the definition of the space $\text{Vect}_{\mathbb{R}}(S^1)$ and its dual. In fact, a closer analysis shows that there are obstructions to making the argument rigorous.

It turns out to be easier to work, at least in the first instance, with a space $\text{Vect}_{\mathbb{C}}(S^1)$ of *complex* vector fields on S^1. The precise definition of $\text{Vect}_{\mathbb{C}}(S^1)$ is as follows. For any $a > 1$, we consider the space V_a of holomorphic vector fields on the annulus $A_a = \{z \in \mathbb{C} \mid 1 < |z| < a\}$, provided with the compact-open topology. Then, $\text{Vect}_{\mathbb{C}}(S^1)$ is the direct limit of the V_a as a tends to 1.

Changing coordinates from θ to $z = e^{\sqrt{-1}\theta}$, (13) becomes

$$(15) \qquad \frac{d}{dz}(g\sqrt{\varphi}) = -\frac{\sqrt{-1}\psi}{2z\sqrt{\varphi}}.$$

We assume that, for some $a > 1$, φ has no zeros in A_a. Because of the presence of square roots, we must work on the double cover $\tilde{A}_a$ of A_a. Let z_0 be any fixed point of $\tilde{A}_a$. Then the solution of (15) can be written

$$(16) \qquad g(z) = -\frac{\sqrt{-1}}{2\sqrt{\varphi(z)}} \int_{\gamma_{z_0}^z} \frac{\psi(w)}{w\sqrt{\varphi(w)}} dw + \frac{\sqrt{\varphi(z_0)}}{\sqrt{\varphi(z)}} g(z_0),$$

where $\gamma_{z_0}^z$ is a path in $\tilde{A}_a$ from z_0 to z. Let z_1 be the other point in $\tilde{A}_a$ with the same image z in A_a as z_0, let $\gamma_{z_0}^{z_1}$ be a path in $\tilde{A}_a$ from z_0 to z_1 of winding number 1 about the origin, and let γ_z be its image in A_a (see Fig. 3). For (16) to define a genuine function on A_a, we must have $g(z_0) = g(z_1)$, i.e.

$$(17) \qquad g(z_0) = -\frac{\sqrt{-1}}{2\sqrt{\varphi(z_1)}} \int_{\gamma_{z_0}^{z_1}} \frac{\psi(w)}{w\sqrt{\varphi(w)}} dw + \frac{\sqrt{\varphi(z_0)}}{\sqrt{\varphi(z_1)}} g(z_0).$$

There are two cases to consider. If φ has even winding number about the origin, then $\sqrt{\varphi(z_0)} = \sqrt{\varphi(z_1)}$ and (17) forces

$$(18) \qquad \int_{\gamma_{z_0}^{z_1}} \frac{\psi(w)}{w\sqrt{\varphi(w)}} dw = 0.$$

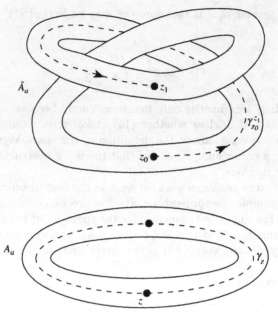

Fig. 3 Paths in A_a and in its double cover $\tilde{A}_a$

This cannot hold for all ψ, since $\mathrm{Vect}_{\mathbf{C}}(S^1)$ is dense in the space of continuous vector fields on S^1. Hence, ω is not surjective in this case.

If φ has odd winding number, so that $\sqrt{\varphi(z_0)} = -\sqrt{\varphi(z_1)}$, then (17) has a unique solution for $g(z_0)$. Reinterpreting the resulting formula on A_a, we see that in this case (15) has the unique solution

$$(19) \qquad r(\psi)(z) = -\frac{\sqrt{-1}}{4\sqrt{\varphi(z)}} \int_{\gamma_z} \frac{\psi(w)}{w\sqrt{\varphi(w)}} dw.$$

This is the rigorous version of (14).

Passing to the direct limit as a tends to 1, one finds that (19) defines a continuous linear map $r : \mathrm{Vect}_{\mathbf{C}}(S^1) \to \mathrm{Vect}_{\mathbf{C}}(S^1)$. Finally, one shows that $\mathrm{Vect}_{\mathbf{C}}(S^1)$ is isomorphic to its (continuous) dual, so r may be regarded as a map $\mathrm{Vect}_{\mathbf{C}}(S^1)^* \to \mathrm{Vect}_{\mathbf{C}}(S^1)$. This gives a genuine Lie bialgebra structure on $\mathrm{Vect}_{\mathbf{C}}(S^1)$, except that its cocommutator does not take values in the algebraic tensor product $\mathrm{Vect}_{\mathbf{C}}(S^1) \otimes \mathrm{Vect}_{\mathbf{C}}(S^1)$ but only in a suitable completion of it.

Returning now to the real case, we can of course regard $\mathrm{Vect}_{\mathbf{R}}(S^1) \subset \mathrm{Vect}_{\mathbf{C}}(S^1)$ as the vector fields which are real valued on the circle, but the question is whether $r(\mathrm{Vect}_{\mathbf{R}}(S^1)) \subset \mathrm{Vect}_{\mathbf{R}}(S^1)$. In fact, this is *never* the case, so the construction does not yield any Lie bialgebra structures on $\mathrm{Vect}_{\mathbf{R}}(S^1)$. This is easiest to see when φ has a zero of order greater than 1 on S^1, say at z_0. In that case, it is obvious from (12) that $\omega(g)(z_0) = 0$ for all g.

Hence, if $\psi(z_0) \neq 0$, then $r(\psi) \notin \text{Vect}_{\mathbb{R}}(S^1)$. If all the zeros have order 1, we argue as follows. The integral in (19) depends continuously on z. Hence, if $r(\psi) \in \text{Vect}_{\mathbb{R}}(S^1)$, then $r(\psi)(z)$ is continuous even at a zero z_0 of φ, so the integral in (18) must vanish. As we noted above, this cannot happen for all ψ. $\Diamond$

C The classical R-matrix Although we have described a number of special properties of Poisson–Lie groups G whose tangent Lie bialgebras $\mathfrak{g}$ are coboundary, we have not yet identified the geometric object on G which corresponds to the classical r-matrix $r \in \mathfrak{g} \otimes \mathfrak{g}$. To do so, we shall make the slightly stronger assumption that r satisfies the CYBE and that $r + \sigma(r)$ is non-degenerate. One says that $\mathfrak{g}$ is *factorizable* in this case (the reason for this terminology will appear in a moment).

As in the previous subsection, we may regard r and $-\sigma(r)$ as linear maps $\mathfrak{g}^* \to \mathfrak{g}$, and, by 2.1.4, these linear maps are actually Lie algebra homomorphisms (note that $-\sigma(r)$ defines the same Lie bialgebra structure on $\mathfrak{g}$ as r because $r + \sigma(r)$ is $\mathfrak{g}$-invariant). Assuming that G and its double G^* are connected and simply-connected, these homomorphisms lift to group homomorphisms S and S^σ from G^* to G. Let $T : G^* \to G$ be defined by $T(x) = S(x)S^\sigma(x)^{-1}$; since $r + \sigma(r)$ is non-degenerate, T is a local diffeomorphism near the identity elements of G^* and G. One says that G is *factorizable* if T is a (global) diffeomorphism, since in this case every element $g \in G$ admits the factorization $g = g_+ g_-^{-1}$, where $g_+ = S(T^{-1}(g))$, $g_- = S^\sigma(T^{-1}(g))$.

PROPOSITION 2.2.8 *Let G be a factorizable Poisson–Lie group and define the classical R-matrix $R : G \times G \to G \times G$ by*

$$R(g, h) = (h_- g h_-^{-1}, (h_- g h_-^{-1})_+^{-1} h (h_- g h_-^{-1})_+).$$

Then, R is a Poisson diffeomorphism if G is equipped with the Poisson structure of G^*, using T to identify G and G^* (see above). Moreover, R is a solution of the quantum Yang–Baxter equation (QYBE):

$$R_{12} R_{13} R_{23} = R_{23} R_{13} R_{12}. \quad \blacksquare$$

In this equation, R_{12} means the map $G \times G \times G \to G \times G \times G$ given by

$$R_{12}(g, h, k) = (R(g, h), k),$$

and the meaning of R_{13} and of R_{23} is similar. Thus, the QYBE asserts the equality of two composites of three maps from $G \times G \times G$ to itself.

Example 2.2.9 Consider the group SU_{n+1} with the standard Poisson structure described in Subsection 1.4C, and let G be its connected, simply-connected double Poisson–Lie group. Then, $G \cong SL_{n+1}(\mathbb{C})$ as a Lie group. It is easy to check that the factorization $g = g_+ g_-^{-1}$ is exactly the Gram–Schmidt process, with $g_+ \in SU_{n+1}$, $g_- \in B$, where B is the group of upper triangular matrices in $SL_{n+1}(\mathbb{C})$ with positive (real) diagonal entries. Thus, G is factorizable, and the map R defined in 2.2.8 is a solution of the QYBE. $\Diamond$

Example 2.2.10 Let G be a connected, simply-connected complex Lie group whose Lie algebra $\mathfrak{g}$ is simple. Give $\mathfrak{g}$ the Lie bialgebra structure defined in 1.3.8, and give G the corresponding complex Poisson–Lie group structure. With the notation of that example, let $N_\pm$ and H be the complex Lie subgroups of G with Lie algebras $\mathfrak{n}_\pm$ and $\mathfrak{h}$. It is clear from 1.3.8 that the dual of G is the subgroup of $G \times G$ given by

$$G^* = \{(hn_+, n_- h^{-1}) \in G \times G \mid n_\pm \in N_\pm, \ h \in H\}.$$

The maps S and S^σ are the projections onto the two factors, so

$$T(hn_+, n_- h^{-1}) = n_+ h^2 (n_-)^{-1}.$$

Hence, T is a 2 to 1 map onto the 'big cell' $N_+ H N_- \subset G$. Thus, although $\mathfrak{g}$ is factorizable, G is not. However, the factorization $g = g_+ g_-^{-1}$ is still valid for elements g in the big cell, and, because g_+ and g_- are well defined up to a sign, it is clear that the R-matrix in 2.2.8 is still uniquely determined. But it is not clear whether it satisfies the QYBE. $\Diamond$

Remark We shall meet the QYBE again when we discuss the quantization of Poisson–Lie groups. In that case, the role of G is played by a representation of the quantized group, and R is a linear map instead of a Poisson diffeomorphism. The reason for the terminology 'quantum Yang–Baxter equation' is easy to see in that situation. For, if we look for solutions of the QYBE of the form $R = \mathrm{id} + hr$, where h is a 'small' parameter, then the QYBE is satisfied identically up to order h, and is satisfied to order h^2 if and only if r satisfies the CYBE. $\triangle$

2.3 Classical integrable systems

We conclude this chapter with a brief discussion of the relationship between the CYBE and the theory of classical hamiltonian systems, where it plays the role of an *integrability condition*.

A Complete integrability We begin by recalling the notion of a completely integrable hamiltonian system. If a hamiltonian system is modelled by a Poisson manifold M (the phase space of the system) and a hamiltonian $\mathcal{H} \in$

$C^\infty(M)$, its time-evolution is given by following the integral curves of the hamiltonian vector field $X_\mathcal{H}$ on M corresponding to $\mathcal{H}$. We recall that

$$X_\mathcal{H}(f) = \{\mathcal{H}, f\}$$

for all $f \in C^\infty(M)$. It follows that

$$\frac{df}{dt} = \{\mathcal{H}, f\}.$$

(This equation is shorthand for $(d/dt)f(m(t)) = \{\mathcal{H}, f\}(m(t))$, where $m(t)$ is any integral curve of $X_\mathcal{H}$.) In particular, an observable f is *conserved*, or a *constant of the motion*, if $\{\mathcal{H}, f\} \equiv 0$. More generally:

DEFINITION 2.3.1 *Two observables f_1, f_2 on a Poisson manifold M are in involution if $\{f_1, f_2\} \equiv 0$.*

Note that every integral curve $m(t)$ of $X_\mathcal{H}$ lies entirely in some symplectic leaf of M. Hence, we might as well assume that M is symplectic, and we do so from now on.

A hamiltonian system is completely integrable if it has the maximum number of conserved quantities in involution.

DEFINITION 2.3.2 *The dynamical system defined on a $2m$-dimensional symplectic manifold M by a hamiltonian $\mathcal{H} \in C^\infty(M)$ is completely integrable if there exist m independent conserved quantities $f_1, \ldots, f_m \in C^\infty(M)$ in involution.*

Remarks [1] The meaning of 'independent' is that the set of critical values of the map $\boldsymbol{f} = (f_1, \ldots, f_m) : M \to \mathbb{R}^m$ is of (Lebesgue) measure zero (a point $c \in \mathbb{R}^m$ is a critical value of $\boldsymbol{f}$ if $c = \boldsymbol{f}(p)$ for some p such that $\boldsymbol{f}'_p : T_p(M) \to T_c(\mathbb{R}^m)$ is not surjective). Intuitively, this means that there should be no functional relation betwen the f_i.

[2] Usually, the hamiltonian $\mathcal{H}$ is taken to be one of the f_i. In any case, $\mathcal{H}$, and any other conserved quantity in involution with the f_i, is 'functionally dependent' on the f_i. $\triangle$

The importance of complete integrability, as the terminology implies, is that systems with this property can be 'solved' completely. In fact, one of the most important results about such systems asserts that if $\mathbf{b} \in \mathbb{R}^m$ is a regular value of $\boldsymbol{f}$, and if the level submanifold $\boldsymbol{f}^{-1}(\mathbf{b})$ is compact, then each of its connected components can be identified with an m-dimensional torus $T^m = S^1 \times \cdots \times S^1$ by a diffeomorphism which carries the flow of $X_\mathcal{H}$ on $\boldsymbol{f}^{-1}(\mathbf{b})$ to a rotation at constant speed in each factor of T^m.

B Lax pairs One of the principal tools used to construct and solve completely integrable systems is the Lax pair.

DEFINITION 2.3.3 *A Lax pair for a hamiltonian system on a symplectic manifold M is a pair (L, P) of C^∞ functions on M with values in a real Lie algebra $\mathfrak{g}$ such that the equations of motion of the system are equivalent to*

$$(20) \qquad \frac{dL}{dt} = [L, P],$$

where $[\ ,\ ]$ is the Lie bracket on $\mathfrak{g}$.

It is easy to produce conserved quantities for systems which admit a Lax pair. In fact, let f be a polynomial function of degree n on $\mathfrak{g}$ which is invariant under the coadjoint action of $\mathfrak{g}$. Then, $f(L, \ldots, L)$ is conserved:

$$
\begin{aligned}
\frac{d}{dt} f(L, \ldots, L) &= f(dL/dt, L, \ldots, L) + \cdots + f(L, L, \ldots, dL/dt) \\
&= f([L, P], L, \ldots, L) + \cdots + f(L, L, \ldots, [L, P]) \\
&= \mathrm{ad}_P(f)(L, L, \ldots, L) \\
&= 0.
\end{aligned}
$$

To produce conserved quantities of this type in involution, we assume further that

$$(21) \qquad \{L, L\} = [r, L_1 + L_2]$$

for some $r \in \mathfrak{g} \otimes \mathfrak{g}$. Here, $L_1 = L \otimes 1$, $L_2 = 1 \otimes L$, and the Poisson bracket is extended to $\mathfrak{g}$-valued functions in the obvious way: if $\{X_s\}$ is a basis of $\mathfrak{g}$, and

$$
f_1 = \sum_s f_1^s X_s, \qquad f_2 = \sum_s f_2^s X_s,
$$

where $f_1^s, f_2^s \in C^\infty(M)$, then

$$
\{f_1, f_2\} = \sum_{s,t} \{f_1^s, f_2^t\}\, X_s \otimes X_t.
$$

Assume for simplicity that $\mathfrak{g} = sl_{n+1}(\mathbb{R})$ for some $n \geq 1$ (we could dispense with this assumption by taking a representation $\mathfrak{g} \to sl_{n+1}(\mathbb{R})$). Then, the functions

$$
f_k(X) = \mathrm{trace}(X^k), \qquad k = 1, 2, \ldots
$$

are obviously invariant polynomial functions on $\mathfrak{g}$. Moreover, the observables $f_k(L)$ and $f_l(L)$ are in involution for any k, l. Indeed,

$$
\begin{aligned}
\{f_k(L), f_l(L)\} &= (\mathrm{trace} \otimes \mathrm{trace})\{L^k, L^l\} \\
&= (\mathrm{trace} \otimes \mathrm{trace}) \sum_{i=0}^{k-1} \sum_{j=0}^{l-1} (L_1^{k-i-1} L_2^{l-j-1})\{L, L\}(L_1^i L_2^j) \\
&= (\mathrm{trace} \otimes \mathrm{trace}) \sum_{i=0}^{k-1} \sum_{j=0}^{l-1} (L_1^{k-i-1} L_2^{l-j-1})[r, L_1 + L_2](L_1^i L_2^j) \\
&= kl(\mathrm{trace} \otimes \mathrm{trace})(L_1^{k-1}[r, L_1 + L_2]L_2^{l-1}),
\end{aligned}
$$

where the second equality used the derivation property of $\{\ ,\ \}$, the third equality used (21), and the last equality used the invariance property of the trace. The last expression vanishes because, for example,

$$k(\text{trace} \otimes 1)(L_1^{k-1}[r, L_1]L_2^{l-1}) = k(\text{trace} \otimes 1)(L_1^{k-1}[r, L_1]).L_2^{l-1}$$
$$= (\text{trace} \otimes 1)([r, L_1^k]).L_2^{l-1}$$
$$= 0.$$

Remark Equation (21) imposes restrictions on the element $r \in \mathfrak{g} \otimes \mathfrak{g}$. In fact, it is easy to show that the skew-symmetry property and Jacobi identity satisfied by the Poisson bracket $\{\ ,\ \}$ are equivalent to

$$[r_{12} + r_{21},\, L_1 + L_2] = 0$$

and

$$[\,[[r, r]],\, L_1 + L_2 + L_3] = 0,$$

respectively. Note that these conditions will hold if $r_{12} + r_{21}$ and $[[r, r]]$ are $\mathfrak{g}$-invariant elements of $\mathfrak{g} \otimes \mathfrak{g}$ and $\mathfrak{g} \otimes \mathfrak{g} \otimes \mathfrak{g}$. In view of 2.1.2, this is a strong indication of the close relationship between coboundary Lie bialgebras and integrable hamiltonian systems. $\triangle$

C Integrable systems from r-matrices We now reverse the preceding discussion and give a general procedure for constructing completely integrable hamiltonian systems starting from an r-matrix.

Let $\mathfrak{g}$ be a finite-dimensional real Lie algebra and let $(\ ,\ )$ be a non-degenerate invariant symmetric bilinear form on $\mathfrak{g}$. Let $r \in \mathfrak{g} \otimes \mathfrak{g}$ be a skew-symmetric solution of the MCYBE (5). We may regard r as a linear map $\mathfrak{g}^* \to \mathfrak{g}$, and hence as a linear map $\rho : \mathfrak{g} \to \mathfrak{g}$, identifying $\mathfrak{g}^*$ with $\mathfrak{g}$ by using the inner product $(\ ,\ )$.

PROPOSITION 2.3.4 *With the above notation, let $r \in \mathfrak{g} \otimes \mathfrak{g}$ be a skew-symmetric solution of the MCYBE and define, for $X, Y \in \mathfrak{g}$,*

$$(22) \qquad\qquad [X, Y]_r = [\rho(X), Y] + [X, \rho(Y)].$$

Then $[\ ,\ ]_r$ is a Lie bracket on $\mathfrak{g}$.

PROOF Skew-symmetry is obvious, and the Jacobi identity is a straightforward verification. ∎

Remark Since r satisfies the MCYBE, it defines via (1) a Lie bialgebra structure on $\mathfrak{g}$, and hence a Lie algebra structure on $\mathfrak{g}^*$. Identifying $\mathfrak{g}$ with $\mathfrak{g}^*$ using the inner product, we get another Lie algebra structure on $\mathfrak{g}$. This is the Lie bracket defined in the proposition. $\triangle$

To avoid confusion, we denote by $\mathfrak{g}_r$ the vector space $\mathfrak{g}$ with the Lie bracket $[\,.\;]_r$, by ad_r the adjoint action of $\mathfrak{g}_r$ on itself, and by ad_r^* the corresponding coadjoint action. Note that ad_r^* can be expressed in terms of ad^*:

$$(23) \qquad (\mathrm{ad}_r)_X^*(\xi) = \mathrm{ad}_{\rho(X)}^*(\xi) + \rho^*(\mathrm{ad}_X^*(\xi)).$$

We also denote by G (resp. G_r) a connected Lie group with Lie algebra $\mathfrak{g}$ (resp. $\mathfrak{g}_r$).

By 2.3.4 and 1.1.3, we have two Poisson-Lie group structures on the vector space $\mathfrak{g}^*$ (viewed as an additive group), namely

$$(24) \qquad \{f_1, f_2\} = \langle [df_1(\xi), df_2(\xi)], \xi \rangle$$

and

$$(25) \qquad \{f_1, f_2\}_r = \langle [df_1(\xi), df_2(\xi)]_r, \xi \rangle$$

for $f_1, f_2 \in C^\infty(\mathfrak{g}^*)$.

The next result shows that the functions f on $\mathfrak{g}^*$ which are invariant under the coadjoint action of G give examples of observables in involution. Note that the infinitesimal expression of the invariance of f is

$$(26) \qquad \langle df(\xi), \mathrm{ad}_X^*(\xi) \rangle = 0 \qquad (X \in \mathfrak{g}, \ \xi \in \mathfrak{g}^*).$$

PROPOSITION 2.3.5 *The C^∞ functions on $\mathfrak{g}^*$ which are invariant under the coadjoint action of G are in involution with respect to both Poisson brackets $\{\,.\,\}$ and $\{\,,\,\}_r$ on $\mathfrak{g}^*$.*

PROOF Suppose that $f_1, f_2 \in C^\infty(\mathfrak{g}^*)$ are invariant under the coadjoint action of G. Then,

$$\{f_1, f_2\}(\xi) = \langle [df_1(\xi), df_2(\xi)], \xi \rangle = \langle df_2(\xi), \mathrm{ad}_{df_1(\xi)}^*(\xi) \rangle = 0$$

by (26).

On the other hand,

$$\{f_1, f_2\}_r(\xi) = -\langle [\rho(df_1(\xi)), df_2(\xi)], \xi \rangle - \langle [df_1(\xi), \rho(df_2(\xi))], \xi \rangle$$
$$= -\langle df_2(\xi), \mathrm{ad}_{\rho(df_1(\xi))}^*(\xi) \rangle + \langle df_1(\xi), \mathrm{ad}_{\rho(df_2(\xi))}^*(\xi) \rangle$$
$$= 0. \quad \blacksquare$$

This proposition. and the definition of complete integrability, suggest that we consider a dynamical system on $\mathfrak{g}^*$ with the Poisson bracket $\{\,,\,\}_r$, using as hamiltonian a function $\mathcal{H}$ satisfying the invariance condition (26). The equations of motion are then, for any $f \in C^\infty(\mathfrak{g}^*)$,

$$\frac{d}{dt} f(\xi) = \{\mathcal{H}, f\}_r(\xi) = \langle df(\xi), (\mathrm{ad}_r)^*_{d\mathcal{H}(\xi)}(\xi) \rangle$$
$$= \langle df(\xi), \mathrm{ad}^*_{\rho(d\mathcal{H}(\xi))}(\xi) \rangle + \langle df(\xi), \rho^*(\mathrm{ad}^*_{d\mathcal{H}(\xi)}(\xi)) \rangle$$

by (23). But the second term on the right-hand side vanishes because

$$\langle df(\xi), \rho^*(\mathrm{ad}^*_{d\mathcal{H}(\xi)}(\xi)) \rangle = \langle \rho(df(\xi)), \mathrm{ad}^*_{d\mathcal{H}(\xi)}(\xi) \rangle$$
$$= \langle [d\mathcal{H}(\xi), \rho(df(\xi))], \xi \rangle$$
$$= -\langle d\mathcal{H}(\xi), \mathrm{ad}^*_{\rho(df(\xi))}(\xi) \rangle$$
$$= 0,$$

by the invariance of $\mathcal{H}$. Hence, the equations of motion are

(27) $$\frac{d}{dt} f(\xi) = \langle df(\xi), \mathrm{ad}^*_{\rho(d\mathcal{H}(\xi))}(\xi) \rangle.$$

Note that (27) implies that the integral curves of the equations of motion lie on the coadjoint orbits of G on $\mathfrak{g}^*$. By the choice of Poisson bracket, they also lie on the coadjoint orbits of G_r.

We shall now show that the equations of motion (27) can be written in Lax form. Let $L : \mathfrak{g}^* \to \mathfrak{g}$ be the canonical map associated to the chosen inner product on $\mathfrak{g}$, and let $P(\xi) = \rho(d\mathcal{H}(\xi))$. Note that $L(\xi) = (\xi \otimes 1)(t)$, where $t \in \mathfrak{g} \otimes \mathfrak{g}$ is the Casimir element (not to be confused with the time t!). Then,

$$\frac{d}{dt} L(\xi) = (\mathrm{ad}^*_{P(\xi)}(\xi) \otimes 1)(t) = (\xi \otimes 1)[P(\xi) \otimes 1, t] = -(\xi \otimes 1)[1 \otimes P(\xi), t],$$

by the $\mathrm{ad}_\mathfrak{g}$-invariance of t, and hence

$$\frac{d}{dt} L(\xi) = [(\xi \otimes 1)(t), P(\xi)] = [L, P](\xi).$$

This proves the first part of

PROPOSITION 2.3.6 *Let* $\mathcal{H}$ *be a smooth function on* $\mathfrak{g}^*$ *which is invariant under the coadjoint action of* G, *and let* $r \in \mathfrak{g} \otimes \mathfrak{g}$ *be a skew-symmetric solution of the MCYBE. Then, the hamiltonian system on* $\mathfrak{g}^*$ *with Poisson bracket* $\{\ ,\ \}_r$ *and hamiltonian* $\mathcal{H}$ *admits a Lax pair* (L, P). *Moreover,*

$$\{L, L\}_r = [r, L \otimes 1 + 1 \otimes L].$$

PROOF Let $\{X_s\}$ be a basis of $\mathfrak{g}$ and let $\{X^s\}$ be the dual basis with respect to the inner product on $\mathfrak{g}$. Suppose that

$$r = \sum_{s,t} r^{st} X^s \otimes X_t = -\sum_{s,t} r^{ts} X_s \otimes X^t.$$

Then,

$$\rho(X_s) = \sum_t r^{st} X_t, \quad \rho(X^s) = -\sum_t r^{ts} X^t.$$

Hence, since

$$L(\xi) = \sum_s L^s(\xi) X_s,$$

where $L^s(\xi) = \langle X^s, \xi \rangle$, we have

$$
\begin{aligned}
\{L, L\}_r(\xi) &= \sum_{s,t} \{L^s, L^t\}_r(\xi)(X_s \otimes X_t) \\
&= \sum_{s,t} \langle [dL^s(\xi), dL^t(\xi)]_r, \xi \rangle (X_s \otimes X_t) \\
&= \sum_{s,t} \langle [X^s, X^t]_r, \xi \rangle (X_s \otimes X_t).
\end{aligned}
$$

Hence, $\{L, L\}_r(\xi)$ is the sum of two terms:

$$(28) \qquad \sum_{s,t,u} \langle r^{us}[X^u, X^t], \xi \rangle (X_s \otimes X_t) + \sum_{s,t,u} \langle r^{ut}[X^s, X^u], \xi \rangle (X_s \otimes X_t).$$

On the other hand,

$$
\begin{aligned}
[r, L_1] &= \sum_{s,t,u} r^{st}[X^s \otimes X_t, \langle X^u, \xi \rangle X_u \otimes 1] \\
(29) \qquad &= \sum_{s,t,u} r^{st} \langle X^u, \xi \rangle [X^s, X_u] \otimes X_t.
\end{aligned}
$$

Now, the $\mathfrak{g}$-invariance of $t = \sum_u X^u \otimes X_u$ implies that

$$\sum_u ([X^s, X^u] \otimes X_u + X^u \otimes [X^s, X_u]) = 0.$$

Applying $\xi \otimes 1$ gives

$$\sum_u \langle X^u, \xi \rangle [X^s, X_u] = -\sum_u \langle [X^s, X^u], \xi \rangle X_u.$$

Hence,

$$[r, L_1] = -\sum_{s,t,u} r^{st} \langle [X^s, X^u], \xi \rangle (X_u \otimes X_t).$$

Interchanging the roles of s and u in the last sum, we see that $[r, L_1]$ is equal to the second sum on the right-hand side of (28). Similarly, $[r, L_2]$ is equal to the first sum on the right-hand side of (28). This proves that

$$\{L, L\}_r = [r, L_1 + L_2]. \quad \blacksquare$$

D Toda systems We illustrate these constructions with an example of Toda lattice type.

Example 2.3.7 Let $\mathfrak{g}$ be a finite-dimensional complex simple Lie algebra of rank n, and take the standard inner product on $\mathfrak{g}$. We use the notation in the Appendix.

Let $\mathfrak{g}^{\mathbb{R}}$ be a *split* real form of $\mathfrak{g}$ (if $\mathfrak{g} = sl_{n+1}(\mathbb{C})$, then $\mathfrak{g}^{\mathbb{R}} = sl_{n+1}(\mathbb{R})$). The elements

(30)
$$\{H_i\}_{i=1,\ldots,n} \cup \{X_\alpha^\pm\}_{\alpha \in \Delta^+}$$

form a basis of $\mathfrak{g}^{\mathbb{R}}$ over $\mathbb{R}$. Let $\mathfrak{h}^{\mathbb{R}}$ be the $\mathbb{R}$-linear span of the H_i.

We take the skew solution

$$r = \frac{1}{2} \sum_{\alpha \in \Delta^+} d_\alpha X_\alpha^+ \wedge X_\alpha^-$$

of the MCYBE (see 2.1.7); note that $r \in \mathfrak{g}^{\mathbb{R}} \otimes \mathfrak{g}^{\mathbb{R}}$. Using the formulas in the Appendix, the associated linear map $\rho : \mathfrak{g}^{\mathbb{R}} \to \mathfrak{g}^{\mathbb{R}}$ is easily found to be

$$\rho(H_i) = 0, \quad \rho(X_\alpha^\pm) = \mp \frac{1}{2} X_\alpha^\pm,$$

and the Lie bracket $[\ ,\ ]_r$ to be

(31)
$$[H_i, H_j]_r = 0, \quad [H_i, X_\alpha^\pm]_r = -\frac{1}{2}(\alpha, \alpha_i) X_\alpha^\pm,$$
$$[X_\alpha^\pm, X_\beta^\pm]_r = \pm C_{\alpha\beta} X_{\alpha+\beta}^\pm, \quad [X_\alpha^+, X_\beta^-]_r = 0,$$

for some constants $C_{\alpha\beta}$.

To describe the Poisson bracket $\{\ ,\ \}_r$ on $(\mathfrak{g}^{\mathbb{R}})^*$, we introduce the coordinate functions

$$h_i(\xi) = \langle H_i, \xi \rangle, \quad x_\alpha^\pm(\xi) = \langle X_\alpha^\pm, \xi \rangle$$

for $\xi \in (\mathfrak{g}^{\mathbb{R}})^*$. We find that

$$\{h_i, h_j\}_r = 0, \quad \{h_i, x_\alpha^\pm\}_r = -\frac{1}{2}(\alpha, \alpha_i) x_\alpha^\pm,$$
$$\{x_\alpha^\pm, x_\beta^\pm\}_r = \pm C_{\alpha\beta} x_{\alpha+\beta}^\pm, \quad \{x_\alpha^+, x_\beta^-\}_r = 0.$$

Our example is essentially a symplectic leaf (i.e. a coadjoint orbit) of this system. In fact, we take the linear submanifold $\mathfrak{g}_T^* \subset (\mathfrak{g}^{\mathbb{R}})^*$ defined by the equations

$$x_\alpha^+ = x_\alpha^- \quad \text{if} \quad \alpha \in \Pi, \quad x_\alpha^\pm = 0 \quad \text{if} \quad \alpha \in \Delta^+ \backslash \Pi.$$

We must show that $\mathfrak{g}_T^*$ is preserved by the coadjoint action of the group $G_r^{\mathbb{R}}$ with Lie algebra $\mathfrak{g}_r^{\mathbb{R}}$. For this, it suffices to show that $(\mathrm{ad}_r)_X^*(\xi) \in \mathfrak{g}_T^*$ for all $X \in \mathfrak{g}^{\mathbb{R}}$, $\xi \in \mathfrak{g}_T^*$, and it is enough to check this when X runs through the basis (30) of $\mathfrak{g}^{\mathbb{R}}$. The verification is straightforward, using the relations (31) and the fact that a sum of two positive roots cannot be simple.

Note that the discussion in this section implies that if we take as hamiltonian the restriction to $\mathfrak{g}_T^*$ of any $G^{\mathbb{R}}$-invariant function on $(\mathfrak{g}^{\mathbb{R}})^*$, the resulting system is completely integrable. Indeed, this follows from 2.3.6 and the fact that there are $n = \frac{1}{2}\dim(\mathfrak{g}_T^*)$ independent $G^{\mathbb{R}}$-invariant polynomials on $\mathfrak{g}^{\mathbb{R}}$.

One natural choice for the hamiltonian is the manifestly $G^{\mathbb{R}}$-invariant function

$$\mathcal{H}(\xi) = \frac{1}{2}(\xi, \xi).$$

This function also admits a simple interpretation in terms of the Lax pair. Indeed, the Lax operator L for this problem is

$$L(\xi) = (\xi \otimes 1)(t) = \sum_{i=1}^{n} \left(h_i(\xi)\check{H}_i + d_i x_i(\xi)(X_i^+ + X_i^-) \right),$$

where $\{\check{H}_i\}$ is the basis of $\mathfrak{h}^{\mathbb{R}}$ dual to $\{H_i\}$ with respect to the inner product. It is clear that $\mathcal{H} = \frac{1}{2}f_2(L)$, where f_2 is the $G^{\mathbb{R}}$-invariant quadratic polynomial on $\mathfrak{g}^{\mathbb{R}}$ given by the inner product.

To interpret this hamiltonian, let us introduce coordinates on $\mathfrak{g}_T^*$. The most obvious choice is the set of functions h_i and $x_i \equiv x_{\alpha_i}^{\pm}$, $i = 1, \ldots, n$. Their Poisson brackets are given by

$$(32) \qquad \{h_i, h_j\}_r = 0, \quad \{h_i, x_j\}_r = -\frac{1}{2}(\alpha_i, \alpha_j)x_j, \quad \{x_i, x_j\}_r = 0.$$

If we introduce new coordinates $p_1, \ldots, p_n, q_1, \ldots, q_n$ on the open subset of $\mathfrak{g}_T^*$ where the x_i are positive by

$$h_i = p_i, \quad x_i = \exp\left[-\frac{1}{2}\sum_{j=1}^{n}(\alpha_i, \alpha_j)q_j \right],$$

we find that the p_i and q_i satisfy the canonical commutation relations

$$(33) \qquad \{p_i, p_j\}_r = 0, \quad \{p_i, q_j\}_r = \delta_{i,j}, \quad \{q_i, q_j\}_r = 0.$$

In these canonical coordinates, the hamiltonian is given by

$$(34) \qquad \mathcal{H}(\xi) = \frac{1}{2}\sum_{i,j=1}^{n} B_{ij}p_i p_j + \sum_{i=1}^{n} \exp\left[-\frac{1}{2}\sum_{j=1}^{n}(\alpha_i, \alpha_j)q_j \right],$$

where B is the inverse of the symmetrized Cartan matrix $(d_j^{-1}a_{ij}) = (H_i, H_j)$ of $\mathfrak{g}$. Since the quadratic form $\sum_{i,j} B_{ij}p_ip_j$ is positive definite, there is a linear change of coordinates

$$P_i = \sum_j C_{ij}p_j, \quad Q_i = \sum_j (C^{-1})_{ji}q_j$$

such that the P_i and Q_i still satisfy the canonical commutation relations (33), and

$$\sum_{i,j} B_{ij}p_ip_j = \sum_i P_i^2.$$

Thus, $\mathcal{H}$ may be regarded as the hamiltonian for a system of n particles with positions Q_i and momenta P_i interacting via a potential given by the second term on the right-hand side of (34).

The simplest case, where $\mathfrak{g}^{\mathbb{R}} = sl_2(\mathbb{R})$, can easily be solved by *ad hoc* methods. We leave it to the reader to check that every integral curve of $\mathcal{H} = h^2 + x^2$ is of the form

$$x = \frac{2Ex_0}{(E + h_0)e^{\frac{1}{2}Et} + (E - h_0)e^{-\frac{1}{2}Et}},$$

$$h = E\left(\frac{(E + h_0)e^{\frac{1}{2}Et} - (E - h_0)e^{-\frac{1}{2}Et}}{(E + h_0)e^{\frac{1}{2}Et} + (E - h_0)e^{-\frac{1}{2}Et}}\right),$$

where E^2 is the constant value of $\mathcal{H}$ along the curve, and x_0 and h_0 are the coordinates at time $t = 0$.

The solution in the general case can be reduced to a 'factorization' problem, but we shall not pursue this further. $\diamond$

Bibliographical notes

2.1 The CYBE first arose in the literature on inverse scattering theory; see Faddeev & Takhtajan (1987) for a thorough exposition. The algebraic results described in this section are due mainly to Drinfel'd (1983b) and Semenov-Tian-Shansky (1983, 1984).

2.2 The main references here are again Drinfel'd (1983b) and Semenov-Tian-Shansky (1983, 1984). See also Lu & Weinstein (1990).

The idea for the construction of coboundary Lie bialgebra structures on the Lie algebra of vector fields on the circle described in 2.2.7 is due to Witten (1988a), and its rigorous formulation to Beggs & Majid (1990).

For the classical R-matrix, see Weinstein & Xu (1992), where the proof of 2.2.8 can be found.

The QYBE was first encountered by C. N. Yang (1967), where it played a crucial role in his computation of the eigenfunctions of a one-dimensional fermion gas with delta function interactions. A little later, the QYBE arose in the solution, by R. J. Baxter (1972), of the eight vertex model in statistical mechanics. The terminology 'quantum Yang–Baxter equation' was apparently coined by L. D. Faddeev around 1980. (Strictly speaking, it was the quantum Yang–Baxter equation 'with spectral parameters' which played a role in the work of Yang and Baxter – see Subsection 7.5C.)

2.3 For the notion of complete integrability for dynamical systems, see Abraham & Marsden (1978). For a more detailed discussion of its relation to the classical Yang–Baxter equation, see Semenov-Tian-Shansky (1983, 1984, 1985).

For further information on systems of Toda lattice type, see Kostant (1979), Reyman & Semenov-Tian-Shansky (1979, 1981), Adler & van Moerbecke (1980) and Symes (1980a,b).

3

Solutions of the
classical Yang–Baxter equation

In Chapter 2, we saw the importance of describing the solutions r of the classical Yang–Baxter equation with values in $\mathfrak{g} \otimes \mathfrak{g}$, where $\mathfrak{g}$ is a Lie algebra. On the one hand, solutions of the CYBE lead to integrable dynamical systems. On the other hand, if the symmetric part of r is $\mathfrak{g}$-invariant, then r gives rise to a Lie bialgebra structure on $\mathfrak{g}$. Moreover, in certain cases, for example if $\mathfrak{g}$ is a finite-dimensional complex simple Lie algebra, every Lie bialgebra structure on $\mathfrak{g}$ arises in this way. In this chapter, we give several results which parametrize and, in some cases, explicitly describe, all the solutions of the CYBE for certain Lie algebras $\mathfrak{g}$.

We begin in Section 3.1 by classifying all the solutions of the CYBE when $\mathfrak{g}$ is a finite-dimensional complex simple Lie algebra. It turns out that rather different methods are required, depending on whether the solutions one is looking for are skew-symmetric or not. In the latter case, the classification is completely explicit; in particular, we shall see that the generic solution depends on $\frac{1}{2}n(n-1)$ parameters, where $n = \operatorname{rank}(\mathfrak{g})$. In the skew case, the classification is rather less explicit; it depends on the classification of the quasi-Frobenius Lie subalgebras of $\mathfrak{g}$, i.e. those subalgebras which admit a non-degenerate 2-cocycle (with values in $\mathbb{C}$).

In Section 3.2, we study the case where $\mathfrak{g}$ is a Lie algebra of functions on a space U with values in a Lie algebra $\mathfrak{a}$. The solutions of the CYBE may then be regarded as functions $r(u_1, u_2)$ on $U \times U$ with values in $\mathfrak{a} \otimes \mathfrak{a}$. The variables u_1 and u_2 are called *spectral parameters* in the physics literature. If U is an open subset of the complex plane, $\mathfrak{g}$ is a complex simple Lie algebra and r is a meromorphic function on $U \times U$ then, under a mild additional hypothesis, it turns out that every solution $r(u_1, u_2)$ is equivalent, in a certain natural sense, to a function which is elliptic, trigonometric (i.e. a rational function of e^{u_1} and e^{u_2}) or rational. In the elliptic and trigonometric cases, we give explicit descriptions of all the solutions.

Throughout the remainder of this chapter, $\mathfrak{g}$ denotes a finite-dimensional complex simple Lie algebra. We use freely the notation established in the Appendix.

3.1 Constant solutions of the CYBE

A The parameter space of non-skew solutions We begin with the classification of the non-skew-symmetric solutions of the CYBE with values in $\mathfrak{g} \otimes \mathfrak{g}$, which is due to Belavin & Drinfel'd (1982, 1984a,b). As we remarked in Subsection 2.1B, this is equivalent to solving the system of equations

$$(1) \qquad r_{12} + r_{21} = t, \qquad [[r, r]] = 0,$$

where t is the Casimir element of $\mathfrak{g} \otimes \mathfrak{g}$ corresponding to a non-degenerate invariant bilinear form $(\ , \ )$ on $\mathfrak{g}$ (this can be any non-zero multiple of the standard bilinear form defined in the Appendix). Note that, if σ is any automorphism of $\mathfrak{g}$, then $(\sigma \otimes \sigma)(r)$ satisfies (1) if r does. We shall say that two solutions of (1) are *equivalent* if they can be obtained from each other by applying an automorphism in this way.

To describe the parameter space of equivalence classes of solutions of (1), we make the following definition. Fix a set Π of simple roots of $\mathfrak{g}$.

DEFINITION 3.1.1 *A quadruple* $(\Pi_1, \Pi_0, \tau, r^0)$, *where* $\Pi_1, \Pi_0 \subset \Pi$, τ *is a bijection* $\Pi_1 \to \Pi_0$, *and* $r^0 \in \mathfrak{h} \otimes \mathfrak{h}$, *is said to be* admissible *if it satisfies the following conditions:*

(i) $(\tau(\alpha), \tau(\beta)) = (\alpha, \beta)$ *for all* α, $\beta \in \Pi_1$ *(we also use* $(\ , \ )$ *to denote the inner product on* $\mathfrak{h}^*$ *induced by that on* $\mathfrak{h}$*);*

(ii) for every $\alpha \in \Pi_1$, *there exists* $m \in \mathbb{N}$ *such that* $\alpha, \tau(\alpha), \ldots, \tau^{m-1}(\alpha) \in \Pi_1$ *but* $\tau^m(\alpha) \notin \Pi_1$;

(iii) $r_{12}^0 + r_{21}^0 = t_0$, *the component of* t *in* $\mathfrak{h}^{\otimes 2} \subset \mathfrak{g}^{\otimes 2} = (\mathfrak{n}_- \oplus \mathfrak{h} \oplus \mathfrak{n}_+)^{\otimes 2}$;

(iv) $(\tau(\alpha) \otimes 1)(r^0) + (1 \otimes \alpha)(r^0) = 0$ *for all* $\alpha \in \Pi_1$.

We shall also say that a triple (Π_1, Π_0, τ) satisfying conditions (i) and (ii) is *admissible*.

For a given admissible triple (Π_1, Π_0, τ), the solutions r^0 of the system of inhomogeneous linear equations (iii) and (iv) form an affine space whose dimension may be computed as follows. Since the simple roots form a basis of $\mathfrak{h}^*$, r^0 is completely determined by the matrix of complex numbers $(r_{\alpha\beta}^0)$ defined by

$$r_{\alpha\beta}^0 = (\alpha \otimes \beta)(r^0) \qquad (\alpha, \beta \in \Pi).$$

Conditions (iii) and (iv) are equivalent to the following equations:

$$(2) \qquad r_{\alpha\beta}^0 + r_{\beta\alpha}^0 = (\alpha, \beta), \quad \text{if } \alpha, \beta \in \Pi,$$

$$(3) \qquad r_{\tau(\alpha)\beta}^0 + r_{\beta\alpha}^0 = 0, \quad \text{if } \alpha \in \Pi_1, \ \beta \in \Pi.$$

Now, every simple root $\gamma \in \Pi$ can be expressed uniquely in the form $\gamma = \tau^{-m}(\alpha)$ for some integer $m \geq 0$ and some $\alpha \in \Pi\backslash\Pi_1$ (if $\gamma \notin \Pi_1$, take $m = 0$; if $\gamma \in \Pi_1$, use condition (ii)). If $\delta = \tau^{-l}(\beta)$, with $l \geq 0$ and $\beta \in \Pi\backslash\Pi_1$, it follows from (2) and (3) that

$$
r_{\gamma\delta}^0 - r_{\alpha\beta}^0 = \begin{cases} (\tau^{-1}(\alpha) + \cdots + \tau^{l-m}(\alpha)\,,\, \beta) & \text{if } m > l, \\ -(\alpha\,,\, \beta + \tau^{-1}(\beta) + \cdots + \tau^{m-l+1}(\beta)) & \text{if } m < l, \\ 0 & \text{if } m = l, \end{cases}
$$

and that, conversely, if we prescribe $r_{\alpha\beta}^0$ for $\alpha, \beta \in \Pi\backslash\Pi_1$ arbitrarily, subject only to (2), and then determine r^0 by using the preceding formulas, then conditions (iii) and (iv) are satisfied. Hence, for a given admissible triple (Π_1, Π_0, τ), the (complex) dimension of the space of solutions of (iii) and (iv) is $\frac{1}{2}k(k-1)$, where $k = |\Pi\backslash\Pi_1|$.

The parameter space of solutions of (1) is essentially the space of admissible quadruples:

THEOREM 3.1.2 *The parameter space of equivalence classes of solutions of (1) is the quotient of the space of admissible quadruples by the natural action of the group of automorphisms of the Dynkin diagram of* $\mathfrak{g}$. *It has a stratification into finitely many strata, indexed by the orbits of the group of diagram automorphisms of* $\mathfrak{g}$ *on the set of admissible triples. The stratum corresponding to the triple* (Π_1, Π_0, τ) *is of complex dimension* $\frac{1}{2}k(k-1)$, *where* $k = |\Pi\backslash\Pi_1|$. $\blacksquare$

Remarks [1] A more precise statement of the last part of the theorem is that the stratum corresponding to the admissible triple (Π_1, Π_0, τ) is the quotient of a complex affine space of dimension $\frac{1}{2}k(k-1)$ by an action of the group of diagram automorphisms of $\mathfrak{g}$ which fix (Π_1, Π_0, τ).

[2] Since the invariant bilinear form on $\mathfrak{g}$ is unique up to scalar multiples, it follows that the parameter space of non-skew-symmetric solutions of the CYBE is the cone over the parameter space of solutions of (1) corresponding to a fixed invariant bilinear form.

[3] There is a unique stratum of maximal dimension, corresponding to $\Pi_1 = \Pi_0 = \emptyset$. In this case, conditions (i), (ii) and (iv) in 3.1.1 are vacuous, and the most obvious solution of (iii) is $r^0 = \frac{1}{2}t_0$. This corresponds to the standard Lie bialgebra structure on $\mathfrak{g}$. The most general solution of (iii) is obtained by adding to the standard solution an arbitrary element of $\wedge^2(\mathfrak{h})$. This shows, as predicted by 3.1.2, that the dimension of this stratum is $\frac{1}{2}n(n-1)$, where $n = \text{rank}(\mathfrak{g})$. $\triangle$

B Description of the solutions We now turn to the explicit description of the solution of (1) corresponding to a point $(\Pi_1, \Pi_0, \tau, r^0)$ of the parameter space.

For any subset $\Sigma \subset \Pi$, let $\mathfrak{g}_\Sigma$ be the 'diagram' subalgebra of $\mathfrak{g}$ generated (as a Lie algebra) by the $X_\alpha^\pm$, H_α for $\alpha \in \Sigma$. Explicitly, $\mathfrak{g}_\Sigma$ is spanned (as a vector space) by the H_α for $\alpha \in \Sigma$ and the root spaces $\mathfrak{g}^\alpha$ for $\alpha \in \Delta_\Sigma$, where Δ_Σ is the set of roots whose expression in terms of the simple roots only involves roots from Σ.

If (Π_1, Π_0, τ) is an admissible triple, there is an isomorphism of Lie algebras $\varphi : \mathfrak{g}_{\Pi_1} \to \mathfrak{g}_{\Pi_0}$ such that

$$\varphi(X_\alpha^\pm) = X_{\tau(\alpha)}^\pm, \qquad \varphi(H_\alpha) = H_{\tau(\alpha)},$$

for all $\alpha \in \Pi_1$ (the existence of φ follows from condition (i) in 3.1.1). Choose elements $E_{\pm\alpha} \in \mathfrak{g}^{\pm\alpha}$ such that

$$(E_\alpha, E_{-\alpha}) = 1 \quad \text{if } \alpha \in \Delta^+,$$

$$\varphi(E_\alpha) = E_{\tau(\alpha)} \quad \text{if } \alpha \in \Delta_{\Pi_1}.$$

Finally, introduce a partially defined ordering $<$ on Π by saying that $\alpha < \beta$ if there is a positive integer m such that $\beta = \tau^m(\alpha)$ (notice that, for this to make sense, we must have $\alpha \in \Pi_1$, $\tau(\alpha), \ldots, \tau^{m-1}(\alpha) \in \Pi_0 \cap \Pi_1$, and $\beta \in \Pi_0$).

THEOREM 3.1.3 *If $(\Pi_1, \Pi_0, \tau, r^0)$ is an admissible quadruple, then, with the above notation,*

$$r = r^0 + \sum_{\alpha \in \Delta^+} E_{-\alpha} \otimes E_\alpha + \sum_{\alpha, \beta \in \Delta^+, \alpha < \beta} E_{-\alpha} \wedge E_\beta$$

is a solution of (1) (here, $E_{-\alpha} \wedge E_\beta$ means $E_{-\alpha} \otimes E_\beta - E_\beta \otimes E_{-\alpha}$).

Conversely, every solution of (1) is equivalent to a solution of this type. ∎

We have nothing to add to the detailed proof available in Belavin & Drinfel'd (1982, 1984a).

C Examples We now make 3.1.2 and 3.1.3 explicit in the cases of $sl_2(\mathbb{C})$ and $sl_3(\mathbb{C})$.

Example 3.1.4 If $\mathfrak{g} = sl_2(\mathbb{C})$, we take $\mathfrak{h}$ (resp. $\mathfrak{n}_+$) to be the diagonal matrices (resp. the strictly upper triangular matrices). Then, there is a unique simple root α, so either $\Pi_0 = \Pi_1 = \emptyset$ or $\Pi_0 = \Pi_1 = \{\alpha\}$. But the second choice does not satisfy condition (ii) in 3.1.1. Taking the inner product to be the trace form

$$(X, Y) = \text{trace}(XY),$$

we obtain the unique solution

$$r = \frac{1}{4} H \otimes H + X^- \otimes X^+$$

in the notation of 2.1.8. Passing to the skew-symmetric part, we see that this corresponds to the standard structure 2.1.8(i) (with $\lambda = -\frac{1}{2}$). Note that the non-trivial triangular Lie bialgebra structure 2.1.8(ii) is not included in the Belavin–Drinfel'd classification. $\diamondsuit$

Example 3.1.5 The simplest examples not treated in Section 2.1 appear when $\mathfrak{g} = sl_3(\mathbb{C})$. Taking $\mathfrak{h}$, $\mathfrak{n}_+$ and the inner product to be as described in the previous example, and writing an element of $\mathfrak{h}$ as $\mathrm{diag}(x_1, x_2, x_3)$, the positive roots are $\alpha = x_1 - x_2$, $\beta = x_2 - x_3$, $\alpha + \beta = x_1 - x_3$. Thus, $\Pi = \{\alpha, \beta\}$. Writing E_{ij} for the elementary matrix with 1 in the ith row and jth column and zeros elsewhere, we take

$$E_\alpha = E_{12}, \quad E_{-\alpha} = E_{21}, \quad E_\beta = E_{23}, \quad E_{-\beta} = E_{32},$$

$$E_{\alpha+\beta} = E_{13}, \quad E_{-\alpha-\beta} = E_{31}, \quad H_\alpha = E_{11} - E_{12}, \quad H_\beta = E_{22} - E_{33}.$$

Then,

$$t_0 = \frac{2}{3} H_\alpha \otimes H_\alpha + \frac{1}{3} H_\alpha \otimes H_\beta + \frac{1}{3} H_\beta \otimes H_\alpha + \frac{2}{3} H_\beta \otimes H_\beta.$$

There are three admissible triples:

(a) $\Pi_0 = \Pi_1 = \emptyset$. Condition (iv) in 3.1.1 is vacuous and the most general solution of (iii) is

$$r^0 = \frac{1}{2} t_0 + \lambda H_\alpha \wedge H_\beta$$

for $\lambda \in \mathbb{C}$. The resulting r-matrix is

$$r = \frac{1}{2} t_0 + \lambda H_\alpha \wedge H_\beta + E_{-\alpha} \otimes E_\alpha + E_{-\beta} \otimes E_\beta + E_{-\alpha-\beta} \otimes E_{\alpha+\beta}.$$

(b) $\Pi_0 = \{\alpha\}$, $\Pi_1 = \{\beta\}$. Conditions (iii) and (iv),

$$r_{12}^0 + r_{21}^0 = t_0,$$

$$(\alpha \otimes 1)(r^0) + (1 \otimes \beta)(r^0) = 0,$$

have the unique solution

$$r^0 = \frac{1}{3} H_\alpha \otimes H_\alpha + \frac{1}{3} H_\beta \otimes H_\alpha + \frac{1}{3} H_\beta \otimes H_\beta.$$

The resulting r-matrix is

$$r = \tfrac{1}{3} H_\alpha \otimes H_\alpha + \tfrac{1}{3} H_\beta \otimes H_\alpha + \tfrac{1}{3} H_\beta \otimes H_\beta + E_{-\alpha} \otimes E_\alpha + E_{-\beta} \otimes E_\beta$$
$$+ E_{-\alpha-\beta} \otimes E_{\alpha+\beta} + E_{-\beta} \otimes E_\alpha - E_\alpha \otimes E_{-\beta}.$$

(c) $\Pi_0 = \{\beta\}$, $\Pi_1 = \{\alpha\}$. The r-matrix in this case is obtained by interchanging α and β in (b).

There is a unique non-trivial diagram automorphism of $sl_3(\mathbb{C})$, which interchanges α and β. Thus, the solution of type (c) is equivalent to the solution of type (b), and the solution of type (a) with parameter λ is equivalent to that with parameter $-\lambda$. The parameter space of solutions of (1) is thus the union of the upper half-space

$$\{\lambda \in \mathbb{C} \mid \Im(\lambda) > 0 \text{ or } \lambda \geq 0\}$$

together with a single point ($\Im(\lambda)$ denotes the imaginary part of a complex number λ). ◊

D Skew solutions and quasi-Frobenius Lie algebras We now consider briefly the problem of constructing *skew-symmetric* solutions of the classical Yang–Baxter equation. In view of 2.2.4 and 2.2.6, this problem is equivalent to that of finding pairs $(\mathfrak{g}_0, \omega)$, where $\mathfrak{g}_0$ is a Lie subalgebra of $\mathfrak{g}$ and ω is a non-degenerate 2-cocycle on $\mathfrak{g}_0$. In general, a Lie algebra $\mathfrak{f}$ which admits a non-degenerate 2-cocycle ω is called *quasi-Frobenius*; it is *Frobenius* if ω is a coboundary, i.e. if $\omega(X, Y) = \mathcal{L}([X, Y])$ for some $\mathcal{L} \in \mathfrak{f}^*$ and all $X, Y \in \mathfrak{f}$.

Since any scalar multiple of a skew solution of the CYBE is also a solution, it is convenient to weaken the notion of equivalence of solutions slightly compared with that used for solutions of the MCYBE. We shall say that two skew solutions r_1 and r_2 of the CYBE are equivalent if r_2 is equivalent in the old sense to a *non-zero scalar multiple* of r_1.

PROPOSITION 3.1.6 *Let $\mathfrak{g}$ be a finite-dimensional Lie algebra. There is a (non-zero) skew-symmetric solution of the CYBE with values in $\mathfrak{g} \otimes \mathfrak{g}$ associated to every (non-zero) quasi-Frobenius Lie subalgebra $(\mathfrak{f}, \omega)$ of $\mathfrak{g}$. The solutions associated to $(\mathfrak{f}_1, \omega_1)$ and $(\mathfrak{f}_2, \omega_2)$ are equivalent if and only if there is an inner automorphism α of $\mathfrak{g}$ such that $\alpha(\mathfrak{f}_1) = \mathfrak{f}_2$ and $\alpha^*(\omega_2) - \omega_1$ is a coboundary.* ∎

Thus, we have the following algorithm for constructing (equivalence classes of) skew solutions of the CYBE:

(a) classify the quasi-Frobenius Lie subalgebras $\mathfrak{f}$ of $\mathfrak{g}$ up to inner automorphisms (such subalgebras must, of course, be even dimensional);

(b) for each such $\mathfrak{f}$, compute $H^2_{\text{Lie}}(\mathfrak{f})$, the Lie algebra cohomology of $\mathfrak{f}$ (the space of 2-cocycles modulo the space of coboundaries);

(c) classify the orbits of the action of the normalizer $N(\mathfrak{f})$ of $\mathfrak{f}$ on $H^2_{\text{Lie}}(\mathfrak{f})$ (note that the centralizer $C(\mathfrak{f})$ acts trivially on $H^2_{\text{Lie}}(\mathfrak{f})$);

(d) determine whether a cohomology class in each orbit has a representative cocycle which is non-degenerate.

Example 3.1.7 The only non-trivial even-dimensional Lie subalgebra of $\mathfrak{g} = sl_2(\mathbb{C})$ up to conjugacy is the Borel subalgebra $\mathfrak{b}$ consisting of the upper triangular matrices. In fact, the Lie algebra cohomology $H^2_{\mathrm{Lie}}(\mathfrak{b}) = 0$, so $\mathfrak{b}$ is Frobenius and there is, up to equivalence, a unique solution associated to it. It is easy to see that this solution is $r = H \wedge X^+$ in the notation of 2.1.8(ii). $\Diamond$

Example 3.1.8 The quasi-Frobenius Lie subalgebras of $\mathfrak{g} = sl_3(\mathbb{C})$ are of the following types.

(i) Up to conjugacy, the only six-dimensional Lie subalgebra of $sl_3(\mathbb{C})$ is

$$\mathfrak{p} = \begin{pmatrix} * & * & * \\ 0 & * & * \\ 0 & * & * \end{pmatrix}$$

(the $*$'s represent arbitrary complex numbers, subject only to the condition that the trace is zero). Since $H^2_{\mathrm{Lie}}(\mathfrak{p}) = 0$, $\mathfrak{p}$ is Frobenius and there is a unique skew solution of the CYBE associated to it.

(ii) The four-dimensional subalgebra

$$\mathfrak{r} = \begin{pmatrix} * & * & * \\ 0 & * & 0 \\ 0 & 0 & * \end{pmatrix}$$

is Frobenius. We have $H^2_{\mathrm{Lie}}(\mathfrak{r}) = \mathbb{C}$, and the quotient $N(\mathfrak{r})/C(\mathfrak{r})$ of the normalizer of $\mathfrak{r}$ in $SL_3(\mathbb{C})$ by its centralizer is $\mathbb{Z}_2$. Thus, there is a one-parameter family of solutions associated to $\mathfrak{r}$.

(iii) The two-parameter family of four-dimensional subalgebras

$$\mathfrak{q}_{a_1,a_2,a_3} = * \begin{pmatrix} a_1 & 0 & 0 \\ 0 & a_2 & 0 \\ 0 & 0 & a_3 \end{pmatrix} + \begin{pmatrix} 0 & * & * \\ 0 & 0 & * \\ 0 & 0 & 0 \end{pmatrix}$$

are quasi-Frobenius. We have

$$H^2_{\mathrm{Lie}}(\mathfrak{q}_{a_1,a_2,a_3}) = \begin{cases} \mathbb{C} & \text{if } (a_1, a_2, a_3) = (0, 1, -1) \text{ or } (1, 1, -2), \\ 0 & \text{otherwise,} \end{cases}$$

and $N(\mathfrak{q}_{a_1,a_2,a_3})/C(\mathfrak{q}_{a_1,a_2,a_3}) = \mathbb{C}^\times$ for all (a_1, a_2, a_3). There are two solutions associated to each of $\mathfrak{q}_{0,1,-1}$ and $\mathfrak{q}_{1,1,-2}$, and one to $\mathfrak{q}_{a_1,a_2,a_3}$ in each of the remaining cases.

(iv) The two-dimensional non-abelian Lie algebra embeds in $sl_3(\mathbb{C})$ in the following ways.

(a) the one-parameter family of subalgebras

$$\mathfrak{b}_\lambda = * \begin{pmatrix} \lambda & 0 & 0 \\ 0 & \lambda - 1 & 0 \\ 0 & 0 & 1 - 2\lambda \end{pmatrix} + \begin{pmatrix} 0 & * & 0 \\ 0 & 0 & 0 \\ 0 & 0 & 0 \end{pmatrix}, \quad (\lambda \in \mathbb{C});$$

(b) the subalgebra

$$\mathfrak{b}^{(0)} = * \begin{pmatrix} 1 & 0 & 0 \\ 0 & 0 & 0 \\ 0 & 0 & -1 \end{pmatrix} + * \begin{pmatrix} 0 & 1 & 0 \\ 0 & 0 & 1 \\ 0 & 0 & 0 \end{pmatrix};$$

(c) the subalgebra

$$\mathfrak{b}^{(1)} = * \begin{pmatrix} 2 & 0 & 0 \\ 0 & -1 & 1 \\ 0 & 0 & -1 \end{pmatrix} + \begin{pmatrix} 0 & 0 & * \\ 0 & 0 & 0 \\ 0 & 0 & 0 \end{pmatrix};$$

(d) the subalgebra

$$\mathfrak{b}^{(2)} = * \begin{pmatrix} 2 & 0 & 0 \\ 0 & -1 & 1 \\ 0 & 0 & -1 \end{pmatrix} + * \begin{pmatrix} 0 & 1 & 1 \\ 0 & 0 & 0 \\ 0 & 0 & 0 \end{pmatrix}.$$

In each case, $H^2_{\text{Lie}} = 0$ and there is a unique associated solution.

(v) The two-dimensional abelian Lie algebra has $H^2_{\text{Lie}} = \mathbb{C}$. In fact, 0 is the only 2-coboundary, so H^2_{Lie} is simply the space of all skew bilinear forms. In particular, the zero class in H^2_{Lie} has no non-degenerate representative. The possible embeddings are as follows:

(a) the Cartan subalgebra

$$\mathfrak{h} = \begin{pmatrix} * & 0 & 0 \\ 0 & * & 0 \\ 0 & 0 & * \end{pmatrix};$$

the group $N(\mathfrak{h})/C(\mathfrak{h})$ is trivial, so this gives a one-parameter family of solutions;

(b) the subalgebra

$$\mathfrak{h}^{(1)} = * \begin{pmatrix} 1 & 0 & 0 \\ 0 & 1 & 0 \\ 0 & 0 & -2 \end{pmatrix} + \begin{pmatrix} 0 & * & 0 \\ 0 & 0 & 0 \\ 0 & 0 & 0 \end{pmatrix};$$

this time, $N(\mathfrak{h}^{(1)})/C(\mathfrak{h}^{(1)}) = \mathbb{C}^{\times}$, and there is a unique associated solution;

(c) the subalgebras

$$\mathfrak{h}^{(0,1)} = \begin{pmatrix} 0 & * & * \\ 0 & 0 & 0 \\ 0 & 0 & 0 \end{pmatrix}$$

and

$$\mathfrak{h}^{(1,1)} = * \begin{pmatrix} 0 & 1 & 0 \\ 0 & 0 & 1 \\ 0 & 0 & 0 \end{pmatrix} + \begin{pmatrix} 0 & 0 & * \\ 0 & 0 & 0 \\ 0 & 0 & 0 \end{pmatrix},$$

for which $N(\mathfrak{h}^{(0,1)})/C(\mathfrak{h}^{(0,1)}) = N(\mathfrak{h}^{(1,1)})/C(\mathfrak{h}^{(1,1)}) = \mathbb{C}^{\times} \times \mathbb{C}^{\times}$; there is a unique solution associated to each subalgebra. $\Diamond$

3.2 Solutions of the CYBE with spectral parameters

A Classification of the solutions There is a generalization of the classical Yang–Baxter equation, which is particularly important in physical applications of the theory, in which the r-matrix r is a function on $U \times U$, for some set U, with values in $\mathfrak{g} \otimes \mathfrak{g}$, satisfying

(4)
$$\begin{aligned}[r_{12}(u_1, u_2), r_{13}(u_1, u_3)] + [r_{12}(u_1, u_2), r_{23}(u_2, u_3)] \\ + [r_{13}(u_1, u_3), r_{23}(u_2, u_3)] = 0.\end{aligned}$$

This is usually called the CYBE with *spectral parameters* u_1 and u_2, but we shall refer to it simply as the CYBE in this section. Thus, the previous section was concerned with *constant* solutions.

In this section, we shall consider solutions of (4) for which $\mathfrak{g}$ is a finite-dimensional complex simple Lie algebra, as in the previous section. We shall further assume that

(a) U is an open subset of the complex plane, and r is a meromorphic function on $U \times U$, and

(b) $r(u_1, u_2)$ is non-degenerate for at least one point $(u_1, u_2) \in U \times U$ (see Subsection 2.2B; we shall simply say that r is *non-degenerate* if this condition is satisfied).

It follows from 2.2.5 that skew-symmetric *constant* solutions cannot be non-degenerate.

Note that (4) and conditions (a) and (b) are preserved if we replace r by an *equivalent* r-matrix, as described in the following definition.

DEFINITION 3.2.1 *Let r and $\tilde{r}$ be two solutions of the CYBE which are meromorphic on open sets $U \times U$ and $\tilde{U} \times \tilde{U}$, respectively. Then, r and $\tilde{r}$ are said to be equivalent if there exists a holomorphic map $\alpha : \tilde{U} \to \mathrm{Aut}(\mathfrak{g})$ and a non-constant biholomorphic function $f : \tilde{U} \to U$ such that*

$$\tilde{r}(\tilde{u}_1, \tilde{u}_2) = (\alpha(\tilde{u}_1) \otimes \alpha(\tilde{u}_2))(r(f(\tilde{u}_1), f(\tilde{u}_2))).$$

Note that the group $\mathrm{Aut}(\mathfrak{g})$ of Lie algebra automorphisms of $\mathfrak{g}$ is a complex Lie group, so this definition makes sense.

We have already met examples of solutions of the CYBE (4) in Section 2.1. In 2.1.9 we encountered the rational solution

$$(5) \qquad\qquad r(u_1, u_2) = \frac{t}{u_1 - u_2},$$

and in 2.1.10 another rational solution

$$(6) \qquad\qquad r(u_1, u_2) = r_{\mathfrak{g}} + \left(\frac{u_1}{u_2 - u_1} \right) t,$$

where $r_{\mathfrak{g}}$ is the standard r-matrix for $\mathfrak{g}$ found in 2.1.6 (we have passed to the quotient of the affine Lie algebra by its centre to obtain a solution with values in $\mathfrak{g} \otimes \mathfrak{g}$).

Note that the solution in (5) is a function of $u_1 - u_2$ only, while that in (6) depends only on u_1/u_2, and, hence, making the change of variable $u_1 \mapsto e^{u_1}$, $u_2 \mapsto e^{u_2}$, is equivalent to a solution which is a function of $u_1 - u_2$ only. In general, we have:

PROPOSITION 3.2.2 *Every solution of the CYBE satisfying conditions (a) and (b) is equivalent to one which can be continued meromorphically to the whole complex plane and which depends only on the difference of the spectral parameters.* ∎

We shall assume from now on that $r(u_1, u_2) = r(u_1 - u_2)$ for some meromorphic function $r(u)$ on $\mathbb{C}$ with values in $\mathfrak{g} \otimes \mathfrak{g}$. Thus, we have to solve

$$(7) \qquad \begin{aligned} [r_{12}(u_1 - u_2), r_{13}(u_1 - u_3)] &+ [r_{12}(u_1 - u_2), r_{23}(u_2 - u_3)] \\ &+ [r_{13}(u_1 - u_3), r_{23}(u_2 - u_3)] = 0. \end{aligned}$$

Non-degenerate meromorphic solutions of (7) have the following regularity properties.

PROPOSITION 3.2.3 *Let $r(u)$ be a non-degenerate meromorphic solution of (7). Then, up to equivalence,*

(i) r has a simple pole at $u = 0$ with residue the Casimir element t (for some choice of non-degenerate invariant bilinear form on $\mathfrak{g}$). Moreover, all the poles of r are simple;

(ii) r *satisfies the* unitarity *condition*

$$r_{12}(u) + r_{21}(-u) = 0;$$

(iii) there is no proper subalgebra $\mathfrak{g}_0$ *of* $\mathfrak{g}$ *such that* $r(u) \in \mathfrak{g}_0 \otimes \mathfrak{g}_0$ *for all* u. ∎

The central result for the analysis of solutions of the CYBE is:

THEOREM 3.2.4 *Let* r(u) *be a non-degenerate meromorphic solution of the CYBE (7). Then, the set of poles of* r *form a discrete subgroup* Γ *of the additive group of* $\mathbb{C}$. *There is a homomorphism* $\chi : \Gamma \to \mathrm{Aut}(\mathfrak{g})$ *such that*

$$(8) \qquad r(u + \gamma) = (\chi(\gamma) \otimes 1)(r(u)) = (1 \otimes \chi(\gamma))^{-1}(r(u))$$

for all $u \in \mathbb{C}$, $\gamma \in \Gamma$.

There are three possibilities, as follows.

(i) $\mathrm{rank}(\Gamma) = 2$: *this is possible only when* $\mathfrak{g} \cong sl_{n+1}(\mathbb{C})$ *for some* $n \geq 1$, *and then* r(u) *is an elliptic function of* u *with period lattice* $(n + 1)\Gamma$; *moreover, for each* $\gamma \in \Gamma$, $\chi(\gamma)$ *is an automorphism of* $\mathfrak{g}$ *of finite order, and the automorphisms* $\{\chi(\gamma)\}_{\gamma \in \Gamma}$ *have no common fixed vector.*

(ii) $\mathrm{rank}(\Gamma) = 1$: *then* r(u) *is equivalent to a rational function of* e^{cu} *for some* $c \in \mathbb{C}^\times$ *(such functions are called trigonometric).*

(iii) $\Gamma = 0$: *then* r(u) *is equivalent to a rational function of* u.

OUTLINE OF PROOF To show that, up to equivalence, only elliptic, trigonometric or rational solutions are possible, one proceeds as follows. Writing $u = u_1 - u_2$, $v = u_2 - u_3$, the CYBE becomes

$$(9) \qquad [r_{12}(u) - r_{23}(v), r_{13}(u + v)] = [r_{23}(v), r_{12}(u)].$$

We regard (9) as an inhomogeneous system of linear equations for $r_{13}(u+v)$, with $r_{12}(u)$ and $r_{23}(v)$ playing the role of coefficients. One shows that, at least for sufficiently small u, v, the associated homogeneous system

$$(10) \qquad [r_{12}(u) - r_{23}(v), s_{13}] = 0$$

has only the trivial solution $s = 0$. Assuming this for a moment, it follows that r(u + v) is a rational function of r(u) and r(v). One now appeals to the generalization to vector-valued functions, due to Myrberg (1922), of the classical theorem of Weierstrass, according to which a meromorphic function $f : \mathbb{C} \to \mathbb{C}$, which satisfies a functional equation of the form

$$P(f(u), f(v), f(u + v)) = 0$$

for some polynomial P, is necessarily elliptic, trigonometric or rational.

To prove the assertion about the system (10), it is enough by continuity to consider the case $u = v = 0$ when, by 3.2.3, (10) becomes

$$[t_{12} - t_{23}, s_{13}] = 0.$$

It is easy to see that this equation is equivalent to

$$[X \otimes 1, s] = [1 \otimes X, s]$$

for all $X \in \mathfrak{g}$. Taking $X \in \mathfrak{h}$, this implies that

$$s = \sum_{i=1}^{n} s_i \otimes H_i + \sum_{\alpha \in \Delta^+} (\lambda_\alpha^+ X_\alpha^+ \otimes X_\alpha^+ + \lambda_\alpha^- X_\alpha^- \otimes X_\alpha^-)$$

for some $s_i \in \mathfrak{h}$, $\lambda_\alpha^\pm \in \mathbb{C}$. Taking $X = X_\beta^\pm$, and considering terms whose second factor lies in $\mathfrak{h}$, one sees that $\lambda_\beta^\pm = 0$ for all $\beta \in \Delta^+$, and then that $s = 0$. ∎

In the remainder of this section, we shall describe how solutions of the CYBE of the three possible types are constructed.

B Elliptic solutions Let $\{\gamma_1, \gamma_2\}$ be a basis of the lattice Γ in 3.2.4(i) and set $\chi_i = \chi(\gamma_i)$. Thus, χ_1 and χ_2 are commuting finite order automorphisms of $\mathfrak{g}$ with no common fixed vector. By using the classification of finite order automorphisms of complex simple Lie algebras, one shows that there is an isomorphism of Lie algebras $\mathfrak{g} \cong sl_{n+1}(\mathbb{C})$ under which χ_1 and χ_2 go over into conjugation by the matrices

$$Z = \begin{pmatrix} 1 & 0 & 0 & \cdots & 0 \\ 0 & \epsilon & 0 & \cdots & 0 \\ 0 & 0 & \epsilon^2 & \cdots & 0 \\ \vdots & \vdots & \vdots & \ddots & \vdots \\ 0 & 0 & 0 & 0 & \epsilon^{n-1} \end{pmatrix} \quad \text{and} \quad X = \begin{pmatrix} 0 & 1 & 0 & \cdots & 0 \\ 0 & 0 & 1 & \cdots & 0 \\ \vdots & \vdots & \vdots & \ddots & \vdots \\ 0 & 0 & 0 & \cdots & 1 \\ 1 & 0 & 0 & \cdots & 0 \end{pmatrix},$$

respectively, where $\epsilon = \exp[2\pi\sqrt{-1}/(n+1)]$. We assume that $\chi_1 = Z$ and $\chi_2 = X$ from now on.

Note that the matrices $X^j Z^k$ for $(j, k) \neq (0, 0)$ form a basis of $sl_{n+1}(\mathbb{C})$ consisting of simultaneous eigenvectors of χ_1 and χ_2, with eigenvalues ϵ^{-j} and ϵ^{-k}, respectively. Let φ_{jk} be the unique meromorphic function on $\mathbb{C}$ such that

(i) $\varphi_{jk}(u + \gamma_1) = \epsilon^j \varphi_{jk}(u)$, $\varphi_{jk}(u + \gamma_2) = \epsilon^k \varphi_{jk}(u)$,

(ii) φ_{jk} has simple poles at the points of Γ, no other poles, and residue 1 at $u = 0$.

(Condition (i) comes from (8) and (ii) from 3.2.3.)

PROPOSITION 3.2.5 *With φ_{jk} as above, the function*

$$(11) \qquad r(u) = \frac{1}{n+1} \sum_{j,k=1}^{n+1} \epsilon^{-jk} \varphi_{jk}(u) \, X^{-j} Z^{-k} \otimes X^j Z^k$$

is a non-degenerate solution of the CYBE with values in $sl_{n+1}(\mathbb{C})^{\otimes 2}$. It has simple poles at the points of the lattice $\Gamma = \mathbb{Z}.\gamma_1 + \mathbb{Z}.\gamma_2$.
    Conversely, non-degenerate meromorphic solutions of the CYBE with values in $\mathfrak{g}^{\otimes 2}$ and whose poles form a lattice of rank 2 exist only when $\mathfrak{g} \cong sl_{n+1}(\mathbb{C})$ for some $n \geq 1$, and every such solution is equivalent to one of the form (11). ∎

    Thus, for each $n \geq 1$, there is, up to equivalence, exactly one non-degenerate elliptic solution of the CYBE with values in $sl_{n+1}(\mathbb{C}) \otimes sl_{n+1}(\mathbb{C})$.

C Trigonometric solutions It turns out that the classification of the non-degenerate trigonometric solutions of the CYBE is very similar to that of the constant solutions described in the previous section. This may explained by the following observations.
    Note first that, by 3.2.3(i), every solution is of the form

$$(12) \qquad r(u) = \frac{t}{e^u - 1} + \tilde{r}(u),$$

where $\tilde{r}(u)$ is holomorphic. A direct computation shows that, if $\tilde{r}$ is *constant*, then r satisfies the CYBE if and only if $\tilde{r}$ satisfies (1).
    Secondly, we have seen in 2.1.10 (see also (6)) that trigonometric solutions of the CYBE arise by considering Lie bialgebra structures on the affine Lie algebra associated to $\mathfrak{g}$. It is well known that the structure of affine Lie algebras resembles closely that of the finite-dimensional complex simple Lie algebras.

    To describe the classification of trigonometric solutions, we must first recall certain results about the finite order automorphisms of $\mathfrak{g}$ (see Helgason (1978)). Let σ be an automorphism of the Dynkin diagram of $\mathfrak{g}$. We may view σ as an element of the quotient group $\mathrm{Aut}(\mathfrak{g})/\mathrm{Inn}(\mathfrak{g})$ of the automorphism group of $\mathfrak{g}$ by the subgroup of inner automorphisms. An automorphism γ of finite order in the coset of σ modulo $\mathrm{Inn}(\mathfrak{g})$ is called a *Coxeter automorphism* of the pair $(\mathfrak{g}, \sigma)$ if

(a) its fixed point subalgebra is abelian;
(b) γ has minimal order among the elements of the coset of σ which satisfy (a).

The order h_σ of γ is called the *Coxeter number* of the pair $(\mathfrak{g}, \sigma)$. Let $\epsilon = \exp[2\pi\sqrt{-1}/h_\sigma]$ and let $\mathfrak{g}_{\sigma,j}$ be the ϵ^j-eigenspace of γ on $\mathfrak{g}$. Then,

$$\mathfrak{g} = \bigoplus_{j=0}^{h_\sigma-1} \mathfrak{g}_{\sigma,j}.$$

The abelian subalgebra $\mathfrak{g}_{\sigma,0}$ will play a similar role to that of the Cartan subalgebra, and we denote it by $\mathfrak{h}_\sigma$ from now on. Let $t_{\sigma,j}$ be the projection of t onto the part $\mathfrak{g}_{\sigma,j} \otimes \mathfrak{g}_{\sigma,-j}$ of $\mathfrak{g} \otimes \mathfrak{g}$. Then,

$$t = \sum_{j=0}^{h_\sigma-1} t_{\sigma,j}.$$

For any $\alpha \in \mathfrak{h}_\sigma^*$, and any $j = 0, 1, \ldots, h_\sigma - 1$, let

$$\mathfrak{g}_{\sigma,j}^\alpha = \{X \in \mathfrak{g}_{\sigma,j} \mid [H, X] = \alpha(H)X \text{ for all } H \in \mathfrak{h}_\sigma\}.$$

Then, $\dim(\mathfrak{g}_{\sigma,j}^\alpha) \leq 1$ for $\alpha \neq 0$, and

$$\mathfrak{g}_{\sigma,j} = \bigoplus_{\alpha \in \mathfrak{h}_\sigma^*} \mathfrak{g}_{\sigma,j}^\alpha.$$

Let $\Pi_\sigma = \{\alpha \in \mathfrak{h}_\sigma^* \mid \mathfrak{g}_{\sigma,1}^\alpha \neq 0\}$. Then, $0 \notin \Pi_\sigma$. The elements of Π_σ are called the *simple weights* of $(\mathfrak{g}, \sigma)$. Unlike the simple roots of $\mathfrak{g}$, they are not linearly independent, but satisfy a single linear relation with positive integer coefficients.

The pair $(\mathfrak{g}, \sigma)$ can be recovered from its *Dynkin diagram*, which is constructed from the simple weights in the same way as the Dynkin diagram of $\mathfrak{g}$ is constructed from the simple roots. Namely, the Dynkin diagram of $(\mathfrak{g}, \sigma)$ is a graph with one vertex for each simple weight, the vertices corresponding to weights α and β being joined by $4(\alpha, \beta)^2/(\alpha, \alpha)(\beta, \beta)$ bonds, with an arrow pointing toward vertex β if $(\alpha, \alpha) > (\beta, \beta)$.

To describe the parameter space of (equivalence classes) of trigonometric solutions of the CYBE, we make the following definition, which is analogous to 3.1.1.

DEFINITION 3.2.6 *A quintuple* $(\Pi_1, \Pi_0, \tau, r^0, \sigma)$, *where* σ *is an automorphism of the Dynkin diagram of* $\mathfrak{g}$, $\Pi_0, \Pi_1 \subset \Pi_\sigma$, $\tau : \Pi_1 \to \Pi_0$ *is a bijection, and* $r^0 \in \mathfrak{h}_\sigma \otimes \mathfrak{h}_\sigma$, *is said to be* admissible *if it satisfies conditions (i)–(iv) of 3.1.1 (in condition (iii), t should be replaced by* $t_{\sigma,0}$). *The quadruple* $(\Pi_1, \Pi_0, \tau, \sigma)$ *is* admissible *if it satisfies conditions (i) and (ii).*

As in the case of constant solutions, the set of admissible quintuples has a decomposition into finitely many strata, indexed by the admissible quadruples, the stratum associated to $(\Pi_1, \Pi_0, \tau, \sigma)$ being a complex affine space of dimension $\frac{1}{2}k(k-1)$, where $k = |\Pi_\sigma \backslash \Pi_1|$.

One must take into account the action of the group of automorphisms of $\mathfrak{g}$ on these data. Let $\mathrm{Aut}^\gamma(\mathfrak{g})$ be the centralizer of the Coxeter automorphism γ in $\mathrm{Aut}(\mathfrak{g})$, and $\mathrm{Inn}^\gamma(\mathfrak{g})$ the subgroup generated by γ and the inner automorphisms given by conjugation by elements of $\exp(\mathfrak{h})$. Then, $\mathrm{Inn}^\gamma(\mathfrak{g})$ is a normal subgroup of $\mathrm{Aut}^\gamma(\mathfrak{g})$, and $\mathrm{Aut}^\gamma(\mathfrak{g})/\mathrm{Inn}^\gamma(\mathfrak{g})$ is isomorphic to the group of automorphisms of the Dynkin diagram of $(\mathfrak{g}, \sigma)$.

PROPOSITION 3.2.7 *The equivalence classes of non-degenerate trigonometric solutions of the CYBE are in one-to-one correspondence with the union, over the set of automorphisms σ of the Dynkin diagram of $\mathfrak{g}$, of the quotients of the spaces of admissible quintuples $(\Pi_1, \Pi_0, \tau, r^0, \sigma)$ by the natural action of the group of automorphisms of the Dynkin diagram of $(\mathfrak{g}, \sigma)$.* ∎

We shall now describe the solution of the CYBE associated to an admissible quintuple $(\Pi_1, \Pi_0, \tau, r^0, \sigma)$. Let $\mathfrak{g}_{\Pi_i}$ be the subalgebra of $\mathfrak{g}$ generated by the subspaces $\mathfrak{g}_{\sigma,1}^\alpha$ for $\alpha \in \Pi_i$. Then, $\mathfrak{g}_{\Pi_i}$ is the sum of certain of the $\mathfrak{g}_{\sigma,j}^\alpha$. Hence, there is a canonical projection map $P : \mathfrak{g} \to \mathfrak{g}_{\Pi_1}$ such that $P(\mathfrak{g}_{\sigma,j}^\alpha) = 0$ if $\mathfrak{g}_{\sigma,j}^\alpha \not\subset \mathfrak{g}_{\Pi_1}$. One shows that there is an isomorphism $\theta : \mathfrak{g}_{\Pi_1} \to \mathfrak{g}_{\Pi_0}$ of Lie algebras such that $\theta(\mathfrak{g}_{\sigma,1}^\alpha) = \mathfrak{g}_{\sigma,1}^{\tau(\alpha)}$ for all $\alpha \in \Pi_1$. Moreover, the operator $\theta \circ P : \mathfrak{g} \to \mathfrak{g}$ is nilpotent, so it makes sense to define $f : \mathfrak{g} \to \mathfrak{g}$ by

$$f = \frac{\theta \circ P}{1 - \theta \circ P} = \theta \circ P + (\theta \circ P)^2 + \cdots .$$

THEOREM 3.2.8 *For any admissible quintuple $(\Pi_1, \Pi_0, \tau, r^0, \sigma)$,*

$$r(u) = r^0 + \frac{1}{e^u - 1} \sum_{j=0}^{h_\sigma - 1} e^{ju/h_\sigma} t_{\sigma,j} - \sum_{j=1}^{h_\sigma - 1} e^{ju/h_\sigma} (f \otimes 1)(t_{\sigma,j})$$

$$+ \sum_{j=1}^{h_\sigma - 1} e^{-ju/h_\sigma} (1 \otimes f)(t_{\sigma,-j})$$

is a non-degenerate solution of the CYBE with poles at the points $2\pi\sqrt{-1}\mathbb{Z}$. Moreover,

$$r(u + 2\pi\sqrt{-1}) = (\gamma \otimes 1)(r(u)) = (1 \otimes \gamma)^{-1}(r(u)).$$

Conversely, every non-degenerate meromorphic solution of the CYBE whose poles form a lattice of rank 1 is equivalent to a solution of this form. ∎

Remark The role played by $\mathfrak{g}$ in the proof of 3.1.3 is played in that of 3.2.8 by the *loop algebra* $L(\mathfrak{g})$ of Laurent polynomial maps $\mathbb{C}^\times \to \mathfrak{g}$. △

Example 3.2.9 For each automorphism σ of the Dynkin diagram of $\mathfrak{g}$, there is a *standard* solution, corresponding to $\Pi_0 = \Pi_1 = \emptyset$, $r^0 = \frac{1}{2}t_{\sigma,0}$, namely

$$r(u) = \frac{1}{2}t_{\sigma,0} + \frac{1}{e^u - 1} \sum_{j=0}^{h_\sigma - 1} e^{ju/h_\sigma} t_j. \quad \diamond$$

Example 3.2.10 If $\mathfrak{g} = sl_2(\mathbb{C})$, the identity is the only diagram automorphism of $\mathfrak{g}$, and the Coxeter automorphism γ is conjugation by the element

$$\begin{pmatrix} 1 & 0 \\ 0 & -1 \end{pmatrix}$$

of $SL_2(\mathbb{C})$ of order $h_1 = 2$. Thus, $\mathfrak{h}_1$ is the usual Cartan subalgebra of $sl_2(\mathbb{C})$, and $\mathfrak{g}_{1,1}$ consists of the matrices

$$\begin{pmatrix} 0 & * \\ * & 0 \end{pmatrix}.$$

The simple weights are $\pm\alpha$, where α is the usual simple root of $sl_2(\mathbb{C})$. The Dynkin diagram of $(sl_2(\mathbb{C}), 1)$ is

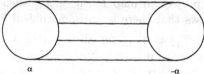

$\qquad\qquad\qquad\qquad \alpha \qquad\qquad\qquad\qquad\qquad -\alpha$

There are three admissible quadruples:

(i) $\Pi_0 = \Pi_1 = \emptyset$: then, in the notation of 2.1.7, we have $\mathfrak{g}_{\Pi_0} = \mathfrak{g}_{\Pi_1} = 0$, $f = 0$, $t_{1,0} = \frac{1}{2}H \otimes H$, $t_{1,1} = X^+ \otimes X^- + X^- \otimes X^+$, $r^0 = \frac{1}{2}t_{1,0}$, and we obtain

$$r(u) = \frac{1}{4}\coth\left(\frac{1}{2}u\right) H \otimes H + \frac{X^+ \otimes X^- + X^- \otimes X^+}{2\sinh(\frac{1}{2}u)}.$$

(ii) $\Pi_0 = \{\alpha\}$, $\Pi_1 = \{-\alpha\}$: then, $\mathfrak{g}_{\Pi_0} = \mathbb{C}.X^+$, $\mathfrak{g}_{\Pi_1} = \mathbb{C}.X^-$. Choose θ such that $\theta(X^-) = X^+$. Then, f is given by

$$f(X^-) = X^+, \quad f(X^+) = f(H) = 0.$$

We obtain

$$r(u) = \frac{1}{4}\coth\left(\frac{1}{2}u\right) H \otimes H + \frac{X^+ \otimes X^- + X^- \otimes X^+}{2\sinh(\frac{1}{2}u)} + 2\sinh\left(\frac{1}{2}u\right) X^- \otimes X^+.$$

(iii) $\Pi_0 = \{-\alpha\}$, $\Pi_1 = \{\alpha\}$: the solution in this case is obtained by interchanging X^+ and X^- in (ii).

There is a unique non-trivial diagram automorphism of $(\mathfrak{g}, 1)$, which interchanges α and $-\alpha$. Thus, the solution of type (iii) is equivalent to that of type (ii), and the parameter space of solutions consists of two points. $\Diamond$

D Rational solutions From 3.2.3(i) and 3.2.4, it follows that every non-degenerate rational solution is equivalent to one of the form

$$(13) \qquad r(u_1, u_2) = \frac{t}{u_1 - u_2} + \tilde{r}(u_1, u_2),$$

where $\tilde{r}$ is a polynomial in u_1 and u_2 with values in $\mathfrak{g} \otimes \mathfrak{g}$. As in the trigonometric case, a computation shows that, if $\tilde{r}$ is constant, then (13) holds if and only if $\tilde{r}$ is a skew-symmetric solution of the CYBE. The following result, conjectured by Drinfel'd, shows that in one sense the general case is not much more complicated.

PROPOSITION 3.2.11 *Every non-degenerate rational solution of the CYBE is equivalent to a solution of the form (13), in which $\tilde{r}$ is a polynomial of degree at most one in each of u_1 and u_2.* ∎

This result is a consequence of a prescription for constructing all non-degenerate rational solutions of the CYBE due to Stolin (1991a–d). We shall only outline his results.

Let $\mathbb{C}[[u^{-1}]]$ be the algebra of formal power series in an indeterminate u with complex coefficients, and let $\mathbb{C}((u^{-1}))$ be its field of fractions. We introduce the Lie algebra of formal power series $\mathfrak{g}((u^{-1})) = \mathfrak{g} \otimes \mathbb{C}((u^{-1}))$ and its subalgebras $\mathfrak{g}[[u^{-1}]] = \mathfrak{g} \otimes \mathbb{C}[[u^{-1}]]$ and $\mathfrak{g}[u] = \mathfrak{g} \otimes \mathbb{C}[u]$. We define an inner product on $\mathfrak{g}((u^{-1}))$ by

$$(f, g) = \mathrm{res}_0(f(u), g(u)) \qquad (f, g \in \mathfrak{g}((u^{-1}))),$$

where 'res$_0$' is the u^{-1} coefficient, and the inner product (,) on the right-hand side is that on $\mathfrak{g}$. Note that $u^{-1}\mathfrak{g}[[u^{-1}]]$ is a lagrangian subspace of $\mathfrak{g}((u^{-1}))$, i.e. it is equal to its orthogonal complement.

PROPOSITION 3.2.12 *There is a natural one-to-one correspondence between non-degenerate rational solutions of the CYBE and subspaces $W \subset \mathfrak{g}((u^{-1}))$ such that*

(i) W is a Lie subalgebra of $\mathfrak{g}((u^{-1}))$;

(ii) $W \oplus \mathfrak{g}[u] = \mathfrak{g}((u^{-1}))$ and $W \supset u^{-N}\mathfrak{g}[[u^{-1}]]$ for some $N \geq 1$;

(iii) W is lagrangian.

Moreover, the solutions associated to subspaces W_1 and W_2 are equivalent if and only if $W_2 = \alpha(u).W_1$ for some $\alpha \in \mathrm{Aut}(\mathfrak{g})[u]$.

OUTLINE OF PROOF By using the inner product, we may identify the vector space dual of $\mathfrak{g}[u]$ with $u^{-1}\mathfrak{g}[[u^{-1}]]$. We associate to any rational solution of the form (13) the graph $W(r)$ of $\tilde{r}$ thought of as a map $\mathfrak{g}[u]^* \to \mathfrak{g}[u]$. That W is a Lie subalgebra follows from the fact that r satisfies the CYBE, and that W is lagrangian follows from the skew property $\tilde{r}_{12}(u_1, u_2) = -\tilde{r}_{21}(u_2, u_1)$

of $\tilde{r}$. Conversely, any W such that $W \oplus \mathfrak{g}\,[u] = \mathfrak{g}\,((u^{-1}))$ is the graph of a map $\mathfrak{g}\,[u]^* \to \mathfrak{g}\,[u]$, and the second condition in (ii) is a 'continuity' condition which guarantees that this map comes from an element of $\mathfrak{g}\,[u] \otimes \mathfrak{g}\,[u]$. ∎

To simplify the presentation, we shall restrict ourselves from now on to the case $\mathfrak{g} = sl_{n+1}(\mathbb{C})$.

PROPOSITION 3.2.13 *Every subspace $W \subseteq sl_{n+1}(\mathbb{C})((u^{-1}))$ satisfying the conditions of 3.2.12 is contained in $g^{-1}.sl_{n+1}(\mathbb{C})[[u^{-1}]].g$ for some invertible matrix $g \in GL_{n+1}(\mathbb{C}((u^{-1})))$.* ∎

Since applying an element of $SL_{n+1}(\mathbb{C}[u])$ to W gives an equivalent solution of the CYBE, and since $sl_{n+1}(\mathbb{C})[[u^{-1}]]$ is preserved by the adjoint action of $GL_{n+1}(\mathbb{C}[[u^{-1}]])$, we are in the situation of the

BIRKHOFF FACTORIZATION THEOREM 3.2.14 *Every $g \in GL_{n+1}(\mathbb{C}((u^{-1})))$ can be written as a product*

$$g = g^- \cdot \begin{pmatrix} u^{k_1} & 0 & \cdots & 0 \\ 0 & u^{k_2} & \cdots & 0 \\ \vdots & \vdots & \ddots & \vdots \\ 0 & 0 & \cdots & u^{k_{n+1}} \end{pmatrix} \cdot g^+,$$

where $g^+ \in GL_{n+1}(\mathbb{C}[u])$, $g^- \in GL_{n+1}(\mathbb{C}[[u^{-1}]])$, and $k_1, k_2, \ldots, k_{n+1}$ are integers satisfying $k_1 \le k_2 \le \cdots \le k_{n+1}$. Moreover, $(k_1, k_2, \ldots, k_{n+1})$ is uniquely determined by g. ∎

It follows that we may restrict ourselves to subspaces W which are contained in $g^{-1}.sl_{n+1}(\mathbb{C})[[u^{-1}]].g$, where g is a diagonal matrix of powers of u. In fact, it is easy to see that condition (ii) in 3.2.12 implies that $|k_i - k_j| \le 1$ for all i, j, so that g is of the form

$$d_k = \begin{pmatrix} 1 & & & & & \\ & \ddots & & & & \\ & & 1 & & & \\ & & & u & & \\ & & & & \ddots & \\ & & & & & u \end{pmatrix},$$

where k is the number of ones. By applying the non-trivial diagram automorphism of $sl_{n+1}(\mathbb{C})$ if necessary, we may assume that $0 \le k \le [(n+1)/2]$. If

$$W \subseteq d_k^{-1}.sl_{n+1}(\mathbb{C})[[u^{-1}]].d_k,$$

we say that W, or the corresponding solution of the CYBE, is of *class k*.

Define Lie subalgebras $\mathfrak{p}_k$ of $sl_{n+1}(\mathbb{C})$, for $0 \le k \le n+1$, by $\mathfrak{p}_0 = sl_{n+1}(\mathbb{C})$, and, for $k > 0$,

$$\mathfrak{p}_k = \begin{pmatrix} * & * \\ 0 & * \end{pmatrix},$$

where the zero matrix is $(n+1-k) \times k$.

PROPOSITION 3.2.15 *There is a one-to-one correspondence between non-degenerate rational solutions of the CYBE of class k with values in sl_{n+1}* $(\mathbb{C})^{\otimes 2}$, *where $1 \leq k \leq [(n+1)/2]$, and pairs $(\mathfrak{f}, \omega)$, where $\mathfrak{f}$ is a Lie subalgebra of $sl_{n+1}(\mathbb{C})$ and ω is a 2-cocycle on $\mathfrak{f}$ which is non-degenerate on $\mathfrak{f} \cap \mathfrak{p}_k$.*

Moreover, the solutions associated to pairs $(\mathfrak{f}_1, \omega_1)$ and $(\mathfrak{f}_2, \omega_2)$ are equivalent if and only if there is an inner automorphism α of $sl_{n+1}(\mathbb{C})$ such that $\alpha(\mathfrak{f}_1) = \mathfrak{f}_2$ and $\alpha^*(\omega_2) - \omega_1$ is a coboundary.

Finally, solutions of class 0 are equivalent to solutions (13) in which $\tilde{r}$ is constant.

OUTLINE OF PROOF The statement about class 0 solutions follows from 3.1.6 and the remarks at the beginning of this subsection. We assume from now on that $k > 0$.

Note that $sl_{n+1}(\mathbb{C}[[u^{-1}]])/u^{-2}sl_{n+1}(\mathbb{C}[[u^{-1}]]) \cong sl_{n+1}(\mathbb{C}[\epsilon]/(\epsilon^2))$; we denote this Lie algebra by $sl_{n+1}(\epsilon)$. As a Lie algebra,

$$sl_{n+1}(\epsilon) \cong sl_{n+1}(\mathbb{C})\tilde{\oplus}\mathrm{ad},$$

the semidirect product in which $sl_{n+1}(\mathbb{C})$ acts on its adjoint representation ad, thought of as an abelian Lie algebra. Note that the inner product on $sl_{n+1}(\mathbb{C}((u^{-1})))$ induces one on $sl_{n+1}(\epsilon)$.

Given W of class k, note that, since W is lagrangian,

$$d_k.W.d_k^{-1} \supset u^{-2}sl_{n+1}(\mathbb{C})[[u^{-1}]].$$

Let W_ϵ be the projection of $d_k.W.d_k^{-1}$ to $sl_{n+1}(\epsilon)$, and let $\mathfrak{f}$ be its further projection to $sl_{n+1}(\mathbb{C})$. Then, $W_\epsilon \subset \mathfrak{f} \oplus \epsilon\,\mathrm{ad}$, and, since W_ϵ is easily seen to be lagrangian, it follows that $W_\epsilon \supset \epsilon\mathfrak{f}^\perp$. Hence, W_ϵ is determined by giving a subspace $V_\epsilon \subset (\mathfrak{f} \oplus \epsilon\,\mathrm{ad})/\epsilon\mathfrak{f}^\perp \cong \mathfrak{f} \oplus \epsilon(\mathrm{ad}/\mathfrak{f}^\perp) \cong \mathfrak{f} \oplus \epsilon\mathfrak{f}^*$. Since W_ϵ projects onto $\mathfrak{f}$, it follows that V_ϵ does also, and hence that V_ϵ is the graph of a map $f : \mathfrak{f} \to \mathfrak{f}^*$. Moreover, since W is a subalgebra of $sl_{n+1}(\mathbb{C})((u^{-1}))$, it follows that W_ϵ is a subalgebra of $sl_{n+1}(\epsilon)$ and that V_ϵ is a subalgebra of $\mathfrak{f} \oplus \mathfrak{f}^*$. The latter condition is equivalent to

$$f([x,y]) = \mathrm{ad}_x(f(y)) - \mathrm{ad}_y(f(x))$$

for all $x, y \in \mathfrak{f}$. If we regard f as an element of $\mathfrak{f}^*\otimes\mathfrak{f}^*$, this in turn is equivalent to f being a 2-cocycle.

Finally, the condition that W is of class k implies that

$$W_\epsilon \oplus (\mathfrak{p}_k \oplus \epsilon\mathfrak{p}_k^\perp) = sl_{n+1}(\epsilon).$$

It is not difficult to see that this is equivalent to requiring that f is non-degenerate on $\mathfrak{f} \cap \mathfrak{p}_k$ and $\mathfrak{f} + \mathfrak{p}_k = sl_{n+1}(\epsilon)$. ∎

Example 3.2.16 For $sl_2(\mathbb{C})$, the only possible choice of $\mathfrak{f}$ is $sl_2(\mathbb{C})$ itself. This gives the class 0 solution

$$r(u_1, u_2) = \frac{t}{u_1 - u_2} + \frac{1}{2} H \wedge X^+$$

and the class 1 solution

$$r(u_1, u_2) = \frac{t}{u_1 - u_2} + \frac{u_2}{2} H \otimes X^+ - \frac{u_1}{2} X^+ \otimes H.$$

(We are taking the inner product on $sl_2(\mathbb{C})$ to be the trace form.) ◊

Example 3.2.17 For $sl_3(\mathbb{C})$, the class 0 solutions can be read off from 3.1.8. In the class 1 case, there are two choices for $\mathfrak{f}$, namely $sl_3(\mathbb{C})$ itself and

$$\mathfrak{p}_1^t = \begin{pmatrix} * & 0 & 0 \\ * & * & * \\ * & * & * \end{pmatrix}.$$

In both cases, $H^2_{\mathrm{Lie}}(\mathfrak{f}) = 0$, so the associated solution is unique up to equivalence. Using the notation of 3.1.5, the associated solutions are, respectively,

$$r(u_1, u_2) = \frac{t}{u_1 - u_2} + \frac{u_2}{3}(H_\alpha + 2H_\beta) \otimes X^+_{\alpha+\beta} - \frac{u_1}{3} X^+_{\alpha+\beta} \otimes (H_\alpha + 2H_\beta)$$

$$+ u_2 X^+_\alpha \otimes X^+_\beta - u_1 X^+_\alpha \otimes X^+_\beta + \frac{1}{3}(H_\alpha - H_\beta) \wedge X^-_\alpha + X^+_\alpha \wedge X^+_{\alpha+\beta},$$

$$r(u_1, u_2) = \frac{t}{u_1 - u_2} + \frac{u_1}{3}(H_\alpha + 2H_\beta) \otimes X^+_{\alpha+\beta} - \frac{u_2}{3} X^+_{\alpha+\beta} \otimes (H_\alpha + 2H_\beta)$$

$$+ \frac{1}{3} X^+_\beta \wedge (H_\alpha + 2H_\beta) + \frac{1}{3} X^-_\alpha \wedge (2H_\alpha + H_\beta),$$

the Casimir element again being computed for the trace form. ◊

Bibliographical notes

3.1 The classification of the non-skew-symmetric constant solutions of the CYBE (or, equivalently, of the skew-symmetric constant solutions of the MCYBE) for a complex simple Lie algebra is due to Belavin & Drinfel'd (1982, 1984a,b), where complete details of the proofs may be found. The case of skew solutions of the CYBE is due to Stolin (1991a,c,d).

3.2 The classification theorem 3.2.4, as well as the explicit description of the elliptic and trigonometric solutions of the CYBE, are again due to Belavin & Drinfel'd (1982, 1984a,b) although, as in the case of constant solutions, many examples had appeared earlier in the physics literature. In particular, all the elliptic solutions had been found by Belavin (1981) by taking the 'classical limit' of the solutions of the quantum Yang–Baxter equation associated to the '$\mathbb{Z}_n$-symmetric' lattice model in statistical mechanics (see Section 7.5).

The rational solutions were classified by Stolin (1991a,b,d).

4

Quasitriangular Hopf algebras

The space $\mathcal{F}(M)$ of C^∞ functions on a smooth manifold M forms a commutative algebra under pointwise addition and multiplication. If G is a Lie group, the multiplication and inverse maps of G endow $\mathcal{F}(G)$ with extra structure, namely the *comultiplication* map $\mathcal{F}(G) \to \mathcal{F}(G) \otimes \mathcal{F}(G)$ and the *antipode*, a map $\mathcal{F}(G) \to \mathcal{F}(G)$ (one must define the tensor product appropriately here). These maps satisfy certain compatibility conditions reflecting the associativity of multiplication in G and the defining properties of the inverse. Abstracting these properties leads to the notion of a *Hopf algebra* (which need not be commutative as an algebra).

In fact, to any Lie group G one can associate a second Hopf algebra, the *universal enveloping algebra* $U(\mathfrak{g})$ of the Lie algebra $\mathfrak{g}$ of G. The universal enveloping algebra is not commutative (unless G is), but is always cocommutative, which means that the comultiplication of $U(\mathfrak{g})$ takes values in the symmetric part of $U(\mathfrak{g}) \otimes U(\mathfrak{g})$.

Before the advent of quantum groups, very few Hopf algebras which are neither commutative nor cocommutative were known (apart from tensor products of function algebras and enveloping algebras), but the theory of quantum groups provides a large class of natural examples.

We summarize the main facts about Hopf algebras in Section 4.1. This rather lengthy section should probably not be read linearly, but referred to when necessary.

The comultiplication map $A \to A \otimes A$ of an arbitrary Hopf algebra A allows one to define the tensor product $V \otimes W$ of two representations V and W of A. If A is not cocommutative, $V \otimes W$ and $W \otimes V$ are not necessarily isomorphic as representations of A, although they are isomorphic for the special class of *quasitriangular* Hopf algebras, discussed in Section 4.2. Such Hopf algebras contain a distinguished invertible element $\mathcal{R} \in A \otimes A$, called the *universal R-matrix*, from which the isomorphism $V \otimes W \to W \otimes V$ is constructed. Most of the quantum groups we shall encounter (or their duals) are quasitriangular, and many of the applications of the theory of quantum groups make essential use of their universal R-matrices. The crucial property of $\mathcal{R}$ in these applications is that it satisfies the *quantum Yang–Baxter equation*

$$\mathcal{R}_{12}\mathcal{R}_{13}\mathcal{R}_{23} = \mathcal{R}_{23}\mathcal{R}_{13}\mathcal{R}_{12}.$$

4.1 Hopf algebras

A Definitions An algebra with unit over a commutative ring k (always assumed to have a unit itself) is a k-module A together with a way of multiplying two elements a_1, $a_2 \in A$ to give their product $a_1.a_2 \in A$. The multiplication should be bilinear over k and associative. The unit element 1 of A has the property $a.1 = 1.a = a$ for all $a \in A$. We shall now reformulate this in terms of commutative diagrams.

DEFINITION 4.1.1 *An algebra over a commutative ring k is a k-module A equipped with k-module maps $\mu^A : A \underset{k}{\otimes} A \to A$, the multiplication, and $\imath^A : A \to A$, the unit, such that the following diagrams commute:*

$$
\begin{array}{ccc}
A \otimes k & \xrightarrow{\mathrm{id} \otimes \imath} & A \otimes A \\
\cong \downarrow & & \downarrow \mu \\
A & \xrightarrow[\mathrm{id}]{} & A
\end{array}
\qquad
\begin{array}{ccc}
k \otimes A & \xrightarrow{\imath \otimes \mathrm{id}} & A \otimes A \\
\cong \downarrow & & \downarrow \mu \\
A & \xrightarrow[\mathrm{id}]{} & A
\end{array}
$$

$$
\begin{array}{ccc}
A \otimes A \otimes A & \xrightarrow{\mu \otimes \mathrm{id}} & A \otimes A \\
\mathrm{id} \otimes \mu \downarrow & & \downarrow \mu \\
A \otimes A & \xrightarrow[\mu]{} & A
\end{array}
$$

(we omit the superscript on μ, $\imath$ and the subscript on the tensor product when this will not lead to confusion).

Of course, in terms of the more familiar description of an algebra given earlier, we have $\imath(\lambda) = \lambda 1$, $\mu(a_1 \otimes a_2) = a_1.a_2$. The first two diagrams express the properties of the unit element and the third the associativity of multiplication.

The condition that an algebra be *commutative* can similarly be expressed by the commutativity of the diagram

$$
\begin{array}{ccc}
A \otimes A & \xrightarrow{\sigma} & A \otimes A \\
\mu \downarrow & & \downarrow \mu \\
A & \xrightarrow[\mathrm{id}]{} & A
\end{array}
$$

where $\sigma : A \otimes A \to A \otimes A$ is the k-linear map such that $\sigma(a_1 \otimes a_2) = a_2 \otimes a_1$ for a_1, $a_2 \in A$. In general, if we set $\mu_{\mathrm{op}} = \mu \circ \sigma$, then $(A, \imath, \mu_{\mathrm{op}})$ is also an algebra, called the *opposite* of A and denoted by A_{op}. Note that k itself is a commutative algebra over k.

If A and B are algebras over k, their tensor product $A \underset{k}{\otimes} B$ is also an algebra. The product operation

$$(a_1 \otimes b_1).(a_2 \otimes b_2) = a_1.a_2 \otimes b_1.b_2, \qquad (a_1, a_2 \in A, \ b_1, b_2 \in B)$$

in $A \otimes B$ can be expressed as the composite

$$(A \otimes B) \otimes (A \otimes B) \xrightarrow{\sigma_{23}} A \otimes A \otimes B \otimes B \xrightarrow{\mu^A \otimes \mu^B} A \otimes B,$$

where σ_{23} is σ applied to the second and third factors of the tensor product, and the unit of $A \otimes B$ is the composite

$$k \xrightarrow{\cong} k \otimes k \xrightarrow{\iota^A \otimes \iota^B} A \otimes B.$$

We leave it to the reader to give a diagrammatic formulation of the definition of a homomorphism of algebras.

The purpose of the reformulation of the definition of an algebra we have given is that it leads naturally to the 'dual' notion.

DEFINITION 4.1.2 *A coalgebra over a commutative ring k is a k-module A equipped with k-module maps $\Delta^A : A \to A \otimes A$, the comultiplication, and $\epsilon^A : A \to k$, the counit, such that the following diagrams commute:*

$$
\begin{array}{ccc}
A \otimes k & \xleftarrow{\mathrm{id} \otimes \epsilon} & A \otimes A \\
{\scriptstyle \cong} \big\uparrow & & \big\uparrow {\scriptstyle \Delta} \\
A & \xleftarrow[\mathrm{id}]{} & A
\end{array}
\qquad
\begin{array}{ccc}
k \otimes A & \xleftarrow{\epsilon \otimes \mathrm{id}} & A \otimes A \\
{\scriptstyle \cong} \big\uparrow & & \big\uparrow {\scriptstyle \Delta} \\
A & \xleftarrow[\mathrm{id}]{} & A
\end{array}
$$

$$
\begin{array}{ccc}
A \otimes A \otimes A & \xleftarrow{\Delta \otimes \mathrm{id}} & A \otimes A \\
{\scriptstyle \mathrm{id} \otimes \Delta} \big\uparrow & & \big\uparrow {\scriptstyle \Delta} \\
A \otimes A & \xleftarrow[\Delta]{} & A
\end{array}
$$

The commutativity of the third diagram is usually referred to as the *coassociativity* of A.

A coalgebra A is called *cocommutative* if the following diagram commutes:

$$
\begin{array}{ccc}
A \otimes A & \xleftarrow{\sigma} & A \otimes A \\
{\scriptstyle \Delta} \big\uparrow & & \big\uparrow {\scriptstyle \Delta} \\
A & \xleftarrow[\mathrm{id}]{} & A
\end{array} \ ;
$$

in other words, if $\Delta(A)$ is contained in the symmetric part of $A \otimes A$. In general, if we set $\Delta^{\mathrm{op}} = \sigma \circ \Delta$, then $(A, \epsilon, \Delta^{\mathrm{op}})$ is also a coalgebra, called the *opposite* of A and denoted by A^{op}. Note that k itself is a cocommutative coalgebra over k, with $\epsilon^k = \mathrm{id}_k$ and $\Delta^k : k \to k \otimes k$ the natural isomorphism.

If A and B are coalgebras, then $A \otimes B$ is a coalgebra with

$$\Delta^{A \otimes B}(a \otimes b) = \Delta_{13}^A(a)\Delta_{24}^B(b), \quad \epsilon^{A \otimes B}(a \otimes b) = \epsilon^A(a)\epsilon^B(b);$$

i.e. if $\Delta^A(a) = \sum_i a_i \otimes a_i'$ and $\Delta^B(b) = \sum_j b_j \otimes b_j'$, then

$$\Delta^{A \otimes B}(a \otimes b) = \sum_{i,j} a_i \otimes b_j \otimes a_i' \otimes b_j'.$$

If A and B are coalgebras, a k-module map $\varphi : A \to B$ is a *homomorphism of coalgebras* if

$$(\varphi \otimes \varphi) \circ \Delta^A = \Delta^B \circ \varphi, \quad \epsilon^B \circ \varphi = \epsilon^A.$$

A Hopf algebra has compatible algebra and coalgebra structures and one extra structure map.

DEFINITION 4.1.3 *A Hopf algebra over a commutative ring k is a k-module A such that*

(i) A is both an algebra and a coalgebra over k;

(ii) the comultiplication $\Delta : A \to A \otimes A$ and the counit $\epsilon : A \to k$ are homomorphisms of algebras;

(iii) the multiplication $\mu : A \otimes A \to A$ and the unit $\imath : k \to A$ are homomorphisms of coalgebras;

(iv) A is equipped with a bijective k-module map $S^A : A \to A$, called the antipode, such that the following diagrams commute:

$$
\begin{array}{ccc}
A \otimes A & \xrightarrow{S \otimes \mathrm{id}} & A \otimes A \\
\Delta \uparrow & & \downarrow \mu \\
A & \xrightarrow{\imath \circ \epsilon} & A
\end{array}
\qquad
\begin{array}{ccc}
A \otimes A & \xrightarrow{\mathrm{id} \otimes S} & A \otimes A \\
\Delta \uparrow & & \downarrow \mu \\
A & \xrightarrow{\imath \circ \epsilon} & A
\end{array}
$$

If A and B are Hopf algebras, a k-module map $\varphi : A \to B$ is a *homomorphism of Hopf algebras* if it is a homomorphism of both the algebra and coalgebra structures of A.

A *Hopf ideal* of a Hopf algebra A over k is a two-sided ideal I of A as an algebra such that

$$\Delta(I) \subseteq I \otimes A + A \otimes I, \quad \epsilon(I) = 0, \quad S(I) \subseteq I.$$

The quotient k-module A/I then inherits a Hopf algebra structure from A in the obvious way. The kernel of any homomorphism of Hopf algebras is a Hopf ideal.

The tensor product $A \otimes B$ of two Hopf algebras is a Hopf algebra with the algebra and coalgebra structures defined above, and with antipode

$$S^{A \otimes B} = S^A \otimes S^B.$$

If A is a Hopf algebra, there are three associated Hopf algebras A_{op}, A^{op} and $A^{\mathrm{op}}_{\mathrm{op}}$. Their algebra and coalgebra structures have already been defined, and their antipodes are $(S^A)^{-1}$, $(S^A)^{-1}$ and S^A, respectively.

Remarks [1] Conditions (ii) and (iii) in the definition are equivalent (in the presence of the other conditions). Omitting condition (iv) in the definition gives the notion of a *bialgebra*. If a bialgebra has an antipode satisfying (iv), then it is unique.

[2] A homomorphism of Hopf algebras $\varphi : A \to B$ automatically satisfies $S^B \circ \varphi = \varphi \circ S^A$ (see Hazewinkel (1978), Proposition 37.1.10).

[3] The antipode S is automatically an *anti-automorphism* of A, i.e. $S : A \to A^{\mathrm{op}}_{\mathrm{op}}$ is an isomorphism of Hopf algebras. Thus, S^2 is an automorphism of A. It can be shown that, if A is either commutative or cocommutative, then S^2 is the identity map (see Abe (1980), Theorem 2.1.4), but we shall encounter many Hopf algebras in this book for which S^2 is non-trivial.

[4] The defining property of the antipode S can be interpreted in terms of the *convolution* algebra structure on the set of k-module maps $A \to A$. If f, $g \in \mathrm{End}(A)$, their convolution is the map $f \bullet g : A \to A$ defined by

$$f \bullet g = \mu \circ (f \otimes g) \circ \Delta.$$

The two commutative diagrams in part (iv) of 4.1.3 assert that S *is the inverse of the identity map for the convolution product*. The last sentence in Remark [1] above is a consequence of this observation.

[5] In the literature on Hopf algebras, a slightly weaker definition is used: the antipode is not required to be bijective. It is known that the antipode of every Hopf algebra in this weaker sense is necessarily bijective if the Hopf algebra is either commutative or cocommutative. Indeed, it seems to be bijective in essentially all naturally occurring examples. $\triangle$

There are several variants of the definition we have given. In some situations, A and $A \otimes A$ (as well as the ground ring k) are equipped with topologies for which the k-module structure maps of A are continuous. In that case, one usually replaces the algebraic tensor product $A \otimes A$ by a suitable completion, and requires that the Hopf algebra structure maps are continuous.

The most important example of such a situation for us is where the ground ring is $k[[h]]$, the ring of formal power series in an indeterminate h over a field k. Every $k[[h]]$-module V has the *h-adic topology*, which is characterized by requiring that $\{h^n V \mid n \geq 0\}$ is a base of the neighbourhoods of 0 in V, and that translations in V are continuous. It is easy to see that, for modules equipped with this topology, every $k[[h]]$-module map is automatically continuous.

DEFINITION 4.1.4 *A* topological Hopf algebra over $k[[h]]$, *the ring of formal power series in an indeterminate h over a field k, is a complete $k[[h]]$-module A equipped with $k[[h]]$-linear maps $\imath$, μ, ϵ, Δ and S satisfying the axioms of 4.1.3, but with the algebraic tensor products replaced by their completions in the h-adic topology.*

We study topological Hopf algebras further in Subsection 4.1D.

A second variant of 4.1.3 applies to the case where A is a graded k-module. This means that there are k-submodules A_n of A, for $n \in \mathbb{N}$, such that

$$A = \bigoplus_{n \in \mathbb{N}} A_n.$$

One says that A is a *graded Hopf algebra* if the usual axioms are satisfied, except that σ is replaced by its graded version

$$\sigma(a \otimes a') = (-1)^{mn} a' \otimes a, \quad \text{if} \quad a \in A_m, a' \in A_n$$

(thus changing the algebra structure on $A \otimes A$), and that the structure maps of A are required to preserve the graded structure, i.e. $\imath(k) \subset A_0$, $\epsilon(A_n) = 0$ if $n > 0$, $S(A_n) = A_n$, $\mu(A_m \otimes A_n) \subset A_{m+n}$ and $\Delta(A_n) \subset \oplus_{p+q=n} A_p \otimes A_q$. As we shall see in a moment, the graded case preceded the usual case historically, even though the latter is a special case of the former (by regarding an ordinary Hopf algebra as being concentrated in degree zero).

B Examples There are several natural constructions which associate Hopf algebras to groups of various kinds, and we shall discuss some of these now.

Example 4.1.5 The simplest case is that of a finite group G. The *group algebra* $k[G]$ of G over a commutative ring k is the free k-module with basis G and algebra structure obtained by extending linearly the product on G. There is a Hopf algebra structure on $k[G]$ defined by extending linearly the formulas $\imath(1) = e$ (where e is the identity element of G), $\epsilon(g) = 1$, $\Delta(g) = g \otimes g$ and $S(g) = g^{-1}$. Note that $k[G]$ is always cocommutative, but is commutative only if G is commutative.

The second Hopf algebra associated to G is the set $\mathcal{F}(G)$ of k-valued functions on G. The k-module and algebra structures of $\mathcal{F}(G)$ are defined pointwise, and the counit and antipode are given by $\epsilon(f) = f(e)$ and $S(f)(g) = f(g^{-1})$, respectively. To define the comultiplication, note that $\mathcal{F}(G \times G) \cong$

$\mathcal{F}(G) \otimes \mathcal{F}(G)$ as algebras over k: if f_1, $f_2 \in \mathcal{F}(G)$, the isomorphism takes $f_1 \otimes f_2$ to the function on $G \times G$ given by $(g_1, g_2) \mapsto f_1(g_1)f_2(g_2)$. We set $\Delta(f)(g_1, g_2) = f(g_1 g_2)$. Note that $\mathcal{F}(G)$ is always commutative, but is cocommutative only if G is commutative. $\diamond$

Remark Because of the first of these examples, an element a of an arbitrary Hopf algebra A is called *group-like* if $\Delta(a) = a \otimes a$. $\triangle$

Example 4.1.6 Let G be an affine algebraic group over a field k (i.e. an affine variety over k such that the group operations are morphisms of varieties), and let $\mathcal{F}(G)$ be its algebra of regular functions. Then it is still true that $\mathcal{F}(G \times G) \cong \mathcal{F}(G) \otimes \mathcal{F}(G)$, and one can define a Hopf algebra structure on $\mathcal{F}(G)$ exactly as in 4.1.5. For example, if $G = GL_n(k)$, then $\mathcal{F}(G)$ is the commutative algebra with generators x_{ij}, $1 \leq i, j \leq n$, and D^{-1}, and the single relation

$$D^{-1}.\det X = 1,$$

where X is the $n \times n$ matrix with entries x_{ij}. The comultiplication is defined on generators by

$$\Delta(x_{ij}) = \sum_{k=1}^{n} x_{ik} \otimes x_{kj},$$

and the remaining structure maps are given by $\epsilon(x_{ij}) = \delta_{i,j}$, $S(x_{ij}) = (X^{-1})_{ij}$. It follows from the multiplicative property of the determinant that D^{-1} is group-like and that $S(D^{-1}) = D$. $\diamond$

Example 4.1.7 Let G be a compact topological group and let $\mathcal{F}(G)$ be the algebra of continuous real- (or complex-) valued functions on G. If we use the same definitions as in the previous two examples, we do not get a Hopf algebra in the sense of 4.1.3 because the comultiplication takes values in $\mathcal{F}(G \times G)$, which is 'bigger' than $\mathcal{F}(G) \otimes \mathcal{F}(G)$. One solution is to modify our definition of a Hopf algebra, taking into account the natural topology on $\mathcal{F}(G)$ (defined by the sup norm). Since the algebraic tensor product $\mathcal{F}(G) \otimes \mathcal{F}(G)$ is dense in $\mathcal{F}(G \times G)$, we can regard $\mathcal{F}(G \times G)$ as a completion of $\mathcal{F}(G) \otimes \mathcal{F}(G)$.

Alternatively, we can replace $\mathcal{F}(G)$ by a smaller algebra. In fact, let $\rho : G \to GL_n(\mathbb{R})$ be any continuous representation of G on $\mathbb{R}^n$, and let $\rho_{ij} = x_{ij} \circ \rho$ be the *matrix elements* of ρ. Then the ρ_{ij}, as ρ runs through all finite-dimensional representations of G, generate a subalgebra $\text{Rep}(G)$ of $\mathcal{F}(G)$, called the algebra of *representative functions* of G. We know by the Peter–Weyl theorem that $\text{Rep}(G)$ is dense in $\mathcal{F}(G)$. Also, $\text{Rep}(G \times G)$ is isomorphic to the algebraic tensor product $\text{Rep}(G) \otimes \text{Rep}(G)$, and it is easy to see that the usual maps make $\text{Rep}(G)$ into a genuine Hopf algebra. $\diamond$

Example 4.1.8 Let $\mathfrak{g}$ be a Lie algebra over a field k. The *universal enveloping algebra* $U(\mathfrak{g})$ of $\mathfrak{g}$ is the quotient of the tensor algebra $T(\mathfrak{g}) = \oplus_{n \geq 0} \mathfrak{g}^{\otimes n}$ by the two-sided ideal generated by the elements $x \otimes y - y \otimes x - [x, y]$ for all x, $y \in \mathfrak{g}$. The Poincaré–Birkhoff–Witt theorem asserts that, if $\{x_i\}_{i \in I}$ is any basis of $\mathfrak{g}$, where the index set I is totally ordered, the set of monomials $\{x_{i_1} x_{i_2} \ldots x_{i_k}\}$, where $k \geq 1$ and $i_1 \leq i_2 \leq \cdots \leq i_k$, is a basis of $U(\mathfrak{g})$. It follows that the composite of the natural map $\mathfrak{g} \to T(\mathfrak{g})$ with the canonical projection $T(\mathfrak{g}) \to U(\mathfrak{g})$ gives an embedding of $\mathfrak{g}$ into $U(\mathfrak{g})$, and we usually identify $\mathfrak{g}$ with its image under this map. Since $\mathfrak{g}$ clearly generates $U(\mathfrak{g})$ as an algebra, to define a Hopf structure on $U(\mathfrak{g})$ it is enough to give the structure maps on elements of $\mathfrak{g}$:

$$\Delta(x) = x \otimes 1 + 1 \otimes x, \quad S(x) = -x, \quad \epsilon(x) = 0 \quad (x \in \mathfrak{g}).$$

In fact, $\mathfrak{g}$ can be characterized as the set of *primitive* elements of $U(\mathfrak{g})$, i.e. the set of elements x for which $\Delta(x) = x \otimes 1 + 1 \otimes x$.

Of course, one must check that these formulas respect the algebra structure: for example,

$$
\begin{aligned}
\Delta(xy - yx) &= \Delta(x)\Delta(y) - \Delta(y)\Delta(x) \\
&= (x \otimes 1 + 1 \otimes x)(y \otimes 1 + 1 \otimes y) - (y \otimes 1 + 1 \otimes y)(x \otimes 1 + 1 \otimes x) \\
&= (xy - yx) \otimes 1 + 1 \otimes (xy - yx) \\
&= [x, y] \otimes 1 + 1 \otimes [x, y] \\
&= \Delta([x, y]).
\end{aligned}
$$

Note that $U(\mathfrak{g})$ is cocommutative, since $\Delta(\mathfrak{g})$ is obviously contained in the symmetric part of $U(\mathfrak{g}) \otimes U(\mathfrak{g})$, which is a subalgebra of $U(\mathfrak{g}) \otimes U(\mathfrak{g})$. But $U(\mathfrak{g})$ is commutative only if $\mathfrak{g}$ is commutative. $\Diamond$

Example 4.1.9 Let V be a graded vector space over a field k of characteristic $\neq 2$. If V is concentrated in even degrees, its *symmetric algebra* $\mathrm{Sym}(V)$ is the quotient of the tensor algebra $T(V)$ by the two-sided ideal generated by $v \otimes v' - v' \otimes v$ for all v, $v' \in V$. If V is concentrated in odd degrees, its *exterior algebra* $\wedge(V)$ is the quotient of $T(V)$ by the two-sided ideal generated by $v \otimes v' + v' \otimes v$ for all v, $v' \in V$.

The symmetric and exterior algebras are multiplicative, in the sense that, for any two (appropriately graded) vector spaces V and W, there are canonical isomorphisms

$$\mathrm{Sym}(V \oplus W) \cong \mathrm{Sym}(V) \otimes \mathrm{Sym}(W), \quad \wedge(V \oplus W) \cong \wedge(V) \otimes \wedge(W)$$

of graded algebras. Hence, the diagonal map $V \to V \oplus V$, $v \to (v, v)$, induces comultiplications on $\mathrm{Sym}(V)$ and $\wedge(V)$. If we set $\epsilon(v) = 0$, $S(v) = -v$, for $v \in V$, then $\mathrm{Sym}(V)$ and $\wedge(V)$ become graded Hopf algebras over k.

Note that $\mathrm{Sym}(V)$ and $\wedge(V)$ are commutative and cocommutative (in the graded sense). Conversely, it can be shown that every graded Hopf algebra A

over a field of characteristic $\neq 2$ which is both commutative and cocommutative, and for which $\dim(A_0) = 1$ and $\dim(A_n) < \infty$ for all n, is isomorphic to the tensor product of a symmetric algebra and an exterior algebra. $\Diamond$

Example 4.1.10 Let G be a connected, finite-dimensional Lie group, and let $H_*(G, k)$ be the homology of G with coefficients in a field k of characteristic zero. Then, $H_*(G, k)$ is a graded Hopf algebra: the multiplication in $H_*(G, k)$ (called the Pontryagin product) is induced by the product $G \times G \to G$ (note that $H_*(G \times G, k) \cong H_*(G, k) \otimes H_*(G, k)$ as graded algebras by the Künneth theorem); the comultiplication is induced by the diagonal map $G \to G \times G$; the antipode by the inversion map $G \to G$; and the counit is the projection onto the degree zero part, which is one-dimensional because G is connected.

Moreover, the algebra $H_*(G, k)$ is graded commutative and cocommutative, and finite dimensional because G is finite dimensional. Hence, the result stated in the previous example implies that $H_*(G, k)$ is isomorphic to the tensor product of a symmetric algebra and an exterior algebra. But, since $H_*(G, k)$ itself (and not just its graded components) is finite dimensional, the symmetric part must be trivial, whence Hopf's theorem that $H_*(G, k)$ is an exterior algebra on odd-dimensional generators.

This example provided the original motivation for the study of Hopf algebras, as well as the terminology. $\Diamond$

C Representations of Hopf algebras If A is an algebra over a commutative ring k, a k-module V is called a *left A-module* if there is a k-module map $\lambda_V : A \otimes V \to V$ such that the following diagrams commute:

$$
\begin{array}{ccc}
A \otimes A \otimes V & \xrightarrow{\mu \otimes \mathrm{id}_V} & A \otimes V \\
{\scriptstyle \mathrm{id}_A \otimes \lambda} \downarrow & & \downarrow {\scriptstyle \lambda} \\
A \otimes V & \xrightarrow{\quad \lambda \quad} & V
\end{array}
\qquad
\begin{array}{ccc}
k \otimes V & \xrightarrow{\imath \otimes \mathrm{id}_V} & A \otimes V \\
{\scriptstyle \cong} \downarrow & & \downarrow {\scriptstyle \lambda} \\
V & \xrightarrow{\quad \mathrm{id}_V \quad} & V
\end{array}
$$

(as usual, we omit the subscript if this is unambiguous). Equivalently, $a \mapsto \lambda(a \otimes \cdot)$ should be a homomorphism of algebras from A into the endomorphisms of V. We shall sometimes write $\lambda(a \otimes v)$ as $a.v$ if the action is understood. Right A-modules are defined similarly.

If A is a coalgebra over a commutative ring k, a k-module V is called a *right A-comodule* if there is a k-module map $\rho_V : V \to V \otimes A$ such that the following diagrams commute:

$$
\begin{array}{ccc}
V \otimes A \otimes A & \xleftarrow{\mathrm{id}_V \otimes \Delta} & V \otimes A \\
{\scriptstyle \rho \otimes \mathrm{id}_A} \uparrow & & \uparrow {\scriptstyle \rho} \\
V \otimes A & \xleftarrow{\quad \rho \quad} & V
\end{array}
\qquad
\begin{array}{ccc}
V \otimes k & \xleftarrow{\mathrm{id}_V \otimes \epsilon} & V \otimes A \\
{\scriptstyle \cong} \uparrow & & \uparrow {\scriptstyle \rho} \\
V & \xleftarrow{\quad \mathrm{id}_V \quad} & V
\end{array}
$$

Left A-comodules are defined similarly.

If V is itself an algebra or a coalgebra, it is natural to require that these module and comodule structures should respect the extra structure on V. Assume that A is a bialgebra. An algebra V is a *left A-module algebra* if it is a left A-module and

$$a.(vw) = \sum_i (a_i.v)(a^i.w), \quad a.1 = \epsilon_A(a)1$$

for all $a \in A$, $v, w \in V$, where $\Delta_A(a) = \sum_i a_i \otimes a^i$. A coalgebra V is a *left A-module coalgebra* if it is a left A-module and

$$\Delta_V(a.v) = \sum_{i,j} a_i.v_j \otimes a^i.v^j, \quad \epsilon_V(a.v) = \epsilon_A(a)\epsilon_V(v)$$

for all $a \in A$, $v \in V$, where $\Delta_V(v) = \sum_j v_j \otimes v^j$. We leave it to the reader to formulate these definitions diagrammatically, and to define the notions of a right A-comodule algebra and coalgebra.

Example 4.1.11 Let k be a field of characteristic zero, and let G be an affine algebraic group over k. Let $\varphi : G \to GL(V)$ be a representation of G on a finite-dimensional k-vector space V (in the usual sense). Choosing a basis $\{v_i\}$ of V defines an isomorphism $GL(V) \cong GL_n(k)$, and allows us to regard φ as a homomorphism $\varphi : G \to GL_n(k)$. One says that φ is *rational* if the pull-backs $\varphi_{ij} = x_{ij} \circ \varphi$ of the coordinate functions x_{ij} on $GL_n(k)$ are regular functions on G (cf. 4.1.6); it is easy to see that this criterion is independent of the choice of basis of V. For such a representation, the map $\rho : V \to V \otimes \mathcal{F}(G)$ defined by

$$\rho(v_i) = \sum_{j=1}^n v_j \otimes \varphi_{ji}$$

is also independent of the chosen basis of V, and defines a right $\mathcal{F}(G)$-comodule structure on V. $\Diamond$

A *representation* of a Hopf algebra A over a commutative ring k is a left A-module (A being regarded simply as an algebra). The terms subrepresentation, irreducible representation, etc., have their usual meanings. A *corepresentation* of A is a right A-comodule. We shall use the terms (co)representation and (co)module interchangeably.

Remark It is only necessary to consider left A-modules, for if $\rho : V \otimes A \to V$ is a right A-module, then $\lambda = \rho \circ (\mathrm{id}_V \otimes S)$ is a left A-module, and conversely. A similar remark applies, of course, to comodules. $\triangle$

Example 4.1.12 The *trivial representation* of a Hopf algebra A on a k-module V is given by

$$\lambda(a \otimes v) = \epsilon(a)v.$$

More generally, an element v of an arbitrary representation of A is said to be *invariant* under the action of A if the preceding equation holds for all $a \in A$.

Similarly, the trivial corepresentation is defined using the unit $\imath$ of A. $\Diamond$

Example 4.1.13 The *adjoint representation* of a Hopf algebra A on itself is given by

$$\mathrm{ad}(a \otimes a') = \sum_i a_i a' S(a^i),$$

where $\Delta(a) = \sum_i a_i \otimes a^i$. The reader will easily verify that this makes A into a left A-module algebra, and that it reduces, in the case of a group algebra or a universal enveloping algebra, to the action of a group on itself by conjugation or to the usual adjoint representation of a Lie algebra, respectively. $\Diamond$

Example 4.1.14 The *regular representation* of a Hopf algebra A on itself is the multiplication μ of A; it makes A into a left A-module coalgebra. Similarly, the *regular corepresentation* is the comultiplication Δ of A ; it makes A into a right A-comodule algebra. $\Diamond$

Although the definition of a representation of a Hopf algebra A only makes use of its algebra structure, the remaining structure maps still play an important role in representation theory. In fact, we have already seen that the counit allows one to define the trivial representation of A. More significantly, the comultiplication allows one to take the *tensor product* of representations: if V and W are representations of A, then $V \otimes W$ is naturally a representation of $A \otimes A$,

$$(a_1 \otimes a_2).(v \otimes w) = a_1.v \otimes a_2.w,$$

and we make $V \otimes W$ into a representation of A by

$$a.(v \otimes w) = \Delta(a).(v \otimes w).$$

Finally, the antipode allows one to turn the dual $V^* = \mathrm{Hom}_k(V, k)$ of a representation V of A into a representation of A

$$(1) \qquad \langle a.\xi, v \rangle = \langle \xi, S(a).v \rangle,$$

where $a \in A$, $v \in V$, $\xi \in V^*$ ($\langle \ , \ \rangle$ is the natural pairing between V and V^*). One checks without difficulty that the canonical k-module map $V^* \otimes V \to k$ commutes with the action of A. If k is a field and V is finite dimensional, the same is true for the canonical map $k \to V \otimes V^*$, which takes $1 \in k$ to $\sum_i v_i \otimes v^i$ if $\{v_i\}$ is a basis of V and $\{v^i\}$ the dual basis on V^* (this remains true for any commutative ring k if V is finitely generated and projective over k, by the 'dual basis lemma' – see 5.1.3). However, the canonical k-module maps $V \otimes V^* \to k$ and $k \to V^* \otimes V$ do *not* commute with the action of A in general. Note finally that, since S is a coalgebra anti-automorphism of A, if V and W are representations of A, the canonical isomorphism of k-modules $(V \otimes W)^* \cong W^* \otimes V^*$ also commutes with A.

If V and W are representations of A, we make $\mathrm{Hom}_k(V, W)$ into a representation of A by setting

$$a.f(v) = \sum_i a_i.f(S(a^i).v),$$

for $f \in \mathrm{Hom}_k(V, W)$, $v \in V$, where $\Delta(a) = \sum_i a_i \otimes a^i$. If W is the trivial representation, this is of course the dual V^*. We leave it to the reader to check that, if k is a field (say), the canonical vector space isomorphism $\mathrm{Hom}_k(V, W) \cong W \otimes V^*$ is actually an isomorphism of representations of A (but this is not true for the vector space isomorphism $\mathrm{Hom}_k(V, W) \cong V^* \otimes W$!).

Once again, the reader will easily verify that, in the case of group algebras and universal enveloping algebras, these definitions reduce to the usual ones for groups and Lie algebras, respectively.

Remark One can define another action of A on $\mathrm{Hom}_k(V, W)$ by replacing S by S^{-1} in the above formulas. In particular, if $W = k$ this gives a representation V' of A. To distinguish V^* from V', we call them the left and right duals of V, respectively. It is clear that, if $\dim(V) < \infty$, the canonical isomorphisms of k-modules $V^{*'} \cong V \cong V'^*$ commute with the action of A.

There is a simple condition on A which guarantees that $V^* \cong V'$ as representations of A for all V: namely, that the automorphism S^2 of A is *inner*. In fact, if u is an invertible element of A such that $S^2(a) = uau^{-1}$ for all $a \in A$, the map $V^* \to V'$ given by $\xi \mapsto u^{-1}.\xi$ commutes with the action of A (the dot denotes the action of A on V^*). In particular, if $\dim(V) < \infty$, $V^{**} \cong V \cong V''$ in this case. We shall see in the next section a large class of Hopf algebras which have this property, and in Chapter 12 some which do not. $\triangle$

D Topological Hopf algebras and duality If A is a coalgebra (assumed to be defined over a field k for simplicity), its vector space dual A^* is naturally an algebra. In fact, the comultiplication Δ of A induces a map $\Delta^* : (A \otimes A)^* \to A^*$, and the multiplication of A^* is obtained by restricting Δ^* to the subspace $A^* \otimes A^*$ of $(A \otimes A)^*$. The unit of A^* is the dual of the counit of A. That these maps do define an algebra structure on A follows from the fact that the commutative diagrams in 4.1.1 are obtained from those in 4.1.2 by 'reversing the arrows'.

Moreover, if $\rho : V \to V \otimes A$ is a right A-comodule, one defines a left A^*-module structure on V by

$$\lambda(\alpha \otimes v) = \sum_i \langle \alpha, a^i \rangle v_i,$$

where $\alpha \in A^*$, $v \in V$ and $\rho(v) = \sum v_i \otimes a^i$.

In the finite-dimensional case, one passes from algebras to coalgebras (and from modules to comodules) in a similar way. Combining this with the previous remarks, one sees that the dual A^* of any finite-dimensional Hopf algebra A is a Hopf algebra in a natural way (the antipode of the dual is the dual of the antipode), and one has a canonical isomorphism of Hopf algebras $(A^*)^* \cong A$. Moreover, the categories of representations of A and of corepresentations of A^* are canonically equivalent.

Example 4.1.15 If G is a finite group and k is a field, the pairing

$$\mathcal{F}(G) \times k[G] \to k,$$

given, for $f \in \mathcal{F}(G)$, $g_i \in G$, $\lambda_i \in k$, by

$$\left\langle f, \sum_i \lambda_i g_i \right\rangle = \sum_i \lambda_i f(g_i),$$

induces isomorphisms of Hopf algebras $\mathcal{F}(G)^* \cong k[G]$ and $k[G]^* \cong \mathcal{F}(G)$. $\Diamond$

In the infinite-dimensional case, however, the dual of a Hopf algebra A cannot in general be given a Hopf algebra structure by the above procedure, since the dual of the multiplication of A might not take values in the subspace $A^* \otimes A^*$ of $(A \otimes A)^*$.

Example 4.1.16 Let $\mathfrak{g}$ be a Lie algebra over a field k of characteristic zero, and let $\{x_1, x_2, \ldots, x_d\}$ be any basis of $\mathfrak{g}$ (we assume that $\dim(\mathfrak{g}) = d < \infty$ for simplicity). For any $\boldsymbol{\lambda} = (\lambda_1, \lambda_2, \ldots, \lambda_d) \in \mathbf{N}^d$, set

$$x_{\boldsymbol{\lambda}} = \frac{x_1^{\lambda_1}}{\lambda_1!} \frac{x_2^{\lambda_2}}{\lambda_2!} \cdots \frac{x_d^{\lambda_d}}{\lambda_d!}.$$

Then, by the Poincaré–Birkhoff–Witt theorem, $\{x_{\boldsymbol{\lambda}}\}_{\boldsymbol{\lambda} \in \mathbf{N}^d}$ is a basis of $U(\mathfrak{g})$, and it is not difficult to show that

$$(2) \qquad \Delta(x_{\boldsymbol{\lambda}}) = \sum_{\boldsymbol{\mu} + \boldsymbol{\nu} = \boldsymbol{\lambda}} x_{\boldsymbol{\mu}} \otimes x_{\boldsymbol{\nu}},$$

where the addition of elements of $\mathbf{N}^d$ is performed componentwise (see Dixmier (1977), Proposition 2.7.2).

Now define $\xi^\lambda \in U(\mathfrak{g})^*$ by $\xi^\lambda(x_\mu) = \delta_{\lambda\mu}$. Then (2) implies that

(3) $$\xi^\lambda.\xi^\mu = \xi^{\lambda+\mu}.$$

Let $\xi_1, \ldots, \xi_d$ be indeterminates, and let $k[[\xi_1, \ldots, \xi_d]]$ be the algebra of formal power series in $\xi_1, \ldots, \xi_d$. It is clear that the assignment $\xi^\lambda \mapsto \xi_1^{\lambda_1} \xi_2^{\lambda_2} \ldots \xi_d^{\lambda_d}$ defines an algebra isomorphism $U(\mathfrak{g})^* \to k[[\xi_1, \ldots, \xi_d]]$.

Assume now that $\mathfrak{g}$ is abelian. Then,

$$x_\mu.x_\nu = \binom{\mu+\nu}{\mu} x_{\mu+\nu},$$

where

$$\binom{\mu+\nu}{\mu} = \prod_{i=1}^{d} \binom{\mu_i + \nu_i}{\mu_i}.$$

Hence, the comultiplication on $U(\mathfrak{g})^*$ should be given by

$$\Delta_{U^*}(\xi^\lambda) = \sum_{\mu+\nu=\lambda} \binom{\mu+\nu}{\mu} \xi^\mu \otimes \xi^\nu.$$

Identifying $U(\mathfrak{g})^*$ with $k[[\xi_1, \ldots, \xi_d]]$, this means that Δ_{U^*} is defined by taking the ξ_i to be primitive and extending to an algebra homomorphism. As expected, however, Δ_{U^*} does not take values in the algebraic tensor product $k[[\xi_1, \ldots, \xi_d]]^{\otimes 2}$, but only in its 'completion' $k[[\xi_1, \ldots, \xi_d, \xi_1', \ldots, \xi_d']]$. Thus, $U(\mathfrak{g})^*$ is a Hopf algebra, provided we are willing to interpret $U(\mathfrak{g})^* \otimes U(\mathfrak{g})^*$ as an appropriately completed tensor product. $\Diamond$

Despite these difficulties, there is a way to define the dual of any Hopf algebra, finite dimensional or not. We set $A^\circ = \{\alpha \in \dot{A}^* \mid \mu^*(\alpha) \in A^* \otimes A^*\}$. It is easy to see that $A^* \otimes A^*$ is a subalgebra of $(A \otimes A)^*$; hence, A° is a subalgebra of A^*. One proves that $\mu^*(A^\circ) \subset A^\circ \otimes A^\circ$, that $\Delta^*(A^\circ \otimes A^\circ) \subset A^\circ$, and that $S^*(A^\circ) \subset A^\circ$. Hence, with the obvious structure maps, A° is a Hopf algebra, called the *Hopf dual* of A.

This construction has a simple representation-theoretic interpretation. In fact, it can be shown that A° is exactly the space spanned by the matrix elements of all finite-dimensional representations of A. This suggests the following generalization of the Hopf dual. Suppose that we have a set Σ of finite-dimensional representations of A which contains the trivial representation and is closed under taking (left) duals, direct sums and tensor products. Let A_Σ° be the space spanned by the matrix elements of all representations in Σ. Then, A_Σ° is a Hopf subalgebra of A°, and every Hopf subalgebra of A° is of this form.

One can also give a topological interpretation of $A°$ (and similarly of $A^°_\Sigma$). Let

$$\mathcal{I} = \{\ker(\rho) \mid \rho : A \to \mathrm{End}(V) \text{ is a finite-dimensional left } A\text{-module}\}.$$

Then, for any $a \in A$, $\{I + a \mid I \in \mathcal{I}\}$ forms a base of neighbourhoods of a for the $\mathcal{I}$-*adic topology* on A. If we define a topology on $A \otimes A$ in the same way using the family of ideals $A \otimes I + J \otimes A$, where $I, J \in \mathcal{I}$, and give the field k the discrete topology, then all the structure maps of A are continuous. Moreover, $A°$ is the set of continuous linear maps $A \to k$.

Example 4.1.17   Let G be a connected, simply-connected affine algebraic group over $\mathbb{C}$, and let e be the identity element of G. The elements of the Lie algebra $\mathfrak{g}$ of G may be regarded as left-invariant vector fields on G; more generally, any $X \in U(\mathfrak{g})$ defines a left-invariant differential operator on G. The pairing $U(\mathfrak{g}) \times \mathcal{F}(G) \to \mathbb{C}$ given by $(X, f) \mapsto (X.f)(e)$ induces linear maps $U(\mathfrak{g}) \to \mathcal{F}(G)$ and $\mathcal{F}(G) \to U(\mathfrak{g})^*$. Both maps are injective, and it is easy to see that their images are contained in $\mathcal{F}(G)°$ and $U(\mathfrak{g})°$, respectively.

Inside $\mathcal{F}(G)°$, we have the evaluation maps $f \mapsto f(g)$ ($f \in \mathcal{F}(G)$, $g \in G$), and these, together with $U(\mathfrak{g})$, generate $\mathcal{F}(G)°$ as an algebra (see Abe (1980), Theorem 4.3.13.). For an equivalent formulation of this result, note that G acts on $\mathfrak{g}$ by the adjoint action, and that this extends to an action on $U(\mathfrak{g})$ by Hopf algebra automorphisms. Extending by linearity gives $U(\mathfrak{g})$ a left $\mathbb{C}[G]$-module algebra and coalgebra structure. Then, $\mathcal{F}(G)°$ is isomorphic to the semi-direct product $U(\mathfrak{g}) \natural \mathbb{C}[G]$, which is $U(\mathfrak{g}) \otimes \mathbb{C}[G]$ as a coalgebra, and with algebra structure and antipode defined by

$$(a \otimes x)(b \otimes y) = \sum_i a(x_i.b) \otimes x'_i y, \quad S(a \otimes x) = \sum_i S(x_i).S(a) \otimes S(x'_i),$$

where $\Delta(x) = \sum_i x_i \otimes x'_i$ and the dot denotes the action of $\mathbb{C}[G]$ on $U(\mathfrak{g})$.

On the other hand, the map $\mathcal{F}(G) \to U(\mathfrak{g})°$ is an isomorphism of Hopf algebras if $\mathfrak{g}$ is semisimple. $\diamond$

As this example suggests, it is often convenient to bypass these delicate duality questions by working with the weaker notion of a *non-degenerate pairing* of two Hopf algebras $(A, \iota^A, \mu^A, \epsilon^A, \Delta^A, S^A)$ and $(B, \iota^B, \mu^B, \epsilon^B, \Delta^B, S^B)$. This is a pairing $\langle\, ,\, \rangle : A \times B \to k$, bilinear over the ground ring k, which is non-degenerate in the sense that $\langle a, b \rangle = 0$ for all $a \in A$ (resp. for all $b \in B$) implies that $b = 0$ (resp. $a = 0$), and which satisfies the following conditions:

$$\langle \iota^A(a), b \rangle = \langle a, \epsilon^B(b) \rangle, \quad \langle \epsilon^A(a), b \rangle = \langle a, \iota^B(b) \rangle,$$

$$\langle \mu^A(a_1 \otimes a_2), b \rangle = \langle a_1 \otimes a_2, \Delta^B(b) \rangle, \quad \langle \Delta^A(a), b_1 \otimes b_2 \rangle = \langle b, \mu^B(b_1 \otimes b_2) \rangle,$$

$$\langle S^A(a), b \rangle = \langle a, S^B(b) \rangle,$$

for all $a, a_1, a_2 \in A$, $b, b_1, b_2 \in B$. In the second pair of equations, the pairing is extended to one between $A \otimes A$ and $B \otimes B$ in the obvious way:

$$\langle a_1 \otimes a_2, b_1 \otimes b_2 \rangle = \langle a_1, b_1 \rangle \langle a_2, b_2 \rangle.$$

It is clear that, if A and B are finite dimensional and k is a field, such a pairing induces injective homomorphisms of Hopf algebras $A \to B^*$ and $B \to A^*$; in this case, the existence of such a pairing is thus equivalent to the statement that each of A and B is the dual of the other.

E Integration on Hopf algebras A (left-invariant) *integral* on a Hopf algebra A over a field k is a linear functional H on A such that

(4) $$\langle \alpha \otimes H, \Delta(a) \rangle = \langle \alpha, 1 \rangle \langle H, a \rangle$$

for all $a \in A$, $\alpha \in A^*$. It can be shown that, if an integral exists, it is unique up to a scalar multiple. Right-invariant and bi-invariant integrals are defined in the obvious way.

An integral H on A is said to be *normalized* if $\langle H, 1 \rangle = 1$.

Example 4.1.18 Let G be a compact topological group and let $\mathrm{Rep}(G)$ be its algebra of real-valued representative functions (see 4.1.7). Let $d\mu$ be a left-invariant Haar measure on G, and set

$$\langle H, f \rangle = \int_G f \, d\mu$$

for $f \in \mathrm{Rep}(G)$. Then, (4) follows from left invariance of the measure, and the normalization condition is equivalent to $d\mu$ being normalized in the usual sense, that the integral of the constant function 1 over G is equal to 1. Note that the integral is positive definite, in the sense that for any $f \in \mathrm{Rep}(G)$ the integral of f^2 is positive unless f is identically zero.

The integral in this example can also be expressed algebraically. If $\rho : G \to \mathrm{End}(V)$ is an irreducible representation of G on a finite-dimensional complex vector space V, we may choose a positive-definite hermitian form $(\ ,\ )$ on V which is invariant under G:

$$(g.v_1, g.v_2) = (v_1, v_2) \qquad (g \in G, \ v_1, v_2 \in V).$$

If $\{v_i\}$ is an orthonormal basis of V with respect to $(\ ,\ )$, the matrix elements of ρ are the functions $\rho_{ij} : G \to \mathbb{C}$ defined by

$$\rho_{ij}(g) = (g.v_j, v_i).$$

We recall that the Schur orthogonality relations tell us that

$$\langle H, \rho_{ij}\overline{\rho_{i'j'}}\rangle = \frac{\delta_{i,i'}\delta_{j,j'}}{\dim(\rho)},$$

and that, if σ is another finite-dimensional irreducible representation of G, with matrix elements σ_{kl}, then, if ρ and σ are not equivalent,

$$\langle H, \rho_{ij}\overline{\sigma_{kl}}\rangle = 0,$$

for all i, j, k, l.

Let Σ be the set of matrix elements $\{\rho_{ij}\}$, as ρ runs through the set of finite-dimensional irreducible representations of G (one from each equivalence class). It follows from the Schur orthogonality relations that Σ is linearly independent. On the other hand, any product of matrix elements $\rho_{ij}\sigma_{kl}$ is a matrix element of $\rho \otimes \sigma$, and therefore belongs to the linear span of Σ, every finite-dimensional representation of G being completely reducible. Thus, Σ is a basis of $\mathrm{Rep}(G)$.

Further, by the orthogonality relations again,

$$\langle H, \rho_{ij}\rangle = \begin{cases} 1 & \text{if } \rho \text{ is trivial,} \\ 0 & \text{otherwise.} \end{cases}$$

Since Σ is a basis of $\mathrm{Rep}(G)$, these formulas uniquely determine a linear map $H : \mathrm{Rep}(G) \to \mathbb{R}$, and it is easy to see that H is a normalized integral. $\Diamond$

This example has a converse which is a classical example of *Tannaka–Krein duality*.

Example 4.1.19 Let A be a commutative Hopf algebra over $\mathbb{R}$. Then the set G of algebra homomorphisms $A \to \mathbb{R}$ is a group with product $g_1.g_2 = (g_1 \otimes g_2) \circ \Delta$ and inverse $g^{-1} = g \circ S$. Suppose now that A is equipped with a positive-definite left-invariant integral, and assume that G separates the points of A (i.e. if a and a' are distinct points of A, there exists $g \in G$ such that $g(a) \neq g(a')$). Then, G is a compact topological group in the finite-open topology (for which the sets $N(a_1, \ldots, a_n, \epsilon, g) = \{g' \in G \mid |g'(a_i) - g(a_i)| < \epsilon\}$, for $a_1, \ldots, a_n \in A$, $\epsilon > 0$, form a basis of the neighbourhoods of $g \in G$). Moreover, the map which assigns to an element $a \in A$ the evaluation of a representative function at a is an isomorphism of Hopf algebras $A \to \mathrm{Rep}(G)$.

If, in addition, A is finitely generated as an algebra, then G can be given the structure of a compact Lie group, and conversely. $\Diamond$

We shall discuss Tannaka–Krein duality in more detail in Section 5.1.

F Hopf ∗-algebras If A is an associative algebra with 1 over $\mathbb{C}$, a ∗-structure on A is a conjugate-linear map $* : A \to A$, written $a \to a^*$, such that

$$(a^*)^* = a, \qquad \text{(involutive)}$$
$$(a_1 a_2)^* = a_2^* a_1^*, \qquad \text{(anti-multiplicative)}$$
$$1^* = 1, \qquad \text{(unital)}$$

for all a, a_1, $a_2 \in A$. A homomorphism $\varphi : A \to B$ between ∗-algebras is a *∗-homomorphism* if

$$\varphi(a^*) = \varphi(a)^*, \qquad (a \in A).$$

If A and B are ∗-algebras, then so is $A \otimes B$:

$$(a \otimes b)^* = a^* \otimes b^*, \qquad (a \in A, \ b \in B).$$

The canonical example is, of course, the algebra $\mathcal{B}(V)$ of bounded linear operators on a Hilbert space V, the ∗-structure being given by

$$\langle T(v), w \rangle = \langle v, T^*(w) \rangle,$$

where $\langle \, , \, \rangle$ is the inner product on V.

A *Hopf ∗-algebra* is a Hopf algebra A over $\mathbb{C}$ equipped with a ∗-algebra structure which is compatible with the coalgebra structure of A in the sense that the comultiplication $\Delta : A \to A \otimes A$ and counit $\epsilon : A \to \mathbb{C}$ are ∗-homomorphisms ($A \otimes A$ being given its natural ∗-algebra structure, and $\mathbb{C}$ the ∗-structure given by complex-conjugation). The pair $(A, *)$ is also sometimes called a *real form* of A (see below). These two conditions imply that

$$S(a)^* = S^{-1}(a^*), \qquad (a \in A).$$

In fact, they imply that $* \circ S^{-1} \circ *$ satisfies the axioms 4.1.3(iv) for the antipode, so the assertion follows from the uniqueness of the antipode (see Remark [2] following 4.1.3).

Two real forms $(A, *)$ and $(A, \dagger)$ are said to be *equivalent* if there is a Hopf algebra automorphism α of A such that

$$\alpha(a^\dagger) = \alpha(a)^*$$

for all $a \in A$.

If A is a finite-dimensional Hopf ∗-algebra, then the dual Hopf algebra A^* (with a different meaning of * !) is also a Hopf ∗-algebra. In fact, if $\alpha \in A^*$, one defines

$$\alpha^*(a) = \overline{\alpha(S(a)^*)},$$

the bar denoting complex-conjugation.

A representation ρ of A on a separable complex Hilbert space V is *unitary* with respect to a given $*$-struciture on A, or simply a $*$-*representation*, if $\rho : A \to \mathcal{B}(V)$ is a homomorphism of $*$-algebras, i.e. if

$$\langle \rho(a)(v_1), v_2 \rangle = \langle v_1, \rho(a^*)(v_2) \rangle \qquad (v_1, v_2 \in V).$$

A corepresentation $\sigma : V \to A \otimes V$ is unitary if, given an orthonormal basis $\{v_i\}$ of V, so that

$$\sigma(v_i) = \sum_j a_{ij} \otimes v_j$$

for some $a_{ij} \in A$, we have $S(a_{ij}) = a_{ji}^*$.

Example 4.1.20 Let G be a complex algebraic group with Lie algebra $\mathfrak{g}$. Then, there is a one-to-one correspondence between the following three types of objects:

(i) real forms of G;

(ii) Hopf $*$-structures on $U(\mathfrak{g})$;

(iii) Hopf $*$-structures on $\mathcal{F}(G)$.

Indeed, a real form of G determines a real form $\mathfrak{g_R}$ of $\mathfrak{g}$, i.e. a real Lie subalgebra of $\mathfrak{g}$ such that $\mathfrak{g} = \mathfrak{g_R} \oplus \sqrt{-1}\mathfrak{g_R}$. Conjugation with respect to $\mathfrak{g_R}$ defines a conjugate-linear map $* : \mathfrak{g} \to \mathfrak{g}$, which extends uniquely to a Hopf $*$-structure on $U(\mathfrak{g})$. Conversely, given a Hopf $*$-structure on $U(\mathfrak{g})$, the compatibility with the comultiplication implies that $*$ preserves the space $\mathfrak{g}$ of primitive elements of $U(\mathfrak{g})$, and the fixed points of $*$ on $\mathfrak{g}$ define a real form.

Similarly, to define the Hopf $*$-structure on $\mathcal{F}(G)$ corresponding to a real form $G_\mathbf{R}$ of G, take a regular function on G, restrict it to $G_\mathbf{R}$, take the pointwise complex conjugate of this restriction, and then extend the resulting function on $G_\mathbf{R}$ to a regular function on G. Conversely, given a Hopf $*$-structure on $\mathcal{F}(G)$, every $*$-homomorphism $\mathcal{F}(G) \to \mathbf{C}$ is of the form $f \to f(x)$ for some $x \in G_\mathbf{R}$.

The Hopf $*$-structures on $\mathcal{F}(G)$ and $U(\mathfrak{g})$ corresponding to a given real form of G are, of course, dual to each other.

If $G = SL_2(\mathbf{C})$, there are exactly two real forms (up to equivalence):

(a) The compact real form

$$SU_2 = \left\{ \begin{pmatrix} a & b \\ c & d \end{pmatrix} \in SL_2(\mathbf{C}) \ \middle| \ d = \bar{a}, \ c = -\bar{b} \right\}.$$

If we regard the matrix entries a, b, c and d as generators of $\mathcal{F}(SL_2(\mathbf{C}))$, the corresponding $*$-structure is given by $d = a^*$, $c = -b^*$.

(b) The non-compact real form

$$SL_2(\mathbb{R}) = \left\{ \begin{pmatrix} a & b \\ c & d \end{pmatrix} \in SL_2(\mathbb{C}) \;\middle|\; \bar{a} = a,\; \bar{b} = b,\; \bar{c} = c,\; \bar{d} = d \right\},$$

for which the corresponding $*$-structure on $\mathcal{F}(SL_2(\mathbb{C}))$ is given by $a^* = a$, $b^* = b$, $c^* = c$, $d^* = d$. $\diamondsuit$

4.2 Quasitriangular Hopf algebras

There are several results in the literature which assert, under some additional technical hypotheses, that every cocommutative Hopf algebra is isomorphic to the universal enveloping algebra of a Lie algebra (and, dually, that every commutative Hopf algebra is isomorphic to the algebra of functions on a group). The theory of quantum groups provides a large family of Hopf algebras which are neither commutative nor cocommutative. However, many of these examples (or their duals) are 'almost cocommutative' in a certain sense. In this section, we give a number of ways of constructing almost cocommutative Hopf algebras, and we describe some of their properties.

A Almost cocommutative Hopf algebras A Hopf algebra is almost cocommutative if its comultiplication and its opposite comultiplication differ by conjugation by a distinguished invertible element of $A \otimes A$.

DEFINITION 4.2.1 *A Hopf algebra A over a commutative ring k is said to be* almost cocommutative *if there exists an invertible element $\mathcal{R} \in A \otimes A$ such that*

$$(5) \qquad\qquad \Delta^{\mathrm{op}}(a) = \mathcal{R}\Delta(a)\mathcal{R}^{-1}$$

for all $a \in A$.

Remarks [1] It is obvious that if A is commutative and almost cocommutative, then it is actually cocommutative. For example, the algebra $\mathcal{F}(G)$ of functions on a non-abelian group G (in any of the categories we have considered) is not almost cocommutative.

[2] If $\lambda_V : A \to \mathrm{End}_k(V)$ and $\lambda_W : A \to \mathrm{End}_k(W)$ are representations of an almost cocommutative Hopf algebra $(A, \mathcal{R})$, the tensor products $V \otimes W$ and $W \otimes V$ are isomorphic as representations of A. In fact, it is easy to check that, if $\sigma : V \otimes W \to W \otimes V$ is the flip of the two factors, then $\sigma \circ (\lambda_V \otimes \lambda_W)(\mathcal{R})$ is an isomorphism $V \otimes W \to W \otimes V$ which commutes with the action of A. We shall develop this remark much further in the next chapter.

[3] If A is a topological Hopf algebra over $k[[h]]$ (see 4.1.4), one says that A is *topologically almost cocommutative* if there exists an element $\mathcal{R}$ in the h-adic completion of $A \otimes A$ with the properties in 4.2.1. $\triangle$

If A is almost cocommutative, then $\mathcal{R}$ is of course unique up to right multiplication by an element of the centralizer C of $\Delta(A)$ in $A \otimes A$. Obviously, $C \supseteq Z \otimes Z$, where Z is the centre of A. In the other direction, we have

PROPOSITION 4.2.2 *Let $\varphi : A \otimes A \to A$ be the k-module map defined by* $\varphi(a_1 \otimes a_2) = a_1 S(a_2)$. *Then, $\varphi(C) \subseteq Z$.*

PROOF Give A the natural left $(A \otimes A)$-module structure:

$$(a_1 \otimes a_2) * a = a_1 a S(a_2).$$

Note that $\varphi(x) = x * 1$ for any $x \in A \otimes A$, and that $\Delta(a) * 1 = \epsilon(a)1$ by the commutative diagrams in 4.1.3. More generally, if $c \in C$ and $a \in A$,

$$
\begin{aligned}
\Delta(a) * \varphi(c) = \Delta(a) * (c * 1) &= (\Delta(a)c) * 1 \\
&= (c\Delta(a)) * 1 = c * (\Delta(a) * 1) \\
&= \epsilon(a)(c * 1) = \epsilon(a)\varphi(c).
\end{aligned}
$$

Now define, for fixed $c \in C$, a k-module map $\psi : A \otimes A \otimes A \to A$ by

$$\psi(a_1 \otimes a_2 \otimes a_3) = a_1 \varphi(c) S(a_2) a_3.$$

We compute $\psi((\Delta \otimes \mathrm{id})\Delta(a)) = \psi((\mathrm{id} \otimes \Delta)\Delta(a))$ for any $a \in A$. Writing $\Delta(a) = \sum_i a_i \otimes a^i$, we have

$$
\begin{aligned}
\psi((\Delta \otimes \mathrm{id})\Delta(a)) &= \sum_i (\Delta(a_i) * \varphi(c))\, a^i = \sum_i \varphi(c)\, \epsilon(a_i)\, a^i \\
&= \varphi(c)\, (\epsilon \otimes \mathrm{id})\Delta(a) = \varphi(c)\, a, \\
\psi((\mathrm{id} \otimes \Delta)\Delta(a)) &= \sum_i a_i \varphi(c).\mu(S \otimes \mathrm{id})\Delta(a^i) = \sum_i a_i \epsilon(a^i)\varphi(c) \\
&= (\mathrm{id} \otimes \epsilon)\Delta(a).\varphi(c) = a\varphi(c).
\end{aligned}
$$

Thus, $\varphi(c)$ commutes with every element $a \in A$. ∎

Applying the flip map σ to both sides of (5), we see that $\mathcal{R}_{21}\mathcal{R} \in C$, and hence that $\varphi(\mathcal{R}_{21}\mathcal{R}) \in Z$ (here, $\mathcal{R}_{21} = \sigma(\mathcal{R})$). The next result will allow us to refine this observation.

PROPOSITION 4.2.3 *Let $u = \mu(S \otimes \mathrm{id})(\mathcal{R}_{21})$. Then, u is an invertible element of A and, for all $a \in A$,*

$$(6) \qquad\qquad S^2(a) = uau^{-1}.$$

PROOF We show first that

$$(7) \qquad\qquad ua = S^2(a)u$$

for all $a \in A$. Write

$$(\Delta \otimes \mathrm{id})\Delta(a) = \sum_i a_i \otimes a_i' \otimes a_i''.$$

Equation (5) implies that

$$(\mathcal{R} \otimes 1)\left(\sum_i a_i \otimes a_i' \otimes a_i''\right) = \left(\sum_i a_i' \otimes a_i \otimes a_i''\right)(\mathcal{R} \otimes 1).$$

Writing $\mathcal{R} = \sum_j r_j \otimes r^j$, we have

$$\sum_{i,j} r_j a_i \otimes r^j a_i' \otimes a_i'' = \sum_{i,j} a_i' r_j \otimes a_i r^j \otimes a_i''.$$

Applying $\mathrm{id} \otimes S \otimes S^2$ to both sides and reversing the order of the factors, we obtain

$$\sum_{i,j} S^2(a_i'')S(r^j a_i')r_j a_i = \sum_{i,j} S^2(a_i'')S(a_i r^j)a_i' r_j,$$

which we rewrite as

(8) $$\sum_{i,j} S(a_i' S(a_i'')) S(r^j) r_j a_i = \sum_{i,j} S^2(a_i'') S(r^j) S(a_i) a_i' r_j.$$

Now,

$$\sum_i S(a_i) a_i' \otimes a_i'' = (\mu \otimes \mathrm{id})(S \otimes \mathrm{id} \otimes \mathrm{id})((\Delta \otimes \mathrm{id})\Delta(a))$$

$$= (\epsilon \otimes \mathrm{id})\Delta(a) = 1 \otimes a.$$

Hence,

$$\sum_{i,j} S(r^j)S(a_i) a_i' r_j \otimes a_i'' = \sum_j S(r^j) r_j \otimes a = u \otimes a,$$

so the right-hand side of (8) is equal to $S^2(a)u$. Similarly,

$$\sum_i a_i \otimes a_i' S(a_i'') = a \otimes 1,$$

from which we deduce that the left-hand side of (8) is equal to ua. This proves (7).

To complete the proof, we show that $v = \mu(S^{-1} \otimes \mathrm{id})(\mathcal{R}_{21}^{-1})$ is the inverse of u. Indeed, let $\mathcal{R}^{-1} = \sum_k s_k \otimes s^k$. Then,

$$uv = \sum_k u S^{-1}(s^k) s_k = \sum_k S(s^k) u s_k$$

$$= \sum_{j,k} S(r^j s^k) r_j s_k = \mu(S \otimes \mathrm{id})(\mathcal{R}_{21}\mathcal{R}_{21}^{-1})$$

$$= 1$$

(the second equality follows from (6)). Taking $a = v$ in (7) gives $S^2(v)u = 1$. Since u has both a left and a right inverse, it is invertible. ∎

Similar arguments show that, if

$$u_2 = \mu(S \otimes \mathrm{id})(\mathcal{R}^{-1}), \quad u_3 = \mu(\mathrm{id} \otimes S^{-1})(\mathcal{R}_{21}), \quad u_4 = \mu(\mathrm{id} \otimes S)(\mathcal{R}^{-1}),$$

then $S^2(a) = u_i a u_i^{-1}$ for all $a \in A$ and $i = 2, 3, 4$. Setting $u_1 = u$, we deduce

COROLLARY 4.2.4 *The elements $u_i \in A$ commute with each other and the elements $u_i u_j^{-1}$ are in the centre of A, for $i, j = 1, 2, 3, 4$.* ∎

Remarks [1] It can be shown further that

$$u_1^{-1} = \varphi((S^{-1} \otimes S)(\mathcal{R}_{21}^{-1})), \quad u_2^{-1} = \varphi((S^{-1} \otimes S^{-1})(\mathcal{R})),$$
$$u_3^{-1} = \varphi(\mathcal{R}_{21}^{-1}), \qquad\qquad u_4^{-1} = \varphi(\mathcal{R}).$$

Moreover, $S(u_1) = u_2^{-1}$, $S(u_2) = u_1^{-1}$, $S(u_3) = u_4^{-1}$, $S(u_4) = u_3^{-1}$.

[2] The central element $\varphi(\mathcal{R}_{21}\mathcal{R})$ of A encountered above is equal to $u_3 u_4^{-1}$.

[3] If $(S \otimes S)(\mathcal{R}) = \mathcal{R}$, then $u_1 = \mu(S \otimes \mathrm{id})(\mathcal{R}_{21}) = \mu(\mathrm{id} \otimes S^{-1})(\mathcal{R}_{21}) = u_3$, and similarly $u_2 = u_4$. If $\mathcal{R}_{21} = \mathcal{R}^{-1}$, a similar argument shows that $u_1 = u_2$ and $u_3 = u_4$. △

Proposition 4.2.3 has a number of interesting consequences for the representation theory of A. For example, from the discussion at the end of Subsection 4.1D, we immediately deduce

COROLLARY 4.2.5 *The left and right duals of any representation of an almost cocommutative Hopf algebra are canonically isomorphic.* ∎

Proposition 4.2.3 also enables one to define a new notion of dimension for a representation $\rho : A \rightarrow \mathrm{End}_k(V)$ when A is almost cocommutative and, more generally, of the trace of any k-module map $f : V \rightarrow V$. We recall from Subsection 4.1C that there is a canonical isomorphism $\mathrm{End}_k(V) \cong V \otimes V^*$ of representations of A. The trace of f would normally be defined as the image of f under the canonical pairing $V \otimes V^* \rightarrow k$, but this is unnatural from a Hopf algebra viewpoint because this pairing does not commute with the action of A in general. However, if A is almost cocommutative and V is finitely generated and projective as a k-module (for example, if k is a field and $\dim(V) < \infty$), 4.2.3 and the discussion in Subsection 4.1C imply that $V \cong V^{**}$ as representations of A. We may therefore think of f as an element of $V^{**} \otimes V^*$ and use the canonical pairing $V^{**} \otimes V^* \rightarrow k$, which does commute with the A-action. We leave it to the reader to check that the image of f under this map is $\mathrm{trace}(\rho(u) \circ f)$. One drawback of this new trace, however, is that, if $\tilde{V}$ is another representation of A and $\tilde{f} \in \mathrm{End}_k(\tilde{V})$, the new trace of $f \otimes \tilde{f} \in \mathrm{End}_k(V \otimes \tilde{V})$ is not in general equal to the product of the new traces of f and $\tilde{f}$, because u is not in general group-like. In Subsection 4.2C, we shall see how to circumvent this difficulty for a large class of almost cocommutative Hopf algebras.

B Quasitriangular Hopf algebras It is clear that the element $\mathcal{R}$ in (5) cannot be arbitrary since we know that A^{op} is a Hopf algebra. In fact,

$$(\Delta^{\mathrm{op}} \otimes \mathrm{id})\Delta^{\mathrm{op}}(a) = \mathcal{R}_{12}.(\Delta \otimes \mathrm{id})(\mathcal{R}).(\Delta \otimes \mathrm{id})\Delta(a).(\Delta \otimes \mathrm{id})(\mathcal{R})^{-1}.\mathcal{R}_{12}^{-1},$$

$$(\mathrm{id} \otimes \Delta^{\mathrm{op}})\Delta^{\mathrm{op}}(a) = \mathcal{R}_{23}.(\mathrm{id} \otimes \Delta)(\mathcal{R}).(\mathrm{id} \otimes \Delta)\Delta(a).(\mathrm{id} \otimes \Delta)(\mathcal{R})^{-1}.\mathcal{R}_{23}^{-1}.$$

Hence, a sufficient condition for coassociativity of Δ^{op} is

$$(9) \qquad \mathcal{R}_{12}(\Delta \otimes \mathrm{id})(\mathcal{R}) = \mathcal{R}_{23}(\mathrm{id} \otimes \Delta)(\mathcal{R}).$$

Similarly, a sufficient condition for the properties $(\epsilon \otimes \mathrm{id}) \circ \Delta^{\mathrm{op}} = \mathrm{id}$ and $(\mathrm{id} \otimes \epsilon) \circ \Delta^{\mathrm{op}} = \mathrm{id}$ of the counit to hold is that $(\epsilon \otimes \mathrm{id})(\mathcal{R}) = (\mathrm{id} \otimes \epsilon)(\mathcal{R}) = 1$.

It will turn out to be convenient to make a slightly stronger assumption.

DEFINITION 4.2.6 *An almost cocommutative Hopf algebra* $(A, \mathcal{R})$ *is said to be*

(i) coboundary if $\mathcal{R}$ *satisfies* (9), $\mathcal{R}_{21} = \mathcal{R}^{-1}$ *and* $(\epsilon \otimes \epsilon)(\mathcal{R}) = 1$;

(ii) quasitriangular if

$$(10) \qquad (\Delta \otimes \mathrm{id})(\mathcal{R}) = \mathcal{R}_{13}\mathcal{R}_{23},$$

$$(11) \qquad (\mathrm{id} \otimes \Delta)(\mathcal{R}) = \mathcal{R}_{13}\mathcal{R}_{12};$$

(iii) triangular if it is quasitriangular, and, in addition, $\mathcal{R}_{21} = \mathcal{R}^{-1}$.

If A *is quasitriangular, the element* $\mathcal{R}$ *is called the* universal R-matrix *of* $(A, \mathcal{R})$.

Remarks [1] If $(A, \mathcal{R})$ is quasitriangular, so is $(A, \mathcal{R}_{21}^{-1})$.

[2] If $(A, \mathcal{R})$ is quasitriangular, so are $(A_{\mathrm{op}}, \mathcal{R}_{21})$, $(A^{\mathrm{op}}, \mathcal{R}_{21})$ and $(A_{\mathrm{op}}^{\mathrm{op}}, \mathcal{R})$.

[3] Examples of non-cocommutative quasitriangular Hopf algebras are given in Subsection 4.2F. However, those of most importance for us are studied in Subsection 6.4D and Section 8.3.

[4] The notion of a topologically quasitriangular (or coboundary, or triangular) Hopf algebra is defined in the obvious way (see Remark [3] following 4.2.1). $\triangle$

If A is a quasitriangular Hopf algebra, the central elements of A constructed in the previous subsection reduce essentially to the single element $uS(u)$.

In fact, it follows from the remarks following 4.2.5 that $u_1 = u_3 = u$ and $u_2 = u_4 = S(u)^{-1}$. Hence, the central elements $u_i u_j^{-1}$ are either trivial or equal to $uS(u)$ or its inverse. Moreover, using 4.2.3 and the following proposition, it is not difficult to show that

$$(12) \qquad \Delta(u) = (\mathcal{R}_{21}\mathcal{R}_{12})^{-1}(u \otimes u) = (u \otimes u)(\mathcal{R}_{21}\mathcal{R}_{12})^{-1}.$$

The next proposition also implies that any triangular Hopf algebra is coboundary.

PROPOSITION 4.2.7 *Let $(A, \mathcal{R})$ be a quasitriangular Hopf algebra. Then,*

$$(13) \qquad \mathcal{R}_{12}\mathcal{R}_{13}\mathcal{R}_{23} = \mathcal{R}_{23}\mathcal{R}_{13}\mathcal{R}_{12},$$

$$(14) \qquad (\epsilon \otimes \mathrm{id})(\mathcal{R}) = 1 = (\mathrm{id} \otimes \epsilon)(\mathcal{R}),$$

$$(15) \qquad (S \otimes \mathrm{id})(\mathcal{R}) = \mathcal{R}^{-1} = (\mathrm{id} \otimes S^{-1})(\mathcal{R}),$$

$$(16) \qquad (S \otimes S)(\mathcal{R}) = \mathcal{R}.$$

PROOF By (5), we have

$$\mathcal{R}_{12}\mathcal{R}_{13}\mathcal{R}_{23} = \mathcal{R}_{12}.(\Delta \otimes \mathrm{id})(\mathcal{R}) = (\Delta^{\mathrm{op}} \otimes \mathrm{id})(\mathcal{R}).\mathcal{R}_{12}.$$

Applying σ_{12} to both sides of (10), we find that $(\Delta^{\mathrm{op}} \otimes \mathrm{id})(\mathcal{R}) = \mathcal{R}_{23}\mathcal{R}_{13}$, proving (13).

Next, (14) follows immediately by applying $\epsilon \otimes \mathrm{id} \otimes \mathrm{id}$ and $\mathrm{id} \otimes \mathrm{id} \otimes \epsilon$ to both sides of (10) and (11), respectively.

To prove the first equation in (15), we compute

$$\begin{aligned}
\mathcal{R}.(S \otimes \mathrm{id})(\mathcal{R}) &= (\mu \otimes \mathrm{id})(\mathrm{id} \otimes S \otimes \mathrm{id})(\mathcal{R}_{13}\mathcal{R}_{23}) \\
&= (\mu \otimes \mathrm{id})(\mathrm{id} \otimes S \otimes \mathrm{id})(\Delta \otimes \mathrm{id})(\mathcal{R}) \\
&= (\mu(\mathrm{id} \otimes S)\Delta \otimes \mathrm{id})(\mathcal{R}) \\
&= (\epsilon \otimes \mathrm{id})(\mathcal{R}) = 1
\end{aligned}$$

(the second equality follows from (10)). The second equation in (15) follows by applying the same argument to the quasitriangular Hopf algebra $(A^{\mathrm{op}}, \mathcal{R}_{21})$. Finally, (16) follows immediately from (15). ∎

Remark Let A be a Hopf algebra and let $\mathcal{R} \in A \otimes A$. Define $f : A^* \to A$ by

$$f(\alpha) = \langle \alpha \otimes \mathrm{id}, \mathcal{R} \rangle,$$

i.e. if $\mathcal{R} = \sum_j r_j \otimes r_j'$, then $f(\alpha) = \sum_j \langle \alpha, r_j \rangle r_j'$. Then:

(a) Equation (10) and the first equation in (14) are equivalent to f being a homomorphism of algebras;

(b) if A is finite dimensional (so that A^* inherits a coalgebra structure from the algebra structure of A), (11) and the second equation in (14) are equivalent to $f : A^* \to A^{\mathrm{op}}$ being a homomorphism of coalgebras.

The verification is straightforward. For example, (10) implies that

$$
\begin{aligned}
f(\alpha\beta) &= \langle \alpha\beta \otimes \mathrm{id}, \mathcal{R} \rangle \\
&= \langle \alpha \otimes \beta \otimes \mathrm{id}, (\Delta \otimes \mathrm{id})(\mathcal{R}) \rangle \\
&= \langle \alpha \otimes \beta \otimes \mathrm{id}, \mathcal{R}_{13}\mathcal{R}_{23} \rangle \\
&= \langle \alpha \otimes \mathrm{id}, \mathcal{R} \rangle \, \langle \beta \otimes \mathrm{id}, \mathcal{R} \rangle \\
&= f(\alpha)f(\beta),
\end{aligned}
$$

for all α, $\beta \in A^*$. $\triangle$

We have already encountered (13), called the *quantum Yang–Baxter equation*, or QYBE, in 2.2.8. Note, however, that the R-matrices which arise in the study of Hopf algebras are linear maps of k-modules, while those in 2.2.8 are nonlinear.

We remarked in Subsection 2.2C that the CYBE can be regarded as the 'classical limit' of the QYBE. We shall have much more to say about this in Chapter 6, where we study the relation between the terms triangular, quasitriangular and coboundary in the classical and quantum situations, as well as the question of when a classical r-matrix occurs as the 'first order part' of a quantum R-matrix.

C Ribbon Hopf algebras and quantum dimension From (12), it follows that if A is a triangular Hopf algebra, the element $u = \mu(S \otimes \mathrm{id})(\mathcal{R}_{21})$ is group-like. From the discussion at the end of Subsection 4.2A, we see that the new notion of trace defined there is multiplicative with respect to tensor products. In fact, if we are willing to modify this notion of trace slightly, we are able to preserve this multiplicative property for a somewhat larger class of quasitriangular Hopf algebras.

DEFINITION 4.2.8 *A ribbon Hopf algebra* $(A, \mathcal{R}, v)$ *is a quasitriangular Hopf algebra* $(A, \mathcal{R})$ *equipped with an invertible central element* v *such that*

$$
\begin{aligned}
v^2 &= uS(u), \quad S(v) = v, \quad \epsilon(v) = 1, \\
\Delta(v) &= (\mathcal{R}_{21}\mathcal{R}_{12})^{-1}(v \otimes v),
\end{aligned}
$$

where $u = \mu(S \otimes \mathrm{id})(\mathcal{R}_{21})$.

Note that, by (12) and the remarks following 4.2.4, if $(A, \mathcal{R})$ is a triangular Hopf algebra, then $(A, \mathcal{R}, 1)$ is a ribbon Hopf algebra. The reason for the picturesque terminology will appear in Section 5.3.

Although not every quasitriangular Hopf algebra is a ribbon Hopf algebra, every quasitriangular Hopf algebra A can be enlarged to a ribbon Hopf algebra by formally adjoining a square root of $uS(u)$. In fact, let $\tilde{A} = A \oplus A$ as a k-module, and define

$$(a_1, a_2)(a'_1, a'_2) = (a_1 a'_1 + a_2 a'_2 uS(u), a_1 a'_2 + a_2 a'_1),$$

$$\Delta(a_1, a_2) = (\Delta(a_1), 0, 0, \Delta(a_2)(\mathcal{R}_{21}\mathcal{R}_{12})^{-1}),$$

$$S(a_1, a_2) = (S(a_1), S(a_2)), \quad \epsilon(a_1, a_2) = (\epsilon(a_1), \epsilon(a_2)).$$

In the second equation, the first component of the right-hand side is in the tensor product of the first factors in $\tilde{A} \otimes \tilde{A} = (A \oplus A) \otimes (A \oplus A)$, and the last component is in the tensor product of the second factors. It is not difficult to check that these definitions make $\tilde{A}$ into a ribbon Hopf algebra containing A as the Hopf subalgebra $A \oplus 0$, and with $v = (0, 1)$.

See 8.3.16 for an important example of a ribbon Hopf algebra.

Let us now return to the question of traces. Let A be a ribbon Hopf algebra, and let $\rho : A \to \mathrm{End}_k(V)$ be a representation of A. We assume for simplicity that V is a free k-module of finite rank. In Subsection 4.2A, we saw that it was natural to consider $\mathrm{trace}(\rho(u) \circ f)$. The crucial property of u here is that S^2 is given by conjugation by u. But, since v is in the centre of A, S^2 is also given by conjugation by the element $g = v^{-1}u$, so we could just as well consider $\mathrm{trace}(\rho(g) \circ f)$. Moreover, (12) implies that g is group-like, so the latter trace will be multiplicative with respect to tensor products.

DEFINITION 4.2.9 *Let $(A, \mathcal{R}, v)$ be a ribbon Hopf algebra and let $\rho : A \to \mathrm{End}_k(V)$ be a representation of A which is a free k-module of finite rank. If $f : V \to V$ is a k-linear map, its* quantum trace *is defined to be*

$$\mathrm{qtr}(f) = \mathrm{trace}(\rho(g)f),$$

where $g = v^{-1}u$ and $u = \mu(S \otimes \mathrm{id})(\mathcal{R}_{21})$. In particular, the quantum dimension *of V is*

$$\mathrm{qdim}(V) = \mathrm{trace}(\rho(g)).$$

Of course, if A is cocommutative, we may take $\mathcal{R} = 1 \otimes 1$, $u = v = g = 1$, and then the quantum trace becomes the ordinary trace.

Remarks [1] As we mentioned above, the quantum trace is multiplicative, i.e. if V_1 and V_2 are representations of A and $f_1 \in \mathrm{End}_k(V_1)$, $f_2 \in \mathrm{End}_k(V_2)$, then the quantum trace of $f_1 \otimes f_2 \in \mathrm{End}_k(V_1 \otimes V_2)$ is $\mathrm{qtr}(f_1 \otimes f_2) = \mathrm{qtr}(f_1)\mathrm{qtr}(f_2)$.

[2] The quantum trace $\mathrm{qtr} : \mathrm{End}_k(V) \to k$ is a homomorphism of representations of A, where $\mathrm{End}_k(V)$ is made into a representation of A as described in Subsection 4.1C, and k is the trivial representation. In fact, qtr is the composite

$$\mathrm{End}_k(V) \cong V \otimes V^* \cong V^{**} \otimes V^* \to k.$$

Each of these maps is a homomorphism of representations of A. (The maps are all canonical except the isomorphism $V \to V^{**}$, which is defined by $v \mapsto v'$, where $\langle v', \xi \rangle = \langle \xi, \rho(g)(v) \rangle$ for all $\xi \in V^*$; that this commutes with A is the statement that S^2 is conjugation by g.) $\triangle$

D The quantum double In Section 2.1, we showed that the double $\mathcal{D}(\mathfrak{g})$ of any Lie bialgebra $\mathfrak{g}$ is a quasitriangular Lie bialgebra. We recall that $\mathcal{D}(\mathfrak{g})$ is the direct sum $\mathfrak{g} \oplus \mathfrak{g}^*$ as a vector space, but not as a Lie algebra. Now, for a direct sum of Lie algebras $\mathfrak{g} \oplus \mathfrak{h}$, $U(\mathfrak{g} \oplus \mathfrak{h})$ is isomorphic as a Hopf algebra to $U(\mathfrak{g}) \otimes U(\mathfrak{h})$. Hence, it is not surprising that we shall need to use a certain 'twisted' tensor product of Hopf algebras to define the quantum double.

Let B and C be Hopf algebras over a commutative ring k, and let $\mathcal{R}$ be an invertible element of $C \otimes B$ such that

(17)
$$(\Delta^C \otimes \mathrm{id})(\mathcal{R}) = \mathcal{R}_{13}\mathcal{R}_{23}, \qquad (\mathrm{id} \otimes \Delta^B)(\mathcal{R}) = \mathcal{R}_{12}\mathcal{R}_{13},$$
$$(\mathrm{id} \otimes S^B)(\mathcal{R}) = \mathcal{R}^{-1}, \qquad (S^C \otimes \mathrm{id})(\mathcal{R}) = \mathcal{R}^{-1}.$$

PROPOSITION 4.2.10 *If $\mathcal{R} \in C \otimes B$ satisfies (17), then $B \otimes C$, with the usual algebra structure and with comultiplication*

$$\Delta(b \otimes c) = \mathcal{R}_{23}\Delta_{13}^B(b)\Delta_{24}^C(c)\mathcal{R}_{23}^{-1},$$

antipode

$$S(b \otimes c) = \mathcal{R}_{21}^{-1}(S^B(b) \otimes S^C(c))\mathcal{R}_{21},$$

and counit

$$\epsilon(b \otimes c) = \epsilon^B(b)\epsilon^C(c),$$

is a Hopf algebra, which we denote by $B \underset{\mathcal{R}}{\otimes} C$.

OUTLINE OF PROOF Let us check, for example, that $\mu(S \otimes \mathrm{id})\Delta = \iota\epsilon$. For $b \in B$, $c \in C$, write

$$\mathcal{R} = \sum_i c_i \otimes b_i, \qquad \mathcal{R}^{-1} = \sum_j c^j \otimes b^j,$$
$$\Delta^B(b) = \sum_\beta b_\beta \otimes b'_\beta, \qquad \Delta^C(c) = \sum_\gamma c_\gamma \otimes c'_\gamma.$$

Then,

$$\mu(S \otimes \mathrm{id})\Delta(b \otimes c) = \mu(\mathcal{R}_{21}^{-1}.(S^B \otimes S^C \otimes \mathrm{id} \otimes \mathrm{id})(\mathcal{R}_{23}\Delta_{13}^B(b)\Delta_{24}^C(c)\mathcal{R}_{23}^{-1}).\mathcal{R}_{21})$$
$$= \sum_{i,j,k,l,\beta,\gamma} b^l S^B(b_\beta)b_k b_i b'_\beta b^j \otimes c^l S^C(c^j)S^C(c_\gamma)S^C(c_i)c_k c'_\gamma.$$

Now,

$$\sum_{i,k} S^C(c_i)c_k \otimes b_k b_i = (S^C \otimes \mathrm{id})(\mathcal{R}.(S^{C^{-1}} \otimes \mathrm{id})(\mathcal{R}))$$

$$= (S^C \otimes \mathrm{id})(\mathcal{R}.\mathcal{R}^{-1}) = 1 \otimes 1.$$

Hence,

$$\mu(S \otimes \text{id})\Delta(b \otimes c) = \sum_{j,l,\beta,\gamma} b^l S^B(b_\beta) b'_\beta b^j \otimes c^l S^C(c^j) S^C(c_\gamma) c'_\gamma$$

$$= \epsilon^B(b)\epsilon^C(c) \sum_{j,l} b^l b^j \otimes c^l S^C(c^j)$$

$$= \epsilon^B(b)\epsilon^C(c)\mathcal{R}_{21}^{-1}.(\text{id} \otimes S^C)(\mathcal{R}_{21}^{-1})$$

$$= \epsilon^B(b)\epsilon^C(c)\mathcal{R}_{21}^{-1}.\mathcal{R}_{21}$$

$$= \epsilon(b \otimes c). \quad \blacksquare$$

Suppose now that A is a finite-dimensional Hopf algebra over a field k. We apply the preceding construction with $B = A^*$, $C = A_{\text{op}}$.

LEMMA 4.2.11 *The canonical element $\mathcal{R} \in A_{\text{op}} \otimes A^*$ associated to the identity map $A \to A$ satisfies the conditions in (17).*

OUTLINE OF PROOF Let us verify, for example, the first of the four equations in (17). Let $\{a_i\}$ be a basis of A and let $\{\alpha^i\}$ be the dual basis of A^*. Thus, $\mathcal{R} = \sum_i a_i \otimes \alpha^i$. Write

$$\Delta(a_i) = \sum_{j,k} \gamma_i^{jk} a_j \otimes a_k.$$

Note that

$$\langle \alpha^j \alpha^k, a_i \rangle = \langle \alpha^j \otimes \alpha^k, \Delta(a_i) \rangle = \gamma_i^{jk}$$

so that

$$\alpha^j \alpha^k = \sum_i \gamma_i^{jk} \alpha^i.$$

Hence,

$$(\Delta \otimes \text{id})(\mathcal{R}) = \sum_i \Delta(a_i) \otimes \alpha^i = \sum_{i,j,k} \gamma_i^{jk} a_j \otimes a_k \otimes \alpha^i$$

$$= \sum_{jk} a_j \otimes a_k \otimes \alpha^j \alpha^k = \mathcal{R}_{13}\mathcal{R}_{23}. \quad \blacksquare$$

Thus, we may define the *quantum double* of A to be

$$\mathcal{D}(A) = (A^* \underset{\mathcal{R}}{\otimes} A_{\text{op}})^*.$$

Note that $\mathcal{D}(A) \cong A \otimes A^*$ as a coalgebra (and, in particular, as a vector space). Further, it is not hard to see that the canonical embeddings

$$A \hookrightarrow \mathcal{D}(A) \hookleftarrow (A^*)^{\text{op}}$$

are homomorphisms of Hopf algebras (cf. 1.4.2).

It will be useful to have a more explicit description of the multiplication in $\mathcal{D}(A)$. The problem is to express a product $\alpha.a$, where $a \in A$ and $\alpha \in (A^*)^{\mathrm{op}}$, as a linear combination of products $a'.\alpha'$. Write

$$(S^{-1} \otimes \mathrm{id})\Delta(a) = \sum_i a_i \otimes a_i' \otimes a_i'',$$

$$\Delta^{\mathrm{op}}(\alpha) = \sum_j \alpha_j \otimes \alpha_j' \otimes \alpha_j''.$$

Then, one finds that

$$\alpha.a = \sum_{i,j} \langle \alpha_j \, , \, a_i \rangle \langle \alpha_j'' \, , \, a_i'' \rangle a_i'.\alpha_j',$$

where $\langle \, , \, \rangle$ denotes the natural pairing between A and its dual.

The importance of quantum doubles is that they have a simple quasitriangular structure.

PROPOSITION 4.2.12 *The double $\mathcal{D}(A)$ of a finite-dimensional Hopf algebra A over a field k is quasitriangular, with universal R-matrix the canonical element of $A \otimes A^* \subset \mathcal{D}(A) \otimes \mathcal{D}(A)$ associated to the identity map $A \to A$.* ∎

The proof is a straightforward, but tedious, computation similar to those given in 4.2.10 and 4.2.11.

If A is infinite dimensional, one must define the duals in an appropriate sense and suitably complete the tensor products. The element $\mathcal{R}$ will then lie in some completion of the algebraic tensor product $A \otimes A^\circ$. Apart from this, the construction of $\mathcal{D}(A)$ and 4.2.12 is still valid. One may even work over any commutative ring. We shall see some examples of doubles of infinite-dimensional Hopf algebras in Section 8.3.

E Twisting There is a way to construct quasitriangular Hopf algebras by starting with a cocommutative Hopf algebra A and 'twisting' it with an element $\mathcal{F} \in A \otimes A$ satisfying certain conditions. This will be used in Chapter 6 to construct quantizations of a large class of Lie bialgebras.

PROPOSITION 4.2.13 *Let $(A, \mu, \imath, \Delta, \epsilon, S)$ be a Hopf algebra over a commutative ring. Let $\mathcal{F}$ be an invertible element of $A \otimes A$ such that*

$$(18) \qquad \mathcal{F}_{12}(\Delta \otimes \mathrm{id})(\mathcal{F}) = \mathcal{F}_{23}(\mathrm{id} \otimes \Delta)(\mathcal{F}),$$

$$(19) \qquad (\epsilon \otimes \mathrm{id})(\mathcal{F}) = 1 = (\mathrm{id} \otimes \epsilon)(\mathcal{F}).$$

Then, $v = \mu(\text{id} \otimes S)(\mathcal{F})$ is an invertible element of A with

$$v^{-1} = \mu(S \otimes \text{id})(\mathcal{F}^{-1}).$$

Moreover, if we define $\Delta^{\mathcal{F}} : A \to A \otimes A$ and $S^{\mathcal{F}} : A \to A$ by

(20) $$\Delta^{\mathcal{F}}(a) = \mathcal{F}\Delta(a)\mathcal{F}^{-1}, \quad S^{\mathcal{F}}(a) = vS(a)v^{-1},$$

then $(A, \mu, \imath, \Delta^{\mathcal{F}}, \epsilon, S^{\mathcal{F}})$ is a Hopf algebra, denoted by $A^{\mathcal{F}}$ and called the twist of A by $\mathcal{F}$. ∎

This result is proved by arguments similar to those used earlier in this section. We leave the details to the reader. The following special case will be of most interest to us.

COROLLARY 4.2.14 *If A is a cocommutative Hopf algebra and $\mathcal{F}$ is an invertible element of $A \otimes A$ satisfying (18) and (19), then $A^{\mathcal{F}}$ is a triangular Hopf algebra with universal R-matrix*

(21) $$\mathcal{R} = \mathcal{F}_{21}\mathcal{F}^{-1}.$$

PROOF It is obvious that $A^{\mathcal{F}}$ is almost cocommutative with universal R-matrix (21), and that $\mathcal{R}_{21} = \mathcal{R}^{-1}$. Let us check that $(\Delta^{\mathcal{F}} \otimes \text{id})(\mathcal{R}) = \mathcal{R}_{13}\mathcal{R}_{23}$. This equation reduces to

$$\mathcal{F}_{12}.(\Delta \otimes \text{id})(\mathcal{F}_{21}).(\text{id} \otimes \Delta)(\mathcal{F})^{-1}.\mathcal{F}_{23}^{-1} = \mathcal{F}_{31}.\mathcal{F}_{13}^{-1}.\mathcal{F}_{32}.\mathcal{F}_{23}^{-1}.$$

Applying the flip σ_{23} to both sides and using the cocommutativity of Δ, the preceding equation is equivalent to

$$\mathcal{F}_{13}.\sigma_{23}(\Delta \otimes \text{id})(\mathcal{F}_{21}) = \mathcal{F}_{21}.\mathcal{F}_{12}^{-1}.\mathcal{F}_{23}.(\text{id} \otimes \Delta)(\mathcal{F})$$
$$= \mathcal{F}_{21}.(\Delta \otimes \text{id})(\mathcal{F}),$$

by (18). It is easy to see, using the cocommutativity of Δ again, that applying σ_{12} to the left-hand side of the preceding equation gives $\mathcal{F}_{23}(\text{id} \otimes \Delta)(\mathcal{F})$, and applying it to the right-hand side gives $\mathcal{F}_{12}(\Delta \otimes \text{id})(\mathcal{F})$. So the result follows on using (18) once again. ∎

Another variant of 4.2.13 is

COROLLARY 4.2.15 *Suppose that A and $\mathcal{F}$ are as in 4.2.13, but assume in addition that A is quasitriangular with universal R-matrix $\mathcal{R}$ and that $\mathcal{F}_{21} = \mathcal{F}^{-1}$ and $\mathcal{F}_{12}\mathcal{F}_{13}\mathcal{F}_{23} = \mathcal{F}_{23}\mathcal{F}_{13}\mathcal{F}_{12}$. Then, $A^{\mathcal{F}}$ is quasitriangular, with universal R-matrix $\mathcal{R}^{\mathcal{F}} = \mathcal{F}^{-1}\mathcal{R}\mathcal{F}^{-1}$.* ∎

Remark If we drop condition (18), then $\Delta^{\mathcal{F}}$ is not coassociative in general. However, it is 'almost coassociative' in the sense that

$$(\text{id} \otimes \Delta^{\mathcal{F}})\Delta^{\mathcal{F}}(a) = \Phi^{-1}.(\Delta^{\mathcal{F}} \otimes \text{id})\Delta^{\mathcal{F}}(a).\Phi,$$

where Φ is the invertible element of $A \otimes A \otimes A$ given by

$$\Phi = \mathcal{F}_{12}.(\Delta \otimes \text{id})(\mathcal{F}).(\text{id} \otimes \Delta)(\mathcal{F})^{-1}.\mathcal{F}_{23}^{-1}.$$

Weakening the coassociativity condition in the definition of a Hopf algebra to almost coassociativity leads to the notion of a *quasi-Hopf algebra*, which we shall study in detail in Chapter 16. $\triangle$

F Sweedler's example As we remarked in the Introduction, before the advent of quantum groups, very few examples of Hopf algebras which are neither commutative nor cocommutative were known (except for those obtained by taking tensor products of the standard commutative and cocommutative examples). However, Sweedler (1969) constructed an interesting four-dimensional example, which we shall use to illustrate some of the concepts introduced in this chapter.

Sweedler's Hopf algebra A is defined as follows (we assume that the ground ring k is a field of characteristic $\neq 2$). As an associative algebra with unit, it has generators g and x with defining relations

$$g^2 = 1, \quad x^2 = 0, \quad gxg = -x,$$

coalgebra structure given by

$$\Delta(g) = g \otimes g, \quad \Delta(x) = x \otimes g + 1 \otimes x,$$
$$\epsilon(g) = 1, \quad \epsilon(x) = 0,$$

and antipode given by

$$S(g) = g, \quad S(x) = -x.$$

The reader should verify that this definition of Δ on the generators of A extends to a homomorphism of algebras $A \rightarrow A \otimes A$, etc. The elements $\{1, x, g, gx\}$ are a basis of A over k. Obviously, A is neither commutative nor cocommutative. The following properties hold.

(a) $A^* \cong A$ *as Hopf algebras.* In fact, let $\{\bar{1}, \bar{g}, \bar{x}, \overline{gx}\}$ be the basis of A^* dual to the basis $\{1, g, x, gx\}$ of A. Setting $\gamma = 1 - \bar{g}$, $\xi = \bar{x} + \overline{gx}$, we have

$$\gamma^2 = \epsilon, \quad \xi^2 = 0, \quad \gamma\xi\gamma = -\xi,$$
$$\Delta(\gamma) = \gamma \otimes \gamma, \quad \Delta(\xi) = \xi \otimes \gamma + \epsilon \otimes \xi,$$
$$\gamma(1) = 1, \quad \xi(1) = 0.$$

For example,

$$\langle \gamma^2, g \rangle = \langle \gamma \otimes \gamma, \Delta(g) \rangle = \langle \gamma, g \rangle^2$$
$$= ((\langle \bar{1}, g \rangle - \langle \bar{g}, g \rangle))^2 = (0 - 1)^2$$
$$= 1,$$
$$\langle g^2, x \rangle = \langle \gamma \otimes \gamma, x \otimes g + 1 \otimes x \rangle$$
$$= \langle \gamma, x \rangle \langle \gamma, g \rangle + \langle \gamma, 1 \rangle \langle \gamma, x \rangle$$
$$= 0,$$
$$\langle \gamma^2, gx \rangle = \langle \gamma \otimes \gamma, (g \otimes g)(x \otimes g + 1 \otimes x) \rangle$$
$$= \langle \gamma, gx \rangle \langle \gamma, 1 \rangle + \langle \gamma, g \rangle \langle \gamma, gx \rangle$$
$$= 0,$$

proving that $\gamma^2 = \epsilon$. The other formulas are established by similar calculations.

(b) *The non-zero group-like elements of A are 1 and g.* In fact, one finds that $a = \lambda 1 + \mu x + \nu g + \rho gx$, with $\lambda, \mu, \nu, \rho \in k$, satisfies $\Delta(a) = a \otimes a$ if and only if

$$\mu = \rho = 0, \quad \lambda \nu = 0, \quad \lambda^2 = \lambda, \quad \nu^2 = \nu,$$

from which the assertion follows. Similarly, the most general element $y \in A$ such that

(22) $$\Delta(y) = y \otimes 1 + g \otimes y$$

is $y = \lambda gx$ for some $\lambda \in k$ (this result will be needed later).

(c) *The element* $(1 + g)x$ *is a left-invariant integral for A, and* $x(1 + g) = (1-g)x$ *is a right-invariant integral.* The verification of this is straightforward. Since these integrals are not scalar multiples of each other, A has no (non-zero) bi-invariant integral.

(d) *A is not isomorphic to a quantum double.* For if $A \cong \mathcal{D}(H)$ for some Hopf algebra H, then H is 2-dimensional, and any such Hopf algebra is both commutative and cocommutative (commutativity is obvious since H is generated as an algebra by a single element, and cocommutativity follows by applying the same argument to H^*). Since $\mathcal{D}(H) \cong H \otimes (H^*)^{op}$ as coalgebras, it follows that $\mathcal{D}(H)$ is cocommutative. But A is not cocommutative.

(e) *A is quasitriangular (in many ways).* In fact, for any $\lambda \in k$,

$$\mathcal{R}_\lambda = \frac{1}{2}(1 \otimes 1 + 1 \otimes g + g \otimes 1 - g \otimes g) + \frac{\lambda}{2}(x \otimes x + x \otimes gx + gx \otimes gx - gx \otimes x)$$

is a universal R-matrix of A, and every universal R-matrix of A is of this form. The former statement can, of course, be verified by direct computation. Now recall from the remark following 4.2.7 that if $\mathcal{R}$ is a universal R-matrix, the map $f : A^* \to A^{\mathrm{op}}$ given by $f(\eta) = \langle \eta \otimes \mathrm{id}, \mathcal{R} \rangle$ is a homomorphism of bialgebras. All such maps are of the following types (identifying A^* with A):

(i) $f_\infty(a) = \epsilon(a)1$, for all $a \in A$;

(ii) $f_\lambda(g) = g$, $f_\lambda(x) = \lambda gx$, for some $\lambda \in k$.

Case (i) would correspond to the universal R-matrix $\mathcal{R}_\infty = 1 \otimes 1$, which is impossible since A is not cocommutative; and case (ii) corresponds to $\mathcal{R}_\lambda$. To see the assertion about bialgebra maps $f : A^* \to A^{\mathrm{op}}$, note that $f(g)$ is group-like, and hence equal to 0, 1 or g by part (b). But $f(g) = 0$ is incompatible with the relation $g^2 = 1$, while $f(g) = 1$ and the relation $gxg = -x$ force $f(x) = 0$, and then we are in case (i). If $f(g) = g$, then $y = f(x)$ satisfies (22), and hence is of the form $y = \lambda gx$ by part (b) again.

(f) *For any $\lambda \in k$, $(A, \mathcal{R}_\lambda)$ is a ribbon Hopf algebra with distinguished central element $v = g$.* In fact, in the notation of Subsection 4.2C, we have $u = S(u) = g$.

(g) Assume now that k is algebraically closed. Then, A has *(up to isomorphism) exactly four indecomposable representations, namely two one-dimensional representations given by*

$$g \mapsto \pm 1, \quad x \mapsto 0,$$

and two two-dimensional representations given by

$$g \mapsto \begin{pmatrix} \pm 1 & 0 \\ 0 & \mp 1 \end{pmatrix}, \quad x \mapsto \begin{pmatrix} 0 & 1 \\ 0 & 0 \end{pmatrix}.$$

To see this, observe that, since $g^2 = 1$, any A-module V splits as the direct sum $V^+ \oplus V^-$ of the ± 1-eigenspaces for g, and that the relation $gxg = -x$ implies that x interchanges V^+ and V^-. Hence, the A-module V is determined by a pair of linear maps $X^\pm : V^\pm \to V^\mp$ such that $X^+ X^- = X^- X^+ = 0$. Choose linear subspaces $W^\pm \subset V^\pm$ such that $\ker(X^\pm) \oplus W^\pm = V^\pm$. Then, V is the direct sum of the A-invariant subspaces $\ker(X^+) \oplus W^-$ and $\ker(X^-) \oplus W^+$. Assume now that V is indecomposable. In that case, it follows that *either X^+ is injective and $X^- = 0$, or X^- is injective and $X^+ = 0$.* By considering $\mathrm{im}(X^\pm)$, a similar argument shows that we can replace 'injective' by 'surjective' in the last statement. Thus, either X^+ is an isomorphism and $X^- = 0$, or X^- is an isomorphism and $X^+ = 0$. It follows easily that we must have $\dim(V^+) = \dim(V^-) = 1$.

Evaluating the universal R-matrix $\mathcal{R}_\lambda$ in the two-dimensional representations gives the two one-parameter families

$$\begin{pmatrix} \pm 1 & 0 & 0 & \lambda \\ 0 & 1 & 0 & 0 \\ 0 & 0 & 1 & 0 \\ 0 & 0 & 0 & \mp 1 \end{pmatrix}$$

of matrix solutions of the quantum Yang–Baxter equation.

Bibliographical notes

4.1 The notion of a Hopf algebra was introduced by Hopf (1941). The basic references are Milnor & Moore (1965) (where the proof of the result stated in 4.1.9 can be found), Quillen (1969), Sweedler (1969) and Abe (1980). For topological Hopf algebras, see Takeuchi (1985) and Abe & Takeuchi (1992). See van Daele (1993) for a discussion of duality in the context of Hopf *-algebras. See Northcott (1968), Chapter 9, for more information on the $\mathcal{I}$-adic (and h-adic) topology. See Hochschild (1965), Section II.3, for the result in 4.1.19.

4.2 The notion of a quasitriangular Hopf algebra, and that of the quantum double, is due to Drinfel'd (1987). Majid (1990g) surveys the properties and applications of quasitriangular Hopf algebras in physics. See Radford (1992, 1993) for a discussion of finite-dimensional quasitriangular Hopf algebras. The more general notion of an almost cocommutative Hopf algebra is discussed in Drinfel'd (1990a) and Reshetikhin (1990b). See Reshetikhin (1990a) for the notion of twisting.

5

Representations
and
quasitensor categories

It is well known that the properties of the operation of taking direct sums (of vector spaces, of group representations, ...) can be abstracted in the notion of an abelian category. Abstracting the properties of the tensor product operation (on vector spaces, on group representations, ...) leads to the notion of a *monoidal category*. Although the basic theory of such categories was developed in the early 1960s by S. MacLane, the subject has attracted the attention of a wide audience only recently, largely because of the discovery of interesting new examples. It is no surprise that many of these are related to quantum groups, for we saw in Chapter 4 that, from an abstract point of view, a Hopf algebra may be regarded as an algebra A equipped with the extra structure necessary to define tensor products (and duals) of representations of A. We describe the basic theory of monoidal categories in Section 5.1. As well as the categories of representations of Hopf algebras, we give a number of geometric examples related to knot theory.

There are several results which assert, roughly speaking, that a monoidal category with some additional properties 'is' the category of representations of some group: this is *Tannaka–Krein duality*. Perhaps the simplest example is the Pontryagin duality for locally compact abelian groups. For such groups G, all representations are direct sums of 1-dimensional representations, or *characters*. The set of characters of G can itself be given a natural structure of a locally compact abelian group $\widehat{G}$. Pontryagin's theorem asserts that G can be recovered completely from $\widehat{G}$: in fact, G is isomorphic to the character group of $\widehat{G}$. At the end of Section 5.1, we give a duality result of this type for a general Hopf algebra.

In many naturally occurring examples of monoidal categories, the tensor product operation is commutative up to isomorphism, and the commutativity isomorphisms are involutive: one then has a *tensor category*. For example, the category of representations of a group obviously has this property. More generally, we saw in Chapter 4 that the category of representations of any triangular Hopf algebra is a tensor category. If we weaken 'triangular' to 'quasitriangular', the tensor product is still commutative, but the commutativity isomorphisms are no longer involutive, in general. The category of representations is then said to be a *quasitensor* (or *braided monoidal*) category. In Section 5.2, we describe the basic properties of tensor and quasitensor categories. Many, but not all, of the monoidal categories encountered in Section 5.1 are quasitensor.

In a semisimple abelian quasitensor category, the tensor product is determined by the decomposition into simple objects of the tensor product of any two simple objects, and hence by a collection of non-negative integers which satisfy certain conditions which reflect the commutativity and associativity properties of the tensor product. Combinatorial data of this kind were encountered in the study of conformal field theories, where they were called *fusion rules*. At the end of Section 5.2, we give a somewhat schematic discussion of the role played by quasitensor categories in physics.

In Section 5.3, we describe an application of the theory of quasitensor categories to the construction of topological invariants of *ribbon tangles*, which are essentially collections of non-intersecting rubber bands in $\mathbb{R}^3$, either closed or open, and with finite width but zero thickness. Ribbon tangles can be organized into a quasitensor category, and the invariants are obtained by constructing a functor from this category to the category of representations of a particular type of quasitriangular Hopf algebra. If we forget the width of the bands, and consider only ribbon tangles with no free ends, we obtain as a special case invariants of links in $\mathbb{R}^3$. The celebrated *Jones polynomial* can be constructed in this way.

5.1 Monoidal categories

In this section, we attempt to axiomatize the properties of the direct sum and tensor product operations in the category of representations of a Hopf algebra. This leads to the notions of abelian and monoidal categories, respectively.

A Abelian categories This subsection is meant only as a summary of the basic definitions and results on abelian categories. For further discussion, the reader is referred to one of the standard sources, such as MacLane (1971).

As we mentioned above, the notion of an abelian category is designed to capture the properties of direct sums of vector spaces, of group representations, etc. In general, a category **C** *admits (finite) direct sums* if, for any finite set of objects $U_1, \ldots, U_n$ of **C**, there is an object U and morphisms $\pi_i : U \to U_i$ such that, for any object V and morphisms $f_i : V \to U_i$, there is a unique morphism $f : V \to U$ such that $\pi_i \circ f = f_i$ for all i. The object U is then unique up to isomorphism; we write $U = \oplus_i U_i$ and $f = \oplus_i f_i$.

A category which admits direct sums is called *additive* if, for any objects U and V, the set $\mathrm{Hom}_{\mathbf{C}}(U, V)$ of morphisms from U to V is an abelian group, and if the compositions

(1) $$\mathrm{Hom}_{\mathbf{C}}(U,V) \times \mathrm{Hom}_{\mathbf{C}}(V,W) \to \mathrm{Hom}_{\mathbf{C}}(U,W)$$

are bi-additive.

To define the stronger concept of an abelian category, we need the notions of the kernel and cokernel of a morphism. If $\mathbf{C}$ is an additive category, a morphism $f \in \mathrm{Hom}_{\mathbf{C}}(U,V)$ is a *monomorphism* (resp. an *epimorphism*) if, for any $g \in \mathrm{Hom}_{\mathbf{C}}(W,U)$, $f \circ g = 0$ implies $g = 0$ (resp. for any $g \in \mathrm{Hom}_{\mathbf{C}}(V,W)$, $g \circ f = 0$ implies $g = 0$). A *kernel* of a morphism $f : U \to V$ is an object $\ker(f)$ and a morphism $\kappa : \ker(f) \to U$ such that $f \circ \kappa = 0$ and, if $g \in \mathrm{Hom}_{\mathbf{C}}(W,U)$ is such that $f \circ g = 0$, then $g = \kappa \circ h$ for some $h \in \mathrm{Hom}_{\mathbf{C}}(W, \ker(f))$. The dual notion of a *cokernel* $\mathrm{coker}(f)$ is defined in the obvious way. The kernel and cokernel of a morphism are unique, up to isomorphism, if they exist.

An additive category $\mathbf{C}$ is *abelian* if

(a) every morphism has a kernel and a cokernel;

(b) every monomorphism is the kernel of its cokernel, and every epimorphism is the cokernel of its kernel;

(c) every morphism is expressible as the composite of an epimorphism followed by a monomorphism.

A morphism in $\mathbf{C}$ which is both mono and epi is an isomorphism. A pair of morphisms

$$U \xrightarrow{f} W \xrightarrow{g} V$$

in $\mathbf{C}$ is a *short exact sequence* if $g \circ f = 0$ and if the morphisms $U \to \ker(g)$ and $\mathrm{coker}(f) \to V$, given by the defining properties of the kernel and cokernel, are isomorphisms.

We shall occasionally need the stronger notion of a *k-linear* category, where k is a commutative ring. This is an abelian category for which the sets of morphisms $\mathrm{Hom}_{\mathbf{C}}(U,V)$ are k-modules, and the compositions (1) are k-bilinear.

The concepts of an irreducible, or completely reducible, representation (of a group, say) have analogues in any abelian category $\mathbf{C}$. In fact, an object U of $\mathbf{C}$ is *simple* if every non-zero monomorphism $V \to U$ is an isomorphism, and every non-zero epimorphism $U \to W$ is an isomorphism. Any non-zero morphism between simple objects is an isomorphism. An object is *semisimple* if it is a (finite) direct sum of simple objects. The category $\mathbf{C}$ itself is called semisimple if every object of $\mathbf{C}$ is semisimple. In a semisimple abelian category, every short exact sequence splits (the meaning of this is clear).

The direct sum operation of an abelian category $\mathbf{C}$ is encoded in its *Grothendieck group* $\mathrm{Gr}(\mathbf{C})$. Let $\mathcal{F}(\mathbf{C})$ be the free abelian group with generators the isomorphism classes of objects of $\mathbf{C}$, and let $[U]$ be the element of $\mathcal{F}(\mathbf{C})$ corresponding to an object U of $\mathbf{C}$. Then, $\mathrm{Gr}(\mathbf{C})$ is the quotient of $\mathcal{F}(\mathbf{C})$ by the subgroup generated by the elements of the form $[W] - [U] - [V]$ for all short exact sequences

$$U \to W \to V$$

in $\mathbf{C}$. Note that, if $\mathbf{C}$ is semisimple, the defining relations of $\mathrm{Gr}(\mathbf{C})$ are simply $[U \oplus V] = [U] + [V]$; moreover, $\mathrm{Gr}(\mathbf{C})$ is generated as an abelian group by the simple objects of $\mathbf{C}$.

We shall see some examples of abelian categories in Subsection 5.1D.

B Monoidal categories We now turn to the axiomatization of the properties of the tensor product of vector spaces, of representations of a Hopf algebra, etc.

DEFINITION 5.1.1 *A monoidal category is a category $\mathbf{C}$ together with a functor $\otimes : \mathbf{C} \times \mathbf{C} \to \mathbf{C}$, written $(U, V) \mapsto U \otimes V$ for objects U and V of $\mathbf{C}$, which satisfies the following conditions:*

(i) there are natural isomorphisms $\alpha_{U,V,W} : U \otimes (V \otimes W) \to (U \otimes V) \otimes W$ such that the diagram

$$U \otimes (V \otimes (W \otimes Z)) \xrightarrow{\alpha} (U \otimes V) \otimes (W \otimes Z) \xrightarrow{\alpha} ((U \otimes V) \otimes W) \otimes Z$$

$$\mathrm{id} \otimes \alpha \downarrow \qquad\qquad\qquad\qquad\qquad\qquad \uparrow \alpha \otimes \mathrm{id}$$

$$U \otimes ((V \otimes W) \otimes Z) \xrightarrow{\qquad\qquad\qquad \alpha \qquad\qquad\qquad} (U \otimes (V \otimes W)) \otimes Z$$

commutes for all U, V, W, Z;

(ii) there is an identity object $\mathbf{1}$ in $\mathbf{C}$ and natural isomorphisms $\rho_U : U \otimes \mathbf{1} \to U$ and $\lambda_U : \mathbf{1} \otimes U \to U$ such that the diagram

$$U \otimes (\mathbf{1} \otimes V) \xrightarrow{\alpha} (U \otimes \mathbf{1}) \otimes V$$

$$\mathrm{id} \otimes \lambda \downarrow \qquad\qquad\qquad \downarrow \rho \otimes \mathrm{id}$$

$$U \otimes V \xrightarrow{\qquad \mathrm{id} \qquad} U \otimes V$$

commutes for all U, V.

An abelian *(resp. k-linear) monoidal category is a monoidal category which is abelian (resp. k-linear) and such that $\otimes$ is a bi-additive functor (resp. a k-bilinear functor).*

A functor $\Phi : \mathbf{C} \to \mathbf{C}'$ *between monoidal categories is a* monoidal functor *if there are natural isomorphisms*

$$\varphi_{U,V} : \Phi(U) \otimes \Phi(V) \to \Phi(U \otimes V)$$

for all U, V, such that $\Phi(\mathbf{1}) = \mathbf{1}'$, and such that the following diagrams commute:

$$
\begin{array}{ccccc}
\Phi(U) \otimes (\Phi(V) \otimes \Phi(W)) & \xrightarrow{\mathrm{id}\otimes\varphi} & \Phi(U) \otimes \Phi(V \otimes W) & \xrightarrow{\varphi} & \Phi(U\otimes(V\otimes W)) \\
\alpha' \downarrow & & & & \downarrow \Phi(\alpha) \\
(\Phi(U) \otimes \Phi(V)) \otimes \Phi(W) & \xrightarrow{\varphi\otimes\mathrm{id}} & \Phi(U \otimes V) \otimes \Phi(W) & \xrightarrow{\varphi} & \Phi((U\otimes V)\otimes W)
\end{array}
$$

$$
\begin{array}{ccc}
\Phi(\mathbf{1}) \otimes \Phi(A) \xrightarrow{\varphi} \Phi(\mathbf{1} \otimes A) & \qquad & \Phi(A) \otimes \Phi(\mathbf{1}) \xrightarrow{\varphi} \Phi(A \otimes \mathbf{1}) \\
\mathrm{id}\downarrow \qquad\qquad \downarrow \Phi(\lambda) & & \mathrm{id}\downarrow \qquad\qquad \downarrow \Phi(\rho) \\
\mathbf{1}' \otimes \Phi(A) \xrightarrow[\lambda']{} \Phi(A) & & \Phi(A) \otimes \mathbf{1}' \xrightarrow[\rho']{} \Phi(A)
\end{array}
$$

Since the subscripts on the structure maps α, φ, etc. appearing in these commutative diagrams are obvious, we have omitted them. For example, the upper left horizontal arrow in the diagram in (i) is $\alpha_{U,V,W\otimes Z}$.

Condition (i) is usually called the *pentagon axiom*, or *associativity constraint*. Its significance is that, according to a theorem of MacLane (1971), it implies that any two iterated tensor products of objects in **C** (in the same order) are isomorphic by a *canonical* isomorphism. The identity object **1** is unique up to isomorphism.

Another theorem of MacLane (1971) asserts that every monoidal category is equivalent to one in which the associativity morphisms α and the morphisms ρ and λ are the identity maps. A monoidal category with this property is called *strict*.

If **C** is an abelian monoidal category, its Grothendieck group $\mathrm{Gr}(\mathbf{C})$ is a ring with product given by

$$[U].[V] = [U \otimes V],$$

and unit [**1**]. The associativity of this product, and the basic property $[\mathbf{1}].[U] = [U].[\mathbf{1}] = [U]$ of the unit, follow from the existence of the isomorphisms $\alpha_{U,V,W}$, ρ_U and λ_U in 5.1.1.

C Rigidity The next step is to incorporate the notion of the dual of a representation into the categorical framework.

If **C** is a monoidal category, an object U^* is said to be a (left) *dual* of an object U in **C** if there are morphisms $\mathrm{ev}_U : U^* \otimes U \to \mathbf{1}$ and $\pi_U : \mathbf{1} \to U \otimes U^*$ such that the following diagrams commute:

$$
\begin{array}{ccc}
U \xrightarrow{\pi\otimes\mathrm{id}} (U \otimes U^*) \otimes U & \qquad & U^* \xrightarrow{\mathrm{id}\otimes\pi} U^* \otimes (U \otimes U^*) \\
\mathrm{id}\uparrow \qquad\qquad \uparrow\alpha & & \mathrm{id}\downarrow \qquad\qquad \downarrow\alpha \\
U \xleftarrow[\mathrm{id}\otimes\mathrm{ev}]{} U \otimes (U^* \otimes U) & & U^* \xleftarrow[\mathrm{ev}\otimes\mathrm{id}]{} (U^* \otimes U) \otimes U^*
\end{array}
$$

(We have suppressed the isomorphisms λ and ρ for simplicity.) If U has a dual object, it is unique up to isomorphism.

Suppose now that every object in $\mathbf{C}$ has a dual object. If $\psi : U \to V$ is a morphism in $\mathbf{C}$, then the composite

$$V^* = V^* \otimes \mathbf{1} \xrightarrow{\text{id} \otimes \pi} V^* \otimes U \otimes U^* \xrightarrow{\text{id} \otimes \psi \otimes \text{id}} V^* \otimes V \otimes U^* \xrightarrow{\text{ev} \otimes \text{id}} \mathbf{1} \otimes U^* = U^*$$

is a morphism $\psi^* : V^* \to U^*$ (we are now abusing notation by omitting the associativity maps). Thus we obtain a contravariant functor $\mathbf{C} \to \mathbf{C}$, called the *dual object functor*, which takes U to U^* and ψ to ψ^*.

DEFINITION 5.1.2 *A monoidal category* $\mathbf{C}$ *is* rigid *if every object in* $\mathbf{C}$ *has a dual object, and if the dual object functor* $\mathbf{C} \to \mathbf{C}$ *is an anti-equivalence of categories.*

Remarks [1] If $\mathbf{C}$ is a rigid monoidal category, it is easy to see that, for all objects U, V and W of $\mathbf{C}$, the map which assigns to a morphism $\psi : U \otimes V \to W$ the morphism $U \to W \otimes V^*$ given by

$$U = U \otimes \mathbf{1} \xrightarrow{\text{id}_U \otimes \pi_V} U \otimes V \otimes V^* \xrightarrow{\psi \otimes \text{id}_{V^*}} W \otimes V^*$$

is bijective; if $\mathbf{C}$ is k-linear, this bijection is an isomorphism of k-modules. In particular, there is a bijection between morphisms $U \otimes V \to \mathbf{1}$ and morphisms $U \to V^*$, and between morphisms $V \to W$ and morphisms $\mathbf{1} \to W \otimes V^*$.

[2] The dual object functor in a rigid monoidal category has a natural monoidal structure. In fact, the composite

$$(V^* \otimes U^*) \otimes (U \otimes V) \xrightarrow{\text{id}_{V^*} \otimes \text{ev}_U \otimes \text{id}_V} V^* \otimes \mathbf{1} \otimes V = V^* \otimes V \xrightarrow{\text{ev}_V} \mathbf{1}$$

corresponds to a morphism $V^* \otimes U^* \to (U \otimes V)^*$, while

$$\mathbf{1} \xrightarrow{\pi_U} U \otimes U^* = U \otimes \mathbf{1} \otimes U^* \xrightarrow{\text{id}_U \otimes \pi_V \otimes \text{id}_{U^*}} (U \otimes V) \otimes (V^* \otimes U^*)$$

corresponds to a morphism $W \to U \otimes V$, where $W^* = V^* \otimes U^*$. It is easy to see that the dual of the latter morphism is the inverse of the former, so that $(U \otimes V)^*$ is naturally isomorphic to $V^* \otimes U^*$.

[3] If U is an object of a rigid category $\mathbf{C}$, U^{**} is *not* isomorphic to U, in general. We shall see examples of this non-reflexivity in Chapter 12 (see Remark [4] following 12.1.11). $\triangle$

D Examples We begin with four examples of monoidal categories of an algebraic nature.

Example 5.1.3 The category of all modules over a commutative ring k is an abelian monoidal category with the obvious associativity maps and identity object k. The same is true for its full subcategory $\mathbf{mod}_k$ consisting of the finitely-generated projective modules. However, only the latter category is rigid (even if k is a field). The dual of a k-module U is $U^* = \mathrm{Hom}_k(U, k)$ as usual, and the map ev_U is the obvious pairing which evaluates an element of U^* on an element of U. To define π_U, we need the 'dual basis lemma' for finitely-generated projective modules, according to which, for any set $\{u_i\}_{i \in I}$ of generators of U, there exists a set of elements $\{\xi^i\}_{i \in I}$ in U^* such that $u = \sum_{i \in I} u_i \xi^i(u)$ for all $u \in U$. We set $\pi_U(1) = \sum_{i \in I} u_i \otimes \xi^i$.

Two special cases are of particular interest to us. If k is a field, $\mathbf{mod}_k$ is simply the category of finite-dimensional vector spaces over k (and the dual basis lemma gives the dual basis in the usual elementary sense). On the other hand, if $k = F[[h]]$ is the ring of formal power series in an indeterminate h over a field F, then, since $F[[h]]$ is a local ring, $\mathbf{mod}_{F[[h]]}$ consists of the free $F[[h]]$-modules of finite rank. We shall usually call such $F[[h]]$-modules 'finite dimensional' from now on. $\Diamond$

Example 5.1.4 If A is a Hopf algebra over a commutative ring k, the category of all representations of A is an abelian monoidal category with the same associativity maps as in the previous example. The identity object is k, regarded as the trivial representation of A. The full subcategory $\mathbf{rep}_A$ consisting of the representations of A on finitely-generated projective k-modules is also rigid abelian monoidal. The dual of a representation U of A is the k-module dual U^* defined in the preceding example, provided with the action of A given by

$$\langle a.\xi, u \rangle = \langle \xi, S(a).u \rangle,$$

for $a \in A$, $u \in U$, $\xi \in U^*$, where S is the antipode of A (see Subsection 4.1C). The forgetful functor $\mathbf{rep}_A \to \mathbf{mod}_k$ is monoidal.

One defines in an analogous way the category $\mathbf{corep}_A$ of A-comodules which are finitely generated and projective as k-modules. It is also rigid abelian monoidal. $\Diamond$

Example 5.1.5 Suppose that A is a Hopf algebra over a commutative ring k, and let $A^{\mathcal{F}}$ be the twist of A by an element $\mathcal{F} \in A \otimes A$ satisfying equations (18) and (19) in Chapter 4 (see Subsection 4.2E). Then $\mathbf{rep}_A$ and $\mathbf{rep}_{A^{\mathcal{F}}}$ are equivalent as monoidal categories. In fact, let $\rho_U : A \to \mathrm{End}(U)$ and $\rho_V : A \to \mathrm{End}(V)$ be representations of A. We leave it to the reader to check that setting $\Phi(U) = U$ and $\varphi_{U,V} = (\rho_U \otimes \rho_V)(\mathcal{F}^{-1})$ defines a monoidal functor $\Phi : \mathbf{rep}_A \to \mathbf{rep}_{A^{\mathcal{F}}}$. $\Diamond$

Example 5.1.6 Let G be a finite group and let $\mathbf{bun}_G$ be the category of finite-dimensional complex vector bundles over G. Thus, an object V of $\mathbf{bun}_G$ is a family $(V_g)_{g \in G}$ of finite-dimensional vector spaces, and a morphism $f : V \to W$ is a collection of linear maps $f_g : V_g \to W_g$. If we define direct sums componentwise, it is easy to see that **bun**$_G$ becomes an abelian category; it is clearly semisimple.

We define the tensor product of two objects of **bun**$_G$ by

$$(V \otimes W)_g = \bigoplus_{g_1 g_2 = g} V_{g_1} \otimes W_{g_2}.$$

If we take the associativity maps

$$(\alpha_{U,V,W})_g : \bigoplus_{g_1 g_2 g_3 = g} U_{g_1} \otimes (V_{g_2} \otimes W_{g_3}) \to \bigoplus_{g_1 g_2 g_3 = g} (U_{g_1} \otimes V_{g_2}) \otimes W_{g_3}$$

to be the identity maps, and the identity object to be

$$(\mathbf{1})_g = \begin{cases} \mathbb{C} & \text{if } g = e, \\ 0 & \text{otherwise,} \end{cases}$$

then **bun**$_G$ becomes an abelian monoidal category.

More generally, let $c : G \times G \times G \to \mathbb{C}^\times$ be any function and take $(\alpha_{U,V,W})_g$ to be $c(g_1, g_2, g_3)$ times the identity map on the component $U_{g_1} \otimes V_{g_2} \otimes W_{g_3}$. Then, the pentagon axiom 5.1.1(i) is equivalent to

$$c(g_1, g_2, g_3 g_4) c(g_1 g_2, g_3, g_4) = c(g_2, g_3, g_4) c(g_1, g_2 g_3, g_4) c(g_1, g_2, g_3)$$

for all $g_1, g_2, g_3, g_4 \in G$, in other words, to c being a 3-cocycle (in the group cohomology of G with coefficients in the trivial module). Thus, for any such cocycle, one obtains another abelian monoidal category.

This example is analogous to the theory of quasi-Hopf algebras, which is the subject of Chapter 16. ◇

We now give three topological examples of monoidal categories which are important for the relation between quantum groups and knot theory discussed at the end of this chapter, and, in more detail, in Chapter 15.

Example 5.1.7 Choose a plane P^0 in $\mathbb{R}^3$, and fix m distinct points $p_1^0, \ldots, p_m^0$ in P^0. Let P^t be the parallel plane obtained by translating P^0 by a distance t perpendicular to P^0, and let $p_1^1, \ldots, p_m^1$ be the points in P^1 corresponding to $p_1^0, \ldots, p_m^0$. A *braid* β *on* m *strands* is the union of m disjoint piecewise smooth curves in the region X between P^0 and P^1, each of which joins some p_i^0 to some p_j^1, so that $j \mapsto i$ is a permutation of $\{1, \ldots, m\}$. We also assume that, for each $0 \leq t \leq 1$, the plane P^t cuts each curve transversely in exactly one point. Two braids β and β' are said to be *equivalent* if there is a diffeomorphism of X which maps each plane P^t, for $0 \leq t \leq 1$, into itself, which fixes each of the points $p_1^0, \ldots, p_m^0, p_1^1, \ldots, p_m^1$, and which maps β to β'.

Although braids are objects in three-dimensional space, it is useful to have a way of representing them in two dimensions. For this, we choose a plane Q perpendicular to P^0, and project each braid β perpendicularly onto Q. By choosing Q suitably, we can assume that the projection consists of n piecewise smooth curves which intersect transversely, and that every line in Q parallel to $P^0 \cap Q$ meets each curve in the projection exactly once. To be able to recover the equivalence class of β from its projection, one must include an extra piece of information at each point where the projected curves cross. Namely, if we think of the direction perpendicular to Q as 'vertical', then the curve which lies below is shown broken in the projection (see Fig. 4). A similar convention will be used in the other topological examples discussed below.

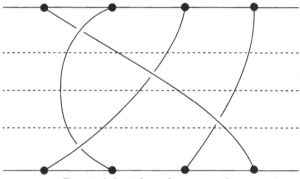

Fig. 4 A braid on four strands

Equivalence classes of braids can be regarded as the morphisms in a category **braid** whose objects are the natural numbers, so that $\mathrm{Hom}(m, n)$ is empty unless $m = n$ in which case it is the set of equivalence classes of braids on m strands. Composition of morphisms is induced by concatenation of braids, i.e. $\beta'\beta$ is obtained by putting β 'on top of' β' and shrinking the vertical coordinate by a factor of two. Moreover, if we define the tensor product $\beta \otimes \beta'$ of two braids by putting them 'side-by-side', with β' to the right of β, then **braid** becomes a monoidal category (0 is the identity object).

It is easy to see that **braid** is rigid. In fact, every object m in **braid** is self-dual, and the morphisms ev_m and π_m are as shown in Fig. 5 (for $m = 3$). $\Diamond$

Remark Although we do not need it at the moment, we should mention that the set of equivalence classes of braids has a natural group structure. The product is defined by concatenation as above, and the inverse of (the equivalence class of) a braid is obtained by reflecting the braid in the plane $P^{1/2}$. The resulting group $\mathcal{B}_m$ has generators $T_1, \ldots, T_{m-1}$ given, with their inverses, by the diagrams

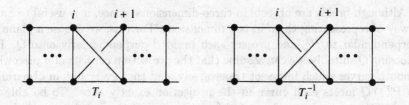

and the expression of an arbitrary braid as a product of these generators can be obtained by dividing the plane projection into horizontal strips so that each strip contains just one crossing point. For example, the braid in Fig. 4 defines the element $T_1^{-1} T_3 T_2 T_1^{-1} \in \mathcal{B}_4$. It is an easy exercise to check that

$$T_i T_j = T_j T_i \qquad \text{if } |i - j| > 1,$$
$$T_i T_j T_i = T_j T_i T_j \qquad \text{if } |i - j| = 1.$$

In fact, these relations define $\mathcal{B}_m$ as an abstract group.

Assigning to a braid the permutation $j \mapsto i$ of its endpoints defines a homomorphism from $\mathcal{B}_m$ to the symmetric group Σ_m on m letters. Its kernel is the *pure braid group* $\mathcal{P}_m$. $\triangle$

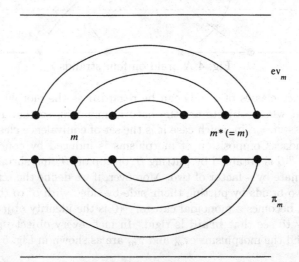

Fig.5 Rigidity of **braid**

Example 5.1.8 A *tangle* is similar to a braid except that the strands may now have both their endpoints on the same plane; one also allows embedded circles, called loops. Equivalence, concatenation and tensor products of tangles are defined in the same way as for braids. In this way, one obtains a monoidal category **tangle** whose objects are the natural numbers and whose morphisms are equivalence classes of tangles. Regarding a braid as a tangle gives a monoidal functor **braid** → **tangle**.

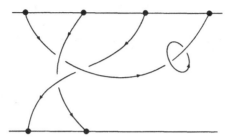

Fig. 6 An element of $\hom(4, 2)$ in **dtangle**

One may also consider *directed tangles*, in which each strand and loop is oriented and concatenation is only allowed when the orientations at the joining points match. To each point $(i, 0, 0)$ (resp. $(i, 0, 1)$), which is an endpoint of a strand, one attaches a sign ϵ_i (resp. ϵ'_i) which is $+1$ or -1 according to whether the oriented tangent vector at the point has negative or positive vertical component.

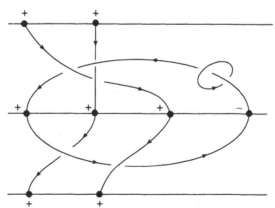

Fig. 7 Concatenation of directed tangles

Equivalence classes of directed tangles (defined in the obvious way) are the morphisms of a monoidal category **dtangle**, whose objects are (finite) words on symbols $+$ and $-$. More precisely, a morphism from $(\epsilon_1, \ldots, \epsilon_m)$ to $(\tilde{\epsilon}_1, \ldots, \tilde{\epsilon}_n)$ is an equivalence class of tangles such that the m ends of the tangle lying in $\mathbb{R} \times \{0\} \times \{1\}$ have signs $\epsilon_1, \ldots, \epsilon_m$ (reading from left to right), while its n ends lying in $\mathbb{R} \times \{0\} \times \{0\}$ have signs $\tilde{\epsilon}_1, \ldots, \tilde{\epsilon}_n$.

The categories of tangles and directed tangles are rigid, with the morphisms ev and π being given by the obvious analogues of Fig. 5 (in the directed case, one must change the signs on passing to the dual).

Forgetting orientation gives a monoidal functor **dtangle** $\rightarrow$ **tangle**. $\Diamond$

Example 5.1.9 A *ribbon tangle* in $\mathbb{R}^3$ is a 'thickened tangle'. More precisely, it is a disjoint union of oriented bands and annuli, where a band is an embedding $[0,1] \times [0,1] \rightarrow \mathbb{R}^3$, an annulus is an embedding $S^1 \times [0,1] \rightarrow \mathbb{R}^3$, and 'orientation' has its usual meaning for surfaces in $\mathbb{R}^3$. The *ends* of a band are the images of $\{0\} \times [0,1]$ and $\{1\} \times [0,1]$. The union of the ends of the bands must be of the form

$$(2) \qquad \bigcup_{i=1}^{m} \left[i - \frac{1}{4}, i + \frac{1}{4} \right] \times \{0\} \times \{0\} \ \cup \ \bigcup_{j=1}^{n} \left[j - \frac{1}{4}, j + \frac{1}{4} \right] \times \{0\} \times \{1\}.$$

Finally, the positively oriented normal must point towards the reader near the end of each band. In pictures, the upper side of each band or annulus is white, the other side shaded.

Fig. 8 A ribbon tangle

Equivalence of ribbon tangles is defined in the obvious way. As in the case of tangles, there is a monoidal category **ribbon** whose objects are the same as in **tangle** and whose morphisms are equivalence classes of ribbon tangles.

The *core* of a ribbon tangle is the union of the images of $[0,1] \times \{\frac{1}{2}\}$ for each band and of $S^1 \times \{\frac{1}{2}\}$ for each annulus. Assigning to a ribbon tangle its core defines a monoidal functor **ribbon** $\rightarrow$ **tangle**.

A *directed ribbon tangle* is a ribbon tangle with a directed core. Directed ribbon tangles are the morphisms in a monoidal category **dribbon** whose objects are the same as in **dtangle**, and with the same rules for concatenation. To each end of each band of a ribbon tangle, one can associate a sign ± 1 by applying the construction in 5.1.8 to its core. Both **ribbon** and **dribbon** are rigid.

For some purposes, it is convenient to consider ribbon tangles with 'free' bands, called *coupons*. We require that any two coupons are disjoint, and that the ends of the coupons do not intersect $\mathbb{R} \times \{0\} \times \{0, 1\}$. The intersection of the bands other than coupons with $\mathbb{R} \times \{0\} \times \{0, 1\}$ is a set of the form (2) as before, and the remaining ends of bands other than coupons are contained in the ends of coupons. We call a collection of bands, annuli and coupons with these properties a *generalized ribbon tangle*. They form a category **ribbon** in the obvious way. Directed generalized ribbon tangles and the category **dribbon** are defined in the same way. ◊

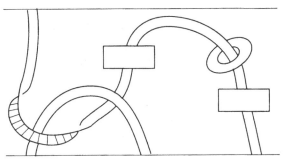

Fig. 9 A generalized ribbon tangle

Example 5.1.10 If X is any set, we can define a category **dribbon**(X) by attaching to each annulus, and to each end of each band, of a ribbon tangle an element of X (a 'colour'). Thus, the objects of **dribbon**(X) are (finite) words, including the empty word, in the elements of the set $X \times \{+, -\}$. A morphism from an object $(x_1, \epsilon_1), \ldots, (x_m, \epsilon_m)$, where the $x_i \in X$ and the $\epsilon_i = \pm$, to an object $(\tilde{x}_1, \tilde{\epsilon}_1), \ldots, (\tilde{x}_n, \tilde{\epsilon}_n)$ is a morphism in **dribbon** from $\epsilon_1, \ldots, \epsilon_m$ to $\tilde{\epsilon}_1, \ldots, \tilde{\epsilon}_n$ (i.e. a ribbon tangle) satisfying the following additional condition. If a band of the ribbon tangle has both ends at the top, i.e. in $\mathbb{R} \times \{0\} \times \{1\}$, these ends being the ith and jth along this line, then $x_i = x_j$; and similarly for bands with both ends in $\mathbb{R} \times \{0\} \times \{0\}$, and for those with one end at the top and one at the bottom.

Colourings of generalized ribbon tangles will be considered in Subsection 5.2B. ◊

E Reconstruction theorems We have seen that, starting from a Hopf algebra A, we can construct a rigid monoidal category. The following result shows essentially that the Hopf algebra can be reconstructed from this categorical data, and gives a rather general instance of Tannaka–Krein duality.

THEOREM 5.1.11 *Let k be a commutative ring, let **C** be an essentially small, k-linear, rigid, abelian, monoidal category, and let $\Phi : \mathbf{C} \to \mathbf{mod}_k$ be a k-linear exact faithful monoidal functor. Then, there exists a Hopf algebra A*

over k and an equivalence of k-linear categories $\mathbf{C} \to \mathbf{corep}_A$ whose composite with the forgetful functor $\mathbf{corep}_A \to \mathbf{mod}_k$ is Φ.

A category is essentially small if it is equivalent to a small category, i.e. one whose class of objects is a set. A functor is faithful if it is injective on sets of morphisms and exact if it takes exact sequences of morphisms to exact sequences.

OUTLINE OF PROOF We describe how the Hopf algebra A is recovered from the categorical data.

If M is a finitely-generated projective k-module, let $\Phi \otimes M$ be the functor $\mathbf{C} \to \mathbf{mod}_k$ given by $U \to \Phi(U) \otimes M$. Let $\mathrm{Nat}(\Phi, \Phi \otimes M)$ be the k-module of natural transformations $\Phi \to \Phi \otimes M$. It can be shown that the functor $M \to \mathrm{Nat}(\Phi, \Phi \otimes M)$ is representable, i.e. there is a k-module A and natural isomorphisms

$$\tau_M : \mathrm{Hom}_k(A, M) \to \mathrm{Nat}(\Phi, \Phi \otimes M).$$

The comultiplication $\Delta : A \to A \otimes A$ and the counit $\epsilon : A \to k$ of A are defined by

$$\Delta = \tau_{A \otimes A}^{-1}((\delta \otimes \mathrm{id}_A)\,\delta), \quad \epsilon = \tau_k^{-1}(e),$$

where $\delta = \tau_A(\mathrm{id}_A)$ and $e : \Phi \to \Phi \otimes k$ is the canonical isomorphism.

Next, define $\Phi^2 : \mathbf{C} \times \mathbf{C} \to \mathbf{mod}_k$ to be the functor $(U, V) \mapsto \Phi(U) \otimes \Phi(V)$ and define $\tau_M^2 : \mathrm{Hom}_k(A \otimes A, M) \to \mathrm{Nat}(\Phi^2, \Phi^2 \otimes M)$ by

$$\tau_M^2(f)_{U,V} = (\mathrm{id}_{\Phi(U)} \otimes \mathrm{id}_{\Phi(V)} \otimes f)(\delta_U^{13} \delta_V^{24}).$$

It can be shown that τ_M^2 is an isomorphism. Let $m_{U,V}$ be the composite

$$\Phi(U) \otimes \Phi(V) \cong \Phi(U \otimes V) \xrightarrow{\delta} \Phi(U \otimes V) \otimes M \cong \Phi(U) \otimes \Phi(V) \otimes M,$$

where the isomorphisms use the fact that Φ is monoidal. Then, m is a natural transformation from Φ^2 to $\Phi^2 \otimes A$, and we define the multiplication $\mu : A \otimes A \to A$ of A to be $(\tau_M^2)^{-1}(m)$.

The unit $\imath : k \to A$ of A is the composite

$$k \cong \Phi(\mathbf{1}) \xrightarrow{\delta_{\mathbf{1}}} \Phi(\mathbf{1}) \otimes A \cong k \otimes A \cong A.$$

Finally, define isomorphisms $\rho_U : \Phi(U)^* \to \Phi(U^*)$ to be the composites

$$\Phi(U)^* \cong \Phi(U)^* \otimes k \xrightarrow{\mathrm{id}_{\Phi(U)^*} \otimes \pi_{\Phi(U)}} \Phi(U)^* \otimes \Phi(U) \otimes \Phi(U^*)$$

$$\downarrow {\mathrm{ev}_{\Phi(U)} \otimes \mathrm{id}_{\Phi(U)^*}}$$

$$k \otimes \Phi(U^*) \cong \Phi(U^*).$$

If $d \in \mathrm{Nat}(\Phi, \Phi \otimes M)$, the morphism $d_U : \Phi(U) \to \Phi(U) \otimes M$ corresponds, via ρ, to a morphism $\bar{d}_U : \Phi(U)^* \to \Phi(U)^* \otimes M$. In view of the canonical isomorphisms

$$\mathrm{Hom}_k(\Phi(U), \Phi(U) \otimes M) \cong \mathrm{Hom}_k(\Phi(U)^*, \Phi(U)^* \otimes M),$$

the assignment $d \mapsto \bar{d}$ defines a map from $\mathrm{Nat}(\Phi, \Phi \otimes M)$ to itself, and hence also a map from $\mathrm{Hom}_k(A, M)$ to itself. These maps are natural in M and k-linear, and so correspond to a k-linear map $S : A \to A$. The invertibility of S follows from the fact that $U \mapsto U^*$ is an anti-equivalence of categories.

It is not difficult to show that $(A, \imath, \mu, \epsilon, \Delta)$ is a Hopf algebra over k, and that $U \mapsto (\Phi(U), \delta_U)$ is an equivalence of categories. ∎

Remark In view of this theorem, it is natural to ask what further conditions on **C** will guarantee that it is equivalent to the category of representations (or corepresentations) of a Hopf algebra of some special type. When k is a field of characteristic zero, Deligne (1991) gives sufficient conditions for **C** to be the category of representations of a proalgebraic group scheme. On the other hand, Gel'fand & Kazhdan (1992) show that Deligne's theorem fails if k has finite characteristic (see also Andersen (1992c)). Their counterexamples are closely related to a category of quantum group representations discussed in Section 11.3. △

5.2 Quasitensor categories

In many monoidal categories, such as the categories of vector spaces and of group representations, the tensor product is commutative up to isomorphism. In this section, we incorporate this symmetry into the categorical framework which we introduced in the preceding section.

A Tensor categories We begin with

DEFINITION 5.2.1 *A* tensor category *is a monoidal category* **C** *which is equipped with natural isomorphisms* $\sigma_{U,V} : U \otimes V \to V \otimes U$, *for all objects U and V of* **C**, *such that*

(i) the diagrams

$$
\begin{array}{ccccc}
U \otimes (V \otimes W) & \xrightarrow{\alpha} & (U \otimes V) \otimes W & \xrightarrow{\sigma} & W \otimes (U \otimes V) \\
{\scriptstyle \mathrm{id}\otimes\sigma}\downarrow & & & & \downarrow{\scriptstyle \alpha} \\
U \otimes (W \otimes V) & \xrightarrow[\alpha]{} & (U \otimes W) \otimes V & \xrightarrow[\sigma\otimes\mathrm{id}]{} & (W \otimes U) \otimes V
\end{array}
$$

$$(U \otimes V) \otimes W \xrightarrow{\sigma \otimes \mathrm{id}} (V \otimes U) \otimes W \xrightarrow{\alpha^{-1}} V \otimes (U \otimes W)$$

$$\alpha^{-1} \downarrow \qquad\qquad\qquad\qquad\qquad\qquad \downarrow \mathrm{id} \otimes \sigma$$

$$U \otimes (V \otimes W) \xrightarrow[\sigma]{} (V \otimes W) \otimes U \xrightarrow[\alpha^{-1}]{} V \otimes (W \otimes U)$$

commute for all U, V, W;

(ii) the diagrams

$$1 \otimes U \xrightarrow{\sigma} U \otimes 1 \qquad U \otimes 1 \xrightarrow{\sigma} 1 \otimes U$$

$$\lambda \downarrow \qquad\qquad \downarrow \rho \qquad\qquad \rho \downarrow \qquad\qquad \downarrow \lambda$$

$$U \xrightarrow[\mathrm{id}]{} U \qquad\qquad U \xrightarrow[\mathrm{id}]{} U$$

commute for all U;

(iii) $\sigma_{V,U} \circ \sigma_{U,V} = \mathrm{id}_{U \otimes V}$ for all U, V.

If $\mathbf{C}$ is abelian (resp. k-linear), σ is required to be bi-additive (resp. k-bilinear).

A monoidal functor $\Phi : \mathbf{C} \to \mathbf{C}'$ between tensor categories is a tensor functor if the diagram

$$\Phi(U) \otimes \Phi(V) \xrightarrow{\varphi} \Phi(U \otimes V)$$

$$\sigma' \downarrow \qquad\qquad\qquad \downarrow \Phi(\sigma)$$

$$\Phi(V) \otimes \Phi(U) \xrightarrow[\varphi]{} \Phi(V \otimes U)$$

commutes for all U, V.

Remark It is easy to see, by using condition (iii), that if one of the diagrams in (i) or (ii) commutes so does the other. We have included these redundant conditions partly to make the definition more symmetric, and partly for convenience later. $\triangle$

Clearly, $\mathbf{mod}_k$ is a tensor category with the obvious commutativity maps $\sigma_{U,V} : U \otimes V \to V \otimes U$ given by $\sigma_{U,V}(u \otimes v) = v \otimes u$; the associativity maps are the identity maps. The same is true (with the same maps) for $\mathbf{rep}_A$ for any *cocommutative* Hopf algebra A, and the forgetful functor $\mathbf{rep}_A \to \mathbf{mod}_k$ is a tensor functor. More generally, we have

Example 5.2.2 Let $(A, \mathcal{R})$ be a *triangular* Hopf algebra over a commutative ring k. Then $\mathbf{rep}_A$ is still a tensor category, but the commutativity maps $\sigma_{U,V}$

are no longer the obvious ones, which would not commute with the action of A in general. If $\rho_U : A \to \text{End}(U)$ and $\rho_V : A \to \text{End}(V)$ are representations of A, the correct choice is

$$\sigma_{U,V} = \sigma \circ (\rho_U \otimes \rho_V)(\mathcal{R}).$$

Let us verify that $\sigma_{U,V}$ is a morphism in $\textbf{rep}_A$. Indeed, if $a \in A$, $u \in U$, $v \in V$,

$$a.(\sigma_{U,V}(u \otimes v)) = \Delta(a)\mathcal{R}_{21}.(v \otimes u).$$

Using the equation $\Delta^{\text{op}}(a) = \mathcal{R}\Delta(a)\mathcal{R}^{-1}$, we obtain $\Delta(a)\mathcal{R}_{21} = \mathcal{R}_{21}\Delta^{\text{op}}(a)$. Hence,

$$\begin{aligned} a.(\sigma_{U,V}(u \otimes v)) &= \mathcal{R}_{21}\Delta^{\text{op}}(a).(v \otimes u) \\ &= \sigma_{U,V}(a.(u \otimes v)). \end{aligned}$$

The commutativity constraint $\sigma_{V,U} \circ \sigma_{U,V} = \text{id}_{U \otimes V}$ follows immediately from $\mathcal{R}_{21}\mathcal{R} = 1$. Finally, going one way round the first hexagon diagram in 5.2.1(i),

$$\alpha_{W,U,V} \circ \sigma_{U \otimes V,W} \circ \alpha_{U,V,W}(u \otimes v \otimes w) = \sigma_{123}(\Delta \otimes \text{id})(\mathcal{R}).(u \otimes v \otimes w),$$

where σ_{123} is the cyclic permutation of the factors. Going the other way round,

$$\begin{aligned} (\sigma_{U,W} \otimes \text{id}_V) &\circ \alpha_{U,W,V} \circ (\text{id}_U \otimes \sigma_{V,W})(u \otimes v \otimes w) \\ &= \sigma_{12}\mathcal{R}_{12}\sigma_{23}\mathcal{R}_{23}.(u \otimes v \otimes w) = \sigma_{123}\mathcal{R}_{13}\mathcal{R}_{23}.(u \otimes v \otimes w). \end{aligned}$$

Thus, the hexagon axiom follows from (10) in Chapter 4 (note that (10) and (11) are equivalent in the triangular case). $\Diamond$

Example 5.2.3 The category $\textbf{bun}_G$ is tensor if and only if G is abelian (we take the associativity maps to be identity maps). Indeed, for any $g \in G$, let $E(g)$ be the object of $\textbf{bun}_G$ such that $E(g)_{g'}$ is $\mathbb{C}$ if $g' = g$, and zero if $g' \neq g$. Then, if g_1 and g_2 are elements of G which do not commute, the only non-zero component of $E(g_1) \otimes E(g_2)$ is the (g_1g_2)th, so $E(g_1) \otimes E(g_2)$ is not isomorphic to $E(g_2) \otimes E(g_1)$.

On the other hand, if G is abelian, we can take $(\sigma_{U,V})_g$ to be the obvious map of vector spaces

$$(U \otimes V)_g = \bigoplus_{h \in G} U_h \otimes V_{h^{-1}g} \to \bigoplus_{h \in G} V_{gh^{-1}} \otimes U_h = (V \otimes U)_g,$$

in which the arrow is the obvious flip map. This is not, however, the only possible choice of $\sigma_{U,V}$. Let us take the simplest non-trivial case when G is the group $\mathbf{Z}_2 = \{0, 1\}$ with two elements. Since $\mathbf{bun}_{\mathbf{Z}_2}$ is semisimple, and since $E(0) = \mathbf{1}$ and $E(1) = X$ (say) are, up to isomorphism, the only simple objects of $\mathbf{bun}_{\mathbf{Z}_2}$, it suffices by condition 5.2.1(ii) to define $\sigma_{X,X}$; note that $X \otimes X \cong \mathbf{1}$. There are two possibilities:

(a) $\sigma_{X,X} = \mathrm{id}_{\mathbf{1}}$;
(b) $\sigma_{X,X} = -\mathrm{id}_{\mathbf{1}}$.

In case (a), $\mathbf{bun}_{\mathbf{Z}_2}$ is the category of $\mathbf{Z}_2$-graded vector spaces; in case (b), it is the category of super vector spaces. $\Diamond$

B Quasitensor categories In view of the notion of a quasitriangular Hopf algebra introduced in 4.2.6, Example 5.2.2 strongly suggests that one should consider the effect of dropping condition (iii) in 5.2.1.

DEFINITION 5.2.4 *A quasitensor category is a monoidal category* $\mathbf{C}$ *equipped with functorial isomorphisms* $\sigma_{U,V} : U \otimes V \to V \otimes U$ *which satisfy conditions (i) and (ii) in 5.2.1.*

Examples 5.2.5 The definition is designed, of course, so that the category of representations of any quasitriangular Hopf algebra $(A, \mathcal{R})$ on free modules of finite rank is a quasitensor category. The commutativity isomorphisms are defined in the same way as in the triangular case (see 5.2.2). One needs (11) from Chapter 4 to verify commutativity of the second hexagon diagram. In fact, one can weaken 'Hopf' to 'quasi-Hopf', as we shall see in Chapter 16.

All of the geometric categories **braid**, **tangle** and **ribbon**, as well as their directed versions, are quasitensor categories (but not tensor categories). The switch map $\sigma_{X,Y}$ in **ribbon**, for example, is shown schematically in Fig. 10. The verification of the axioms of a quasitensor category will be an entertaining exercise for the reader. $\Diamond$

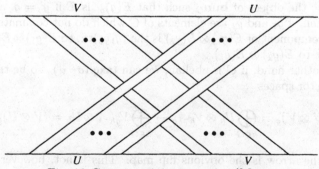

Fig. 10 Commutativity maps in **ribbon**

Remark If **C** is a rigid tensor (or quasitensor) category, then, for all objects U of **C**, there is a morphism $\jmath_U : U \to U^{**}$ corresponding to $\mathrm{ev}_U \circ \sigma_{U,U^*}$: $U \otimes U^* \to \mathbf{1}$ (see the remarks following 5.1.2). If $(A, \mathcal{R})$ is a quasitriangular Hopf algebra and U is an object of **rep**$_A$, we find that

$$\langle \jmath_U(x), \xi \rangle = \langle \xi, u.x \rangle, \qquad (x \in U, \ \xi \in U^*),$$

where $u = \mu(S \otimes \mathrm{id})(\mathcal{R}_{21})$. $\triangle$

The family of isomorphisms $\sigma_{U,V}$ is called the *braiding* of **C** (and quasitensor categories are often called *braided monoidal categories*). In the tensor case, the braiding is often called the *symmetry* of the category. The reason for this terminology is as follows.

Let us assume for simplicity that **C** is strict. Let U, V and W be objects of **C**. There are six possible tensor products of these objects, with each object occurring once, as shown in the following diagram:

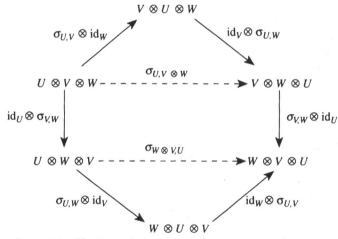

It follows from 5.2.1(i) that the top and bottom triangles commute, and the commutativity of the middle rectangle follows from the naturality of the braiding. If we write σ_1 (resp. σ_2) for the braiding applied to the first two (resp. last two) factors in a triple tensor product, we have shown that

$$\sigma_1 \sigma_2 \sigma_1 = \sigma_2 \sigma_1 \sigma_2.$$

This is the defining relation of the braid group $\mathcal{B}_3$ (cf. 5.1.7); if **C** is actually a tensor category, then $\sigma_1^2 = \sigma_2^2 = 1$, and we have the defining relations of the symmetric group Σ_3. Note that the pure braid group $\mathcal{P}_3$ acts on $U \otimes V \otimes W$, but that a general element $\beta \in \mathcal{B}_3$ takes $U \otimes V \otimes W$ to the tensor product of U, V and W in some other order, determined by the element of Σ_3 associated to β.

C Balancing The morphisms $\jmath_U : U \to U^{**}$, defined in the preceding remark for all objects U of any rigid quasitensor category $\mathbf{C}$, do not, in general, respect the tensor product. However, under the following additional condition, these morphisms can be modified to become monoidal.

DEFINITION 5.2.6 *A rigid quasitensor category $\mathbf{C}$ is* balanced *if it is equipped with an automorphism b_U of U, for every object U of $\mathbf{C}$, such that the following conditions hold:*

(3) $$b_U \otimes b_V = \sigma_{V,U} \circ \sigma_{U,V} \circ b_{U \otimes V},$$

(4) $$b_{U^*} = (b_U)^*,$$

(5) $$b_\mathbf{1} = \mathrm{id}_\mathbf{1}.$$

Of course, if $\mathbf{C}$ is actually a tensor category, we can take $b_U = \mathrm{id}_U$ for all U.

It is easy to see that, in any balanced rigid quasitensor category, the modified morphisms $k_U = \jmath_U \circ b_U^{-1}$ satisfy

(6) $$k_{U \otimes V} = k_U \otimes k_V$$

(we use the identification $(U \otimes V)^{**} \cong U^{**} \otimes V^{**}$ given by Remark [2] following 5.1.2).

Example 5.2.7 Let $(A, \mathcal{R}, v)$ be a ribbon Hopf algebra (see 4.2.8), and let $\mathbf{C} = \mathrm{rep}_A$. If $\rho_U : A \to \mathrm{End}(U)$ is a finite-dimensional representation of A, we set $b_U = \rho_U(v)$. Since v is central, b_U commutes with the action of A on U. Then, (3), (4) and (5), respectively, follow from the following defining properties of v:

$$\Delta(v) = (\mathcal{R}_{21}\mathcal{R}_{12})^{-1}(v \otimes v),$$
$$S(v) = v,$$
$$\epsilon(v) = 1.$$

The morphisms $k_U : U \to U^{**}$ are given by

$$k_U = \rho_U(g),$$

where $g = uv^{-1}$. As we noted in Subsection 4.2C, the element g is group-like, so property (6) is clear in this case. $\Diamond$

D Quasitensor categories and fusion rules Fusion rules are a new kind of algebraic structure, which can be interpreted as giving the 'structure constants' of a quasitensor category.

DEFINITION 5.2.8 *Let Λ be a set equipped with an involution $\lambda \mapsto \lambda^*$ and a distinguished element ω such that $\omega^* = \omega$. A set of fusion rules indexed by Λ is a collection $\{N_{\lambda\mu,\nu}\}_{\lambda,\mu,\nu \in \Lambda}$ of non-negative integers satisfying the following conditions:*

(i) for each λ, $\mu \in \Lambda$, $N_{\lambda\mu,\nu} = 0$ for all except finitely many ν;
(ii) $N_{\lambda\mu,\nu} = N_{\mu\lambda,\nu}$;
(iii) $\sum_{\alpha \in \Lambda} N_{\lambda\alpha,\beta} N_{\mu\nu,\alpha} = \sum_{\alpha \in \Lambda} N_{\lambda\mu,\alpha} N_{\alpha\nu,\beta}$;
(iv) $N_{\lambda\omega,\mu} = N_{\omega\lambda,\mu} = \delta_{\lambda,\mu}$;
(v) $N_{\lambda\mu,\nu} = N_{\mu^*\lambda^*,\nu^*}$;
(vi) $N_{\lambda\mu^*,\omega} = \delta_{\lambda,\mu}$.

Before we see the connection between fusion rules and quasitensor categories, we shall re-interpret the definition. Let $\mathcal{F}$ be the free abelian group with basis a set of formal symbols $\{f_\lambda\}_{\lambda \in \Lambda}$. Define a bi-additive product $\mathcal{F} \times \mathcal{F} \to \mathcal{F}$ by setting

$$f_\lambda . f_\mu = \sum_{\nu \in \Lambda} N_{\lambda\mu,\nu} f_\nu.$$

Condition (i) asserts that this is well defined, and conditions (ii) and (iii) that the product is commutative and associative. Thus, $\mathcal{F}$ is a commutative ring and, by condition (iv), f_ω is its identity element. We call $\mathcal{F}$ the *fusion ring* associated to the fusion rules $N_{\lambda\mu,\nu}$.

To interpret the last two conditions in 5.2.8, extend the involution of Λ to an additive involution $*$ of $\mathcal{F}$ such that $f_\lambda^* = f_{\lambda^*}$, and define an additive map ev : $\mathcal{F} \to \mathbb{Z}$ by setting $\mathrm{ev}(f_\lambda) = \delta_{\lambda,\omega}$. Then, (v) asserts that $\mathrm{ev}(f^*) = \mathrm{ev}(f)$ for all $f \in \mathcal{F}$, and (vi) that $\mathrm{ev}(f_\omega) = 1$. Moreover, setting

$$(f, f') = \mathrm{ev}(f^* f')$$

defines a bi-additive inner product $(,) : \mathcal{F} \times \mathcal{F} \to \mathbb{Z}$ such that

$$(ff', f'') = (f', f^* f'') \qquad (f, f', f'' \in \mathcal{F}),$$

and for which $\{f_\lambda\}$ is an orthonormal basis.

Remark The fusion ring has a natural representation, defined as follows. Let $\mathbf{N}_\lambda$, the *fusion matrices*, be the $\Lambda \times \Lambda$ matrices given by

$$(\mathbf{N}_\lambda)_{\mu\nu} = N_{\lambda\mu,\nu}.$$

Then, condition (iii) is equivalent to

$$\mathbf{N}_\lambda \mathbf{N}_\mu = \sum_{\nu \in \Lambda} N_{\lambda\mu,\nu} \mathbf{N}_\nu,$$

so that $f_\lambda \mapsto \mathbf{N}_\lambda$ is a matrix representation of $\mathcal{F}$. Since $\mathcal{F}$ is commutative,

$$\mathbf{N}_\lambda \mathbf{N}_\mu = \mathbf{N}_\mu \mathbf{N}_\lambda$$

for all $\lambda,\ \mu \in \Lambda$.

Note further that

$$\mathbf{N}_{\lambda^*} = \mathbf{N}_\lambda^t.$$

In fact,

$$(\mathbf{N}_{\lambda^*})_{\mu\nu} = N_{\lambda^*\mu,\nu} = (f_{\lambda^*}f_\mu,\ f_\nu) = (f_\mu,\ f_\lambda f_\nu) = N_{\lambda\mu,\nu} = (\mathbf{N}_\lambda)_{\nu\mu}.$$

In particular, if the involution $*$ is the identity map, the fusion matrices are symmetric. $\triangle$

It is time we gave some examples of fusion rules. The following example is very familiar.

Example 5.2.9 Let G be a compact topological group, and let Λ be the set of isomorphism classes of finite-dimensional irreducible complex representations of G. Since every representation of G is completely reducible (because it admits an invariant inner product, for example), if $U, V \in \Lambda$, we can write

$$U \otimes V \cong \bigoplus_W W^{\oplus N_{UV,W}},$$

for some non-negative integers $N_{UV,W}$. Further, let $V^* = \mathrm{Hom}(V, \mathbb{C})$ be the dual of V (with the usual action of G), and let ω be the one-dimensional trivial representation. Then, $\{N_{UV,W}\}$ is a set of fusion rules. In fact, properties (i) and (iv) in the definition are obvious; (ii) and (iii) follow from the commutativity and associativity of the tensor product; (v) follows from the fact that taking the dual commutes with the tensor product; and (vi) follows from the equality between the multiplicity of the trivial representation in $U \otimes V^*$ and the dimension of the space of G-module maps $V \to U$. $\Diamond$

It is clear that the fusion ring in this example is the representation ring of G, i.e. the Grothendieck ring of the category of finite-dimensional complex representations of G. In this form, the example has the following generalization.

PROPOSITION 5.2.10 *Let k be a commutative ring, and let $\mathbf{C}$ be a rigid, semisimple, k-linear, quasitensor category. Assume that the identity object of $\mathbf{C}$ is simple, and that $\mathrm{Hom}_{\mathbf{C}}(X, X) \cong k$ for all simple objects X of $\mathbf{C}$. Then, the Grothendieck ring $\mathrm{Gr}(\mathbf{C})$ is a fusion ring. Moreover, every fusion ring arises in this way.*

Remark Note that the assumption on $\mathrm{Hom}_{\mathbf{C}}(X, X)$ is satisfied in the category of finite-dimensional representations of any algebra defined over an algebraically closed field (by Schur's lemma).

PROOF Let $\mathbf{C}$ be as in the proposition, and let $\{X_\lambda\}_{\lambda \in \Lambda}$ be the set of (isomorphism classes of) simple objects of $\mathbf{C}$; we suppose that $X_\omega = \mathbf{1}$. Define $*: \Lambda \to \Lambda$ by $X_{\lambda^*} \cong X_\lambda^*$.

Since $\mathbf{C}$ is semisimple,

(7)
$$X_\lambda \otimes X_\mu \cong \bigoplus_\nu X_\nu^{\oplus N_{\lambda\mu,\nu}},$$

for some $N_{\lambda\mu,\nu} \in \mathbf{N}$. Note that $\mathrm{Hom}_{\mathbf{C}}(X_\nu, X_\lambda \otimes X_\mu) \cong k^{N_{\lambda\mu,\nu}}$. Property (i) in 5.2.8 is obvious; (ii) and (iii) follow from the fact that the tensor product is commutative and associative, up to isomorphism; and (iv) follows from the defining property 5.1.1(ii) of the identity object. Finally, (v) and (vi) are consequences of the remarks following 5.1.2, which imply that $(X_\lambda \otimes X_\mu)^* \cong X_\mu^* \otimes X_\lambda^*$ and that there is an isomorphism of k-modules $\mathrm{Hom}_{\mathbf{C}}(\mathbf{1}, X_\lambda \otimes X_\mu^*) \cong \mathrm{Hom}_{\mathbf{C}}(X_\mu, X_\lambda)$.

For the converse, let $\{N_{\lambda\mu,\nu}\}_{\lambda,\mu,\nu \in \Lambda}$ be a set of fusion rules. We define a category $\mathbf{C}$ as follows. The objects of $\mathbf{C}$ are the maps $X: \Lambda \to \mathbf{N}$ such that $X(\lambda) = 0$ for all but finitely many λ. If X and Y are two objects of $\mathbf{C}$, we set
$$\mathrm{Hom}_{\mathbf{C}}(X, Y) = \bigoplus_\lambda \mathrm{Hom}_k(k^{X(\lambda)}, k^{Y(\lambda)}).$$

Composition of morphisms is defined in the obvious way. The direct sum of two objects X and Y is
$$(X \oplus Y)(\lambda) = X(\lambda) + Y(\lambda) \qquad (\lambda \in \Lambda).$$

Then, $X = \oplus_\lambda X_\lambda^{\oplus X(\lambda)}$, where the object X_λ maps λ to one and all other elements of Λ to zero. We define $X_\lambda^* = X_{\lambda^*}$ and $X_\lambda \otimes X_\mu$ by (7), and extend additively to give the dual and tensor product of arbitrary objects of $\mathbf{C}$. The identity object is X_ω.

We leave it to the reader to check that these definitions make $\mathbf{C}$ into a rigid, k-linear, quasitensor category. It is obviously semisimple, with the X_λ being the simple objects. ∎

E Quasitensor categories in quantum field theory We conclude this section with a simplistic account of how quasitensor categories arise in quantum field theory. For a more serious discussion of the physical background, see Haag (1992) and Fröhlich & Kerler (1993), for example.

Traditionally, observables in relativistic quantum field theory are 'operator-valued distributions' $J(x)$ defined on Minkowski spacetime M (thus, $M = \mathbf{R}^n$ equipped with an inner product of signature $(+ - \cdots -)$). This means that, if f is a smooth real-valued function on M of compact support, then
$$J(f) = \int_M J(x) f(x) \, dx$$

should be a bounded operator on some separable Hilbert space. The von Neumann algebra $\mathcal{A}(\mathcal{O})$ of observables localized in a bounded subset $\mathcal{O} \subset M$ then consists of all bounded functions of operators of the form $J(f)$ for which supp$(f) \subset \mathcal{O}$. The local algebras $\mathcal{A}(\mathcal{O}_1)$ and $\mathcal{A}(\mathcal{O}_2)$ associated to spacelike-separated regions $\mathcal{O}_1$ and $\mathcal{O}_2$ commute (recall that $\mathcal{O}_1$ and $\mathcal{O}_2$ are 'spacelike-separated' if $\| x_1 - x_2 \|^2 < 0$ for all $x_1 \in \mathcal{O}_1$, $x_2 \in \mathcal{O}_2$). If S is any, not necessarily bounded, subset of M, one sets

$$\mathcal{A}(S) = \overline{\bigcup_{\mathcal{O}} \mathcal{A}(\mathcal{O})},$$

the union being over all bounded sets $\mathcal{O} \subset S$, and the closure being in the operator norm topology. In particular, $\mathcal{A} = \mathcal{A}(M)$ is the algebra of all localized observables of the theory. Since the theory is to be relativistic, there should be a representation of the Poincaré group $\mathcal{G}$ (i.e. the group of isometries of M) by $*$-automorphisms of $\mathcal{A}$ such that

$$g.\mathcal{A}(S) = \mathcal{A}(g.S)$$

for all $g \in \mathcal{G}$, $S \subset M$.

In order to qualify as a 'local quantum field theory', these data are required to satisfy a number of axioms which imply, among other things, that there is a distinguished $*$-representation $\rho_{\text{vac}} : \mathcal{A} \to \text{End}(\mathcal{H})$ on a Hilbert space $\mathcal{H}$, called the *vacuum representation*, which is both faithful and irreducible. All $*$-representations ρ of $\mathcal{A}$ which have a physical meaning have the following properties:

(i) ρ is unitarily equivalent to ρ_{vac} when restricted to $\mathcal{A}(\mathcal{O}')$, for some bounded region $\mathcal{O}$ (possibly depending on ρ); here, $\mathcal{O}'$ denotes the set of points of M which are spacelike-separated from every point of $\mathcal{O}$. In particular, the Hilbert space of ρ can be identified with $\mathcal{H}$.

(ii) ρ is *covariant*, in the sense that there is a unitary representation π of $\mathcal{G}$ on $\mathcal{H}$ such that

$$\rho(g.A) = \pi(g)\rho(A)\pi(g)^{-1}$$

for all $g \in \mathcal{G}$, $A \in \mathcal{A}$.

Representations satisfying these conditions are called *physical representations*.

The axioms further imply that, if ρ is a physical representation (whose Hilbert space we may as well assume is $\mathcal{H}$), there exists a $*$-endomorphism φ of $\mathcal{A}$ such that

$$\rho \cong \rho_{\text{vac}} \circ \varphi,$$

where $\cong$ means 'unitarily equivalent'. Moreover, the set of $*$-endomorphisms which arise in this way is closed under taking composites. It is this last property which is crucial for us, for it implies that, from any two irreducible physical representations $\rho_1 = \rho_{\text{vac}} \circ \varphi_1$ and $\rho_2 = \rho_{\text{vac}} \circ \varphi_2$, we may construct a third, $\rho_{\text{vac}} \circ \varphi_1 \circ \varphi_2$. We shall write the latter representation as $\rho_1 \overline{\times} \rho_2$. The following properties of $\overline{\times}$ follow from the axioms:

(P1) if ρ and σ are irreducible physical representations, then

$$\rho \overline{\times} \sigma \cong \sigma \overline{\times} \rho\,;$$

(P2) there are non-negative integers $N_{\rho\sigma,\tau}$ such that

$$\rho \overline{\times} \sigma \cong \bigoplus_\tau \tau^{\oplus N_{\rho\sigma,\tau}},$$

where ρ, σ and τ run through the set of irreducible physical representations;

(P3) for each irreducible physical representation ρ, there exists a physical representation ρ^* such that $\rho \overline{\times} \rho^*$ contains the vacuum representation exactly once (i.e. $N_{\rho\rho^*,\rho_{\text{vac}}} = 1$).

These data may be encoded in a category **QFT** whose objects are the physical representations of the theory, the set of morphisms from ρ_1 to ρ_2 being, in the above notation,

$$\text{Hom}_{\textbf{QFT}}(\rho_1, \rho_2) = \{A \in \mathcal{A} \mid A\varphi_1(B) = \varphi_2(B)A \text{ for all } B \in \mathcal{A}\}.$$

The operation $\overline{\times}$ makes **QFT** into a rigid monoidal category: the tensor product of morphisms $A \in \text{Hom}_{\textbf{QFT}}(\rho_1, \rho_2)$, $A' \in \text{Hom}_{\textbf{QFT}}(\rho_1', \rho_2')$ is

$$A \overline{\times} A' = A\varphi_1(A')$$

(it is easy to check that $A \overline{\times} A'$ is in $\text{Hom}_{\textbf{QFT}}(\rho_1 \overline{\times} \rho_1', \rho_2 \overline{\times} \rho_2')$). From property (P1) above, **QFT** is actually quasitensor. Note that **QFT** is also strict.

With this background in place, we can state the following fundamental theorem of Doplicher & Roberts (1990).

THEOREM 5.2.11 *Associated to every local quantum field theory in four or more spacetime dimensions is a compact group G whose representation ring is isomorphic to the fusion ring of the theory.* ∎

The discussion preceding 5.2.11 applies to two-dimensional conformal field theories, if one replaces spacetime M by the circle S^1, thought of as a compactified light ray, and the Poincaré group by the group $\text{PSL}_2(\mathbb{R})$ of Möbius transformations of S^1 ('spacelike-separated' should be interpreted as 'disjoint') (see Fröhlich & Gabbiani (1992)). However, the theorem of Doplicher & Roberts is no longer valid:

Example 5.2.12 One of the simplest conformal field theories is that which describes 'the scaling limit of the Ising model at the critical point'. This has been worked out in detail by Mack & Schomerus (1990). They find three physical representations ρ_0, $\rho_{1/2}$ and ρ_1, with ρ_0 being the vacuum representation. The fusion rules are given by:

$$\rho_{1/2} \overline{\times} \rho_{1/2} \cong \rho_0 \oplus \rho_1, \quad \rho_{1/2} \overline{\times} \rho_1 \cong \rho_1 \overline{\times} \rho_{1/2} \cong \rho_{1/2}, \quad \rho_1 \overline{\times} \rho_1 \cong \rho_0,$$

$$\rho_0 \overline{\times} \rho_s \cong \rho_s \overline{\times} \rho_0 \cong \rho_s, \quad \text{for } s = 0, \frac{1}{2}, 1.$$

The quasitensor category associated to this theory certainly cannot be equivalent to the category of finite-dimensional representations of anything for which the tensor product has its usual meaning on the level of vector spaces. For, comparing dimensions in the above isomorphisms would lead us immediately to the absurd conclusion that $\dim(\rho_{1/2}) = \sqrt{2}$.

Nevertheless, we shall see in 11.3.22 that the fusion ring of the category arising from this example is isomorphic to that of the category of representations of a quantum group, provided with a certain 'truncated' tensor product. Moreover, the quantum dimension of the representation corresponding to $\rho_{1/2}$ (in the sense of 4.2.9) is indeed equal to $\sqrt{2}$! $\Diamond$

Although it does not quite fit into the above discussion of categories and quantum field theories, we include the following example, which shows that a fusion ring is essentially the same thing as a $(1+1)$-dimensional topological quantum field theory (TQFT) in the sense of Witten (1988b) (see also Atiyah (1988)).

Example 5.2.13 A $(1+1)$-dimensional TQFT associates to each compact oriented 1-manifold X (i.e. a disjoint union of circles, possibly empty) a complex vector space V_X, and to each oriented 2-manifold with boundary Y a vector $v_Y \in V_{\partial Y}$. We shall not write down all the axioms which the v_Y and V_X must satisfy, but we note that $V_\emptyset = \mathbb{C}$, that if X^* denotes X with the opposite orientation, then $V_{X^*} = (V_X)^*$, and that $V_{X_1 \coprod X_2} = V_{X_1} \otimes V_{X_2}$. Thus, a surface Y with $\partial Y = X_1^* \coprod X_2$ defines a linear map $v_Y : V_{X_1} \to V_{X_2}$.

It is not hard to see that giving such a theory is equivalent to giving a unital, commutative, associative algebra A together with a linear form ev : $A \to \mathbb{C}$ such that the bilinear form $(a, a') \mapsto \text{ev}(aa')$ on A is non-degenerate (this remark is due to R. H. Dijkgraaf (1989) and G. B. Segal). In fact, A is the vector space associated to a single circle (with a definite orientation), and the multiplication, unit element and linear form correspond to the diagrams shown in Fig. 11.

It is clear from our earlier discussion that the complexification $\mathcal{F}^{\mathbb{C}} = \mathcal{F} \underset{\mathbf{Z}}{\otimes} \mathbb{C}$ of a fusion ring $\mathcal{F}$ for which the involution $*$ is the identity map is an algebra satisfying these conditions. $\Diamond$

Fig. 11 Multiplication, unit element and linear form in a TQFT

5.3 Invariants of ribbon tangles

In this section, we consider ribbon tangles coloured with the set X of representations of a ribbon Hopf algebra A. We construct a functor from **dribbon**(X) to **rep**$_A$. This amounts to associating to each coloured ribbon tangle an isotopy invariant which is a homomorphism between representations of A.

A Isotopy invariants and monoidal functors Let A be a ribbon Hopf algebra over a commutative ring k, and let X be the set of objects of the category **rep**$_A$. To construct a functor **dribbon**$(X) \to$ **rep**$_A$, we need the following topological result, which gives a description of the morphisms in **dribbon**(X) in terms of 'generators and relations'. It is a consequence of classical work due to Reidemeister (1932) and others (see Subsection 15.1A).

PROPOSITION 5.3.1 *Every morphism in* **dribbon**(X) *can be obtained from the morphisms in Fig. 12 by taking compositions and tensor products. The relations between morphisms in* **dribbon**(X) *shown in Fig. 13 hold, and every relation is a consequence of them.* ∎

We can now state the main result of this section.

THEOREM 5.3.2 *Let A be a ribbon Hopf algebra defined over a commutative ring k, with universal R-matrix $\mathcal{R}$ and distinguished central element v, and let X be the set of objects of* **rep**$_A$. *There is a unique monoidal functor $\mathcal{F} :$ **dribbon**$(X) \to$ **rep**$_A$ *such that*

(i) on objects, $\mathcal{F}$ is given by

$$\mathcal{F}((V_1, \epsilon_1), \ldots, (V_n, \epsilon_n)) = V_1^{\epsilon_1} \otimes \cdots \otimes V_n^{\epsilon_n},$$

where $V^+ = V$ and $V^- = V^*$ ($\mathcal{F}$ takes the empty word to k);

(ii) on generating morphisms, $\mathcal{F}$ is given as follows (the notation is as in

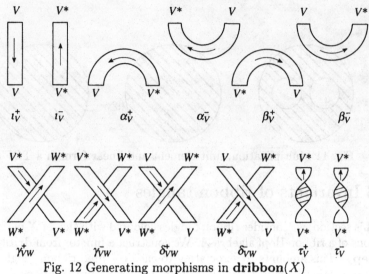

Fig. 12 Generating morphisms in **dribbon**(X)

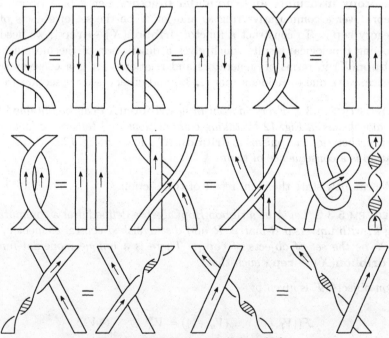

Fig. 13 Relations in **dribbon**(X)

Fig. 12):

$$\mathcal{F}(\iota_V^+) = \mathrm{id}_V, \quad \mathcal{F}(\iota_V^-) = \mathrm{id}_{V^*},$$

$$\mathcal{F}(\alpha_V^+) = \pi_V : k \to V \otimes V^*, \quad \mathcal{F}(\alpha_V^-) = \mathrm{ev}_V : V^* \otimes V \to k,$$

$$\mathcal{F}(\beta_V^+) = \tilde{\pi}_V : k \to V^* \otimes V, \quad \mathcal{F}(\beta_V^-) = \tilde{\mathrm{ev}}_V : V \otimes V^* \to k,$$

$$\mathcal{F}(\gamma_{VW}^+) = \sigma \circ R_{V \cdot W \cdot}, \quad \mathcal{F}(\gamma_{VW}^-) = (R_{W \cdot V \cdot})^{-1} \circ \sigma,$$

$$\mathcal{F}(\delta_{VW}^+) = \sigma \circ R_{VW \cdot}, \quad \mathcal{F}(\delta_{VW}^-) = (R_{W \cdot V})^{-1} \circ \sigma,$$

$$\mathcal{F}(\tau_V^+) = v_{V \cdot}, \quad \mathcal{F}(\tau_V^-) = (v_{V \cdot})^{-1}.$$

Here, R_{VW} is the endomorphism of $V \otimes W$ given by the action of $\mathcal{R}$, v_V the endomorphism of V given by the action of v, and σ the operator which interchanges the order of the factors in a tensor product. The morphisms ev_V and π_V are defined in 5.1.3, and their 'twisted' versions $\tilde{\mathrm{ev}}_V$ and $\tilde{\pi}_V$ are defined as follows: if $\{e_i\}$ is a basis of V and $\{e^i\}$ is the dual basis of V^*,

$$\tilde{\mathrm{ev}}_V(e_i \otimes e^j) = \langle e^j, v^{-1} u e_i \rangle, \quad \tilde{\pi}_V(1) = \sum_i e^i \otimes u^{-1} v e_i.$$

OUTLINE OF PROOF The uniqueness follows immediately from the first part of 5.3.1. To prove existence, we must first check that the endomorphisms on the right-hand side of the formulas in the theorem are actually morphisms in $\mathbf{rep}_A$, i.e. that they commute with the action of A. This is straightforward: for $\alpha_V^\pm$, $\beta_{VW}^\pm$ and $\gamma_{VW}^\pm$, we have carried out most of the necessary computations in 5.2.2, and, for $\tau_V^\pm$, one must use the fact that v lies in the centre of A.

To complete the proof, we must verify that $\mathcal{F}$ is compatible with the relations in Fig. 13. As an example, we do this for the relation shown in the following diagram:

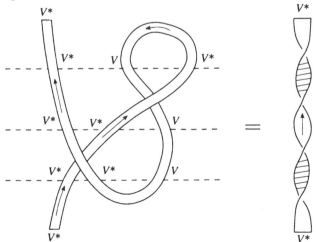

The horizontal lines show how to express the left-hand side of this relation as a composite of tensor products of generating morphisms. Its image under $\mathcal{F}$ is therefore the endomorphism of V^* given by

$$(\mathrm{id}_{V^*} \otimes \mathrm{ev}_V)((R_{V^* V^*})^{-1} \circ \sigma \otimes \mathrm{id}_V)(\mathrm{id}_{V^*} \otimes \sigma \circ R_{V V^*})(\mathrm{id}_{V^*} \otimes \pi_V).$$

Let $\mathcal{R} = \sum_r a_r \otimes b_r$, $\mathcal{R}^{-1} = \sum_s c_s \otimes d_s$. The effect of the preceding endomorphism on a basis vector e_i of V is

$$e_i \mapsto e_i \otimes e^j \otimes e_j \mapsto e^i \otimes b_r e^j \otimes a_r e_j \mapsto c_s b_r e^j \otimes d_s e^i \otimes a_r e_j \mapsto c_s b_r e^j \langle d_s e^i, a_r e_j \rangle$$

(repeated indices are to be summed over). This is the same as the action on V of the element $w = c_s b_r S^{-1}(a_r) d_s \in A$. Since the image under $\mathcal{F}$ of the right-hand side of the relation to be checked is v_V^2, we must show that $w = v^2 = S(u)u$, where $u = \mu(S \otimes \mathrm{id})(\mathcal{R}_{21})$, i.e. that $S(d_s) a_r S(c_s b_r) = S(S(u)u)$. But

$$S(d_s) a_r S(c_s b_r) = S(d_s) a_r S(b_r) S(c_s) = S(d_s) S(u) S(c_s) = S(c_s u d_s),$$

where the second equality is a consequence of the remarks following 4.2.4. Recalling that S^2 is given by conjugation by u, we must show that

$$S(u) = c_s S^2(d_s) = \mu(\mathrm{id} \otimes S^2)(\mathcal{R}^{-1}).$$

But, from (15) in Chapter 4, $(\mathrm{id} \otimes S^2)(\mathcal{R}^{-1}) = (\mathrm{id} \otimes S)(\mathcal{R})$, so our assertion follows from the formula for $S(u)$ used earlier in the proof. ∎

Example 5.3.3 It is interesting to compute the image under $\mathcal{F}$ of an annulus coloured by a representation V. Using the decomposition of the annulus shown below,

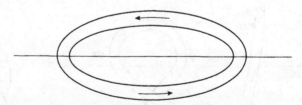

it follows that the image is the composite $k \xrightarrow{\pi_V} V \otimes V^* \xrightarrow{\widetilde{\mathrm{ev}}_V} k$, which is multiplication by the trace of the action of the element $v^{-1}u$ on V. Comparing with 4.2.9 shows that this is exactly the *quantum dimension* of V. ◇

Example 5.3.4 As a generalization of the preceding example, let T be an endomorphism of an object $((V_1, \epsilon_1), \ldots, (V_n, \epsilon_n))$ of **dribbon**(X). Let $\overline{T}$ be the ribbon tangle obtained by 'closing' the coloured ribbon tangle T as shown below. Thus, $\overline{T}$ has no free bands. Then, $\mathcal{F}(\overline{T}) = \text{qtr}(\mathcal{F}(T))$. For the proof, see Reshetikhin & Turaev (1991), page 555. ◇

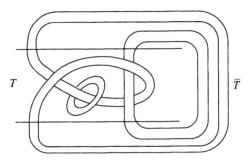

Example 5.3.5 As a simple application of the previous example, consider a single annulus with an ϵ-twist, where $\epsilon = \pm$, coloured with a representation V:

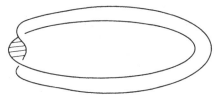

This is the closure of one of the twisted bands shown at the bottom right-hand corner of Fig. 12 (with the same sign convention). Hence, the associated scalar is $\mathcal{F}(\tau_V^\epsilon) = v_V^\epsilon \text{qdim}(V)$, where $v_V^\epsilon = v_V$ or v_V^{-1} according as $\epsilon = +$ or $-$. ◇

Example 5.3.6 As a second application, consider the *Hopf link* $H_{V,W}$, illustrated in the diagram at the top of the next page. It consists of two linked, unknotted, untwisted annuli, the two components being coloured with representations V and W. From this diagram, 5.3.4, and 4.2.7, we obtain

$$\mathcal{F}(H_{V,W}) = \text{qtr}(\gamma_{W^* \cdot V^*}^+ \circ \gamma_{V^* \cdot W^*}^-) = \text{qtr}(\mathcal{R}_{21}\mathcal{R}_{12} : V \otimes W \to V \otimes W). \quad ◇$$

In Chapter 15, we shall need an extension of 5.3.2 to the case of generalized ribbon tangles. With A and X as in 5.3.2, we define a category $\widetilde{\textbf{dribbon}}(X)$ as follows. Its objects are the same as those of $\textbf{dribbon}(X)$, while a morphism from $((V_1, \epsilon_1), \ldots, (V_m, \epsilon_m))$ to $((\tilde{V}_1, \tilde{\epsilon}_1), \ldots, (\tilde{V}_n, \tilde{\epsilon}_n))$ is a generalized ribbon tangle T such that, for any band whose ends both lie in $\mathbb{R} \times \{0\} \times \{0, 1\}$, the same restriction on the colours of its ends applies as in $\textbf{dribbon}(X)$. To completely specify a morphism in $\widetilde{\textbf{dribbon}}(X)$, however, we must give some additional data. Namely, if C is any coupon of T, we assign to each end of a band lying in an end of C a sign $+$ or $-$ according to whether the orientation

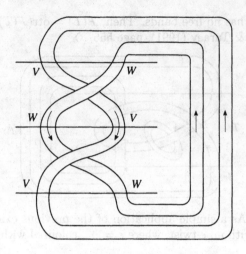

of the band at that end is downwards or upwards, respectively. Let $\epsilon_1, \ldots, \epsilon_r$ (resp. $\tilde{\epsilon}_1, \ldots, \tilde{\epsilon}_s$) be the sequence of signs along the top end (resp. the bottom end) of C. Then, the extra data we need is a homomorphism of A-modules $f : V_1^{\epsilon_1} \otimes \cdots \otimes V_r^{\epsilon_r} \to \tilde{V}_1^{\tilde{\epsilon}_1} \otimes \cdots \otimes \tilde{V}_s^{\tilde{\epsilon}_s}$.

We define a functor $\tilde{\mathcal{F}} : \mathbf{dribbon}(X) \to \mathbf{rep}_A$ as follows. On objects, and on the morphisms involving no coupons, $\tilde{\mathcal{F}}$ is the same as $\mathcal{F}$. To generate all the morphisms in $\mathbf{dribbon}(X)$, including those involving coupons, we must add to the generating morphisms in Fig. 12 those involving a single coupon, as shown in Fig. 14. Then, $\tilde{\mathcal{F}}$ associates to such a morphism the homomorphism $f : V_1^{\epsilon_1} \otimes \cdots \otimes V_r^{\epsilon_r} \to \tilde{V}_1^{\tilde{\epsilon}_1} \otimes \cdots \otimes \tilde{V}_s^{\tilde{\epsilon}_s}$ attached to the coupon.

THEOREM 5.3.7 *There is a unique monoidal functor $\tilde{\mathcal{F}} : \mathbf{dribbon}(X) \to$ $\mathbf{rep}_A$ defined on objects and generating morphisms as described above.* ∎

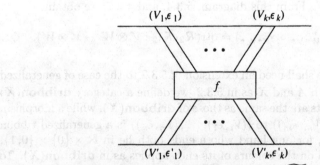

Fig. 14 A generating morphism in $\mathbf{dribbon}(X)$

B Tangle invariants The following result is an analogue of 5.3.2 for coloured tangles instead of coloured ribbon tangles.

THEOREM 5.3.8 *Let* $(A, \mathcal{R}, v)$ *be as in 5.3.2. There is a unique monoidal functor* $\mathcal{G} : \mathbf{dtangle}(X) \to \mathbf{rep}_A$ *such that*

(i) on objects, $\mathcal{G}$ *is given by*

$$\mathcal{F}((V_1, \epsilon_1), \ldots, (V_n, \epsilon_n)) = V_1^{\epsilon_1} \otimes \cdots \otimes V_n^{\epsilon_n},$$

where $V^+ = V$ *and* $V^- = V^*$;

(ii) on the generating morphisms,

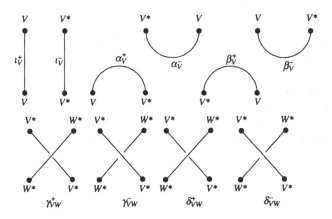

$\mathcal{G}$ *is given as follows:*

$$\mathcal{G}(\iota_V^+) = \mathrm{id}_V, \quad \mathcal{G}(\iota_V^-) = \mathrm{id}_{V^*},$$
$$\mathcal{G}(\alpha_V^+) = \pi_V : k \to V \otimes V^*, \quad \mathcal{G}(\alpha_V^-) = \mathrm{ev}_V : V^* \otimes V \to k,$$
$$\mathcal{G}(\beta_V^+) = \tilde{\pi}_V : k \to V^* \otimes V, \quad \mathcal{G}(\beta_V^-) = \tilde{\mathrm{ev}}_V : V \otimes V^* \to k,$$
$$\mathcal{G}(\gamma_{VW}^+) = \sigma \circ R_{V \cdot W \cdot} \circ (v_{V \cdot} \otimes \mathrm{id}_{W \cdot}),$$
$$\mathcal{G}(\gamma_{VW}^-) = (R_{W \cdot V \cdot})^{-1} \circ \sigma \circ (\mathrm{id}_{V \cdot} \otimes v_{W \cdot}),$$
$$\mathcal{G}(\delta_V^+) = \sigma \circ R_{VW \cdot} \circ (v_V \otimes \mathrm{id}_{W \cdot}),$$
$$\mathcal{G}(\delta_V^-) = (R_{W \cdot V})^{-1} \circ \sigma \circ (\mathrm{id}_V \otimes v_{W \cdot}^{-1}). \quad \blacksquare$$

A *link* is simply a tangle with no free ends. Hence, this theorem gives isotopy invariants of links with values in the algebra $\mathrm{Hom}_k(k, k) \cong k$, i.e. numerical invariants. To give a non-trivial example of such an invariant, we must anticipate a result from the next chapter. In Section 6.4, we construct a topological ribbon Hopf algebra $U_h(sl_2(\mathbb{C}))$ over the algebra $\mathbb{C}[[h]]$ of formal power series in an indeterminate h. It is a 'deformation' of the universal enveloping algebra of $sl_2(\mathbb{C})$ and, like $sl_2(\mathbb{C})$ itself, it has exactly one irreducible representation of each positive dimension. Taking $A = U_h(sl_2(\mathbb{C}))$

and X to consist of a single 'colour', namely the two-dimensional irreducible representation of $U_h(sl_2(\mathbb{C}))$, 5.3.8 gives a link invariant with values in $\mathbb{C}[[h]]$. It turns out that its values actually lie in the subalgebra consisting of the Laurent polynomials in $t = e^{h/2}$. If the normalization of the invariant is adjusted so that its value on a single unknotted circle is 1, it becomes the *Jones polynomial*, discovered by V. F. R. Jones (1985).

We shall study link invariants in much more detail in Chapter 15.

C Central elements In the statement of 5.3.2, we only allowed colourings by finite-dimensional representations of A. This restriction was necessary to be able to define the image of an annulus under $\mathcal{F}$, which involves taking a (quantum) trace. However, it can be relaxed as far as the bands are concerned. In particular, we may colour the bands with the (left) regular representation of A (which is infinite dimensional in most interesting cases).

Let us consider ribbon tangles with just one band: if the tangle contains annuli, they must be coloured with finite-dimensional representations of A. The image under $\mathcal{F}$ of such a ribbon tangle is a map $f : A \to A$, which, by the nature of its construction, is of the form $f(a) = za$ for some $z \in A$. Obviously, this map commutes with the action of A on itself if and only if z lies in the centre of A. More generally, if we consider ribbon tangles with n bands, we obtain elements of the centralizer of $\Delta^{(n)}(A)$ in $A^{\otimes n}$.

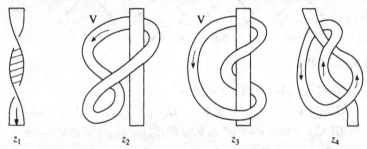

For example, the central elements of A corresponding to the four ribbon tangles shown above are given, respectively, by

$$z_1 = b_r S^2(a_s) b_s a_r, \qquad\qquad z_2 = \mathrm{trace}_V(a_r b_s S(b_t) a_t) b_r a_s,$$
$$z_3 = \mathrm{trace}_V(b_q a_r b_s a_p S(b_t) a_t) a_q b_r a_s a_p, \qquad z_4 = a_p b_q a_r S(b_t) a_t b_p a_q b_r,$$

where $\mathcal{R} = a_r \otimes b_r$ (repeated indices are to be summed over). We leave it to the reader to check that $z_1 = (uS(u))^{-1}$.

Bibliographical notes

5.1 For the general background in category theory, which is assumed known in this section, see MacLane (1971). For monoidal categories, see Saavedra

Rivano (1972), Deligne & Milne (1982), Joyal & Street (1986), Yetter (1990) and Deligne (1991). The reconstruction theorem 5.1.11 is due to Ulbrich (1990).

5.2 For fusion rules in quantum and conformal field theory, see, for example, Verlinde (1988), Witten (1990), Gawedski (1991), Kerler (1992a,b) and Fröhlich & Kerler (1993).

5.3 The results of this subsection are taken from Reshetikhin (1990b) and Reshetikhin & Turaev (1990, 1991). See also Rosso (1992c). For other constructions of central elements in quasitriangular Hopf algebras, see Drinfel'd (1990a).

6
Quantization of Lie bialgebras

The most important examples of quantum groups are deformations of universal enveloping algebras and of algebras of functions on groups. The purpose of this chapter is to lay the foundations of the deformation theory of Hopf algebras, paying particular attention to the first of these examples. Quantized algebras of functions will be discussed in the next chapter.

The obstructions to the existence and uniqueness of deformations of a given class of mathematical objects are generally described by some cohomology theory of the objects concerned. In Section 6.1, we describe the appropriate theory for Hopf algebras, which is a combination of the Hochschild cohomology theory of associative algebras and the dual theory for coalgebras. This cohomology theory implies, among other things, that, for semisimple algebraic groups and Lie algebras over fields of characteristic zero, every deformation of the algebra of functions has unchanged coalgebra structure and every deformation of the universal enveloping algebra has unchanged algebra structure.

In the next two sections, we concentrate on deformations of enveloping algebras, which are called *quantized universal enveloping algebras*, or simply QUE algebras, and on deformations of the algebra of functions on a Lie group. We argued in Chapter 1 that to quantize a Lie group or Lie algebra one should equip it with extra structure, namely a Poisson–Lie group structure or Lie bialgebra structure, respectively. Incorporating this into the universal enveloping algebra leads to the notion of a *co-Poisson–Hopf algebra*. We show that every deformation of a universal enveloping algebra induces on it a co-Poisson–Hopf structure, so the Lie bialgebra structure is in fact 'built in' to the deformation, and can be obtained from it by taking an appropriate 'classical limit'.

In Section 6.2, we show that every finite-dimensional Lie bialgebra can be quantized. The proof is completely algebraic. We also outline an analytic proof of the same result in the triangular case, which uses the ideas of Moyal described in Section 1.6, and perhaps provides some justification for the use of the term 'quantization' in this context.

After describing some general properties of QUE algebras in Section 6.3, we give in Section 6.4 the standard quantization of $sl_2(\mathbb{C})$ (with its standard Lie bialgebra structure). We show explicitly that its algebra structure is unchanged, and that it is quasitriangular as a Hopf algebra. The importance of this example is that a large class of QUE algebras, namely the quantum

Kac–Moody algebras, can be constructed from it. We describe these algebras in Section 6.5, and show further that, by suitably 'twisting' these standard quantizations, one obtains a quantization of the 'generic' Lie bialgebra structure on any finite-dimensional complex simple Lie algebra.

6.1 Deformations of Hopf algebras

A Definitions The notion of a deformation of an associative algebra was introduced informally in Section 1.6. We now give a more precise definition adapted to the case of Hopf algebras.

DEFINITION 6.1.1 *A deformation of a Hopf algebra* $(A, \imath, \mu, \epsilon, \Delta, S)$ *over a field k is a topological Hopf algebra* $(A_h, \imath_h, \mu_h, \epsilon_h, \Delta_h, S_h)$ *over the ring $k[[h]]$ of formal power series in an indeterminate h over k, such that*

(i) A_h is isomorphic to $A[[h]]$ as a $k[[h]]$-module;

(ii) $\mu_h \equiv \mu \pmod{h}$, $\Delta_h \equiv \Delta \pmod{h}$.

Two deformations A_h and A'_h are equivalent if there is an isomorphism $f_h : A_h \to A'_h$ of Hopf algebras over $k[[h]]$ which is the identity (mod h).

Remarks [1] The notation $A[[h]]$ means the algebra of formal power series in h with coefficients in A:

$$(1) \qquad a_h = a_0 + a_1 h + a_2 h^2 + \cdots \qquad (a_0, a_1, a_2, \ldots \in A).$$

Note that, unless A is finite-dimensional, $A[[h]]$ is 'bigger' than $A \otimes_k k[[h]]$: in fact, it is the completion of the latter in the h-adic topology (see Subsection 4.1A).

[2] The definition does not mention the unit, counit and antipode of A_h. In fact, it can be shown that any deformation is equivalent to one in which the unit and counit are obtained simply by extending $k[[h]]$-linearly those of A. Further, any deformation of A as a *bialgebra* (i.e. forgetting the antipode) is automatically a Hopf algebra.

[3] If $f_h : A_h \to A'_h$ is an equivalence, we shall write $A'_h = f_h * A_h$. Note that $g_h * (f_h * A_h) = (g_h f_h) * A_h$.

[4] The simplest deformation of A, called the *null deformation*, is obtained simply by extending $k[[h]]$-linearly the structure maps of A. Any deformation equivalent to the null deformation is said to be *trivial*. $\triangle$

Since μ_h and Δ_h are $k[[h]]$-module maps, they are determined by giving their values on elements (1) for which $a_1 = a_2 = \cdots = 0$. Condition (ii) requires that, for $a, a' \in A$,

$$(2) \qquad \mu_h(a \otimes a') = \mu(a \otimes a') + \mu_1(a \otimes a')h + \mu_2(a \otimes a')h^2 + \cdots,$$

$$(3) \qquad \Delta_h(a) = \Delta(a) + \Delta_1(a)h + \Delta_2(a)h^2 + \cdots,$$

for some k-module maps $\mu_i : A \otimes A \to A$, $\Delta_i : A \to A \otimes A$.

Similarly, an equivalence of deformations $f_h : A_h \to A'_h$ can be written on elements of A as

$$f_h(a) = a + hf_1(a) + h^2 f_2(a) + \cdots.$$

Note that

$$f_h^{-1}(a) = a - hf_1(a) + \cdots.$$

The associativity condition for μ_h is

$$(4) \qquad \mu_h(\mu_h(a_1 \otimes a_2) \otimes a_3) = \mu_h(a_1 \otimes \mu_h(a_2 \otimes a_3))$$

for all $a_1, a_2, a_3 \in A$. Writing both sides of this equation as formal power series in h, we get an infinite number of conditions on the components μ_i of μ_h; the component of order h gives

$$(5) \qquad \mu_1(a_1 a_2 \otimes a_3) + \mu_1(a_1 \otimes a_2)a_3 = a_1\mu_1(a_2 \otimes a_3) + \mu_1(a_1 \otimes a_2 a_3).$$

Similarly, the coassociativity condition for Δ_h is

$$(6) \qquad (\Delta_h \otimes \mathrm{id})\Delta_h(a) = (\mathrm{id} \otimes \Delta_h)\Delta_h(a)$$

for all $a \in A$, the component of order h in which is

$$(7) \qquad (\Delta \otimes \mathrm{id})\Delta_1(a) + (\Delta_1 \otimes \mathrm{id})\Delta(a) = (\mathrm{id} \otimes \Delta)\Delta_1(a) + (\mathrm{id} \otimes \Delta_1)\Delta(a).$$

Finally, the condition that Δ_h is a homomorphism of algebras is

$$(8) \qquad \Delta_h(\mu_h(a_1 \otimes a_2)) = (\mu_h \otimes \mu_h)\Delta_h^{13}(a_1)\Delta_h^{24}(a_2),$$

the term of order h in which gives

$$(9) \qquad \begin{aligned} \Delta(\mu_1(a_1 \otimes a_2)) + \Delta_1(a_1 a_2) &= (\mu \otimes \mu_1 + \mu_1 \otimes \mu)\Delta^{13}(a_1)\Delta^{24}(a_2) \\ &\quad + \Delta_1(a_1)\Delta(a_2) + \Delta(a_1)\Delta_1(a_2). \end{aligned}$$

A pair of k-module maps (μ_1, Δ_1) is called a *deformation (mod h^2)* of A if it satisfies (5), (7) and (9). (Deformations (mod h^2) are usually called *infinitesimal deformations* in the literature on deformation theory.) More generally, a $2n$-tuple $(\mu_1, \ldots, \mu_n, \Delta_1, \ldots, \Delta_n)$ of k-module maps is a deformation (mod h^{n+1}) of A if (4), (6) and (8) hold (mod h^{n+1}). Thus, any deformation of A

induces a deformation (mod h^{n+1}) for all n. However, a deformation (mod h^{n+1}) does not in general extend to a genuine deformation.

We shall see shortly that (5), (7) and (9) admit a cohomological interpretation, but first we examine in a little more detail the notion of equivalence of deformations. The conditions for a $k[[h]]$-module isomorphism $f_h : A_h \to A'_h$ to be an equivalence of deformations are

$$\mu'_h = f_h \mu_h (f_h^{-1} \otimes f_h^{-1}), \quad \Delta'_h = (f_h \otimes f_h) \Delta_h f_h^{-1}.$$

Working to first order as usual, the equations become

(10) $\qquad \mu'_1(a_1 \otimes a_2) = \mu_1(a_1 \otimes a_2) + f_1(a_1 a_2) - a_1 f_1(a_2) - f_1(a_1) a_2,$

(11) $\qquad \Delta'_1(a) = \Delta_1(a) - \Delta(f_1(a)) + (f_1 \otimes \mathrm{id} + \mathrm{id} \otimes f_1)\Delta(a).$

Two deformations (mod h^2) of A, say (μ_1, Δ_1) and (μ'_1, Δ'_1), are said to be *equivalent* if there is a k-module map $f_1 : A \to A$ satisfying (10) and (11).

B Cohomology theory The calculations in the previous section suggest the following definition of Hopf algebra cohomology, which is modelled on the Hochschild cohomology of associative algebras.

For $i, j \geq 1$, define the space of (i, j)-cochains of a Hopf algebra A to be $C^{ij} = \mathrm{Hom}_k(A^{\otimes i}, A^{\otimes j})$ and define k-module maps $d'_{ij} : C^{ij} \to C^{i+1\,j}$ and $d''_{ij} : C^{ij} \to C^{i\,j+1}$ as follows (we shall usually omit the subscripts on d'_{ij} and d''_{ij}): for $\gamma \in C^{ij}$,

$$(d'\gamma)(a_1 \otimes \cdots \otimes a_{i+1}) = \Delta^{(j)}(a_1).\gamma(a_2 \otimes \cdots \otimes a_{i+1})$$

(12) $$+ \sum_{r=1}^{i} (-1)^r \gamma(a_1 \otimes \cdots \otimes a_{r-1} \otimes a_r a_{r+1} \otimes a_{r+2} \otimes \cdots \otimes a_{i+1})$$

$$+ (-1)^{i+1} \gamma(a_1 \otimes \cdots \otimes a_i).\Delta^{(j)}(a_{i+1}),$$

$$(d''\gamma)(a_1 \otimes \cdots \otimes a_i) = (\mu^{(i)} \otimes \gamma)(\Delta_{1\,i+1}(a_1)\Delta_{2\,i+2}(a_2)\ldots\Delta_{i\,2i}(a_i))$$

(13) $$+ \sum_{r=1}^{j} (-1)^r (\mathrm{id}^{\otimes r-1} \otimes \Delta \otimes \mathrm{id}^{\otimes j-r})(\gamma(a_1 \otimes \cdots \otimes a_i))$$

$$+ (-1)^{j+1}(\gamma \otimes \mu^{(i)})(\Delta_{1\,i+1}(a_1)\Delta_{2\,i+2}(a_2)\ldots\Delta_{i\,2i}(a_i)).$$

Here, $\mu^{(i)} : A^{\otimes i} \to A$ and $\Delta^{(j)} : A \to A^{\otimes j}$ are the iterated multiplication and comultiplication maps:

$$\mu^{(i)}(a_1 \otimes \cdots \otimes a_i) = a_1.\ \ldots\ .a_i,$$

$$\Delta^{(j)}(a) = (\mathrm{id} \otimes \cdots \otimes \mathrm{id} \otimes \Delta) \cdots (\mathrm{id} \otimes \Delta)\Delta(a)$$

(with $j-1$ Δs). The meaning of the notation $\Delta_{k\,i+k}(a)$ is obvious, namely the element of $A^{\otimes 2i}$ which has 1 in all positions except the kth and the $(i+k)$th, where it has the components of $\Delta(a)$.

The following result can be proved by direct computation.

DEFINITION-PROPOSITION 6.1.2 $(C^{..}, d', d'')$ is a bicomplex, i.e. one has

$$d' \circ d' = d'' \circ d'' = d' \circ d'' + d'' \circ d' = 0.$$

Set $d = d'_{ij} + (-1)^i d''_{ij}$ and $C^n = \oplus_{i+j=n+1} C^{ij}$. Then, $d : C^n \to C^{n+1}$ and $(C^{.}, d)$ is a complex whose cohomology groups (actually k-modules) we denote by $H^{.}(A, A)$. ∎

To see what this definition means, suppose first that $f_1 \in C^{11}$. Then,

$$d'(f_1)(a_1 \otimes a_2) = a_1 f_1(a_2) + f_1(a_1)a_2 - f_1(a_1 a_2),$$
$$d''(f_1)(a) = (f_1 \otimes \mathrm{id} + \mathrm{id} \otimes f_1)\Delta(a) - \Delta(f_1(a)).$$

Comparing with (10) and (11), we see that a deformation (mod h^2) (μ_1, Δ_1) is trivial if and only if it is a 2-coboundary.

Next, for any $\mu_1 \in C^{21}$ and $\Delta_1 \in C^{12}$,

$$d'(\mu_1)(a_1 \otimes a_2 \otimes a_3) = a_1 \mu_1(a_2 \otimes a_3) - \mu_1(a_1 a_2 \otimes a_3)$$
$$+ \mu_1(a_1 \otimes a_2 a_3) - \mu_1(a_1 \otimes a_2)a_3,$$
$$d''\mu_1(a_1 \otimes a_2) = (\mu \otimes \mu_1 + \mu_1 \otimes \mu)\Delta^{13}(a_1)\Delta^{24}(a_2) - \Delta(\mu_1(a_1 \otimes a_2))),$$
$$d'(\Delta_1)(a_1 \otimes a_2) = \Delta_1(a_1)\Delta(a_2) - \Delta_1(a_1 a_2) + \Delta(a_1)\Delta_1(a_2),$$
$$d''(\Delta_1)(a) = (\mathrm{id} \otimes \Delta_1)\Delta(a) - (\Delta \otimes \mathrm{id})\Delta_1(a)$$
$$+ (\mathrm{id} \otimes \Delta)\Delta_1(a) - (\Delta_1 \otimes \mathrm{id})\Delta(a).$$

Comparing with (5), (7) and (9) shows that (μ_1, Δ_1) is a deformation (mod h^2) of A if and only if it is a 2-cocycle.

We have proved the first part of

PROPOSITION 6.1.3 Let A be a Hopf algebra over a field k.

(i) There is a natural bijection between $H^2(A, A)$ and the set of equivalence classes of deformations (mod h^2) of A.

(ii) If $H^2(A, A) = 0$, every deformation of A is trivial.

PROOF We know from part (i) that, if $H^2(A, A) = 0$, then every deformation A_h of A is trivial (mod h^2); we must prove that it is actually trivial.

Write the multiplication and comultiplication maps of A_h as in (2) and (3). Choose a 1-cochain $f^{(1)} : A \to A$ such that $df^{(1)} = (\mu_1, \Delta_1)$. Let $A'_h = (\mathrm{id} + h f^{(1)}) * A_h$. Then, $\mu'_1 = \Delta'_1 = 0$. Repeating the calculations which led to (5), (7) and (9) (but now looking at the h^2 terms) shows that (μ'_2, Δ'_2) is again a 2-cocycle. Choose a 1-cochain $f^{(2)}$ such that $df^{(2)} = (\mu'_2, \Delta'_2)$ and define a deformation $A''_h = (\mathrm{id} + h^2 f^{(2)}) * A'_h$. Then $\mu''_1 = \mu''_2 = \Delta''_1 = \Delta''_2 = 0$. Let $f^{(1)}, f^{(2)}, f^{(3)}, \ldots$ be the sequence of 1-cochains obtained by iterating this process indefinitely, and set

$$A_h^{(\infty)} = (\cdots (\mathrm{id} + h^n f^{(n)}) \cdots (\mathrm{id} + h^3 f^{(3)})(\mathrm{id} + h^2 f^{(2)})(\mathrm{id} + h f^{(1)})) * A_h.$$

Note that the infinite product makes sense as an element of $A[[h]]$. Since $A_h^{(\infty)}$ is the null deformation, this completes the proof. ∎

Example 6.1.4 If A is the group algebra $k[G]$ of a finite group G over a field k of characteristic zero, one can show (by a spectral sequence argument) that $H^2(A, A) = 0$. Thus, $k[G]$ is non-deformable as a Hopf algebra. ◊

The cohomology theory of Hopf algebras that we have defined is modelled, as we have said, on the Hochschild cohomology of algebras and coalgebras. In fact, the cohomology $H_{\text{alg}}(A, A)$ of A as an algebra is that of the complex $(C^{\cdot 1}, d')$, and its cohomology $H_{\text{coalg}}(A, A)$ as a coalgebra is that of the complex $(C^{1 \cdot}, d'')$. Moreover, the proof of 6.1.3 contains that of the following corollary.

COROLLARY 6.1.5 *If $H_{\text{alg}}^2(A, A) = 0$ (resp. if $H_{\text{coalg}}^2(A, A) = 0$), every deformation of A is isomorphic to $A[[h]]$ as an algebra (resp. as a coalgebra).* ∎

The 'space' of genuine deformations of a Hopf algebra A is generally smaller than the space $H^2(A, A)$ of its deformations (mod h^2). In fact, if (μ_1, Δ_1) extends to a genuine deformation of A, or just a deformation (mod h^3), then equating h^2 terms in (4), (6) and (8) gives

$$d'\mu_2(a_1 \otimes a_2 \otimes a_3) = \mu_1(\mu_1(a_1 \otimes a_2) \otimes a_3) - \mu_1(a_1 \otimes \mu_1(a_2 \otimes a_3)),$$

$$-d''\Delta_2(a) = (\text{id} \otimes \Delta_1)\Delta_1(a) - (\Delta_1 \otimes \text{id})\Delta_1(a),$$

$$d'\Delta_2 + d''\mu_2 = \Delta_1(\mu_1(a_1 \otimes a_2)) - (\mu_1 \otimes \mu_1)\Delta^{13}(a_1)\Delta^{24}(a_2) - \Delta_1(a_1)\Delta_1(a_2)$$

$$-(\mu \otimes \mu_1 + \mu_1 \otimes \mu)(\Delta_1^{13}(a_1)\Delta^{24}(a_2) + \Delta^{13}(a_1)\Delta_1^{24}(a_2)),$$

respectively. It is not difficult to check that the right-hand sides of these equations define, for any 2-cocycle (μ_1, Δ_1), a 3-cocycle, whose cohomology class depends only on the cohomology class of (μ_1, Δ_1). One therefore has a quadratic map $Sq : H^2(A, A) \to H^3(A, A)$ such that a deformation (mod h^2) of A extends to a deformation (mod h^3) if and only if its image under Sq is zero.

In general, if one has a deformation (mod h^n) of A, the obstruction to extending it to a deformation (mod h^{n+1}) is again an element of $H^3(A, A)$. In particular, we obtain

PROPOSITION 6.1.6 *If $H^3(A, A) = 0$, every deformation (mod h^2) of A extends to a genuine deformation of A.* ∎

Unfortunately, for most of the Hopf algebras of interest to us, $H^3(A, A)$ is non-zero, and the question of whether a given deformation (mod h^2) extends is often non-trivial.

C Rigidity theorems We shall be particularly interested in deformations A_h of the universal enveloping algebra $U(\mathfrak{g})$ of a Lie algebra $\mathfrak{g}$, and of the algebra $\mathcal{F}(G)$ of regular functions on an algebraic group G. For these Hopf algebras A, $H^2(A, A)$ is (fortunately!) non-zero in general, so non-trivial deformations are possible.

DEFINITION 6.1.7 *A Hopf algebra deformation of the universal enveloping algebra $U(\mathfrak{g})$ of a Lie algebra $\mathfrak{g}$ is called a* quantized universal enveloping algebra, *or simply a QUE algebra.*

A Hopf algebra deformation of the algebra $\mathcal{F}(G)$ of regular functions on an algebraic group G is called a quantized function algebra, *or simply a QF algebra.*

QUE algebras in general will be discussed in Section 6.3, and the most important examples are in Sections 6.4 and 6.5. QF algebras are the subject of Chapter 7. The justification for the adjective 'quantized' will be discussed in the next section.

The following result shows that, in constructing QUE algebras or QF algebras in the semisimple case, one may assume at the outset that the algebra or coalgebra structure is unchanged (although it is often not convenient to do this).

THEOREM 6.1.8 *Let G be a reductive algebraic group over a field k of characteristic zero, and let $\mathfrak{g}$ be its Lie algebra. Let $\wedge^m(\mathfrak{g})^{\mathfrak{g}}$ be the subspace of $\wedge^m(\mathfrak{g})$ consisting of the elements which are invariant under the adjoint action of $\mathfrak{g}$.*

(i) For all $m \geq 1$,

$$H^m(\mathcal{F}(G), \mathcal{F}(G)) \cong \wedge^m(\mathfrak{g})/\wedge^m(\mathfrak{g})^{\mathfrak{g}}, \qquad H^m_{\mathrm{coalg}}(\mathcal{F}(G), \mathcal{F}(G)) = 0.$$

Hence, every deformation of $\mathcal{F}(G)$ is isomorphic to $\mathcal{F}(G)[[h]]$ as a coalgebra.

(ii) If $\mathfrak{g}$ is semisimple,

$$H^2(U(\mathfrak{g}), U(\mathfrak{g})) \cong \wedge^2(\mathfrak{g}), \quad H^3(U(\mathfrak{g}), U(\mathfrak{g})) \cong \wedge^3(\mathfrak{g})/\wedge^3(\mathfrak{g})^{\mathfrak{g}},$$
$$H^2_{\mathrm{alg}}(U(\mathfrak{g}), U(\mathfrak{g})) = 0.$$

Hence, every deformation of $U(\mathfrak{g})$ is isomorphic to $U(\mathfrak{g})[[h]]$ as an algebra. ∎

Remarks [1] Part (ii) does not hold for arbitrary reductive $\mathfrak{g}$, as the example of the Heisenberg Lie algebra shows. In fact, let $\mathfrak{g}$ be the three-dimensional abelian Lie algebra, let $\{X, Y, Z\}$ be a basis of $\mathfrak{g}$ and define a Lie bracket on $\mathfrak{g}[[h]]$ by

$$[X, Y] = hZ, \quad [X, Z] = [Y, Z] = 0.$$

With this Lie algebra structure, $U(\mathfrak{g}[[h]])$ is a deformation of $U(\mathfrak{g})$ (when completed in the h-adic topology), but is non-commutative while $U(\mathfrak{g})[[h]]$ is commutative.

[2] In spite of this theorem, for most of the QUE algebras we shall study later in the book, no *explicit* equivalence with a deformation with unchanged algebra structure is known. $\triangle$

6.2 Quantization

A (Co-) Poisson–Hopf algebras We argued in Section 1.6 that a 'quantization' of a Poisson manifold M might be interpreted as a deformation $\mathcal{F}_h(M)$ of the Poisson algebra $\mathcal{F}(M)$ of C^∞ functions on M, such that the 'first order part' of the commutator of two elements of $\mathcal{F}_h(M)$ is equal to the Poisson bracket of their classical limits. If G is a Poisson–Lie group, then $\mathcal{F}(G)$ is also a Hopf algebra (modulo questions of completeness, which we ignore for the moment), and the two structures are compatible. In fact, at the group level the compatibility condition is

$$\{f_1 \circ m, \, f_2 \circ m\}_{G \times G} = \{f_1, \, f_2\}_G \circ m,$$

where $f_1, f_2 \in \mathcal{F}(G)$, $m : G \times G \to G$ is the multiplication in G and $\{ \, , \, \}_{G \times G}$ and $\{ \, , \, \}_G$ are the Poisson brackets of $G \times G$ and G. Since $f_i \circ m = \Delta(f_i)$, the preceding equation becomes

$$\{\Delta(f_1), \, \Delta(f_2)\}_{G \times G} = \Delta(\{f_1, \, f_2\}).$$

This suggests the following definition.

DEFINITION 6.2.1 *A* Poisson algebra *over a commutative ring k is a commutative algebra A over k equipped with a skew-symmetric k-module map $\{ \, , \, \}_A : A \otimes A \to A$, the* Poisson bracket, *such that:*

(i) the Jacobi identity

$$\{a_1, \{a_2, a_3\}_A\}_A + \{a_3, \{a_1, a_2\}_A\}_A + \{a_2, \{a_3, a_1\}_A\}_A = 0$$

holds for all $a_1, a_2, a_3 \in A$;

(ii) the Leibniz identity

$$\{a_1 a_2, \, a\}_A = \{a_1, \, a\}_A a_2 + a_1 \{a_2, \, a\}_A$$

holds for all $a_1, a_2, a \in A$.

A *Poisson–Hopf algebra over a commutative ring k is a Poisson algebra $(A, \{ \, , \, \}_A)$ over k, which is also a Hopf algebra $(A, \imath, \mu, \epsilon, \Delta, S)$ over k, the two structures being compatible in the sense that*

(iii) for all a_1, $a_2 \in A$,

$$\{\Delta(a_1),\, \Delta(a_2)\}_{A\otimes A} = \Delta(\{a_1,\, a_2\}_A).$$

The Poisson bracket $\{\ ,\ \}_{A\otimes A}$ on $A \otimes A$ is defined by

$$\{a_1 \otimes a_1',\, a_2 \otimes a_2'\}_{A\otimes A} = \{a_1,\, a_2\}_A \otimes a_1'a_2' + a_1a_2 \otimes \{a_1',\, a_2'\}_A.$$

To discuss the corresponding structures on universal enveloping algebras, we must dualize 6.2.1.

DEFINITION 6.2.2 *A co-Poisson algebra over a commutative ring k is a co-commutative coalgebra* (A, ϵ, Δ) *equipped with a skew-symmetric k-module map $\delta : A \to A \otimes A$, the Poisson co-bracket, satisfying the following conditions:*

(i) *the composite*

$$A \xrightarrow{\delta} A \otimes A \xrightarrow{\delta \otimes \mathrm{id}} A \otimes A \otimes A \xrightarrow{c.p.} A \otimes A \otimes A$$

is zero, where 'c.p.' means the sum over cyclic permutations of the factors in the triple tensor product (this is the co-Jacobi identity);

(ii) *the co-Leibniz identity*

$$(\Delta \otimes \mathrm{id})\delta = (\mathrm{id} \otimes \delta)\Delta + \sigma_{23}(\delta \otimes \mathrm{id})\Delta$$

holds.

A *co-Poisson–Hopf algebra is a co-Poisson algebra* $(A, \epsilon, \Delta, \delta)$ *which is also a Hopf algebra* $(A, \iota, \mu, \epsilon, \Delta, S)$, *the two structures being compatible in the sense that*

(iii) *for all a_1, $a_2 \in A$,*

$$\delta(a_1a_2) = \delta(a_1)\Delta(a_2) + \Delta(a_1)\delta(a_2).$$

Since Poisson–Hopf structures on $\mathcal{F}(G)$ correspond to Poisson–Lie group structures on G, which in turn correspond to Lie bialgebra structures on the Lie algebra of G, it is natural to expect

PROPOSITION 6.2.3 *Let $\mathfrak{g}$ be a Lie algebra over a field k of characteristic zero. If its universal enveloping algebra $U(\mathfrak{g})$ has a co-Poisson structure δ, making it a co-Poisson–Hopf algebra, then $\delta(\mathfrak{g}) \subset \mathfrak{g} \otimes \mathfrak{g}$ and $\delta|_{\mathfrak{g}}$ is a Lie bialgebra structure on $\mathfrak{g}$. Conversely, any Lie bialgebra structure $\delta : \mathfrak{g} \to \mathfrak{g}\otimes\mathfrak{g}$ extends uniquely to a Poisson co-bracket on $U(\mathfrak{g})$, which makes $U(\mathfrak{g})$ into a co-Poisson–Hopf algebra.*

PROOF Let $\delta : U(\mathfrak{g}) \to U(\mathfrak{g})\otimes U(\mathfrak{g})$ be a Poisson co-bracket on $U(\mathfrak{g})$. To show that $\delta(\mathfrak{g}) \subset \mathfrak{g} \otimes \mathfrak{g}$, let $x \in \mathfrak{g}$ and write $\delta(x) = \sum_i a_i \otimes a_i'$, where $a_i, a_i' \in U(\mathfrak{g})$.

We may assume that the a_i' are linearly independent. By condition (ii) in 6.2.2,

$$\sum_i \Delta(a_i) \otimes a_i' = 1 \otimes \delta(x) + a \otimes \delta(1) + \sigma_{23}(\delta(x) \otimes 1 + \delta(1) \otimes x).$$

Taking $a_1 = a_2 = 1$ in (iii) gives $\delta(1) = 0$, so

$$\sum_i \Delta(a_i) \otimes a_i' = \sum_i (a_i \otimes 1 + 1 \otimes a_i) \otimes a_i'.$$

It follows that the a_i are primitive elements of $U(\mathfrak{g})$. Hence, $\delta(\mathfrak{g}) \subset \mathfrak{g} \otimes U(\mathfrak{g})$. Since δ is skew-symmetric,

$$\delta(\mathfrak{g}) \subset (\mathfrak{g} \otimes U(\mathfrak{g})) \cap (U(\mathfrak{g}) \otimes \mathfrak{g}) = \mathfrak{g} \otimes \mathfrak{g}.$$

To prove that $\delta|_{\mathfrak{g}}$ is a 1-cocycle, let x_1, $x_2 \in \mathfrak{g}$ and compute

$$\begin{aligned}
\delta([x_1, x_2]) &= \delta(x_1 x_2 - x_2 x_1) \\
&= [\Delta(x_1), \delta(x_2)] - [\Delta(x_2), \delta(x_1)] \\
&= x_1.\delta(x_2) - x_2.\delta(x_1),
\end{aligned}$$

where the dot denotes the adjoint representation of $\mathfrak{g}$ in each factor of $\mathfrak{g} \otimes \mathfrak{g}$ (the second equality follows from 6.2.2(iii)).

Finally, the Jacobi identity for δ^* is easily seen to be equivalent to the co-Jacobi identity for δ.

The converse is proved in a similar way, using 6.2.2(iii) to extend δ from $\mathfrak{g}$ to $U(\mathfrak{g})$. ∎

B Quantization The following definition formalizes the discussion given in Section 1.6.

DEFINITION 6.2.4 *Let A be a commutative Poisson–Hopf algebra over a field k of characteristic zero, and let $\{\ ,\ \}$ be its Poisson bracket. A quantization of A is a Hopf algebra deformation A_h of A such that*

$$\{x_1, x_2\} \equiv \frac{a_1 a_2 - a_2 a_1}{h} \quad (\mathrm{mod}\ h),$$

if a_1, $a_2 \in A_h$ reduce to x_1, $x_2 \in A$ (mod h).

A quantization of an algebraic Poisson–Lie group $(G, \{\ ,\ \})$ is a quantization $\mathcal{F}_h(G)$ of the algebra $\mathcal{F}(G)$ of regular functions on G, regarded as a Poisson algebra, and $(G, \{\ ,\ \})$ is called the classical limit *of $\mathcal{F}_h(G)$.*

Before giving the dual definition, which will be the most important for us, we pause to give an interesting example.

Example 6.2.5 Take the Poisson–Lie group to be the dual space $\mathfrak{g}^*$ of a real Lie algebra $\mathfrak{g}$, equipped with the trivial Lie bracket and the Lie–Poisson structure (see 1.1.3). The algebra $\mathcal{F}(\mathfrak{g}^*)$ of polynomial functions on $\mathfrak{g}^*$ is a commutative (and cocommutative) Poisson–Hopf algebra. Let $\mathfrak{g}_h$ be the Lie algebra obtained by multiplying the Lie bracket of $\mathfrak{g}$ by h, and let $U(\mathfrak{g}_h)$ be (the h-adic completion of) its universal enveloping algebra. It is easy to see that $U(\mathfrak{g}_h)$ is a quantization of $\mathcal{F}(\mathfrak{g}^*)$. For example, if $X, Y \in \mathfrak{g}$ are regarded as linear functions on $\mathfrak{g}^*$, their commutator in $U(\mathfrak{g}_h)$ is $h[X, Y]$, where $[\ ,\ ]$ is the Lie bracket in $\mathfrak{g}$, while

$$\frac{h[X, Y]}{h} = [X, Y]$$

is their Poisson bracket in $\mathcal{F}(\mathfrak{g}^*)$ (see 1.1.3).

It is clear that $U(\mathfrak{g}_h)$ is isomorphic to $U(\mathfrak{g})[[h]]$ as a Hopf algebra, at least over the quotient field of $\mathbb{R}[[h]]$. Thus, in a sense, $U(\mathfrak{g})$ is a deformation of $\mathcal{F}(\mathfrak{g}^*)$. Now, according to the well-known Kirillov–Kostant orbit principle, the irreducible unitary representations of $\mathfrak{g}$ should correspond to its coadjoint orbits. Recalling that these coadjoint orbits are exactly the symplectic leaves of $\mathfrak{g}^*$ (see 1.1.3), we might expect that there should be a generalization of the orbit method according to which the irreducible unitary representations of a quantization of a Poisson–Lie group G should correspond to its symplectic leaves. In Chapter 13 we shall see that, if G is a compact simple Lie group with the standard Poisson structure described in Subsection 1.4C, then G has a 'standard' quantization for which this expectation is fulfilled. $\Diamond$

We now turn to the dual picture.

DEFINITION 6.2.6 *Let A be a cocommutative co-Poisson–Hopf algebra over a field k of characteristic zero, and let δ be its Poisson co-bracket. A quantization of A is a Hopf algebra deformation A_h of A such that*

$$\delta(x) \equiv \frac{\Delta_h(a) - \Delta_h^{\mathrm{op}}(a)}{h} \quad (mod\ h),$$

where $x \in A$ and a is any element of A_h such that $x \equiv a$ (mod h).

A quantization of a Lie bialgebra $(\mathfrak{g}, \delta)$ is a quantization $U_h(\mathfrak{g})$ of its universal enveloping algebra $U(\mathfrak{g})$ equipped with the co-Poisson–Hopf structure from 6.2.3. Conversely, $(\mathfrak{g}, \delta)$ is called the classical limit of the QUE algebra $U_h(\mathfrak{g})$.

Remarks [1] Note that the condition in this definition makes sense because $U(\mathfrak{g})$ is cocommutative, so that $\Delta_h(a) \equiv \Delta_h^{\mathrm{op}}(a)$ (mod h). Note also that this condition only depends on the behaviour of Δ_h up to order h, so it makes sense to discuss deformations (mod h^n) for all $n \geq 2$. Similar remarks apply, of course, to quantizations of Poisson–Hopf algebras.

[2] The Lie algebra $\mathfrak{g}$ is uniquely determined by the QUE algebra $U_h(\mathfrak{g})$, being the set of primitive elements of $U_h(\mathfrak{g})/hU_h(\mathfrak{g})$. △

The following result shows that a quantization of a Lie bialgebra is the same thing as a QUE algebra.

PROPOSITION 6.2.7 *Let $\mathfrak{g}$ be a Lie algebra over a field k of characteristic zero and let $U_h(\mathfrak{g})$ be a Hopf algebra deformation of $U(\mathfrak{g})$. Define $\delta : U(\mathfrak{g}) \to U(\mathfrak{g}) \otimes U(\mathfrak{g})$ by*

$$\delta(x) \equiv \frac{\Delta_h(a) - \Delta_h^{\mathrm{op}}(a)}{h} \quad (mod\ h),$$

where $x \in U(\mathfrak{g})$, and $a \in U_h(\mathfrak{g})$ reduces to x (mod h). Then $(U(\mathfrak{g}), \delta)$ is a co-Poisson–Hopf algebra.

PROOF It is easy to see that δ is well-defined. We shall check that it satisfies 6.2.2(i), leaving the other two conditions to the reader.

We have

$$(\delta \otimes \mathrm{id})\delta(x) = \frac{1}{h^2} \Big\{ (\Delta_h \otimes \mathrm{id})\Delta_h(a) - (\Delta_h^{\mathrm{op}} \otimes \mathrm{id})\Delta_h(a) -$$

$$(\Delta_h \otimes \mathrm{id})\Delta_h^{\mathrm{op}}(a) + (\Delta_h^{\mathrm{op}} \otimes \mathrm{id})\Delta_h^{\mathrm{op}}(a) \Big\} \quad (mod\ h).$$

Write

$$(\Delta_h \otimes \mathrm{id})\Delta_h(a) = \sum_i a_i \otimes a_i' \otimes a_i''.$$

Then,

$$(\Delta_h^{\mathrm{op}} \otimes \mathrm{id})\Delta_h(a) = \sum_i a_i' \otimes a_i \otimes a_i''$$

$$(\Delta_h \otimes \mathrm{id})\Delta_h^{\mathrm{op}}(a) = \sum_i a_i' \otimes a_i'' \otimes a_i$$

$$(\Delta_h^{\mathrm{op}} \otimes \mathrm{id})\Delta_h^{\mathrm{op}}(a) = \sum_i a_i'' \otimes a_i' \otimes a_i.$$

The result is now clear. ∎

This result has an interesting cohomological interpretation. In fact, let $\mathfrak{g}$ be a Lie algebra over a field of charcteristic zero, and let $r \in \wedge^2(\mathfrak{g})$. The map $\delta : \mathfrak{g} \to \mathfrak{g} \otimes \mathfrak{g}$ given by $\delta(x) = (\mathrm{ad}_x \otimes \mathrm{id} + \mathrm{id} \otimes \mathrm{ad}_x)(r)$ extends uniquely to a map $\delta : U(\mathfrak{g}) \to U(\mathfrak{g}) \otimes U(\mathfrak{g})$, using the formula in 6.2.2(iii). Then, $\Delta_1 = \delta$ is a deformation (mod h^2) of $U(\mathfrak{g})$. Indeed, (7) is equivalent to

$$((\Delta \otimes \mathrm{id})\Delta(a)).((\Delta \otimes \mathrm{id})(r) + r \otimes 1) = ((\mathrm{id} \otimes \Delta)\Delta(a)).((\mathrm{id} \otimes \Delta)(r) + 1 \otimes r).$$

The terms involving a on the two sides are equal by the coassociativity of Δ, and those involving r are both equal to the result of applying the operation 'c.p.' to $r \otimes 1$.

By 6.1.6 and 6.1.8, the obstruction to extending this to a deformation (mod h^3) is an element of $\wedge^3(\mathfrak{g}) / \wedge^3 (\mathfrak{g})^{\mathfrak{g}}$. Explicit computation shows that this element is exactly $[[r, r]]$, the 'left-hand side of the classical Yang–Baxter equation'. In view of 2.1.2, we see that the obstruction vanishes precisely when δ is a Lie bialgebra structure on $\mathfrak{g}$.

C Existence of quantizations The question of the existence of quantizations of Lie bialgebras was settled, only recently, by Reshetikhin (1992a).

THEOREM 6.2.8 *Every finite-dimensional Lie bialgebra* $(\mathfrak{g}, \mathfrak{g}^*)$ *over a field* k *of characteristic zero admits a quantization.*

OUTLINE OF PROOF Let $H(X, Y)$, for X, $Y \in \mathfrak{g}$, be the Baker–Campbell–Hausdorff series for $\mathfrak{g}$:

$$H(X,Y) = X + Y + \frac{1}{2}[X,Y] + \frac{1}{12}[X,[X,Y]] + \frac{1}{12}[Y,[Y,X]]$$
$$- \frac{1}{24}[X,[Y,[X,Y]]] + \cdots$$

(see Bourbaki (1972), Chap. 2, §6, no. 4). We shall need the following properties of H:

(i) $H = \sum_{n=1}^{\infty} H_n$, where H_n is a finite linear combination of terms of the form

$$[Z_1, [Z_2, [\cdots [Z_{n-1}, Z_n] \cdots]]],$$

where each $Z_i = X$ or Y;

(ii) $H(X, H(Y, Z)) = H(H(X,Y), Z)$ (associativity);

(iii) $H(X, -X) = 0$, $H(X, 0) = H(0, X) = X$.

The first property implies that $(X, Y) \mapsto H(hX, hY)$ is a well-defined k-bilinear map $\mathfrak{g} \times \mathfrak{g} \to \mathfrak{g}[[h]]$. We denote its natural extension to a $k[[h]]$-linear map $\mathfrak{g}[[h]] \underset{k[[h]]}{\otimes} \mathfrak{g}[[h]] \to \mathfrak{g}[[h]]$ by H_h.

Let $\mathcal{F}(\mathfrak{g})$ be the algebra of polynomial functions on $\mathfrak{g}$, and let $\mathcal{F}_h(\mathfrak{g}) = \mathcal{F}(\mathfrak{g})[[h]]$ as a $k[[h]]$-algebra. Then, properties (ii) and (iii) of H imply that $\mathcal{F}_h(\mathfrak{g})$ has a Hopf algebra structure with comultiplication, antipode and counit given by

$$\Delta_h(f)(X \otimes Y) = f(H_h(X, Y)),$$
$$S_h(f) = -f, \quad \epsilon_h(f) = 0,$$

for $f \in \mathfrak{g}^* \subset \mathcal{F}(\mathfrak{g})$, X, $Y \in \mathfrak{g}$.

Let $\bar{h}$ be a second indeterminate, independent of h. The crux of the proof is to show that there exist a pair of Hopf algebras $(\text{Sym}_h^{\bar{h}}(\mathfrak{g}), \text{Sym}_{\bar{h}}^h(\mathfrak{g}^*))$ over $k[[h, \bar{h}]]$ with the following properties:

(i) $\text{Sym}_h^{\bar{h}}(\mathfrak{g}) \cong \text{Sym}(\mathfrak{g}) \underset{k}{\otimes} k[[h, \bar{h}]]$ and $\text{Sym}_{\bar{h}}^h(\mathfrak{g}^*) \cong \text{Sym}(\mathfrak{g}^*) \underset{k}{\otimes} k[[h, \bar{h}]]$ as $k[[h, \bar{h}]]$-modules ($\text{Sym}(\mathfrak{g})$ denotes the symmetric algebra of $\mathfrak{g}$);

(ii) $\text{Sym}_h^{\bar{h}}(\mathfrak{g}) \cong \mathcal{F}_h(\mathfrak{g}^*) \underset{k[[h]]}{\otimes} k[[h, \bar{h}]]$ and $\text{Sym}_{\bar{h}}^h(\mathfrak{g}^*) \cong \mathcal{F}_{\bar{h}}(\mathfrak{g}) \underset{k[[\bar{h}]]}{\otimes} k[[h, \bar{h}]]$ as coalgebras over $k[[h, \bar{h}]]$;

(iii) there exists a non-degenerate pairing of Hopf algebras

$$\langle\, , \,\rangle : \text{Sym}_h^{\bar{h}}(\mathfrak{g}) \times \text{Sym}_{\bar{h}}^h(\mathfrak{g}^*) \to k[[h, \bar{h}]]$$

such that the restriction of $\langle\, , \,\rangle$ to $\mathfrak{g} \times \mathfrak{g}^* \subset \text{Sym}_h^{\bar{h}}(\mathfrak{g}) \times \text{Sym}_{\bar{h}}^h(\mathfrak{g}^*)$ is equal to the canonical pairing $\mathfrak{g} \times \mathfrak{g}^* \to k \pmod{(h, \bar{h})}$. (The notion of a non-degenerate pairing of Hopf algebras was introduced in Subsection 4.1D.)

Note that the existence of the pairing and the known coalgebra structure of $\text{Sym}_h^{\bar{h}}(\mathfrak{g})$ and $\text{Sym}_{\bar{h}}^h(\mathfrak{g}^*)$ enable one to determine completely the algebra structure of $\text{Sym}_{\bar{h}}^h(\mathfrak{g}^*)$ and $\text{Sym}_h^{\bar{h}}(\mathfrak{g})$. From this, one deduces that $\text{Sym}_h^{\bar{h}}(\mathfrak{g})$ admits a $k[\bar{h}][[h]]$-form, i.e. a Hopf algebra $U_h^{\bar{h}}(\mathfrak{g})$ over $k[\bar{h}][[h]]$ such that

$$\text{Sym}_h^{\bar{h}}(\mathfrak{g}) \cong U_h^{\bar{h}}(\mathfrak{g}) \underset{k[\bar{h}][[h]]}{\otimes} k[[h, \bar{h}]]$$

as Hopf algebras over $k[[h, \bar{h}]]$. The point is that $U_h^{\bar{h}}(\mathfrak{g})$ can be specialized in $\bar{h}$: we set

$$U_h(\mathfrak{g}) = U_h^{\bar{h}}(\mathfrak{g}) \underset{k[\bar{h}][[h]]}{\otimes} k[[h]]$$

via the homomorphism $k[\bar{h}][[h]] \to k[[h]]$ of $k[[h]]$-modules given by sending $\bar{h}$ to 1. This gives the desired quantization of $(\mathfrak{g}, \mathfrak{g}^*)$. ∎

We shall describe briefly another approach to the existence question due to Drinfel'd (1983b), since it establishes a connection between quantum groups and the deformation-theoretic approach to the quantization of dynamical systems discussed in Section 1.6.

THEOREM 6.2.9 *Let $\mathfrak{g}$ be a finite-dimensional real Lie algebra, and let $r \in \mathfrak{g} \otimes \mathfrak{g}$ be a skew-symmetric solution of the classical Yang–Baxter equation*

$$[r_{12}, r_{13}] + [r_{12}, r_{23}] + [r_{13}, r_{23}] = 0.$$

Then, there exists a deformation $U_h(\mathfrak{g})$ of $U(\mathfrak{g})$ whose classical limit is $\mathfrak{g}$ with the Lie bialgebra structure defined by r. Moreover, $U_h(\mathfrak{g})$ is a triangular Hopf algebra and is isomorphic to $U(\mathfrak{g})[[h]]$ as an algebra over $\mathbb{R}[[h]]$.

PROOF We start with the trivial deformation $U(\mathfrak{g})[[h]]$ and twist it with an invertible element $\mathcal{F}_h \in U(\mathfrak{g})[[h]] \otimes U(\mathfrak{g})[[h]]$ (completed tensor product) satisfying (18) and (19) in 4.2.13. By 4.2.14, $U(\mathfrak{g})[[h]]^{\mathcal{F}_h}$ is a triangular Hopf algebra with universal R-matrix $\mathcal{R}_h = \mathcal{F}_h^{21}\mathcal{F}_h^{-1}$. (Here, and elsewhere in this chapter, we shall usually write $\mathcal{F}_h^{12}$ instead of $(\mathcal{F}_h)_{12}$, etc.) In order that $U(\mathfrak{g})[[h]]^{\mathcal{F}_h}$ be a quantization of the Lie bialgebra structure on $\mathfrak{g}$ defined by r, it suffices, by 6.2.7, that

$$\mathcal{F}_h \equiv 1 \otimes 1 \pmod{h} \quad \text{and} \quad \frac{\mathcal{F}_h - 1 \otimes 1}{h} \equiv -\frac{1}{2}r \pmod{h}.$$

Note that, by 2.2.6, we may assume without loss of generality that r is non-degenerate. By 2.2.2, r then defines a left-invariant symplectic structure on the connected, simply-connected Lie group G with Lie algebra $\mathfrak{g}$. Let $\{\,,\,\}$ be the corresponding Poisson bracket on $C^\infty(G)$. The existence of $\mathcal{F}_h$ follows from

LEMMA 6.2.10 *There exist left-invariant, bilinear, bi-differential operators*

$$D_k : C^\infty(G) \times C^\infty(G) \to C^\infty(G),$$

for $k \geq 2$, such that the product $*_h$ on $C^\infty(G)[[h]]$ defined by

$$f_1 *_h f_2 = f_1 f_2 + \frac{1}{2}h\{f_1,\,f_2\} + \sum_{k=2}^{\infty} h^k D_k(f_1, f_2),$$

for f_1, $f_2 \in C^\infty(G)$, is associative with the constant function 1 as unit element.

Indeed, if we identify $\mathfrak{g}$ with the left-invariant vector fields on G, then left-invariant differential operators $C^\infty(G) \to C^\infty(G)$ are identified with elements of $U(\mathfrak{g})$, and bi-differential operators $C^\infty(G) \times C^\infty(G) \to C^\infty(G)$ with elements of $U(\mathfrak{g}) \otimes U(\mathfrak{g})$. Note that, if X^L is the left-invariant differential operator corresponding to $X \in U(\mathfrak{g})$ (and similarly for elements of $U(\mathfrak{g}) \otimes U(\mathfrak{g})$), then

(14) $$X^L(f_1 f_2) = \Delta(X)^L(f_1, f_2)$$

for f_1, $f_2 \in C^\infty(G)$.

Let $\tilde{\mathcal{F}}_k \in U(\mathfrak{g}) \otimes U(\mathfrak{g})$ correspond to D_k and set

$$\tilde{\mathcal{F}}_h = 1 + \frac{1}{2}hr + \sum_{k=2}^{\infty} h^k \tilde{F}_k \in (U(\mathfrak{g}) \otimes U(\mathfrak{g}))[[h]].$$

The associativity of the product $*_h$ is equivalent to

(15) $$\tilde{\mathcal{F}}_h^L(f_1, \tilde{\mathcal{F}}_h^L(f_2, f_3)) = \tilde{\mathcal{F}}_h^L(\tilde{\mathcal{F}}_h^L(f_1, f_2), f_3),$$

which, in view of (14), is equivalent to

$$(\mathrm{id} \otimes \Delta)(\tilde{\mathcal{F}}_h).\tilde{\mathcal{F}}_h^{23} = (\Delta \otimes \mathrm{id})(\tilde{\mathcal{F}}_h).\tilde{\mathcal{F}}_h^{12}.$$

Hence, $\mathcal{F}_h = \tilde{\mathcal{F}}_h^{-1}$ satisfies (18) in 4.2.13. Similarly, the unital property of $*_h$ is equivalent to (19) in the same proposition. ∎

To prove 6.2.10, one first linearizes the problem by essentially realizing G as a coadjoint orbit (of a larger group), and then applies the method used in Section 1.6 to construct the Moyal product on $C^\infty(\mathbb{R}^2)$. Actually, we shall only construct the product $*_h$ on the space $C_e^\infty(G)$ of germs of C^∞ functions at the identity element e of G, but this is sufficient for our purposes.

Let ω be the 2-cocycle on $\mathfrak{g}$ defined by r (recall that r is non-degenerate), and let

$$0 \to \mathbb{R} \to \hat{\mathfrak{g}} \to \mathfrak{g} \to 0$$

be the central extension of $\mathfrak{g}$ defined by ω. Thus, $\hat{\mathfrak{g}} = \mathfrak{g} \oplus \mathbb{R}$ as a vector space, and the Lie bracket is given by

$$[(x, a), (y, b)] = ([x, y], \omega(x, y))$$

for x, $y \in \mathfrak{g}$, a, $b \in \mathbb{R}$. Let $\hat{G}$ be the connected, simply-connected Lie group with Lie algebra $\hat{\mathfrak{g}}$. Define $\xi_0 \in \hat{\mathfrak{g}}^*$ by $\xi_0(x, a) = a$, and let $\mathcal{H} \subset \hat{\mathfrak{g}}^*$ be the hyperplane $\xi_0 + \mathfrak{g}^*$. Define $\alpha : G \to \mathcal{H}$ by

$$\alpha(g) = (\mathrm{Ad}_{\hat{G}})^*_{g^{-1}}(\xi_0).$$

Note that the kernel of the central extension $\hat{G}$ acts trivially on $\hat{\mathfrak{g}}^*$, and hence there is an induced action of G on $\hat{\mathfrak{g}}^*$. It is easy to see that α is a local diffeomorphism in a neighbourhood of the identity: in fact, its derivative there is $\alpha'_e(X)(Y) = \omega(X, Y)$, so the assertion follows from the non-degeneracy of ω.

To construct the desired product on $C_e^\infty(G)$, it suffices to construct an associative unital product $*^{\mathcal{H}}$ on $C^\infty(\mathcal{H})$ of the form in 6.2.10 which is invariant under the coadjoint action of G, for we can then use α to obtain a product on $C_e^\infty(G)$ with the desired properties. We shall make use of the Baker–Campbell–Hausdorff map $H_h : \hat{\mathfrak{g}}[[h]] \times \hat{\mathfrak{g}}[[h]] \to \hat{\mathfrak{g}}[[h]]$ introduced earlier. Note that, by property (i) above, it follows that H_h is invariant under the adjoint action of $\hat{G}$:

$$H_h(\hat{g}.\hat{X}, \hat{g}.\hat{X}) = \hat{g}.H_h(\hat{X}, \hat{Y}).$$

Identifying $X \in \mathfrak{g}$ with $\hat{X} = (X, 0) \in \hat{\mathfrak{g}}$, we define the $*^{\mathcal{H}}$ product of φ_1, $\varphi_2 \in C^\infty(\mathcal{H})$ by

$$(\varphi_1 *^{\mathcal{H}} \varphi_2)(\xi) = \int_{\mathfrak{g} \times \mathfrak{g}} \hat{\varphi}_1(X)\hat{\varphi}_2(Y) \exp\left[\frac{\sqrt{-1}}{h} \langle H_h(X, Y), \xi \rangle\right] dX\, dY,$$

where

$$\hat{\varphi}(X) = \left(\frac{1}{2\pi}\right)^{\dim(\mathfrak{g})} \int_{\mathcal{H}} \varphi(\eta) \exp[-\sqrt{-1}\langle X, \eta\rangle] \, d\eta$$

is the Fourier transform of φ. This product is invariant under the coadjoint action of G because H_h is invariant under the adjoint action of $\hat{G}$. Furthermore, since

$$H_h(X, Y) \equiv h(X + Y) + \frac{h^2}{2}[X, Y] \pmod{h^3},$$

it follows that

$$(\varphi_1 *^{\mathcal{H}} \varphi_2)(\xi) \equiv \varphi_1(\xi)\varphi_2(\xi) + \frac{1}{2}h\{\varphi_1, \varphi_2\}_{\mathcal{H}}(\xi) \pmod{h^2},$$

where $\{\ ,\ \}_{\mathcal{H}}$ is the Poisson bracket on $\mathcal{H}$ defined by ω (regarded as an element of $\mathfrak{g}^* \otimes \mathfrak{g}^*$).

As for associativity, a direct computation shows that

$$(\varphi_1 *^{\mathcal{H}} (\varphi_2 *^{\mathcal{H}} \varphi_3))(\xi)$$

$$= \left(\frac{1}{2\pi}\right)^{3\dim(\mathfrak{g})} \int_{\mathfrak{g}\times\mathfrak{g}\times\mathfrak{g}} \int_{\mathcal{H}\times\mathcal{H}\times\mathcal{H}} \exp\left[-\sqrt{-1}(\langle Y, \eta\rangle + \langle Z, \zeta\rangle + \langle W, \theta\rangle)\right]$$

$$\times \exp\left[\frac{\sqrt{-1}}{h}\langle H_h(Y, H_h(Z, W)), \xi\rangle\right]$$

$$\times \varphi_1(\eta)\varphi_2(\zeta)\varphi_3(\theta) \, dY \, dZ \, dW \, d\eta \, d\zeta \, d\theta,$$

while $((\varphi_1 *^{\mathcal{H}} \varphi_2) *^{\mathcal{H}} \varphi_3)(\xi)$ is given by the same formula but with the second exponential in the integral replaced by $\exp[\frac{\sqrt{-1}}{h}\langle H_h(H_h(Y, Z), W), \xi\rangle]$. Thus, the associativity of $*^{\mathcal{H}}$ follows from the associativity property (ii) of H_h. Finally, the unital property of $*^{\mathcal{H}}$ follows immediately from the unital property (iii) of H_h. ∎

Remark This calculation is a generalization of that in Subsection 1.6C. There, $\mathfrak{g}$ is the two-dimensional abelian Lie algebra, and $\hat{\mathfrak{g}}$ is the Heisenberg Lie algebra. The Baker–Campbell–Hausdorff map for $\hat{\mathfrak{g}}$ is given by

$$H_h(\hat{X}, \hat{Y}) = h(\hat{X} + \hat{Y}) + \frac{h^2}{2}[\hat{X}, \hat{Y}],$$

so that

$$h^{-1}H_h(X, Y) - X - Y = h(x_1 y_2 - x_2 y_1)$$

if $X = (x_1, x_2)$, $Y = (y_1, y_2) \in \mathbb{R}^2$. △

There is yet another approach to the problem of quantizing Lie bialgebras which is closer in spirit to the origins of quantum groups in inverse scattering theory. In this approach, the problem is to construct, for a given matrix solution $r \in \text{End}(V \otimes V)$ of the classical Yang–Baxter equation, where V is a vector space over k, a solution $R_h \in \text{End}(V \otimes V)[[h]]$ of the quantum Yang–Baxter equation such that $R_h \equiv 1 + hr \pmod{h^2}$. In Chapter 7, we shall describe, given such a solution of the QYBE, a way to construct a bialgebra over $k[[h]]$. It turns out that many of the 'standard' quantizations can be obtained this way, although in general there is no guarantee that this construction leads to a QUE algebra.

There is, however, a relation between the R-matrix approach and the definition of quantization we have given, at least in the case of triangular Lie bialgebra structures on $gl_n(\mathbb{R})$.

PROPOSITION 6.2.11 *Let $R_h \in (gl_n(\mathbb{R}) \otimes gl_n(\mathbb{R}))[[h]]$ be a solution of the QYBE such that $R_h \equiv 1 \pmod{h}$ and $R_h^{21} R_h = 1$. Then, there exists an element $F_h \equiv 1 - \frac{1}{2}hr \pmod{h^2}$ in $(U(gl_n(\mathbb{R})) \otimes U(gl_n(\mathbb{R})))[[h]]$ satisfying (18) and (19) in 4.2.13 such that R_h is the image of $F_h^{21} F_h^{-1}$ under the canonical map $\varphi : U(gl_n(\mathbb{R}))[[h]] \to gl_n(\mathbb{R})[[h]]$ (φ is the unique homomorphism of associative $\mathbb{R}[[h]]$-algebras which is the identity on $gl_n(\mathbb{R})[[h]]$). Moreover, F_h is uniquely determined by R_h up to transformations of the form*

$$F_h \mapsto \Delta(f_h).F_h.(f_h \otimes f_h)^{-1},$$

where $f_h \in U(gl_n(\mathbb{R}))[[h]]$ and $\varphi(f_h) = 1$. ∎

We give a concrete application of 6.2.9 and 6.2.11 in Subsection 6.4F.

6.3 Quantized universal enveloping algebras

In this section, we discuss some general properties of QUE algebras. Examples will be given in the following two sections.

A Cocommutative QUE algebras The next result shows, by contrast with 6.1.8, that, although in many cases one may restrict attention to QUE algebras which have the 'usual' algebra structure, the coalgebra structure must be changed.

PROPOSITION 6.3.1 *Let A_h be a cocommutative QUE algebra over $k[[h]]$, where k is a field of characteristic zero. Then A_h is isomorphic as a topological Hopf algebra over $k[[h]]$ to $U(\mathfrak{g}_h)$, where $\mathfrak{g}_h$ is a deformation of a Lie algebra $\mathfrak{g}$.*

The meaning of the last sentence should be clear: $\mathfrak{g}_h$ is isomorphic as a $k[[h]]$-module to $\mathfrak{g}[[h]]$, and $\mathfrak{g}_h/h\mathfrak{g}_h \cong \mathfrak{g}$ as Lie algebras over k. It is understood that $U(\mathfrak{g}_h)$ means the h-adic completion of the universal enveloping algebra of $\mathfrak{g}_h$.

OUTLINE OF PROOF Suppose that $A_h \cong U(\mathfrak{g})[[h]]$ as a $k[[h]]$-module, where $\mathfrak{g}$ is a Lie algebra over k. Let $\varphi_h : A_h \to A_h$ be the composite $\varphi_h = \mu_h \circ \Delta_h$. Note that φ preserves $I_h = \ker(\epsilon_h) \subset A_h$, and that, for all $n \geq 0$, φ (mod h) has eigenvalue 2^n on the image in I_h/hI_h of the nth symmetric power $\mathrm{Sym}^n(\mathfrak{g})$. It follows that

$$A_h = \bigoplus_{n=0}^{\infty} A_h^{(n)},$$

where $A_h^{(0)} = k[[h]].1$ and, for $n \geq 1$, $A_h^{(n)}$ is the generalized eigenspace of φ_h on I_h with eigenvalue 2^n (the direct sum is the h-adic completion of the algebraic direct sum).

Observe that $A_h^{(1)}$ is the set of primitive elements of A_h. Indeed, from the cocommutativity of Δ_h, it follows that φ_h is a coalgebra homomorphism and hence that $\varphi_h(A_h^{(1)}) \subset A_h^{(1)} \otimes A_h^{(0)} + A_h^{(0)} \otimes A_h^{(1)}$ for any $a \in A_h^{(1)}$. Thus, $\Delta_h(a) = b \otimes 1 + 1 \otimes b$ for some $b \in A_h^{(1)}$. But then $b = (\epsilon_h \otimes \mathrm{id})\Delta_h(a) = a$.

Let $\mathfrak{g}_h = A_h^{(1)}$. Then $\mathfrak{g}_h$ is a Lie algebra deformation of $\mathfrak{g}$ and $A_h \cong U(\mathfrak{g}_h)$ as Hopf algebras over $k[[h]]$. ∎

In particular, if $\mathfrak{g}$ is a complex simple Lie algebra, every cocommutative deformation of $U(\mathfrak{g})$ is trivial, since every deformation of $\mathfrak{g}$ as a Lie algebra is trivial (this follows from Whitehead's lemma – see Jacobson (1962)).

B Quasitriangular QUE algebras It is natural to ask which QUE algebras have as their classical limits coboundary, quasitriangular or triangular Lie bialgebras (see Section 2.1). We shall say that a QUE algebra $U_h(\mathfrak{g})$ is *coboundary, quasitriangular* or *triangular* if there is an element $\mathcal{R}_h \in U_h(\mathfrak{g}) \otimes U_h(\mathfrak{g})$ such that $(U_h(\mathfrak{g}), \mathcal{R}_h)$ has these properties as a topological Hopf algebra and $\mathcal{R}_h \equiv 1 \otimes 1$ (mod h) (see 4.2.6).

PROPOSITION 6.3.2 *Let* $(U_h(\mathfrak{g}), \mathcal{R}_h)$ *be a coboundary QUE algebra and define* $r \in U(\mathfrak{g}) \otimes U(\mathfrak{g})$ *by*

$$r \equiv \frac{\mathcal{R}_h - 1 \otimes 1}{h} \quad (\text{mod } h).$$

Then $r \in \mathfrak{g} \otimes \mathfrak{g}$ *and the classical limit of* $U_h(\mathfrak{g})$ *is the coboundary Lie bialgebra* $(\mathfrak{g}, \delta)$ *defined by* r, *i.e.*

$$\delta(x) = (\mathrm{ad}_x \otimes \mathrm{id} + \mathrm{id} \otimes \mathrm{ad}_x)(r)$$

for $x \in \mathfrak{g}$. *Moreover, if* $U_h(\mathfrak{g})$ *is quasitriangular or triangular, so is* $(\mathfrak{g}, \delta)$.

OUTLINE OF PROOF From the coboundary conditions

$$\mathcal{R}_h^{12}(\Delta_h \otimes \mathrm{id})(\mathcal{R}_h) = \mathcal{R}_h^{23}(\mathrm{id} \otimes \Delta)(\mathcal{R}_h), \quad \mathcal{R}_h^{21}\mathcal{R}_h = 1 \otimes 1$$

on $\mathcal{R}_h$, it follows immediately that

(16) $$r_{12} + (\Delta \otimes \mathrm{id})(r) = r_{23} + (\mathrm{id} \otimes \Delta)(r),$$

(17) $$r_{12} = -r_{21}.$$

We show that $r \in \mathfrak{g} \otimes \mathfrak{g}$. The rest of the proof in the coboundary case is completely straightforward. We leave the triangular and quasitriangular cases to the reader.

We assume that $\mathfrak{g}$ is finite dimensional for simplicity. Let $\{x_1, x_2, \ldots, x_d\}$ be any basis of $\mathfrak{g}$, and let

$$x_{\boldsymbol{\lambda}} = \frac{x_1^{\lambda_1}}{\lambda_1!} \frac{x_2^{\lambda_2}}{\lambda_2!} \cdots \frac{x_d^{\lambda_1}}{\lambda_d!}, \qquad (\boldsymbol{\lambda} \in \mathbb{N}^d)$$

be the basis of $U(\mathfrak{g})$ introduced in 4.1.16. Recall that

$$\Delta(x_{\boldsymbol{\lambda}}) = \sum_{\boldsymbol{\mu}+\boldsymbol{\nu}=\boldsymbol{\lambda}} x_{\boldsymbol{\mu}} \otimes x_{\boldsymbol{\nu}},$$

where the addition of elements of $\mathbb{N}^d$ is performed componentwise.

Let

$$r = \sum_{\boldsymbol{\lambda},\boldsymbol{\mu}} r_{\boldsymbol{\lambda}\boldsymbol{\mu}} \, x_{\boldsymbol{\lambda}} \otimes x_{\boldsymbol{\mu}}.$$

From (17), we have $r_{\boldsymbol{\lambda}\boldsymbol{\mu}} = -r_{\boldsymbol{\mu}\boldsymbol{\lambda}}$. Next, (16) is equivalent to the following equations:

(18)
$$r_{\boldsymbol{\lambda}0} = r_{0\boldsymbol{\lambda}} = 0 \quad \text{if } \boldsymbol{\lambda} \neq \mathbf{0} = (0,\ldots,0),$$
$$r_{\boldsymbol{\lambda}+\boldsymbol{\mu}\,\boldsymbol{\nu}} = r_{\boldsymbol{\lambda}\,\boldsymbol{\mu}+\boldsymbol{\nu}} \quad \text{if } \boldsymbol{\lambda}, \boldsymbol{\mu}, \boldsymbol{\nu} \text{ are all non-zero.}$$

Let $\epsilon_i = (0, \ldots, 1, \ldots, 0)$, with 1 in the ith place. We must show that $r_{\boldsymbol{\lambda}\boldsymbol{\mu}} = 0$ unless $\boldsymbol{\lambda} = \epsilon_i$ and $\boldsymbol{\mu} = \epsilon_j$ for some i, j. Suppose, on the contrary, that $\boldsymbol{\lambda} = \epsilon_i + \boldsymbol{\lambda}'$, where $\boldsymbol{\lambda}' \neq \mathbf{0}$, and let $\boldsymbol{\mu} \neq \mathbf{0}$. Then, using (18) and the skewness of r, we have

$$r_{\boldsymbol{\lambda}\boldsymbol{\mu}} = r_{\epsilon_i + \boldsymbol{\lambda}'\,\boldsymbol{\mu}} = r_{\epsilon_i\,\boldsymbol{\lambda}'+\boldsymbol{\mu}}$$
$$= r_{\epsilon_i + \boldsymbol{\mu}\,\boldsymbol{\lambda}'} = -r_{\boldsymbol{\lambda}'\,\epsilon_i + \boldsymbol{\mu}}$$
$$= -r_{\boldsymbol{\lambda}\boldsymbol{\mu}}. \ \blacksquare$$

C QUE duals and doubles We have seen in 4.1.16 that the vector space dual $U(\mathfrak{g})^*$ of the universal enveloping algebra $U(\mathfrak{g})$ of a Lie algebra $\mathfrak{g}$ can be identified with an algebra of formal power series, and that it has a natural Hopf algebra structure, provided we interpret the tensor product $U(\mathfrak{g})^* \otimes U(\mathfrak{g})^*$ in a suitably completed sense. In the quantum case, Drinfel'd (1987) gives the following definition.

DEFINITION 6.3.3 *A quantum formal series Hopf algebra (or QFSH algebra) is a topological Hopf algebra B_h over $k[[h]]$, where k is a field, such that B_h is isomorphic as a $k[[h]]$-module to Map $(I, k[[h]])$ for some set I, and $B_h/hB_h \cong k[[\xi_1, \xi_2, \ldots]]$ as a topological algebra, for some (possibly infinite) sequence of indeterminates $\xi_1, \xi_2, \ldots$.*

As in the classical case, when appropriately defined, the dual of a QUE algebra is a QFSH algebra, and vice versa. But Drinfel'd also constructs a QFSH algebra *inside* any QUE algebra A_h. In fact, define $\Delta_n : A_h \to A_h^{\otimes n}$ by

$$\Delta_n(a) = (\mathrm{id} - \iota_h \epsilon_h)^{\otimes n} \Delta_h^{(n)}(a).$$

Then,

$$B_h = \{a \in A \mid \Delta_n(a) \equiv 0 \pmod{h^n} \text{ for all } n \geq 0\}$$

is a QFSH algebra. The dual of B_h is a new QUE algebra, called the *QUE dual* of A_h.

If the classical limit of A_h is a Lie bialgebra $\mathfrak{g}$ over k, that of the QUE dual of A_h is the dual of $\mathfrak{g}$ in the sense of Lie bialgebras (see Section 1.4).

Moreover, if we replace the dual A_h^* of A_h by its QUE dual in the construction of the quantum double in Section 4.2, then the double of A_h becomes a QUE algebra whose classical limit is the classical double $\mathcal{D}(\mathfrak{g})$ of $\mathfrak{g}$.

Explicit examples of QUE duals and doubles will be discussed in Section 8.3.

D The square of the antipode In 1.3.13, we showed that the composite D of the cocommutator $\delta : \mathfrak{g} \to \mathfrak{g} \otimes \mathfrak{g}$ of a Lie bialgebra $\mathfrak{g}$ with the Lie bracket $[\,,\,] : \mathfrak{g} \otimes \mathfrak{g} \to \mathfrak{g}$ is a derivation of $\mathfrak{g}$ as a Lie bialgebra. We shall now show that D is closely related to the square of the antipode of any quantization of $\mathfrak{g}$.

PROPOSITION 6.3.4 *Let $\mathfrak{g}$ be a Lie bialgebra over a field k, let $U_h(\mathfrak{g})$ be a quantization of $\mathfrak{g}$, and let S_h be the antipode of $U_h(\mathfrak{g})$. Then, $h^{-1}(S_h^2 - \mathrm{id})$ (mod h) is the unique derivation of $U(\mathfrak{g})$ as an associative algebra whose restriction to $\mathfrak{g}$ is equal to $-D/2$.*

Since the antipode S of $U(\mathfrak{g})$ is given by $S(x) = -x$ for all $x \in \mathfrak{g}$, it follows that $S_h^2 \equiv \mathrm{id} \pmod{h}$, so the statement makes sense. Formally, it is clear that $h^{-1}(S_h^2 - \mathrm{id})$ (mod h) is a derivation of $U(\mathfrak{g})$ since it is the 'derivative' $(dS_h^2/dh)_{h=0}$ of a 'one-parameter family' of automorphisms.

PROOF We make use of the *convolution product* $f \bullet_h g$ of $k[[h]]$-module maps $f, g : U_h(\mathfrak{g}) \to U_h(\mathfrak{g})$, defined by

$$f \bullet_h g = \mu_h(f \otimes g)\Delta_h,$$

where μ_h and Δ_h are the multiplication and comultiplication of $U_h(\mathfrak{g})$, respectively (cf. Remark [4] in Subsection 4.1A). Using the Hopf algebra axioms, it is easy to show that $\bullet_h$ gives the set of $k[[h]]$-module maps $U_h(\mathfrak{g}) \to U_h(\mathfrak{g})$ the structure of an associative algebra over $k[[h]]$, for which $\imath_h\epsilon_h$ is the identity and S_h is the inverse of the identity map.

We claim that

$$(S_h^2 - \mathrm{id}) \bullet_h S_h = -S_h\mu_h(\mathrm{id} \otimes S_h)(\Delta_h - \Delta_h^{\mathrm{op}}).$$

In fact, if $a \in U_h(\mathfrak{g})$ and $\Delta_h(a) = \sum_i a_i \otimes a_i'$,

$$((S_h^2 - \mathrm{id}) \bullet_h S_h)(a) = \sum_i S_h^2(a_i)S_h(a_i') - \imath_h\epsilon_h(a)$$

$$= S_h\left(\sum_i a_i'S_h(a_i)\right) - S_h\imath_h\epsilon_h(a)$$

$$= S_h\left(\sum_i a_i'S_h(a_i) - S_h(a_i)a_i'\right).$$

The second equality follows from the fact that S_h is an anti-homomorphism, and the third from the identity $S_h \bullet_h \mathrm{id} = \imath_h\epsilon_h$. This proves our assertion.

Passing to the quotient (mod h), we obtain

$$\mathcal{D} \bullet S = -S\mu(\mathrm{id} \otimes S)\delta,$$

where $\mathcal{D} \equiv h^{-1}(S_h^2 - \mathrm{id})$ (mod h) $: U(\mathfrak{g}) \to U(\mathfrak{g})$ and $\bullet$ is the convolution product for $U(\mathfrak{g})$. Applying the right-hand side to an element $x \in \mathfrak{g}$ and writing $\delta(x) = \sum_j x_j \otimes x_j'$, we find that

$$-S\mu(\mathrm{id} \otimes S)(x) = \sum_j x_j'x_j.$$

On the other hand,

$$(\mathcal{D} \bullet S)(x) = \mathcal{D}(x)S(1) + \mathcal{D}(1)S(x) = \mathcal{D}(x)$$

since we clearly have $\mathcal{D}(1) = 0$. Finally,

$$D(x) = \sum_j [x_j, x_j'] = -2\sum_j x_j'x_j$$

because δ is skew-symmetric.

This proves that $\mathcal{D}(x) = -D(x)/2$ for $x \in \mathfrak{g}$. If we show that $\mathcal{D}$ is a derivation, it will follow that the equation holds for all $x \in U(\mathfrak{g})$.

Suppose then that x, $y \in U(\mathfrak{g})$, and choose a, $b \in U_h(\mathfrak{g})$ such that $x \equiv a$ and $y \equiv b \pmod{h}$. Then, $\pmod{h}$ we have

$$\mathcal{D}(xy) - \mathcal{D}(x)y - x\mathcal{D}(y) \equiv \frac{S_h^2(ab) - ab}{h} - \frac{(S_h^2(a) - a)b}{h} - \frac{a(S_h^2(b) - b)}{h}$$

$$\equiv \frac{(S_h^2(a) - a)(S_h^2(b) - b)}{h}$$

$$\equiv 0. \quad \blacksquare$$

6.4 The basic example

We now describe a quantization of the standard Lie bialgebra structure on $sl_2(\mathbb{C})$ defined in 1.3.8 (one could replace $\mathbb{C}$ by any field of characteristic zero). Many of the results in this section will be generalized later in the book.

A Construction of the standard quantization The 'standard' Lie bialgebra structure

$$\delta(H) = 0, \quad \delta(X^+) = X^+ \wedge H, \quad \delta(X^-) = X^- \wedge H$$

on $sl_2(\mathbb{C})$ is defined by the r-matrix $r = X^+ \wedge X^-$ (cf. 2.1.8). Let $\mathfrak{b}^\pm$ be the subalgebras of $sl_2(\mathbb{C})$ spanned by $\{H, X^\pm\}$, and let $\mathfrak{h} = \mathfrak{b}^+ \cap \mathfrak{b}^-$. Obviously, $\delta(\mathfrak{b}^\pm) \subset \mathfrak{b}^\pm \otimes \mathfrak{b}^\pm$, so $\mathfrak{b}^\pm$ are actually sub-Lie bialgebras of $sl_2(\mathbb{C})$. Our strategy is to quantize the Lie bialgebras $(\mathfrak{b}^\pm, \delta|_{\mathfrak{b}^\pm})$ in such a way that the quantizations 'fit together' to give a quantization of $(sl_2(\mathbb{C}), \delta)$.

We look for a quantization of $\mathfrak{b}^+$ which is isomorphic to $U(\mathfrak{b}^+)[[h]]$ as an algebra. Since $\delta(H) = 0$, the choice

$$(19) \qquad\qquad \Delta_h(H) = H \otimes 1 + 1 \otimes H$$

satisfies the condition in 6.2.6. To guess $\Delta_h(X^+)$, note that $U(\mathfrak{b}^+)$ is a graded vector space, with $\deg(H) = 0$, $\deg(X^+) = 1$, and that its comultiplication Δ preserves the grading. To preserve this grading, we try

$$(20) \qquad\qquad \Delta_h(X^+) = X^+ \otimes f + g \otimes X^+,$$

where f, $g \in U(\mathfrak{h})[[h]]$. The condition $\Delta_h \equiv \Delta \pmod{h}$ forces f, $g \equiv 1 \pmod{h}$.

It is easy to check that, for any choice of f, g, the Δ_h defined by (19) and (20) extends to an algebra homomorphism

$$\Delta_h : U(\mathfrak{b}^+)[[h]] \to U(\mathfrak{b}^+)[[h]] \otimes U(\mathfrak{b}^+)[[h]],$$

and that Δ_h is coassociative if f and g are group-like for Δ_h. The tensor product on the right-hand side is understood to be completed in the h-adic topology (see Subsection 4.1D) – in this case, the completion is simply $(U(\mathfrak{b}^+) \otimes U(\mathfrak{b}^+))[[h]]$.

LEMMA 6.4.1 *The group-like elements f of $U(\mathfrak{h})[[h]]$ such that $f \equiv 1$ (mod h) are precisely those of the form $f = e^{h\mu H}$, where $\mu \in \mathbb{C}[[h]]$.*

PROOF First, we compute

$$
\begin{aligned}
\Delta_h(e^{h\mu H}) &= \sum_{n=0}^{\infty} \frac{(h\mu)^n}{n!}(H \otimes 1 + 1 \otimes H)^n \\
&= \sum_{n=0}^{\infty} \sum_{m=0}^{n} \binom{n}{m} \frac{(h\mu)^n}{n!} H^m \otimes H^{n-m} \\
&= \sum_{m=0}^{\infty} \sum_{k=0}^{\infty} \frac{(h\mu)^{m+k}}{m!k!} H^m \otimes H^k \\
&= e^{h\mu H} \otimes e^{h\mu H}.
\end{aligned}
$$

For the converse, let

$$
f = 1 + \sum_{n=1}^{\infty} f_n H^n \in U(\mathfrak{h})[[h]]
$$

be group-like, where $f_n \in \mathbb{C}[[h]]$. A computation similar to that above shows that $f_{m+n} = f_m f_n$ for all $m, n \geq 1$. Thus, if $f_1 = h\mu$, then $f = e^{h\mu H}$. ∎

We can therefore assume that

$$
\Delta_h(X^+) = X^+ \otimes e^{h\mu H} + e^{h\nu H} \otimes X^+
$$

for some $\mu, \nu \in \mathbb{C}[[h]]$. In fact, by replacing X^+ by $e^{-h\nu H} X^+$, we may assume that $\nu = 0$. Definition 6.2.6 then requires that $\mu \equiv 1$ (mod h). The simplest choice is $\mu = 1$, giving

(21) $$\Delta_h(X^+) = X^+ \otimes e^{hH} + 1 \otimes X^+.$$

(In fact, up to automorphisms of $\mathbb{C}[[h]]$ of the form $h \to h + O(h^2)$, this is the only possible choice.) It is now easy to check that, if we set

(22) $$S_h(H) = -H, \quad S_h(X^+) = -X^+ e^{-hH}, \quad \epsilon_h(H) = \epsilon_h(X^+) = 0,$$

then Δ_h, S_h and ϵ_h extend to algebra homomorphisms (an anti-homomorphism in the case of S_h) and satisfy the Hopf algebra axioms.

Of course, $(\mathfrak{b}^-, \delta|_{\mathfrak{b}^-})$ may be quantized in the same way. As above, there is a choice in the definition of $\Delta_h(X^-)$, and the most convenient turns out to be

(23) $$\Delta_h(X^-) = X^- \otimes 1 + e^{-hH} \otimes X^-.$$

The antipode and counit are then

(24)
$$S_h(H) = -H, \quad S_h(X^-) = -e^{hH}X^-,$$
$$\epsilon_h(H) = \epsilon_h(X^-) = 0.$$

We take the quantization $U_h(sl_2(\mathbb{C}))$ to be $U(sl_2(\mathbb{C}))[[h]]$ as a $\mathbb{C}[[h]]$-module, with the coalgebra structure defined by (19)–(24). We must choose the algebra structure so that Δ_h is a homomorphism. We compute

$$\Delta_h[X^+, X^-] = [X^+ \otimes e^{hH} + 1 \otimes X^+, \; X^- \otimes 1 + e^{-hH} \otimes X^-]$$
$$= [X^+, X^-] \otimes e^{hH} + e^{-hH} \otimes [X^+, X^-] + X^+ e^{-hH} \otimes e^{hH} X^-$$
(25)
$$- e^{-hH} X^+ \otimes X^- e^{hH}.$$

To simplify this, we use

LEMMA 6.4.2 *Let A be an algebra over $\mathbb{C}[[h]]$, let a, b, $c \in A$ and assume that a commutes with c. Then,*

(i) $e^{h(a+c)} = e^{ha}e^{hc}$;

(ii) if $[a, b] = bc$, then

$$e^{ha}be^{-ha} = be^{hc}.$$

PROOF The first part is well known (and easy to prove). For the second, we have $ab = b(a + c)$, and hence $a^n b = b(a + c)^n$ by an obvious induction. Hence,

$$e^{ha}b = \sum_{n=0}^{\infty} \frac{h^n}{n!} a^n b = \sum_{n=0}^{\infty} \frac{h^n}{n!} b(a + c)^n$$
$$= be^{h(a+c)} = be^{hc}e^{ha}. \quad \blacksquare$$

It follows from this lemma that the last two terms on the right-hand side of (25) cancel, so that

(26) $$\Delta_h[X^+, X^-] = [X^+, X^-] \otimes e^{hH} + e^{-hH} \otimes [X^+, X^-].$$

Comparing with (19), we see that the classical relation $[X^+, X^-] = H$ must be changed. Recalling that $e^{\pm hH}$ are group-like, it is easy to see that taking $[X^+, X^-]$ to be any multiple of $e^{hH} - e^{-hH}$ will satisfy (26). To get the correct classical limit, we take

$$[X^+, X^-] = \frac{e^{hH} - e^{-hH}}{e^h - e^{-h}}.$$

We summarize our computations in

DEFINITION–PROPOSITION 6.4.3 *Quantum $sl_2(\mathbb{C})$ is the topological Hopf algebra $U_h(sl_2(\mathbb{C}))$ over $\mathbb{C}[[h]]$ defined as follows.*

Let $P = \mathbb{C}\{H, X^+, X^-\}$ be the algebra of non-commutative polynomials in three generators H, X^+ and X^-, let I be the two-sided ideal of $P[[h]]$ generated by

$$(27) \qquad [H, X^+] - 2X^+, \quad [H, X^-] + 2X^-, \quad [X^+, X^-] - \frac{e^{hH} - e^{-hH}}{e^h - e^{-h}},$$

and let $\bar{I}$ be the closure of I in the h-adic topology. Then, $U_h(sl_2(\mathbb{C})) = P[[h]]/\bar{I}$ as an algebra over $\mathbb{C}[[h]]$.

There are homomorphisms of $\mathbb{C}[[h]]$-algebras

$$\Delta_h : U_h(sl_2(\mathbb{C})) \to U_h(sl_2(\mathbb{C})) \otimes U_h(sl_2(\mathbb{C})),$$

$$\epsilon_h : U_h(sl_2(\mathbb{C})) \to \mathbb{C}[[h]], \quad S_h : U_h(sl_2(\mathbb{C})) \to U_h(sl_2(\mathbb{C})),$$

(an anti-homomorphism in the case S_h) given on generators by (19) and (21)–(24), which define the structure of a topological Hopf algebra on $U_h(sl_2(\mathbb{C}))$.

Moreover, $U_h(sl_2(\mathbb{C}))$ is a QUE algebra whose classical limit is the Lie bialgebra structure on $sl_2(\mathbb{C})$ defined by $r = X^+ \wedge X^-$.

Remarks [1] All $\mathbb{C}[[h]]$-modules which appear are understood to have the h-adic topology, and all tensor products to be the completion of the algebraic tensor products.

[2] To show that Δ_h (say) extends to an algebra homomorphism

$$\Delta_h : U_h(sl_2(\mathbb{C})) \to U_h(sl_2(\mathbb{C})) \otimes U_h(sl_2(\mathbb{C})),$$

note that it extends trivially to a homomorphism $P[[h]] \to P[[h]] \otimes P[[h]]$, and that

$$(28) \qquad \Delta_h(I) \subset P[[h]] \otimes I + I \otimes P[[h]]$$

(it is enough to observe that $\Delta_h(H)$, $\Delta_h(X^+)$ and $\Delta_h(X^-)$ are contained in the right-hand side of (28)). Since Δ_h is $\mathbb{C}[[h]]$-linear, it follows that $\bar{I}$ is a Hopf ideal in $P[[h]]$:

$$\Delta_h(\bar{I}) \subset P[[h]] \otimes \bar{I} + \bar{I} \otimes P[[h]].$$

[3] The square of the antipode of $U_h(sl_2(\mathbb{C}))$ is conjugation by e^{hH}:

$$S_h^2(a) = e^{hH} a e^{-hH}$$

for all $a \in U_h(sl_2(\mathbb{C}))$. (To prove this, it is enough to verify that the equation holds when $a = H$, X^+ and X^-.) Taking 6.3.4 into account, this agrees with one of the examples in 1.3.16.

[4] To complete the proof that $U_h(sl_2(\mathbb{C}))$ is a QUE algebra, we must show that it is isomorphic to $U(sl_2(\mathbb{C}))[[h]]$ as a $\mathbb{C}[[h]]$-module. This is discussed in the next subsection. Note that the completeness of $U_h(sl_2(\mathbb{C}))$ follows from that of $P[[h]]$ and the fact that $\bar{I}$ is closed.

[5] We shall often abbreviate the second paragraph in 6.4.3 by saying that $U_h(sl_2(\mathbb{C}))$ is 'topologically generated by H, X^+ and X^- with defining relations $[H, X^+] = 2X^+$, etc.' $\triangle$

B Algebra structure We begin by introducing a quantum analogue of the Casimir element

$$\bar{\Omega} = \frac{1}{4}(\bar{H} + 1)^2 + \bar{X}^- \bar{X}^+ = \frac{1}{4}(\bar{H} - 1)^2 + \bar{X}^+ \bar{X}^-$$

of $U(sl_2(\mathbb{C}))$. (Here, and in the remainder of this section, we use bars to distinguish elements of $U(sl_2(\mathbb{C}))$ from those of $U_h(sl_2(\mathbb{C}))$.)

DEFINITION 6.4.4 *The quantum Casimir element of $U_h(sl_2(\mathbb{C}))$ is*

$$\Omega = \left(\frac{\sinh \frac{1}{2} h(H + 1)}{\sinh h} \right)^2 + X^- X^+$$

$$= \left(\frac{\sinh \frac{1}{2} h(H - 1)}{\sinh h} \right)^2 + X^+ X^-.$$

The verification that the two expressions for Ω are equal is straightforward. It is clear that $\Omega \equiv \bar{\Omega} \pmod{h}$.

PROPOSITION 6.4.5 *The quantum Casimir element is a topological generator of the centre of $U_h(sl_2(\mathbb{C}))$.*

This means that the $\mathbb{C}[[h]]$-subalgebra of $U_h(sl_2(\mathbb{C}))$ generated by Ω is dense in the centre of $U_h(sl_2(\mathbb{C}))$ in the h-adic topology. To prove that Ω lies in the centre, it is of course enough to show that it commutes with the generators H, X^+ and X^-. We leave this to the reader, as similar but more involved computations will be carried out in the proof of the next proposition, which also implies that Ω generates the centre.

PROPOSITION 6.4.6 *There is an isomorphism*

$$\varphi : U_h(sl_2(\mathbb{C})) \to U(sl_2(\mathbb{C}))[[h]]$$

of algebras over $\mathbb{C}[[h]]$ such that

$$\varphi(H) = \bar{H}, \quad \varphi(X^-) = \bar{X}^-,$$

$$\varphi(X^+) = 2 \left(\frac{\cosh h(\bar{H} - 1) - \cosh 2h\sqrt{\bar{\Omega}}}{((\bar{H} - 1)^2 - 4\bar{\Omega}) \sinh^2 h} \right) \bar{X}^+.$$

Note that the formula for $\varphi(X^+)$ makes sense because, if u and v are indeterminates,

$$\frac{\cosh u - \cosh v}{u^2 - v^2}$$

can be written as a formal power series $f(u^2, v^2)$. Note also that φ is equal to the identity (mod h).

PROOF To show that φ extends to a homomorphism of algebras, we must check that

(29) $$\varphi(X^+)\bar{X}^- - \bar{X}^-\varphi(X^+) = \frac{\sinh h\bar{H}}{\sinh h}$$

(the other relations are obvious). From the formula for $\bar{\Omega}$,

$$\varphi(X^+)\bar{X}^- = -\left(\frac{\cosh h(\bar{H} - 1) - \cosh 2h\sqrt{\bar{\Omega}}}{2\sinh^2 h}\right).$$

On the other hand, since $\bar{X}^-(\bar{H} - 1) = (\bar{H} + 1)\bar{X}^-$, it follows that

$$\bar{X}^- f(h^2(\bar{H} - 1)^2, 4h^2\bar{\Omega}) = f(h^2(\bar{H} + 1)^2, 4h^2\bar{\Omega})\bar{X}^-,$$

and hence that

$$\bar{X}^-\varphi(X^+) = -\left(\frac{\cosh h(\bar{H} + 1) - \cosh 2h\sqrt{\bar{\Omega}}}{2\sinh^2 h}\right).$$

Subtracting gives (29).

We prove that φ is an isomorphism by constructing its inverse. First we compute

$$\varphi(\Omega) = \frac{\sinh^2 \frac{1}{2}h(\bar{H} + 1)}{\sinh^2 h} + \bar{X}^-\varphi(X^+)$$

$$= \frac{\sinh^2 \frac{1}{2}h(\bar{H} + 1)}{\sinh^2 h} - \left(\frac{\cosh h(\bar{H} + 1) - \cosh 2h\sqrt{\bar{\Omega}}}{2\sinh^2 h}\right)$$

$$= \frac{\sinh^2 h\sqrt{\bar{\Omega}}}{\sinh^2 h}.$$

Let

$$g(h, u) = \frac{\sinh^2 h\sqrt{u}}{\sinh^2 h} \in \mathbb{C}[[h, u]],$$

where u is an indeterminate. Since

$$g = u\left(\frac{h^2}{\sinh^2 h} + O(u)\right),$$

and since $h^2/\sinh^2 h$ is an invertible element of $\mathbb{C}[[h]]$, it follows that there exists a power series $\tilde{g} \in \mathbb{C}[[h, u]]$ such that

$$g(h, \tilde{g}(h, u)) = \tilde{g}(h, g(h, u)) = u.$$

Define $\psi(\bar{H}) = H$, $\psi(\bar{X}^-) = X^-$ and

$$\psi(\bar{X}^+) = \frac{1}{2}\left(\frac{((H-1)^2 - 4\tilde{g}(h,\Omega))\sinh^2 h}{\cosh h(H-1) - \cosh 2h\sqrt{\tilde{g}(h,\Omega)}}\right) X^+.$$

It is easy to see that $\psi(\varphi(X^+)) = X^+$ and that $\varphi(\psi(\bar{X}^+)) = \bar{X}^+$, so to complete the proof it suffices to show that ψ extends to an algebra homomorphism $\psi : U(sl_2(\mathbb{C}))[[h]] \to U_h(sl_2(\mathbb{C}))$. For this, we compute

$$\psi(\bar{X}^+)X^- - X^-\psi(\bar{X}^+)$$

$$= \frac{1}{2}\left(\frac{((H-1)^2 - 4\tilde{g}(h,\Omega))\sinh^2 h}{\cosh h(H-1) - \cosh 2h\sqrt{\tilde{g}(h,\Omega)}}\right) X^+ X^-$$

$$- \frac{1}{2}\left(\frac{((H+1)^2 - 4\tilde{g}(h,\Omega))\sinh^2 h}{\cosh h(H+1) - \cosh 2h\sqrt{\tilde{g}(h,\Omega)}}\right) X^- X^+$$

using $[H, X^-] = -2X^-$, which equals

$$\frac{1}{2}\left(\frac{((H-1)^2 - 4\tilde{g}(h,\Omega))\sinh^2 h}{\cosh h(H-1) - \cosh 2h\sqrt{\tilde{g}(h,\Omega)}}\right)\left(\Omega - \frac{\sinh^2 \frac{1}{2}h(H-1)}{\sinh^2 h}\right)$$

$$- \frac{1}{2}\left(\frac{((H+1)^2 - 4\tilde{g}(h,\Omega))\sinh^2 h}{\cosh h(H+1) - \cosh 2h\sqrt{\tilde{g}(h,\Omega)}}\right)\left(\Omega - \frac{\sinh^2 \frac{1}{2}h(H+1)}{\sinh^2 h}\right).$$

Now,

$$\cosh h(H-1) - \cosh 2h\sqrt{\tilde{g}(h,\Omega)}$$

$$= 2\left(\sinh^2 \frac{1}{2}h(H-1) - \sinh^2 h\sqrt{\tilde{g}(h,\Omega)}\right)$$

$$= 2\left(\sinh^2 \frac{1}{2}h(H-1) - \Omega\sinh^2 h\right).$$

Thus,

$$\psi(\bar{X}^+)X^- - X^-\psi(\bar{X}^+)$$

$$= \frac{1}{4}((H+1)^2 - 4\tilde{g}(h,\Omega)) - \frac{1}{4}((H-1)^2 - 4\tilde{g}(h,\Omega))$$

$$= H. \quad \blacksquare$$

This computation completes the proof of 6.4.5. For we know that Ω generates the centre of $U(sl_2(\mathbb{C}))[[h]]$, and hence so does $g(h, \bar{\Omega}) = \bar{\Omega}(1 + O(h^2))$. It follows that $\psi(g(h, \bar{\Omega})) = \Omega$ generates the centre of $U_h(sl_2(\mathbb{C}))$. ∎

It also completes 6.4.3. For, using the isomorphism φ, we can identify $U_h(sl_2(\mathbb{C}))$ with $U(sl_2(\mathbb{C}))[[h]]$ as an algebra, and in particular as a $\mathbb{C}[[h]]$-module. Since, on the generators H, X^+ and X^-, $\varphi \equiv$ id, $\epsilon_h \equiv \epsilon$, $\Delta_h \equiv \Delta$ and $S_h \equiv S \pmod h$, it follows that $U_h(sl_2(\mathbb{C}))$ is a deformation of $U(sl_2(\mathbb{C}))$ in the sense of 6.1.1. And of course Δ_h was constructed so as to give the standard Lie bialgebra structure on $sl_2(\mathbb{C})$ in the classical limit. ∎

C PBW basis In the structure and representation theory of Lie algebras, an important role is played by the Poincaré–Birkhoff–Witt theorem, which associates to any (ordered) basis of a Lie algebra a basis of its universal enveloping algebra. The following is an analogue of this result for $U_h(sl_2(\mathbb{C}))$:

PROPOSITION 6.4.7 *The monomials* $(X^-)^r H^s (X^+)^t$, *for* r, s, $t \in \mathbb{N}$, *form a topological basis of* $U_h(sl_2(\mathbb{C}))$.

PROOF This can be deduced from the isomorphism 6.4.6, together with the classical PBW theorem. However, the following direct argument, which makes use of the comultiplication of $U_h(sl_2(\mathbb{C}))$, is instructive. The fact that the given monomials span is easy; the point is to prove linear independence.

First, the generators H, $X^\pm \in U_h = U_h(sl_2(\mathbb{C}))$ are all non-zero, for we shall exhibit in 6.4.10 representations of U_h in which they act as non-zero operators. To prove, say, that $(X^+)^t \neq 0$ for all t, note that U_h is a $\mathbb{Z}$-graded algebra with $\deg(X^\pm) = \pm 1$. One now observes that the term of degree $(1, 1, \ldots, 1)$ in $\Delta_h^{(t)}(X^+)^t$ is

$$\prod_{s=1}^{t} \left(\frac{e^{sh} - e^{-sh}}{e^h - e^{-h}} \right) X^+ \otimes e^{hH} X^+ \otimes e^{2hH} X^+ \otimes \cdots \otimes e^{(t-1)hH} X^+.$$

This is non-zero by the previous observation, and hence so is $\Delta_h^{(t)}(X^+)^t$.

If there were a linear relation

$$\sum_s c_s H^s = 0$$

in U_h, with $c_s \in \mathbb{C}[[h]]$, then applying the left-hand side to the vector $v_0^{(m)}$ in the representation V_m defined in 6.4.10 shows that

$$\sum_s c_s m^s = 0$$

for all $m \geq 0$. This implies that all the c_s are zero.

Suppose now that we have a linear relation between monomials $H^s(X^+)^t$. Taking into account the grading, we can assume the relation is of the form

$$\sum_{s=0}^{S} c_s H^s (X^+)^t = 0,$$

where $c_S \neq 0$. Applying Δ_h, using the linear independence of the powers of H, and looking for the term of maximal degree in the second factor, we see that the term $c_S H^S \otimes (X^+)^t$ must itself be zero. But this contradicts previous results.

Finally, suppose that there is a linear relation between the monomials $(X^-)^r H^s (X^+)^t$. Applying Δ_h and looking for the term of maximal degree in the second factor, we see that we can assume that only one value of t occurs in the relation. Thus, we consider a relation of the form

$$\sum_{s=0}^{S} c_s (X^-)^r H^s (X^+)^t = 0$$

where $c_S \neq 0$. Applying $\Delta_h^{(2)}$, looking for the term of minimal degree in the first factor and maximal degree in the third factor, we obtain

$$\sum_{s=0}^{S} c_s \Delta_h^{(2)} (H^s) = 0.$$

The term $H^S \otimes 1 \otimes 1$ occurs on the left-hand side with coefficient c_S, contradicting the linear independence of the powers of H. ∎

D Quasitriangular structure The classical limit of $U_h(sl_2(\mathbb{C}))$ is a quasi-triangular Lie bialgebra structure on $sl_2(\mathbb{C})$, so we might expect, in view of 6.2.7, that $U_h(sl_2(\mathbb{C}))$ is itself a quasitriangular QUE algebra (cf. 6.3.2). We now show that this is indeed the case.

First, we recall the notion of *q-binomial coefficients*. If q is an indeterminate, and $m \geq n \in \mathbb{N}$, define

$$[n]_q = \frac{q^n - q^{-n}}{q - q^{-1}},$$

$$[n]_q! = [n]_q.[n-1]_q. \cdots .[2]_q.[1]_q,$$

$$\begin{bmatrix} m \\ n \end{bmatrix}_q = \frac{[m]_q!}{[n]_q![m-n]_q!}.$$

These symbols are well-defined elements of the field $\mathbb{Q}(q)$ of rational functions of q over $\mathbb{Q}$; in fact, it is obvious that $[n]_q \in \mathbb{Z}[q, q^{-1}]$. Moreover, taking $q = e^h$, it is clear that $[n]_{e^h} \equiv n \pmod{h}$.

PROPOSITION 6.4.8 *The Hopf algebra* $U_h(sl_2(\mathbb{C}))$ *is topologically quasitriangular with universal R-matrix*

$$(30) \qquad \mathcal{R}_h = \sum_{n=0}^{\infty} R_n(h) e^{\frac{1}{2}h(H \otimes H)}(X^+)^n \otimes (X^-)^n,$$

where

$$R_n(h) = \frac{q^{\frac{1}{2}n(n+1)}(1 - q^{-2})^n}{[n]_q!}, \qquad (q = e^h).$$

Note that $\mathcal{R}_h$ is a well defined invertible element of the completion of the tensor product $U_h(sl_2(\mathbb{C})) \otimes U_h(sl_2(\mathbb{C}))$, since $R_n(h) \equiv 2^n h^n / n! \pmod{h^{n+1}}$.

PROOF First we verify that

$$(31) \qquad (\Delta_h \otimes \mathrm{id})(\mathcal{R}_h) = \mathcal{R}_h^{13} \mathcal{R}_h^{23}.$$

The left-hand side of (31) can be written as a sum,

$$\sum_{m,n \geq 0} Q_{m,n}(X^+)^m \otimes (X^+)^n \otimes (X^-)^{m+n},$$

where $Q_{m,n} \in U(\mathfrak{h})[[h]]$ is given by

$$Q_{m,n} = q^{-mn} \begin{bmatrix} m+n \\ n \end{bmatrix}_q R_{m+n}(h)(\Delta_h \otimes \mathrm{id})(e^{\frac{1}{2}h(H \otimes H)}).(1 \otimes e^{mhH} \otimes 1).$$

The term involving $(X^+)^m \otimes (X^+)^n \otimes (X^-)^{m+n}$ on the right-hand side of (31) is $R_m(h)R_n(h)$ multiplied by

$$e^{\frac{1}{2}h(H \otimes 1 \otimes H)}((X^+)^m \otimes 1 \otimes (X^-)^m) e^{\frac{1}{2}h(1 \otimes H \otimes H)}(1 \otimes (X^+)^n \otimes (X^-)^n).$$

By induction on m, we find that

$$[1 \otimes H \otimes H, (X^+)^m \otimes 1 \otimes (X^-)^m] = -2m(X^+)^m \otimes H \otimes (X^-)^m,$$

and hence, by 6.4.2,

$$e^{-\frac{1}{2}h(1 \otimes H \otimes H)}((X^+)^m \otimes 1 \otimes (X^-)^m)e^{\frac{1}{2}h(1 \otimes H \otimes H)}$$
$$= (1 \otimes e^{mhH} \otimes 1)((X^+)^m \otimes 1 \otimes (X^-)^m).$$

On the other hand,

$$(\Delta_h \otimes \mathrm{id})(e^{\frac{1}{2}h(H \otimes H)}) = e^{\frac{1}{2}h(\Delta_h(H) \otimes H)}$$
$$= e^{\frac{1}{2}h(H \otimes 1 \otimes H + 1 \otimes H \otimes H)}.$$

Thus, to verify (31), it suffices to show that

$$q^{-mn} \begin{bmatrix} m+n \\ n \end{bmatrix}_q R_{m+n}(h) = R_m(h)R_n(h),$$

and this is an easy computation.

The relation

$$(\mathrm{id} \otimes \Delta_h)(\mathcal{R}_h) = \mathcal{R}_h^{13}\mathcal{R}_h^{12}$$

is verified in a similar way.

To show that

$$\Delta_h^{\mathrm{op}}(a) = \mathcal{R}_h\Delta_h(a)\mathcal{R}_h^{-1}$$

for all $a \in U_h(sl_2(\mathbf{C}))$, note that both sides of this equation define algebra homomorphisms $U_h(sl_2(\mathbf{C})) \to U_h(sl_2(\mathbf{C})) \otimes U_h(sl_2(\mathbf{C}))$, so it suffices to check that

$$\Delta_h^{\mathrm{op}}(a)\mathcal{R}_h = \mathcal{R}_h\Delta_h(a)$$

when $a = H$, X^+ and X^-. When $a = X^+$, for example, we must show that

$$(X^+ \otimes 1 + e^{hH} \otimes X^+). \sum_{n=0}^{\infty} R_n(h)e^{\frac{1}{2}h(H \otimes H)}(X^+)^n \otimes (X^-)^n$$

$$= \left\{ \sum_{n=0}^{\infty} R_n(h)e^{\frac{1}{2}h(H \otimes H)}(X^+)^n \otimes (X^-)^n \right\} .(X^+ \otimes e^{hH} + 1 \otimes X^+).$$

We find that the difference between the left- and right-hand sides is

$$\sum_{n=0}^{\infty} e^{\frac{1}{2}h(H \otimes H)} \left\{ R_n(h)(1 \otimes (e^{-hH} - q^{2n}e^{hH})) \right.$$

$$\left. + R_{n+1}(h)[n+1]_q \left(1 \otimes \left(\frac{q^n e^{hH} - q^{-n}e^{-hH}}{q - q^{-1}} \right) \right) \right\} ((X^+)^{n+1} \otimes (X^-)^n).$$

The coefficient of $(X^+)^{n+1} \otimes (X^-)^n$ vanishes provided that

$$[n+1]_q R_{n+1}(h) = q^n(q - q^{-1})R_n(h).$$

This relation is easily verified. ∎

Remark It is clear that the second part of the above computation could have been used to actually compute $\mathcal{R}_h$, given the general form of $\mathcal{R}_h$ in the statement of the proposition. This suggests that there may be a uniqueness theorem for the universal R-matrix; we shall discuss such a result in Subsection 8.3E. △

From 4.2.7, we deduce

COROLLARY 6.4.9 *The element* $\mathcal{R}_h \in U_h(sl_2(\mathbb{C})) \otimes U_h(sl_2(\mathbb{C}))$ *(completed tensor product) defined in (30) satisfies the quantum Yang–Baxter equation*

$$\mathcal{R}_h^{12}\mathcal{R}_h^{13}\mathcal{R}_h^{23} = \mathcal{R}_h^{23}\mathcal{R}_h^{13}\mathcal{R}_h^{12}. \quad \blacksquare$$

The universal R-matrix $\mathcal{R}_h$ can be derived by 'almost' realizing $U_h(sl_2(\mathbb{C}))$ as a quantum double and using 4.2.12 (cf. the computation of the r-matrix for the standard Lie bialgebra structure on complex simple Lie algebras in Subsection 2.1D). We have preferred to give a direct proof in the sl_2 case, partly to show that the proof is completely elementary, and partly because it will be convenient to have the R-matrix in the sl_2 case available before we discuss the general case. However, we shall return to the quantum double method in Chapter 8, when we investigate the structure of the quantization of the standard Lie bialgebra structure on an arbitrary complex simple Lie algebra.

Remark Note that, by 4.2.3, if $u_h = \mu_h(S_h \otimes \mathrm{id})(\mathcal{R}_h^{21})$, then $S_h^2(a) = u_h a u_h^{-1}$ for all $a \in U_h(sl_2(\mathbb{C}))$. On the other hand, it is easy to see directly that $S_h^2(a) = e^{hH}ae^{-hH}$ when a is one of the generators X^+, X^-, H, and hence for all a. It follows that $v_h = e^{-hH}u_h = u_h e^{-hH}$ lies in the centre of $U_h(sl_2(\mathbb{C}))$. Moreover, since e^{hH} is obviously group-like, it follows that $(U_h(sl_2(\mathbb{C})), \mathcal{R}_h, v_h)$ is a topological ribbon Hopf algebra (see Subsection 4.2C). △

E Representations Thanks to 6.4.6, it is easy to describe the quasitensor category **rep**$_A$ when $A = U_h(sl_2(\mathbb{C}))$. We recall from Section 4.1 that **rep**$_A$ consists of the representations of a Hopf algebra A defined over a commutative ring k which are finitely generated and projective as k-modules. In this situation, $k = \mathbb{C}[[h]]$ is a local ring, so that **rep**$_A$ consists precisely of the representations which are free and of finite rank as $\mathbb{C}[[h]]$-modules. We shall simply call such representations *finite dimensional* from now on.

If V_h is a representation of $U_h(sl_2(\mathbb{C}))$, then V_h/hV_h is a representation of $sl_2(\mathbb{C})$. Conversely, if V is a representation of $sl_2(\mathbb{C})$, then $V[[h]]$ is a representation of $U_h(sl_2(\mathbb{C}))$, by composing the obvious action of $U(sl_2(\mathbb{C}))[[h]]$ on $V[[h]]$ with the isomorphism $U_h(sl_2(\mathbb{C})) \to U(sl_2(\mathbb{C}))[[h]]$ constructed in 6.4.6. Moreover, these two operations are mutually inverse, at least for finite-dimensional representations, because it is known that such representations of $sl_2(\mathbb{C})$ admit no non-trivial deformations (see Jacobson (1962)). Finally, irreducible representations of $sl_2(\mathbb{C})$ correspond to indecomposable representations of $U_h(sl_2(\mathbb{C}))$. Of course, (non-zero) representations V_h of $U_h(sl_2(\mathbb{C}))$ which are free as $\mathbb{C}[[h]]$-modules are never irreducible, since hV_h is a proper subrepresentation.

After these observations, it is easy to give a complete description of the objects of **rep**$_A$ when $A = U_h(sl_2(\mathbb{C}))$.

PROPOSITION 6.4.10 *Let V_m be the free $\mathbb{C}[[h]]$-module of rank $m+1$ with basis $\{v_0^{(m)}, \ldots, v_m^{(m)}\}$. Then the formulas*

(32)
$$H.v_r^{(m)} = (m - 2r)v_r^{(m)},$$
$$X^+.v_r^{(m)} = [m - r + 1]_q v_{r-1}^{(m)}, \quad X^-.v_r^{(m)} = [r + 1]_q v_{r+1}^{(m)},$$

where $q = e^h$ and $v_{-1}^{(m)} = v_{m+1}^{(m)} = 0$, define an indecomposable representation of $U_h(sl_2(\mathbb{C}))$ on V_m. Conversely, every finite-dimensional indecomposable representation of $U_h(sl_2(\mathbb{C}))$ is isomorphic to one of the V_m.

PROOF The fact that (32) does define a representation of $U_h(sl_2(\mathbb{C}))$ is a straightforward calculation, using the relations (27). Now, $\bar{V}_m = V_m/hV_m$ has basis $\{\bar{v}_0^{(m)}, \ldots, \bar{v}_m^{(m)}\}$, where $v_r^{(m)} \equiv \bar{v}_r^{(m)} \pmod{h}$, with action of $sl_2(\mathbb{C})$ given by

$$\bar{H}.\bar{v}_r^{(m)} = (m - 2r)\bar{v}_r^{(m)}, \quad \bar{X}^+.\bar{v}_r^{(m)} = (m - r + 1)\bar{v}_{r-1}^{(m)}, \quad \bar{X}^-.\bar{v}_r^{(m)} = (r + 1)\bar{v}_{r+1}^{(m)}.$$

Thus, $\bar{V}_m$ is the $(m + 1)$-dimensional irreducible representation of $sl_2(\mathbb{C})$. Proposition 6.4.10 therefore follows from the preceding discussion, together with the well known fact that the $\bar{V}_m$ are, up to isomorphism, all of the finite-dimensional irreducible representations of $sl_2(\mathbb{C})$. ∎

The next result completes the description of $\mathbf{rep}_A$ when $A = U_h(sl_2(\mathbb{C}))$, by determining its tensor structure. It is not, of course, a consequence of 6.4.6, since $U_h(sl_2(\mathbb{C}))$ is not isomorphic to $U(sl_2(\mathbb{C}))[[h]]$ as a coalgebra (the latter being cocommutative, the former non-cocommutative).

PROPOSITION 6.4.11 *As representations of $U_h(sl_2(\mathbb{C}))$, we have*

$$V_m \otimes V_n \cong \bigoplus_{l=0}^{\min\{m,n\}} V_{m+n-2l}.$$

PROOF Note that the representation V_n contains, up to scalar multiples, a unique eigenvector of H which is annihilated by X^+, and that its H-eigenvalue is n. Since $V_m \otimes V_n$ is obviously a direct sum of indecomposables, it suffices to show that $V_m \otimes V_n$ contains, for each $l = 0, 1, \ldots, \min\{m, n\}$, an eigenvector $v^{(l)}$ of H with eigenvalue $m + n - 2l$ such that $v^{(l)}$ is annihilated by X^+, and that, up to scalar multiples, these are the only eigenvectors of H which are annihilated by X^+.

Let $\{v_i^{(m)}\}_{i=0,\ldots,m}$ and $\{v_j^{(n)}\}_{j=0,\ldots,n}$ be bases of V_m and V_n, as in 6.4.10. It is obvious that the eigenvalues of H on $V_m \otimes V_n$ are $m + n - 2l$ for $l = 0, 1, \ldots, m + n$, and that any vector with eigenvalue $m + n - 2l$ is of the form

$$v = \sum_{r=0}^{l} c_r v_{l-r}^{(m)} \otimes v_r^{(n)}.$$

for some $c_r \in \mathbb{C}[[h]]$. Using (21) for $\Delta_h(X^+)$, it is easy to see that $X^+.v = 0$ if and only if

$$q^{n-2r}[m - l + r + 1]_q c_r + [n - r]_q c_{r+1} = 0, \quad r = 0, \ldots, l - 1.$$

If $l > \min\{m, n\}$, then $[n - r]_q = 0$ or $[m - l + r + 1]_q = 0$ for some $r = 0, \ldots, l$, and then $c_r = 0$ for all r. On the other hand, if $l \leq \min\{m, n\}$, these equations imply that v is a multiple of the vector

$$v^{(l)} = \sum_{r=0}^{l} (-1)^r q^{r(n-r+1)} \frac{\left[\begin{matrix} m-l+r \\ m-l \end{matrix} \right]_q}{\left[\begin{matrix} n \\ r \end{matrix} \right]_q} v_{l-r}^{(m)} \otimes v_r^{(n)}.$$

This proves the proposition. ∎

Remarks [1] The coefficients in the summation on the right-hand side of the last equation are examples of *quantum Clebsch–Gordan coefficients*. The most general such coefficient expresses an arbitrary eigenvector of H in the component V_{m+n-2l} of $V_m \otimes V_n$ as a linear combination of tensor products of the standard basis vectors in V_m and V_n. Formulas in the general case can be found in Kirillov & Reshetikhin (1989).

[2] There is a much quicker proof of 6.4.11 using characters (see Subsection 10.1C), but we feel that it is instructive to give a completely direct argument. △

It follows from this result that $V_m \otimes V_n \cong V_n \otimes V_m$ as representations of $U_h(sl_2(\mathbb{C}))$. In fact, by 5.2.2, the isomorphism is the composite with the flip map $\sigma : V_m \otimes V_n \to V_n \otimes V_m$ of the operator on $V_m \otimes V_n$ corresponding to the universal R-matrix $\mathcal{R}_h$ of $U_h(sl_2(\mathbb{C}))$ given in (30). Note that $\mathcal{R}_h$ gives a well-defined operator on $V_m \otimes V_n$, since the elements $(X^{\pm})^k$ act as zero on V_m and V_n if $k > m, n$.

Example 6.4.12 Let us compute this operator in the simplest case $m = n = 1$. The representation V_1 has a basis $\{v_0, v_1\}$ with action

$$H.v_0 = v_0, \quad X^+.v_0 = 0, \quad X^-.v_0 = v_1,$$
$$H.v_1 = -v_1, \quad X^+.v_1 = v_0, \quad X^-.v_1 = 0.$$

From (30), we find that, with respect to the basis

$$\{v_0 \otimes v_0, v_1 \otimes v_0, v_0 \otimes v_1, v_1 \otimes v_1\}$$

of $V_1 \otimes V_1$, the universal R-matrix acts as

$$e^{-\frac{1}{2}h} \begin{pmatrix} e^h & 0 & 0 & 0 \\ 0 & 1 & 0 & 0 \\ 0 & e^h - e^{-h} & 1 & 0 \\ 0 & 0 & 0 & e^h \end{pmatrix}.$$

Note that, by 6.4.9, this is a matrix solution of the quantum Yang–Baxter equation.

We shall see on a number of occasions later in this book that this simple calculation has profound implications. $\diamond$

F A non-standard quantization We showed in 2.1.8 that there are, up to isomorphism and scalar multiples, exactly three Lie bialgebra structures on $sl_2(\mathbb{C})$. Apart from the trivial structure, and the standard structure which we have quantized in this section, there is a 'non-standard' structure given by the r-matrix $r^{ns} = H \wedge X^+$.

Since r^{ns} is a skew-symmetric solution of the CYBE, 6.2.9 shows that there is a quantization $U_h^{ns}(sl_2(\mathbb{C}))$ of the non-standard structure which is isomorphic to $U(sl_2(\mathbb{C}))[[h]]$ as an algebra, and with comultiplication given by $\Delta_h = \mathcal{F}_h.\Delta.\mathcal{F}_h^{-1}$, where Δ is the comultiplication of $U(sl_2(\mathbb{C}))$, extended $\mathbb{C}[[h]]$-linearly to $U(sl_2(\mathbb{C}))[[h]]$, and $\mathcal{F}_h$ is a certain invertible element of $(U(sl_2(\mathbb{C})) \otimes U(sl_2(\mathbb{C})))[[h]]$. (Strictly, we are applying 6.2.9 to the Lie bialgebra structure on $sl_2(\mathbb{R})$ defined by r^{ns}, and then complexifying the result.) Ohn (1992) finds that

$$\mathcal{F}_h = \exp\left(-\tfrac{1}{2}\Delta(H) + \tfrac{1}{2}\left(H\tfrac{\sinh(hX^+)}{hX^+} \otimes e^{-hX^+} + e^{hX^+} \otimes H\tfrac{\sinh(hX^+)}{hX^+}\right)\tfrac{h\Delta(X^+)}{\sinh(h\Delta(X^+))}\right).$$

From the proof of 6.2.9, we know that $U_h^{ns}(sl_2(\mathbb{C}))$ is a triangular QUE algebra with universal R-matrix $\mathcal{R}_h^{ns} = \mathcal{F}_h^{21}\mathcal{F}_h^{-1}$. In the two-dimensional representation ρ_h^t of $U_h^{ns}(sl_2(\mathbb{C}))$, one finds that

$$(33) \qquad (\rho_h^t \otimes \rho_h^t)(\mathcal{R}_h^{ns}) = \begin{pmatrix} 1 & h & -h & h^2 \\ 0 & 1 & 0 & h \\ 0 & 0 & 1 & -h \\ 0 & 0 & 0 & 1 \end{pmatrix}.$$

As we noted at the end of Section 6.1, it is often inconvenient to assume that a deformation has the usual algebra structure, even when this is possible, and in this case the explicit formula for the comultiplication of $U_h^{ns}(sl_2(\mathbb{C}))$ obtained from 6.2.9 would be very cumbersome. However, it is straightforward to check that the following formulas also define a quantization of the non-standard Lie bialgebra structure on $sl_2(\mathbb{C})$:

$$[X^+.X^-] = H, \quad [H.X^+] = \frac{4}{h}(1 - e^{-2hX^+}), \quad [H,X^-] = -2X^- - hH^2,$$

$$\Delta_h(X^+) = X^+ \otimes 1 + 1 \otimes X^+,$$

$$\Delta_h(X^-) = X^- \otimes 1 + e^{-2hX^+} \otimes X^-.$$

$$\Delta_h(H) = H \otimes 1 + e^{-2hX^+} \otimes H,$$

$$S_h(X^+) = -X^+, \quad S_h(X^-) = -e^{2hX^+}X^-, \quad S_h(H) = -e^{2hX^+}H,$$

$$\epsilon_h(X^+) = \epsilon_h(X^-) = \epsilon_h(H) = 0.$$

Note that the algebra structure has been changed, although by 6.1.8 it must be isomorphic to the usual one.

Finally, let us consider the R-matrix approach to the quantization of the classical r-matrix $r^{\mathrm{ns}} = H \wedge X^+$. This is very easy to carry out, thanks to the following amusing observation (which may be verified by direct computation).

PROPOSITION 6.4.13 *Let* $\mathfrak{g}$ *be a Lie algebra consisting of* $m \times m$ *matrices over a field* k *of characteristic zero. Assume that* $r \in \mathfrak{g} \otimes \mathfrak{g}$ *satisfies* $r^3 = 0$, *where the cube is computed using the associative algebra structure of* $M_{m^2}(k) \cong M_m(k) \otimes M_m(k)$. *Then, the* $m^2 \times m^2$ *matrix* $R = \exp(r)$ *is a solution of the QYBE.* ∎

In our case, if we regard H and X^+ as 2×2 matrices (in the natural representation of $sl_2(\mathbb{C})$), then $HX^+ = -X^+H = X^+$ and $(X^+)^2 = 0$, so it is clear that $(H \wedge X^+)^3 = 0$, and hence that $R_h^{\mathrm{ns}} = \exp(hr^{\mathrm{ns}})$ is a solution of the QYBE for which $R_h^{\mathrm{ns}} = 1 + hr^{\mathrm{ns}} + \mathrm{O}(h^2)$. In fact, R_h^{ns} is exactly the matrix (33).

However, if r^{ns} is viewed as an element of $U(\mathfrak{g}) \otimes U(\mathfrak{g})$, $\exp(hr^{\mathrm{ns}})$ does *not* satisfy the QYBE, and so is not the universal R-matrix of any QUE algebra. (This can be verified by computing the operator defined by $\exp(hr^{\mathrm{ns}})$ in the three-dimensional indecomposable representation of $U_h^{\mathrm{ns}}(sl_2(\mathbb{C}))$.)

6.5 Quantum Kac–Moody algebras

In this section, we quantize the standard Lie bialgebra structure 1.3.8 on an arbitrary symmetrizable Kac–Moody algebra $\mathfrak{g}$.

A The standard quantization Let A be a symmetrizable generalized Cartan matrix, and let $\mathfrak{g} = \mathfrak{g}'(A)$ be the associated Kac–Moody algebra. We recall from the Appendix that $\mathfrak{g}$ is generated by certain distinguished subalgebras isomorphic to $sl_2(\mathbb{C})$. In fact, using the notation in the Appendix, $\mathfrak{g}$ is generated by elements H_i, X_i^+, X_i^-, $i = 1, \ldots, n$, satisfying certain relations which include

$$[H_i, X_i^\pm] = \pm 2X_i^\pm, \quad [X_i^+, X_i^-] = H_i.$$

When $\mathfrak{g}$ is given the standard Lie bialgebra structure δ defined in 1.3.8, the subalgebra $\mathfrak{g}_i$ spanned by H_i, X_i^+ and X_i^- is a sub-Lie bialgebra:

$$\delta(H_i) = 0, \quad \delta(X_i^\pm) = d_i X_i^\pm \wedge H_i.$$

Thus, it is reasonable to try to quantize $\mathfrak{g}$ by 'gluing together' the quantizations of $sl_2(\mathbb{C})$ obtained in the previous section. This procedure uniquely specifies the action of the coalgebra structure maps on the generators of the desired quantization of $\mathfrak{g}$, as well as the 'quantum analogues' of all the defining relations of $\mathfrak{g}$ except the Serre relations

$$\sum_{k=0}^{1-a_{ij}} (-1)^k \binom{1-a_{ij}}{k} (X_i^\pm)^k X_j^\pm (X_i^\pm)^{1-a_{ij}-k} = 0 \qquad (i \neq j).$$

The quantum analogues of these relations must be chosen so that the comultiplication extends to an algebra homomorphism. This leads to

DEFINITION–PROPOSITION 6.5.1 *Let $\mathfrak{g} = \mathfrak{g}'(A)$ be the Lie algebra associated to a symmetrizable generalized Cartan matrix $(a_{ij})_{i,j=1,\ldots,n}$ as in the Appendix. Let $U_h(\mathfrak{g})$ be the algebra over $\mathbb{C}[[h]]$ topologically generated by elements H_i, X_i^+, X_i^-, $i = 1, \ldots, n$, and with the following defining relations:*

$$[H_i, H_j] = 0, \quad [H_i, X_j^\pm] = \pm a_{ij} X_j^\pm,$$

(34)
$$X_i^+ X_j^- - X_j^- X_i^+ = \delta_{i,j} \frac{e^{d_i h H_i} - e^{-d_i h H_i}}{e^{d_i h} - e^{-d_i h}},$$

$$\sum_{k=0}^{1-a_{ij}} (-1)^k \begin{bmatrix} 1 - a_{ij} \\ k \end{bmatrix}_{e^{d_i h}} (X_i^\pm)^k X_j^\pm (X_i^\pm)^{1-a_{ij}-k} = 0 \qquad (i \neq j).$$

Then, $U_h(\mathfrak{g})$ is a topological Hopf algebra over $\mathbb{C}[[h]]$ with comultiplication defined by

(35)
$$\Delta_h(H_i) = H_i \otimes 1 + 1 \otimes H_i,$$

$$\Delta_h(X_i^+) = X_i^+ \otimes e^{d_i h H_i} + 1 \otimes X_i^+, \quad \Delta_h(X_i^-) = X_i^- \otimes 1 + e^{-d_i h H_i} \otimes X_i^-,$$

antipode defined by

$$S_h(H_i) = -H_i, \quad S_h(X_i^+) = -X_i^+ e^{-d_i h H_i}, \quad S_h(X_i^-) = -e^{d_i h H_i} X_i^-,$$

and counit defined by

$$\epsilon_h(H_i) = \epsilon_h(X_i^\pm) = 0.$$

Moreover, $U_h(\mathfrak{g})$ is a quantization of the standard Lie bialgebra structure on $\mathfrak{g}$ defined in 1.3.8.

For the precise meaning of the statement that $U_h(\mathfrak{g})$ is topologically generated by the elements H_i, $X_i^\pm$, see Remark [5] following 6.4.3. As in that case, all tensor products are to be understood in the h-adically completed sense.

PROOF We must first show that the formulas for Δ_h, S_h and ϵ_h do define algebra homomorphisms (or rather an anti-homomorphism in the case of the antipode). Let us compute, for example,

$$[\Delta_h(X_i^+), \Delta_h(X_j^-)] = [X_i^+ \otimes e^{d_i h H_i} + 1 \otimes X_i^+, X_j^- \otimes 1 + e^{-d_j h H_j} \otimes X_j^-]$$

$$= [X_i^+, X_j^-] \otimes e^{d_i h H_i} + e^{-d_j h H_j} \otimes [X_i^+, X_j^-]$$

$$+ X_i^+ e^{-d_j h H_j} \otimes e^{d_i h H_i} X_j^- - e^{-d_j h H_j} X_i^+ \otimes X_j^- e^{d_i h H_i}.$$

By 6.4.2, the sum of the last two terms is

$$\left(e^{d_j a_{ji} h} - e^{d_i a_{ij} h}\right) e^{-d_j h H_j} X_i^+ \otimes e^{d_i h H_i} X_j^-,$$

which vanishes because the matrix $(d_i a_{ij})$ is symmetric. The sum of the first two terms is

$$\delta_{i,j} \left(\frac{e^{d_i h H_i} \otimes e^{d_i h H_i} - e^{-d_i h H_i} \otimes e^{-d_i h H_i}}{e^{d_i h} - e^{-d_i h}} \right) = \delta_{i,j} \Delta_h \left(\frac{e^{d_i h H_i} - e^{-d_i h H_i}}{e^{d_i h} - e^{-d_i h}} \right),$$

as we want.

The most difficult relations to check are, of course, the 'quantum Serre relations'

$$(36) \qquad \sum_{k=0}^{1-a_{ij}} (-1)^k \begin{bmatrix} 1 - a_{ij} \\ k \end{bmatrix}_{e^{d_i h}} \Delta_h(X_i^{\pm})^k \Delta_h(X_j^{\pm}) \Delta_h(X_i^{\pm})^{1-a_{ij}-k} = 0,$$

for $i \neq j$. For this, we shall need certain facts about q-binomial coefficients:

LEMMA 6.5.2 *Let q be an indeterminate. Then:*

(i) $\begin{bmatrix} r \\ k \end{bmatrix}_q = q^{-k} \begin{bmatrix} r-1 \\ k \end{bmatrix}_q + q^{r-k} \begin{bmatrix} r-1 \\ k-1 \end{bmatrix}_q$ if $r \geq k \geq 0$;

(ii) $\sum_{k=0}^{r} (-1)^k \begin{bmatrix} r \\ k \end{bmatrix}_q q^{-(r-1)k} = 0$ if $r > 0$.

PROOF Part (i) is proved by direct computation:

$$\begin{bmatrix} r \\ k \end{bmatrix}_q - q^{-k} \begin{bmatrix} r-1 \\ k \end{bmatrix}_q = \frac{[r-1]_q!}{[k]_q! [r-k]_q!} \left(\frac{(q^r - q^{-r}) - q^{-k}(q^{r-k} - q^{-r+k})}{q - q^{-1}} \right)$$

$$= q^{r-k} \frac{[r-1]_q!}{[k]_q! [r-k]_q!} \frac{q^k - q^{-k}}{q - q^{-1}}$$

$$= q^{r-k} \begin{bmatrix} r-1 \\ k-1 \end{bmatrix}_q.$$

As for (ii), we have, using (i),

$$\sum_{k=0}^{r} (-1)^k \begin{bmatrix} r \\ k \end{bmatrix}_q q^{-(r-1)k} = \sum_{k=0}^{r} (-1)^k q^{-(r-1)k} \left(q^{-k} \begin{bmatrix} r-1 \\ k \end{bmatrix}_q + q^{r-k} \begin{bmatrix} r-1 \\ k-1 \end{bmatrix}_q \right),$$

where the first term in the brackets on the right-hand side is understood to be zero if $k = r$, and the second to be zero if $k = 0$. Hence, if $r > 0$ the right-hand side is equal to

$$\sum_{k=0}^{r-1} (-1)^k q^{-rk} \begin{bmatrix} r-1 \\ k \end{bmatrix}_q + \sum_{k=1}^{r} (-1)^k q^{-r(k-1)} \begin{bmatrix} r-1 \\ k-1 \end{bmatrix}_q,$$

which obviously vanishes (change k to $k+1$ in the second sum). ∎

The following result, which follows immediately from part (i) by induction on r, will be important later in the book.

COROLLARY 6.5.3 $\begin{bmatrix} r \\ k \end{bmatrix}_q \in \mathbf{Z}[q, q^{-1}]$ if $r \geq k \geq 0$. ∎

Returning to the proof of 6.5.1, we shall establish (36) with the $+$ sign, the other case being similar. It will be convenient to abbreviate the formulas by writing X_i for X_i^+, q_i for $e^{d_i h}$, and K_i for $e^{d_i h H_i}$. Note that

$$K_i X_j K_i^{-1} = q_i^{a_{ij}} X_j$$

by 6.4.2.

The first step is to compute $\Delta(X_i)^r$; we find, by induction on r using 6.5.2(i),

$$
(37) \qquad \Delta(X_i)^r = \sum_{k=0}^{n} q_i^{-k(r-k)} \begin{bmatrix} r \\ k \end{bmatrix}_{q_i} X_i^k \otimes K_i^k X_i^{r-k}.
$$

Hence, the left-hand side of (36) is equal to

$$
\sum_{k=0}^{1-a_{ij}} \sum_{t=0}^{k} \sum_{s=0}^{1-a_{ij}-k} (-1)^k q^{-t(k-t)-s(1-a_{ij}-k-s)} \begin{bmatrix} 1 - a_{ij} \\ k \end{bmatrix}_{q_i} \begin{bmatrix} k \\ t \end{bmatrix}_{q_i} \begin{bmatrix} 1-a_{ij}-k \\ s \end{bmatrix}_{q_i}
$$
$$
\times (X_i^t \otimes K_i^t X_i^{k-t})(X_j \otimes K_j + 1 \otimes X_j)(X_i^s \otimes K_i^s X_i^{1-a_{ij}-k-s}).
$$

We show that the sum of the terms which have X_j in the second factor of the tensor product is zero (the argument for the remaining terms is similar). The sum of these terms is

$$
\sum_{k=0}^{1-a_{ij}} \sum_{t=0}^{k} \sum_{s=0}^{1-a_{ij}-k} (-1)^k q_i^{s(2t+s-k-1)-t(k-t)} \begin{bmatrix} 1 - a_{ij} \\ k \end{bmatrix}_{q_i} \begin{bmatrix} k \\ t \end{bmatrix}_{q_i} \begin{bmatrix} 1 - a_{ij} - k \\ s \end{bmatrix}_{q_i}
$$
$$
\times X_i^{s+t} \otimes K_i^{s+t} X_i^{k-t} X_j X_i^{1-a_{ij}-k-s}.
$$

Setting $m = s + t$, $p = k - t$, and noting the identity

$$
\begin{bmatrix} 1 - a_{ij} \\ p+t \end{bmatrix}_{q_i} \begin{bmatrix} p+t \\ t \end{bmatrix}_{q_i} \begin{bmatrix} 1 - a_{ij} - p - t \\ m - t \end{bmatrix}_{q_i} = \begin{bmatrix} 1 - a_{ij} - m \\ p \end{bmatrix}_{q_i} \begin{bmatrix} 1-a_{ij} \\ m \end{bmatrix}_{q_i} \begin{bmatrix} m \\ t \end{bmatrix}_{q_i},
$$

we obtain

$$
\sum_{m=0}^{1-a_{ij}} \sum_{p=0}^{1-a_{ij}-m} \left(\sum_{t=0}^{m} (-1)^t \begin{bmatrix} m \\ t \end{bmatrix}_{q_i} q_i^{-t(m-1)} \right) (-1)^p q^{m(m-t-1)}
$$
$$
\times \begin{bmatrix} 1 - a_{ij} - m \\ p \end{bmatrix}_{q_i} \begin{bmatrix} 1 - a_{ij} \\ m \end{bmatrix}_{q_i} X_i^m \otimes K_i^m X_i^p X_j X_i^{1-a_{ij}-m-p}.
$$

By 6.4.2(ii), the sum over t vanishes if $m > 0$, while the sum of the terms with $m = 0$ is equal to

$$\sum_{p=0}^{1-a_{ij}} \begin{bmatrix} 1 - a_{ij} \\ p \end{bmatrix}_{q_i} 1 \otimes X_i^p X_j X_i^{1-a_{ij}-p},$$

which is zero by the relations (34) in $U_h(\mathfrak{g})$.

The proof that Δ_h respects the remaining relations in (34) is easy, as are the corresponding arguments for S_h and ϵ_h.

It remains to show that $U_h(\mathfrak{g})$ is a quantization of the standard Lie bialgebra structure on $\mathfrak{g}$. Since $\Delta_h(H_i)$ is symmetric and

$$\frac{\Delta_h(X_i^{\pm}) - \Delta_h^{\mathrm{op}}(X_i^{\pm})}{h} \equiv d_i\, X_i^{\pm} \wedge H_i \pmod{h},$$

$U_h(\mathfrak{g})$ certainly has the correct classical limit. For the proof that $U_h(\mathfrak{g})$ is isomorphic to $U(\mathfrak{g})[[h]]$ as a $\mathbb{C}[[h]]$-module, see Tanisaki (1990). ∎

Using the rigidity theorem 6.1.8, we deduce

COROLLARY 6.5.4 *If $\mathfrak{g}$ is a finite-dimensional complex simple Lie algebra, then $U_h(\mathfrak{g})$ is isomorphic to $U(\mathfrak{g})[[h]]$ as an algebra over $\mathbb{C}[[h]]$, by an isomorphism equal to the identity (mod h).* ∎

Remarks [1] According to Drinfel'd (1987) and Shnider (1991, 1992), it follows from the cohomology theory of Hopf algebras outlined in Subsection 4.1A that the quantization $U_h(\mathfrak{g})$ can be characterized, up to changes in the deformation parameter of the form $h \mapsto h + \mathrm{O}(h^2)$, by the existence of a quantum analogue of the Cartan involution. More precisely, a quantization A_h of $\mathfrak{g}$ is equivalent to the standard one if there exists a commutative Hopf subalgebra H_h of A_h and an involution θ_h of A_h which preserves H_h, such that the isomorphism $A_h/hA_h \cong U(\mathfrak{g})$ takes H_h/hH_h isomorphically onto $U(\mathfrak{h})$, where $\mathfrak{h}$ is the Cartan subalgebra of $\mathfrak{g}$, and such that θ_h is an algebra automorphism and a coalgebra anti-automorphism, and reduces to the usual Cartan involution (mod h). For $U_h(\mathfrak{g})$, we can take H_h to be the subalgebra topologically generated by the H_i, and θ_h to be given on the generators by the same formulas as in the classical case:

$$\theta_h(X_i^{\pm}) = -X_i^{\mp}, \quad \theta_h(H_i) = -H_i.$$

[2] We shall discuss the structure of $U_h(\mathfrak{g})$ when $\dim(\mathfrak{g}) < \infty$ in more detail in Chapter 8. We show, in particular, that $U_h(\mathfrak{g})$ is quasitriangular, and we compute its universal R-matrix.

[3] To extend the quantization of $\mathfrak{g} = \mathfrak{g}'(A)$ to one of $\mathfrak{g}(A)$, one adjoins commuting primitive elements D_i satisfying the same relations with the generators H_j, $X_j^{\pm}$ of $U_h(\mathfrak{g})$ as in the classical case (see the Appendix). △

The following refinement of 6.5.4 is proved in Drinfel'd (1990a).

PROPOSITION 6.5.5 *Let $\mathfrak{g}$ be a finite-dimensional complex simple Lie algebra, let $U_h(\mathfrak{h})$ be the subalgebra of $U_h(\mathfrak{g})$ topologically generated by the H_i, and let $\mathfrak{h}$ be the Lie subalgebra generated by the corresponding elements of $\mathfrak{g}$. Then, there is an isomorphism of algebras $\varphi : U_h(\mathfrak{g}) \rightarrow U(\mathfrak{g})[[h]]$ over $\mathbb{C}[[h]]$ such that $\varphi \equiv \mathrm{id} \pmod{h}$ and $\varphi|_{\mathfrak{h}} = \mathrm{id}$.* ∎

As in Subsection 6.4E, this implies

COROLLARY 6.5.6 *If $\mathfrak{g}$ is a finite-dimensional complex simple Lie algebra, the assignment $V \mapsto V[[h]]$ is a one-to-one correspondence between the finite-dimensional irreducible representations of $\mathfrak{g}$ and the indecomposable representations of $U_h(\mathfrak{g})$ which are free and of finite rank as $\mathbb{C}[[h]]$-modules.* ∎

B The centre Another consequence of the rigidity theorem 6.1.8 is

PROPOSITION 6.5.7 *If $\mathfrak{g}$ is a finite-dimensional complex simple Lie algebra, the centre $Z_h(\mathfrak{g})$ of $U_h(\mathfrak{g})$ is canonically isomorphic to $Z(\mathfrak{g})[[h]]$, where $Z(\mathfrak{g})$ is the centre of $U(\mathfrak{g})$.*

PROOF The existence of the isomorphism in question is an immediate consequence of 6.1.8. To see that it is canonical, note that any two isomorphisms $U_h(\mathfrak{g}) \cong U(\mathfrak{g})[[h]]$ which are the identity (mod h) differ by an automorphism of $U(\mathfrak{g})[[h]]$ which is a deformation of the identity automorphism of $U(\mathfrak{g})$. But, since $H^1_{\mathrm{Lie}}(\mathfrak{g}, U(\mathfrak{g})) = 0$, any such deformation is trivial. This means that the isomorphisms $U_h(\mathfrak{g}) \cong U(\mathfrak{g})[[h]]$ differ by an inner automorphism, and therefore have the same effect on $Z_h(\mathfrak{g})$. ∎

Remarks [1] The vanishing of the Lie algebra cohomology group $H^1_{\mathrm{Lie}}(\mathfrak{g}, U(\mathfrak{g}))$ follows from the fact that the adjoint representation of $\mathfrak{g}$ on $U(\mathfrak{g})$ is a direct sum of finite-dimensional irreducible representations, and that, for each such representation V, we have $H^1_{\mathrm{Lie}}(\mathfrak{g}, V) = 0$. See Jacobson (1962) and Dixmier (1977).

[2] A proof of 6.5.7 which does not use 6.1.8 has been given by Tanisaki (1990). △

C Multiparameter quantizations We can use the 'twisting' construction described in Subsection 4.2E to obtain from the single deformation $U_h(\mathfrak{g})$ of $U(\mathfrak{g})$ a family of deformations depending on $\binom{n}{2}$ parameters, where n is the rank of $\mathfrak{g}$.

In fact, let $\mathfrak{g}$ be a symmetrizable Kac–Moody algebra, let $\mathfrak{h}$ be a Cartan subalgebra of $\mathfrak{g}$, let $r \in \wedge^2(\mathfrak{h})$, and set $\mathcal{F}_h = \exp(\frac{1}{2}hr) \in U_h(\mathfrak{g}) \otimes U_h(\mathfrak{g})$. Since $\mathfrak{h}$ is commutative, the equation

$$\mathcal{F}_h^{12} \mathcal{F}_h^{13} \mathcal{F}_h^{23} = \mathcal{F}_h^{23} \mathcal{F}_h^{13} \mathcal{F}_h^{12}$$

holds trivially, and since r is skew we have $\mathcal{F}_h^{21} = \mathcal{F}_h^{-1}$. Hence, all the conditions in the statement of 4.2.15 are satisfied, and we may define as there a new Hopf algebra $U_h(\mathfrak{g})^{\mathcal{F}_h}$ with the same algebra structure and counit as $U_h(\mathfrak{g})$, but with comultiplication

$$\Delta_h^{\mathcal{F}_h}(a) = \mathcal{F}_h.\Delta_h(a).\mathcal{F}_h^{-1}.$$

Since $\mathcal{F}_h \equiv 1 \otimes 1 \pmod{h}$, we clearly have

PROPOSITION 6.5.8 *If $\mathfrak{g}$ is any symmetrizable Kac–Moody algebra and $r \in \wedge^2(\mathfrak{h})$, the Hopf algebra $U_h(\mathfrak{g})^{\mathcal{F}_h}$ defined above is a QUE algebra.* ∎

To see the significance of this result, suppose that $\mathfrak{g}$ is finite dimensional and simple. In Section 3.1 we saw that the set of Lie bialgebra structures on $\mathfrak{g}$ has a stratification into finitely many pieces, with a single stratum of maximal dimension which is an affine space of dimension $\binom{n}{2}$. More precisely, any Lie bialgebra structure in this maximal stratum is defined by a skew solution of the MCYBE of the form $r + r^0$, where $r^0 \in \wedge^2(\mathfrak{g})$ defines the standard structure and $r \in \wedge^2(\mathfrak{h})$ is arbitrary (the formula for r^0 is given in 2.1.7). Since

$$\mathcal{F}_h \equiv 1 \otimes 1 + \frac{1}{2}hr \pmod{h^2},$$

it is clear that $U_h(\mathfrak{g})^{\mathcal{F}_h}$ is a quantization of the Lie bialgebra structure defined by $r + r^0$. Thus, *we have described explicitly a quantization of every Lie bialgebra structure in the stratum of maximal dimension.*

Bibliographical notes

6.1 For general background in the deformation theory of algebras, see Gerstenhaber & Schack (1988). For the particular case of Hopf algebras, see Gerstenhaber & Schack (1990, 1992), Shnider (1991, 1992) and Gerstenhaber, Giaquinto & Schack (1992), where proofs of 6.1.8 may be found.

6.2 Most of the results of this section are in Drinfel'd (1987). Theorem 6.2.8 is due to Reshetikhin (1992a). Theorem 6.2.9 is taken from Drinfel'd (1983b), although the proof given here is a little more explicit. The existence and uniqueness of quantizations of Lie bialgebras are also discussed in Drinfel'd (1992a).

6.3 This section is an elaboration of parts of Drinfel'd (1987).

6.4 The algebra $U_h(sl_2(\mathbb{C}))$ was discovered by Kulish & Sklyanin (1982b), although its Hopf structure was only understood later. Most of the statements in this section are special cases of more general results for quantum Kac–Moody algebras discussed elsewhere in the book, although in the general case such direct proofs are not usually possible. The explicit formulas in 6.4.6 are taken (with some modification) from Jimbo (1985). The non-standard R-matrix (33) was found by Zakrzewski (1991a), and the associated deformation of $U(sl_2(\mathbb{C}))$ by Lazarev & Movshev (1991b) and Ohn (1992).

6.5 The defining relations of $U_h(\mathfrak{g})$ appeared first in Drinfel'd (1987) and in Jimbo (1985) (in a slightly different form – see Chapter 9), and now appear in nearly every paper on quantum groups. Multiparameter quantizations were constructed by Sudbery (1990), Artin, Schelter & Tate (1991), and others, although it seems to have been Reshetikhin (1990a) who recognized that they could be obtained from the standard quantization by twisting.

A quantization of the infinite general linear Lie algebra gl_∞, and of its central extension A_∞, which is an infinite rank Kac–Moody algebra, has been constructed by Levendorskii & Soibelman (1991b). Classically, A_∞ is important because all affine Lie algebras can be embedded in it, although it is not clear whether this continues to hold in the quantum situation.

Other examples of QUE algebras will be introduced later in the book. We mention in particular the Yangians studied in Chapter 12.

7

Quantized function algebras

Partly because of their wealth of applications, and partly as a matter of personal taste, we shall concentrate most of our attention in this book on quantized universal enveloping algebras. However, the picture of the subject which would emerge would be too biased if we did not spend some time discussing the dual approach in terms of quantized function algebras. Accordingly, we devote the present chapter and Chapter 13 to this subject.

We begin in Section 7.1 with an *ad hoc* construction, due to Yu. Kobyzev and Yu. I. Manin, of a quantization $\mathcal{F}_h(SL_2(\mathbb{C}))$ of the algebra of polynomial functions on $SL_2(\mathbb{C})$. The quantization is obtained by allowing the matrix entries which generate the classical function algebra to become noncommuting, and demanding that the usual action of $SL_2(\mathbb{C})$ on the spaces of row and column vectors survives the quantization unchanged. These spaces are themselves allowed to become non-commutative, in a way analogous to the deformation of the torus group to the Heisenberg group.

It turns out that the multiplicative relations between the generators of the quantized function algebra can be expressed in terms of a solution of the quantum Yang–Baxter equation. This suggests that, conversely, given a solution of the QYBE one might be able to construct deformations of function algebras from it. We discuss this approach in Section 7.2.

We also show in Section 7.1 that $\mathcal{F}_h(SL_2(\mathbb{C}))$ is exactly the dual Hopf algebra of the quantization of $U(sl_2(\mathbb{C}))$ constructed in the previous chapter. This leads us to define the (standard) quantization $\mathcal{F}_h(G)$ of the algebra of functions on an arbitrary complex simple Lie group G with Lie algebra $\mathfrak{g}$ to be the dual of the QUE algebra $U_h(\mathfrak{g})$ defined in Section 6.5. One is then faced with the problem of giving an 'explicit' description of $\mathcal{F}_h(G)$. This has been satisfactorily accomplished only in the $SL_{n+1}(\mathbb{C})$ case, although there are candidates for the orthogonal and symplectic cases, which are constructed from solutions of the QYBE as described above. We discuss the present state of affairs in Section 7.3.

There are also multiparameter deformations of $GL_{n+1}(\mathbb{C})$ which have attracted a good deal of attention, although most of them are accounted for by the twisting construction described in Subsection 4.2E. There remains a two-parameter family of deformations: of these, one parameter is a simple scaling of the underlying Lie bialgebra structure, while the other parameter arises from the fact that $GL_{n+1}(\mathbb{C})$ has one-dimensional centre.

In Section 7.4, we outline a theory of 'differential calculus on the quantum general linear group'. More precisely, we describe a deformation of the de Rham complex of $GL_{n+1}(\mathbb{C})$ which is compatible with the deformation of the algebra of functions defined in Section 7.3.

As we mentioned in the Introduction to this book, the quantum Yang–Baxter equation first appeared in the physics literature. In Section 7.5, we describe briefly the theory of two-dimensional lattice models in statistical mechanics, where the QYBE plays the role of an integrability condition. In fact, even the defining relations of $\mathcal{F}_h(G)$, in the form described in Section 7.2, appear explicitly in this theory.

7.1 The basic example

We begin with an elementary discussion of the Hopf algebra structure of the space of regular functions on $GL_2(\mathbb{C})$ and $SL_2(\mathbb{C})$.

A Definition If we write a general element of the vector space $M_2(\mathbb{C})$ of 2×2 complex matrices in the form

$$(1) \qquad\qquad T = \begin{pmatrix} a & b \\ c & d \end{pmatrix},$$

we can think of the entries a, b, c and d of T as complex-valued functions on $M_2(\mathbb{C})$; in fact, the algebra $\mathcal{F}(M_2(\mathbb{C}))$ is simply the polynomial algebra $\mathbb{C}[a, b, c, d]$. Matrix multiplication endows $\mathcal{F}(M_2(\mathbb{C}))$ with a bialgebra structure given by

$$(2) \qquad\qquad \Delta(T) = T \,\square\, T, \qquad \epsilon(T) = 1,$$

which means

$$\Delta(a) = a \otimes a + b \otimes c, \qquad \Delta(b) = a \otimes a + b \otimes d,$$
$$\Delta(c) = c \otimes a + d \otimes c, \qquad \Delta(d) = c \otimes b + d \otimes d,$$
$$\epsilon(a) = \epsilon(d) = 1, \qquad \epsilon(b) = \epsilon(c) = 0.$$

(See Subsection 7.1C for the general definition of $\square$.)

As we showed in 4.1.6, one obtains the Hopf algebra $\mathcal{F}(GL_2(\mathbb{C}))$ of regular functions on the general linear group $GL_2(\mathbb{C})$ by localizing at the determinant $\det(T) = ad - bc$ (i.e. by 'formally adjoining' the inverse of $\det(T)$). A straightforward calculation shows that $\det(T)$ is a group-like element of $\mathcal{F}(M_2(\mathbb{C}))$. If one sets

$$\Delta(\det(T)^{-1}) = \det(T)^{-1} \otimes \det(T)^{-1}, \qquad \epsilon(\det(T)^{-1}) = 1,$$

and defines the antipode by $S(T) = T^{-1}$, i.e.

$$S(a) = \det(T)^{-1}d, \quad S(b) = -\det(T)^{-1}b,$$
$$S(c) = -\det(T)^{-1}c, \quad S(d) = \det(T)^{-1}a,$$

and $S(\det(T)) = \det(T)^{-1}$, then $\mathcal{F}(GL_2(\mathbb{C}))$ becomes a Hopf algebra.

When we regard a, b, c and d as functions on the special linear group $SL_2(\mathbb{C})$, they are no longer independent because of the determinant condition

(3) $$\det(T) = 1.$$

Indeed, it is easy to show that the algebra $\mathcal{F}(SL_2(\mathbb{C}))$ of regular functions on $SL_2(\mathbb{C})$ is the quotient

$$\mathcal{F}(SL_2(\mathbb{C})) = \mathcal{F}(M_2(\mathbb{C}))/I$$

of $\mathcal{F}(M_2(\mathbb{C}))$ by the two-sided ideal I generated by $\det(T) - 1$. Writing $D = \det(T)$ for simplicity, we have

$$\Delta(D-1) = (D-1)\otimes 1 + D\otimes(D-1), \quad \epsilon(D-1) = 0, \quad S(D-1) = -D^{-1}(D-1),$$

from which it follows that I is a Hopf ideal of $\mathcal{F}(M_2(\mathbb{C}))$ and hence that $\mathcal{F}(SL_2(\mathbb{C}))$ is a Hopf algebra (cf. Subsection 4.1A).

If V is a representation of $SL_2(\mathbb{C})$, the map $SL_2(\mathbb{C}) \times V \to V$ which gives the action of $SL_2(\mathbb{C})$ on V induces an algebra homomorphism

$$\mathrm{Sym}(V^*) \to \mathcal{F}(SL_2(\mathbb{C})) \otimes \mathrm{Sym}(V^*),$$

where $\mathrm{Sym}(V^*)$ is the symmetric algebra of V^*, i.e. the algebra of polynomial functions on V. For example, if V is the natural representation on the space of 2×1 column vectors, in which case $\mathrm{Sym}(V^*) = \mathbb{C}[x, y]$ is a polynomial algebra on two generators, the homomorphism is

(4) $$\begin{pmatrix} x \\ y \end{pmatrix} \mapsto \begin{pmatrix} a & b \\ c & d \end{pmatrix} \square \begin{pmatrix} x \\ y \end{pmatrix},$$

i.e. $x \mapsto a \otimes x + b \otimes y$, $y \mapsto c \otimes x + d \otimes y$.

We look for a deformation of $\mathcal{F}(SL_2(\mathbb{C}))$ which is generated by elements a, b, c and d which satisfy certain relations which reduce (mod h) to those defining $\mathcal{F}(SL_2(\mathbb{C}))$. Our strategy is first to deform $\mathrm{Sym}(V^*)$ and then to require that the map (4) remains a homomorphism of algebras.

As the deformation of $\mathrm{Sym}(V^*)$ we take

$$\mathrm{Sym}_h(V^*) = \mathbb{C}\{x, y\}[[h]]/(xy - e^{-h}yx).$$

Here, $\mathbb{C}\{x, y\}$ is the free (non-commutative) associative algebra over $\mathbb{C}$ with generators x and y, and $(xy - e^{-h}yx)$ is the closure in the h-adic topology of the two-sided ideal in $\mathbb{C}\{x, y\}[[h]]$ generated by the element $xy - e^{-h}yx$. This is analogous to passing from the torus to the Heisenberg group (cf. Remark [1] following 6.1.8).

We set

$$\mathcal{F}_h(SL_2(\mathbb{C})) = \mathbb{C}\{a, b, c, d\}[[h]]/I_h,$$

where I_h is an ideal of relations to be determined (the meaning of $\mathbb{C}\{a, b, c, d\}$ is clear). The condition that (4) defines a homomorphism of algebras $\mathrm{Sym}_h(V^*) \to \mathcal{F}_h(SL_2(\mathbb{C})) \otimes \mathrm{Sym}_h(V^*)$ is

$$(a \otimes x + b \otimes y)(c \otimes x + d \otimes y) = e^{-h}(c \otimes x + d \otimes y)(a \otimes x + b \otimes y).$$

Multiplying out and using the relation $xy = e^{-h}yx$, we find

$$(ac - e^{-h}ca) \otimes x^2 + (ad - da + e^h bc - e^{-h}cb) \otimes xy + (bd - e^{-h}db) \otimes y^2 = 0.$$

This will be satisfied if

$$(5) \qquad ac = e^{-h}ca, \quad ad - da + e^h bc - e^{-h}cb = 0, \quad bd = e^{-h}db.$$

In fact, these relations are forced because it is not difficult to show that the monomials $\{x^m y^n\}_{m,n \in \mathbb{N}}$ form a topological basis of $\mathrm{Sym}_h(V^*)$. We shall not prove this since the purpose of this discussion is only to motivate the definition of $\mathcal{F}_h(SL_2(\mathbb{C}))$.

Only half the necessary relations are contained in (5), for we can just as well require that the action of $SL_2(\mathbb{C})$ on the space of 1×2 row vectors also survives the deformation. This amounts to requiring that the relations (5) continue to hold when the matrix T is replaced by its transpose, or

$$(6) \qquad ab = e^{-h}ba, \quad ad - da + e^h cb - e^{-h}bc = 0, \quad cd = e^{-h}dc.$$

Combining (5) and (6) gives

$$(e^{-h} + e^h)(bc - cb) = 0,$$

which implies that b and c commute. Then, (5) and (6) are together equivalent to

$$(7) \qquad \begin{array}{l} ac = e^{-h}ca, \quad bd = e^{-h}db, \quad ab = e^{-h}ba, \quad cd = e^{-h}dc, \\ bc = cb, \quad ad - da + (e^h - e^{-h})bc = 0. \end{array}$$

Note that, if we set $h = 0$, the relations (7) simply say that a, b, c and d commute with each other, and thus generate the algebra $\mathcal{F}(M_2(\mathbb{C}))$. We shall write

$$\mathcal{F}_h(M_2(\mathbb{C})) = \mathbb{C}\{a, b, c, d\}[[h]]/J_h,$$

where J_h is the closure of the two-sided ideal generated by the relations (7) (i.e. by the elements $ac - e^{-h}ca$, etc.). It is easy to check that the usual comultiplication map (2) extends to a homomorphism of algebras $\mathcal{F}_h(M_2(\mathbb{C})) \to \mathcal{F}_h(M_2(\mathbb{C})) \otimes \mathcal{F}_h(M_2(\mathbb{C}))$ (completed tensor product), and thus defines a (topological) bialgebra structure on $\mathcal{F}_h(M_2(\mathbb{C}))$.

To complete the definition of $\mathcal{F}_h(SL_2(\mathbb{C}))$, we need an analogue of the determinant relation (3). A straightforward computation shows that $ad - bc$ is not a group-like element of $\mathcal{F}_h(SL_2(\mathbb{C}))$, but that

$$\text{qdet}(T) = ad - e^{-h}bc$$

is group-like.

We have finally arrived at

DEFINITION–PROPOSITION 7.1.1 *Let I_h is the closure of the ideal in the $\mathbb{C}[[h]]$-algebra $\mathbb{C}\{a, b, c, d\}[[h]]$ generated by the relations (7) and*

$$(8) \qquad\qquad ad - e^{-h}bc = 1.$$

The quantized algebra of functions on $SL_2(\mathbb{C})$ *is*

$$\mathcal{F}_h(SL_2(\mathbb{C})) = \mathbb{C}\{a, b, c, d\}[[h]]/I_h.$$

It is a topological Hopf algebra over $\mathbb{C}[[h]]$ with comultiplication, antipode and counit defined by

$$\Delta_h \begin{pmatrix} a & b \\ c & d \end{pmatrix} = \begin{pmatrix} a & b \\ c & d \end{pmatrix} \boxdot \begin{pmatrix} a & b \\ c & d \end{pmatrix},$$

$$S_h \begin{pmatrix} a & b \\ c & d \end{pmatrix} = \begin{pmatrix} a & b \\ c & d \end{pmatrix}^{-1} = \begin{pmatrix} d & -e^h b \\ -e^{-h}c & a \end{pmatrix},$$

$$\epsilon_h \begin{pmatrix} a & b \\ c & d \end{pmatrix} = \begin{pmatrix} 1 & 0 \\ 0 & 1 \end{pmatrix}.$$

Remarks [1] It is easy to verify, using the relations (7) and (8), that the *quantum determinant* qdet(T) lies in the centre of $\mathcal{F}_h(M_2(\mathbb{C}))$; in fact, it generates the centre topologically.

[2] One might ask why, in the derivation of the defining relations (7) and (8) of $\mathcal{F}_h(SL_2(\mathbb{C}))$, we assumed that (4) was not deformed. An 'explanation' for the success of this assumption lies in the duality between $\mathcal{F}_h(SL_2(\mathbb{C}))$ and the quantized universal enveloping algebra $U_h(sl_2(\mathbb{C}))$, discussed below, and the fact that the formulas in 6.4.12 which define the action of $U_h(sl_2(\mathbb{C}))$ on its two-dimensional indecomposable representation are *identical* to those which define the two-dimensional representation of $sl_2(\mathbb{C})$.

[3] It is also natural to ask for the most general deformation of $\mathcal{F}(SL_2(\mathbb{C}))$ that one can construct by the above procedure. Ewen, Ogievetsky & Wess (1991) find a two-parameter family of deformations of $\mathcal{F}(M_2(\mathbb{C}))$, which are discussed in Subsection 7.3D, and a 'strange' algebra whose defining relations are given in Subsection 7.2A. $\triangle$

Before proving 7.1.1, we describe briefly an alternative approach to the definition of $\mathcal{F}_h(SL_2(\mathbb{C}))$ from which the quantum determinant $\mathrm{qdet}(T) = ad - e^{-h}bc$ appears more naturally. For this, we consider a deformation of the exterior algebra $\wedge(V^*)$. Classically,

$$\wedge(V^*) = \mathbb{C}\{\xi, \eta\}/(\xi^2, \eta^2, \xi\eta + \eta\xi),$$

and if one takes the deformation to be

$$\wedge_h(V^*) = \mathbb{C}\{\xi, \eta\}[[h]]/(\xi^2, \eta^2, \xi\eta + e^h \eta\xi),$$

then one finds that the relations (6) are the conditions for the natural action of $SL_2(\mathbb{C})$ on $\wedge(V^*)$ to survive the deformation unchanged. But now it is natural to define the quantum determinant by $\mathrm{qdet}(T) = ad - e^{-h}bc$ because

$$(a \otimes \xi + b \otimes \eta)(c \otimes \xi + d \otimes \eta) = \mathrm{qdet}(T) \otimes \xi\eta.$$

B A basis of $\mathcal{F}_h(SL_2(\mathbb{C}))$ What remains to be proved in 7.1.1 is that

$$\mathcal{F}_h(SL_2(\mathbb{C})) \cong \mathcal{F}(SL_2(\mathbb{C}))[[h]]$$

as $\mathbb{C}[[h]]$-modules. This is a consequence of the following lemma.

LEMMA 7.1.2 The set $\mathcal{X} = \{a^\alpha b^\beta c^\gamma,\ d^\delta b^\beta c^\gamma,\ \alpha, \beta, \gamma, \delta \geq 0\}$ is a topological basis of $\mathcal{F}_h(SL_2(\mathbb{C}))$.

This means that the elements of $\mathcal{X}$ are linearly independent over the quotient field $\mathbb{C}((h))$ of $\mathbb{C}[[h]]$, and that their $\mathbb{C}[[h]]$-linear span is dense in $\mathcal{F}_h(SL_2(\mathbb{C}))$ in its h-adic topology.

PROOF The linear span of the set $\mathcal{Y}$ of finite products

$$(9) \qquad a^{\alpha_1} d^{\delta_1} b^{\beta_1} c^{\gamma_1} a^{\alpha_2} d^{\delta_2} b^{\beta_2} c^{\gamma_2} \ldots \quad (\alpha_1, \delta_1, \ldots \geq 0)$$

is obviously dense in $\mathcal{F}_h(SL_2(\mathbb{C}))$. By using the relations

$$ab = e^{-h}ba, \quad ac = e^{-h}ca, \quad da = e^h bc + 1,$$

it is clear that every such product can be written as a linear combination of products in which $\alpha_i = 0$ for $i > 1$. Similarly, using the relations

$$bd = e^{-h}db, \quad cd = e^{-h}dc,$$

every product of the latter kind can be written as a linear combination of products for which $\alpha_i = \delta_i = 0$ for $i > 1$. Since b and c commute, it follows that $\mathcal{F}_h(SL_2(\mathbb{C}))$ is spanned (topologically) by the products $a^\alpha d^\delta b^\beta c^\gamma$. If α and δ are both positive, write

$$a^\alpha d^\delta b^\beta c^\gamma = a^{\alpha-1}.ad.d^{\delta-1}b^\beta c^\gamma$$

and then use the relation $ad = e^{-h}bc + 1$. Iterating this procedure, one sees that $a^\alpha d^\delta b^\beta c^\gamma$ can be written as a linear combination of elements of $\mathcal{X}$.

To prove that the elements of $\mathcal{X}$ are linearly independent, we formalize the preceding argument by defining a $\mathbb{C}[[h]]$-linear 'straightening map'

$$\chi : \mathbb{C}\{a, b, c, d\}[[h]] \to \mathbb{C}\{a, b, c, d\}[[h]]$$

with the following properties:

(i) $\chi(u) = u$ for all $u \in \mathcal{X}$;

(ii) $\chi(I_h) = 0$, where I_h is the ideal of relations defining $\mathcal{F}_h(SL_2(\mathbb{C}))$.

Evidently, this will complete the proof of the lemma, since $\mathcal{X}$ is certainly linearly independent when regarded as a subset of $\mathbb{C}\{a, b, c, d\}[[h]]$.

It suffices to straighten each of the elements of $\mathcal{Y}$, since they are obviously a topological basis of $\mathbb{C}\{a, b, c, d\}[[h]]$. To accomplish this, we 'work from left to right'; more precisely, we define a linear map ψ from $\mathbb{C}\{a, b, c, d\}[[h]]$ to itself as follows:

$$\psi(a^\alpha b^\beta c^\gamma y) = a^\alpha b^\beta c^\gamma y,$$
$$\psi(a^\alpha b^\beta c^\gamma ay) = e^{(\beta+\gamma)h}a^{\alpha+1}b^\beta c^\gamma y,$$
$$\psi(a^\alpha b^{\beta+1} c^\gamma by) = a^\alpha b^\beta c^\gamma y,$$
$$\psi(a^\alpha b^\beta c^\gamma dy) = e^{-(\beta+\gamma)h}(a^{\alpha-1}b^\beta c^\gamma + e^{-h}a^{\alpha-1}b^{\beta+1}c^{\gamma+1})y,$$
$$\psi(d^\delta b^\beta c^\gamma y) = d^\delta b^\beta c^\gamma y,$$
$$\psi(d^\delta b^\beta c^\gamma ay) = e^{(\beta+\gamma)h}(d^{\delta-1}b^\beta c^\gamma + e^h d^{\delta-1}b^{\beta+1}c^{\gamma+1})y,$$
$$\psi(d^\delta b^\beta c^\gamma by) = d^\delta b^{\beta+1}c^\gamma y,$$
$$\psi(d^\delta b^\beta c^\gamma dy) = e^{-(\beta+\gamma)h}d^{\delta+1}b^\beta c^\gamma y.$$

Here, y is any element of $\mathcal{Y}$; it is clear that every element of $\mathcal{Y}$ can be written in exactly one of the eight forms on the left-hand sides of these equations. Thus, ψ is well defined. It is clear that $\psi(x) = x$ if $x \in \mathcal{X}$, and that applying ψ iteratively takes any element of $\mathcal{Y}$ into $\mathcal{X}$ (indeed, this is the argument of the first paragraph). Thus, we may define, for $y \in \mathcal{Y}$,

$$\chi(y) = \lim_{n \to \infty} \psi^n(y).$$

It is now clear that χ satisfies condition (i) above, and that this implies that, to check condition (ii), it suffices to show that $\chi(xr) = 0$ whenever $x \in \mathcal{X}$ and r is one of the generators in (7) and (8) of the ideal I_h. This can be verified by direct calculation. For example,

$$\chi(a^\alpha b^\beta c^\gamma ad) = \psi^2(a^\alpha b^\beta c^\gamma ad) = a^\alpha b^\beta c^\gamma + e^{-h} a^\alpha b^{\beta+1} c^{\gamma+1},$$
$$\chi(a^\alpha b^\beta c^\gamma da) = \psi^2(a^\alpha b^\beta c^\gamma da) = a^\alpha b^\beta c^\gamma + e^{h} a^\alpha b^{\beta+1} c^{\gamma+1},$$
$$\chi(a^\alpha b^\beta c^\gamma bc) = \psi(a^\alpha b^\beta c^\gamma bc) = a^\alpha b^{\beta+1} c^{\gamma+1},$$

and hence

$$\chi(a^\alpha b^\beta c^\gamma (ad - da - (e^{-h} - e^{h})bc)) = 0. \quad \blacksquare$$

An analogous, but simpler, argument shows that

$$\{a^\alpha b^\beta c^\gamma,\ d^\delta b^\beta c^\gamma,\ \alpha, \beta, \gamma, \delta \geq 0\}$$

is a basis of $\mathcal{F}(SL_2(\mathbb{C}))$ (over $\mathbb{C}$). This proves that $\mathcal{F}_h(SL_2(\mathbb{C}))$ is a genuine Hopf algebra deformation of $\mathcal{F}(SL_2(\mathbb{C}))$.

C The R-matrix formulation Let $M_{m,n}(\mathcal{A}_h)$ be the $\mathbb{C}[[h]]$-module of $m \times n$ matrices with entries in $\mathcal{A}_h = \mathbb{C}\{a, b, c, d\}[[h]]$. There are natural maps

$$\odot : M_{m,n}(\mathcal{A}_h) \otimes M_{r,s}(\mathcal{A}_h) \to M_{mr,ns}(\mathcal{A}_h)$$

given by

$$(A \odot B)_{ij,kl} = A_{ik} B_{jl}.$$

We shall need the following simple properties of $\odot$:

$$(A \odot 1)(B \odot 1) = AB \odot 1, \quad (1 \odot A)(1 \odot B) = 1 \odot AB,$$
$$(A \odot 1)(1 \odot B) = A \odot B, \quad (1 \odot A)(B \odot 1) = \sigma(A \odot B)\sigma,$$

where σ is the flip map. However, because $\mathcal{A}_h$ is not commutative, it is *not* true in general that $(A \odot B)(C \odot D)$ is equal to $AC \odot BD$.

We also define

$$\square : M_{m,p}(\mathcal{A}_h) \otimes M_{p,n}(\mathcal{A}_h) \to M_{m,n}(\mathcal{A}_h \otimes \mathcal{A}_h)$$

by

$$(A \square B)_{ij} = \sum_{k=1}^{p} A_{ik} \otimes B_{kj}.$$

For example, if

$$T = \begin{pmatrix} a & b \\ c & d \end{pmatrix} \in M_{2,2}(\mathcal{A}_h) = M_2(\mathcal{A}_h),$$

then

$$T \odot 1 = \begin{pmatrix} a & 0 & b & 0 \\ 0 & a & 0 & b \\ c & 0 & d & 0 \\ 0 & c & 0 & d \end{pmatrix}, \quad 1 \odot T = \begin{pmatrix} a & b & 0 & 0 \\ c & d & 0 & 0 \\ 0 & 0 & a & b \\ 0 & 0 & c & d \end{pmatrix}$$

(the rows and columns are numbered in the order 11, 12, 21, 22). The defining relations of $\mathcal{F}_h(M_2(\mathbb{C}))$ admit a simple formulation in terms of these matrices:

PROPOSITION 7.1.3 *The defining relations (8) of $\mathcal{F}_h(M_2(\mathbb{C}))$ are equivalent to*

$$(T \odot 1)(1 \odot T)R_h = R_h(1 \odot T)(T \odot 1),$$

or to

$$\check{R}_h(T \odot T) = (T \odot T)\check{R}_h,$$

where

$$R_h = \begin{pmatrix} e^h & 0 & 0 & 0 \\ 0 & 1 & 0 & 0 \\ 0 & e^h - e^{-h} & 1 & 0 \\ 0 & 0 & 0 & e^h \end{pmatrix}$$

and $\check{R}_h = R_h\sigma$, with σ being the flip operator. ∎

The proof is an easy computation.

Remark Similarly, it is easy to check that the defining relations of $\mathrm{Sym}_h(V^*)$ and of $\wedge_h(V^*)$ can be written as

$$(1 - e^{-h}\check{R})(X \odot X) = 0 \quad \text{and} \quad (1 + e^h\check{R})(\Xi \odot \Xi) = 0,$$

respectively. In the first formula, for example, we think of X as the 2×1 matrix $\begin{pmatrix} x \\ y \end{pmatrix}$, so that $X \odot X$ is the 4×1 matrix

$$\begin{pmatrix} x^2 \\ xy \\ yx \\ y^2 \end{pmatrix}. \quad \triangle$$

D Duality We saw in Subsection 3.1D that the algebra $\mathcal{F}(SL_2(\mathbb{C}))$ of regular functions on $SL_2(\mathbb{C})$ is the dual Hopf algebra, in an appropriate sense, of the universal enveloping algebra $U(sl_2(\mathbb{C}))$. This suggests that the dual of a deformation of $U(sl_2(\mathbb{C}))$ should be a deformation of $\mathcal{F}(SL_2(\mathbb{C}))$. We shall now show that this is the case for the deformation of $\mathcal{F}(SL_2(\mathbb{C}))$ described in 7.1.1 and that of $U(sl_2(\mathbb{C}))$ described in Section 6.4.

As in Subsection 4.1D, we define the dual Hopf algebra $U_h(sl_2(\mathbb{C}))^\circ$ to be the set of $\mathbb{C}[[h]]$-linear maps $U_h(sl_2(\mathbb{C})) \to \mathbb{C}[[h]]$ which are continuous when $U_h(sl_2(\mathbb{C}))$ is given the $\mathcal{I}$-adic topology, where $\mathcal{I}$ is the set of ideals $\ker \rho$, for any representation ρ of $U_h(sl_2(\mathbb{C}))$ on a free $\mathbb{C}[[h]]$-module of finite rank.

THEOREM 7.1.4 *As a topological Hopf algebra over* $\mathbb{C}[[h]]$, $U_h(sl_2(\mathbb{C}))^\circ$ *is isomorphic to* $\mathcal{F}_h(SL_2(\mathbb{C}))$.

Remark The dual $U_h(sl_2(\mathbb{C}))^\circ$ is *not* the same as the QFSH dual of $U_h(sl_2(\mathbb{C}))$ as a QUE algebra, defined in Subsection 6.3C. In fact, the QFSH dual is the quotient of the algebra of formal power series $\mathbb{C}\{\{a-1, b, c, d-1\}\}[[h]]$ by the relations in (7) and (8). $\triangle$

OUTLINE OF PROOF We shall simplify the notation a little by writing q for e^h, $\mathcal{F}_h$ for $\mathcal{F}_h(SL_2(\mathbb{C}))$ and U_h for $U_h(sl_2(\mathbb{C}))$ throughout this proof. As above, we write $\mathcal{A}_h = \mathbb{C}\{a, b, c, d\}[[h]]$, so that $\mathcal{F}_h \cong \mathcal{A}_h/I_h$, where I_h is the (closure of the) ideal generated by the relations in (7) and (8).

Let $\rho : U_h \to \mathrm{End}(V)$ be the two-dimensional indecomposable U_h-module. Writing

$$\rho(x) = \begin{pmatrix} A(x) & B(x) \\ C(x) & D(x) \end{pmatrix} \qquad (x \in U_h),$$

the matrix elements A, B, C, D lie in U_h°. We set $\Phi(a) = A$, $\Phi(b) = B$, $\Phi(c) = C$ and $\Phi(d) = D$. Then, Φ obviously extends to an algebra homomorphism $\Phi : \mathcal{A}_h \to U_h^\circ$. We must show that Φ is surjective, and that $\ker(\Phi) = I_h$.

To show that Φ is surjective, we have to show that A, B, C, D generate U_h° as a topological algebra. This follows from 6.4.11, which implies that every finite-dimensional indecomposable representation of $U_h(sl_2(\mathbb{C}))$ occurs as a direct summand of an iterated tensor product of copies of V.

To see that $\ker(\Phi) \supset I_h$, we must show that A, B, C, D satisfy the relations in (7) and (8). Let us verify, for example, that

$$(10) \qquad\qquad AD - e^{-h}BC = 1.$$

To compute the left-hand side, note that, by 6.4.7,

$$(11) \qquad\qquad \{(X^-)^r H^s (X^+)^t\}_{r,s,t \in \mathbb{N}}$$

is a topological basis of U_h. It is almost obvious that

$$A((X^-)^r H^s (X^+)^t) = \begin{cases} 1 & \text{if } r = t = 0, \\ 0 & \text{otherwise,} \end{cases}$$

$$B((X^-)^r H^s (X^+)^t) = \begin{cases} 1 & \text{if } r = 0 \text{ and } t = 1, \\ 0 & \text{otherwise,} \end{cases}$$

$$C((X^-)^r H^s (X^+)^t) = \begin{cases} 1 & \text{if } r = 1 \text{ and } t = 0, \\ 0 & \text{otherwise,} \end{cases}$$

$$D((X^-)^r H^s (X^+)^t) = \begin{cases} (-1)^s, & \text{if } r = t = 0, \\ 1 & \text{if } r = t = 1, \\ 0 & \text{otherwise.} \end{cases}$$

Since, for $x \in U_h$,

$$(AD)(x) = (A \otimes D)(\Delta_h(x)),$$

we must compute the comultiplication Δ_h on the basis elements (11). Using the formulas (19), (21) and (23) in Chapter 6, one shows, by induction on k, that

$$\Delta_h(H)^k = \sum_{r=0}^{k} \binom{k}{r} H^r \otimes H^{k-r},$$

$$\Delta_h(X^+)^k = \sum_{r=0}^{k} e^{-r(k-r)h} \begin{bmatrix} k \\ r \end{bmatrix}_{e^h} (X^+)^r \otimes e^{rhH}(X^+)^{k-r},$$

$$\Delta_h(X^-)^k = \sum_{r=0}^{k} e^{-r(k-r)h} \begin{bmatrix} k \\ r \end{bmatrix}_{e^h} e^{-(k-r)hH}(X^-)^r \otimes (X^-)^{k-r}.$$

Using these formulas, we find that

$$(AD)((X^-)^r H^s (X^+)^t) = \begin{cases} 1 & \text{if } r = s = t = 0, \\ 2^s e^{-h} & \text{if } r = t = 1, \\ 0 & \text{otherwise.} \end{cases}$$

Similarly,

$$(BC)((X^-)^r H^s (X^+)^t) = \begin{cases} 2^s & \text{if } r = t = 1, \\ 0 & \text{otherwise.} \end{cases}$$

Hence, $AD - e^{-h}BC$ vanishes on all basis elements (11) except the identity element 1, on which it is equal to one. This proves (10).

We now know that Φ induces a surjective homomorphism

$$\Phi : \mathcal{F}_h \to U_h^\circ.$$

To prove that it is injective, it is enough to show that the homomorphisms

$$\Phi^{(n)} : \mathcal{F}_h/h^n \mathcal{F}_h \to U_h^\circ/h^n U_h^\circ$$

are injective for all $n > 0$: the sufficiency follows from the completeness of $\mathcal{F}_h$ as a $\mathbb{C}[[h]]$-module, which in turn follows from the completeness of $\mathcal{A}_h$ and the fact that the ideal I_h is closed. Let $\mathcal{A}_h^{(n,k)}$ be the $\mathbb{C}[[h]]/(h^n)$-submodule of

$$\mathcal{A}_h^{(n)} = \mathcal{A}_h/h^n \mathcal{A}_h \cong \mathbb{C}\{a,b,c,d\} \otimes \mathbb{C}[[h]]/(h^n)$$

consisting of the polynomials in a, b, c, d of total degree at most k. Thus,

$$\mathbb{C}[[h]]/(h^n) = \mathcal{A}_h^{(n,0)} \subset \mathcal{A}_h^{(n,1)} \subset \mathcal{A}_h^{(n,2)} \subset \cdots \subset \mathcal{A}_h^{(n)} = \bigcup_{k=0}^{\infty} \mathcal{A}_h^{(n,k)}.$$

Similarly, we filter $\mathcal{F}_h^{(n)} = \mathcal{F}_h/h^n\mathcal{F}_h$ by setting $\mathcal{F}_h^{(n,k)}$ equal to the $\mathbb{C}[[h]]/(h^n)$-submodule spanned by the matrix elements of representations of U_h of dimension at most $k+1$.

By 6.4.11, $\Phi^{(n)}(\mathcal{A}_h^{(n,k)}) \subset \mathcal{F}_h^{(n,k)}$, and it clearly suffices to show that the induced map

$$\Phi^{(n,k)} : \mathcal{A}_h^{(n,k)}/\mathcal{A}_h^{(n,k)} \cap (\mathcal{A}_h^{(n,k-1)} + I_h) \to \mathcal{F}_h^{(n,k)}/\mathcal{F}_h^{(n,k-1)}$$

is injective for all n and k. It follows, by using 6.4.11 again, that $\Phi^{(n,k)}$ is surjective and $\mathcal{F}_h^{(n,k)}/\mathcal{F}_h^{(n,k-1)}$ is a free $\mathbb{C}[[h]]/(h^n)$-module of rank $(k+1)^2$. On the other hand, from 7.1.2 it follows that $\mathcal{A}_h^{(n,k)} \cap (\mathcal{A}_h^{(n,k-1)} + I_h)$ is spanned by

$$\{b^\beta c^\gamma\}_{\beta+\gamma=k} \cup \{a^\alpha b^\beta c^\gamma\}_{\alpha>0,\, \alpha+\beta+\gamma=k} \cup \{d^\delta b^\beta c^\gamma\}_{\delta>0,\, \delta+\beta+\gamma=k}.$$

Since this set has $(k+1)^2$ elements, $\Phi^{(n,k)}$ must be injective.

Finally, it is a tautology that Φ is a homomorphism of coalgebras. For this amounts to showing that

$$\begin{pmatrix} A & B \\ C & D \end{pmatrix} \square \begin{pmatrix} A & B \\ C & D \end{pmatrix} = \Delta_h^\circ \begin{pmatrix} A & B \\ C & D \end{pmatrix},$$

where Δ_h° is the comultiplication of U_h°, or that

$$\begin{pmatrix} A(x) & B(x) \\ C(x) & D(x) \end{pmatrix} \begin{pmatrix} A(y) & B(y) \\ C(y) & D(y) \end{pmatrix} = \begin{pmatrix} A(xy) & B(xy) \\ C(xy) & D(xy) \end{pmatrix}$$

for all $x, y \in U_h$. But this is what it means for

$$x \mapsto \begin{pmatrix} A(x) & B(x) \\ C(x) & D(x) \end{pmatrix}$$

to be a representation of U_h. $\blacksquare$

Let $C_{rs}^{(m)}$, $0 \le r, s \le m$, be the matrix elements of the representation V_m of $U_h(sl_2(\mathbb{C}))$ constructed in 6.4.10 with respect to its basis $\{v_r^{(m)}\}_{r=0,\ldots,m}$:

$$x.v_s^{(m)} = \sum_{r=0}^m C_{rs}^{(m)} v_r^{(m)}.$$

The preceding proof implies

COROLLARY 7.1.5 *The matrix elements* $C_{rs}^{(m)}$*, for* $0 \le r, s \le m$*, are a topological basis of* $\mathcal{F}_h(SL_2(\mathbb{C}))$*.* $\blacksquare$

E Representations In the classical case, the algebra $\mathcal{F}(SL_2(\mathbb{C}))$ is commutative, and hence its irreducible representations are all one-dimensional. The quantized algebra $\mathcal{F}_h(SL_2(\mathbb{C}))$, on the other hand, is non-commutative, and, at first sight, it seems unlikely to have any one-dimensional representations at all. In fact, it is easy to see that there is a family of one-dimensional representations parametrized by the set of invertible elements of $\mathbb{C}[[h]]$; in addition, there are infinite-dimensional indecomposable representations in the quantum case, as the following result shows.

PROPOSITION 7.1.6 *The algebra* $\mathcal{F}_h(SL_2(\mathbb{C}))$ *has two families of indecomposable representations, each parametrized by the invertible elements* $t \in \mathbb{C}[[h]]$, *which can be described as follows:*

(a) a family τ_t *of representations on* $\mathbb{C}[[h]]$ *given by*

$$\tau_t(a) = t, \quad \tau_t(b) = \tau_t(c) = 0, \quad \tau_t(d) = t^{-1};$$

(b) a family π_t *of representations on the space* $\mathbb{C}[[h]]_0^{\mathbb{N}}$ *of sequences of elements of* $\mathbb{C}[[h]]$, *all but finitely many of whose terms are zero, given by*

$$\pi_t(a)(e_k) = (1 + e^{-kh})e_{k-1}, \quad \pi_t(d)(e_{k-1}) = (1 - e^{-kh})e_k,$$
$$\pi_t(b)(e_k) = -e^{-(k+1)h}t^{-1}e_k, \quad \pi_t(c)(e_k) = e^{-kh}t\,e_k,$$

where e_k *is the sequence which has 1 in the kth place and zeros elsewhere, and* $e_{-1} = 0$.

PROOF An easy calculation. ∎

Remarks [1] The formulas for π_t can be made to look a little more symmetrical by rescaling the basis $\{e_k\}$. In fact, it is easy to see that the formulas for $\pi_t(a)$ and $\pi_t(d)$ can be put in the form

$$\pi_t(a)(e_k) = (1 - e^{-2kh})^{1/2}e_{k-1}, \quad \pi_t(d)(e_{k-1}) = (1 - e^{-2kh})^{1/2}e_k,$$

(with no change in $\pi_t(b)$ or $\pi_t(c)$), provided one is willing to extend scalars from $\mathbb{C}[[h]]$ to $\mathbb{C}[[h^{1/2}]]$. In this form, we shall meet the representations π_t again in Chapter 13.

[2] Note the resemblance between this result and the parametrization of the symplectic leaves in $SU(2)$ with its standard Poisson–Lie group structure, described in Section 1.4. This correspondence has already been hinted at in 6.2.5, and will be made more precise and 'explained' in Chapter 13. △

7.2 R-matrix quantization

In this section and the next, we shall reverse the order of events in the previous section by defining a Hopf algebra, or at least a bialgebra, starting from an arbitrary R-matrix. In some cases, these Hopf algebras will turn out to be quantizations of algebras of functions on complex simple groups.

A From R-matrices to bialgebras Let R be a matrix in $\mathrm{End}_k(k^m \otimes k^m) \cong \mathrm{End}_k(k^m) \otimes \mathrm{End}_k(k^m)$, where k^m denotes the free k-module of rank m, and k is any commutative ring with unit. We write $\check{R} = R\sigma$, where σ is the flip operator. We also choose an element $q \in k^\times$ (the set of invertible elements of k). Motivated by 7.1.3 and the remarks following it, we make

DEFINITION 7.2.1 (i) $\mathrm{Sym}_{R,q}$ is the associative k-algebra generated by elements $x_1, \ldots, x_m$ with defining relations

$$(1 - q^{-1}\check{R})(X \odot X) = 0,$$

where X is the $m \times 1$ matrix with entries $x_1, \ldots, x_m$.

(ii) $\wedge_{R,q}$ is the associative k-algebra generated by the elements $\xi_1, \ldots, \xi_m$ with defining relations
$$(1 + q\check{R})(\Xi \odot \Xi) = 0,$$
where Ξ is the $m \times 1$ matrix with entries $\xi_1, \ldots, \xi_m$.

(iii) $\mathcal{F}_R(M_m)$ is the associative k-algebra generated by the elements t_{ij}, where $1 \le i, j \le m$, with defining relations

$$(12) \qquad\qquad \check{R}(T \odot T) = (T \odot T)\check{R},$$

where $T = (t_{ij})$.

Remark We shall be particularly interested in the case $k = \mathbb{C}[[h]]$. In that case, one should replace 'generated' by 'topologically generated' in all three parts of the definition. The meaning of 'topologically generated' is explained in 6.4.3. $\triangle$

As in the basic example considered in the preceding section, we have

PROPOSITION 7.2.2 For any matrix $R \in \mathrm{End}_k(k^m \otimes k^m)$, $\mathcal{F}_R(M_m)$ is a bialgebra with comultiplication and counit determined by

$$\Delta(T) = T \boxdot T \quad and \quad \epsilon(T) = 1,$$

respectively.

Moreover, for any $q \in k^{\times}$, there exist unique left k-comodule algebra structures

$$\operatorname{Sym}_{R,q} \to \mathcal{F}_R(M_m) \otimes \operatorname{Sym}_{R,q}, \qquad \wedge_{R,q} \to \mathcal{F}_R(M_m) \otimes \wedge_{R,q}$$

given on generators by

$$X \mapsto T \,\square\, X, \quad \Xi \mapsto T \,\square\, \Xi.$$

PROOF We first verify that Δ is compatible with the defining relations (12). Now,

$$\begin{aligned}
\Delta(\check{R}(T \odot T)) &= \check{R}(T \,\square\, T) \odot (T \,\square\, T) \\
&= \check{R}(T \odot T) \,\square\, (T \odot T) \\
&= (T \odot T)\check{R} \,\square\, (T \odot T) \\
&= (T \odot T) \,\square\, \check{R}(T \odot T) \\
&= (T \odot T) \,\square\, (T \odot T)\check{R}
\end{aligned}$$

(the third and fifth equalities follow from (12)). By a similar argument,

$$\Delta((T \odot T)\check{R}) = (T \odot T) \,\square\, (T \odot T)\check{R}.$$

(We recommend that the reader write out the preceding computation in components to be convinced that it is correct.) The verification for ϵ, and the second part of the proposition, are proved by similar calculations, and are left to the reader. ∎

Although $\mathcal{F}_R(M_m)$ is a well-defined bialgebra for any R, one usually assumes that R is a solution of the quantum Yang–Baxter equation (QYBE):

$$(13) \qquad\qquad R_{12}R_{13}R_{23} = R_{23}R_{13}R_{12}.$$

This is a reflection of the associativity of multiplication in $\mathcal{F}_R(M_m)$. In fact, writing $T_1 = T \odot 1 \odot 1$, $T_2 = 1 \odot T \odot 1$, $T_3 = 1 \odot 1 \odot T$, one can pass from the product $T_1 T_2 T_3$ to the product $T_3 T_2 T_1$ in two ways: either

$$T_1 T_2 T_3 \longrightarrow T_1 T_3 T_2 \longrightarrow T_3 T_1 T_2 \longrightarrow T_3 T_2 T_1,$$

or

$$T_1 T_2 T_3 \longrightarrow T_2 T_1 T_3 \longrightarrow T_2 T_3 T_1 \longrightarrow T_3 T_2 T_1.$$

The first sequence is effected by conjugation by $R_{12}R_{13}R_{23}$, the second by conjugation by $R_{23}R_{13}R_{12}$. *We assume throughout the remainder of this section that R satisfies the QYBE.*

The simplest example is, of course, $R = 1$ and $q = 1$, in which case $\mathrm{Sym}_{R,q}$, $\wedge_{R,q}$ and $\mathcal{F}_R(M_m)$ become $\mathrm{Sym}((k^m)^*)$, $\wedge((k^m)^*)$ and $\mathcal{F}(M_m(k))$, respectively. We shall be mainly interested in the case $k = \mathbb{C}[[h]]$, and we would like $\mathcal{F}_R(M_m)$ to be a deformation of $\mathcal{F}(M_m(\mathbb{C}))$, in the sense that $\mathcal{F}_R(M_m)/h\mathcal{F}_R(M_m) \cong \mathcal{F}(M_m(\mathbb{C}))$ as bialgebras. It is thus natural to assume that

$$(14) \qquad\qquad R \equiv 1 \pmod{h},$$

since then the defining relations of $\mathcal{F}_R(M_m)$ will become those of $\mathcal{F}(M_m(\mathbb{C}))$ when we set $h = 0$. Conditions (13) and (14) are satisfied if $m = 2$ and R is the matrix in 7.1.3; as we saw in the preceding section, $\mathcal{F}_R(M_m)$ is then a (non-trivial) deformation of $\mathcal{F}(M_2(\mathbb{C}))$. Ewen, Ogievetsky & Wess (1991) consider the example

$$(15) \qquad\qquad R = \begin{pmatrix} 1 & 0 & 0 & 0 \\ e^{-h} - e^h & 1 & 0 & 0 \\ e^h - e^{-h} & 0 & 1 & 0 \\ e^h - e^{-h} & -1 & 1 & 1 \end{pmatrix}.$$

This is a solution of the QYBE, first found by Demidov *et al.* (1990), but clearly $R \not\equiv 1 \pmod{h}$. Thus, the algebra $\mathcal{F}_R(M_m)$ associated to this 'strange' R-matrix is not a deformation of $\mathcal{F}(M_2(\mathbb{C}))$. (Ewen *et al.* assert that this strange algebra has a 'classical limit', but that is not in the sense meant in this book: to obtain this limit, they use a change of basis which becomes singular when $h = 0$.)

Unfortunately, however, even for a matrix satisfying (13) and (14), $\mathcal{F}_R(M_m)$ is not necessarily a deformation of $\mathcal{F}(M_m(\mathbb{C}))$, and there seems to be no simple criterion on R which guarantees that it is. The question has been discussed extensively by Gerstenhaber & Schack (1990, 1992) and by Gerstenhaber, Giaquinto & Schack (1992). We close this subsection with a simple example of what can go wrong.

Example 7.2.3 Take $m = 2$ and let R be a diagonal matrix :

$$R = \begin{pmatrix} \alpha & 0 & 0 & 0 \\ 0 & \beta & 0 & 0 \\ 0 & 0 & \gamma & 0 \\ 0 & 0 & 0 & \delta \end{pmatrix},$$

where α, β, γ, $\delta \in \mathbb{C}[[h]]$ are $\equiv 1 \pmod{h}$. Any matrix of this form trivially satisfies the QYBE, and, since $R \equiv 1 \pmod{h}$, one intuitively expects that 'the relations in $\mathcal{F}_R(M_m)$ go over into those of $\mathcal{F}(M_2(\mathbb{C}))$ as h tends to zero'. But the relations (12) imply that

$$\alpha t_{11} t_{12} = \gamma t_{12} t_{11}, \qquad \alpha t_{12} t_{11} = \beta t_{11} t_{12}.$$

These equations imply the relation $(\alpha^2 - \beta\gamma)t_{11}t_{12} = 0$, so that, if $\alpha^2 \neq \beta\gamma$, $\mathcal{F}_R(M_m)$ is not a deformation of $\mathcal{F}(M_2(\mathbb{C}))$. (This argument can easily be made precise by noting that the algebras $\mathcal{F}_R(M_m)$ and $\mathcal{F}(M_2(\mathbb{C}))$ are graded by the total degree in the generators t_{ij} and considering their Poincaré series.)
$\diamond$

B From bialgebras to Hopf algebras: the quantum determinant The next step is to construct a Hopf algebra which is related to $\mathcal{F}_R(M_m)$ in the same way as $\mathcal{F}(GL_2(\mathbb{C}))$ is related to $\mathcal{F}(M_2(\mathbb{C}))$. Thus, we must construct an appropriate analogue of the determinant $\det(T)$. To do this, it will be necessary to make an additional assumption on R.

Note first that $\wedge_{R,q}$ is a graded algebra with $\deg(\xi_i) = 1$:

$$\wedge_{R,q} = \bigoplus_{i=0}^{\infty} \wedge^i_{R,q}, \qquad \wedge^i_{R,q} \cdot \wedge^j_{R,q} \subseteq \wedge^{i+j}_{R,q}.$$

We shall say that R is *Frobenius* if, for some $n \geq 1$,

(FR1) $\wedge^n_{R,q} \cong k$ and $\wedge^i_{R,q} = 0$ for $i > n$;

(FR2) the multiplication map $\wedge^i_{R,q} \otimes \wedge^{n-i}_{R,q} \to \wedge^n_{R,q} \cong k$ is a non-degenerate pairing for all $i \leq n$.

Of course, if $R = 1$ and $q = 1$ then $n = m$, but we allow $n \neq m$ in the general case.

DEFINITION–PROPOSITION 7.2.4 *If $R \in \mathrm{End}_k(k^m \otimes k^m)$ is Frobenius, there exists an element* $\mathrm{qdet}(T) \in \mathcal{F}_R(M_m)$, *called the* quantum determinant, *with the following properties:*

(i) $\delta_R(\omega) = \mathrm{qdet}(T) \otimes \omega$ for all $\omega \in \wedge^n_{R,q}$, where δ_R denotes the left k-comodule algebra structure given by 7.2.2;

(ii) $\Delta(\mathrm{qdet}(T)) = \mathrm{qdet}(T) \otimes \mathrm{qdet}(T)$ and $\epsilon(\mathrm{qdet}(T)) = 1$.

PROOF By its definition, $\delta_R(\wedge^1_{R,q}) \subseteq \mathcal{F}_R(M_m) \otimes \wedge^1_{R,q}$, so, since δ_R is an algebra homomorphism, $\delta_R(\wedge^n_{R,q}) \subseteq \mathcal{F}_R(M_m) \otimes \wedge^n_{R,q}$. Part (i) now follows from (FR1).

Since δ_R is a k-comodule structure, $(\mathrm{id} \otimes \delta_R)\delta_R = (\Delta \otimes \mathrm{id})\delta_R$. Applying both sides to a non-zero element of $\wedge^n_{R,q}$ and using (FR2) shows that $\mathrm{qdet}(T)$ is group-like. The second equation in part (ii) follows in the same way. ∎

Example 7.2.5 We consider the 'strange' R-matrix (15), and take $q = 1$ (this choice will be justified below). We find that $\wedge_{R,q}$ has defining relations

$$\xi_1^2 = 0, \quad \xi_1\xi_2 + \xi_2\xi_1 = 0, \quad \xi_2^2 + \xi_1\xi_2 = 0.$$

It follows that $\wedge_{R,q}^2 = \mathbb{C}.(\xi_1\xi_2)$, and that $\wedge_{R,q}^i = 0$ for $i > 2$. Thus, R is Frobenius. Moreover,

$$\delta_R(\xi_1\xi_2) = (t_{11} \otimes \xi_1 + t_{12} \otimes \xi_2)(t_{21} \otimes \xi_1 + t_{22} \otimes \xi_2)$$
$$= (t_{11}t_{22} - t_{12}t_{21} - t_{11}t_{21}) \otimes \xi_1\xi_2.$$

Thus, $\mathrm{qdet}(T) = t_{11}t_{22} - t_{12}t_{21} - t_{11}t_{21}$. With a little more effort, one can show that $\mathrm{qdet}(T)$ lies in the centre of $\mathcal{F}_R(M_2)$. $\Diamond$

We should like to construct a Hopf algebra by localizing $\mathcal{F}_R(M_m)$ at $\mathrm{qdet}(T)$, in the same way as $\mathcal{F}(GL_2(\mathbb{C}))$ was obtained from $\mathcal{F}(M_2(\mathbb{C}))$. Since we are now dealing with non-commutative rings, however, the notion of localization is a little more subtle. In general, if $\mathcal{M}$ is a ring with unit and $\mathcal{S}$ is a multiplicatively closed subset of $\mathcal{M}$, one can define the localization of $\mathcal{M}$ at $\mathcal{S}$ only if the *Ore condition* is satisfied:

if $m \in \mathcal{M}$, $s \in \mathcal{S}$, there exist $m' \in \mathcal{M}$, $s' \in \mathcal{S}$ such that $ms' = m's$.

Apart from this problem, the localization is constructed in essentially the same way as in the commutative case (see McConnell & Robson (1987), for example). Note that the Ore condition is satisfied trivially if $\mathcal{M}$ is commutative or, more generally, if $\mathcal{S}$ is contained in the centre of $\mathcal{M}$.

To ensure that the Ore condition is satisfied in our case, when $\mathcal{M} = \mathcal{F}_R(M_m)$ and $\mathcal{S} = \{\mathrm{qdet}(T)^r\}_{r\geq 1}$, we make an additional assumption on R. We say that R is *Hecke* if

(16) $$\check{R}^2 = (q - q^{-1})\check{R} + 1.$$

Of course, this holds in the trivial case $R = 1$, $q = 1$, and it is easy to check that the R-matrix of 7.1.3 is Hecke with $q = e^h$. The 'strange' R-matrix (15) is also Hecke, with $q = 1$ (this justifies the choice of q in 7.2.5).

LEMMA 7.2.6 *Let* $R \in \mathrm{End}_k(k^m \otimes k^m)$ *be Frobenius–Hecke. Then, there exists a matrix* $Q \in GL_m(k)$ *such that*

$$\mathrm{qdet}(T).T.Q^{-1} = Q^{-1}.T.\mathrm{qdet}(T). \quad \blacksquare$$

This result is due to Lyubashenko (1986) and Gurevich (1991); see also Tsygan (1993). It clearly implies the Ore condition. For the R-matrix in 7.1.3, we have $Q = 1$.

Thus, if $R \in \mathrm{End}_k(k^m \otimes k^m)$ is Frobenius–Hecke, we may define $\mathcal{F}_{R,q}(GL_m)$ to be the localization of $\mathcal{F}_R(M_m)$ at $\mathrm{qdet}(T)$. Using 7.2.4(ii), it is easy to see that $\mathcal{F}_{R,q}(GL_m)$ inherits a bialgebra structure from $\mathcal{F}_R(M_m)$. It follows from the equation

$$\Delta(T) = T \,\square\, T$$

and the Hopf algebra axioms that, if $\mathcal{F}_{R,q}(GL_m)$ can be made into a Hopf algebra, its antipode must be given by $S(T) = T^{-1}$. We must therefore prove that T is invertible in the algebra of $m \times m$ matrices with entries in $\mathcal{F}_{R,q}(GL_m)$.

Fix $0 \neq \Omega \in \wedge^n_{R,q}$ and let $\omega_1, \ldots, \omega_m$ be the basis of $\wedge^{n-1}_{R,q}$ which is right dual to the basis $\xi_1, \ldots, \xi_m$ of $\wedge^1_{R,q}$, i.e.

(17) $$\xi_i . \omega_j = \delta_{i,j} \Omega.$$

Define elements $\tilde{t}_{ij} \in \mathcal{F}_R(M_m)$ by

$$\delta_R(\omega_i) = \sum_{j=1}^m \tilde{t}_{ji} \otimes \omega_j.$$

Applying δ_R to both sides of (17), we find that

$$T.\tilde{T} = \text{qdet}(T).1,$$

where $\tilde{T} = (\tilde{t}_{ij})$. Thus, T has a right inverse in $\mathcal{F}_{R,q}(GL_m)$. But it is easy to see that the map $t_{ij} \mapsto t_{ij}$ induces an isomorphism of algebras

$$\mathcal{F}_{R^{-1}, q^{-1}}(GL_m) \cong \mathcal{F}_{R,q}(GL_m)_{\text{op}},$$

and so T also has a left inverse in $\mathcal{F}_{R,q}(GL_m)$.

It is now straightforward to prove

PROPOSITION 7.2.7 *Let R be a Frobenius–Hecke R-matrix. Then, $\mathcal{F}_{R,q}(GL_m)$ is a Hopf algebra with comultiplication, counit and antipode given by*

$$\Delta(T) = T \,\square\, T, \quad \Delta(\text{qdet}(T)^{-1}) = \text{qdet}(T)^{-1} \otimes \text{qdet}(T)^{-1},$$
$$\epsilon(T) = 1, \quad \epsilon(\text{qdet}(T)^{-1}) = 1,$$
$$S(T) = T^{-1}, \quad S(\text{qdet}(T)^{-1}) = \text{qdet}(T). \quad \blacksquare$$

One can now define $\mathcal{F}_{R,q}(SL_m)$ to be the quotient of $\mathcal{F}_{R,q}(GL_m)$ by the two-sided ideal I generated by $\text{qdet}(T) - 1$. It is clear from 7.2.4(ii) and the formulas in 7.2.7 that I is actually a Hopf ideal, and hence that $\mathcal{F}_{R,q}(SL_m)$ is a Hopf algebra.

C Solutions of the QYBE As the results of this section indicate, it is very desirable to write down all the solutions $R \in \text{End}_k(k^m \otimes k^m)$ of the quantum Yang–Baxter equation (13). Thinking of R as an $m^2 \times m^2$ matrix, the QYBE is a system of m^6 cubic equations in m^4 unknowns. The problem of solving these equations can be simplified somewhat by noting that the QYBE has a number of 'symmetries'. Thus, it is clear that, if R is a solution of (13), so is

(i) cR, for any $c \in k^\times$,

(ii) $(A \otimes A)R(A \otimes A)^{-1}$, for any $A \in GL_m(k)$,

(iii) R_{21},

(iv) the transpose of R.

Let $\mathcal{G}$ be the group of linear automorphisms of $\text{End}_k(k^m \otimes k^m)$ generated by the transformations (i)–(iv). Then, the problem is to describe the space of orbits of $\mathcal{G}$ on the space of solutions of the QYBE.

Unfortunately, although there are a large number of examples of solutions of the QYBE in the literature, little is known about the description of this orbit space, in general. Experimental evidence suggests that, if $k = \mathbb{C}$, the diagonal solutions will form a dense open subset of the orbit space, which will therefore have complex dimension $m^2 - 1$, but this does not seem to have been proved (the loss of one dimension is due to the possibility of an overall scaling). However, if $m = 2$, Hietarinta (1993b) has given a complete solution to the problem. The diagonal solutions form a dense open set, but the orbit space is highly singular, with several 'strata' of lower dimension.

It would also be interesting to know which solutions of the QYBE arise from quantum groups. The answer, of course, will depend on how one interprets the question. Any solution which arises by evaluating the universal R-matrix of a quantized universal enveloping algebra in a (finite-dimensional) representation will lie on a curve of solutions passing through the identity matrix. It is known that the flip operator σ, which is clearly a solution of the QYBE, does not have this property. In fact, it is an isolated point in the space of all solutions of the QYBE (this remark is due to M. Gerstenhaber, S. Majid & S. D. Schack).

7.3 Examples of quantized function algebras

A The general definition To define a quantized algebra of functions $\mathcal{F}_h(G)$ on an arbitrary complex simple Lie group G, note that we have already constructed a quantization $U_h(\mathfrak{g})$ of the Lie algebra $\mathfrak{g}$ of G in Section 6.5, so we may turn 7.1.4 into a definition. Classically, the dual of $U(\mathfrak{g})$ is the algebra of regular functions on the *simply-connected* group G (cf. 4.1.17).

DEFINITION 7.3.1 *Let G be a connected, simply-connected, finite-dimensional, complex simple Lie group G with Lie algebra $\mathfrak{g}$. The (standard) quantized algebra of functions $\mathcal{F}_h(G)$ on G is the Hopf algebra dual of the QUE algebra $U_h(\mathfrak{g})$ defined in 6.5.1.*

By the Hopf dual we mean the set of continuous $\mathbb{C}[[h]]$-linear maps $U_h(\mathfrak{g}) \to \mathbb{C}[[h]]$, where the topology on $U_h(\mathfrak{g})$ is defined using the kernels of the finite-dimensional representations of $U_h(\mathfrak{g})$ (see Subsection 4.1D). Since, by 6.5.4, $U_h(\mathfrak{g}) \cong U(\mathfrak{g})[[h]]$ as topological algebras over $\mathbb{C}[[h]]$, and since $\mathcal{F}(G) \cong U(\mathfrak{g})^\circ$ as Hopf algebras, it follows that $\mathcal{F}_h(G) \cong \mathcal{F}(G)[[h]]$ as topological coalgebras. In particular, $\mathcal{F}_h(G)$ is a Hopf algebra deformation of $\mathcal{F}(G)$ in the sense of 6.1.1.

Furthermore, the classical limit of the QF algebra $\mathcal{F}_h(G)$ is the Poisson–Lie group structure on G whose tangent Lie bialgebra structure $\delta : \mathfrak{g} \to \mathfrak{g} \otimes \mathfrak{g}$ is the 'first order part' of the deformation $U_h(\mathfrak{g})$ of $U(\mathfrak{g})$ (see Chapter 1). Indeed, let a, $b \in \mathcal{F}_h(G)$ and let a_0, $b_0 \in \mathcal{F}(G)$ be their $h = 0$ limit. We have to prove that

$$\frac{[a,b](X)}{h} \equiv \langle d(\{a_0, b_0\})_e, X_0 \rangle \pmod{h}$$

for all $X \in U_h(\mathfrak{g})$ which are equal to $X_0 \in \mathfrak{g} \pmod h$. From Subsection 1.3A, we see that the right-hand side is equal to $\langle a_0 \otimes b_0, \delta(X_0) \rangle$, while the left-hand side is

$$\frac{\langle a \otimes b - b \otimes a, \Delta_h(X) \rangle}{h} = \left\langle a \otimes b, \frac{\Delta_h(X) - \Delta_h^{\mathrm{op}}(X)}{h} \right\rangle$$
$$\equiv \langle a_0 \otimes b_0, \delta(X_0) \rangle \pmod{h}$$

as we want.

In the remainder of this section, we attempt to give a more explicit description of $\mathcal{F}_h(G)$ in the cases in which G is a classical group of matrices.

B The quantum special linear group To construct a deformation of the algebra of functions on $SL_{n+1}(\mathbb{C})$, it is natural, in view of the developments in Section 7.2, to consider a Hopf algebra of the form $\mathcal{F}_{R,q}(SL_{n+1})$ for some invertible element $q \in \mathbb{C}[[h]]$ and some $R \in \mathrm{End}_{\mathbb{C}}(\mathbb{C}^{n+1} \otimes \mathbb{C}^{n+1})[[h]]$. We take $q = e^h$ and

$$(18) \qquad R = e^h \sum_i E_{ii} \otimes E_{ii} + \sum_{i \neq j} E_{ii} \otimes E_{jj} + (e^h - e^{-h}) \sum_{i<j} E_{ij} \otimes E_{ji},$$

where $0 \leq i, j \leq n$ and the E_{ij} are the elementary matrices (i.e. E_{ij} has 1 in the ith row and jth column and zeros elsewhere). If $n = 1$, this is the R-matrix in 7.1.3. One can check directly that R satisfies the QYBE for all m (but see the remarks below); and it is easy to see that (R, q) satisfies the Hecke condition (16).

Furthermore, R is Frobenius. Indeed, we find that the defining relations of $\wedge_{R,q}$ are

$$\xi_0^2 = \xi_1^2 = \cdots = \xi_n^2 = 0, \quad \xi_i \xi_j + q \xi_j \xi_i = 0, \quad \text{for } i < j.$$

It follows that

$$\wedge_{R,q}^{n+1} = \mathbb{C}[[h]].(\xi_0\xi_1\ldots\xi_n), \quad \wedge_{R,q}^i = 0 \text{ for } i > n+1.$$

Computing $\delta_R(\xi_0\xi_1\ldots\xi_n)$, one obtains for the quantum determinant:

$$(19) \qquad \text{qdet}(T) = \sum_{w\in\Sigma_{n+1}} (-e^{-h})^{\ell(w)} t_{0\,w(0)} t_{1\,w(1)} \cdots t_{n\,w(n)},$$

where $\ell(w)$ is the length of an element w of the group Σ_{n+1} of permutations of $\{0, 1, \ldots, n\}$, i.e. the minimum number of adjacent transpositions $(i, i+1)$ of which w is a product. It follows from 7.2.4 that $\text{qdet}(T)$ is a group-like element of $\mathcal{F}_R(M_{n+1})$.

Since R is Frobenius–Hecke, one can define the Hopf algebra $\mathcal{F}_{R,q}(SL_{n+1})$, and we can state

THEOREM 7.3.2 *Let R be the matrix defined in (18), and let $q = e^h$. Then, the Hopf algebra $\mathcal{F}_{R,q}(SL_{n+1})$ is isomorphic to $\mathcal{F}_h(SL_{n+1}(\mathbb{C}))$.* ∎

The proof of this result is analogous to that of 7.1.4.

Remarks [1] The R-matrix (18) will be 'derived' in Section 8.3, where we show that $U_h(sl_{n+1}(\mathbb{C}))$ is topologically quasitriangular. By 6.5.6, the finite-dimensional representations of $U_h(sl_{n+1}(\mathbb{C}))$ are all deformations of representations of $sl_{n+1}(\mathbb{C})$. The R-matrix (18) is obtained, up to a scalar multiple, by evaluating the universal R-matrix of $U_h(sl_{n+1}(\mathbb{C}))$ in the natural representation of dimension $n + 1$. In particular, it necessarily satisfies the QYBE.

[2] It follows from its defining relations that the quantum determinant lies in the centre of $\mathcal{F}_R(M_{n+1})$, and that it has the alternative expression

$$\text{qdet}(T) = \sum_{w\in\Sigma_{n+1}} (-e^{-h})^{\ell(w)} t_{w(0)\,0} t_{w(1)\,1} \cdots t_{w(n)\,n}.$$

[3] The antipode of $\mathcal{F}_{R,q}(SL_{n+1})$ is given by

$$S(t_{ij}) = (-e^{-h})^{i-j}\text{qdet}(T_{ji}),$$

where T_{ji} is the submatrix of T obtained by omitting its jth row and ith column, and its quantum determinant is computed using (19) (with n replaced by $n-1$). △

C The quantum orthogonal and symplectic groups A uniform description of the R-matrix appropriate to the orthogonal and symplectic groups can be given as follows. Fix an integer $m \geq 1$ and, for $r = 1, \ldots, m$, let $\lambda_r \in \mathbb{C}$, $\sigma_r \in \{+1, -1\}$. Let J be the diagonal matrix

$$
J = \begin{pmatrix}
\sigma_1 e^{\lambda_1 h} & 0 & \cdots & 0 \\
0 & \sigma_2 e^{\lambda_2 h} & \cdots & 0 \\
\vdots & \vdots & \ddots & \vdots \\
0 & 0 & \cdots & \sigma_m e^{\lambda_m h}
\end{pmatrix},
$$

and let

$$
R = e^h \sum_{i \neq i'} E_{ii} \otimes E_{ii} + \sum_{\substack{i \neq j \\ i \neq j'}} E_{ii} \otimes E_{jj} + e^{-h} \sum_{i \neq i'} E_{i'i'} \otimes E_{ii}
$$

$$
+ (e^h - e^{-h}) \sum_{i < j} E_{ij} \otimes E_{ji} - (e^h - e^{-h}) \sum_{i < j} e^{(\lambda_j - \lambda_i)h} \sigma_i \sigma_j E_{ij} \otimes E_{i'j'}
$$

(20)

$$
+ E_{\frac{m+1}{2} \frac{m+1}{2}} \otimes E_{\frac{m+1}{2} \frac{m+1}{2}}.
$$

Here, $i' = m + 1 - i$, $1 \leq i \leq m$, and the last term is present only if m is odd. The values of the parameters λ_r and σ_r are as follows:

(a) if $G = SO_{2n+1}(\mathbb{C})$, then $m = 2n + 1$, $\sigma_r = 1$ for all r, and

$$
(\lambda_1, \ldots, \lambda_{2n+1}) = \left(n - \frac{1}{2}, n - \frac{3}{2}, \ldots, \frac{1}{2}, 0, -\frac{1}{2}, \ldots, -n + \frac{1}{2} \right);
$$

(b) if $G = SO_{2n}(\mathbb{C})$, then $m = 2n$, $\sigma_r = 1$ for all r, and

$$
(\lambda_1, \ldots, \lambda_{2n}) = (n - 1, n - 2, \ldots, 1, 0, 0, -1, \ldots, -n + 1);
$$

(c) if $G = Sp_n(\mathbb{C})$, then $m = 2n$, $\sigma_r = 1$ for $r \leq n$, $\sigma_r = -1$ for $r > n$, and

$$
(\lambda_1, \ldots, \lambda_{2n}) = (n, n - 1, \ldots, 1, -1, \ldots, -n).
$$

DEFINITION–PROPOSITION 7.3.3 *For R as above, let* Fun_R *be the quotient of* $\mathcal{F}_R(M_m)$ *by the relations*

$$
TJT^t J^{-1} = JT^t J^{-1} T = 1
$$

(T^t is the transpose of T). Then, Fun_R is a Hopf algebra over $\mathbb{C}[[h]]$ with comultiplication, counit and antipode given by

$$
\Delta(T) = T \odot T, \quad \epsilon(T) = 1, \quad S(T) = JT^t J^{-1},
$$

respectively. ∎

We write Fun_R as $\mathrm{Fun}_h(SO_{2n+1}(\mathbb{C}))$, $\mathrm{Fun}_h(SO_{2n}(\mathbb{C}))$ and $\mathrm{Fun}_h(Sp_n(\mathbb{C}))$ in the three cases (a), (b) and (c), respectively. Formally at least, it is clear that their defining relations go over into those of $\mathcal{F}(SO_{2n+1}(\mathbb{C}))$, $\mathcal{F}(SO_{2n}(\mathbb{C}))$ and $\mathcal{F}(Sp_n(\mathbb{C}))$, respectively, in the limit $h = 0$.

The reason for the change of notation from $\mathcal{F}_h$ to Fun_h is twofold. First, one would *not* expect that $\text{Fun}_h(SO_m(\mathbb{C}))$ is isomorphic to $U_h(so_m(\mathbb{C}))^\circ$, for in the classical case the dual of $U(so_m(\mathbb{C}))$ is the algebra of functions on the simply-connected group with Lie algebra $so_m(\mathbb{C})$, i.e. $\mathcal{F}(\text{Spin}_m(\mathbb{C}))$, which is 'larger' than $\mathcal{F}(SO_m(\mathbb{C}))$. Put another way, the $h = 0$ limit of $\text{Fun}_h(SO_m(\mathbb{C}))$ is generated by the matrix elements of the natural m-dimensional representation of $SO_m(\mathbb{C})$, but not all irreducible representations of $so_m(\mathbb{C})$ occur inside the tensor algebra of the natural representation (for example, the spin representation(s) do(es) not).

Secondly, although one would expect that $\text{Fun}_h(Sp_n(\mathbb{C})) \cong \mathcal{F}_h(Sp_n(\mathbb{C}))$, there does not seem to be a proof of this in the literature. (However, a pairing between $U_h(sp_n(\mathbb{C}))$ and $\text{Fun}_h(Sp_n(\mathbb{C}))$ is constructed by Reshetikhin, Takhtajan & Faddeev (1990) and Hayashi (1992b).)

Remark One difficulty with the algebras $\text{Fun}_h(G)$ for the orthogonal and symplectic groups is that they are not quotients, at least in any obvious way, of the quantized algebra of functions $\mathcal{F}_h(SL_m(\mathbb{C}))$ (nor are the quantized universal enveloping algebras $U_h(\mathfrak{g})$, in any obvious way, subalgebras of $U_h(sl_m(\mathbb{C}))$). The reason for this is that the Dynkin diagrams of the orthogonal and symplectic groups are not subdiagrams of the Dynkin diagram of any special linear group. Takeuchi (1989) defines non-standard quantum analogues of the orthogonal and symplectic groups which are 'quantum subgroups' of quantum linear groups. $\triangle$

D Multiparameter quantized function algebras An obvious way to construct multiparameter deformations of $\mathcal{F}(G)$, where G is, for the moment, an arbitrary complex simple Lie group, is to take the dual of the deformations $U_h(\mathfrak{g})^{\mathcal{F}_h}$ of $U(\mathfrak{g})$ constructed in Subsection 6.5C. Since $U_h(\mathfrak{g})^{\mathcal{F}_h}$ has the same algebra structure as $U_h(\mathfrak{g})$, its dual is isomorphic to the standard quantized function algebra $\mathcal{F}_h(G)$ as a coalgebra. Moreover, in view of 4.2.15, the R-matrix which one expects to give the defining relations of the dual can be obtained from that which defines the relations in $\mathcal{F}_h(G)$ and the image of the element $\mathcal{F}_h$ in the natural representation of G (at least for classical G).

Another possibility which has attracted attention applies to reductive rather than semisimple groups. In the case of $GL_{n+1}(\mathbb{C})$, this leads to a two-parameter family of quantizations, as follows.

Fix λ, $\mu \in \mathbb{C}$ and define $R^{\lambda,\mu} \in \text{End}_{\mathbb{C}}(\mathbb{C}^{n+1} \otimes \mathbb{C}^{n+1})[[h]]$ by

$$R^{\lambda,\mu} = e^{(\lambda+\mu)h/2} \sum_i E_{ii} \otimes E_{ii} + e^{(\lambda-\mu)h/2} \sum_{i<j} E_{ii} \otimes E_{jj}$$

$$+ e^{(\mu-\lambda)h/2} \sum_{i>j} E_{ii} \otimes E_{jj} + (e^{(\lambda+\mu)h/2} - e^{-(\lambda+\mu)h/2}) \sum_{i<j} E_{ij} \otimes E_{ji},$$

where $0 \leq i, j \leq n$. One can check directly that $R^{\lambda,\mu}$ is a solution of the QYBE for all λ and μ, and that it is Hecke with $q = e^{(\lambda+\mu)h/2}$.

PROPOSITION 7.3.4 *The R-matrix $R^{\lambda,\mu}$ defined in (21) is Frobenius–Hecke with $q = e^{(\lambda+\mu)h/2}$. Its quantum determinant is*

$$\text{qdet}(T) = \sum_{w \in \Sigma_{n+1}} (-e^{-\lambda h})^{\ell(w)} t_{0\,w(0)} t_{1\,w(1)} \cdots t_{n\,w(n)}$$

$$= \sum_{w \in \Sigma_{n+1}} (-e^{-\mu h})^{\ell(w)} t_{w(0)\,0} t_{w(1)\,1} \cdots t_{w(n)\,n}.$$

The Hopf algebra $\mathcal{F}_{R^{\lambda,\mu},q}(GL_{n+1})$ is a deformation of $\mathcal{F}(GL_{n+1}(\mathbb{C}))$. Its antipode is given by

$$S(t_{ij}) = (-e^{-\lambda h})^{i-j} \text{qdet}(T_{ji}) \text{qdet}(T)^{-1}$$

$$= (-e^{-\mu h})^{i-j} \text{qdet}(T)^{-1} \text{qdet}(T_{ji}),$$

where T_{ji} is the matrix obtained from T by removing its jth row and ith column. ■

Remarks [1] If $\lambda = \mu = 1$, then $\mathcal{F}_{R^{\lambda,\mu},q}(GL_{n+1})$ is the standard deformation of $\mathcal{F}(GL_{n+1}(\mathbb{C}))$ defined in Subsection 7.3B. The case $\lambda = 0$ and μ arbitrary is the subject of Dipper & Donkin (1991). The case $\lambda + \mu = 0$ is discussed in Gerstenhaber & Schack (1992) in the case $n = 1$. In the general case, 7.3.4 is due to Takeuchi (1990a).

[2] The quantum determinant $\text{qdet}(T)$ does not lie in the centre of $\mathcal{F}_{R^{\lambda,\mu},q}(GL_{n+1})$ unless $\lambda = \mu$: in fact,

$$t_{ij} \text{qdet}(T) = e^{(\lambda-\mu)(i-j)h} \text{qdet}(T) t_{ij}.$$

[3] It is easy to identify the corresponding construction in the context of enveloping algebras. In fact, $U(gl_{n+1}(\mathbb{C}))$ is obtained from $U(sl_{n+1}(\mathbb{C}))$ by adjoining a primitive central element (equal to the identity matrix in the natural representation of $gl_{n+1}(\mathbb{C})$). To obtain a two-parameter family of deformations of $U(gl_{n+1}(\mathbb{C}))$, one first replaces h by λh in the defining relations of $U_h(sl_{n+1}(\mathbb{C}))$ and then adjoins a primitive central element c. The comultiplication is unchanged on the generators H_i, and on the $X_i^{\pm}$ it becomes

$$\Delta_h(X_i^+) = X^+ \otimes Ce^{\lambda hH} + C^{-1} \otimes X^+,$$

$$\Delta_h(X^-) = X^- \otimes C^{-1} + Ce^{-\lambda hH} \otimes X^-,$$

where $C = e^{(\mu-\lambda)hc}$.

[4] One can combine the two-parameter deformation of $GL_{n+1}(\mathbb{C})$ with the $\frac{1}{2}n(n-1)$-parameter deformation of $\mathcal{F}(SL_{n+1}(\mathbb{C}))$ discussed at the beginning

of this subsection to give a family of deformations of $\mathcal{F}(GL_{n+1}(\mathbb{C}))$ depending on $\frac{1}{2}n(n-1)+1$ parameters (modulo rescalings of h). This has been worked out by Artin, Schelter & Tate (1991). $\triangle$

7.4 Differential calculus on quantum groups

Our aim in this section is to construct an analogue for 'quantum' general linear groups of the (algebraic) de Rham complex of $GL_m(k)$, where k is any commutative ring with unit. It is natural to define at the same time the de Rham complex of the 'quantum' plane which is the natural representation of quantum $GL_m(k)$.

We work in the setting of Section 7.2. Thus, we fix an invertible element $q \in k$ and a matrix $R \in \operatorname{End}_k(k^m \otimes k^m)$, and we assume throughout this section that R satisfies the quantum Yang–Baxter equation (13) and the Hecke relation (16); we assume also that $q^2 + 1$ is not a zero divisor. The algebras $\operatorname{Sym}_{R,q}$ and $\mathcal{F}_{R,q}(GL_m)$ defined in Section 7.2 are thought of as the 'algebras of functions' on the 'quantum' plane and the 'quantum' general linear group. As in the classical case, their de Rham complexes will be generated, as differential graded algebras, by $\operatorname{Sym}_{R,q}$ and $\mathcal{F}_{R,q}(GL_m)$.

We recall that a *differential graded algebra* is a graded associative algebra $A = \oplus_{i=0}^{\infty} A^i$ over k equipped with k-linear maps $d : A^i \to A^{i+1}$ such that $d^2 = 0$ and

$$d(a.b) = d(a)b + (-1)^{\deg(a)} a d(b)$$

for any two homogeneous elements $a, b \in A$.

A The de Rham complex of the quantum plane We begin by defining a complex $\Omega^*(\operatorname{Sym}_{R,q})$ which, when $R = 1$ and $q = 1$, becomes the usual de Rham complex of k^m. It will be convenient to define an analogous complex for $\wedge_{R,q}$ at the same time, which can be thought of as the de Rham complex of the 'odd' quantum plane (in the language of 'supergeometry').

DEFINITION 7.4.1 (i) $\Omega^*(\operatorname{Sym}_{R,q})$ *is the associative algebra over k with generators* $x_1, \ldots, x_m$ *and* $dx_1, \ldots, dx_m$, *and defining relations*

$$X \odot X = q^{-1} \check{R}(X \odot X),$$
$$X \odot dX = q \check{R}(dX \odot X),$$
$$dX \odot dX = -q \check{R}(dX \odot dX).$$

(ii) $\Omega^*(\wedge_{R,q})$ *is the associative k-algebra with generators* $\xi_i, d\xi_i, 1 \le i \le m$, *and defining relations*

$$\Xi \odot \Xi = -q\check{R}(\Xi \odot \Xi),$$
$$\Xi \odot d\Xi = q^{-1}\check{R}(d\Xi \odot \Xi),$$
$$d\Xi \odot d\Xi = q^{-1}\check{R}(d\Xi \odot d\Xi).$$

The notation is as in Section 7.2. For example, X is the $m \times 1$ matrix with entries $x_1, \ldots, x_m$, and $X \odot X$ is the $m^2 \times 1$ matrix with entries $x_1^2, \ldots, x_1 x_m, x_2 x_1, \ldots, x_m^2$.

PROPOSITION 7.4.2 $\Omega^*(\mathrm{Sym}_{R,q})$ *(resp.* $\Omega^*(\wedge_{R,q})$*) has a unique structure of a differential graded algebra with grading given by* $\deg(x_i) = 0$, $\deg(dx_i) = 1$ *(resp.* $\deg(\xi_i) = 1$, $\deg(d\xi_i) = 2$*) and differential d such that $d(x_i) = dx_i$ (resp. $d(\xi_i) = d\xi_i$).*

OUTLINE OF PROOF We give the proof for $\mathrm{Sym}_{R,q}$; that for $\wedge_{R,q}$ is similar.

It is clear that $\Omega^*(\mathrm{Sym}_{R,q})$ is a graded algebra. To show that d is compatible with its defining relations, note that applying d to the difference between the left- and right-hand sides of the first relation in 7.4.1(i) gives

$$(1 - q^{-1}\check{R})(dX \odot X + X \odot dX) = (1 - q^{-1}\check{R})(1 + q\check{R})(dX \odot X) = 0,$$

where the first equality uses the second relation in 7.4.1(i) and the second equality the Hecke relation (16). The rest of the proof is easy, for it is clear that applying d to the second relation in 7.4.1(i) gives the third relation, both sides of which are annihilated by d. ∎

Example 7.4.3 For the R-matrix (18) (and with $q = e^h \in \mathbb{C}[[h]]$), one finds that $\Omega^*(\mathrm{Sym}_{R,q})$ has defining relations

$$x_i.dx_i = q^2 dx_i.x_i, \quad (dx_i)^2 = 0, \quad \text{for all } i,$$

$$\left. \begin{array}{l} x_i.x_j = q^{-1}x_j.x_i, \quad x_i.dx_j = qdx_j x_i, \\ dx_i.x_j = q^{-1}x_j.dx_i - (q - q^{-1})dx_j.x_i, \\ dx_i.dx_j = -qdx_j.dx_i, \end{array} \right\} \quad \text{for } i < j,$$

and that $\Omega^*(\wedge_{R,q})$ *has defining relations*

$$\xi_i.d\xi_i = d\xi_i.\xi_i, \quad (\xi_i)^2 = 0, \quad \text{for all } i,$$

$$\left. \begin{array}{l} \xi_i.\xi_j = -q\xi_j.\xi_i, \quad \xi_i.d\xi_j = q^{-1}d\xi_j\xi_i, \\ d\xi_i.\xi_j = q\xi_j.d\xi_i - (q - q^{-1})d\xi_j.\xi_i, \\ d\xi_i.d\xi_j = q^{-1}d\xi_j.d\xi_i, \end{array} \right\} \quad \text{for } i < j. \quad \Diamond$$

B The de Rham complex of the quantum $m \times m$ matrices Recall
that, if (A, d_A) and (B, d_B) are differential graded algebras over k, the tensor
product algebra $A \underset{k}{\otimes} B$ is a differential graded algebra with grading given by

$$(A \otimes B)^i = \bigoplus_{j=0}^{i} A^j \otimes B^{i-j},$$

and differential

$$d_{A \otimes B}(a \otimes b) = d_A(a) \otimes b + (-1)^{\deg(a)} a \otimes d_B(b),$$

for homogeneous elements $a \in A$, $b \in B$. Recall also that a homomorphism
of differential graded algebras $f : A \to B$ is a homomorphism of algebras for
which $f(A^i) \subseteq B^i$ for all i and such that f commutes with the differentials
of A and B.

In the classical situation, pulling back differential forms via the natural
representation $GL_m(k) \times k^m \to k^m$ gives a homomorphism of differential
graded algebras

$$(22) \qquad \Omega^*(\mathrm{Sym}((k^m)^*)) \to \mathcal{F}(GL_m(k)) \otimes \Omega^*(\mathrm{Sym}((k^m)^*)).$$

The complex $\Omega^*(\mathcal{F}_R(M_m))$ must be defined so that it contains $\mathcal{F}_R(M_m)$ as
a subalgebra (the quantum 0-forms!), and such that there is an analogue of
the homomorphism (22). In the literature, the latter property is expressed
by saying that the resulting 'differential calculus' is *covariant*.

The appropriate definition is the following.

DEFINITION–PROPOSITION 7.4.4 $\Omega^*(\mathcal{F}_R(M_m))$ *is the associative algebra
over k with generators t_{ij} and dt_{ij} for $1 \le i, j \le m$, and defining relations*

$$\check{R}(T \odot T) = (T \odot T)\check{R},$$
$$T \odot dT = \check{R}(dT \odot T)\check{R},$$
$$dT \odot dT = -\check{R}(dT \odot dT)\check{R}.$$

It has a unique structure of a differential graded algebra with $\deg(t_{ij}) = 0$,
$\deg(dt_{ij}) = 1$ *and* $d(t_{ij}) = dt_{ij}$.

OUTLINE OF PROOF As in the proof of 7.4.2, the main point is to show that
applying d to the first relation in 7.4.4 gives the second relation. Now,

$$\begin{aligned}
d[\check{R}, T \odot T] &= [\check{R}, dT \odot T + T \odot dT] \\
&= [\check{R}, dT \odot T + \check{R}(dT \odot T)\check{R}] \\
&= [\check{R}, dT \odot T] + ((q - q^{-1})\check{R} + 1)(dT \odot T)\check{R} \\
&\qquad\qquad - \check{R}(dT \odot T)((q - q^{-1})\check{R} + 1) \\
&= 0,
\end{aligned}$$

where the first equality used the second relation in 7.4.4, and the third equality used the Hecke relation (16). ∎

As we mentioned above, the definitions of $\Omega^*(\mathrm{Sym}_{R,q})$ and $\Omega^*(\mathcal{F}_R(M_m))$ have been chosen so that

PROPOSITION 7.4.5 *There exist unique homomorphisms of differential graded algebras*

$$\Omega^*(\mathrm{Sym}_{R,q}) \to \Omega^*(\mathcal{F}_R(M_m)) \otimes \Omega^*(\mathrm{Sym}_{R,q}),$$
$$\Omega^*(\wedge_{R,q}) \to \Omega^*(\mathcal{F}_R(M_m)) \otimes \Omega^*(\wedge_{R,q}),$$

such that

$$x_i \mapsto \sum_j t_{ij} \otimes x_j, \quad \xi_i \mapsto \sum_j t_{ij} \otimes \xi_j. \quad \blacksquare$$

The proof of this result rests on a computation similar to, though rather more difficult than, that used to prove 7.4.4.

Example 7.4.6 For the R-matrix in (18), one finds that the defining relations of $\Omega^*(\mathcal{F}_R(M_m))$ are as follows:

$$t_{ij}dt_{ij} = q^2 dt_{ij}t_{ij}, \quad (dt_{ij})^2 = 0,$$

for all i, j,

$$t_{ij}.t_{kj} = q^{-1}t_{kj}.t_{ij}, \quad t_{ij}.dt_{kj} = qdt_{kj}t_{ij},$$
$$dt_{ij}.t_{kj} = q^{-1}t_{kj}.dt_{ij} - (q - q^{-1})dt_{kj}.t_{ij},$$
$$dt_{ij}.dt_{kj} = -qdt_{kj}.dt_{ij},$$

for $i < k$ and all j,

$$t_{ij}.t_{il} = q^{-1}t_{il}.t_{ij}, \quad t_{ij}.dt_{il} = qdt_{il}t_{ij},$$
$$dt_{ij}.t_{il} = q^{-1}t_{il}.dt_{ij} - (q - q^{-1})dt_{il}.t_{ij},$$
$$dt_{ij}.dt_{il} = -qdt_{il}.dt_{ij},$$

for $j < l$ and all i, and

$$[t_{il}, t_{kj}] = 0, \quad [t_{ij}, t_{kl}] = -(q - q^{-1})t_{kj}t_{il},$$
$$t_{ij}.dt_{kl} = dt_{kl}.t_{ij}, \quad [t_{kj}, dt_{il}] = [t_{il}, dt_{kj}] + (q - q^{-1})dt_{kl}t_{ij},$$
$$t_{kl}dt_{ij} = dt_{ij}t_{kl} + (q - q^{-1})(dt_{kj}t_{il} + dt_{il}t_{kj}) + (q - q^{-1})^2dt_{kl}t_{ij},$$
$$dt_{ij}dt_{kl} = -dt_{kl}dt_{ij}, \quad dt_{jk}dt_{il} + dt_{il}dt_{jk} = -(q - q^{-1})dt_{kl}dt_{ij},$$

for $i < k$ and $j < l$. ◇

Remark When R is the R-matrix in (18) and k is a field, the cohomology of the complex $\Omega^*(\mathcal{F}_R(M_m))$ has been computed by P. Feng. If q is not a root of unity,

$$H^*(\Omega^*(\mathcal{F}_R(M_m))) \cong k,$$

concentrated in degree zero, while if q is a primitive ℓth root of unity, where ℓ is odd and ≥ 3, then

$$\{(dt_{11})^{\sigma_{11}}(t_{11})^{a_{11}}(dt_{12})^{\sigma_{12}}(t_{12})^{a_{12}}\ldots(dt_{mm})^{\sigma_{mm}}(t_{mm})^{a_{mm}}\},$$

where $\sigma_{ij} = 0$ or 1, $a_{ij} \in \ell\mathbb{N}$, is a basis of $H^*(\Omega^*(\mathcal{F}_R(M_m)))$ (the grading is the obvious one). We shall encounter such 'root of unity phenomena' frequently in later chapters. $\triangle$

C The de Rham complex of the quantum general linear group In this subsection, we assume that R is Frobenius. As we saw in Subsection 7.2B, this enables one to define $\mathcal{F}_{R,q}(GL_m)$, the algebra of functions on the 'quantum general linear group', to be the localization of $\mathcal{F}_R(M_m)$ at the quantum determinant qdet(T). In the same way, we should like to define $\Omega^*(\mathcal{F}_{R,q}(GL_m))$ to be the localization of $\Omega^*(\mathcal{F}_R(M_m))$ at qdet$(T) \in \Omega^0(\mathcal{F}_R(M_m))$. For this, we need the following analogue of 7.2.6, which is due to Tsygan (1993).

LEMMA 7.4.7 *There exists an invertible element $\beta \in k$ such that*

$$\mathrm{qdet}(T).dT.Q^{-1} = \beta q^2 Q^{-1}.dT.\mathrm{qdet}(T),$$

where $Q \in GL_m(k)$ is defined in 7.2.6. ∎

Together with 7.2.6, this lemma clearly implies that the Ore condition is satisfied for the multiplicative subset $\{\mathrm{qdet}(T)^r\}_{r\geq 1}$ of the ring $\Omega^*(\mathcal{F}_R(M_m))$, so we may define $\Omega^*(\mathcal{F}_{R,q}(GL_m))$ to be the localization of $\Omega^*(\mathcal{F}_R(M_m))$ at the quantum determinant.

PROPOSITION 7.4.8 $\Omega^*(\mathcal{F}_{R,q}(GL_m))$ *is a Hopf algebra with comultiplication, counit and antipode determined by*

$$\Delta(T) = T \,\square\, T, \quad \Delta(dT) = dT \,\square\, T + T \,\square\, dT,$$
$$\epsilon(T) = 1, \quad \epsilon(dT) = 0,$$
$$S(T) = T^{-1}, \quad S(dT) = d(S(T)).$$

Moreover, the differential graded algebra structure of $\Omega^*(\mathcal{F}_R(M_m))$ extends uniquely to $\Omega^*(\mathcal{F}_{R,q}(GL_m))$, and then Δ and ϵ (resp. S) are homomorphisms (resp. is an anti-homomorphism) of differential graded algebras (k is concentrated in degree zero and has the zero differential).

OUTLINE OF PROOF The main point is to show that there is a homomorphism of algebras $S : \Omega^*(\mathcal{F}_{R,q}(GL_m)) \to \Omega^*(\mathcal{F}_{R,q}(GL_m))_{\mathrm{op}}$ given on the generators by the formulas in the proposition. Recalling from Subsection 7.2B that there is an isomorphism of algebras $(\mathcal{F}_{R,q})_{\mathrm{op}} \cong \mathcal{F}_{R^{-1},q^{-1}}$, we have to show that, if we replace T by T^{-1} and $\check{R}$ by $(R^{-1})\check{} = \sigma\check{R}^{-1}\sigma$, the defining relations in

7.4.4 continue to hold. We do this for the second relation, as the other two are similar but easier.

We start with the second relation in 7.4.4 in the form

$$(T^{-1} \odot 1)\check{R}(dT \odot 1) = (1 \odot dT)\check{R}^{-1}(1 \odot T^{-1}).$$

Multiplying both sides on the left by $\sigma(1 \odot T^{-1})$ and on the right by $(T^{-1} \odot 1)\sigma$, and using the relation $(1 \odot A)(B \odot 1) = \sigma(A \odot B)\sigma$, we get

$$(T^{-1} \odot T^{-1})\sigma\check{R}(dT.T^{-1} \odot 1)\sigma = \sigma(1 \odot T^{-1}.dT)\check{R}^{-1}\sigma(T^{-1} \odot T^{-1}).$$

Using the first relation in 7.4.4, the left-hand side of this equation is equal to

$$(T^{-1} \odot T^{-1})\sigma(T \odot T)\check{R}(T \odot T)^{-1}(dT.T^{-1} \odot 1)\sigma$$
$$= (T^{-1} \odot T^{-1})\sigma(T \odot T)\check{R}(1 \odot T^{-1})(T^{-1}.dT.T^{-1} \odot 1)\sigma$$
$$= (T^{-1} \odot T^{-1})\sigma(T \odot T)\check{R}\sigma(T^{-1} \odot T^{-1}.dT.T^{-1})$$
$$= (T^{-1} \odot T^{-1})(1 \odot T)(T \odot 1)\sigma\check{R}\sigma(T^{-1} \odot T^{-1}.dT.T^{-1})$$
$$= \sigma\check{R}\sigma(T^{-1} \odot T^{-1}.dT.T^{-1}).$$

Similarly, the right-hand side is equal to $(T^{-1}.dT.T^{-1} \odot T^{-1})\sigma\check{R}^{-1}\sigma$. Noting that $d(T^{-1}) = -T^{-1}.dT.T^{-1}$ completes the proof. ∎

D Invariant forms on quantum GL_m Finally, we consider briefly the problem of defining right-invariant forms in $\Omega^*(\mathcal{F}_{R,q}(GL_m))$ (left-invariant forms are defined similarly). We begin by formulating the right-invariance condition algebraically in the classical situation.

Suppose then that G is a Lie group, and let ω be an n-form on G. Then, ω is right-invariant if $R_g^*(\omega) = \omega$ for all $g \in G$, where $R_g : G \to G$ is right translation by g. Let $m : G \times G \to G$ be the multiplication map of G, and define maps $\rho_g : G \to G \times G$ by $\rho_g(h) = (h, g)$. Then, $R_g = m \circ \rho_g$ so right-invariance is equivalent to $\rho_g^*(m^*(\omega)) = \omega$ for all $g \in G$. Note that m^* is precisely the comultiplication Δ in the Hopf algebra $\Omega^*(G)$, and that ρ_g^* kills all bigraded components of $\Delta(\omega)$ except that of degree $(n, 0)$. Suppose that

$$\Delta(\omega)^{n,0} = \sum_\alpha \omega_\alpha \otimes f_\alpha, \quad (\omega_\alpha \in \Omega^n, \ f_\alpha \in \Omega^0).$$

The right-invariance condition is thus $\sum_\alpha f_\alpha(g)\omega_\alpha = \omega$, for all $g \in G$. Assuming, as we may, that the ω_α are linearly independent, this implies that the f_α are constant, and hence that

$$(23) \qquad\qquad \Delta(\omega)^{n,0} = \omega \otimes 1.$$

Returning to the quantum situation, an element $\omega \in \Omega^n(\mathcal{F}_{R,q}(GL_m))$ is called *right-invariant* if it satisfies (23), where Δ is the comultiplication defined in 7.4.8. It is clear that the set $\Omega^*_{inv}(\mathcal{F}_{R,q}(GL_m))$ of such forms is a differential graded subalgebra of $\Omega^*(\mathcal{F}_{R,q}(GL_m))$ (in the obvious sense). We state without proof

PROPOSITION 7.4.9 *The entries of the matrix* $\Omega = dT.T^{-1}$ *are right-invariant 1-forms and generate* $\Omega^*_{\mathrm{inv}}(\mathcal{F}_{R,q}(GL_m))$ *subject to the relations*

$$(\Omega \odot 1)\check{R}(\Omega \odot 1) = -\check{R}(\Omega \odot 1)\check{R}(\Omega \odot 1)\check{R}. \quad \blacksquare$$

7.5 Integrable lattice models

In this section, we shall indicate briefly the origin and meaning of the relations (12) and of the quantum Yang–Baxter equation (13) in one area of the theory of integrable systems, namely two-dimensional lattice models in statistical mechanics.

A Vertex models We consider a collection of 'atoms' located at the vertices of a 2-dimensional lattice. Each atom interacts only with its nearest neighbours with an energy which depends on the 'state' of the bonds joining each pair of neighbouring atoms. If the possible states of the bonds are labelled by the elements of a finite set $\{1, \ldots, n\}$, we write the interaction energy of an atom as $\mathcal{E}_{ij}^{kl}$ if the states of the bonds connected to the atom are as shown in Fig. 15.

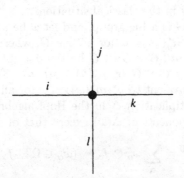

Fig. 15 Labelling of the bonds

Usually, $\mathcal{E}_{ij}^{kl}$ will depend on parameters other than the bond states, such as the values of external electric or magnetic fields, but we shall assume that this function does not depend on the location of the vertex itself. A knowledge of the $\mathcal{E}_{ij}^{kl}$ specifies the model.

A *state* of the lattice is an assignment of a state to each bond, and the energy $\mathcal{E}$ of such a state is simply the sum of the $\mathcal{E}_{ij}^{kl}$ over all the atoms in the lattice. For an infinite lattice, this sum will usually diverge, so one works with finite lattices, say an $M \times N$ rectangular array of atoms, keeping in mind that one is interested ultimately in what happens when M and N tend to infinity. For finite lattices, one must impose *boundary conditions* on the states of the bonds at the edges; we shall always use periodic boundary conditions, as indicated in Fig. 16.

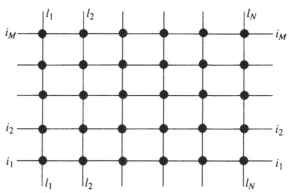

Fig. 16 Periodic boundary conditions

The statistical mechanical properties of such a system are encoded in its *partition function*

$$(24) \qquad Z = \sum_{\text{states}} \exp(-\beta \mathcal{E}(\text{state})),$$

where $\beta = 1/kT$ is the reciprocal of Boltzmann's constant k times the temperature T. The principle is that $\frac{1}{Z} e^{-\beta \mathcal{E}}$ is the probability that the system is in a particular state, so that the average value of any physical quantity Q is given by

$$\frac{\sum_{\text{states}} Q e^{-\beta \mathcal{E}}}{\sum_{\text{states}} e^{-\beta \mathcal{E}}}.$$

For example, the average energy of the system is

$$\langle \mathcal{E}(\text{system}) \rangle = kT^2 \frac{\partial}{\partial T} \ln Z.$$

Thus, one would like an 'explicit' formula for Z; if this can be found, one says that the model is *exactly solvable*.

Note that

$$\exp(-\beta \mathcal{E}(\text{state})) = \prod_{\text{vertices}} \exp(-\beta \mathcal{E}_{ij}^{kl}).$$

The quantities in the product are called *Boltzmann weights*; we denote them by

$$R_{ij}^{kl} = \exp(-\beta \mathcal{E}_{ij}^{kl}).$$

B Transfer matrices To evaluate (24), we first sum over all possible states of the bonds in the first row, excluding the two free bonds (which we temporarily allow to have different states):

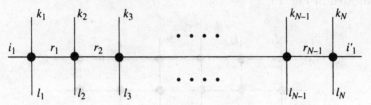

Thus, we form

(25)
$$T^{i'_1 l_1 \cdots l_N}_{i_1 k_1 \cdots k_N} = \sum_{r_1, \ldots, r_{N-1}} R^{r_1 l_1}_{i_1 k_1} R^{r_2 l_2}_{r_1 k_2} \cdots R^{i'_1 l_N}_{r_{N-1} k_N}.$$

Before we lose control of the indices, we introduce an auxiliary N-dimensional vector space V with basis $\{v_1, \ldots, v_N\}$, and define $R \in \text{End}(V \otimes V)$ by

$$R(v_i \otimes v_j) = \sum_{k,l} R^{kl}_{ij} v_k \otimes v_l,$$

and $T \in \text{End}(V \otimes V^{\otimes N})$ by

$$T(v_{i_1} \otimes v_{k_1} \otimes \cdots \otimes v_{k_N}) = \sum_{i'_1, l_1, \ldots, l_N} T^{i'_1 l_1 \cdots l_N}_{i_1 k_1 \cdots k_N} v_{i'_1} \otimes v_{l_1} \otimes \cdots \otimes v_{l_N}.$$

Then, (25) can be written

(26)
$$T = R_{01} R_{02} \ldots R_{0N},$$

where, as usual, R_{ij} means R acting in the ith and jth factors of the tensor product $V \otimes V^{\otimes N}$, the first factor being labelled 0 and the rest $1, \ldots, N$.

Recalling that periodic boundary conditions are assumed, we have $i_1 = i'_1$ and the result of summing over the states of all bonds in the first row is

$$(\text{trace}_V(T))^{l_1 \cdots l_N}_{k_1 \cdots k_N}.$$

The endomorphism $\text{trace}_V(T)$ is called the *row-to-row transfer matrix*.

The result of summing the contributions to (24) over all states of bonds in the first two rows is therefore

$$(\text{trace}_V(T))^{l_1 \cdots l_N}_{k_1 \cdots k_N} (\text{trace}_V(T))^{k_1 \cdots k_N}_{j_1 \cdots j_N}.$$

Summing over the vertical bonds joining the first two rows thus gives

$$((\text{trace}_V(T))^2)^{l_1 \cdots l_N}_{j_1 \cdots j_N}.$$

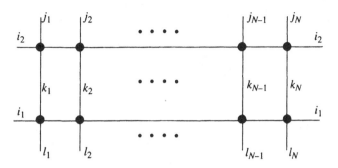

Continuing in this way, it is clear that

$$((\mathrm{trace}_V(T))^M)^{l_1\cdots l_N}_{l'_1\cdots l'_N}$$

is the result of summing over all bond states except the free vertical bonds on the bottom and top rows. Imposing periodic boundary conditions $l_i = l'_i$ and summing over these free bonds, we finally obtain

PROPOSITION 7.5.1 *The partition function Z of a vertex model can be expressed in terms of its transfer matrix T by*

$$Z = \mathrm{trace}_{V^{\otimes N}}((\mathrm{trace}_V(T))^M). \quad \blacksquare$$

Remark The way that the partition function is built up row by row using the transfer matrices is reminiscent of the way, discussed in Subsection 5.3A, that an endomorphism of a representation of a ribbon Hopf algebra is associated to a ribbon tangle by breaking up the ribbon tangle into 'elementary tangles'. We shall see in Section 15.2 that this similarity is not accidental. $\triangle$

It follows from this proposition that, if κ_N is the largest eigenvalue (assumed to be simple) of $\mathrm{trace}_V(T) \in \mathrm{End}(V^{\otimes N})$, then for large M we have the asymptotic estimate

$$Z \sim \kappa_N^M.$$

Thus, one would like to find the eigenvalues of $\mathrm{trace}_V(T)$, or at least the largest. This is complicated, of course, by the fact that the size of the matrix tends to infinity with N. It would help if we knew a large family of operators commuting with $\mathrm{trace}_V(T)$, for then we could restrict attention to their common invariant subspaces. We shall now give a condition which guarantees the existence of such operators.

C Integrability Recall that the vertex energies $\mathcal{E}^{kl}_{ij}$, and hence also R and T, may depend on the values of external parameters. We shall write an element of this parameter space as λ, μ, ν, etc.

DEFINITION 7.5.2 A vertex model is integrable if, for each μ, ν, there is a λ such that

$$(27) \qquad R_{12}(\lambda)R_{13}(\mu)R_{23}(\nu) = R_{23}(\nu)R_{13}(\mu)R_{12}(\lambda)$$

as endomorphisms of $V \otimes V \otimes V$.

In some examples, λ, μ and ν are complex numbers, and it turns out that the parametrization can be chosen so that (27) holds with $\lambda = \mu - \nu$:

$$(28) \quad R_{12}(\lambda - \mu)R_{13}(\lambda - \nu)R_{23}(\mu - \nu) = R_{23}(\mu - \nu)R_{13}(\lambda - \nu)R_{12}(\lambda - \mu).$$

This is called the quantum Yang–Baxter equation with *spectral parameters*. It should be compared to the CYBE with spectral parameters studied in Section 3.2, to which it reduces if

$$R = 1 + hr,$$

where h is a vanishingly small quantity.

The importance of the integrability condition is explained by the following result and its corollary.

PROPOSITION 7.5.3 Consider an integrable vertex model, and let λ, μ and ν be as in 7.5.2. Then,

$$R(\lambda)(1 \otimes T(\mu))(T(\nu) \otimes 1) = (T(\nu) \otimes 1)(1 \otimes T(\mu))R(\lambda)$$

as endomorphisms of $V \otimes V \otimes V^{\otimes N}$, where $R(\lambda)$ acts on the first two factors of the tensor product.

COROLLARY 7.5.4 For an integrable vertex model for which $R(\lambda)$ is invertible for all λ, $\mathrm{trace}_V(T(\mu))$ commutes with $\mathrm{trace}_V(T(\nu))$ for all μ, ν.

PROOF OF 7.5.4 Simply note that

$$(T(\nu) \otimes 1)(1 \otimes T(\mu)) = R(\lambda)(1 \otimes T(\mu))(T(\nu) \otimes 1)R(\lambda)^{-1}.$$

Taking $\mathrm{trace}_{V \otimes V}$ on both sides gives

$$\mathrm{trace}_V(T(\nu))\mathrm{trace}_V(T(\mu)) = \mathrm{trace}_V(T(\mu))\mathrm{trace}_V(T(\nu)). \quad \blacksquare$$

PROOF OF 7.5.3 We label the first two factors in $V \otimes V \otimes V^{\otimes N}$ as 0 and $\bar{0}$ and the remainder as $1, \ldots, N$. Then, from (26), we have

$$T_0 = R_{01}R_{02} \ldots R_{0N}, \quad T_{\bar{0}} = R_{\bar{0}1}R_{\bar{0}2} \ldots R_{\bar{0}N}.$$

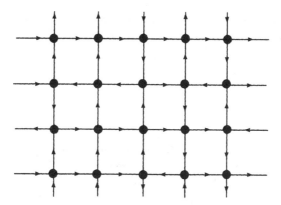

Fig. 17 A possible arrangement of dipoles for the six vertex model

Thus,

$$R(\lambda)(1 \otimes T(\mu))(T(\nu) \otimes 1)$$
$$= R_{0\bar{0}}(\lambda)T_{\bar{0}}(\mu)T_0(\nu)$$
$$= R_{0\bar{0}}(\lambda)R_{\bar{0}1}(\mu)R_{\bar{0}2}(\mu)\dots R_{\bar{0}N}(\mu)R_{01}(\nu)R_{02}(\nu)\dots R_{0N}(\nu)$$
$$= R_{0\bar{0}}(\lambda)R_{\bar{0}1}(\mu)R_{01}(\nu)R_{\bar{0}2}(\mu)\dots R_{\bar{0}N}(\mu)R_{02}(\nu)\dots R_{0N}(\nu)$$
$$= R_{01}(\nu)R_{\bar{0}1}(\mu)R_{0\bar{0}}(\lambda)R_{\bar{0}2}(\mu)\dots R_{0N}(\nu),$$

where the penultimate equality uses the fact that $R_{01}(\nu)$ commutes with $R_{\bar{0}i}(\mu)$ for $i = 2,\dots,N$, and the last equality uses the QYBE. Bringing $R_{02}(\nu)$ to the left in the same way, using the QYBE to re-order the product

$$R_{0\bar{0}}(\lambda)R_{\bar{0}2}(\mu)R_{02}(\nu),$$

and then using the fact that $R_{02}(\nu)$ commutes with $R_{\bar{0}1}(\mu)$, we obtain

$$R_{01}(\nu)R_{02}(\nu)R_{\bar{0}1}(\mu)R_{\bar{0}2}(\mu)R_{0\bar{0}}(\lambda)\dots R_{0N}(\nu).$$

Bringing $R_{03}(\nu),\dots,R_{0N}(\nu)$ successively to the left in the same way, we eventually obtain

$$R_{01}(\nu)\dots R_{0N}(\nu)R_{\bar{0}1}(\mu)\dots R_{\bar{0}N}(\mu)R_{0\bar{0}}(\lambda) = T_0(\nu)T_{\bar{0}}(\mu)R_{0\bar{0}}(\lambda)$$
$$= (T(\nu) \otimes 1)(1 \otimes T(\mu))R(\lambda)$$

as we want. ∎

D Examples We consider two examples, in both of which there are just two possible states for each bond. We can imagine that two nearest neighbour atoms interact by means of an electron cloud which can be either 'close' to one atom or to the other. The two possibilities can be represented by means of an arrow (or dipole) pointing (say) from the region of low charge density to the region of high charge density (see Fig. 17).

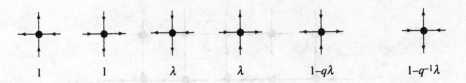

1   1 λ λ $1-q\lambda$ $1-q^{-1}\lambda$

Six vertex model (XXX model) In this, perhaps the simplest, model, one assumes that all Boltzmann weights are zero except those which correspond to configurations which have two arrows pointing towards the atom and two pointing away (a possible arrangement of dipoles on the whole lattice is shown in Fig. 17). There are six such configurations (hence the name of the model), and we take their Boltzmann weights to be as shown above (the spectral parameter λ is an arbitrary complex number; q is an auxiliary non-zero parameter). The auxiliary vector space has dimension 2, and we label its basis by $\{1, 2\}$, where 1 corresponds to arrows which point up or to the left, and 2 to arrows with the opposite orientation. Then, the R-matrix is

$$R(q, \lambda) = \begin{pmatrix} 1 & 0 & 0 & 0 \\ 0 & \lambda & 1 - q\lambda & 0 \\ 0 & 1 - q^{-1}\lambda & \lambda & 0 \\ 0 & 0 & 0 & 1 \end{pmatrix}.$$

This model is integrable, for a direct calculation shows that (27) holds if

(29) $$\lambda - \mu + \nu + \lambda\mu\nu = (q + q^{-1})\lambda\nu$$

(the value of q is the same for all three R-matrices).

There are two interesting special cases. If we take $\lambda = \mu = \nu = q^{-1}$, then (29) is satisfied and so $R(q, q^{-1})$ is a *constant* solution of the QYBE for each q. In fact, if we replace q by e^h, then $qR(q, q^{-1})$ is exactly the R-matrix which defines the standard deformation $\mathcal{F}_h(M_2(\mathbb{C}))$ (see 7.1.3).

On the other hand, if we take $q = 1$ and set

$$\tilde{\lambda} = \frac{\lambda}{\lambda - 1}, \quad \tilde{\mu} = \frac{\mu}{\mu - 1}, \quad \tilde{\nu} = \frac{\nu}{\nu - 1},$$

we find that (29) is equivalent to

$$\tilde{\lambda} = \tilde{\mu} - \tilde{\nu}.$$

Thus, $R(1, \frac{\lambda}{\lambda - 1})$ is a rational solution of the QYBE (28), and the model is accordingly called the *rational six vertex model*. We shall see in Chapter 12 that this R-matrix is associated to a two-dimensional representation of the *Yangian* of $sl_2(\mathbb{C})$.

Eight vertex model (XYZ model) In this model, one relaxes the conditions on the Boltzmann weights by allowing them to be non-zero for the eight configurations in which the number of arrows pointing towards each atom is even. One also assumes that the weights are unchanged when all the arrows are reversed. The following weights lead to an integrable model:

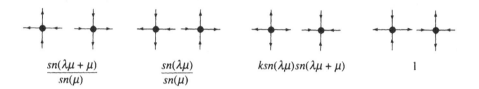

$$\frac{sn(\lambda\mu + \mu)}{sn(\mu)} \qquad \frac{sn(\lambda\mu)}{sn(\mu)} \qquad ksn(\lambda\mu)sn(\lambda\mu + \mu) \qquad 1$$

Here, sn is a jacobian elliptic function and k is its modulus. The R-matrix is

$$R(k,\lambda,\mu) = \begin{pmatrix} \frac{sn(\lambda\mu+\mu)}{sn(\mu)} & 0 & 0 & ksn(\lambda\mu)sn(\lambda\mu + \mu) \\ 0 & \frac{sn(\lambda\mu)}{sn(\mu)} & 1 & 0 \\ 0 & 1 & \frac{sn(\lambda\mu)}{sn(\mu)} & 0 \\ ksn(\lambda\mu)sn(\lambda\mu + \mu) & 0 & 0 & \frac{sn(\lambda\mu+\mu)}{sn(\mu)} \end{pmatrix}.$$

Again, there are a number of interesting special cases. As $\mu \to 0$, we get the rational six vertex solution above (up to a scalar multiple and a change of sign of the parameter). As $k \to 0$, the elliptic function sn becomes the trigonometric function sin and we have the *trigonometric six vertex model*.

Bibliographical notes

7.1 Our discussion of $\mathcal{F}_h(SL_2(\mathbb{C}))$ is taken from Manin (1988). The proof of 7.1.2 is adapted from Meister (1992), and that of 7.1.4 from Vaksman & Soibelman (1988).

7.2 The most important references for this section and the next are Manin (1988), Reshetikhin, Takhtajan & Faddeev (1990) and Tsygan (1993); see also Gurevich (1991). See Majid (1990g) for another method of passing from bialgebras to Hopf algebras. Lambe & Radford (1993) contains an extensive discussion of various algebraic structures related to the QYBE.

For a catalogue of solutions of the QYBE, pre-dating quantum groups, see Kulish & Sklyanin (1982a).

7.3 Theorem 7.3.2 is stated in Drinfel'd (1987) and in Reshetikhin, Takhtajan & Faddeev (1990), although neither reference contains a complete proof. One can give a proof along the same lines as that of 7.1.4, but the details are more unpleasant. Takeuchi (1990a) gives a more general result.

Our discussion of the orthogonal and symplectic cases is again based on Reshetikhin, Takhtajan & Faddeev (1990). Further details may be found in Hayashi (1992b). The former reference also defines quantum orthogonal and symplectic vector spaces on which the quantum orthogonal and symplectic groups act, generalizing their classical realizations as groups which act on a vector space preserving a non-degenerate symmetric or skew-symmetric bilinear form.

To define $\text{Fun}_h(G)$ when G is an exceptional group, the first step is to compute a suitable R-matrix. When G is of type G_2, the image of the universal R-matrix of $U_h(\mathfrak{g})$ in the deformation of the seven-dimensional representation of G is computed in Kirillov & Reshetikhin (1989).

Multiparameter quantizations have been studied by a number of authors, including Sudbery (1990), Takeuchi (1990a) and Artin, Schelter & Tate (1991).

7.4 A 'quantum' analogue of differential calculus was first introduced by Woronowicz (1989), in the context of 'compact matrix pseudogroups' (see Chapter 13). In the form we have presented them, the results are due to Tsygan (1993), where further details and omitted proofs may be found. Examples 7.4.3 and 7.4.6 can be found in Maltsiniotis (1990), Wess & Zumino (1991) and elsewhere.

There is, of course, a complex of 'differential forms' associated to any k-algebra, namely its Hochschild complex (see Chapter 4), but this is different from the de Rham complex we have described. For quantized function algebras, the Hochschild complex has been studied in detail by Feng & Tsygan (1991).

7.5 This section is essentially an extract from Baxter (1982). In Baxter's book, the QYBE is called the 'star-triangle relation', because of a graphical interpretation which we do not discuss in this book (although it appears, unannounced, in Section 15.2). See also Faddeev (1980, 1984), and Majid (1990g) for more details on the connection with Hopf algebras.

8

Structure of QUE algebras: the universal R-matrix

The Weyl group W of a complex simple Lie algebra $\mathfrak{g}$ acts as a group of reflections on the Cartan subalgebra $\mathfrak{h}$ of $\mathfrak{g}$. It is well known that this action extends to an action of a finite covering $\tilde{W}$ of W as a group of Lie algebra automorphisms of $\mathfrak{g}$, and hence to an action on the universal enveloping algebra $U(\mathfrak{g})$ by Hopf algebra automorphisms. It is important for a number of reasons. For example, it enables one to define root vectors in $\mathfrak{g}$ associated to every root, only the simple ones being 'given' as generators of $\mathfrak{g}$. This in turn enables one to formulate and prove results such as the Poincaré–Birkhoff–Witt theorem, which is crucial for the representation theory of $\mathfrak{g}$.

In this chapter, we discuss the analogues and applications of these results in the quantum situation. The algebra $U(\mathfrak{h})[[h]]$ of formal power series in h with coefficients in the classical universal enveloping algebra $U(\mathfrak{h})$ is a subalgebra of $U_h(\mathfrak{g})$, but the obvious action of W on $U(\mathfrak{h})[[h]]$ only extends to an action on $U_h(\mathfrak{g})$ of an infinite covering of W, called the *braid group* $\mathcal{B}_{\mathfrak{g}}$ of $\mathfrak{g}$ (if $\mathfrak{g} = sl_{n+1}(\mathbb{C})$, then $\mathcal{B}_{\mathfrak{g}}$ is Artin's braid group on $n+1$ strands). However, one can still define root vectors in $U_h(\mathfrak{g})$ associated to every root of $\mathfrak{g}$: in fact, for any given root, there is one root vector corresponding to each choice of a decomposition of the longest element w_0 of W (which takes positive roots to negative roots) as a product of a minimal number of simple reflections. Using these root vectors, one can construct a (topological) basis of $U_h(\mathfrak{g})$ analogous to the Poincaré–Birkhoff–Witt basis of $U(\mathfrak{g})$. This basis depends, however, on the choice of the decomposition of w_0, for, contrary to the classical situation, the different choices for a given root vector are not even proportional to each other. (We show in Chapter 14 how to construct a basis independent of such choices.)

Another difference between the classical and quantum situations is that the braid group action on $U_h(\mathfrak{g})$ preserves its algebra structure, but not its coalgebra structure, so that it is not obvious how to find the action of the comultiplication on non-simple root vectors. This problem can be solved by means of another interpretation of the braid group action. In fact, there is a Hopf algebra $\tilde{U}_h(\mathfrak{g})$, called the *quantum Weyl group* of $U_h(\mathfrak{g})$, which contains $U_h(\mathfrak{g})$ as a Hopf subalgebra, and is such that the braid group action is given by conjugation by certain elements $\tilde{w}_h$ of $\tilde{U}_h(\mathfrak{g})$. This is analogous to the fact that the classical action of $\tilde{W}$ on $\mathfrak{g}$ is given by the adjoint action of certain elements of the group G with Lie algebra $\mathfrak{g}$. The incompatibility of the braid group action with the coalgebra structure of $U_h(\mathfrak{g})$ is reflected in the failure of the elements $\tilde{w}_h$ to be group-like: the extent of this failure can be precisely described, and this enables one to compute the comultiplication on arbitrary root vectors. The incompatibility also means that there is no finite-dimensional Hopf subalgebra of $\tilde{U}_h(\mathfrak{g})$ which has the group algebra of W as its classical limit: one can only define the quantum analogue of the 'crossed product' of $U(\mathfrak{g})$ with W.

One of the most important properties of $U_h(\mathfrak{g})$ is that it is quasitriangular. Indeed, we show that $U_h(\mathfrak{g})$ can 'almost' be realized as a quantum double (the classical analogue of this statement was proved in 1.4.3). Its universal R-matrix can therefore be computed by the method of Subsection 4.2D. To perform this calculation, one needs an explicit basis of $U_h(\mathfrak{g})$ and to know the action of the comultiplication in this basis. As we have seen, the quantum Weyl group is precisely the tool which enables one to carry this out.

Note Unless indicated otherwise, $\mathfrak{g}$ denotes in this chapter a finite-dimensional complex simple Lie algebra, and $U_h(\mathfrak{g})$ denotes the QUE algebra defined in 6.5.1. We denote $U_h(\mathfrak{g})$ by U_h if this will not lead to confusion.

8.1 The braid group action

A The braid group Let $A = (a_{ij})_{1 \le i \le n}$ be a symmetrizable generalized Cartan matrix, and let $\mathfrak{g} = \mathfrak{g}'(A)$ be the associated Kac–Moody algebra (see the Appendix). Let m_{ij} be equal to 2, 3, 4, 6, or infinity, if $a_{ij}a_{ji}$ is equal to 0, 1, 2, 3, or is at least 4, respectively. We begin with

DEFINITION 8.1.1 *The braid group* $\mathcal{B}_{\mathfrak{g}}$ *associated to* $\mathfrak{g}$ *has generators* T_i, $1 \le i \le n$, *and defining relations*

$$T_i T_j T_i T_j \ldots = T_j T_i T_j T_i \ldots$$

for all $i \ne j$ *such that* $m_{ij} < \infty$, *where there are* m_{ij} T's *on each side of the equation.*

Note that, if $\mathfrak{g}$ is finite dimensional, the products $a_{ij}a_{ji}$, for $i \ne j$, are all equal to 0, 1, 2 or 3.

Remarks [1] The Weyl group W of $\mathfrak{g}$ is generated by the simple reflections s_i:

$$(1) \qquad s_i(H_j) = H_j - a_{ji}H_i$$

(in this formula, the H_i are the standard generators of $\mathfrak{g}$ – see the Appendix). Its defining relations are those in 8.1.1 (with T_i replaced by s_i), together with

$s_i^2 = 1$. Hence, W is a quotient of $\mathcal{B}_\mathfrak{g}$. On the other hand, there is a surjective homomorphism $\mathcal{B}_\mathfrak{g} \to \mathbb{Z}$ given by $T_i \mapsto 1$ for all i; in particular, $\mathcal{B}_\mathfrak{g}$ is an infinite group.

[2] If $\mathfrak{g}$ is of type A_n, the relations become

$$T_i T_{i+1} T_i = T_{i+1} T_i T_{i+1},$$
$$T_i T_j = T_j T_i \quad \text{if } |i - j| > 1.$$

These are the defining relations of Artin's braid group on $n + 1$ strands (see the remark following 5.1.7). △

For any integer $r \geq 0$, and for $1 \leq i \leq n$, define

$$(X_i^+)^{(r)} = \frac{(X_i^+)^r}{[r]_{q_i}!}, \quad (X_i^-)^{(r)} = \frac{(X_i^-)^r}{[r]_{q_i}!}.$$

The q-factorial $[r]_q!$ is defined in Subsection 6.4D.

THEOREM 8.1.2 *There is an action of the braid group $\mathcal{B}_\mathfrak{g}$ by algebra automorphisms of $U_h(\mathfrak{g})$ defined on the standard generators (see 6.5.1) as follows:*

$$T_i(X_i^+) = -X_i^- e^{d_i h H_i}, \quad T_i(X_i^-) = -e^{-d_i h H_i} X_i^+, \quad T_i(H_j) = H_j - a_{ji} H_i,$$

$$T_i(X_j^+) = \sum_{r=0}^{-a_{ij}} (-1)^{r-a_{ij}} e^{-r d_i h} (X_i^+)^{(-a_{ij}-r)} X_j^+ (X_i^+)^{(r)} \quad \text{if } i \neq j,$$

$$T_i(X_j^-) = \sum_{r=0}^{-a_{ij}} (-1)^{r-a_{ij}} e^{r d_i h} (X_i^-)^{(r)} X_j^- (X_i^-)^{(-a_{ij}-r)} \quad \text{if } i \neq j. \ \blacksquare$$

This can be proved by direct verification. Of course, it is enough to consider the case when $\mathfrak{g}$ has rank 1 or 2.

Remarks [1] Note that the action of T_i on the H_j is the 'same' as that of the simple reflections $s_i \in W$ on the corresponding generators of $\mathfrak{g}$.

[2] The inverses of the automorphisms T_i are given by

$$T_i^{-1}(X_i^+) = -e^{-d_i h H_i} X_i^-, \ T_i^{-1}(X_i^-) = -X_i^+ e^{d_i h H_i}, \ T_i^{-1}(H_j) = H_j - a_{ji} H_i,$$

$$T_i^{-1}(X_j^+) = \sum_{r=0}^{-a_{ij}} (-1)^{r-a_{ij}} e^{-r d_i h} (X_i^+)^{(r)} X_j^+ (X_i^+)^{(-a_{ij}-r)} \quad \text{if } i \neq j,$$

$$T_i^{-1}(X_j^-) = \sum_{r=0}^{-a_{ij}} (-1)^{r-a_{ij}} e^{r d_i h} (X_i^-)^{(-a_{ij}-r)} X_j^- (X_i^-)^{(r)} \quad \text{if } i \neq j.$$

[3] The T_i are *not* Hopf algebra automorphisms of U_h. We shall see in the next section, however, that they are 'almost' Hopf algebra automorphisms, in the same sense as U_h is 'almost' cocommutative.

[4] The T_i can be expressed in terms of the Hopf algebra adjoint representation of U_h and that of the opposite coalgebra U_h^{op} (see 4.1.13):

$$T_i(X_j^+) = \mathrm{ad}^{\mathrm{op}}_{-(X_i^+)^{(-a_{ij})}}(X_j^+), \quad T_i(X_j^-) = \mathrm{ad}_{-(X_i^-)^{(-a_{ij})}}(X_j^-).$$

[5] In the sequel, we shall make use of the $\mathbb{C}$-algebra anti-automorphism ω and the $\mathbb{C}$-algebra automorphism ϕ of U_h defined on generators by:

$$\omega(X_i^+) = X_i^-, \quad \omega(X_i^-) = X_i^+, \quad \omega(H_i) = H_i, \quad \omega(h) = -h,$$
$$\phi(X_i^+) = X_i^-, \quad \phi(X_i^-) = X_i^+, \quad \phi(H_i) = H_i, \quad \phi(h) = -h.$$

The action of $\mathcal{B}_{\mathfrak{g}}$ is compatible with these involutions, in the sense that:

$$\omega \circ T_i = T_i \circ \omega, \quad \phi \circ T_i = T_i^{-1} \circ \phi. \quad \triangle$$

Recall from the Appendix that any element $w \in W$ can be written as a product $w = s_{i_1} s_{i_2} \ldots s_{i_k}$ of simple reflections; the expression is *reduced* if its length k is minimal among all such expressions for w, and this minimal length is denoted by $\ell(w)$.

The following classical result shows that the natural surjective homomorphism $\mathcal{B}_{\mathfrak{g}} \to W$ which sends T_i to s_i has a canonical section (which is not a homomorphism). See Bourbaki (1968), Chap. 4, § 1.5, Proposition 5.

PROPOSITION 8.1.3 *Let* $w = s_{i_1} s_{i_2} \ldots s_{i_{\ell(w)}}$ *be a reduced expression of an element* $w \in W$. *Then, the automorphism* $T_w = T_{i_1} T_{i_2} \ldots T_{i_{\ell(w)}}$ *of* $U_h(\mathfrak{g})$ *depends only on* w, *and not on the choice of reduced expression.* ∎

B Root vectors and the PBW basis The classical Poincaré–Birkhoff–Witt theorem associates to any (ordered) basis of a Lie algebra a basis of its universal enveloping algebra. In the case of a quantized universal enveloping algebra such as $U_h(\mathfrak{g})$, one has no underlying Lie algebra, much less a basis of it, so it is not immediately clear how to obtain an analogue of the PBW theorem for $U_h(\mathfrak{g})$. The most natural basis of the classical enveloping algebra $U(\mathfrak{g})$ is the PBW basis associated to the basis of $\mathfrak{g}$ given by a basis of the Cartan subalgebra $\mathfrak{h} \subset \mathfrak{g}$ together with a root vector in each root space of $\mathfrak{g}$. In $U_h(\mathfrak{g})$, one has an analogue of the basis of $\mathfrak{h}$, namely $\{H_i\}_{i=1,\ldots,n}$, but one is only given 'root vectors' $X_i^{\pm}$ associated to the simple roots α_i (and their negatives). Thus, to construct a basis of $U_h(\mathfrak{g})$ of PBW type, one must define analogues of root vectors associated to the non-simple roots of $\mathfrak{g}$. As we recall in the Appendix, if $\mathfrak{g}$ is finite dimensional, non-simple root vectors of $\mathfrak{g}$ can be defined by using the action of (a finite covering of) the Weyl group W on $\mathfrak{g}$. This is based on the fact that, if

(2) $w_0 = s_{i_1} s_{i_2} \ldots s_{i_N}$

is a reduced decomposition of the longest element of W, then every positive root occurs exactly once in the following set:

$$(3) \qquad \beta_1 = \alpha_{i_1}, \ \beta_2 = s_{i_1}(\alpha_{i_2}), \ldots, \ \beta_N = s_{i_1} s_{i_2} \ldots s_{i_{N-1}}(\alpha_{i_N}).$$

In the quantum case, the action of the Weyl group is replaced by that of the braid group $\mathcal{B}_{\mathfrak{g}}$, and we make

DEFINITION 8.1.4 *Let* $\mathfrak{g}$ *be a finite-dimensional complex simple Lie algebra. Fix a reduced decomposition (2) of the longest element* $w_0 \in W$, *and define* $\beta_1, \ldots, \beta_N$ *as in (3). Define elements* $X_{\beta_r}^\pm \in U_h(\mathfrak{g})$ *as follows:*

$$X_{\beta_r}^\pm = T_{i_1} T_{i_2} \ldots T_{i_{r-1}}(X_{i_r}^\pm).$$

The $X_{\beta_r}^+$ *(resp.* $X_{\beta_r}^-$*) are called the positive (resp. negative) root vectors of* $U_h(\mathfrak{g})$.

If $\mathfrak{g}$ is an infinite-dimensional Kac–Moody algebra, it is no longer true that every root lies in the Weyl group orbit of a simple root, and the above procedure for defining root vectors fails. (See Beck (1993), however, for the case when $\mathfrak{g}$ is affine.) *For the remainder of this chapter, we assume that* $\mathfrak{g}$ *is finite dimensional.*

One important difference between the classical and quantum situations is that, in the classical case, the root vectors defined in this way are independent, up to sign, of the choice of the reduced decomposition (2), whereas, in the quantum case, the root vectors corresponding to different reduced decompositions of w_0 are not even proportional to each other, in general.

Example 8.1.5 If $\mathfrak{g}$ is of type A_2, there are two choices for the reduced decomposition of w_0:

(i) $w_0 = s_1 s_2 s_1$: then the positive root vectors are

$$X_1^+, \ T_1(X_2^+) = -X_1^+ X_2^+ + e^{-h} X_2^+ X_1^+, \ T_1 T_2(X_1^+) = X_2^+ \, ;$$

(ii) $w_0 = s_2 s_1 s_2$: then the positive root vectors are

$$X_2^+, \ T_2(X_1^+) = -X_2^+ X_1^+ + e^{-h} X_1^+ X_2^+, \ T_2 T_1(X_2^+) = X_1^+ \, . \quad \Diamond$$

Let U_h^+, U_h^0 and U_h^- be the $\mathbb{C}[[h]]$-subalgebras of U_h topologically generated by the X_i^+, by the H_i and by the X_i^-, respectively.

PROPOSITION 8.1.6 *(i) Let* $w \in W$ *be such that* $w(\alpha_i) \in \Delta^+$ *for some simple root* α_i. *Then,* $T_w(X_i^\pm) \in U_h^\pm$. *If* $w(\alpha_i) = \alpha_j$ *is simple, then* $T_w(X_i^\pm) = X_j^\pm$. *(ii) In particular, for any choice of reduced decomposition of* w_0, *the positive (resp. negative) root vectors lie in* U_h^+ *(resp.* U_h^-*).*

(iii) The braid group action preserves U_h^0 and induces on it the classical action of the Weyl group. ∎

The meaning of the third part is that, if $\beta = \sum_i k_i \alpha_i \in Q$, and if we define $H_\beta = \sum_i k_i H_i$, then $T_w(H_\beta) = H_{w(\beta)}$ for all $w \in W$. Note that, if $\beta = \beta_r \in \Delta^+$ is a positive root (in the notation of (3)), the vectors $\{H_{\beta_r}, X_{\beta_r}^\pm\}$ satisfy the defining relations of $U_{d_{i_r}, h}(sl_2(\mathbb{C}))$: this follows by applying the automorphism $T_{i_1} T_{i_2} \ldots T_{i_{r-1}}$ to the relations satisfied by $\{H_{i_r}, X_{i_r}^\pm\}$.

Now that we have quantum analogues of all the root vectors of $\mathfrak{g}$ at our disposal, we can use them to construct a topological basis of U_h. Fix a choice of positive and negative root vectors as in 8.1.4. Define, for $\boldsymbol{r} = (r_1, \ldots, r_N) \in \mathbb{N}^N$,

$$(X^+)^{\boldsymbol{r}} = (X_{\beta_N}^+)^{r_N} (X_{\beta_{N-1}}^+)^{r_{N-1}} \ldots (X_{\beta_1}^+)^{r_1},$$
$$(X^-)^{\boldsymbol{r}} = \omega(X^+)^{\boldsymbol{r}} = (X_{\beta_1}^-)^{r_1} (X_{\beta_2}^-)^{r_2} \ldots (X_{\beta_N}^-)^{r_N},$$

and for $\boldsymbol{s} = (s_1, \ldots, s_n) \in \mathbb{N}^n$,

$$H^{\boldsymbol{s}} = H_1^{s_1} H_2^{s_2} \ldots H_n^{s_n}.$$

PROPOSITION 8.1.7 *The elements $(X^-)^{\boldsymbol{r}}$, $H^{\boldsymbol{s}}$ and $(X^+)^{\boldsymbol{t}}$, for $\boldsymbol{r}, \boldsymbol{t} \in \mathbb{N}^N$, $\boldsymbol{s} \in \mathbb{N}^n$, form topological bases of U_h^-, U_h^0 and U_h^+, respectively, and the products $(X^-)^{\boldsymbol{r}} H^{\boldsymbol{s}} (X^+)^{\boldsymbol{t}}$ form a topological basis of U_h.* ∎

This result has the following immediate

COROLLARY 8.1.8 *Multiplication defines an isomorphism of $\mathbb{C}[[h]]$-modules*

$$U_h^- \otimes U_h^0 \otimes U_h^+ \to U_h$$

(completed tensor products). ∎

Proposition 8.1.7 was first proved by Rosso (1989) when $\mathfrak{g} = sl_{n+1}(\mathbb{C})$, using a definition of the root vectors due to Jimbo (1986a), which did not of course use the braid group action. Lusztig (1990a) deals with the case in which $\mathfrak{g}$ is simply laced, and Lusztig (1990b) considers the general case. Berger (1992) gives a proof using the 'diamond lemma' in ring theory (see Bergman (1978)). S. Z. Levendorskii has given a proof using the next result, which is also used in the computation of the universal R-matrix of $U_h(\mathfrak{g})$ (see Section 8.3). Although we shall make use of 8.1.7 in the proof of this result, Levendorskii gives an independent proof by making use of a weak version of 8.1.7, which in turn is established by the argument used to prove 6.4.7.

PROPOSITION 8.1.9 *If* $0 \leq r < s \leq N$, *we have*

$$X_{\beta_r}^+ X_{\beta_s}^+ - e^{h(\beta_r, \beta_s)} X_{\beta_s}^+ X_{\beta_r}^+ = \sum c(t_{r+1}, \ldots, t_{s-1})(X_{\beta_{s-1}}^+)^{t_{s-1}} \ldots (X_{\beta_{r+1}}^+)^{t_{r+1}},$$

where the sum is over all $t_{r+1}, \ldots, t_{s-1} \in \mathbb{N}$, *and all but finitely many of the* $c(t_{r+1}, \ldots, t_{s-1}) \in \mathbb{C}[[h]]$ *are zero.*

PROOF By 8.1.7, we know that there is an expression of the form

$$X_{\beta_r}^+ X_{\beta_s}^+ = \sum_{t_1, \ldots, t_N = 0}^{\infty} c(t_1, \ldots, t_N)(X_{\beta_N}^+)^{t_N} \ldots (X_{\beta_1}^+)^{t_1}.$$

Suppose that there is a $p < r$ and a non-zero coefficient $c(0, \ldots, 0, t_p, \ldots, t_N)$ with $t_p \neq 0$; and choose p minimal with this property. Applying $(T_{i_1} \ldots T_{i_p})^{-1}$ to both sides and using 8.1.6(i), we see that every term lies in U_h^+ except those with $t_p \neq 0$, which give a term of the form $E T_{i_p}^{-1}(X_{i_p}^{t_p})$, where $E \in U_h^+$ is non-zero. Since $T_{i_p}^{-1}(X_{i_p}^{t_p}) = (-e^{-d_{i_p} h H_{i_p}} X_{i_p}^-)^{t_p}$, this contradicts 8.1.7. Similar arguments show that $c(t_1, \ldots, t_N) = 0$ unless $t_1 = \cdots = t_{r-1} = t_{s+1} = \cdots = t_N = 0$, and that, if these equalities hold, then $c(t_1, \ldots, t_N) = 0$ unless t_r and t_s are both non-zero. For weight reasons, we then have the equation in the statement of the proposition, except that the coefficient $e^{h(\beta_r, \beta_s)}$ must be replaced by an as yet unknown scalar κ, say. To determine κ, apply $(T_{i_1} \ldots T_{i_r})^{-1}$ to both sides. The right-hand side is in U_h^+, while the left-hand side is

$$-e^{-d_{i_r} h H_{i_r}} X_{i_r}^- \tilde{E} + \kappa \tilde{E} e^{-d_{i_r} h H_{i_r}} X_{i_r}^-,$$

where $\tilde{E} = T_{i_{r+1}} \ldots T_{i_{s-1}}(X_{i_s}^+) \in U_h^+$ has weight $s_{i_r}(\beta_s)$:

$$e^{d_{i_r} h H_{i_r}} \tilde{E} e^{-d_{i_r} h H_{i_r}} = e^{h(\beta_r, s_{i_r}(\beta_s))} \tilde{E}.$$

Thus, the left-hand side becomes

$$-e^{-d_{i_r} h H_{i_r}} (X_{i_r}^- \tilde{E} - \kappa e^{h(\beta_r, s_{i_r}(\beta_s))} \tilde{E} X_{i_r}^-).$$

Writing $\tilde{E}$ as a linear combination of monomials in the X_j^+, $j = 1, \ldots, n$, we see that $\tilde{E} X_{i_r}^- - X_{i_r}^- \tilde{E} \in U_h^0.U_h^+$. It follows by using 8.1.7 again that $\kappa e^{h(\beta_r, s_{i_r}(\beta_s))} = 1$, and hence that

$$\kappa = e^{-h(\beta_r, s_{i_r}(\beta_s))} = e^{-h(s_{i_r}(\beta_r), \beta_s)} = e^{h(\beta_r, \beta_s)},$$

as required. ∎

8.2 The quantum Weyl group

Classically, the action of (a finite covering of) the Weyl group on $\mathfrak{g}$ is given by the adjoint action of certain elements of the group G with Lie algebra $\mathfrak{g}$. In this section, we show that essentially the same result holds in the quantum case, the role of the group elements being played by certain elements of $\mathcal{F}_h^*$, the dual of the quantized algebra of functions on G. Conjugation by invertible elements of $\mathcal{F}_h^*$ makes sense, because $\mathcal{F}_h^*$ is an algebra and there is a natural embedding $U_h \to \mathcal{F}_h^*$.

A The sl_2 case We recall from 7.1.5 that the matrix elements $C_{rs}^{(m)}$ of the representations V_m of $U_h = U_h(sl_2(\mathbb{C}))$ are a topological basis of the dual Hopf algebra $\mathcal{F}_h$ of U_h. It therefore makes sense to define a linear functional $w_h \in \mathcal{F}_h^*$, the full $\mathbb{C}[[h]]$-module dual of $\mathcal{F}_h$, by

(4)
$$w_h(C_{rs}^{(m)}) = \begin{cases} (-1)^r e^{\frac{1}{4}hm^2 + rh} & \text{if } r + s = m, \\ 0 & \text{otherwise.} \end{cases}$$

Since $\mathcal{F}_h^*$ is an algebra, the following result makes sense.

LEMMA 8.2.1 *In $\mathcal{F}_h^*$, we have*
$$w_h X^+ w_h^{-1} = -e^h X^-, \quad w_h X^- w_h^{-1} = -e^{-h} X^+, \quad w_h H w_h^{-1} = -H.$$

PROOF Let us prove the first formula. We have
$$\langle w_h X^+, C_{rs}^{(m)} \rangle = \langle w_h \otimes X^+, \Delta_{\mathcal{F}_h}(C_{rs}^{(m)}) \rangle$$
$$= \sum_t \langle w_h, C_{rt}^{(m)} \rangle \langle X^+, C_{ts}^{(m)} \rangle.$$

Now, by 6.4.10,
$$\langle X^+, C_{rs}^{(m)} \rangle = [m - s + 1]_q \delta_{r,s-1}, \quad \langle X^-, C_{rs}^{(m)} \rangle = [s + 1]_q \delta_{r,s+1},$$

where $q = e^h$. Hence,
$$\langle w_h X^+, C_{rs}^{(m)} \rangle = \sum_t (-1)^r e^{\frac{1}{4}hm^2 + rh} \delta_{r+t,m} [m - s + 1]_q \delta_{t,s-1}$$
$$= (-1)^r e^{\frac{1}{4}hm^2 + rh} [m - s + 1]_q \delta_{r+s,m+1}.$$

On the other hand,
$$-e^h \langle X^- w_h, C_{rs}^{(m)} \rangle = -e^h \sum_t \langle X^-, C_{rt}^{(m)} \rangle \langle w_h, C_{ts}^{(m)} \rangle$$
$$= -e^h \sum_t [t + 1]_q \delta_{r,t+1} (-1)^t e^{\frac{1}{4}hm^2 + th} \delta_{s+t,m}$$
$$= -e^h [r]_q (-1)^{r-1} e^{\frac{1}{4}hm^2 + (r-1)h} \delta_{r+s,m+1}.$$

Thus, $w_h X^+ = -e^h X^- w_h$. ∎

Now define $\tilde{w}_h \in \mathcal{F}_h^*$ to be the product

$$(5) \qquad \tilde{w}_h = w_h \exp\left[\frac{hH^2}{4}\right] = \exp\left[\frac{hH^2}{4}\right] w_h$$

(the latter equality follows from 8.2.1).

PROPOSITION 8.2.2 *Let T be the generator of the braid group of $sl_2(\mathbb{C})$. Then, for all $a \in U_h$, we have*

$$T(a) = \tilde{w}_h a \tilde{w}_h^{-1}.$$

PROOF It suffices to check this when a is one of the generators of U_h. It is obvious when $a = H$. By 6.4.2,

$$\exp\left[\frac{hH^2}{4}\right] X^\pm \exp\left[\frac{-hH^2}{4}\right] = X^\pm \exp[h(H \pm 1)].$$

The result now follows from 8.2.1 and the definition of T in 8.1.2. For example,

$$\begin{aligned}
\tilde{w}_h X^+ \tilde{w}_h^{-1} &= e^{hH^2/4} w_h X^+ w_h^{-1} e^{-hH^2/4} \\
&= -e^h e^{hH^2/4} X^- e^{-hH^2/4} \\
&= -e^h X^- e^{h(H-1)} \\
&= -X^- e^{hH} \\
&= T(X^+)
\end{aligned}$$

(the second equation follows from 8.2.1, and the third equation from 6.4.2). ∎

B The relation with the universal R-matrix Although the elements $C_{rs}^{(m)}$ are linearly independent, there are many multiplicative relations between them. In fact, in the Clebsch–Gordan decomposition 6.4.11 of the tensor product $V_m \otimes V_n$ as a direct sum of indecomposable representations V_p, we can express any basis vector $v_t^{(p)}$ of any V_p which occurs as a linear combination of tensor products $v_r^{(m)} \otimes v_s^{(n)}$ of basis vectors in V_m and V_n, and vice versa. By considering the action of U_h on both sides, one then obtains relations of the form

$$(6) \qquad C_{rr'}^{(m)} C_{ss'}^{(n)} = \sum_{t,t'} c_{rst,r's't'}^{mnp} C_{tt'}^{(p)}.$$

These relations allow one to compute the action on $\tilde{w}_h$ of the comultiplication Δ_h dual to the multiplication on $\mathcal{F}_h$.

PROPOSITION 8.2.3 For the element $w_h \in \mathcal{F}_h^*$ defined in (4), we have

$$(7) \qquad \Delta_h(w_h) = \mathcal{R}_h^{-1}(w_h \otimes w_h),$$

where $\mathcal{R}_h$ is the universal R-matrix of U_h defined in 6.4.8.

OUTLINE OF PROOF It suffices to show that the two sides of (7) agree on any tensor product $C_{rr'}^{(m)} \otimes C_{ss'}^{(n)}$. The left-hand side gives $w_h(C_{rr'}^{(m)} C_{ss'}^{(n)})$, which can be computed using (6). The right-hand side is

$$(\mathcal{R}_h^{-1})_{rs,r's'}^{m,n} w_h(C_{rr'}^{(m)}) w_h(C_{ss'}^{(n)}),$$

where

$$\mathcal{R}_h^{-1}.(v_r^{(m)} \otimes v_s^{(n)}) = \sum_{rr'ss'} (\mathcal{R}_h^{-1})_{rs,r's'}^{m,n} v_{r'}^{(m)} \otimes v_{s'}^{(n)},$$

which can be computed using the formula for the R-matrix in 6.4.8. See Kirillov & Reshetikhin (1989) for the details. ∎

Similar calculations give the action of the counit and antipode on w_h:

$$S_h(w_h) = w_h e^{hH}, \quad \epsilon_h(w_h) = 1.$$

Remarks [1] This result shows that T is 'almost' a Hopf algebra automorphism of U_h. In fact, for $a \in U_h$,

$$
\begin{aligned}
\Delta_h(T(a)) &= \Delta_h(\tilde{w}_h a \tilde{w}_h^{-1}) \\
(8) \qquad &= \mathcal{R}_h^{-1}(w_h \otimes w_h)\Delta_h(e^{hH^2/4})\Delta_h(a)\Delta_h(e^{-hH^2/4})(w_h \otimes w_h)^{-1}\mathcal{R}_h \\
&= \tilde{\mathcal{R}}_h^{-1}.(T \otimes T)(\Delta_h(a)).\tilde{\mathcal{R}}_h,
\end{aligned}
$$

where

$$
\begin{aligned}
\tilde{\mathcal{R}}_h &= (e^{hH^2/4} \otimes e^{hH^2/4}).\Delta_h(e^{-hH^2/4}).\mathcal{R}_h \\
(9) \qquad &= e^{-\frac{1}{2}h(H \otimes H)}\mathcal{R}_h \\
&= \sum_{n=0}^{\infty} \frac{(1 - e^{-2h})}{[n]_{q_i}!} e^{\frac{1}{2}hn(n+1)}(X^+)^n \otimes (X^-)^n,
\end{aligned}
$$

by 6.4.8.

[2] One can 'explain' 8.2.3 as follows. Recall that the Poisson bracket on $\mathcal{F} = \mathcal{F}(SL_2(\mathbb{C}))$ induced by the quantization U_h is given by $\{f, g\} \equiv h^{-1}(f_h g_h - g_h f_h) \pmod{h}$, where $f_h, g_h \in \mathcal{F}_h$ have classical limits $f, g \in \mathcal{F}$. Now, if $w_h \in \tilde{U}_h$ were group-like, we should have $w_h(f_h g_h) = w_h(f_h)w_h(g_h)$, and hence $\{f, g\}(w) = 0$ for all f, g, where $w \in W$ is the classical limit of w_h. But we saw in Section 1.4 that the Poisson bracket on $SL_2(\mathbb{C})$ vanishes only

at the points of the diagonal (Cartan) subgroup of $SL_2(\mathbb{C})$ (strictly speaking, we discussed SU_2 rather than $SL_2(\mathbb{C})$ in Section 1.4, but the same result holds). Since the Poisson bracket is determined by the classical r-matrix, it is natural that the extent to which w_h fails to be group-like is measured by the universal (quantum) R-matrix. $\triangle$

It follows from 8.2.3 and the above remarks that the subalgebra $\tilde{U}_h$ of $\mathcal{F}_h^*$ topologically generated by U_h and w_h (or, equivalently, by U_h and $\tilde{w}_h$) is a topological Hopf algebra. Moreover, every finite-dimensional representation of U_h, being a direct sum of indecomposables, is naturally a representation of $\tilde{U}_h$. The topological Hopf algebra $\tilde{U}_h$ is called the *quantum Weyl group* of U_h. Note that, unlike the classical situation, in which the group algebra of the Weyl group is itself a Hopf algebra, 8.2.3 shows that neither the subalgebra of $\mathcal{F}_h^*$ topologically generated by $\tilde{w}_h$, nor that topologically generated by w_h, is a topological Hopf algebra.

Another difference from the classical situation is that neither w_h nor $\tilde{w}_h$ is an involution. The corresponding fact in the quantum situation is

PROPOSITION 8.2.4 *The square w_h^2 of the quantum Weyl group element w_h lies in the centre of U_h. In fact,*

$$w_h^2 = e^{-hH} u_h \sigma,$$

where $u_h = \mu_h(S_h \otimes \mathrm{id})(\mathcal{R}_h^{21})$ and σ is the central element of $\mathcal{F}_h^*$ given by

$$\sigma(C_{rs}^{(m)}) = (-1)^m C_{rs}^{(m)}$$

for all r, $s = 0, \ldots, m$, $m \geq 0$ (we write $\mathcal{R}_h^{21}$ instead of $(\mathcal{R}_h)_{21}$ for typographical convenience).

OUTLINE OF PROOF The first statement is immediate from 8.2.1. The fact that $e^{-hH} u_h$ lies in the centre of U_h was shown in the remark at the end of Subsection 6.4D. The formula for w_h^2 is proved by a direct computation. ∎

C The general case Turning now to the algebra $U_h = U_h(\mathfrak{g})$ for an arbitrary finite-dimensional complex simple Lie algebra $\mathfrak{g}$, we use the fact that U_h is generated topologically by the subalgebras $U_{h,1}, \ldots, U_{h,n}$ of $U_h(sl_2(\mathbb{C}))$, where $U_{h,i}$ is the subalgebra topologically generated by H_i, X_i^+ and X_i^-; note that $U_{h,i} \cong U_{d_i h}(sl_2(\mathbb{C}))$. The inclusions $U_{h,i} \to U_h$ induce projections $\mathcal{F}_h \to \mathcal{F}_{h,i}$ of the dual Hopf algebras, and hence the element $w_{h,i} \in \mathcal{F}_{h,i}^*$ constructed in Subsection 8.2A may be regarded as an element of $\mathcal{F}_h^*$. We set

$$\tilde{w}_{h,i} = w_{h,i} e^{\frac{1}{4} d_i h H_i^2}.$$

Then, we have

PROPOSITION 8.2.5 *For all $a \in U_h$ and $i = 1, \ldots, n$,*

$$T_i(a) = \tilde{w}_{h,i} a \tilde{w}_{h,i}^{-1}. \quad \blacksquare$$

The proof is by explicit computation. See Kirillov & Reshetikhin (1990a) for the details.

We also have the following generalization of 8.2.5.

PROPOSITION 8.2.6 *For $i = 1, \ldots, n$,*

$$\Delta_h(\tilde{w}_{h,i}) = \tilde{\mathcal{R}}_{h,i}^{-1}.(\tilde{w}_{h,i} \otimes \tilde{w}_{h,i}),$$

where

$$\tilde{\mathcal{R}}_{h,i} = \sum_{n=0}^{\infty} \frac{(1 - q_i^{-2})^n}{[n]_{q_i}!} q_i^{\frac{1}{2}n(n+1)} (X_i^+)^n \otimes (X_i^-)^n$$

and $q_i = e^{d_i h}$. $\blacksquare$

As in the sl_2 case, this result implies that the subalgebra $\tilde{U}_h$ of $\mathcal{F}_h^*$ topologically generated by U_h and $w_{h,1}, \ldots, w_{h,n}$ is a topological Hopf algebra. It is called the *quantum Weyl group* of U_h.

Remarks [1] One can show that the $\tilde{w}_{h,i}$ satisfy the braid relations in 8.1.1. Together with 8.2.5, this gives another proof of 8.1.2.

[2] Levendorskii & Soibelman (1990) define an element $\bar{w}_h \in \mathcal{F}_h^*$ (in the sl_2 case) directly as a matrix element of an infinite-dimensional representation of $\mathcal{F}_h$. In fact, in the notation of 7.1.6, one sets $\tilde{w}_h(f)$ equal to the 00 matrix element of $f \in \mathcal{F}_h$ acting in the representation π_{-1}. $\triangle$

8.3 The quasitriangular structure

In this section, we show that $U_h(\mathfrak{g})$ is 'almost' a quantum double, deduce that it is topologically quasitriangular and compute its universal R-matrix.

A The quantum double construction We recall from Subsection 4.2D that the double $\mathcal{D}(A)$ of any finite-dimensional Hopf algebra A is a Hopf algebra isomorphic to $A \otimes (A^*)^{\mathrm{op}}$ as a coalgebra, and that it is quasitriangular with the element of $A \otimes (A^*)^{\mathrm{op}} \subset \mathcal{D}(A) \otimes \mathcal{D}(A)$ corresponding to the identity map $A \to A$ being the universal R-matrix. As explained in Subsection 6.3C, when A is a QUE algebra, the tensor products must be understood in the completed sense, and A^* must be the QUE dual of A.

The crucial observation for the proof that $U_h = U_h(\mathfrak{g})$ is topologically quasitriangular is the following result, which shows that U_h is 'almost' a quantum double. We write $U_h^{\geq 0} = U_h^0.U_h^+$, $U_h^{\leq 0} = U_h^-.U_h^0$. It is obvious that $U_h^{\leq 0}$, U_h^0 and $U_h^{\geq 0}$ are Hopf subalgebras of U_h.

PROPOSITION 8.3.1 *There is a surjective homomorphism of topological Hopf algebras*

$$\mathcal{D}(U_h^{\geq 0}) \to U_h.$$

The image of the universal R-matrix of $\mathcal{D}(U_h^{\geq 0})$ under the homomorphism in 8.3.1 will clearly be a universal R-matrix for U_h. This not only proves that U_h is topologically quasitriangular, but also gives an explicit formula for its universal R-matrix.

Remark In fact, there is an isomorphism of topological Hopf algebras

$$\mathcal{D}(U_h^{\geq 0}) \to U_h \,\hat{\otimes}\, U_h^0$$

(completed tensor product) and the homomorphism in the proposition is obtained by composing this isomorphism with $\mathrm{id} \otimes \epsilon_h$, where ϵ_h is the counit of U_h^0. $\triangle$

As explained above, one of our first tasks is to describe the QUE dual of $U_h^{\geq 0}$:

PROPOSITION 8.3.2 *The QUE dual of $U_h^{\geq 0}$, with the opposite comultiplication, is isomorphic to $U_h^{\leq 0}$ as a topological Hopf algebra over $\mathbb{C}[[h]]$.*

We prove 8.3.1 and 8.3.2 in detail in the $sl_2(\mathbb{C})$ case to bring out the main ideas of the quantum double method, and then outline the general case where the quantum Weyl group is used.

B The sl_2 case In this subsection, $\mathfrak{g} = sl_2(\mathbb{C})$. By 8.1.7, the monomials $H^s(X^+)^t$, $s, t \geq 0$, are a topological basis of $U_h^{\geq 0}$. We recall from Subsection 6.3C that the QUE dual of $U_h^{\geq 0}$ is the Hopf algebra dual of a QFSH subalgebra $U_{h,\geq 0}$ of $U_h^{\geq 0}$.

PROPOSITION 8.3.3 *The QFSH subalgebra $U_{h,\geq 0}$ of $U_h^{\geq 0}$ defined in Subsection 6.3C is the subalgebra topologically generated, in the sense of formal power series over $\mathbb{C}[[h]]$, by $\tilde{H} = hH$ and $\tilde{X}^+ = hX^+$.*

PROOF By the discussion in Subsection 6.3 C, an element $a \in U_h^{\geq 0}$ lies in the QFSH subalgebra $U_{h,\geq 0}$ provided that $\Delta_n(a) \equiv 0 \pmod{h^n}$ for all $n \geq 1$, where $\Delta_n = \pi^{\otimes n} \Delta_h^{(n)}$ and $\pi : U_h^{\geq 0} \to U_h^{\geq 0}$ is the $\mathbb{C}[[h]]$-module projection which sends 1 to 0 and fixes each monomial $H^s(X^+)^t$ for which $s + t > 0$. Consider an element $a = f_{s,t}(h)H^s(X^+)^t \in U_h^{\geq 0}$, where $s + t > 0$ and $f_{s,t}(h) \in \mathbb{C}[[h]]$. Applying Δ_{s+t}, we obtain a sum of terms with exactly one H or one X^+ in each position in the tensor product. In particular, the term

$$H \otimes \cdots \otimes H \otimes X^+ \otimes \cdots \otimes X^+$$

occurs with coefficient $s!t!f_{s+t}(h)$. It follows that a lies in $U_{h,\geq 0}$ if and only if $f_{s,t}(h)$ is divisible by h^{s+t}. ∎

Note that

$$(10) \qquad \Delta_h(\tilde{H}) = \tilde{H} \otimes 1 + 1 \otimes \tilde{H}, \quad \Delta_h(\tilde{X}^+) = \tilde{X}^+ \otimes e^{\tilde{H}} + 1 \otimes \tilde{X}^+,$$

and that $e^{\tilde{H}}$ is *not* in the h-adic completion of the subalgebra of $U_h^{\geq 0}$ generated by $\tilde{H}$.

To describe the QUE dual of $U_h^{\geq 0}$, let ξ and η be the linear functionals on $U_{h,\geq 0}$ which are equal to one on $\tilde{X}^+$ and $\tilde{H}$, respectively, and equal to zero on all other monomials $\tilde{H}^s(\tilde{X}^+)^t$. Then, the QUE dual of $U_h^{\geq 0}$ is topologically generated by ξ and η.

LEMMA 8.3.4 *Let* $\langle \ , \ \rangle$ *be the natural pairing between* $U_{h,\geq 0}$ *and its dual. Then,*

$$\langle \xi^s \eta^t , \ \tilde{H}^{s'}(\tilde{X}^+)^{t'} \rangle = \delta_{s,s'} \delta_{t,t'} s! \, q^{-\frac{1}{2}t(t-1)} \, [t]_q!,$$

where $q = e^h$.

PROOF Consider first the case $t = 0$. From (10), it is clear that

$$\langle \xi^s , \ \tilde{H}^{s'}(\tilde{X}^+)^{t'} \rangle = \langle \xi \otimes \cdots \otimes \xi , \Delta_h^{(s)}(\tilde{H}^{s'}(\tilde{X}^+)^{t'}) \rangle$$

can be non-zero only if $t' = 0$, and that in this case it is equal to the coefficient of $\tilde{H} \otimes \cdots \otimes \tilde{H}$ in

$$\Delta_h^{(s)}(\tilde{H}^{s'}) = (\tilde{H} \otimes 1 \otimes \cdots \otimes 1 + 1 \otimes \tilde{H} \otimes \cdots \otimes 1 + \cdots + 1 \otimes 1 \otimes \cdots \otimes \tilde{H})^{s'}.$$

This vanishes unless $s = s'$, in which case it is equal to $s!$.

Similarly, $\langle \eta^t , \ \tilde{H}^{s'}(\tilde{X}^+)^{t'} \rangle$ clearly vanishes unless $s' = 0$ and $t' = t$, in which case it is equal to the coefficient of the term in the product

$$\Delta_h^{(t)}(\tilde{X}^+)^t = (\tilde{X}^+ \otimes e^{\tilde{H}} \otimes \cdots \otimes e^{\tilde{H}} + 1 \otimes \tilde{X}^+ \otimes e^{\tilde{H}} \otimes \cdots e^{\tilde{H}} + \cdots + 1 \otimes \cdots \otimes 1 \otimes \tilde{X}^+)^t$$

which has exactly one $\tilde{X}^+$ in each factor. Thus, we are considering an expression

$$(a_1 + a_2 + \cdots + a_t)^t$$

in which $a_i a_j = q^2 a_j a_i$ if $i < j$, and we require the coefficient c_t of $a_1 a_2 \ldots a_t$, all products being written so that the a_i occur in increasing order. An obvious induction on t shows that

$$c_{t+1} = c_t(1 + q^{-2} + q^{-4} + \cdots + q^{-2t}),$$

from which we find that

$$c_t = q^{-\frac{1}{2}t(t-1)}[t]_q!.$$

To prove the formula in the general case, we must consider the term in $\Delta_h(\tilde{H}^{s'}(\tilde{X}^+)^{t'})$ which has s $\tilde{H}$'s (and no $\tilde{X}^+$'s) in the first factor, and t $\tilde{X}^+$'s (and no $\tilde{H}$'s) in the second factor. Such a term exists only if $s' = s$ and $t' = t$, in which case it is simply $\tilde{H}^s \otimes (\tilde{X}^+)^t$. It follows that

$$\langle \xi^s \eta^t , \tilde{H}^{s'}(\tilde{X}^+)^{t'}\rangle = \langle \xi^s , \tilde{H}^{s'}\rangle \langle \eta^t , (\tilde{X}^+)^{t'}\rangle. \quad \blacksquare$$

We can now describe the Hopf algebra structure of the QUE dual of $U_h^{\geq 0}$.

LEMMA 8.3.5 *In the QUE dual of* $U_h^{\geq 0}$, *we have*

(a) $[\xi , \eta] = -\eta$,

(b) $\Delta_h(\xi) = \xi \otimes 1 + 1 \otimes \xi$, $\qquad \Delta_h(\eta) = \eta \otimes e^{-2h\xi} + 1 \otimes \eta$.

PROOF For (a), we must compute $\xi\eta$ and $\eta\xi$. The only non-zero pairings are

$$\langle \xi\eta , \tilde{H}\tilde{X}^+\rangle = \langle \xi \otimes \eta , (\tilde{H} \otimes 1 + 1 \otimes \tilde{H})(\tilde{X}^+ \otimes e^{\tilde{H}} + 1 \otimes \tilde{X}^+)\rangle$$
$$= \langle \xi \otimes \eta , \tilde{H} \otimes \tilde{X}^+\rangle = 1,$$
$$\langle \eta\xi , \tilde{H}\tilde{X}^+\rangle = \langle \eta , \tilde{X}^+\rangle \langle \xi , \tilde{H}e^{\tilde{H}}\rangle = 1,$$
$$\langle \eta\xi , \tilde{X}^+\rangle = \langle \eta , \tilde{X}^+\rangle \langle \xi , e^{\tilde{H}}\rangle = 1.$$

The formula $\xi\eta - \eta\xi = -\eta$ is now clear.

To prove the second formula in (b) (the proof of the first is similar but easier), note that, since η vanishes on all monomials $\tilde{H}^s(\tilde{X}^+)^t$ except that for which $s = 0$ and $t = 1$, it follows that $\Delta_h(\eta)$ vanishes on all tensor products of such monomials except two:

$$\langle \Delta_h(\eta) , 1 \otimes \tilde{X}^+\rangle = \langle \eta , \tilde{X}^+\rangle = 1,$$
$$\langle \Delta_h(\eta) , \tilde{X}^+ \otimes \tilde{H}^s\rangle = \langle \eta , \tilde{X}^+\tilde{H}^s\rangle = \langle \eta , (\tilde{H} - 2h)^s\tilde{X}^+\rangle = (-2h)^s. \quad \blacksquare$$

We can now prove 8.3.2 in the $sl_2(\mathbb{C})$ case. In fact, it is clear from the preceding lemma that the map

$$H \mapsto 2\xi, \qquad X^- \mapsto \eta$$

defines an isomorphism of Hopf algebras from $U_h^{\leq 0}$ to the QUE dual of $U_h^{\geq 0}$ with the opposite comultiplication.

Remark At first sight, this result seems paradoxical, since the classical analogue $U^{\leq 0}$ of $U_h^{\leq 0}$, being non-commutative, is certainly not isomorphic to the classical analogue $((U^{\geq 0})^*)^{\mathrm{op}}$ of $((U_h^{\geq 0})^*)^{\mathrm{op}}$, which is commutative. However, we are taking the *QUE dual* here, i.e. the dual of the QFSH algebra $U_{h,\geq 0}$ generated by $\tilde{H}$ and $\tilde{X}^+$. Since

$$[\tilde{H}, \tilde{X}^+] = 2h\tilde{X}^+, \quad \Delta_h(\tilde{X}^+) = \tilde{X}^+ \otimes e^{\tilde{H}} + 1 \otimes \tilde{X}^+,$$

the classical limit of $U_{h,\geq 0}$ is the commutative algebra of formal power series in two variables $\tilde{H}$ and $\tilde{X}^+$, with the non-cocommutative coalgebra structure (10). Hence, the classical limit of $U_{h,\geq 0}^*$ is cocommutative but non-commutative, and of course $U^{\leq 0}$ has the same properties. $\triangle$

We are finally in a position to describe the algebra structure in $\mathcal{D}(U_h^{\geq 0})$.

LEMMA 8.3.6 *The following relations hold in* $\mathcal{D}(U_h^{\geq 0})$:

$$[X^+, \eta] = \frac{e^{hH} - e^{-2h\xi}}{h}, \quad [X^+, \xi] = -X^+,$$

$$[H, \eta] = -2\eta, \quad [H, \xi] = 0.$$

OUTLINE OF PROOF We prove the first formula. According to the description of the algebra structure in the double $\mathcal{D}(A)$ given in Subsection 4.2D, if $a \in A$ and $\alpha \in (A^*)^{\mathrm{op}}$, and if

$$(S_h^{-1} \otimes \mathrm{id})\Delta_h^{(2)}(a) = \sum_i a_i \otimes a_i' \otimes a_i'', \quad \Delta_h^{(2)}(\alpha) = \sum_j \alpha_j \otimes \alpha_j' \otimes \alpha_j'',$$

then

$$\alpha.a = \sum_{i,j} \langle \alpha_j, a_j \rangle \langle \alpha_j'', a_j'' \rangle a_j'.\alpha_j'.$$

In our case, we have

$$(S_h^{-1} \otimes \mathrm{id})\Delta_h^{(2)}(X^+) = -e^{-hH}X^+ \otimes e^{hH} \otimes e^{hH} + 1 \otimes X^+ \otimes e^{hH} + 1 \otimes 1 \otimes X^+,$$

$$\Delta_h^{(2)}(\eta) = \eta \otimes 1 \otimes 1 + e^{-2h\xi} \otimes \eta \otimes 1 + e^{-2h\xi} \otimes e^{-2h\xi} \otimes \eta.$$

Each term in $\Delta_h^{(2)}(X^+)$ pairs with exactly one term in $\Delta_h^{(2)}(\eta)$, and we find that

$$\eta X^+ = X^+ \eta - \frac{e^{hH} - e^{-2h\xi}}{h}. \quad \blacksquare$$

We can now prove 8.3.1 in the $sl_2(\mathbb{C})$ case. In fact, the required homomorphism $\mathcal{D}(U_h^{\geq 0}) \to U_h$ is given by

$$X^+ \mapsto X^+, \quad H \mapsto H,$$

(11)

$$\eta \mapsto \left(\frac{e^h - e^{-h}}{h} \right) X^-, \quad \xi \mapsto \frac{1}{2}H.$$

It is now a simple matter to compute the universal R-matrix of U_h. By 8.3.4, the basis of $U_{h,\geq 0}^*$ dual to the basis $\{H^s(X^+)^t\}$ of $U_h^{\geq 0}$ is

$$\left\{ \frac{h^{s+t}q^{\frac{1}{2}t(t-1)}}{s![t]_q!} \, \xi^s \eta^t \right\}.$$

Hence, the universal R-matrix of $\mathcal{D}(U_h^{\geq 0})$ is

$$\sum_{s=0}^{\infty} \sum_{t=0}^{\infty} \frac{h^{s+t}q^{\frac{1}{2}t(t-1)}}{s![t]_q!} (H^s(X^+)^t \otimes \xi^s \eta^t).$$

Applying the homomorphism (11), we find that the universal R-matrix of U_h is

$$\mathcal{R}_h = \sum_{s=0}^{\infty} \sum_{t=0}^{\infty} \frac{1}{s![t]_q!} \left(\frac{h}{2}\right)^s (q - q^{-1})^t q^{\frac{1}{2}t(t-1)} (H^s \otimes H^s)((X^+)^t \otimes (X^-)^t)$$

$$= \exp\left[\frac{h}{2}H \otimes H\right] \sum_{t=0}^{\infty} q^{\frac{1}{2}t(t+1)} \frac{(1-q^{-2})^t}{[t]_q!} (X^+)^t \otimes (X^-)^t.$$

This is the same as the formula in 6.4.8.

Remark The formula for $\mathcal{R}_h$ can also be written as follows:

$$\mathcal{R}_h = \left\{ \sum_{t=0}^{\infty} q^{\frac{1}{2}t(t+1)} \frac{(1-q^{-2})^t}{[t]_q!} (E^t \otimes F^t) \right\} \exp\left[\frac{h}{2}H \otimes H\right],$$

where $E = e^{-hH}X^+$, $F = e^{hH}X^-$. $\triangle$

C The general case For an arbitrary finite-dimensional complex simple Lie algebra $\mathfrak{g}$, one follows the same procedure as in the $sl_2(\mathbb{C})$ case, the additional difficulty arising from the need to use root vectors corresponding to all the positive roots, not just the simple roots, and the fact that the comultiplication in U_h is known *a priori* only on the simple root vectors.

The solution to the latter problem is contained in 8.2.5 and 8.2.7. Let $X_{\beta_1}^+$, $X_{\beta_2}^+, \ldots, X_{\beta_N}^+$ be the root vectors defined in 8.1.4, corresponding to some choice of a reduced decomposition of the longest element of the Weyl group of $\mathfrak{g}$. The first root which is not simple is β_2, and by 8.2.5 and 8.2.6 we have

$$\Delta_h(X_{\beta_2}^+) = \Delta_h(T_{i_1}(X_{i_2}^+))$$

$$= \Delta_h(\tilde{w}_{h,i_1})\Delta_h(X_{i_2}^+)\Delta_h(\tilde{w}_{h,i_1})^{-1}$$

$$= \tilde{\mathcal{R}}_{h,i_1}^{-1}(\tilde{w}_{h,i_1} \otimes \tilde{w}_{h,i_1})(X_{i_2}^+ \otimes e^{d_{i_2}hH_{i_2}} + 1 \otimes X_{i_2}^+)(\tilde{w}_{h,i_1}^{-1} \otimes \tilde{w}_{h,i_1}^{-1})\tilde{\mathcal{R}}_{h,i_1}$$

$$= \tilde{\mathcal{R}}_{h,i_1}^{-1}(X_{\beta_2}^+ \otimes e^{d_{i_2}hs_{i_1}(H_{i_2})} + 1 \otimes X_{\beta_2}^+)\tilde{\mathcal{R}}_{h,i_1}.$$

This procedure can clearly be iterated to compute the effect of the comultiplication on all the root vectors. To state the general result, set

$$
(12) \qquad \begin{aligned}
\tilde{\mathcal{R}}_{h,\beta_r} &= (T_{i_1} \ldots T_{i_{r-1}} \otimes T_{i_1} \ldots T_{i_{r-1}})(\tilde{\mathcal{R}}_{h,i_r}), \\
\tilde{\mathcal{R}}_{h,<\beta_r} &= \tilde{\mathcal{R}}_{h,\beta_{r-1}} \tilde{\mathcal{R}}_{h,\beta_{r-2}} \ldots \tilde{\mathcal{R}}_{h,\beta_1},
\end{aligned}
$$

for every positive root β_r.

PROPOSITION 8.3.7 *For any positive root β, we have*

$$
\Delta_h(X_\beta^+) = \tilde{\mathcal{R}}_{h,<\beta}^{-1}(X_\beta^+ \otimes e^{hH_\beta} + 1 \otimes X_\beta^+)\tilde{\mathcal{R}}_{h,<\beta},
$$

where, if $\beta = \sum_i k_i\alpha_i \in Q$, then $H_\beta = \sum_i d_i k_i H_i$. ∎

To determine the QUE dual of $U_h^{\geq 0}$, we must first describe its QFSH subalgebra $U_{h,\geq 0}$. As in 8.3.3, one finds that $U_{h,\geq 0}$ is generated by the elements $\tilde{H}_i = hH_i$, $i = 1, \ldots, n$, and $\tilde{X}_\beta^+ = hX_\beta^+$ as β runs through the positive roots of $\mathfrak{g}$. Define elements $\xi_1, \ldots, \xi_n, \eta_{\beta_1}, \ldots, \eta_{\beta_N}$ in $U_{h,\geq 0}^*$ by requiring that $\langle \xi_i, \tilde{H}_i \rangle = \langle \eta_{\beta_r}, \tilde{X}_{\beta_r}^+ \rangle = 1$, and ξ_i and η_{β_r} give zero when paired against any other monomial of the form $\tilde{H}_1^{s_1} \ldots \tilde{H}_n^{s_n}(\tilde{X}_{\beta_N}^+)^{t_N} \ldots (\tilde{X}_{\beta_1}^+)^{t_1}$. Then, we have the following generalization of 8.3.4. It is proved by induction, making use of 8.3.7.

LEMMA 8.3.8 *For any $s_1, s_1', \ldots, t_N, t_N' \in \mathbb{N}$, we have*

$$
\langle \xi_1^{s_1} \ldots \xi_n^{s_n} \eta_{\beta_N}^{t_N} \ldots \eta_{\beta_1}^{t_1}, H_1^{s_1'} \ldots H_n^{s_n'}(X_{\beta_N}^+)^{t_N'} \ldots (X_{\beta_1}^+)^{t_1'} \rangle
$$

$$
= \prod_{i=1}^{n}\prod_{r=1}^{N} \delta_{s_i,s_i'}\delta_{t_r,t_r'} s_i! \, q_{\beta_r}^{-\frac{1}{2}t_r(t_r-1)} \, [t_r]_{q_{\beta_r}}!,
$$

where $q_\beta = e^{d_\beta h}$ and $d_\beta = d_i$ if the positive root β is Weyl group conjugate to the simple root α_i. ∎

Using this result, the Hopf algebra structure of $U_{h,\geq 0}^*$ can be determined and 8.3.2 proved in the general case. In fact, one finds that the isomorphism $U_h^{\leq 0} \to U_{h,\geq 0}^*$ is given by

$$
H_i \mapsto \sum_j d_j^{-1}a_{ij}\xi_j, \quad X_i^- \mapsto \eta_{\alpha_i}.
$$

To prove 8.3.1, one shows that the following defines a homomorphism of Hopf algebras $\mathcal{D}(U_h^{\geq 0}) \to U_h$:

$$
(13) \qquad \begin{aligned}
H_i &\mapsto H_i, \quad X_\beta^+ \mapsto X_\beta^+, \\
\xi_i &\mapsto \sum_j (B^{-1})_{ij}H_j, \quad \eta_\beta \mapsto \left(\frac{q_\beta - q_\beta^{-1}}{h}\right)X_\beta^-,
\end{aligned}
$$

where B is the matrix $(d_j^{-1} a_{ij})$.

Finally, the universal R-matrix of $\mathcal{D}(U_h^{\geq 0})$ is, by 8.3.8,

$$\sum_{s_1,\ldots,s_n=0}^{\infty} \sum_{t_1,\ldots,t_N=0}^{\infty} \prod_{i=1}^{n} \prod_{r=1}^{N} \frac{h^{s_i+t_r} q_{\beta_r}^{\frac{1}{2}t_r(t_r-1)}}{s_i! [t_r]_{q_{\beta_r}}!}$$
$$\times \left(H_1^{s_1} \ldots H_n^{s_n} (X_{\beta_N}^+)^{t_N} \ldots (X_{\beta_1}^+)^{t_1} \otimes \xi_1^{s_1} \ldots \xi_n^{s_n} \eta_{\beta_N}^{t_N} \ldots \eta_{\beta_1}^{t_1} \right).$$

Applying the homomorphism (13), we find that the universal R-matrix $\mathcal{R}_h$ of U_h is

$$\exp\left[h \sum_{i,j} (B^{-1})_{ij} H_i \otimes H_j \right] \sum_{t_1,\ldots,t_N=0}^{\infty} \prod_{r=1}^{N} q_{\beta_r}^{\frac{1}{2}t_r(t_r+1)} \frac{(1-q_{\beta_r}^{-2})^{t_r}}{[t_r]_{q_{\beta_r}}!} (X_{\beta_r}^+)^{t_r} \otimes (X_{\beta_r}^-)^{t_r}.$$

The order of the factors in the product is such that the β_r-term appears to the left of the β_s-term if $r > s$.

The expression for $\mathcal{R}_h$ can be written more concisely by making use of the q-exponential

$$(14) \qquad \exp_q(x) = \sum_{k=0}^{\infty} q^{\frac{1}{2}k(k+1)} \frac{x^k}{[k]_q!}.$$

Then, we obtain

THEOREM 8.3.9 *For any finite-dimensional complex simple Lie algebra* $\mathfrak{g}$, $U_h(\mathfrak{g})$ *is topologically quasitriangular with universal R-matrix*

$$\mathcal{R}_h = \exp\left[h \sum_{i,j} (B^{-1})_{ij} H_i \otimes H_j \right] \prod_{\beta} \exp_{q_\beta}[(1 - q_\beta^{-2}) X_\beta^+ \otimes X_\beta^-],$$

where the product is over all the positive roots of $\mathfrak{g}$, *and the order of the terms is such that the* β_r-*term appears to the left of the* β_s-*term if* $r > s$ *(see 8.1.4).* ∎

Remark The r-matrix $r = h^{-1}(\mathcal{R}_h - 1 \otimes 1) \pmod{h}$ which is the classical limit of $\mathcal{R}_h$ is clearly given by

$$r = \sum_{i,j=1}^{n} (B^{-1})_{ij} H_i \otimes H_j + 2 \sum_{r=1}^{N} d_{\beta_r} X_{\beta_r}^+ \otimes X_{\beta_r}^-.$$

This agrees with the formula in 2.1.7, since the inner products of the generators are given by

$$(H_i, H_j) = d_j^{-1} a_{ij}, \quad (X_\beta^+, X_\gamma^-) = \delta_{\beta,\gamma} d_\beta^{-1}$$

(see the Appendix). △

D Multiplicative properties Recalling the 'partial' R-matrices $\tilde{\mathcal{R}}_{h,\beta}$ defined in (12), we deduce from 8.3.9 the following multiplicative formula for the universal R-matrix of $U_h(\mathfrak{g})$ in terms of those of its $U_h(sl_2(\mathbb{C}))$-subalgebras :

PROPOSITION 8.3.10   *The universal R-matrix of $U_h(\mathfrak{g})$ is given by*

$$\mathcal{R}_h = \exp\left[h\sum_{i,j}(B^{-1})_{ij}H_i \otimes H_j\right]\tilde{\mathcal{R}}_{h,\beta_N}\tilde{\mathcal{R}}_{h,\beta_{N-1}}\ldots\tilde{\mathcal{R}}_{h,\beta_1}. \quad\blacksquare$$

We can obtain a third formula for the universal R-matrix of U_h if we recall the remark at the end of Section 8.2, according to which the $\tilde{w}_{h,i}$ satisfy the defining relations of the braid group of $\mathfrak{g}$. As in 8.1.3, this implies that there is a canonical element of the quantum Weyl group of U_h associated to any element of the Weyl group of $\mathfrak{g}$. In particular, if

$$w_0 = s_{i_1}s_{i_2}\ldots s_{i_N}$$

is a reduced expression of the longest element of the Weyl group, we have a canonical element

$$\tilde{w}_{h,0} = \tilde{w}_{h,i_1}\tilde{w}_{h,i_2}\ldots\tilde{w}_{h,i_N}$$

of the quantum Weyl group. Now, by 8.2.6,

$$\begin{aligned}
\Delta_h(\tilde{w}_{h,0}) &= \Delta_h(\tilde{w}_{h,i_1})\Delta_h(\tilde{w}_{h,i_2})\ldots\Delta_h(\tilde{w}_{h,i_N}) \\
&= \tilde{\mathcal{R}}_{h,i_1}^{-1}(\tilde{w}_{h,i_1} \otimes \tilde{w}_{h,i_1})\tilde{\mathcal{R}}_{h,i_2}^{-1}(\tilde{w}_{h,i_2} \otimes \tilde{w}_{h,i_2})\ldots\tilde{\mathcal{R}}_{h,i_N}^{-1}(\tilde{w}_{h,i_N} \otimes \tilde{w}_{h,i_N}) \\
&= \tilde{\mathcal{R}}_h^{-1}(\tilde{w}_{h,0} \otimes \tilde{w}_{h,0}),
\end{aligned}$$

where

$$\tilde{\mathcal{R}}_h = \tilde{\mathcal{R}}_{h,\beta_N}\tilde{\mathcal{R}}_{h,\beta_{N-1}}\ldots\tilde{\mathcal{R}}_{h,\beta_1}.$$

This proves

PROPOSITION 8.3.11   *The universal R-matrix of $U_h(\mathfrak{g})$ is given by*

$$\mathcal{R}_h = \exp\left[h\sum_{i,j}(B^{-1})_{ij}H_i \otimes H_j\right](\tilde{w}_{h,0} \otimes \tilde{w}_{h,0})\Delta_h(\tilde{w}_{h,0})^{-1}. \quad\blacksquare$$

COROLLARY 8.3.12   *The universal R-matrix of $U_h(\mathfrak{g})$ defined in 8.3.9 is independent of the choice of reduced expression of the longest element of the Weyl group of $\mathfrak{g}$.* $\blacksquare$

It would be interesting to find a proof of 8.3.11 and 8.3.12 which does not require one to compute the R-matrices involved.

E Uniqueness of the universal R-matrix The solution to the problem of finding an invertible element $\mathcal{R}_h$ in (some completion of) $U_h \otimes U_h$ such that $\Delta_h^{\mathrm{op}} = \mathcal{R}_h.\Delta_h.\mathcal{R}_h^{-1}$ is certainly not unique (see the discussion preceding 4.2.2). However, the next result shows that any solution $\mathcal{R}_h$, which is of *the same form* as that in 8.3.9, coincides with that solution up to a scalar multiple (cf. the remark following the proof of 6.4.8).

Let $\tilde{U}_h^0$ be the field of fractions of the subalgebra of $U_h^0 \otimes U_h^0$ generated by elements of the form $e^{hH_i} \otimes 1$, $1 \otimes e^{hH_i}$ and $e^{hH_i \otimes H_i}$, $i = 1, \ldots, n$. Let $\tilde{U}_h^\pm$ be the set of formal infinite sums, with coefficients in $\tilde{U}_h^0$, of monomials of the form

$$(X_{\beta_N}^+)^{s_N} \ldots (X_{\beta_1}^+)^{s_1} \otimes (X_{\beta_1}^-)^{t_1} \ldots (X_{\beta_N}^-)^{t_N}$$

such that the total weight of the terms in each monomial appearing in a given formal sum is bounded, i.e. if $m_{\beta,i}$ is the multiplicity of a simple root α_i in a positive root β, then

$$\left| \sum_{\beta \in \Delta^+} (s_\beta - t_\beta) m_{\beta,i} \right| < \text{constant}.$$

It is not hard to see that $\tilde{U}_h^\pm$ is an associative algebra over $\mathbb{C}[[h]]$. Moreover, the R-matrix in 8.3.9 clearly lies in $\tilde{U}_h^\pm$.

PROPOSITION 8.3.13 *If $\mathcal{R}_h' \in \tilde{U}_h^\pm$ is such that $\mathcal{R}_h'.\Delta_h.(\mathcal{R}_h')^{-1} = \Delta_h^{\mathrm{op}}$, then $\mathcal{R}_h'$ is a scalar multiple of the R-matrix in 8.3.9.* ∎

F The centre of U_h If $u_h = \mu_h(S_h \otimes \mathrm{id})(\mathcal{R}_h)_{21}$, where $\mathcal{R}_h$ is the universal R-matrix given by 8.3.9, then by 4.2.4 and the remarks following it, $z_h = u_h S_h(u_h) = S_h(u_h)u_h$ lies in the centre of U_h. This element is closely related to the Casimir element of the classical enveloping algebra U. Note that, since $\mathcal{R}_h \equiv 1 \otimes 1 \pmod{h}$, we have $z_h \equiv 1 \pmod{h}$, so that it makes sense to define

$$C_h = -\frac{1}{h}\ln(z_h).$$

PROPOSITION 8.3.14 *The element $C_h \in U_h(\mathfrak{g})$ corresponds, under the isomorphism in 6.5.5, to the Casimir element of $U(\mathfrak{g})$.*

The square of the antipode S_h, which is given by conjugation by u_h (see 4.2.3), can be computed directly, and this leads to another central element of U_h. In fact, let ρ^* be the unique element of $\mathfrak{h} \subset U_h(\mathfrak{h})$ such that $\alpha_i(\rho^*) = 2d_i$ for $i = 1, \ldots, n$. (If $\sum k_i \alpha_i$ is the sum of the positive roots of $\mathfrak{g}$, then $\rho^* = \sum k_i d_i H_i$.) From the definition of S_h in 6.5.1, it is easy to see that $S_h^2(a) = e^{h\rho^*} a e^{-h\rho^*}$ for all $a \in U_h(\mathfrak{g})$ (it is enough to check this when $a = X_i^\pm$). It follows that $v_h = e^{-h\rho^*} u_h$ is in the centre of U_h. The central element v_h is not really new, however, for we have

PROPOSITION 8.3.15 $z_h = v_h^2$.

COROLLARY 8.3.16 *The topologically quasitriangular Hopf algebra $U_h(\mathfrak{g})$, with the central element v_h described above, is a topological ribbon Hopf algebra.* ∎

We shall prove 8.3.14 and 8.3.15 in Subsection 10.1B.

G Matrix solutions of the quantum Yang–Baxter equation We know from 4.2.7 that $\mathcal{R}_h$ satisfies the quantum Yang–Baxter equation

$$\mathcal{R}_h^{12}\mathcal{R}_h^{13}\mathcal{R}_h^{23} = \mathcal{R}_h^{23}\mathcal{R}_h^{13}\mathcal{R}_h^{12}.$$

If $\rho : U_h \to \mathrm{End}(V)$ is a finite-dimensional representation of U_h, then obviously $R_h^\rho = (\rho \otimes \rho)(\mathcal{R}_h)$ will be a solution of the QYBE with values in $\mathrm{End}(V \otimes V)$, i.e. a 'matrix' solution. Computing R_h^ρ is not easy, however, since one needs to know the action on V of all the root vectors of U_h, not just the simple ones. We outline one such calculation:

Example 8.3.17 Let $\mathfrak{g} = sl_{n+1}(\mathbb{C})$ and let $\rho_h^\natural : U_h \to \mathrm{End}(V_h^\natural)$ be the natural $(n + 1)$-dimensional representation of U_h. This means that $V_h^\natural$ has a basis $\{v_0, v_1, \ldots, v_n\}$ with respect to which the generators of U_h act as follows:

$$(15) \qquad \rho_h^\natural(H_i) = E_{i-1\,i-1} - E_{ii}, \quad \rho_h^\natural(X_i^+) = E_{i-1\,i}, \quad \rho_h^\natural(X_i^-) = E_{i\,i-1}$$

(E_{ij} is the elementary matrix with one in the ith row and jth column and zeros elsewhere). Using 6.5.1, one verifies directly that these formulas do define a representation of U_h; its '$h = 0$ limit' is clearly the natural representation of $sl_{n+1}(\mathbb{C})$ (in fact, the formulas in (15) are exactly the same as in the classical case). From the discussion in Subsection 6.4E, it follows that $V_h^\natural$ is indecomposable.

We must now compute the action of the remaining root vectors $X_\beta^\pm$ on $V_h^\natural$. We take the reduced decomposition of w_0 to be

$$w_0 = s_1 s_2 \ldots s_n s_1 s_2 \ldots s_{n-1} \ldots s_1 s_2 s_1.$$

The positive roots, in the order in which they appear in (3), are

$$(16) \qquad \beta_{01}, \beta_{02}, \ldots, \beta_{0n}, \beta_{12}, \ldots, \beta_{1n}, \ldots, \beta_{n-2\,n-1}, \beta_{n-2\,n}, \beta_{n-1\,n},$$

where $\beta_{ij} = \alpha_{i+1} + \alpha_{i+2} + \cdots + \alpha_j$. Noting that a product of the form

$$\rho_h^\natural(X_{i_1}^+)\rho_h^\natural(X_{i_2}^+)\ldots\rho_h^\natural(X_{i_k}^+)$$

vanishes unless each difference $i_{r+1} - i_r = 1$ (and similarly for the $\rho_h^\natural(X_i^-)$), it is not difficult to see that

$$\rho_h^\natural(X_{\beta_{ij}}^+) = (-1)^{i-j+1}E_{ij}, \quad \rho_h^\natural(X_{\beta_{ij}}^-) = (-1)^{i-j+1}E_{ji}.$$

Hence,

$$(\rho_h^\natural \otimes \rho_h^\natural)(\exp_q[(1 - q^{-2})X_{\beta_{ij}}^+ \otimes X_{\beta_{ij}}^-]) = 1 + (e^h - e^{-h})E_{ij} \otimes E_{ji}.$$

Remembering to write the terms in the *reverse* of the order (16), we see that all the cross terms vanish and so

$$(17) \quad (\rho_h^\natural \otimes \rho_h^\natural)\left(\prod_{\beta \in \Delta^+} \exp_q[(1-q^{-2})X_\beta^+ \otimes X_\beta^-]\right) = 1 + (e^h - e^{-h})\sum_{i<j} E_{ij} \otimes E_{ji}.$$

The action of the remaining factor $\exp[h\sum_{i,j}(B^{-1})_{ij}H_i \otimes H_j]$ is quite easy to compute, since we already know the action of all the H_i. We find that it gives

$$(18) \qquad e^{-h/(n+1)}\left(e^h\sum_i E_{ii} \otimes E_{ii} + \sum_{i \neq j} E_{ii} \otimes E_{jj}\right).$$

Multiplying (17) and (18), we obtain finally that $(\rho_h^\natural \otimes \rho_h^\natural)(\mathcal{R}_h)$ is equal to

$$e^{-h/(n+1)}\left(e^h\sum_i E_{ii} \otimes E_{ii} + \sum_{i \neq j} E_{ii} \otimes E_{jj} + (e^h - e^{-h})\sum_{i<j} E_{ij} \otimes E_{ji}\right)$$

Apart from the unimportant scalar factor $e^{-h/(n+1)}$, this is the same as the R-matrix in equation (18) of Chapter 7. $\diamondsuit$

If $\mathfrak{g}$ is of type B, C or D, the calculation proceeds along similar lines. The main point is that the indecomposable representation of $U_h(\mathfrak{g})$ which is a deformation of the natural representation of $\mathfrak{g}$ is again given, in a suitable basis, by precisely the same formulas as the natural representation of $\mathfrak{g}$. We leave it to the reader to check that this is a consequence of the following observations:

(a) the matrices representing the simple root vectors $\tilde{X}_i^\pm$ of $\mathfrak{g}$ in the standard realization of its natural representation all have square zero (see Varadarajan (1974), Section 4.4, for example);

(b) the eigenvalues of the generators $\tilde{H}_i$ of $\mathfrak{g}$ in the natural representation are all 0, 1 or -1.

One finds that the image of the universal R-matrix $\mathcal{R}_h$ of $U_h(\mathfrak{g})$ is given, up to a scalar multiple, by the R-matrices written down in Subsection 7.3C.

Bibliographical notes

8.1 The braid group action on $U_h(\mathfrak{g})$ was introduced by Lusztig (1988), and its detailed properties were proved in Lusztig (1990a,b, 1993). Further details, and the omitted proofs, of the results in this section may be found in these references.

8.2 The quantum Weyl group was introduced by Vaksman & Soibelman (1988) in the sl_2 case, by Soibelman (1990b) in the sl_{n+1} case, and by Kirillov & Reshetikhin (1989, 1990a), and Soibelman (1991b) in the general case. We have mainly followed Kirillov & Reshetikhin, where the proofs of the results of this section may be found.

8.3 The fact that $U_h(\mathfrak{g})$ is 'almost' a quantum double was pointed out by Drinfel'd (1987), where the formula for the universal R-matrix in the sl_2 case can also be found. The universal R-matrix in the sl_{n+1} case was computed by Rosso (1989), and in the general case by Kirillov & Reshetikhin (1990a) and Levendorskii & Soibelman (1990). Proposition 8.3.13 is due to Khoroshkin & Tolstoy (1992b).

9

Specializations of QUE algebras

Intuitively, one thinks of the quantized universal enveloping algebras $U_h(\mathfrak{g})$ studied in Chapters 6 and 8 as families of Hopf algebras (over $\mathbb{C}$) depending on a parameter h. Strictly speaking, however, this does not make sense, for an algebra defined over the formal power series ring $\mathbb{C}[[h]]$ cannot be specialized to any value of h except $h = 0$. To remedy this situation, we introduce a new algebra $U_q(\mathfrak{g})$, defined over the field of rational functions of an indeterminate q, which is a kind of 'rational' counterpart of the 'formal' object $U_h(\mathfrak{g})$. The main advantage of $U_q(\mathfrak{g})$ over $U_h(\mathfrak{g})$ is that one can specialize q to any transcendental complex number. In fact, with a little more effort, one can construct an 'integral' form of the algebra, defined over $\mathbb{Z}[q, q^{-1}]$, which enables one to specialize q to any non-zero complex number ϵ. Moreover, the definition and properties of $U_q(\mathfrak{g})$ closely resemble those of $U_h(\mathfrak{g})$. In addition, $U_q(\mathfrak{g})$ is in some ways technically simpler than $U_h(\mathfrak{g})$: for example, one no longer has to worry about h-adic topologies, completed tensor products, etc.

We discuss two such integral forms of $U_h(\mathfrak{g})$, called the 'non-restricted' and 'restricted' forms. We shall see that the resulting specializations coincide when ϵ is transcendental, in fact even when ϵ is not a root of unity, but have very different properties when ϵ is a root of unity. In the former case, the properties of the specialization closely resemble those of the classical universal enveloping algebra. On the other hand, it turns out that the specializations at a root of unity provide a characteristic zero analogue of the classical Lie algebras in characteristic p. Indeed, this is more than an analogy: we shall formulate some precise results relating the two theories in this chapter and Chapter 11.

The non-restricted specialization $U_\epsilon(\mathfrak{g})$ is the easiest to define: one simply replaces q by ϵ in the defining relations of $U_q(\mathfrak{g})$. The crucial fact about $U_\epsilon(\mathfrak{g})$, when ϵ is a primitive ℓth root of unity ($\ell > 1$), is that it is finite dimensional over its centre Z_ϵ; in particular, the representation theory of $U_\epsilon(\mathfrak{g})$ reduces essentially to that of a finite-dimensional algebra. We show that Z_ϵ is the algebra of regular functions on a complex algebraic variety $\mathcal{Z}_\epsilon$, and that $\mathcal{Z}_\epsilon$ is a finite (ramified) covering of the 'big cell' in the group G with Lie algebra $\mathfrak{g}$. (Recall that the centre of $U(\mathfrak{g})$ is much smaller: it is a polynomial algebra on rank$(\mathfrak{g})$ generators.) We also exhibit an infinite-dimensional group $\mathcal{G}$ of automorphisms of $U_\epsilon(\mathfrak{g})$ for which the induced action on $\mathcal{Z}_\epsilon$ corresponds, in a precise sense, to the adjoint action of G on itself. We shall see in the next chapter that, conjecturally at least, this leads to a picture of the representation theory of $U_\epsilon(\mathfrak{g})$ which is in some ways analogous to that provided for G by the Kostant–Kirillov 'orbit method'.

The definition of the restricted specialization $U_\epsilon^{\mathrm{res}}(\mathfrak{g})$ mimics that of the Chevalley–Kostant $\mathbb{Z}$-form of the universal enveloping algebra $U(\mathfrak{g})$, which is itself central to the classical characteristic p theory. When ϵ is a primitive ℓth root of unity, we show that $U_\epsilon^{\mathrm{res}}(\mathfrak{g})$ has a remarkable finite-dimensional Hopf subalgebra $U_\epsilon^{\mathrm{fin}}(\mathfrak{g})$ such that the 'quotient' of $U_\epsilon^{\mathrm{res}}(\mathfrak{g})$ by $U_\epsilon^{\mathrm{fin}}(\mathfrak{g})$ is isomorphic to the classical universal enveloping algebra $U(\mathfrak{g})$. Thus, as in the non-restricted case, the representation theory of $U_\epsilon^{\mathrm{res}}(\mathfrak{g})$ reduces to that of a finite-dimensional algebra. Moreover, if $\ell = p$ is an odd prime, then, after taking the quotient $\tilde{U}_\epsilon^{\mathrm{fin}}(\mathfrak{g})$ of $U_\epsilon^{\mathrm{fin}}(\mathfrak{g})$ by a small subalgebra of its centre and reducing (mod p), $U_\epsilon^{\mathrm{fin}}(\mathfrak{g})$ becomes what is usually called the 'restricted' universal enveloping algebra in the characteristic p theory (hence the terminology used for this specialization). Conjecturally, $\tilde{U}_\epsilon^{\mathrm{fin}}(\mathfrak{g})$ has the same representation theory (in characteristic zero) as the restricted enveloping algebra in characteristic p (if p is not too small).

Note Unless indicated otherwise, $\mathfrak{g}$ denotes in this chapter a finite-dimensional complex simple Lie algebra. We use freely the notation in the Appendix.

9.1 Rational forms

A The definition of U_q If $\mathfrak{g}$ is a finite-dimensional complex simple Lie algebra with Cartan matrix (a_{ij}), we let d_i be the coprime positive integers such that the matrix $(d_i a_{ij})$ is symmetric. Let q be an indeterminate, and let $\mathbb{Q}(q)$ be the field of rational functions of q with coefficients in $\mathbb{Q}$. Write $q_i = q^{d_i}$.

The definition of $U_q(\mathfrak{g})$ is modelled on that of $U_h(\mathfrak{g})$ in 6.5.1.

DEFINITION–PROPOSITION 9.1.1 *Let $\mathfrak{g}$ be a finite-dimensional complex simple Lie algebra. Then $U_q(\mathfrak{g})$ is the associative algebra over $\mathbb{Q}(q)$ with generators X_i^+, X_i^-, K_i and K_i^{-1}, $1 \le i \le n$, and the following defining relations:*

$$K_i K_j = K_j K_i, \quad K_i K_i^{-1} = K_i^{-1} K_i = 1,$$

$$(1) \quad K_i X_j^+ K_i^{-1} = q_i^{a_{ij}} X_j^+, \quad K_i X_j^- K_i^{-1} = q_i^{-a_{ij}} X_j^-,$$

$$(2) \quad X_i^+ X_j^- - X_j^- X_i^+ = \delta_{i,j} \frac{K_i - K_i^{-1}}{q_i - q_i^{-1}},$$

$$(3) \quad \sum_{r=0}^{1-a_{ij}} (-1)^r \begin{bmatrix} 1 - a_{ij} \\ r \end{bmatrix}_{q_i} (X_i^\pm)^{1-a_{ij}-r} X_j^\pm (X_i^\pm)^r = 0 \quad \text{if } i \ne j.$$

There is a Hopf algebra structure on $U_q(\mathfrak{g})$ with comultiplication Δ_q, an-

tipode S_q and counit ϵ_q defined on generators as follows:

$$\Delta_q(K_i) = K_i \otimes K_i,$$

$$\Delta_q(X_i^+) = X_i^+ \otimes K_i + 1 \otimes X_i^+, \quad \Delta_q(X_i^-) = X_i^- \otimes 1 + K_i^{-1} \otimes X_i^-,$$

(4) $\qquad S_q(K_i) = K_i^{-1}, \quad S_q(X_i^+) = -X_i^+ K_i^{-1}, \quad S_q(X_i^-) = -K_i X_i^-,$

$$\epsilon_q(K_i) = 1, \quad \epsilon_q(X_i^+) = \epsilon_q(X_i^-) = 0. \quad \blacksquare$$

PROOF The verification that the formulas for Δ_q, S_q and ϵ_q do extend to algebra homomorphisms (an anti-homomorphism in the case of the antipode) is analogous to the case of $U_h(\mathfrak{g})$ treated in Section 6.5. $\blacksquare$

Remarks [1] We shall omit the subscript q on Δ_q, S_q and ϵ_q, and will denote $U_q(\mathfrak{g})$ simply by U_q, whenever this will not lead to confusion.

[2] If we let $q_i = e^{d_i h}$ and $K_i = e^{d_i h H_i}$, the relations in 9.1.1 are consequences of those in 6.5.1. For example, by using 6.4.2, it is clear that relations (1) above are consequences of the relations $[H_i, X_j^\pm] = \pm a_{ij} X_j^\pm$ in 6.5.1.

[3] The same relations and coalgebra maps serve to define $U_q(\mathfrak{g})$ when $\mathfrak{g} = \mathfrak{g}'(A)$ is the Kac–Moody algebra associated to an arbitrary symmetrizable Cartan matrix A (see the Appendix).

[4] There are other rational forms of $U_h(\mathfrak{g})$, which bear the same relation to $U_q(\mathfrak{g})$ as the various complex Lie groups with Lie algebra $\mathfrak{g}$ bear to the adjoint group of $\mathfrak{g}$. The largest of these, the *simply-connected* rational form, is obtained by adjoining to $U_q(\mathfrak{g})$ invertible elements L_i, $i = 1, \ldots, n$, such that $K_i = \prod_j L_j^{a_{ji}}$; the relations in (1) are replaced by

$$L_i X_j^\pm L_i^{-1} = q_i^{\pm \delta_{ij}} X_j^\pm.$$

More generally, if M is any lattice such that $Q \subseteq M \subseteq P$, there is a rational form $U_q^M(\mathfrak{g})$ obtained by adjoining to $U_q(\mathfrak{g})$ the elements $K_\beta = \prod_j L_j^{m_{ij}}$ for every $\beta = \sum_j m_{ij} \lambda_j \in M$, where $\lambda_1, \ldots, \lambda_n$ are the fundamental weights of $\mathfrak{g}$. Thus, $U_q(\mathfrak{g})$ is the *adjoint form* $U_q^Q(\mathfrak{g})$. When $\mathfrak{g} = sl_2(\mathbb{C})$, for example, $P/Q = \mathbb{Z}_2$, so there are only two rational forms, the adjoint form and the simply-connected form. The latter is obtained from the former by adjoining a square root of K_1.

We shall deal only with the adjoint rational form $U_q(\mathfrak{g})$, as is customary in most of the literature, although for some purposes the simply-connected form is more convenient. Note that, as in the case of $U(\mathfrak{g})$, one cannot distinguish adjoint and simply-connected forms of $U_h(\mathfrak{g})$.

[5] One can define a rational form $\mathcal{F}_q$ of the coordinate algebra as follows. Let $\mathcal{I}$ be the set of finite-codimensional two-sided ideals I of U_q such that, for some $r \in \mathbb{N}$ and all $i = 1, \ldots, n$, $\prod_{s=-r}^{r}(K_i - q_i^s) \in I$. Then, $\mathcal{F}_q$ is the set of all $\mathbb{Q}(q)$-linear maps $U_q \to \mathbb{Q}(q)$ whose kernel contains some ideal $I \in \mathcal{I}$. It is not difficult to show that the comultiplication (resp. the antipode) dual to

the multiplication (resp. to the antipode) in U_q maps $\mathcal{F}_q$ into $\mathcal{F}_q \otimes \mathcal{F}_q$ (resp. into $\mathcal{F}_q$), so that $\mathcal{F}_q$ is a well-defined Hopf algebra over $\mathbb{Q}(q)$. $\triangle$

B Some basic properties of U_q As we mentioned above, many of the results we have established for U_h have counterparts for U_q. Most of these translations are obvious, and we shall not write them all down. We content ourselves with two examples, which will be especially important in the sequel.

First, we translate the braid group action on U_h discussed in Section 8.1.

PROPOSITION 9.1.2 *The braid group $\mathcal{B}_\mathfrak{g}$ of $\mathfrak{g}$ acts by $\mathbb{Q}(q)$-algebra automorphisms on U_q as follows:*

$$T_i(X_i^+) = -X_i^- K_i, \quad T_i(X_i^-) = -K_i^{-1} X_i^+, \quad T_i(K_j) = K_j K_i^{-a_{ij}},$$

$$T_i(X_j^+) = \sum_{r=0}^{-a_{ij}} (-1)^{r-a_{ij}} q_i^{-r} (X_i^+)^{(-a_{ij}-r)} X_j^+ (X_i^+)^{(r)} \quad \text{if } i \neq j,$$

$$T_i(X_j^-) = \sum_{r=0}^{-a_{ij}} (-1)^{r-a_{ij}} q_i^{r} (X_i^-)^{(r)} X_j^- (X_i^-)^{(-a_{ij}-r)} \quad \text{if } i \neq j. \blacksquare$$

The divided powers $(X_i^\pm)^{(r)}$ are defined as in Subsection 8.1A.

The formulas in this proposition are obtained in an obvious way from those in 8.1.2. For example,

$$T_i(K_j) = T_i(e^{d_j h H_j}) = e^{d_j h H_j} e^{-d_j a_{ji} h H_i}$$
$$= e^{d_j h H_j} e^{-d_i a_{ij} h H_i} = K_j K_i^{-a_{ij}}$$

(note that we used the fact that $(d_i a_{ij})$ is symmetric).

Root vectors $X_\beta^\pm$ can now be defined exactly as in 8.1.4. We let U_q^+, U_q^0 and U_q^- be the $\mathbb{Q}(q)$-subalgebras of U_q generated by the X_i^+, the $K_i^\pm$ and the X_i^-, respectively, and write $U_q^{\geq 0} = U_q^0 . U_q^+$, $U_q^{\leq 0} = U_q^- . U_q^0$. Further, for $\beta = \sum_i k_i \alpha_i \in Q$, we let $K_\beta = \prod_i K_i^{k_i}$; then, $T_w(K_\beta) = K_{w(\beta)}$ for all $w \in W$.

We have the following analogue of 8.1.7 and 8.1.8:

PROPOSITION 9.1.3 *Multiplication defines an isomorphism of vector spaces over $\mathbb{Q}(q)$*

$$U_q^- \otimes U_q^0 \otimes U_q^+ \to U_q.$$

Moreover:

(i) *the products $(X^+)^{\boldsymbol{t}} = (X_{\beta_N}^+)^{t_N} (X_{\beta_{N-1}}^+)^{t_{N-1}} \ldots (X_{\beta_1}^+)^{t_1}$, where $\boldsymbol{t} = (t_1, \ldots, t_N) \in \mathbb{N}^N$, form a basis of U_q^+;*

(ii) *the products $K^{\boldsymbol{s}} = K_1^{s_1} K_2^{s_2} \ldots K_n^{s_n}$, where $\boldsymbol{s} = (s_1, \ldots, s_n) \in \mathbb{Z}^n$, form a basis of U_q^0;*

(iii) the products $(X^-)^{\mathbf{r}} = (X_{\beta_1}^-)^{r_1}(X_{\beta_2}^-)^{r_2} \ldots (X_{\beta_N}^-)^{r_N}$, *where* $\mathbf{r} = (r_1, \ldots, r_N)$ $\in \mathbb{N}^N$, *form a basis of* U_q^-. ∎

One should not think, however, that every property of U_h translates into a property of U_q, and vice versa. We give three (counter) examples.

First, while $U_h(\mathfrak{g})$ is isomorphic as an algebra to $U(\mathfrak{g})[[h]]$ (if $\dim(\mathfrak{g}) < \infty$), it is not true that $U_q(\mathfrak{g}) \underset{\mathbb{Q}(q)}{\otimes} \mathbb{C}(q)$ is isomorphic to $U(\mathfrak{g}) \underset{\mathbb{C}}{\otimes} \mathbb{C}(q)$. This follows, for example, from the fact that, for any Lie algebra $\mathfrak{g}$ over a field, the only invertible elements of $U(\mathfrak{g})$ are the non-zero scalar multiples of the identity (see Bourbaki (1972), Chap. 1, §2, Exercise 2a). Nevertheless, we shall see in the next chapter that, up to a rather trivial twisting, there is still a one-to-one correspondence between the finite-dimensional representations of $U_q(\mathfrak{g})$ and those of $\mathfrak{g}$.

The second property concerns the twisting just mentioned. Since it will be used later, we formulate it as

PROPOSITION 9.1.4 *For each sequence of signs* $(\sigma_1, \ldots, \sigma_n) \in \{\pm 1\}^n$, *there is an algebra automorphism of* U_q *given by*

$$K_i \mapsto \sigma_i K_i, \quad X_i^+ \mapsto \sigma_i X_i^+, \quad X_i^- \mapsto X_i^-. \quad ∎$$

The proof is an easy verification. Note that, except for the identity automorphism corresponding to $(1, \ldots, 1)$, there are no analogous automorphisms of U_h. For, if there were such automorphisms, mapping H_i to η_i say, then on reducing the equation

$$\sigma_i e^{d_i h H_i} = e^{d_i h \eta_i}$$

(mod h), we obtain that $\sigma_i = 1$. Note also that, again except for the identity, these algebra automorphisms do not respect the coalgebra structure of U_q.

Let us formulate 9.1.4 a little more invariantly. Let Q_2^* be the group of homomorphisms $Q \to \mathbb{Z}_2 = \{\pm 1\}$ (the product is pointwise multiplication). Then, the assertion is that Q_2^* acts as a group of automorphisms on U_q by

$$\sigma.K_\beta = \sigma(\beta)K_\beta, \quad \sigma.X_\alpha^+ = \sigma(\alpha)X_\alpha^+, \quad \sigma.X_\alpha^- = X_\alpha^-$$

for all $\alpha \in \Delta^+$, $\beta \in Q$, $\sigma \in Q_2^*$. Note that W acts on Q_2^*:

$$(w.\sigma)(\beta) = \sigma(w^{-1}(\beta));$$

moreover, the action of Q_2^* on U_q^0 obviously extends to an action of the semidirect product $W \ltimes Q_2^*$.

Finally, the universal R-matrix does not quite translate to U_q. For the q-exponentials appearing in 8.3.9 involve infinite sums, and so are not elements of $U_q \otimes U_q$. Furthermore, there is no element of $U_q \otimes U_q$ corresponding to the factor

$$\exp\left[h\sum_{ij}(B^{-1})_{ij}H_i \otimes H_j\right]$$

in $\mathcal{R}$. We shall see in Subsection 10.1D, however, that even though the universal R-matrix does not exist as an element of $U_q \otimes U_q$, it nevertheless acts on any tensor product of finite-dimensional U_q-modules!

C The Harish Chandra homomorphism and the centre of U_q We now discuss briefly the centre Z_q of U_q. Proofs and further details will be given in the next chapter, after certain necessary facts about the representation theory of U_q have been obtained.

Any element $z \in Z_q$ can, of course, be written as a linear combination of the basis elements of U_q given in 9.1.3. Since z commutes, in particular, with the K_i for all i, it follows that

$$(5) \qquad z = \sum_{\eta \in Q^+}\sum_{\mathbf{r},\mathbf{t}\in\mathrm{Par}(\eta)}(X^-)^{\mathbf{r}}\varphi_{\mathbf{r},\mathbf{t}}(X^+)^{\mathbf{t}},$$

where the $\varphi_{\mathbf{r},\mathbf{t}} \in U_q^0$ and

$$\mathrm{Par}(\eta) = \left\{(m_1,\ldots,m_N) \in \mathbf{N}^N \mid \sum_i m_i\alpha_i = \eta\right\}.$$

The proof of the following result is almost exactly as in the classical case.

DEFINITION–PROPOSITION 9.1.5 *The map $h_q : Z_q \to U_q^0$ defined by $h_q(z) = \varphi_{\mathbf{0},\mathbf{0}}$ is a homomorphism of algebras, called the* Harish Chandra homomorphism. ∎

Any element $\varphi \in U_q^0$ may be regarded as a $\mathbf{Q}(q)$-valued function on the weight lattice P in an obvious way: if $\varphi = \prod_{i=1}^n K_i^{t_i}$, where $t_1, t_2,\ldots, t_n \in \mathbf{Z}$, set

$$\varphi(\lambda) = q^{\sum_i t_i(\alpha_i,\lambda)},$$

and extended to U_q^0 by linearity. Define an automorphism of $\mathbf{Q}(q)$-algebras $\gamma_q : U_q^0 \to U_q^0$ by setting $\gamma_q(K_i) = q_iK_i$. It is clear that

$$\gamma_q(\varphi)(\lambda) = \varphi(\lambda + \rho)$$

for all $\varphi \in U_q^0$, where ρ is half the sum of the positive roots of $\mathfrak{g}$ (see the Appendix).

As we saw in the previous subsection, the semidirect product $W \ltimes Q_2^*$ acts as a group of automorphisms of U_q^0. Let $\tilde{W}$ be the subgroup generated by all conjugates $\sigma W \sigma^{-1}$ of W by elements $\sigma \in Q_2^*$.

THEOREM 9.1.6 *The homomorphism $\gamma_q^{-1} \circ h_q : Z_q \to U_q^0$ is injective, and its image is precisely the set $U_q^{0\,\tilde{W}}$ of fixed points of the action of $\tilde{W}$ on U_q^0.* ∎

Thus, an element $\varphi_{0,0} \in U_q^0$ occurs as the 'U_q^0 part' of an element $z \in Z_q$ as in (5) if and only if $\gamma_q^{-1}(\varphi_{0,0}) \in U_q^{0\tilde{W}}$. In this case, there is an inductive procedure for calculating the remaining coefficients $\varphi_{\mathbf{r},\mathbf{t}}$ in (5), but this, like the proof of 9.1.6, will have to wait until Subsection 10.1B. We content ourselves at this point with the following example.

Example 9.1.7 Let $\mathfrak{g} = sl_2(\mathbb{C})$. Then, $\tilde{W} = W$ and $U_q^{0\tilde{W}}$ consists of the Laurent polynomials in K_1 which are invariant under $K_1 \mapsto K_1^{-1}$. Thus, $U_q^{0\tilde{W}}$ is generated as an algebra over $\mathbb{Q}(q)$ by

$$\varphi = \frac{K_1 + K_1^{-1}}{(q - q^{-1})^2}.$$

It is easy to check that the quantum Casimir element

$$\Omega = \frac{qK_1 + q^{-1}K_1^{-1}}{(q - q^{-1})^2} + X_1^- X_1^+$$

lies in Z_q, and, since $\gamma_q^{-1}(h_q(\Omega)) = \varphi$, it follows from 9.1.6 that Ω generates Z_q as a $\mathbb{Q}(q)$-algebra. ◇

We conclude this subsection with the quantum analogue of Harish Chandra's theorem on the central characters of the classical enveloping algebra U. Note that any weight $\lambda \in P$ defines a homomorphism $U_q^0 \to \mathbb{Q}(q)$ by sending K_i to $q^{(\alpha_i, \lambda)}$ (cf. the discussion following 9.1.5); let $\chi_{q,\lambda} : Z_q \to \mathbb{Q}(q)$ be its composite with the homomorphism $\gamma_q^{-1} \circ h_q$ in 9.1.6.

THEOREM 9.1.8 *Let λ, $\mu \in P$. Then, $\chi_{q,\lambda} = \chi_{q,\mu}$ if and only if $\mu = w(\lambda)$ for some $w \in W$.* ∎

D A geometric realization We conclude this section by outlining an interpretation, due to Beilinson, Lusztig & MacPherson (1990), of $U_q(sl_{n+1}(\mathbb{C}))$ in terms of spaces of flags in vector spaces over finite fields. It is based on a rather general combinatorial construction, which has other interesting applications.

In fact, suppose that Γ is a finite group which acts transitively on a finite set X. Let $\mathbf{F}$ be the free vector space over a field k with basis the set $\mathcal{A}$ of orbits of the diagonal action of Γ on $X \times X$. We define a k-algebra structure on $\mathbf{F}$ as follows. For $\mathcal{O}$, $\mathcal{O}'$, $\mathcal{O}'' \in \mathcal{A}$, choose $(x_1, x_2) \in \mathcal{O}''$ and let $c_{\mathcal{O}, \mathcal{O}'}^{\mathcal{O}''}$ be the number of elements $x \in X$ such that $(x_1, x) \in \mathcal{O}$ and $(x, x_2) \in \mathcal{O}'$ (this is obviously independent of the choice of (x_1, x_2)). Then, one defines a product on $\boldsymbol{F}$ by

$$\mathcal{O}.\mathcal{O}' = \sum_{\mathcal{O}''} c_{\mathcal{O},\mathcal{O}'}^{\mathcal{O}''} \mathcal{O}''.$$

The unit element of $\boldsymbol{F}$ is the diagonal in $X \times X$ (which is a single Γ-orbit).

As an example, let $X = \Gamma$ with the action given by left multiplication. Then, it is easy to see that $\boldsymbol{F}$ is the group algebra $k[\Gamma]$ of Γ. Another example can be found in Subsection 10.2A.

We now show how this construction, with some modifications, can be used to construct U_q. Let V be a vector space of dimension m over a finite field with r elements, and let $\mathrm{Flag}_n(V)$ be the set of $(n+1)$-step filtrations

$$F = (0 = V_0 \subseteq V_1 \subseteq V_2 \subseteq \cdots \subseteq V_{n+1} = V).$$

The finite general linear group $GL(V)$ acts naturally on $\mathrm{Flag}_n(V)$, and its orbits are the fibres of the map which associates to the flag F the $(n+1)$-tuple of integers

$$(\dim(V_1/V_0), \dim(V_2/V_1), \ldots, \dim(V_{n+1}/V_n)).$$

If we have a second flag

$$F' = (0 = V_0' \subseteq V_1' \subseteq V_2' \subseteq \cdots \subseteq V_{n+1}' = V),$$

the subspaces $W_{ij} = V_{i-1} + (V_i \cap V_j')$ form an $(n+1)^2$-step filtration of V:

$$0 = W_{10} \subseteq W_{11} \subseteq \cdots \subseteq W_{1\,n+1} = W_{20} \subseteq \cdots \subseteq W_{n+1\,n+1} = V.$$

Let $A_{ij} = \dim(W_{ij}/W_{i\,j-1})$, $1 \leq i, j \leq n+1$ (we interpret W_{i0} as $W_{i-1\,n+1}$). Obviously, $\sum_{ij} A_{ij} = m$.

PROPOSITION 9.1.9 *The map which associates to a pair of flags (F, F') the matrix $A = (A_{ij})$ is a bijection between the set $\mathcal{O}_{m,n}$ of orbits of the diagonal action of $GL(V)$ on $\mathrm{Flag}_n(V) \times \mathrm{Flag}_n(V)$ and the set $\mathcal{A}_{m,n}$ of all $(n+1) \times (n+1)$ matrices with non-negative integer entries, the sum of whose entries is m.* ∎

Let $\mathcal{O}(A)$ be the orbit in $\mathrm{Flag}(V)_n \times \mathrm{Flag}_n(V)$ corresponding to a matrix $A \in \mathcal{A}_{m,n}$. If A, A' and $A'' \in \mathcal{A}_{m,n}$, choose $(F_1, F_2) \in \mathcal{O}(A'')$ and let $c_{A,A'}^{A''}(r)$ be the number of flags F such that $(F_1, F) \in \mathcal{O}(A)$ and $(F, F_2) \in \mathcal{O}(A')$.

PROPOSITION 9.1.10 *For any A, A', $A'' \in \mathcal{A}_{m,n}$, there exist integers c_0, $c_1, \ldots, c_\ell$, for some $\ell \geq 0$, such that*

$$c_{A,A'}^{A''}(r) = c_0 + c_1 r + \cdots + c_\ell r^\ell$$

whenever r is a prime power. Define $c_{A,A'}^{A''} \in \mathbb{Q}(q)$ by

$$c_{A,A'}^{A''} = c_0 + c_1 q^2 + \cdots + c_\ell q^{2\ell}.$$

Let $\boldsymbol{F}_{m,n}$ be the free vector space over $\mathbb{Q}(q)$ with basis $\mathcal{A}_{m,n}$. Then, the formula

$$A * A' = \sum_{A'' \in \mathcal{A}_{m,n}} c_{A,A'}^{A''} A''$$

defines the structure of an associative algebra over $\mathbb{Q}(q)$ on $\boldsymbol{F}_{m,n}$ with unit element $\sum A$, as A runs over the diagonal matrices in $\mathcal{A}_{m,n}$. ∎

It is convenient to renormalize the structure constants $c_{A,A'}^{A''}$ by multiplying them by certain powers of q. In fact, for $A \in \mathcal{A}_{m,n}$, set $\{A\} = q^{-s} A \in \boldsymbol{F}_{m,n}$, where

$$s = \sum A_{ij} A_{kl},$$

and the sum is over $\{i, j, k, l \mid i \geq k, \, j < l\}$. Then, we have

$$\{A\}.\{A'\} = \sum_{A''} \tilde{c}_{A,A'}^{A''} \{A''\}$$

for some $\tilde{c}_{A,A'}^{A''} \in \mathbb{Q}(q)$.

Let $\tilde{\mathcal{A}}_n$ be the set of $(n+1) \times (n+1)$ matrices with integer entries, the off-diagonal entries being non-negative. If $A \in \tilde{\mathcal{A}}_n$ and $t \in \mathbb{Z}$, set

$$_t A = A + tI \in \tilde{\mathcal{A}}_n,$$

where I is the identity matrix. If A, A', $A'' \in \tilde{\mathcal{A}}_n$ are such that the sum of the entries in A, in A' and in A'' are the same, then, by taking t sufficiently large, we can obviously arrange that $_t A$, $_t A'$ and $_t A'' \in \mathcal{A}_{m,n}$ for some m.

PROPOSITION 9.1.11 *Let q' be an indeterminate independent of q. There is a unique element $\gamma_{A,A'}^{A''}(q, q') \in \mathbb{Q}(q)[q']$ such that*

$$\gamma_{A,A'}^{A''}(q, q^{-t}) = \tilde{c}_{\,_t A, \,_t A'}^{\,_t A''}$$

for all sufficiently large t if the sum of the entries in A, in A' and in A'' are the same, and $\gamma_{A,A'}^{A''}(q, q') = 0$ otherwise.

Let $\boldsymbol{F}_n$ be the free $\mathbb{Q}(q)$-vector space with basis $\tilde{\mathcal{A}}_n$ (we write $[A]$ for the basis element corresponding to $A \in \tilde{\mathcal{A}}_n$ to avoid confusion). Then, the formula

$$[A] \bullet [A'] = \sum_{A'' \in \tilde{\mathcal{A}}_n} \gamma_{A,A'}^{A''}(q, 1) \, [A'']$$

defines the structure of an associative algebra (without unit) over $\mathbb{Q}(q)$ *on* $\boldsymbol{F}_n$. $\blacksquare$

One thinks of $\boldsymbol{F}_n$ as being a kind of limit of the $\boldsymbol{F}_{m,n}$ as $m \to \infty$.

Finally, let $\mathcal{A}_n^0$ be the set of all $(n+1) \times (n+1)$ matrices with non-negative integer entries and all diagonal entries equal to zero. For $A \in \mathcal{A}_n^0$ and integers $r_1, \ldots, r_{n+1}$, consider the formal infinite sum of elements of $\boldsymbol{F}_n$ given by

$$A(r_1, \ldots, r_{n+1}) = \sum_D q^{d_1 r_1 + \cdots + d_{n+1} r_{n+1}} [A + D],$$

where the sum is over all diagonal matrices $D = \mathrm{diag}(d_1, \ldots, d_{n+1})$ with integer entries. Let $\boldsymbol{F}_n^\infty$ be the free $\mathbb{Q}(q)$-vector space with basis the elements $A(r_1, \ldots, r_{n+1})$ above. One shows that the product of two elements of $\boldsymbol{F}_n^\infty$, computed term-by-term using the product in $\boldsymbol{F}_n$, is well defined and in $\boldsymbol{F}_n^\infty$. Thus, $\boldsymbol{F}_n^\infty$ is an associative algebra over $\mathbb{Q}(q)$. It has a unit given by $\sum_D [D]$ as D runs through all the diagonal matrices with integer entries.

THEOREM 9.1.12 *There is an isomorphism* $U_q \to \boldsymbol{F}_n^\infty$ *of algebras over* $\mathbb{Q}(q)$ *given on the generators of* U_q *by*

$$X_i^+ \mapsto \sum_A [A], \quad X_i^- \mapsto \sum_A [A], \quad K_i \mapsto \sum_D q^{d_i - d_{i+1}} [D],$$

where, in the first (resp. second) sum, A runs over all the matrices in $\tilde{\mathcal{A}}_n$ with 1 in the ith row and $(i+1)$th column (resp. in the $(i+1)$th row and ith column) and zeros elsewhere, and in the third sum, $D = \mathrm{diag}(d_1, \ldots, d_{n+1})$ runs over the diagonal matrices in $\tilde{\mathcal{A}}_n$. $\blacksquare$

Related results can be found in Subsection 9.3D.

9.2 The non-restricted specialization

To obtain from U_q a well-defined Hopf algebra by specializing q to an arbitrary non-zero complex number ϵ, one constructs an *integral form* of U_q. This is an $\mathcal{A}$-subalgebra $U_\mathcal{A}$ of U_q, where $\mathcal{A} = \mathbb{Z}[q, q^{-1}]$, such that the natural map

$$U_\mathcal{A} \underset{\mathcal{A}}{\otimes} \mathbb{Q}(q) \to U_q$$

is an isomorphism of $\mathbb{Q}(q)$-algebras. One then defines

$$U_\epsilon = U_\mathcal{A} \underset{\mathcal{A}}{\otimes} \mathbb{C},$$

using the homomorphism $\mathcal{A} \to \mathbb{C}$ taking q to ϵ. One could replace $\mathbb{C}$ by $\mathbb{Q}(\epsilon)$ here, or even by $\mathbb{Z}[\epsilon, \epsilon^{-1}]$, and we shall find it necessary to do this in the next section.

There are two candidates for $U_\mathcal{A}$ which lead to different specializations (with markedly different representation theories) for certain values of ϵ. We shall discuss them in this section and the next.

A The non-restricted integral form Introduce the elements

$$(6) \qquad [K_i \, ; \, m]_{q_i} = \frac{K_i q_i^m - K_i^{-1} q_i^{-m}}{q_i - q_i^{-1}} \in U_q^0 \qquad (m \geq 0),$$

where $q_i = q^{d_i}$.

DEFINITION–PROPOSITION 9.2.1 *The algebra $U_\mathcal{A}$ is the $\mathcal{A}$-subalgebra of U_q generated by the elements X_i^+, X_i^-, $K_i^{\pm 1}$ and $[K_i \, ; \, 0]_{q_i}$ for $1 \leq i \leq n$. With the maps Δ, S and ϵ defined on the first three sets of generators as in 9.1.1, and with*

$$(7) \qquad \Delta_q([K_i \, ; \, 0]_{q_i}) = [K_i \, ; \, 0]_{q_i} \otimes K_i + K_i^{-1} \otimes [K_i \, ; \, 0]_{q_i},$$
$$(8) \qquad S_q([K_i \, ; \, 0]_{q_i}) = -[K_i \, ; \, 0]_{q_i},$$
$$(9) \qquad \epsilon_q([K_i \, ; \, 0]_{q_i}) = 0,$$

$U_\mathcal{A}$ becomes a Hopf algebra. Moreover, $U_\mathcal{A}$ is an integral form of U_q. ∎

Remarks [1] The defining relations for $U_\mathcal{A}$ are as in 9.1.1, but replacing (2) by

$$X_i^+ X_j^- - X_j^- X_i^+ = \delta_{i,j}[K_i \, ; \, 0]$$

and adding the relation

$$(10) \qquad (q_i - q_i^{-1})[K_i \, ; \, 0]_{q_i} = K_i - K_i^{-1}.$$

[2] The elements $[K_i \, ; \, m]_{q_i}$ are in $U_\mathcal{A}$ for all $m \in \mathbb{N}$, for it is easy to see that

$$[K_i \, ; \, m]_{q_i} = [K_i \, ; \, 0]_{q_i} q_i^{-m} + K_i[m]_{q_i}. \qquad \triangle$$

We now define, as above, $U_\epsilon = U_\mathcal{A} \underset{\mathcal{A}}{\otimes} \mathbb{C}$ for any $\epsilon \in \mathbb{C}^\times$. Define U_ϵ^+, U_ϵ^- and U_ϵ^0 in the obvious way.

PROPOSITION 9.2.2 *If $\epsilon^{2d_i} \neq 1$ for all i, then U_ϵ is generated over $\mathbb{C}$ by elements X_i^+, X_i^- and $K_i^{\pm 1}$ with defining relations obtained from those in 9.1.1 by replacing q by ϵ. Moreover, in this case, the products $(X^-)^\mathbf{r} K^\mathbf{s} (X^+)^\mathbf{t}$ in 9.1.3 are a basis of U_ϵ over $\mathbb{C}$.* ∎

One might expect that the specialization $U_1(\mathfrak{g})$ would be isomorphic to the classical universal enveloping algebra $U(\mathfrak{g})$. In fact, this is not quite true. Note that, by (1), the K_i are in the centre of U_1, and by (10) that $K_i^2 = 1$ in U_1. From this, it is easy to deduce

PROPOSITION 9.2.3 *The quotient of $U_1(\mathfrak{g})$ by the ideal generated by the $K_i - 1$ is isomorphic to the classical universal enveloping algebra $U(\mathfrak{g})$.* ∎

On the other hand, the specialization $U_1^+(\mathfrak{g})$ of $U_q^+(\mathfrak{g})$ has generators X_i^+ and defining relations (3), and hence is isomorphic to the universal enveloping algebra of the nilpotent subalgebra $\mathfrak{n}^+$ of $\mathfrak{g}$ generated by the classical analogues of the X_i^+. In particular, since $U(\mathfrak{n}^+)$ is an integral domain, so is $U_1^+(\mathfrak{g})$. Similar remarks apply, of course, to U_1^-, and it is obvious that U_1^0 is an integral domain. This proves the first part of

PROPOSITION 9.2.4 *The rings U_q and $U_{\mathcal{A}}$ are integral domains. If $\epsilon^{2d_i} \neq 1$ for all i, then U_ϵ is also an integral domain.* ∎

B The centre Suppose first that $\epsilon \in \mathbb{C}^\times$ is not a root of unity. Let Z_ϵ denote the centre of U_ϵ, and define $U_\epsilon^{0\,\tilde{W}}$ and the Harish Chandra homomorphism $h_\epsilon : Z_\epsilon \to U_\epsilon^0$ as in Subsection 9.1C, and the homomorphism $\gamma_\epsilon : U_\epsilon^0 \to U_\epsilon^0$ by $\gamma_\epsilon(K_i) = \epsilon^{d_i} K_i$. Further, for any $\lambda \in P$, let $\chi_{\epsilon,\lambda} : Z_\epsilon \to \mathbb{C}$ be the composite of $\gamma_\epsilon^{-1} \circ h_\epsilon : Z_\epsilon \to U_\epsilon^0$ with the homomorphism $U_\epsilon^0 \to \mathbb{C}$ which sends K_i to $\epsilon^{(\alpha_i, \lambda)}$. Then, as in 9.1.6, we have

PROPOSITION 9.2.5 *Suppose that $\epsilon \in \mathbb{C}^\times$ is not a root of unity. Then, the map $\gamma_\epsilon^{-1} \circ h_\epsilon$ is injective and its image is $U_\epsilon^{0\,\tilde{W}}$. Moreover, if $\lambda, \mu \in P$, we have $\chi_{\epsilon,\lambda} = \chi_{\epsilon,\mu}$ if and only if $\mu = w(\lambda)$ for some $w \in W$.* ∎

Note In the remainder of this section, we assume that ϵ is a primitive ℓ^{th} root of unity, where ℓ is odd, and $\ell > d_i$ for all i.

Our aim is to describe the structure of the centre Z_ϵ of U_ϵ. We shall use a different approach from that used in the generic case, for the following reason. One can show that the inductive procedure which, in the generic case, associates to each element $\varphi \in U_q^{0\,\tilde{W}}$ an element $z_\varphi \in Z_q$ (see the discussion following 9.1.6), can be specialized to give a well-defined homomorphism $U_\epsilon^{0\,\tilde{W}} \to Z_\epsilon$. This homomorphism is injective but, unlike the generic case, it is not surjective (see Remark [4] following 9.2.19). Thus, the Harish Chandra approach now captures only a part of the centre.

We begin our study of Z_ϵ with

PROPOSITION 9.2.6 *The elements $(X_\alpha^\pm)^\ell$ for $\alpha \in \Delta^+$, and K_i^ℓ for $i = 1, \ldots, n$, lie in Z_ϵ.*

OUTLINE OF PROOF The second part is obvious, for it is enough to prove that K_i^ℓ commutes with each $X_j^\pm$, and this is immediate from the equation $K_i X_j^\pm K_i^{-1} = \epsilon^{\pm d_i a_{ij}} X_j^\pm$. The first part follows from certain relations in U_ϵ proved in Lusztig (1990b). ∎

For $\alpha \in \Delta^+$, $\beta \in Q$, let $x_\alpha^\pm = (X_\alpha^\pm)^\ell$, $k_\beta = K_\beta^\ell$; we shall often write $x_i^\pm$ for $x_{\alpha_i}^\pm$. Let Z_0 (resp. $Z_0^\pm$, Z_0^0) be the subalgebra of Z_ϵ generated by the $x_\alpha^\pm$ and the $k_i^{\pm 1}$ (resp. by the $x_\alpha^\pm$, by the $k_i^{\pm 1}$).

PROPOSITION 9.2.7 *(i) We have $Z_0^\pm \subset U_\epsilon^\pm$ and $Z_0^0 \subset U_\epsilon^0$.*

(ii) Multiplication defines an isomorphism of algebras

$$Z_0^- \otimes Z_0^0 \otimes Z_0^+ \to Z_0.$$

(iii) Z_0^0 is the algebra of Laurent polynomials in the k_i, and $Z_0^\pm$ is the polynomial algebra with generators the $x_\alpha^\pm$.

(iv) We have $Z_0^\pm = U_\epsilon^\pm \cap Z_\epsilon$.

(v) The subalgebra Z_0 of Z_ϵ is preserved by the action of the braid group automorphisms T_i.

(vi) U_ϵ is a free Z_0-module with basis the set of monomials $(X^-)^r K^s (X^+)^t$ in the statement of 9.1.3 for which $0 \leq r_k, s_i, t_k < \ell$, for $i = 1,\ldots,n$, $k = 1,\ldots,N$.

OUTLINE OF PROOF Parts (i), (ii), (iii) and (vi) follow from 9.2.2 and the analogue of 8.1.6 for U_ϵ. Part (iv) is proved by considering the action of Z_0 on 'diagonal' U_ϵ-modules (see Subsection 11.1A). Part (v) follows from part (iv) and 8.1.6(i). ∎

The preceding proposition shows that U_ϵ is a noetherian Z_0-module. It follows that $Z_\epsilon \subset U_\epsilon$ is finitely generated over Z_0, and hence integral over Z_0. By the Hilbert basis theorem, Z_ϵ is a finitely-generated algebra. Thus, the affine schemes $\mathrm{Spec}(Z_\epsilon)$ and $\mathrm{Spec}(Z_0)$, namely the sets of algebra homomorphisms from Z_ϵ and Z_0 to $\mathbb{C}$, are algebraic varieties. In fact, it is obvious that $\mathrm{Spec}(Z_0)$ is isomorphic to $\mathbb{C}^{2N} \times (\mathbb{C}^\times)^n$, a complex affine space of dimension equal to that of $\mathfrak{g}$ with n codimension one hyperplanes removed. Moreover, the inclusion $Z_0 \hookrightarrow Z_\epsilon$ induces a projection map $\tau : \mathrm{Spec}(Z_\epsilon) \to \mathrm{Spec}(Z_0)$ and, by the Cohen–Seidenberg theorem, τ is a finite (surjective) map. More precisely, we have

PROPOSITION 9.2.8 $\mathrm{Spec}(Z_\epsilon)$ *is a normal affine variety and $\tau : \mathrm{Spec}(Z_\epsilon) \to \mathrm{Spec}(Z_0)$ is a finite map of degree ℓ^n.* ∎

This means that the fibres of τ have at most ℓ^n points, and that the generic fibre has precisely ℓ^n points. The computation of the degree of τ is carried out in Subsection 11.1A; it depends on the representation theory of U_ϵ.

Example 9.2.9 Suppose that $\mathfrak{g} = sl_2(\mathbb{C})$ and that ℓ is odd. Then, Z_ϵ contains the quantum Casimir element

$$\Omega = \frac{K_1 \epsilon + K_1^{-1}\epsilon^{-1}}{(\epsilon - \epsilon^{-1})^2} + X_1^- X_1^+.$$

By an easy induction on r, one shows that

$$(X_1^-)^r(X_1^+)^r = \prod_{s=0}^{r-1}\left(\Omega - \left(\frac{K_1\epsilon^{2s+1} + K_1^{-1}\epsilon^{-2s-1}}{(\epsilon - \epsilon^{-1})^2}\right)\right).$$

Taking $r = \ell$, we obtain the relation

$$(11) \qquad \prod_{r=0}^{\ell-1}\left(\Omega - \left(\frac{K_1\epsilon^{2s+1} + K_1^{-1}\epsilon^{-2s-1}}{(\epsilon - \epsilon^{-1})^2}\right)\right) = x_1^+ x_1^-.$$

It is not difficult to show that the left-hand side is a polynomial in Ω, all of whose coefficients are complex numbers (not involving K_1), except the term independent of Ω, which is equal to

$$-\frac{k_1 + k_1^{-1}}{(\epsilon - \epsilon^{-1})^{2\ell}}.$$

In fact, Z_ϵ is generated over Z_0 by Ω subject to the single relation (11). ◊

We conclude this subsection by discussing the relation between the centre and the Hopf structure of U_ϵ. To compute the effect of the comultiplication on the ℓth powers, we need the following lemma, which is easily proved by induction.

LEMMA 9.2.10 *Suppose that a and b are elements of an associative algebra over $\mathbb{C}$ such that $ab = \epsilon^2 ba$, where $\epsilon \in \mathbb{C}^\times$. Then, for all $\ell \geq 0$,*

$$(a+b)^\ell = \sum_{k=0}^{\ell}\epsilon^{-k(\ell-k)}\begin{bmatrix}\ell\\k\end{bmatrix}_\epsilon a^k b^{\ell-k}.$$

In particular, if ϵ is a primitive ℓth root of unity and ℓ is odd, then $(a+b)^\ell = a^\ell + b^\ell$. ∎

It follows immediately that, if Δ_ϵ denotes the comultiplication of U_ϵ,

$$\Delta(x_i^+) = x_i^+ \otimes k_i + 1 \otimes x_i^+, \quad \Delta(x_i^-) = x_i^- \otimes 1 + k_i^{-1} \otimes x_i^-, \quad \Delta(k_i) = k_i \otimes k_i.$$

Unfortunately, there is no such simple formula for $\Delta(x_\alpha^\pm)$ when α is a non-simple root. Nevertheless, one can prove

PROPOSITION 9.2.11 Z_0 *is a Hopf subalgebra of U_ϵ, as are Z_0^0, $Z_0^{\geq 0}$ and $Z_0^{\leq 0}$.* ∎

The meaning of the notation is clear: $Z_0^{\geq 0} = Z_0^0 Z_0^+$, $Z_0^{\leq 0} = Z_0^0 Z_0^-$.

It follows from 9.2.11 that $\mathrm{Spec}(Z_0)$ inherits a Lie group structure from the Hopf structure of U_ϵ. In fact:

PROPOSITION 9.2.12 *The formula*

$$\{z\,,\,z'\} = \lim_{q \to \epsilon} \frac{zz' - z'z}{\ell(q^\ell - q^{-\ell})}$$

defines a Poisson bracket on Z_0 which gives $\mathrm{Spec}(Z_0)$ the structure of a Poisson–Lie group. ∎

The chosen normalization of the Poisson bracket will enable us, in the next subsection, to identify $\mathrm{Spec}(Z_0)$ with a Poisson–Lie group studied in Chapter 1.

C The quantum coadjoint action In this subsection, we assume that ϵ is a primitive ℓ^{th} root of unity, where ℓ is odd and $\ell > d_i$ for all i. We set $\epsilon_i = \epsilon^{d_i}$.

Define derivations $\underline{x}_i^\pm$ of U_q by

(12) $$\underline{x}_i^\pm(u) = \left[\frac{(X_i^\pm)^\ell}{[\ell]_{q_i}!}\,,\,u \right].$$

Note that, if we specialize to $q = \epsilon$, we obtain at first sight an indeterminate result, since the $(X_i^\pm)^\ell$ are central and $[\ell]_{\epsilon_i}! = 0$. However, we have

PROPOSITION 9.2.13 *On specializing to $q = \epsilon_i$, (12) induces well-defined derivations of U_ϵ. In fact, we have the following explicit formulas:*

$$\underline{x}_i^+(X_j^+) = \frac{1}{2} \sum_{r=1}^{-a_{ij}} \begin{bmatrix} -a_{ij} \\ r \end{bmatrix}_{\epsilon_i} \left((X_i^+)^{(r)} X_j^+ (X_i^+)^{(\ell-r)} - (X_i^+)^{(\ell-r)} X_j^+ (X_i^+)^{(r)} \right),$$

$$\underline{x}_i^+(X_j^-) = \frac{1}{\ell} \delta_{i,j} (\epsilon_i - \epsilon_i^{-1})^{\ell-2} (K_i \epsilon_i - K_i^{-1} \epsilon_i^{-1}) (X_i^+)^{\ell-1},$$

$$\underline{x}_i^+(K_j^{\pm 1}) = \mp \frac{1}{2\ell} a_{ij} (\epsilon_i - \epsilon_i^{-1})^\ell (X_i^+)^\ell K_j^{\pm 1},$$

and $\underline{x}_i^-$ is obtained from $\underline{x}_i^+$ by using

$$T_{w_0} \underline{x}_{\overline{i}}^+ T_{w_0}^{-1} = \underline{x}_i^-\,,$$

where $i \mapsto \overline{i}$ is the permutation of the nodes of the Dynkin diagram of $\mathfrak{g}$ such that $w_0(\alpha_i) = -\alpha_{\overline{i}}$. ∎

We should like to exponentiate these derivations to obtain automorphisms of U_ϵ. To do this, we must enlarge U_ϵ. Let $\hat{Z}_0$ be the algebra of formal power series in the $x_\alpha^\pm$, $\alpha \in \Delta^+$, and the $k_i^{\pm 1}$, $i = 1, \ldots, n$, which define holomorphic functions on $\mathbb{C}^{2N} \times (\mathbb{C}^\times)^n$. Let

$$\hat{U}_\epsilon = U_\epsilon \underset{Z_0}{\otimes} \hat{Z}_0, \qquad \hat{Z}_\epsilon = Z_\epsilon \underset{Z_0}{\otimes} \hat{Z}_0.$$

Then, another direct computation gives

LEMMA 9.2.14 *The series*

$$\exp(t\,\underline{x}_i^{\pm}) = \sum_{n=0}^{\infty} \frac{t^n}{n!} (\underline{x}_i^{\pm})^n$$

converge, for all $t \in \mathbb{C}$, to well-defined automorphisms of the algebra $\hat{U}_\epsilon$ over $\mathbb{C}$. ∎

Let $\mathcal{G}$ be the group of automorphisms of $\hat{U}_\epsilon$ generated by the one-parameter groups of automorphisms appearing in the previous lemma.

PROPOSITION 9.2.15 *The action of $\mathcal{G}$ on $\hat{U}_\epsilon$ preserves the subalgebras $\hat{Z}_\epsilon$ and $\hat{Z}_0$, and (hence) acts by holomorphic automorphisms on the complex algebraic varieties $\mathrm{Spec}(Z_0)$ and $\mathrm{Spec}(Z_\epsilon)$. Moreover, the action of $\mathcal{G}$ on $\hat{U}_\epsilon$ is normalized by that of the braid group transformations T_i.*

OUTLINE OF PROOF For the first statement, it is enough to show that $\hat{Z}_0$ is preserved by the derivations $\underline{x}_\alpha^{\pm}$. This is proved for the $\underline{x}_i^{\pm}$ by direct computation, and the general case is deduced from this by using the braid group action. For the second statement, one proves by a direct computation that the T_i normalize the Lie algebra generated by the derivations $\underline{x}_\alpha^{\pm}$. ∎

The group $\mathcal{G}$ is 'infinite dimensional', and at first sight there seems little prospect of understanding its action on $\mathrm{Spec}(Z_0)$. We shall now see, however, that its orbits are very closely related to the conjugacy classes in the Lie group with Lie algebra $\mathfrak{g}$.

Let G be the adjoint group of $\mathfrak{g}$ (i.e. the connected complex Lie group with Lie algebra $\mathfrak{g}$ and trivial centre), let H be a Cartan subgroup of G, and let $N_{\pm}$ be the unipotent subgroups of G with Lie algebra $\mathfrak{n}_{\pm}$. Note that H is canonically identified with $\mathrm{Spec}(Z_0^0)$ by the pairing

$$(\exp(\eta),\, k_i) = \exp(2\pi\sqrt{-1}\alpha_i(\eta))$$

for $\eta \in \mathfrak{h}$, the Lie algebra of H. The product $G^0 = N_- H N_+$ is well known to be a dense open subset of G (in the complex topology), called the *big cell*.

We define maps

$$\boldsymbol{X}^{\pm} : \mathrm{Spec}(Z_0^{\pm}) \to N^{\pm}, \quad \boldsymbol{K} : \mathrm{Spec}(Z_0^0) \to H$$

and a map

$$\pi = \boldsymbol{X}^- \boldsymbol{K} \boldsymbol{X}^+ : \mathrm{Spec}(Z_0) = \mathrm{Spec}(Z_0^-) \times \mathrm{Spec}(Z_0^0) \times \mathrm{Spec}(Z_0^+) \to G^0$$

as follows. Fix a reduced decomposition $w_0 = s_{i_1} \ldots s_{i_N}$, and let $\bar{X}_{\beta_1}^-, \ldots, \bar{X}_{\beta_N}^-$ be the corresponding negative root vectors of $\mathfrak{g}$ (see the Appendix: we have changed notation to $\bar{X}_\beta^-$ to avoid confusion with the generators of U_ϵ). Let

$$x_{\beta_k}^- = (\epsilon_{i_k} - \epsilon_{i_k}^{-1}) T_{i_1} \ldots T_{i_{k-1}}(x_{i_k}^-) \in Z_0,$$

which we regard as a complex-valued function on $\mathrm{Spec}(Z_0)$. Then, we define maps $\boldsymbol{X}^{\pm} : \mathrm{Spec}(Z_0^{\pm}) \to N_{\pm}$, $\boldsymbol{K} : \mathrm{Spec}(Z_0) \cong H \to H \cong \mathrm{Spec}(Z_0)$ to be the products

$$\boldsymbol{X}^- = \exp(x_{\beta_N}^- \bar{X}_{\beta_N}^-)\exp(x_{\beta_{N-1}}^-, \bar{X}_{\beta_{N-1}}^-) \dots \exp(x_{\beta_1}^-, \bar{X}_{\beta_1}^-),$$

$$\boldsymbol{X}^+ = \exp(T_{w_0}(x_{\beta_N}^-)T_{w_0}(\bar{X}_{\beta_N}^-)) \dots \exp(T_{w_0}(x_{\beta_1}^-)T_{w_0}(\bar{X}_{\beta_1}^-)),$$

$$\boldsymbol{K}(h) = h^2, \quad (h \in H).$$

In the formula for $\boldsymbol{X}^+$, the action of $T_0 = T_{i_1} \dots T_{i_N}$ on $\mathfrak{g}$ is as described in the Appendix.

PROPOSITION 9.2.16 *The product map* $\pi = \boldsymbol{X}^- \boldsymbol{K} \boldsymbol{X}^+ : \mathrm{Spec}(Z_0) \to G^0$ *is independent of the choice of reduced decomposition of* w_0, *and is an (unramified) covering of degree* 2^n. *The set of fixed points of* $\mathcal{G}$ *on* $\mathrm{Spec}(Z_0)$ *is precisely the pre-image under* π *of the identity element of* G. ∎

Example 9.2.17 When $G = \mathrm{PSL}_2(\mathbb{C})$ is the adjoint group of $sl_2(\mathbb{C})$, the map π is given by

$$\pi(x_1^-, x_1^+, k_1) = \begin{pmatrix} 1 & 0 \\ x_1^- & 1 \end{pmatrix} \begin{pmatrix} k_1^2 & 0 \\ 0 & k_1^{-2} \end{pmatrix} \begin{pmatrix} 1 & x_1^+ \\ 0 & 1 \end{pmatrix}. \quad \diamond$$

Since π is a covering map, vector fields on G^0 can be lifted to vector fields on $\mathrm{Spec}(Z_0)$. In particular, we may regard the Chevalley generators $\bar{X}_i^{\pm}$ of $\mathfrak{g}$ as left-invariant vector fields on G^0, and hence as vector fields on $\mathrm{Spec}(Z_0)$. On the other hand, the derivations $\underline{x}_i^{\pm}$ may also be regarded as vector fields on $\mathrm{Spec}(Z_0)$. One has the following remarkable relations, which are proved by direct computation (it is enough to do the cases in which $\mathfrak{g}$ has rank 1 or 2).

THEOREM 9.2.18 *As vector fields on* $\mathrm{Spec}(Z_0)$, *we have*

$$\underline{x}_i^{\pm} = \pm k_i \bar{X}_i^{\pm},$$

where the k_i *are regarded as complex-valued functions on* $\mathrm{Spec}(Z_0)$. ∎

This is the basic link between the quantum object, namely the action of $\mathcal{G}$ on $\mathrm{Spec}(Z_0)$, and the classical action of G on itself by conjugation. The following consequences are almost immediate.

COROLLARY 9.2.19 *(i) The action of* $\mathcal{G}$ *in the tangent space at each of its fixed points induces the coadjoint action of* G *on* $\mathfrak{g}^*$ *(cf. 9.2.20(i)).*

(ii) If Γ *is a conjugacy class in* G, *the connected components of* $\pi^{-1}(\Gamma)$ *are* $\mathcal{G}$-*orbits in* $\mathrm{Spec}(Z_0)$.

(iii) If $\mathcal{O}$ is a $\mathcal{G}$-orbit in $\mathrm{Spec}(Z_0)$, the connected components of $\tau^{-1}(\mathcal{O})$ are $\mathcal{G}$-orbits in $\mathrm{Spec}(Z_\epsilon)$. ∎

Remarks [1] Because of part (i), the action of $\mathcal{G}$ on $\mathrm{Spec}(Z_0)$ is called the *quantum coadjoint action*.

[2] Part (ii) makes sense since every conjugacy class in G intersects G^0 (because, for example, every element of G is conjugate to an element of the Borel subgroup $B_+ = HN_+$).

[3] Corollary 9.2.19 is analogous to the description of the centre of the universal enveloping algebra in characteristic p – see Kac & Weisfeiler (1976).

[4] One can show that the image of the map $U_\epsilon^{0\bar{W}} \to Z_\epsilon$ described at the beginning of Subsection 9.2B is exactly the $\mathcal{G}$-invariant part of Z_ϵ. △

We conclude this section by showing that the Poisson–Lie group structure on $\mathrm{Spec}(Z_0)$ described in the previous subsection is closely related to that on G defined in 1.3.8. In fact, when G is equipped with the Poisson–Lie structure described in that example (or any non-zero multiple of it), it is clear that the subgroup of $G \times G$ given by

$$G^* = \{(n_-t, n_+t^{-1}) \mid n_\pm \in N_\pm, t \in H\}$$

is a dual Poisson–Lie group of G. Define a map $\psi : \mathrm{Spec}(Z_0) \to G^*$ by

$$\psi(u_-, t, u_+) = ((\boldsymbol{X}^-(u_-)t, \boldsymbol{X}^+(u_+)^{-1}t^{-1}).$$

Note that $\pi = \tilde{\pi} \circ \psi$, where $\tilde{\pi} : G^* \to G^0$ is the unramified covering of degree 2^n given by $\tilde{\pi}(b_-, b_+) = b_-(b_+)^{-1}$. We have

THEOREM 9.2.20 *Let G be a complex simple Lie group equipped with the Poisson–Lie group structure defined in 1.3.8, but with the standard bilinear form on the Lie algebra of G replaced by the Killing form. Let G^* be the dual Poisson–Lie group of G defined above. Then:*

(i) the map $\psi : \mathrm{Spec}(Z_0) \to G^*$ is an isomorphism of Poisson–Lie groups;

(ii) the orbits of $\mathcal{G}$ on $\mathrm{Spec}(Z_0)$ are precisely its symplectic leaves. ∎

If we recall from Subsection 1.5B that the symplectic leaves of Poisson–Lie groups are the orbits of the dressing transformations, it is tempting to think that the action of $\mathcal{G}$ on $\mathrm{Spec}(Z_0)$ should factor through the dressing action of G on G^*. However, there does not seem to be any homomorphism $\mathcal{G} \to G$ with this property.

9.3 The restricted specialization

In this section, we discuss the second of the two specializations mentioned in the introduction.

A The restricted integral form The restricted integral form is defined as follows. As usual, $q_i = q^{d_i}$.

DEFINITION–PROPOSITION 9.3.1 *The algebra $U_{\mathcal{A}}^{\mathrm{res}}(\mathfrak{g})$ is the $\mathcal{A}$-subalgebra of $U_q(\mathfrak{g})$ generated by the elements $(X_i^+)^{(r)}$, $(X_i^-)^{(r)}$ and $K_i^{\pm 1}$ for $1 \leq i \leq n$ and $r \geq 1$. It is an integral form of $U_q(\mathfrak{g})$ and a Hopf algebra over $\mathcal{A}$ with comultiplication given by*

$$\Delta_q(K_i) = K_i \otimes K_i,$$

$$\Delta_q((X_i^+)^{(r)}) = \sum_{k=0}^{r} q_i^{-k(r-k)}(X_i^+)^{(k)} \otimes K_i^k(X_i^+)^{(r-k)},$$

$$\Delta_q((X_i^-)^{(r)}) = \sum_{k=0}^{r} q_i^{k(r-k)}(X_i^-)^{(k)} \otimes (X_i^-)^{(r-k)}K_i^{-k},$$

counit given by

$$\epsilon_q(K_i) = 1, \qquad \epsilon_q((X_i^{\pm})^{(r)}) = 0.$$

and antipode given by

$$S_q(K_i) = K_i^{-1},$$

$$S_q((X_i^+)^{(r)}) = (-1)^r q_i^{r(r+1)} K_i^{-r}(X_i^+)^{(r)},$$

$$S_q((X_i^-)^{(r)}) = (-1)^r q_i^{-r(r+1)}(X_i^-)^{(r)} K_i^r. \quad \blacksquare$$

The proof that $U_{\mathcal{A}}^{\mathrm{res}}$ really is an integral form of U_q is non-trivial. It proceeds by constructing an $\mathcal{A}$-basis of $U_{\mathcal{A}}^{\mathrm{res}}$, analogous to the $\mathbb{Q}(q)$-basis of U_q given in 9.1.3. To do this, one first shows that the root vectors $X_{\beta}^{\pm}$ defined in Section 9.1 lie in $U_{\mathcal{A}}^{\mathrm{res}}$. This follows from the following lemma.

LEMMA 9.3.2 *For $i = 1, \ldots, n$, $T_i(U_{\mathcal{A}}^{\mathrm{res}}) = U_{\mathcal{A}}^{\mathrm{res}}$.*

PROOF It suffices to show that T_i and T_i^{-1} take each generator of $U_{\mathcal{A}}^{\mathrm{res}}$ into $U_{\mathcal{A}}^{\mathrm{res}}$. This is a straightforward computation. $\blacksquare$

Fix a reduced expression $w_0 = s_{i_1} \ldots s_{i_N}$ of the longest element of the Weyl group of $\mathfrak{g}$ and, for $r \in \mathbb{N}$, set

$$(X_{\beta_k}^{\pm})^{(r)} = T_{i_1} \ldots T_{i_{k-1}}((X_{i_k}^{\pm})^{(r)}).$$

Let $U_{\mathcal{A}}^{\mathrm{res}\,+}$ (resp. $U_{\mathcal{A}}^{\mathrm{res}\,-}$) be the $\mathcal{A}$-subalgebra of $U_{\mathcal{A}}^{\mathrm{res}}$ generated by the $(X_i^+)^{(r)}$ (resp. by the $(X_i^-)^{(r)}$) for all $i = 1, \ldots, n$, $r \in \mathbb{N}$, and let $U_{\mathcal{A}}^{\mathrm{res}\,0}$ be the $\mathcal{A}$-subalgebra of $U_{\mathcal{A}}^{\mathrm{res}}$ generated by the $K_i^{\pm 1}$ and the

$$\begin{bmatrix} K_i; c \\ r \end{bmatrix}_{q_i} = \prod_{s=1}^{r} \frac{K_i q_i^{c+1-s} - K_i^{-1} q_i^{s-1-c}}{q_i^s - q_i^{-s}}$$

for all $i = 1, \ldots, n$, $c \in \mathbb{Z}$ and $r \in \mathbb{N}$. (It follows from (13) below that these elements of U_q^0 do, in fact, belong to $U_{\mathcal{A}}^{\mathrm{res}}$.) Then, we have the following result, which includes 9.1.3 as a special case.

PROPOSITION 9.3.3 *For any choice of reduced decomposition of w_0,*
(i) the set of products

$$(X_{\beta_N}^+)^{(t_N)}(X_{\beta_{N-1}}^+)^{(t_{N-1})}\ldots(X_{\beta_1}^+)^{(t_1)},$$

where $t_1, t_2, \ldots, t_N \in \mathbb{N}$, form an $\mathcal{A}$-basis of $U_{\mathcal{A}}^{\mathrm{res}\,+}$ and a $\mathbb{Q}(q)$-basis of $U_q^{\mathrm{res}\,+}$;
(ii) the set of products

$$(X_{\beta_1}^-)^{(r_1)}(X_{\beta_2}^-)^{(r_2)}\ldots(X_{\beta_N}^-)^{(r_N)},$$

where $r_1, r_2, \ldots, r_N \in \mathbb{N}$, form an $\mathcal{A}$-basis of $U_{\mathcal{A}}^{\mathrm{res}\,-}$ and a $\mathbb{Q}(q)$-basis of $U_q^{\mathrm{res}\,-}$;
(iii) the set of products

$$\prod_{i=1}^{n} K_i^{\sigma_i} \begin{bmatrix} K_i; 0 \\ s_i \end{bmatrix}_{q_i},$$

where $s_1, \ldots, s_n \in \mathbb{N}$ and each σ_i is zero or one, form an $\mathcal{A}$-basis of $U_{\mathcal{A}}^{\mathrm{res}\,0}$ and a $\mathbb{Q}(q)$-basis of $U_q^{\mathrm{res}\,0}$;
(iv) multiplication defines an isomorphism of $\mathcal{A}$-modules

$$U_{\mathcal{A}}^{\mathrm{res}\,-} \underset{\mathcal{A}}{\otimes} U_{\mathcal{A}}^{\mathrm{res}\,0} \underset{\mathcal{A}}{\otimes} U_{\mathcal{A}}^{\mathrm{res}\,+} \to U_{\mathcal{A}}^{\mathrm{res}}.$$

In particular, $U_{\mathcal{A}}^{\mathrm{res}\pm}$, $U_{\mathcal{A}}^{\mathrm{res}\,0}$ and $U_{\mathcal{A}}^{\mathrm{res}}$ are free $\mathcal{A}$-modules. ∎

Remarks [1] The bases described in this proposition depend on the choice of the reduced expression of w_0. We shall describe in Chapter 14 certain 'canonical' bases which are independent of any such choices.

[2] The importance of the elements $\begin{bmatrix} K_i; c \\ s_i \end{bmatrix}_{q_i}$ derives from the following identity in $U_{\mathcal{A}}^{\mathrm{res}}$:

(13) $$(X_i^+)^{(r)}(X_i^-)^{(s)} = \sum_{r,s \geq t \geq 0} (X_i^-)^{(s-t)} \begin{bmatrix} K_i; 2t - s - r \\ t \end{bmatrix}_{q_i} (X_i^+)^{(r-t)}.$$

This can be proved by induction on r for fixed s, after first proving the case $r = 1$ by induction on s. △

Using the elements introduced above, one can give a description of $U_{\mathcal{A}}^{\mathrm{res}}$ in terms of generators and relations. For simplicity, we restrict ourselves to the case where $\mathfrak{g}$ is simply laced, i.e. $d_i = 1$ for all i (thus, $\mathfrak{g}$ is of type A, D or E). We number the nodes of the Dynkin diagram of $\mathfrak{g}$ as in Bourbaki (1968), Chaps 4, 5 and 6, except in the D_n case, when we compose the numbering in Planche V with the permutation $i \mapsto n - i + 1$. For any $\alpha \in \Delta^+$, let $g(\alpha)$ be the largest index of any simple root which occurs in α, let c_α be the coefficient of that simple root, and let $h(\alpha) = c_\alpha^{-1}\mathrm{height}(\alpha)$.

THEOREM 9.3.4 *Suppose that $\mathfrak{g}$ is of type A, D or E. Then $U_{\mathcal{A}}^{\text{res}}$ is isomorphic to the associative $\mathcal{A}$-algebra with generators $(X_\alpha^\pm)^{(r)}$, $K_i^{\pm 1}$ and $\left[\begin{array}{c} K_i;c \\ t \end{array}\right]_q$, for $\alpha \in \Delta^+$, $i = 1, \ldots, n$, $c \in \mathbb{Z}$ and $r, t \in \mathbb{N}$, and the following defining relations:*

the generators $K_i^{\pm 1}$, $\left[\begin{array}{c} K_i;0 \\ t_i \end{array}\right]_q$ commute with each other,

$$K_i K_i^{-1} = K_i^{-1} K_i = 1, \quad \left[\begin{array}{c} K_i;0 \\ 0 \end{array}\right]_q = 1,$$

$$\sum_{0 \le t \le s} (-1)^t q^{r(s-t)} \left[\begin{array}{c} r+t-1 \\ t \end{array}\right]_q K_i^t \left[\begin{array}{c} K_i;0 \\ r \end{array}\right]_q \left[\begin{array}{c} K_i;0 \\ s-t \end{array}\right]_q$$
$$= \left[\begin{array}{c} r+s \\ r \end{array}\right]_q \left[\begin{array}{c} K_i;0 \\ r+s \end{array}\right]_q, \quad r > 0, \, s \ge 0,$$

$$\left[\begin{array}{c} K_i;-c \\ r \end{array}\right]_q = \sum_{0 \le s \le r} (-1)^s q^{c(r-s)} \left[\begin{array}{c} c+s-1 \\ s \end{array}\right]_q K_i^s \left[\begin{array}{c} K_i;0 \\ r-s \end{array}\right]_q, \quad r \ge 0, \, c > 0,$$

$$\left[\begin{array}{c} K_i;c \\ r \end{array}\right]_q = \sum_{0 \le s \le r} q^{c(r-s)} \left[\begin{array}{c} c \\ s \end{array}\right]_q K_i^{-s} \left[\begin{array}{c} K_i;0 \\ r-s \end{array}\right]_q, \quad s \ge 0, \, c \ge 0,$$

$$(X_{\alpha_i}^+)^{(r)} (X_{\alpha_j}^-)^{(s)} = (X_{\alpha_j}^-)^{(s)} (X_{\alpha_i}^+)^{(r)}, \text{ if } i \ne j,$$

$$(X_{\alpha_i}^+)^{(r)} (X_{\alpha_i}^-)^{(s)} = \sum_{0 \le t \le r,s} (X_{\alpha_i}^-)^{(s-t)} \left[\begin{array}{c} K_i;2t-s-r \\ t \end{array}\right]_q (X_{\alpha_i}^+)^{(r-t)},$$

$$K_i (X_{\alpha_j}^\pm)^{(r)} K_i^{-1} = q^{\pm a_{ij} r} (X_{\alpha_j}^\pm)^{(r)},$$

$$\left[\begin{array}{c} K_i;c \\ r \end{array}\right]_q (X_{\alpha_j}^\pm)^{(s)} = (X_{\alpha_j}^\pm)^{(s)} \left[\begin{array}{c} K_i;c \pm a_{ij}s \\ r \end{array}\right]_q,$$

$$(X_\alpha^\pm)^{(r)} (X_\alpha^\pm)^{(s)} = \left[\begin{array}{c} r+s \\ r \end{array}\right]_q (X_\alpha^\pm)^{(r+s)}, \quad (X_\alpha^\pm)^{(0)} = 1,$$

$$(X_{\alpha_i}^\pm)^{(r)} (X_\alpha^\pm)^{(s)} = (X_\alpha^\pm)^{(s)} (X_{\alpha_i}^\pm)^{(r)} \text{ if } (\alpha, \alpha_i) = 0, \, i < g(\alpha), \, h(\alpha) \in \mathbb{Z},$$

$$(X_{\alpha'}^+)^{(r)} (X_\alpha^+)^{(s)} = \sum_{r,s \ge t \ge 0} q^{t+(r-t)(s-t)} (X_\alpha^+)^{(s-t)} (X_{\alpha+\alpha'}^+)^{(t)} (X_{\alpha'}^+)^{(r-t)},$$

$$(X_\alpha^-)^{(r)} (X_{\alpha'}^-)^{(s)} = \sum_{r,s \ge t \ge 0} q^{-t-(r-t)(s-t)} (X_{\alpha'}^-)^{(s-t)} (X_{\alpha+\alpha'}^-)^{(t)} (X_\alpha^-)^{(r-t)},$$

$$q^{rs} (X_{\alpha'}^\pm)^{(r)} (X_{\alpha+\alpha'}^\pm)^{(s)} = (X_{\alpha+\alpha'}^\pm)^{(s)} (X_{\alpha'}^\pm)^{(r)},$$

$$q^{rs}(X_{\alpha+\alpha'}^{\pm})^{(r)}(X_{\alpha}^{\pm})^{(s)} = (X_{\alpha}^{\pm})^{(s)}(X_{\alpha+\alpha'}^{\pm})^{(r)},$$

where, in the last four relations, we assume that $(\alpha, \alpha') = -1$ and either

$$\alpha' = \alpha_i \text{ and } i < g(\alpha)$$

or

$$\text{height}(\alpha') = \text{height}(\alpha) + 1 \text{ and } g(\alpha') = g(\alpha). \quad \blacksquare$$

As in the non-restricted case, one can now define, for any $\epsilon \in \mathbb{C}^\times$, the *restricted specializations*

$$U_{\epsilon,\mathbb{Z}}^{\text{res}} = U_{\mathcal{A}}^{\text{res}} \underset{\mathcal{A}}{\otimes} \mathbb{Z}[\epsilon, \epsilon^{-1}], \quad U_{\epsilon}^{\text{res}} = U_{\mathcal{A}}^{\text{res}} \underset{\mathcal{A}}{\otimes} \mathbb{Q}(\epsilon)$$

(it is important for later results to work over $\mathbb{Z}[\epsilon, \epsilon^{-1}]$ or $\mathbb{Q}(\epsilon)$ rather than $\mathbb{C}$). Hopf algebras $U_{\epsilon}^{\text{res}\pm}$, $U_{\epsilon,\mathbb{Z}}^{\text{res}\pm}$, $U_{\epsilon}^{\text{res}\,0}$ and $U_{\epsilon,\mathbb{Z}}^{\text{res}\,0}$ are defined in a similar way.

If ϵ is not a root of unity, it is clear that the restricted and non-restricted specializations coincide,

$$U_\epsilon \cong U_\epsilon^{\text{res}} \underset{\mathbb{Q}(\epsilon)}{\otimes} \mathbb{C},$$

since in that case $[r]_\epsilon \neq 0$ for all $r \geq 1$. But if ϵ is a root of unity, these two algebras are not isomorphic, as will become clear later in this section.

Note From now until the end of Subsection 9.3C, we assume that ϵ is a primitive ℓth root of unity, where ℓ is odd and greater than d_i for all i.

The following result is the counterpart of 9.2.6 in the restricted case.

PROPOSITION 9.3.5 *In $U_{\epsilon,\mathbb{Z}}^{\text{res}}$, we have*

(i) $(X_\alpha^\pm)^\ell = 0$ *for all* $\alpha \in \Delta^+$;

(ii) K_i^ℓ *is central and* $K_i^{2\ell} = 1$ *for* $i = 1, \ldots, n$.

PROOF Part (i) is obvious, since $(X_\alpha^\pm)^\ell = (X_\alpha^\pm)^{(\ell)}[\ell]_q!$ in $U_{\mathcal{A}}^{\text{res}}$, and $[\ell]_\epsilon! = 0$. The first assertion in (ii) is also obvious, and the second follows on specializing the relation

$$\prod_{r=1}^{\ell} (K_i q_i^{1-r} - K_i^{-1} q_i^{r-1}) = \prod_{r=1}^{\ell} (q_i^r - q_i^{-r}) \cdot \begin{bmatrix} K_i; 0 \\ \ell \end{bmatrix}_{q_i}$$

in $U_{\mathcal{A}}^{\text{res}}$. $\quad \blacksquare$

Remarks [1] It is clear from its definition that, if $0 \leq s_i < \ell$, $\begin{bmatrix} K_i;0 \\ s_i \end{bmatrix}_{\epsilon_i}$ lies in the $\mathbb{Q}(\epsilon)$-subalgebra of U_ϵ^{res} generated by $K_i^{\pm 1}$, and that, for all $s_i \in \mathbb{N}$,

$$\begin{bmatrix} K_i; 0 \\ s_i + \ell \end{bmatrix}_{\epsilon_i} = \begin{bmatrix} K_i; 0 \\ s_i \end{bmatrix}_{\epsilon_i} \begin{bmatrix} K_i; 0 \\ \ell \end{bmatrix}_{\epsilon_i}.$$

It follows that $U_\epsilon^{\mathrm{res},0}$ is generated as a $\mathbb{Q}(\epsilon)$-algebra by the $K_i^{\pm 1}$ and the $\left[\begin{smallmatrix} K_i;0 \\ \ell \end{smallmatrix}\right]_{\epsilon_i}$.

[2] We define the restricted integral form and specialization of the coordinate algebra $\mathcal{F}_q$ as follows (cf. Remark [5] at the end of Subsection 9.1A). Let $U_\mathcal{A}^*$ be the set of all $\mathbb{Q}[q,q^{-1}]$-linear maps $U_\mathcal{A}^{\mathrm{res}} \underset{\mathcal{A}}{\otimes} \mathbb{Q}[q,q^{-1}] \to \mathbb{Q}[q,q^{-1}]$, and let $\mathcal{F}_\mathcal{A} = \mathcal{F}_q \cap U_\mathcal{A}^*$. We define

$$\mathcal{F}_\epsilon = \mathcal{F}_\mathcal{A} \underset{\mathbb{Q}[q,q^{-1}]}{\otimes} \mathbb{Q}(\epsilon).$$

Taking $\epsilon = 1$, one can show that

$$\mathcal{F}_\mathbb{Q} = \mathcal{F}_\mathcal{A} \underset{\mathbb{Q}[q,q^{-1}]}{\otimes} \mathbb{Q}$$

is the coordinate algebra of the simply-connected complex algebraic group with Lie algebra $\mathfrak{g}$. $\triangle$

B A remarkable finite-dimensional Hopf algebra In this subsection, we show that $U_\epsilon^{\mathrm{res}}$ can be 'factorized' into a kind of 'product' of the classical universal enveloping algebra and a very interesting finite-dimensional Hopf algebra. This has important implications for the representation theory of $U_\epsilon^{\mathrm{res}}$, as we shall see in Chapter 11. We define this finite-dimensional Hopf algebra in this subsection, and describe the factorization in the next.

The origin of the factorization is the following property of ϵ-binomial coefficients, which is easily deduced from the identity

$$\prod_{k=0}^{m-1}(1 + \epsilon^{2k}\xi) = \sum_{k=0}^{m} \epsilon^{k(m-1)} \begin{bmatrix} m \\ k \end{bmatrix}_\epsilon \xi^k,$$

in $\mathbb{Q}(\epsilon)[\xi]$, ξ being an indeterminate, and the observation that

$$\prod_{k=0}^{\ell-1}(1 + \epsilon^{2k}\xi) = 1 + \xi^\ell.$$

LEMMA 9.3.6 *Let ϵ be a primitive ℓth root of unity, where ℓ is odd and greater than one. Let r, $s \in \mathbb{N}$, and write $r = r_0 + \ell r_1$, $s = s_0 + \ell s_1$, where $0 \le r_0, s_0 < \ell$, $r_1, s_1 \ge 0$, $r \ge s$. Then,*

$$\begin{bmatrix} r \\ s \end{bmatrix}_\epsilon = \begin{bmatrix} r_0 \\ s_0 \end{bmatrix}_\epsilon \begin{pmatrix} r_1 \\ s_1 \end{pmatrix},$$

where $\begin{pmatrix} r_1 \\ s_1 \end{pmatrix}$ is an ordinary binomial coefficient. ∎

Now, in $U_\mathcal{A}^{\mathrm{res}}$ we have the following evident identity (already noted in 9.3.4):

$$(X_i^\pm)^{(r)}(X_i^\pm)^{(s)} = \begin{bmatrix} r+s \\ r \end{bmatrix}_{q_i} (X_i^\pm)^{(r+s)}.$$

Using the lemma, we deduce that, in U_ϵ^{res},

$$(X_i^\pm)^{(r)} = (X_i^\pm)^{(r_0)} \frac{((X_i^\pm)^{(\ell)})^{r_1}}{r_1!}.$$

It follows that U_ϵ^{res} is generated as a $\mathbb{Q}(\epsilon)$-algebra by the $X_i^\pm$, $(X_i^\pm)^{(\ell)}$, $K_i^{\pm 1}$ and $\begin{bmatrix} K_i;0 \\ r_i \end{bmatrix}$, where $0 \le r_i < \ell$ and $i = 1,\ldots,n$. The factorization will essentially be into the part generated by the $X_i^\pm$ and the remainder consisting of the ℓth divided powers.

More precisely, let U_ϵ^{fin} (resp. $U_{\epsilon,\mathbb{Z}}^{\text{fin}}$) be the $\mathbb{Q}(\epsilon)$-subalgebra of U_ϵ^{res} (resp. the $\mathbb{Z}[\epsilon,\epsilon^{-1}]$-subalgebra of $U_{\epsilon,\mathbb{Z}}^{\text{res}}$) generated by the $X_i^\pm$, $K_i^{\pm 1}$ and $\begin{bmatrix} K_i;0 \\ r_i \end{bmatrix}_{\epsilon_i}$, where $0 \le r_i < \ell$, $i = 1,\ldots,n$. Define subalgebras $U_{\epsilon,\mathbb{Z}}^{\text{fin}-}$, $U_{\epsilon,\mathbb{Z}}^{\text{fin}\,0}$, and $U_{\epsilon,\mathbb{Z}}^{\text{fin}+}$ of $U_{\epsilon,\mathbb{Z}}^{\text{fin}}$ in the obvious way. The following result is analogous to 9.3.3 (although it is not a consequence of it).

PROPOSITION 9.3.7 *Suppose that* $\mathfrak{g}$ *is of type* A, D *or* E, *and that* ϵ *is a primitive ℓth root of unity, where ℓ is odd and greater than one. Then, for any choice of reduced decomposition of* w_0,
(i) $U_{\epsilon,\mathbb{Z}}^{\text{fin}+}$ *is a free* $\mathbb{Z}[\epsilon,\epsilon^{-1}]$-*module of rank* ℓ^N *with basis the set of products*

$$(X_{\beta_N}^+)^{(t_N)}(X_{\beta_{N-1}}^+)^{(t_{N-1})} \ldots (X_{\beta_1}^+)^{(t_1)},$$

where $0 \le t_1, t_2, \ldots, t_N < \ell$;
(ii) $U_{\epsilon,\mathbb{Z}}^{\text{fin}-}$ *is a free* $\mathbb{Z}[\epsilon,\epsilon^{-1}]$-*module of rank* ℓ^N *with basis the set of products*

$$(X_{\beta_1}^-)^{(r_1)}(X_{\beta_2}^-)^{(r_2)} \ldots (X_{\beta_N}^-)^{(r_N)},$$

where $0 \le r_1, r_2, \ldots, r_N < \ell$;
(iii) $U_{\epsilon,\mathbb{Z}}^{\text{fin}\,0}$ *is a free* $\mathbb{Z}[\epsilon,\epsilon^{-1}]$-*module of rank* $(2\ell)^n$ *with basis the set of products*

$$\prod_{i=1}^n K_i^{\sigma_i} \begin{bmatrix} K_i;0 \\ s_i \end{bmatrix}_\epsilon,$$

where $0 \le s_i < \ell$ *and each* σ_i *is zero or one*;
(iv) *multiplication defines an isomorphism of* $\mathbb{Z}[\epsilon,\epsilon^{-1}]$-*modules*

$$U_{\epsilon,\mathbb{Z}}^{\text{fin}-} \otimes U_{\epsilon,\mathbb{Z}}^{\text{fin}\,0} \otimes U_{\epsilon,\mathbb{Z}}^{\text{fin}+} \to U_{\epsilon,\mathbb{Z}}^{\text{fin}}. \quad\blacksquare$$

A similar, but more complicated, result holds when $\mathfrak{g}$ is not simply laced.

Remark One shows in a similar way that the Hopf algebra U_ϵ^{fin} has finite dimension $2^n \ell^{2N+n}$ over $\mathbb{Q}(\epsilon)$. The basis is as in the preceding proposition. In this case, however, the elements in part (iii) may be replaced by the products

$$\prod_{i=1}^n K_i^{s_i},$$

where $0 \leq s_i < 2\ell$, since the $\begin{bmatrix} K_i;0 \\ s_i \end{bmatrix}_\epsilon$ for $0 \leq s_i < \ell$ clearly lie in the $\mathbb{Q}(\epsilon)$-subalgebra of $U_\epsilon^{\mathrm{fin}}$ generated by the $K_i^{\pm 1}$, and since $K_i^{2\ell} = 1$ (we are still assuming that $\mathfrak{g}$ is simply laced). $\triangle$

One can also describe $U_\epsilon^{\mathrm{fin}}$ in terms of generators and relations.

PROPOSITION 9.3.8 *Suppose that $\mathfrak{g}$ is of type A, D or E. Then, $U_\epsilon^{\mathrm{fin}}$ is isomorphic to the associative $\mathbb{Q}(\epsilon)$-algebra with generators $X_\alpha^\pm$ and $K_i^{\pm 1}$, for $\alpha \in \Delta^+$, $i = 1, \ldots, n$, and the following defining relations:*

$$K_i K_j = K_j K_i, \quad K_i K_i^{-1} = K_i^{-1} K_i = 1,$$

$$X_{\alpha_i}^+ X_{\alpha_j}^- - X_{\alpha_j}^- X_{\alpha_i}^+ = \delta_{i,j} \frac{K_i - K_i^{-1}}{\epsilon - \epsilon^{-1}},$$

$$K_i X_{\alpha_j}^\pm K_i^{-1} = \epsilon^{\pm a_{ij}} X_{\alpha_j}^\pm,$$

$$X_{\alpha_i}^\pm X_\alpha^\pm = X_\alpha^\pm X_{\alpha_i}^\pm, \quad \text{if } (\alpha, \alpha_i) = 0, \ i < g(\alpha), \ h(\alpha) \in \mathbb{Z},$$

$$(X_\alpha^\pm)^\ell = 0, \quad K_i^{2\ell} = 1,$$

$$X_{\alpha'}^+ X_\alpha^+ = \epsilon X_\alpha^+ X_{\alpha'}^+ + \epsilon X_{\alpha+\alpha'}^+, \quad X_\alpha^- X_{\alpha'}^- = \epsilon^{-1} X_{\alpha'}^- X_\alpha^- + \epsilon^{-1} X_{\alpha+\alpha'}^-,$$

$$\epsilon X_{\alpha'}^\pm X_{\alpha+\alpha'}^\pm = X_{\alpha+\alpha'}^\pm X_{\alpha'}^\pm, \quad \epsilon X_{\alpha+\alpha'}^\pm X_\alpha^\pm = X_\alpha^\pm X_{\alpha+\alpha'}^\pm,$$

where, in the last three relations, we assume that $(\alpha, \alpha') = -1$, $i < g(\alpha)$ and either

$$\alpha' = \alpha_i$$

or

$$\mathrm{height}(\alpha') = \mathrm{height}(\alpha) + 1 \text{ and } g(\alpha') = g(\alpha). \quad \blacksquare$$

We conclude our discussion of $U_\epsilon^{\mathrm{fin}}$ with

PROPOSITION 9.3.9 *The finite-dimensional Hopf algebra $U_\epsilon^{\mathrm{fin}}$ is quasitriangular, with universal R-matrix $\mathcal{R}_\epsilon$ given by*

$$\frac{1}{\ell^n} \left(\sum_{\beta,\gamma \in Q_\ell} \epsilon^{-(\beta,\gamma)} K_\beta \otimes K_\gamma \right) \prod_{\alpha \in \Delta^+} \left(\sum_{r=0}^{\ell-1} \frac{(1 - \epsilon_\alpha^{-2})^r}{[r]_{\epsilon_\alpha}!} \epsilon_\alpha^{r(r+1)/2} (X_\alpha^+)^r \otimes (X_\alpha^-)^r \right).$$

Here, $\epsilon_\alpha = \epsilon^{d_i}$ if α is in the Weyl group orbit of the simple root α_i; $K_\beta = \prod_i K_i^{m_i}$ if $\beta = \sum_i m_i \alpha_i$; and

$$Q_\ell = \left\{ \sum_{i=1}^n m_i \alpha_i \ \Big| \ 0 \leq m_i < \ell \text{ for all } i \right\}. \quad \blacksquare$$

This formula is due to Rosso (1992c), who proved it by using the quantum double method (see Section 8.3). We shall omit the details, but we note that the formula cannot be deduced directly from the formula for the universal R-matrix of $U_h(\mathfrak{g})$ obtained in Chapter 8 since, as we observed at the end of Subsection 9.1B, that formula cannot be translated to U_q^{res}, and hence cannot be specialized.

It is unlikely that U_ϵ^{res} is quasitriangular in the strict sense, since the quantum double method would suggest that any potential R-matrix would be an infinite sum of elements of $U_\epsilon^{\text{res}} \otimes U_\epsilon^{\text{res}}$. See, however, Subsection 10.1D.

C A Frobenius map in characteristic zero The definition of U_ϵ^{res} mimics that of the Chevalley–Kostant $\mathbf{Z}$-form $U_{\mathbf{Z}}$ of the classical universal enveloping algebra U (over $\mathbb{C}$). We recall that $U_{\mathbf{Z}}$ is the subring of U generated by the divided powers

$$(\bar{X}^{\pm})^{(r)} = \frac{(\bar{X}^{\pm})^r}{r!}, \quad r \in \mathbf{N}.$$

(Here, and elsewhere in this section, we use a bar to distinguish the generators of U from those of U_ϵ^{res}.)

The following result gives a simple relation between $U_{\mathbf{Z}}$ and $U_{\mathcal{A}}^{\text{res}}$. Let

$$U_1^{\text{res}} = U_{\mathcal{A}}^{\text{res}} \underset{\mathcal{A}}{\otimes} \mathbf{Q}$$

be the Hopf algebra over $\mathbf{Q}$ obtained by specializing q to 1, and let $\tilde{U}_1^{\text{res}}$ be the quotient of U_1^{res} by the ideal generated by the $K_i - 1$ (it is clear that the K_i are central in U_1^{res}). The following result is the analogue for the restricted specialization of 9.2.3.

PROPOSITION 9.3.10 *There is an isomorphism of Hopf algebras over* $\mathbf{Q}$

$$U_{\mathbf{Q}} = U_{\mathbf{Z}} \underset{\mathbf{Z}}{\otimes} \mathbf{Q} \overset{\cong}{\to} \tilde{U}_1^{\text{res}}$$

which takes $\bar{X}_i^{\pm}$ *to* $X_i^{\pm}$. ∎

Suppose now that $p \in \mathbf{N}$ is an odd prime, and let $\mathbb{F}_p$ be the field with p elements. The Hopf algebra

$$U_{\mathbb{F}_p} = U_{\mathbf{Z}} \underset{\mathbf{Z}}{\otimes} \mathbb{F}_p$$

over $\mathbb{F}_p$ is the hyperalgebra of the algebraic group $G_{\mathbb{F}_p}$ over $\mathbb{F}_p$ associated to $\mathfrak{g}$. This is clearly analogous to the construction of $U_{\epsilon,\mathbf{Z}}^{\text{res}}$ from $U_{\mathcal{A}}^{\text{res}}$. The next result shows that this is more than an analogy.

Let $\tilde{U}_{\epsilon,\mathbf{Z}}^{\text{res}}$ be the quotient of $U_{\epsilon,\mathbf{Z}}^{\text{res}}$ by the ideal generated by the $K_i^\ell - 1$, and let $U_{\mathbb{F}_p}^{\text{fin}}$, the *restricted enveloping algebra*, be the subalgebra of $U_{\mathbb{F}_p}$ generated by the $\bar{X}_i^{\pm}$ (we shall see the reason for the term 'restricted' in Subsection 11.2B).

PROPOSITION 9.3.11 *Suppose that $\ell = p$ is an odd prime. Then, as Hopf algebras over* $\mathbb{F}_p$,

$$U_{\mathbb{F}_p} \cong \tilde{U}^{\text{res}}_{\epsilon,\mathbb{Z}} \underset{\mathbb{Z}[\epsilon,\epsilon^{-1}]}{\otimes} \mathbb{F}_p, \qquad U^{\text{fin}}_{\mathbb{F}_p} \cong \tilde{U}^{\text{fin}}_{\epsilon,\mathbb{Z}} \underset{\mathbb{Z}[\epsilon,\epsilon^{-1}]}{\otimes} \mathbb{F}_p. \quad \blacksquare$$

Note that, since ϵ is a primitive pth root of unity in this situation, there is a well-defined ring homomorphism $\mathbb{Z}[\epsilon, \epsilon^{-1}] \to \mathbb{F}_p$ which takes ϵ to 1.

Recall that there is a homomorphism of Hopf algebras over $\mathbb{F}_p$

$$\text{Fr} : U_{\mathbb{F}_p} \to U_{\mathbb{F}_p},$$

called the *Frobenius map*, such that

(14) $$\text{Fr}((\bar{X}^{\pm}_i)^{(r)}) = \begin{cases} (\bar{X}^{\pm}_i)^{(r/p)} & \text{if } p \text{ divides } r, \\ 0 & \text{otherwise}, \end{cases}$$

(see Jantzen (1987), for example). Moreover, the kernel of Fr is the two-sided ideal of $U_{\mathbb{F}_p}$ generated by the augmentation ideal of $U^{\text{fin}}_{\mathbb{F}_p}$ (the kernel of its counit).

The following result provides an analogous map for U^{res}_ϵ (in characteristic zero!). *For the remainder of this subsection, we assume, in addition to the usual conditions on ℓ, that if $\mathfrak{g}$ is of type G_2, then ℓ is not divisible by three.*

THEOREM 9.3.12 *(i) There is a unique homomorphism*

$$\text{Fr}_\epsilon : U^{\text{res}}_\epsilon \to U_{\mathbb{Q}} \underset{\mathbb{Q}}{\otimes} \mathbb{Q}(\epsilon)$$

of Hopf algebras over $\mathbb{Q}(\epsilon)$ *such that* $\text{Fr}_\epsilon(K_i) = 1$ *and*

$$\text{Fr}_\epsilon((X^{\pm}_i)^{(r)}) = \begin{cases} (\bar{X}^{\pm}_i)^{(r/\ell)} & \text{if } \ell \text{ divides } r, \\ 0 & \text{otherwise}. \end{cases}$$

(ii) The kernel of Fr_ϵ *is the two-sided ideal of* U^{res}_ϵ *generated by the augmentation ideal of* U^{fin}_ϵ *(the kernel of its counit).*

(iii) Fr_ϵ *restricts to a homomorphism of* $\mathbb{Z}[\epsilon, \epsilon^{-1}]$-*modules*

$$\text{Fr}_\epsilon : U^{\text{res}}_{\epsilon,\mathbb{Z}} \to U_{\mathbb{Z}} \underset{\mathbb{Z}}{\otimes} \mathbb{Z}[\epsilon, \epsilon^{-1}]. \quad \blacksquare$$

Remarks [1] Suppose that $\ell = p$ is an odd prime. Since $\text{Fr}_\epsilon(K_i) = 1$, there is an induced homomorphism

$$\text{Fr}_\epsilon : \tilde{U}^{\text{res}}_{\epsilon,\mathbb{Z}} \to U_{\mathbb{Z}} \underset{\mathbb{Z}}{\otimes} \mathbb{Z}[\epsilon, \epsilon^{-1}].$$

By 9.3.11, if we tensor over $\mathbb{F}_p$, the source and target algebras both become $U_{\mathbb{F}_p}$, and clearly Fr_ϵ becomes Fr. This gives the first proof, *working directly at the algebra level*, of the existence of a homomorphism $U_{\mathbb{F}_p} \to U_{\mathbb{F}_p}$ satisfying (14). The classical proof goes via the algebraic group $G_{\mathbb{F}_p}$ over $\mathbb{F}_p$ associated to $\mathfrak{g}$: taking the pth power of each of the matrix entries of $G_{\mathbb{F}_p}$ (in a suitable matrix representation) defines an endomorphism of the coordinate algebra $\mathbb{F}_p[G_{\mathbb{F}_p}]$; one checks that the dual of this map preserves the hyperalgebra $U_{\mathbb{F}_p} \subset \mathbb{F}_p[G_{\mathbb{F}_p}]^*$, and that it is given by the formulas in (14).

[2] Parts (i) and (ii) of the theorem make precise the 'factorization' property mentioned at the beginning of Subsection 9.3B.

[3] Dualizing the Frobenius map Fr_ϵ, we obtain an embedding

$$\mathcal{F}_\mathbb{Q} \underset{\mathbb{Q}}{\otimes} \mathbb{Q}(\epsilon) \subset \mathcal{F}_\epsilon$$

of Hopf algebras over $\mathbb{Q}(\epsilon)$. Thus, the classical coordinate algebra $\mathcal{F}_\mathbb{Q}$ appears as a Hopf subalgebra of the quantum coordinate algebra $\mathcal{F}_\epsilon$.

[4] Despite the analogies between $U_\epsilon^{\mathrm{res}}$ and $U_{\mathbb{F}_p}$, there are some differences. These are reflections of the differences between 9.3.6 and its characteristic p analogue. In fact, let $r, s \in \mathbb{N}$ with $r \geq s$, and write

$$r = \sum_{k=0}^{\infty} r_k p^k, \quad s = \sum_{k=0}^{\infty} s_k p^k,$$

where $0 \leq r_k, s_k < p$ (of course, all but finitely many of the r_k and s_k are zero). Then, we have the well-known congruence property

$$(15) \qquad \binom{r}{s} \equiv \binom{r_0}{s_0}\binom{r_1}{s_1}\binom{r_2}{s_2} \cdots \pmod{p}.$$

(Note that, even if $r \geq s$, it may happen that $r_k < s_k$ for some k; in this case, the binomial coefficient $\binom{r_k}{s_k}$ is understood to be zero.)

Although formula (15) closely resembles that in 9.3.6, note that there are only two factors on the right-hand side of the formula in 9.3.6, whereas there is no bound (independent of r and s) on the number of factors on the right-hand side of (15). This is reflected in the fact that $U_\epsilon^{\mathrm{res}}$ is finitely generated, but $U_{\mathbb{F}_p}$ is not (to generate the latter algebra, one needs all of the divided powers $\bar{X}_i^\pm, (\bar{X}_i^\pm)^{(p)}, (\bar{X}_i^\pm)^{(p^2)}, \dots$). Note also that, whereas all the factors on the right-hand side of (15) are of the same kind, those on the right-hand side of the formula in 9.3.6 are not, one being an ϵ-binomial coefficient and one an ordinary binomial coefficient. This is reflected in the fact that Fr is a map from $U_{\mathbb{F}_p}$ to itself, whereas Fr_ϵ is a map between two different algebras. In particular, while in the classical case one can consider the iterates Fr^k of the Frobenius map, in the quantum situation this is meaningless. $\triangle$

D The quiver approach We conclude this section by describing a combinatorial realization of the subalgebra $U_\epsilon^{\mathrm{res}\,\geq 0}$ of $U_\epsilon^{\mathrm{res}}$, for certain (algebraic) values of ϵ. We assume throughout this subsection that $\mathfrak{g}$ is simply laced, i.e. that the Dynkin diagram Γ of $\mathfrak{g}$ has only single bonds.

Fix an orientation of Γ, i.e. attach an arrow $\overset{i}{\bullet} \longrightarrow \overset{j}{\bullet}$ to each edge $\overset{i}{\bullet}\underline{\quad}\overset{j}{\bullet}$ of Γ. A *representation* ρ of Γ over a field $\mathbb{F}$ is a collection of finite-dimensional vector spaces V_i over $\mathbb{F}$, one for each vertex (or node) of Γ, and $\mathbb{F}$-linear maps $T_{ij} : V_i \to V_j$, one for each oriented edge $\overset{i}{\bullet}\longrightarrow\overset{j}{\bullet}$ of Γ. The *dimension* of ρ is the n-tuple $\dim(\rho) = (\dim(V_1), \dim(V_2), \ldots, \dim(V_n))$. We shall often identify $\dim(\rho)$ with the element $\sum_i \dim(V_i)\,\alpha_i \in Q$.

If ρ' is another representation of Γ, given by vector spaces V_i' and linear maps T_{ij}', a *morphism* from ρ to ρ' is a collection of $\mathbb{F}$-linear maps $\phi_i : V_i \to V_i'$ such that the diagram

$$
\begin{array}{ccc}
V_i & \xrightarrow{\;T_{ij}\;} & V_j \\
\phi_i \downarrow & & \downarrow \phi_j \\
V_i' & \xrightarrow[\;T_{ij}'\;]{} & V_j'
\end{array}
$$

commutes for every oriented edge $\overset{i}{\bullet}\longrightarrow\overset{j}{\bullet}$ of Γ. Subrepresentations, quotient representations, direct sums of representations, etc., are defined in the obvious way. This makes the category of representations of Γ and morphisms between them into an abelian category $\mathbf{rep}_\Gamma$.

The simplest examples of representations of Γ, apart from the zero representation for which $V_i = 0$ for all i, are the representations e_i for which

$$
V_j = \begin{cases} \mathbb{F} & \text{if } j = i, \\ 0 & \text{otherwise.} \end{cases}
$$

This representation is obviously irreducible, and every irreducible representation of Γ is isomorphic to some e_i. The category $\mathbf{rep}_\Gamma$ is not semisimple, however, for the representations e_{ij}, defined, for every oriented edge $\overset{i}{\bullet}\longrightarrow\overset{j}{\bullet}$ of Γ, by

$$
V_k = \begin{cases} \mathbb{F} & \text{if } k = i \text{ or } j, \\ 0 & \text{otherwise,} \end{cases} \qquad T_{ij} = \mathrm{id},
$$

are clearly indecomposable but not irreducible. In general, one has the following fundamental result of Gabriel (1972).

THEOREM 9.3.13 *A representation ρ of Γ is indecomposable if and only if $\dim(\rho) \in \Delta^+$. Conversely, for every $\alpha \in \Delta^+$, there is, up to isomorphism, a unique indecomposable representation ρ of Γ such that $\dim(\rho) = \alpha$.* ∎

Remark The oriented Dynkin graphs Γ are examples of *quivers*. In general, a quiver Υ is a set of vertices joined by (oriented) arrows, but possibly with more than one arrow joining each pair of vertices:

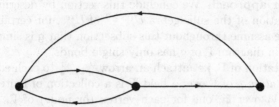

One can then define an abelian category **rep$_{\Upsilon}$** of representations of **Υ** exactly as above. It is known that every finite-dimensional representation of **Υ** can be written uniquely as a sum of indecomposables (in particular, this holds for Γ).

According to another fundamental result of Gabriel (1972), the oriented Dynkin graphs are characterized as the quivers **Υ** for which **rep$_{\Upsilon}$** has only finitely-many indecomposables. $\triangle$

Suppose from now on that **$\mathbb{F}$** is a finite field with m elements. Let $\text{Rep}(\Gamma)$ be the complex vector space with basis the set of isomorphism classes of representations of Γ, and define a bilinear product on $\text{Rep}(\Gamma)$ by

$$\rho'\rho'' = \sum_{\rho} c^{\rho}_{\rho',\rho''}\,\rho,$$

where $c^{\rho}_{\rho',\rho''}$ is the number of subrepresentations of ρ isomorphic to ρ'' for which the quotient representation is isomorphic to ρ'. This gives $\text{Rep}(\Gamma)$ the structure of an associative algebra with unit element (the zero representation), called the *Hall algebra* associated to Γ.

As an exercise, the reader may like to verify the following relations in $\text{Rep}(\Gamma)$:

$$e_{ij} = e_i e_j - e_j e_i, \quad e_i e_{ij} = m e_{ij} e_i, \quad e_{ij} e_i = m e_j e_{ij}.$$

It follows that

$$e_i^2 e_j - (m+1)e_i e_j e_i + m e_j e_i^2 = 0,$$
$$e_i e_j^2 - (m+1)e_j e_i e_j + m e_j^2 e_i = 0.$$

We note the similarity between these identities and some of the defining relations of $U_q(\mathfrak{g})$ in 9.1.1.

To make this observation precise, let

$$\widetilde{\text{Rep}(\Gamma)} = U^{\text{res }0}_{m^{1/2}} \underset{\mathbb{C}}{\otimes} \text{Rep}(\Gamma)$$

(as a complex vector space), and define an associative algebra structure on $\widetilde{\text{Rep}(\Gamma)}$ by requiring that $U^{\text{res }0}_{m^{1/2}}$ and $\text{Rep}(\Gamma)$ are subalgebras, and that

$$K_i \rho K_i^{-1} = m^{\frac{1}{2}\sum_j a_{ij} r_j}\,\rho$$

for $\rho \in \mathrm{Rep}(\Gamma)$, where (a_{ij}) is the Cartan matrix of $\mathfrak{g}$ and $\dim(\rho) = (r_1, \ldots, r_n)$. Then, we have the following result of Ringel (1990).

THEOREM 9.3.14 *Choose integers* $k_1, \ldots, k_n$ *such that* $k_i - k_j = 1$ *whenever* $\overset{i}{\bullet} \longrightarrow \overset{j}{\bullet}$ *is an oriented edge of* Γ. *Then, there is an algebra isomorphism*

$$U_{m^{1/2}}^{\mathrm{res}\,\geq 0} \to \widetilde{\mathrm{Rep}(\Gamma)}$$

which takes $K_i^{\pm 1}$ *to* $K_i^{\pm 1}$ *and* X_i^+ *to* $K_i^{k_i} e_i$ *for all* i. ∎

Note that, since $m^{1/2}$ is not a root of unity, $U_{m^{1/2}}^{\mathrm{res}\,\geq 0}$ is generated by the $K_i^{\pm 1}$ and the X_i^+, so the isomorphism in the theorem is uniquely determined.

Theorem 9.3.14 is important for the construction of 'canonical' bases in Chapter 14.

9.4 Automorphisms and real forms

The main purpose of this section is to classify the real forms of the Hopf algebras $U_\epsilon(\mathfrak{g})$. This is necessary for the discussion of unitary representations of quantum groups in Chapters 10 and 13. In the case in which ϵ is a primitive ℓth root of unity, we assume that ℓ is odd and $\ell > 2d_i$ for all i.

A Automorphisms We begin by determining all the automorphisms of $U_\epsilon(\mathfrak{g})$ as a Hopf algebra. Recall that a *diagram automorphism* of $\mathfrak{g}$ is a permutation μ of the set $\{1, \ldots, n\}$ of vertices of the Dynkin diagram of $\mathfrak{g}$ such that $a_{\mu(i)\mu(j)} = a_{ij}$ for all i, j.

PROPOSITION 9.4.1 *Let* $c_1, \ldots, c_n \in \mathbb{C}^\times$, *and let* μ *be a diagram automorphism of* $\mathfrak{g}$. *Then, there is a unique Hopf algebra automorphism* ρ *of* $U_\epsilon(\mathfrak{g})$ *such that*

$$\rho(K_i) = K_{\mu(i)}, \quad \rho(X_i^\pm) = c_i^{\pm 1} X_{\mu(i)}^\pm,$$

and every Hopf algebra automorphism of $U_\epsilon(\mathfrak{g})$ *is of this form.* ∎

We shall comment on the proof of 9.4.1 after the statement of the next result.

B Real forms The definition of a real form (or ∗-structure) on a Hopf algebra may be found in Subsection 4.1F.

PROPOSITION 9.4.2 *The Hopf algebra* $U_\epsilon(\mathfrak{g})$ *has real forms if and only if* $|\epsilon| = 1$, $\epsilon \in \mathbb{R}$ *or* $\epsilon \in \sqrt{-1}\mathbb{R}$.

(i) If $|\epsilon| = 1$, the Hopf $*$-structures on $U_\epsilon(\mathfrak{g})$ are of the form

$$K_i^* = K_{\mu(i)}, \quad (X_i^\pm)^* = c_i^{\pm 1} X_{\mu(i)}^\pm,$$

where μ is a diagram automorphism of $\mathfrak{g}$ such that $\mu^2 = \mathrm{id}$ and the non-zero complex numbers c_i satisfy $c_{\mu(i)}\bar{c}_i = 1$. A second $*$-structure, given by data c_i' and μ', is equivalent to this one if and only if μ and μ' are conjugate in the group of diagram automorphisms.

(ii) If $\epsilon \in \mathbb{R}$, the Hopf $*$-structures on $U_\epsilon(\mathfrak{g})$ are of the form

$$K_i^* = K_{\mu(i)}, \quad (X_i^+)^* = c_i X_{\mu(i)}^- K_{\mu(i)}, \quad (X_i^-)^* = c_i^{-1} K_{\mu(i)}^{-1} X_{\mu(i)}^+,$$

where μ is a diagram automorphism of $\mathfrak{g}$ such that $\mu^2 = \mathrm{id}$ and the non-zero complex numbers c_i satisfy $c_{\mu(i)} = \bar{c}_i$. A second $*$-structure, given by data c_i' and μ', is equivalent to this one if and only if there is a diagram automorphism ν such that $\nu\mu = \mu'\nu$ and $c_i' c_{\nu(i)}^{-1} > 0$ for all i such that $\mu(i) = i$.

(iii) If $\epsilon \in \sqrt{-1}\mathbb{R}$, then $U_\epsilon(\mathfrak{g})$ has real forms only if $\mathfrak{g} \cong \mathrm{so}(2n + 1, \mathbb{C})$. In this case, any Hopf $*$-structure on $U_\epsilon(\mathfrak{g})$ is of the form

$$K_i^* = K_i, \quad (X_i^+)^* = c_i X_i^- K_i, \quad (X_i^-)^* = (-1)^{d_i} c_i^{-1} K_i^{-1} X_i^+,$$

where the non-zero complex numbers c_i satisfy $c_i = (-1)^{d_i} \bar{c}_i$. A second $*$-structure, given by data c_i', is equivalent to this one if and only if $c_i' c_i^{-1} > 0$ for all i. ∎

The main point of the proofs of 9.4.1 and 9.4.2 is to show that any map $\eta : U_\epsilon(\mathfrak{g}) \to U_\epsilon(\mathfrak{g})$ such that $(\eta \otimes \eta) \circ \Delta = \Delta \circ \eta$, and $\epsilon \circ \eta = \eta$ takes K_i to K_{γ_i} for some $\gamma_i \in \sum_j \mathbb{N}.\alpha_j$; note that automorphisms and $*$-structures both enjoy this property. If η is bijective, the γ_i must be simple roots, and from this point the proofs are straightforward.

Example 9.4.3 The Hopf algebra automorphisms of $U_\epsilon(sl_2(\mathbb{C}))$, for any $\epsilon \in \mathbb{C}^\times$, are of the form

$$K_1 \mapsto K_1, \quad X_1^\pm \mapsto c^{\pm 1} X_1^\pm$$

for some $c \in \mathbb{C}^\times$. Up to equivalence, there are exactly five $*$-structures:

$$
\begin{array}{llll}
\epsilon \in \mathbb{R} : & K_1^* = K_1, & (X_1^+)^* = X_1^- K_1, & (X_1^-)^* = K_1^{-1} X_1^+ ; \\
\epsilon \in \mathbb{R} : & K_1^* = K_1, & (X_1^+)^* = -X_1^- K_1, & (X_1^-)^* = -K_1^{-1} X_1^+ ; \\
|\epsilon| = 1 : & K_1^* = K_1, & (X_1^+)^* = -X_1^+, & (X_1^-)^* = -X_1^- ; \\
\epsilon \in \sqrt{-1}\mathbb{R} : & K_1^* = K_1, & (X_1^+)^* = \sqrt{-1} X_1^- K_1, & (X_1^-)^* = \sqrt{-1} K_1^{-1} X_1^+ ; \\
\epsilon \in \sqrt{-1}\mathbb{R} : & K_1^* = K_1, & (X_1^+)^* = -\sqrt{-1} X_1^- K_1, & (X_1^-)^* = -\sqrt{-1} K_1^{-1} X_1^+
\end{array}
$$

The first three are usually called quantum su_2, quantum $su_{1,1}$ and quantum $sl_2(\mathbb{R})$, respectively (if we let $\epsilon \to 1$ through the appropriate ranges, we obtain

*-structures on $sl_2(\mathbb{C})$, and the sets of elements $x \in sl_2(\mathbb{C})$ for which $x^* = -x$ are su_2, $su_{1,1}$ and $sl_2(\mathbb{R})$, respectively). Note that, whereas $su_{1,1}$ and $sl_2(\mathbb{R})$ are isomorphic as real Lie algebras (cf. 4.1.20), there is no question of their quantum versions being isomorphic, as they are defined for different values of ϵ. Note also that the last two *-structures have no classical limit, since $1 \notin \sqrt{-1}\mathbb{R}$. $\diamond$

Remark Masuda *et al.* (1990a) (and a number of other authors) find only the first three of the real forms because they work with the simply-connected version U_ϵ^P of U_ϵ (cf. Remark [4] in Subsection 9.1A). This is obtained by adjoining to U_ϵ a square root L_1 of K_1. For any *-structure on U_ϵ^P, $L_1^* = L_1$, but then it is clear that the last two *-structures above are incompatible with the relations $L_1 X_1^\pm L_1^{-1} = \epsilon^{\pm 1} X_1^\pm$ in U_ϵ^P. $\triangle$

Bibliographical notes

9.1 The rational form U_q was first introduced by Jimbo (1985). Theorem 9.1.8 was proved by Rosso (1990b); the explicit description of the centre in 9.1.6 is due to De Concini & Kac (1990) and Joseph & Letzter (1992) (we have followed De Concini & Kac). See these papers, as well as Lusztig (1988, 1993), for the omitted proofs.

9.2 The results of this section can be found in De Concini & Kac (1990) and De Concini, Kac & Procesi (1992a). See Kac & Weisfeiler (1976) for the analogous classical theory in characteristic p.

9.3 The restricted integral form of U_q was introduced by Lusztig (1988). The basic properties of U_ϵ^{res} are in Lusztig (1989). The Poincaré–Birkhoff–Witt bases, the description of $U_\mathcal{A}$ and its specializations, and the characteristic zero Frobenius map, including the description of the finite-dimensional Hopf algebra U_ϵ^{fin} and its relation with the classical characteristic p theory, were obtained in Lusztig (1990a) in the simply-laced case, and in Lusztig (1990b) in the general case. Full details of the proofs of all the results stated in Subsections 9.3A–C may be found in these papers, and in Lusztig (1993). See Jantzen (1987) for the analogous classical theory in characteristic p.

See Feng & Tsygan (1991) and Ginzburg & Kumar (1992) for information on the cohomology of $U_\epsilon^{\text{res}}(\mathfrak{g})$ when ϵ is not, or is, a root of unity, respectively.

For the basic properties of quivers, see Gabriel (1972) and Bernstein, Gel'fand & Ponomarev (1973). Their connection with quantum groups was noticed by Ringel (1990). We have followed the discussion in Lusztig (1990d). See Ringel (1992, 1993) for further information.

9.4 The classification of the automorphisms and real forms of U_ϵ is due to

Twietmeyer (1992) in the general case. In fact, he classifies the automorphisms and real forms of $U_\epsilon(\mathfrak{g})$ when $\mathfrak{g}$ is an arbitrary symmetrizable Kac–Moody algebra. Several authors had previously considered the sl_2 case. See Lyubashenko (1992) for a different approach, based on category-theoretic considerations.

10
Representations of QUE algebras: the generic case

This chapter and the next are devoted to the representation theory of QUE algebras. In the present chapter, we discuss $U_q(\mathfrak{g})$ and its specialization $U_\epsilon(\mathfrak{g})$ when ϵ is not a root of unity, leaving the root of unity case to Chapter 11.

In the former case, it turns out that the representation theory is closely parallel to that of $\mathfrak{g}$ itself, although one sometimes has to work a little harder than in the classical case to establish the basic results. A notable example is the complete reducibility of finite-dimensional $U_\epsilon(\mathfrak{g})$-modules when ϵ is not a root of unity. In the classical case, complete reducibility is most easily established by transcendental methods: one uses Weyl's unitary trick, regarding representations of $\mathfrak{g}$ as representations of the compact Lie group K whose complexified Lie algebra is $\mathfrak{g}$, and then making use of integration over K. In the quantum case a purely algebraic argument must be found.

The basic facts about the representation theory of $U_\epsilon(\mathfrak{g})$ when ϵ is not a root of unity are given in Section 10.1. In Section 10.2, we disuss the quantum analogue of the Frobenius–Schur duality between the representations of the special linear and symmetric groups. The role of the latter in the quantum situation is played by the Hecke algebra.

Note Unless indicated otherwise, $\mathfrak{g}$ denotes in this chapter a finite-dimensional complex simple Lie algebra. We use the notation in the Appendix, and write U_q for $U_q(\mathfrak{g})$ when this will not lead to confusion.

10.1 Classification of finite-dimensional representations

Classically, the finite-dimensional irreducible representations of $\mathfrak{g}$ are parametrized by their highest weights. The main result of this section is that, up to a rather trivial twisting, this continues to hold for $U_q(\mathfrak{g})$.

A Highest weight modules In this subsection and the next, we work with the Hopf algebra $U_q(\mathfrak{g})$ over the field $\mathbb{Q}(q)$ of rational functions of an indeterminate q, defined in 9.1.1. We begin by describing the analogues for $U_q(\mathfrak{g})$ of some of the definitions and results concerning highest weight representations of $\mathfrak{g}$.

By a *weight*, we mean an n-tuple $\boldsymbol{\omega} = (\omega_1, \ldots, \omega_n) \in (\mathbb{Q}(q)^\times)^n$. If $\boldsymbol{\omega}'$ is another weight, we write $\boldsymbol{\omega}' \leq \boldsymbol{\omega}$ if $\omega_i'^{-1}\omega_i = q^{(\alpha_i, \beta)}$ for some $\beta \in Q^+$ and all $i = 1, \ldots, n$.

If V is a (left) U_q-module, its *weight spaces* are all the non-zero $\mathbb{Q}(q)$-linear subspaces of V of the form

$$V_{\boldsymbol{\omega}} = \{v \in V \mid K_i.v = \omega_i v, \ i = 1, \ldots, n\},$$

where $\boldsymbol{\omega}$ is a weight. A *primitive* vector in V is a non-zero vector $v \in V$ such that $X_i^+.v = 0$ for all i, and $v \in V_{\boldsymbol{\omega}}$ for some $\boldsymbol{\omega}$.

A *highest weight U_q-module* is a U_q-module V which contains a primitive vector v such that $V = U_q.v$. It follows immediately from 9.1.3 that, if $v \in V_{\boldsymbol{\omega}}$, then

$$V = \bigoplus_{\boldsymbol{\omega}' \leq \boldsymbol{\omega}} V_{\boldsymbol{\omega}'}$$

and that $\dim_{\mathbb{Q}(q)}(V_{\boldsymbol{\omega}}) = 1$. In particular, $\boldsymbol{\omega}$ is uniquely determined by V; it is called the *highest weight* of V, and v is called a *highest weight vector*.

For any weight $\boldsymbol{\omega}$, define the *Verma module* $M_q(\boldsymbol{\omega})$ to be the quotient of U_q by the left ideal generated by X_i^+ and $K_i - \omega_i 1$, for $i = 1, \ldots, n$. Obviously, $M_q(\boldsymbol{\omega})$ is a highest weight module with highest weight $\boldsymbol{\omega}$ and canonical highest weight vector $v_{\boldsymbol{\omega}}$ equal to the image of $1 \in U_q$. Moreover, every highest weight module with highest weight $\boldsymbol{\omega}$ is isomorphic to a quotient of $M_q(\boldsymbol{\omega})$. It follows from 9.1.3, as in the classical case, that $M_q(\boldsymbol{\omega})$ is a free U_q^--module generated by $v_{\boldsymbol{\omega}}$. Finally, since $\dim(M_q(\boldsymbol{\omega})_{\boldsymbol{\omega}}) = 1$, it follows by the well-known classical argument that $M_q(\boldsymbol{\omega})$ has a unique irreducible quotient $V_q(\boldsymbol{\omega})$, and hence that every irreducible highest weight module is isomorphic to some $V_q(\boldsymbol{\omega})$.

We shall be interested mainly in modules with highest weight $\boldsymbol{\omega}_{\sigma, \lambda}$ of the form

(1) $$\omega_i = \sigma(\alpha_i)\, q^{(\alpha_i, \lambda)},$$

where $\sigma \in Q_2^*$, $\lambda \in P$. (We recall from Subsection 9.1B that Q_2^* is the set of homomorphisms $Q \to \{\pm 1\}$.) Note that the partial ordering we have defined induces, on the weights $\boldsymbol{\omega}_{\sigma,\lambda}$ of a fixed *type* σ, the usual partial ordering on P: $\boldsymbol{\omega}_{\sigma,\lambda}' \leq \boldsymbol{\omega}_{\sigma,\lambda}$ if and only if $\lambda - \lambda' \in Q^+$. In particular, the weights of a U_q-module with highest weight $\boldsymbol{\omega}_{\sigma,\lambda}$ are all of the form $\boldsymbol{\omega}_{\sigma,\mu}$ with $\mu \leq \lambda$. The reason for the importance of such weights is contained in the next proposition.

As in the classical case, we say that a U_q-module is *integrable* if it is the direct sum of its weight spaces, and if the $X_i^\pm$ act locally nilpotently on V, i.e. if, for any $v \in V$, we have $(X_i^+)^r.v = (X_i^-)^r.v = 0$ for sufficiently large r.

PROPOSITION 10.1.1 *The irreducible U_q-module $V_q(\boldsymbol{\omega})$ with highest weight $\boldsymbol{\omega}$ is integrable if and only if $\boldsymbol{\omega} = \boldsymbol{\omega}_{\sigma,\lambda}$ for some $\sigma \in Q_2^*$, $\lambda \in P^+$.* ∎

The proof of this result is very similar to that of its classical analogue (cf. Humphreys (1972), Sect. 21, for example). As in the classical case, the reason for the importance of integrable modules is:

PROPOSITION 10.1.2 *Every finite-dimensional irreducible U_q-module V is integrable and highest weight.*

PROOF The fact that V is highest weight will be proved at the same time as 10.1.7 below. We may thus assume that $V = V_q(\boldsymbol{\omega})$ for some $\boldsymbol{\omega} \in (\mathbb{Q}(q)^\times)^n$; let $v_{\boldsymbol{\omega}}$ be a highest weight vector of $V_q(\boldsymbol{\omega})$. Since the vectors $(X_i^\pm)^r.v_{\boldsymbol{\omega}}$ have distinct K_i-eigenvalues $\omega_i q_i^{\pm 2r}$ for all r, they are linearly independent if non-zero. Thus, the local nilpotence of $X_i^\pm$ follows from the finite dimensionality of V. ∎

Note that the module $V_q(\boldsymbol{\omega}_{\sigma,0})$ is one dimensional, the action on the highest weight vector being

$$X_i^\pm.v_{\boldsymbol{\omega}_{\sigma,0}} = 0, \quad K_i.v_{\boldsymbol{\omega}_{\sigma,0}} = \sigma(\alpha_i)\, v_{\boldsymbol{\omega}_{\sigma,0}}.$$

Moreover, it is easy to see that

$$V_q(\boldsymbol{\omega}_{\sigma,\lambda}) \cong V_q(\boldsymbol{\omega}_{\sigma,0}) \underset{\mathbb{Q}(q)}{\otimes} V_q(\boldsymbol{\omega}_{\mathbf{1},\lambda})$$

as U_q-modules, where $\mathbf{1}(\alpha_i) = 1$ for all i. Thus, it is enough to consider modules of type $\mathbf{1}$. If doing so will not lead to confusion, we shall abuse notation by referring to a weight $\boldsymbol{\omega}_{\mathbf{1},\lambda}$ simply as λ, and $V_q(\boldsymbol{\omega}_{\mathbf{1},\lambda})$ as $V_q(\lambda)$.

If W is any highest weight U_q-module with highest weight $\lambda \in P$, its *character* $\mathrm{ch}(W) \in \mathbb{Z}[P]$ is defined, as in the classical case, by

$$\mathrm{ch}(W) = \sum_{\mu \in P} \dim(W_\mu)\, e^\mu,$$

where e^μ denotes the basis element of the group algebra $\mathbb{Z}[P]$ of P corresponding to $\mu \in P$.

Remark If W is as above, one proves, as in the classical case, that the centre Z_ϵ acts on W by the central character $\chi_{\epsilon,\lambda+\rho}$ (see Subsection 9.2B). It follows from 9.2.5 that if two irreducible highest weight modules with highest weights λ and μ have the same central characters, then $\mu + \rho = w(\lambda + \rho)$ for some Weyl group element $w \in W$. In particular, by 10.1.2, the finite-dimensional irreducible U_q-modules are separated by their central characters. △

Example 10.1.3 There are exactly two irreducible $U_q(sl_2(\mathbb{C}))$-modules of each (finite) dimension $\lambda + 1 \geq 1$, namely the modules $V_q(\boldsymbol{\omega}_{\sigma,\lambda})$, $\sigma = \pm 1$, with basis $\{v_0^{(\lambda)}, \ldots, v_\lambda^{(\lambda)}\}$ on which the action of the generators of U_q is as follows:

$$K_1.v_r^{(\lambda)} = \sigma q^{\lambda - 2r} v_r^{(\lambda)}, \quad X_1^+.v_r^{(\lambda)} = \sigma[\lambda - r + 1]_q v_{r-1}^{(\lambda)}, \quad X_1^-.v_r^{(\lambda)} = [r+1]_q v_{r+1}^{(\lambda)}.$$

The vector $v_0^{(\lambda)}$ is a highest weight vector of $V_q(\boldsymbol{\omega}_{\sigma,\lambda})$. Note that, if $\sigma = 1$, these formulas are the counterparts for U_q of those in 6.4.10 for U_h (recall that, to make the translation, one puts $q = e^h$ and $K_1 = e^{hH}$). Those with $\sigma = -1$ have no analogues for U_h (cf. 9.1.4 and the discussion following it). $\Diamond$

Note that the formulas in this example describing the $U_q(sl_2(\mathbb{C}))$-modules $V_q(\omega_{1,\lambda})$ become, in the limit $q \to 1$, the formulas describing the $(\lambda + 1)$-dimensional irreducible representation of $sl_2(\mathbb{C})$ (one thinks of K_1 as 'q^H'). Our main aim in this subsection is to make this argument rigorous, and to show, for arbitrary $\mathfrak{g}$, that the structure of the $U_q(\mathfrak{g})$-modules $V_q(\lambda)$ with $\lambda \in P^+$ is exactly parallel to that of the corresponding highest weight $\mathfrak{g}$-modules. For this, we must specialize q to 1, and hence must first find an $\mathcal{A}$-submodule of $V_q(\lambda)$ preserved by the action of $U_{\mathcal{A}}^{\text{res}}$ (it makes little difference whether we work with the restricted form $U_{\mathcal{A}}^{\text{res}}$ or the non-restricted form $U_{\mathcal{A}}$ here, cf. 9.2.3 and 9.3.10).

PROPOSITION 10.1.4 *Let V be a highest weight U_q-module with highest weight vector v, and let $V_{\mathcal{A}}^{\text{res}} = U_{\mathcal{A}}^{\text{res}}.v$. Then*

(i) *$V_{\mathcal{A}}^{\text{res}}$ is a $U_{\mathcal{A}}^{\text{res}}$-submodule of V;*

(ii) *$V_{\mathcal{A}}^{\text{res}}$ is an $\mathcal{A}$-form of V, in the sense that the natural map*

$$V_{\mathcal{A}}^{\text{res}} \underset{\mathcal{A}}{\otimes} \mathbb{Q}(q) \to V$$

is an isomorphism of vector spaces over $\mathbb{Q}(q)$;

(iii) *$V_{\mathcal{A}}^{\text{res}}$ is the direct sum of its intersections with the weight spaces of V, and each such intersection is a free $\mathcal{A}$-module of finite rank.* ∎

The proof is straightforward, making use of 9.3.3.

If V is as in 10.1.4, let

$$\overline{V} = V_{\mathcal{A}}^{\text{res}} \underset{\mathcal{A}}{\otimes} \mathbb{Q}$$

be the $\mathbb{Q}$-vector space obtained by specializing q to 1. By part (i) of 10.1.4, $\overline{V}$ is a U_1^{res}-module, where

$$U_1^{\text{res}} = U_{\mathcal{A}}^{\text{res}} \underset{\mathcal{A}}{\otimes} \mathbb{Q}.$$

Moreover, by our assumption on the highest weight of V, the eigenvalues of the K_i on V are powers of q, hence the images of the K_i in U_1^{res}, which we also denote by K_i, act as the identity on $\overline{V}$. Hence, $\overline{V}$ is naturally a $\tilde{U}_1^{\text{res}}$-module, where $\tilde{U}_1^{\text{res}}$ is the quotient of U_1^{res} by the ideal generated by the central elements $K_i - 1$, $i = 1, \ldots, n$.

By 9.3.10, $\overline{V}$ becomes a module for the classical enveloping algebra U (over $\mathbb{Q}$). It is clear that $\overline{V}$ is highest weight as a U-module, with highest weight vector the image $\bar{v}$ of v in $\overline{V}$. Moreover, the highest weight of $\overline{V}$ is the same as that of V. For,

$$(X_i^+ X_i^- - X_i^- X_i^+).v = \left(\frac{K_i - K_i^{-1}}{q_i - q_i^{-1}} \right).v = \left(\frac{q^{(\alpha_i, \lambda)} - q^{-(\alpha_i, \lambda)}}{q_i - q_i^{-1}} \right) v$$

in V_1^{res}, and hence, using bars to distinguish generators of U from those of U_q,

$$\overline{H}_i.\bar{v} = (\overline{X_i^+ X_i^-} - \overline{X_i^- X_i^+}).\bar{v} = \lim_{q \to 1} \left(\frac{q^{(\alpha_i, \lambda)} - q^{-(\alpha_i, \lambda)}}{q_i - q_i^{-1}} \right) \bar{v}$$
$$= d_i^{-1}(\alpha_i, \lambda)\bar{v} = \lambda(\overline{H}_i)\bar{v}.$$

The following is a more precise result:

PROPOSITION 10.1.5 *Let $V_q(\lambda)$ be the irreducible highest weight U_q-module with highest weight $\lambda \in P^+$. Then, $\overline{V_q(\lambda)}$ is the irreducible U-module with highest weight λ. Moreover,*

$$\dim_{\mathbb{Q}(q)}(V_q(\lambda)_\mu) = \dim_{\mathbb{Q}}(\overline{V_q(\lambda)}_\mu)$$

for all $\mu \in P$. In particular, the character of $V_q(\lambda)$ is given by the classical Weyl character formula.

PROOF To prove that $\overline{V_q(\lambda)}$ is irreducible, note that, by 10.1.2, the $X_i^{\pm}$ act locally nilpotently on $\overline{V_q(\lambda)}$, and $\overline{V_q(\lambda)}$ is the direct sum of its weight spaces. Hence, $\overline{V_q(\lambda)}$ is an integrable U-module. But it is well known that every integrable highest weight U-module is irreducible.

The statement about the dimensions of the weight spaces follows from 10.1.4. ∎

COROLLARY 10.1.6 *Every finite-dimensional highest weight U_q-module is irreducible and has highest weight in P^+.*

PROOF By 10.1.1 and 10.1.2, the highest weight λ of W lies in P^+. From the proof of 10.1.5, we see that $\overline{W}$ is a finite-dimensional U-module with the same highest weight. But a finite-dimensional highest weight U-module is irreducible. It follows, as above, that $\mathrm{ch}(W)$ is given by the Weyl character formula. But the same argument applies to the unique irreducible quotient W' of W. Hence, W and W' have the same character. It follows that $W = W'$, and hence that W is irreducible. ∎

An important consequence of this is

THEOREM 10.1.7 *Every finite-dimensional U_q-module is completely reducible.*

OUTLINE OF PROOF We shall prove simultaneously that every finite-dimensional irreducible U_q-module is highest weight.

Suppose first that $\mathfrak{g} = sl_2(\mathbb{C})$. Let V be a finite-dimensional irreducible U_q-module. Let $\widetilde{\mathbb{Q}(q)}$ be the algebraic closure of $\mathbb{Q}(q)$, let $\widetilde{V} = V \underset{\mathbb{Q}(q)}{\otimes} \widetilde{\mathbb{Q}(q)}$, and define $\widetilde{U}_q$ similarly; then, $\widetilde{V}$ is a $\widetilde{U}_q$-module. Since we are now working over an algebraically closed field, there is a K-eigenvector $0 \neq \tilde{v} \in \widetilde{V}$ with eigenvalue $\omega \in \widetilde{\mathbb{Q}(q)}$, say. Then,

$$\widetilde{V} = \bigoplus_{r \in \mathbb{Z}} \widetilde{V}_{\omega q^{2r}}$$

since the right-hand side is a non-zero submodule of $\widetilde{V}$. Since $\dim_{\widetilde{\mathbb{Q}(q)}}(\widetilde{V}) < \infty$, there is a maximal $r = r_0$ (say) such that $\widetilde{V}_{\omega q^{2r_0}} \neq 0$. Replacing $\tilde{v}$ by a non-zero vector in $\widetilde{V}_{\omega q^{2r_0}}$, we may therefore assume that $X_1^+.\tilde{v} = 0$.

Suppose for a contradiction that $\omega \notin \pm q^{\mathbb{Z}}$. Define $\tilde{v}_r = (X_1^-)^r.\tilde{v}/[r]_q!$ for $r \in \mathbb{N}$. Then,

$$K_1.\tilde{v}_r = \omega q^{-2r}\tilde{v}_r,$$

(2)
$$X_1^-.\tilde{v}_r = [r+1]_q\tilde{v}_{r+1}, \quad X_1^+.\tilde{v}_r = \left(\frac{\omega q^{-r+1} - \omega^{-1}q^{r-1}}{q - q^{-1}}\right)\tilde{v}_{r-1}.$$

By the assumption on ω, all the coefficients on the right-hand sides of these equations are non-zero. From the third equation, it follows that $\tilde{v}_r \neq 0$ for all r, and from the first that the $\tilde{v}_r$ are linearly independent. This contradicts $\dim_{\widetilde{\mathbb{Q}(q)}}(\widetilde{V}) < \infty$.

Thus, $\omega \in \pm q^{\mathbb{Z}}$ and, in particular, $\omega \in \mathbb{Q}(q)$. Hence, there is a vector $v \in V$ of weight ω. Then, v is primitive, and V is a highest weight module.

Now suppose, for a contradiction, that V is a finite-dimensional U_q-module which is not completely reducible, and that V has minimal dimension among U_q-modules with this property (we are still in the sl_2 case). An argument with Jordan–Hölder series shows that V belongs to a short exact sequence

(3)
$$0 \to V_q(\mu) \to V \to V_q(\lambda) \to 0.$$

In fact, by a classical argument (see, for example, Serre (1965), LA 6.5–6.6), one may even assume that $\lambda = 0$. Thus, we have a short exact sequence

$$0 \to V_q(\mu) \to V \to \mathbb{Q}(q) \to 0,$$

where $\mathbb{Q}(q)$ is the trivial U_q-module. Let $\{v_0^{(\mu)}, v_1^{(\mu)}, \ldots, v_\mu^{(\mu)}\}$ be the basis of $V_q(\mu)$ as in 10.1.3, and denote the images of these vectors in V by the same symbols. Let v' be any vector in V such that $V = V_q(\mu) \oplus \mathbb{Q}(q).v'$

as $\mathbb{Q}(q)$-vector spaces. Then, $K_1.v' = v' + \sum_{r=0}^{\mu} a_r v_r$ for some $a_r \in \mathbb{Q}(q)$, $r = 1, \ldots, \mu$. If μ is odd, the vector

$$(4) \qquad v' + \sum_{r=0}^{\mu} \frac{a_r}{1 - q^{\mu - 2r}} v_r$$

is an eigenvector of K_1 of eigenvalue 1, so we may as well assume that $K_1.v' = v'$. In the case where μ is even, the same conclusion holds, since the relation $K_1 X_1^+ K_1^{-1} = q^2 X_1^+$ forces $a_{\mu/2} = 0$. The remaining relations now force $X_1^{\pm}.v' = 0$ (the even case again causes a little more difficulty than the odd case). Thus, v' generates a highest weight submodule V^0 of V of highest weight 0. By 10.1.6, V^0 is irreducible. It follows that $V \cong V_q(\mu) \oplus V_q(0)$, contrary to hypothesis.

We now turn to the case where $\mathfrak{g}$ is arbitrary. We first prove that every finite-dimensional irreducible U_q-module V is highest weight. By considering V as a module for the $U_{q_i}(sl_2(\mathbb{C}))$ subalgebras of U_q generated by $X_i^{\pm}$, $K_i^{\pm 1}$ for $i = 1, \ldots, n$, and using the sl_2 case proved above, it follows that V is the direct sum of its weight spaces. But if $v_\omega \in V$ is a vector of maximal weight ω, then v_ω is primitive, and so V is a highest weight module.

For the proof of complete reducibility, we are reduced, as in the sl_2 case, to considering a short exact sequence of the form (3), in which V is assumed, for a contradiction, not to be completely reducible. We prove that $\lambda < \mu$ (there is now no particular advantage in assuming that $\lambda = 0$). Since V is the direct sum of its weight spaces, there is a vector $v' \in V$ which projects to a highest weight vector in $V(\lambda)$. If $\lambda \not< \mu$, then $X_i^+.v'$ maps to zero in $V_q(\lambda)$, hence is in the image of $V_q(\mu)$, and hence must be zero since it has weight $\lambda + \alpha_i$, which is not $\leq \mu$. Thus, v' is a primitive vector. As in the sl_2 case, 10.1.6 implies that v' generates a submodule of V isomorphic to $V_q(\lambda)$, which contradicts our hypothesis.

By taking duals in (3), it follows that we also have $-w_0(\mu) < -w_0(\lambda)$, i.e. $\mu < \lambda$. This contradicts the conclusion of the preceding paragraph, and completes the proof of complete reducibility. ∎

Remark The argument reducing the proof of complete reducibility to the statement in 10.1.6 is due to A. Borel. △

B The determinant formula In this subsection, we establish some important results about Verma modules and use them to sketch the proof of 9.1.6.

We must first introduce the quantum analogue of another familiar notion from the classical theory. If V is a $\mathbb{Q}(q)$-vector space, a *hermitian form* on V is a bi-additive form $(\, , \,)$ on V such that

$$(v_1 \, , v_2) = \overline{(v_2 \, , v_1)}, \quad (v_1 \, , cv_2) = c(v_1 \, , v_2),$$

for $c \in \mathbb{Q}(q)$, v_1, $v_2 \in V$. Here, the bar denotes the homomorphism of $\mathbb{Q}$-algebras $\mathbb{Q}(q) \to \mathbb{Q}(q)$ such that $\bar{q} = q^{-1}$. If V is a U_q-module, a hermitian form $(\,,\,)$ on V is said to be *contravariant* if

$$(x.v_1 \, , \, v_2) = (v_1 \, , \, \omega(x).v_2)$$

for all v_1, $v_2 \in V$, $x \in U_q$, where ω is the $\mathbb{Q}$-algebra anti-automorphism of U_q such that

$$\omega(X_i^{\pm}) = X_i^{\mp}, \quad \omega(K_i) = K_i^{-1}, \quad \omega(q) = q^{-1}$$

(cf. the automorphism of U_h with the same name defined in Subsection 8.1A).

PROPOSITION 10.1.8 *The Verma module $M_q(\lambda)$, for $\lambda \in P$, has a unique contravariant hermitian form $(\,,\,)$ such that $(v_\lambda, v_\lambda) = 1$, where v_λ is the canonical highest weight vector of $M_q(\lambda)$. Its kernel is the unique maximal proper submodule of $M_q(\lambda)$.*

OUTLINE OF PROOF This is exactly as in the classical case. In fact, by 9.1.3, we have a decomposition of $\mathbb{Q}(q)$-vector spaces

$$U_q = U_q^0 \oplus (U_q^{--} U_q + U_q U_q^{++}),$$

where $U_q^{\pm\pm} = \sum_{i=1}^{n} X_i^{\pm} U_q^{\pm}$. Let π be the projection of U_q onto U_q^0. Then, we set

$$(x_1.v_\lambda \, , \, x_2.v_\lambda) = \pi(\omega(x_2)x_1)$$

for all x_1, $x_2 \in U_q^-$. ∎

The last sentence of 10.1.8 implies

COROLLARY 10.1.9 *Every highest weight U_q-module V admits a contravariant hermitian form. If V is irreducible, the form is non-degenerate.* ∎

It follows immediately from the definitions of $(\,,\,)$ and of ω that distinct weight spaces of $M_q(\lambda)$ are orthogonal:

$$(M_q(\lambda)_\mu \, , \, M_q(\lambda)_\nu) = 0 \quad \text{if } \mu \neq \nu.$$

For $\eta \in Q^+$, let $\det_{q,\eta}(\lambda)$ denote the determinant of the restriction of $(\,,\,)$ to the (finite-dimensional) weight space $M_q(\lambda)_{\lambda-\eta}$, computed in the basis $\{(X^-)^{\mathbf{r}}.v_\lambda\}_{\mathbf{r} \in \text{Par}(\eta)}$. We recall that

$$\text{Par}(\eta) = \{\mathbf{r} = (r_1, \ldots, r_N) \in \mathbb{N}^N \mid \sum_k r_k \alpha_k = \eta\}.$$

PROPOSITION 10.1.10 *For* $\lambda \in P$, $\eta \in Q^+$, *the determinant* $\det_{q,\eta}(\lambda)$ *is given by*

$$\prod_{\beta \in \Delta^+} \prod_{m=0}^{\infty} \left(\frac{(q^{\frac{m}{2}(\beta,\beta)} - q^{-\frac{m}{2}(\beta,\beta)})(q^{(\lambda+\rho,\beta)-\frac{m}{2}(\beta,\beta)} - q^{-(\lambda+\rho,\beta)+\frac{m}{2}(\beta,\beta)})}{(q^{\frac{1}{2}(\beta,\beta)} - q^{-\frac{1}{2}(\beta,\beta)})^2} \right)^{|\mathrm{Par}(\eta - m\beta)|}.$$

∎

The proof of 10.1.10 is almost identical to that in the classical case (see Kac (1983), for example). As in that situation, we deduce

COROLLARY 10.1.11 *The Verma module* $M_q(\lambda)$, $\lambda \in P$, *is irreducible if and only if* $2(\lambda + \rho, \beta) \neq m(\beta, \beta)$ *for all* $\beta \in \Delta^+$, $m \in \mathbb{N}$. ∎

In fact, this is just the condition for $\det_{q,\eta}(\lambda)$ to be non-zero for all $\eta \in Q^+$.

One can also deduce conditions for inclusions of Verma modules, and for the occurrence of irreducible highest weight modules as subquotients of Verma modules. We shall not write down these conditions since they are exactly the same as in the classical case.

We shall now apply these results on the structure of Verma modules to complete the discussion of the centre Z_q of U_q begun in Subsection 9.1C. As we mentioned there, any $z \in Z_q$ can be expressed in the form

$$(5) \qquad z = \sum_{\eta \in Q^+} \sum_{\mathbf{r}, \mathbf{t} \in \mathrm{Par}(\eta)} (X^-)^{\mathbf{r}} \varphi_{\mathbf{r}, \mathbf{t}} (X^+)^{\mathbf{t}},$$

where $\varphi_{\mathbf{r}, \mathbf{t}} \in U_q^0$. The $\mathrm{Par}(\eta) \times \mathrm{Par}(\eta)$ matrix $\varphi_\eta = (\varphi_{\mathbf{r}, \mathbf{t}})$ can be computed by induction on $\eta \in Q^+$, as follows.

We consider the action of z on a Verma module $M_q(\lambda)$, $\lambda \in P$. Since $z.v_\lambda = \varphi_{\mathbf{0},\mathbf{0}}(\lambda) v_\lambda$ and $z \in Z_q$, it follows that z acts as the scalar $\varphi_{\mathbf{0},\mathbf{0}}(\lambda)$ on the whole of $M_q(\lambda)$. For θ, $\eta \in Q^+$, let $G_{\theta,\eta}(\lambda)$ be the endomorphism of the weight space $M_q(\lambda)_{\lambda-\eta}$ induced by the action of the element

$$\sum_{\mathbf{r}, \mathbf{t} \in \mathrm{Par}(\theta)} (X^-)^{\mathbf{r}} \varphi_{\mathbf{r}, \mathbf{t}} (X^+)^{\mathbf{t}}.$$

Note that $G_{\theta,\eta}(\lambda) = 0$ unless $\theta \leq \eta$. Hence, the fact that z acts as the scalar $\varphi_{\mathbf{0},\mathbf{0}}(\lambda)$ on $M_q(\lambda)_{\lambda-\eta}$ translates into the equation

$$(6) \qquad G_{\eta,\eta}(\lambda) + \sum_{\theta < \eta} G_{\theta,\eta}(\lambda) = \varphi_{\mathbf{0},\mathbf{0}}(\lambda).\mathrm{id}.$$

Let $H_\eta(\lambda)$ (resp. $G_{\theta,\eta}(\lambda)$) be the matrix of the contravariant form (,) (resp. of $G_{\theta,\eta}(\lambda)$) in the basis $\{(X^-)^{\mathbf{s}}.v_\lambda\}_{\mathbf{s} \in \mathrm{Par}(\eta)}$ of $M_q(\lambda)_{\lambda-\eta}$. Then, we have

$$(7) \qquad G_{\eta,\eta}(\lambda) = \varphi_\eta(\lambda) H_\eta(\lambda).$$

In fact, since

$$(X^+)^t(X^-)^s.v_\lambda = H_\eta(\lambda)_{t,s}v_\lambda,$$

we have

$$\left(\sum_{t,r\in\text{Par}(\eta)} (X^-)^r\varphi_{r,t}(X^+)^t \right) (X^-)^s.v_\lambda = \sum_{t,r\in\text{Par}(\eta)} (X^-)^r\varphi_{r,t}(\lambda)H_\eta(\lambda)_{t,s}.v_\lambda$$

and hence

$$G_{\eta,\eta}(\lambda)_{r,s} = \sum_t \varphi_\eta(\lambda)_{r,t}H_\eta(\lambda)_{t,s}.$$

Combining (6) and (7), we have

(8) $$\varphi_\eta H_\eta + \sum_{\theta<\eta} G_{\theta,\eta} = \varphi_{0,0}.\text{id}.$$

Assume that φ_θ is known for all $\theta < \eta$. Then, $G_{\theta,\eta}$ is known for all $\theta < \eta$, and hence (8) determines φ_η, provided that H_η is invertible. By 10.1.10, inverting H_η involves introducing denominators of the form

(9) $$\frac{q^{(\rho,\beta)-\frac{m}{2}(\beta,\beta)}K_\beta - q^{-(\rho,\beta)+\frac{m}{2}(\beta,\beta)}K_\beta^{-1}}{q_\beta - q_\beta^{-1}} = \left[K_\beta;\frac{2(\rho,\beta)}{(\beta,\beta)} - m\right]_{q_\beta}$$

for $\beta \in \Delta^+$ ($q_\beta = q^{\frac{1}{2}(\beta,\beta)}$). *A priori*, therefore, the inductive procedure gives elements $\varphi_{r,t}$ lying in $S^{-1}U_q^0$, where S is the set of products of terms as in (9).

The crux of the proof of 9.1.6 is to show that, starting with an arbitrary element $\varphi_{0,0} \in U_q^0$, the inductive procedure leads to elements $\varphi_{r,t} \in U_q^0$ if and only if $\varphi_{0,0} \in \gamma_q(U_q^0{}^{\widetilde{W}})$. When this condition holds, the resulting element $z \in U_q^0$ lies in Z_q because it acts as a scalar on every Verma module.

We shall not prove this last statement, but we give the following example, which may help to clarify the preceding argument.

Example 10.1.12 Suppose that $\mathfrak{g} = sl_2(\mathbb{C})$. The expansion (5) now takes the form

$$z = \varphi_0 + X_1^-\varphi_1 X_1^+ + (X_1^-)^2\varphi_2(X_1^+)^2 + \cdots,$$

where $\varphi_0, \varphi_1,\ldots \in U_q^0$ are now thought of as Laurent polynomials in K_1 with coefficients in $\mathbb{Q}(q)$. The inductive procedure leads to the following equations:
(10)
$$\varphi_0(K_1) = \varphi_0(q^{-2m}K_1) + [m]_q[K_1;1-m]_q\varphi_1(q^{-2m+2}K_1)$$
$$+ [m]_q[m-1]_q[K_1;1-m]_q[K_1;2-m]_q\varphi_2(q^{-2m+4}K_1) + \cdots$$
$$+ [m]_q![K_1;1-m]_q[K_1;2-m]_q\ldots[K_1;0]_q\varphi_m(K_1).$$

We have used (13) from Chapter 9 to compute the elements $G_{\theta,\eta} \in U_q^0$.
Taking $m = 1$, we find

$$\varphi_0(K_1) - \varphi_0(q^{-2}K_1) = \frac{K_1 - K_1^{-1}}{q - q^{-1}}\varphi_1(K_1);$$

in particular, the left-hand side must be divisible by $K_1 - K_1^{-1}$ in U_q^0. The
choice

$$\varphi_0(K_1) = \frac{qK_1 + q^{-1}K_1^{-1}}{(q - q^{-1})^2}$$

has this property and gives $\varphi_1(K_1) = 1$. It now follows from (10) for $m > 1$,
by considering the degrees of the various terms in K_1, that $\varphi_2 = \varphi_3 = \cdots = 0$.
Thus, the central element we obtain is

$$\frac{qK_1 + q^{-1}K_1^{-1}}{(q - q^{-1})^2} + X_1^- X_1^+,$$

the quantum Casimir element (cf. 9.1.7). $\diamond$

The discussion in this subsection can be given for U_h just as well as for U_q,
and we use it to give the

PROOF OF PROPOSITION 8.3.15 We use the notation of Subsection 8.3E.
The equation to be proved is equivalent to

$$S_h(e^{-h\rho^{\bullet}}u_h) = e^{-h\rho^{\bullet}}u_h.$$

In view of the results of this subsection, and those of Subsection 9.1C (which
also apply to U_h), it is enough to show that both sides of this equation act
by the same scalar on any finite-dimensional indecomposable U_h-module V_h.
It suffices to consider the highest weight vector v of V_h, and if this vector has
weight $\lambda \in P$, say, we have, using the formula for the universal R-matrix in
8.3.9,

$$e^{-h\rho^{\bullet}}u_h.v = \exp\left(-h\rho^* - h\sum_{ij}(B^{-1})_{ij}H_iH_j\right).v$$

$$= \exp\left(-h\lambda(\rho^*) - h\sum_{ij}(B^{-1})_{ij}\lambda(H_i)\lambda(H_j)\right)v$$

$$= e^{-h(\lambda,\lambda+2\rho)}v,$$

since $\rho^* \in \mathfrak{h}$ corresponds to $2\rho \in \mathfrak{h}^*$ under the standard invariant bilinear form
$(\ ,\ )$. But $S_h(e^{-h\rho^{\bullet}}u_h)$ acts on V_h as $e^{-h\rho^{\bullet}}u_h$ acts on the (left) dual module
V_h^*, which has highest weight $-w_0(\lambda)$, where w_0 is the longest element of the
Weyl group of $\mathfrak{g}$. Thus, our assertion follows from the computation

$$e^{-h(-w_0(\lambda),-w_0(\lambda)+2\rho)} = e^{-h(\lambda,\lambda-2w_0(\rho))} = e^{-h(\lambda,\lambda+2\rho)},$$

where we have used the invariance of $(\,,\,)$ and the fact that $w_0(\rho) = -\rho$. ∎

We can also prove 8.3.14. Let $\tilde{C}$ be the image of C_h under the canonical isomorphism between the centres of $U_h(\mathfrak{g})$ and $U(\mathfrak{g})[[h]]$ (see 6.5.7). By the preceding proof, $\tilde{C}$ acts by the scalar $(\lambda, \lambda + 2\rho)$ on any $\mathfrak{g}$-module of highest weight λ. But it is well known that the classical Casimir element also has this property and, by the classical analogue of 9.2.5, that an element of the centre of $U(\mathfrak{g})$ is determined by its action on all the highest weight representations of $\mathfrak{g}$. ∎

C Specialization: the non-root of unity case *In this subsection, ϵ denotes a non-zero complex number which is not a root of unity.* We examine the extent to which the results of the previous two subsections carry over from U_q to U_ϵ.

We can define highest weight modules V whose highest weight is any n-tuple of non-zero complex numbers $\boldsymbol{\omega} = (\omega_1, \ldots, \omega_n)$: V should be generated as a U_ϵ-module by a vector $v_{\boldsymbol{\omega}}$ such that $U_\epsilon^+.v_{\boldsymbol{\omega}} = 0$ and $K_i.v_{\boldsymbol{\omega}} = \omega_i v_{\boldsymbol{\omega}}$ for $i = 1, \ldots, n$. Any such module is the direct sum of its weight spaces

$$V_{\boldsymbol{\omega}'} = \{v \in V \mid K_i.v = \omega_i' v, \ i = 1, \ldots, n\}.$$

The Verma module $M_\epsilon(\boldsymbol{\omega})$, defined in the obvious way for any $\boldsymbol{\omega}$, is the universal highest weight module with highest weight $\boldsymbol{\omega}$. By 9.2.2, it is a free U_ϵ^--module with basis $\{(X^-)^{\boldsymbol{t}}.v_{\boldsymbol{\omega}}\}_{\boldsymbol{t} \in \mathbb{N}^N}$. Moreover, it has a unique irreducible quotient $V_\epsilon(\boldsymbol{\omega})$.

We shall usually be interested in the case where $\omega_i = \epsilon^{(\alpha_i, \lambda)}$ for some $\lambda \in P$. We shall then denote $M_\epsilon(\boldsymbol{\omega})$ by $M_\epsilon(\lambda)$, etc. Note that, by the preceding remarks, $\mathrm{ch}(M_\epsilon(\lambda))$ is the same as in the classical case.

Suppose that ϵ is transcendental. Then all the arguments of Subsection 10.1A carry over *verbatim* to U_ϵ. In particular, one can specialize ϵ to 1 to obtain from each highest weight U_ϵ-module V a highest weight U-module $\overline{V}$ (over $\mathbb{Q}$). Then, we have the following result, whose proof is exactly the same as that of 10.1.5.

PROPOSITION 10.1.13 *Assume that $\epsilon \in \mathbb{C}^\times$ is transcendental. Let $V_\epsilon(\lambda)$ be the irreducible highest weight U_ϵ-module with highest weight $\lambda \in P^+$. Then, $\overline{V_\epsilon(\lambda)}$ is the irreducible highest weight U-module with highest weight λ.* ∎

In the same way, complete reducibility also holds when ϵ is transcendental. In fact, we have the following stronger result.

THEOREM 10.1.14 *Suppose that $\epsilon \in \mathbb{C}^\times$ is not a root of unity. Then, every finite-dimensional U_ϵ-module is completely reducible, and every finite-dimensional irreducible U_ϵ-module is highest weight.*

OUTLINE OF PROOF As we have just noted, if ϵ is transcendental, the proof given in the previous subsection carries over *verbatim* to the U_ϵ case. If ϵ

is algebraic (but not a root of unity), the proof we gave for $U_q(sl_2(\mathbb{C}))$ still works for $U_\epsilon(sl_2(\mathbb{C}))$ (note, in particular, that the vector defined in (4) still makes sense when q is replaced by ϵ – this might not be the case if ϵ were a root of unity). The passage to the case of arbitrary $\mathfrak{g}$ relied on 10.1.6, which in turn was deduced from 10.1.5. Unfortunately, however, the latter result does not even make sense when ϵ is algebraic, since one cannot then specialize ϵ to 1. Thus, we must find a proof of the analogue of 10.1.6 for U_ϵ when ϵ is algebraic, but not a root of unity.

We first look at the Verma modules $M_\epsilon(\lambda)$, $\lambda \in P$. If v_μ is a primitive vector in $M_\epsilon(\lambda)$ of weight μ, then we see by considering the action of Z_ϵ that $\chi_{\epsilon,\lambda} = \chi_{\epsilon,\mu}$. By 9.2.5, $\mu + \rho \in W(\lambda + \rho)$. As in the classical case (see Humphreys (1972), for example), one sees by considering Jordan–Hölder series that $\mathrm{ch}(M_\epsilon(\lambda))$ is a linear combination of characters $\mathrm{ch}(V_\epsilon(\mu))$ for $\mu \in P_\lambda = \{\mu \in P \mid \mu \leq \lambda,\ \mu + \rho \in W(\lambda + \rho)\}$, and that these relations can be inverted to obtain an expression for $\mathrm{ch}(V_\epsilon(\lambda))$ as a linear combination of the $\mathrm{ch}(M_\epsilon(\mu))$ for $\mu \in P_\lambda$.

Suppose now that V is any finite-dimensional U_ϵ-module with highest weight λ. As in the proof of 10.1.6, to prove that V is irreducible, it is enough to show that $\mathrm{ch}(V)$ is given by the Weyl character formula. By considering Jordan–Hölder series as before, we see that $\mathrm{ch}(V)$ can be written as a linear combination of the $\mathrm{ch}(V_\epsilon(\mu))$ for $\mu \in P_\lambda$. Hence,

$$\mathrm{ch}(V) = \sum_{w \in W} c_w(\lambda)\, \mathrm{ch}(M_\epsilon(w(\lambda + \rho) - \rho))$$

for some $c_w(\lambda) \in \mathbb{Q}(\epsilon)$.

The next step is to show that

$$\dim_{\mathbb{Q}(\epsilon)}(V_\mu) = \dim_{\mathbb{Q}(\epsilon)}(V_{w\mu})$$

for all $w \in W$. In fact, the proof of this is exactly as in the classical case (see Humphreys (1972), Sects 7.2 and 21.2). It uses complete reducibility under $U_\epsilon(sl_2(\mathbb{C}))$, but this is already known. It now follows as in the classical case that $c_w(\lambda) = \mathrm{sign}(w)$, where $\mathrm{sign}(w) = (-1)^k$ if $w = s_{i_1} s_{i_2} \ldots s_{i_k}$. Finally, since the characters $\mathrm{ch}(M_\epsilon(\lambda))$ are the same as the characters of the classical Verma modules, it follows that $\mathrm{ch}(V)$ is given by the Weyl character formula. ∎

From the proof, we note

COROLLARY 10.1.15 *Suppose that $\epsilon \in \mathbb{C}^\times$ is not a root of unity, and let $\lambda \in P^+$. Then, $\mathrm{ch}(V_\epsilon(\lambda))$ is given by the classical Weyl character formula.* ∎

Remark We shall see later in this chapter that 10.1.14 and its corollary, 10.1.15, are both false in general when ϵ is a root of unity (this holds for both the restricted and non-restricted specializations). △

Let us give an application of 10.1.14 and 10.1.15.

PROPOSITION 10.1.16 *Suppose that $\epsilon \in \mathbb{C}^\times$ is not a root of unity, and let λ, $\mu \in P^+$. Then*

$$V_\epsilon(\lambda) \otimes V_\epsilon(\mu) \cong \bigoplus_{\nu \in P^+} V_\epsilon(\nu)^{\oplus m_\nu},$$

where the multiplicities m_ν are the same as in the decomposition of the tensor product $V(\lambda) \otimes V(\mu)$ of the irreducible representations of $\mathfrak{g}$ with highest weights λ and μ.

PROOF Note first that, if V_ϵ and W_ϵ are any two finite-dimensional U_ϵ-modules, then

$$\mathrm{ch}(V_\epsilon \otimes W_\epsilon) = \mathrm{ch}(V_\epsilon)\mathrm{ch}(W_\epsilon)$$

(this follows from the fact that, if $v \in V_\epsilon$ and $w \in W_\epsilon$ are weight vectors, then the weight of $v \otimes w$ is the sum of the weights of v and w, which in turn follows from the fact that the K_i are group-like). From this remark and 10.1.15, we deduce that

$$\mathrm{ch}(V_\epsilon(\lambda) \otimes V_\epsilon(\mu)) = \mathrm{ch}(V(\lambda) \otimes V(\mu)).$$

Thus, if n_ν are the multiplicities in the classical case, we have, using 10.1.15 again,

$$\sum_{\nu \in P^+} m_\nu \mathrm{ch}(V(\nu)) = \sum_{\nu \in P^+} n_\nu \mathrm{ch}(V(\nu)).$$

Since finite-dimensional $\mathfrak{g}$-modules are determined by their characters, we have

$$\bigoplus_{\nu \in P^+} V(\nu)^{\oplus m_\nu} \cong \bigoplus_{\nu \in P^+} V(\nu)^{\oplus n_\nu}.$$

But m_ν (for example) is the dimension of the space of $\mathfrak{g}$-module homomorphisms from $V(\nu)$ to the left-hand side. Hence, $m_\nu = n_\nu$ for all ν. ∎

Suppose now that $|\epsilon| = 1$. Then, a *contravariant form* on a U_ϵ-module V is a hermitian form $(\ ,\ )$ on V (in the usual sense) such that

$$(x.v_1 \, , \, v_2) = (v_1 \, , \, \omega(x).v_2),$$

where ω is the anti-automorphism of U_ϵ defined by the same formulas as in the U_q case. Arguing as in 10.1.8 and 10.1.9, we see that the Verma module $M_\epsilon(\lambda)$ admits such a form whose kernel is the unique maximal submodule of $M_\epsilon(\lambda)$. It follows that every highest weight U_ϵ-module of type **1** with highest weight $\lambda \in P$ admits a non-zero contravariant form.

Let $\det_{\epsilon,\eta}(\lambda)$ be the determinant of the restriction of the contravariant form on $M_\epsilon(\lambda)$ to the weight space $M_\epsilon(\lambda)_{\lambda-\eta}$, $\eta \in Q^+$, computed in the basis $\{(X^-)^{\boldsymbol{t}}.v_\lambda\}_{\boldsymbol{t} \in \mathrm{Par}(\eta)}$.

PROPOSITION 10.1.17 *Suppose that $|\epsilon| = 1$ and that $\epsilon^{2d_i} \neq 1$ for $i = 1, \ldots, n$. Then, $\det_{\epsilon,\eta}(\lambda)$ is obtained by replacing q by ϵ in the formula in 10.1.10.* ∎

D R-matrices associated to representations of U_q We observed at the end of Subsection 9.1B that the universal R-matrix

$$\mathcal{R}_h = \exp\left(h \sum_{i,j} (B^{-1})_{ij} H_i \otimes H_j\right)$$

$$\times \sum_{t_1,\ldots,t_N=0}^{\infty} \prod_{r=1}^{N} q_{\beta_r}^{\frac{1}{2}t_r(t_r+1)} \frac{(1 - q_{\beta_r}^{-2})^{t_r}}{[t_r]_{q_{\beta_r}}!} (X_{\beta_r}^+)^{t_r} \otimes (X_{\beta_r}^-)^{t_r}$$

of U_h does not translate to give an element of $U_q \otimes U_q$ (the notation is that of Subsection 8.3C). In fact, there are two problems, which must be discussed separately. Let

$$\mathcal{E}_h = \exp\left(h \sum_{i,j} (B^{-1})_{ij} H_i \otimes H_j\right), \quad \tilde{\mathcal{R}}_h = \mathcal{E}_h^{-1}\mathcal{R}_h.$$

The first difficulty is that, although each term in $\tilde{\mathcal{R}}_h$ is meaningful as an element of $U_q \otimes U_q$, the sum is infinite. This problem is easily overcome by taking a suitable completion of U_q. In fact, for any $r \geq 1$, let

$$U_q^{+,r} = \bigoplus_{\{\beta \in Q | \text{height}(\beta) \geq r\}} (U_q^+)_\beta,$$

where $(U_q^+)_\beta = \{u \in U_q^+ \mid K_i u K_i^{-1} = q^{(\beta,\alpha_i)}u \text{ for } i = 1, \ldots, n\}$ is the subspace of U_q^+ consisting of elements of weight β (recall that $\text{height}(\sum_i m_i \alpha_i) = \sum_i m_i$). The appropriate completion of U_q is

$$\hat{U}_q = \varprojlim U_q/U_q U_q^{+,r}.$$

We define the completed tensor product $\hat{U}_q(\mathfrak{g})\hat{\otimes}\hat{U}_q(\mathfrak{g})$ to be $\hat{U}_q(\mathfrak{g} \oplus \mathfrak{g})$, the subspaces $U_q(\mathfrak{g} \oplus \mathfrak{g})^{+,r} \subset U_q^+ \otimes U_q^+$ being defined in the obvious way, using the height function $\text{height}(\sum_i m_i\alpha_i, \sum_j n_j\alpha_j) = \sum_i m_i + \sum_j n_j$. Higher completed tensor products are defined similarly. It is then easy to check that the algebra structure on U_q extends naturally to $\hat{U}_q$, that the coalgebra structure maps extend to the appropriate completions, and that the Hopf algebra axioms are satisfied, provided one replaces $\otimes$ by $\hat{\otimes}$. Moreover, $\tilde{\mathcal{R}}_h$ is obviously a well-defined element of $\hat{U}_q\hat{\otimes}\hat{U}_q$, and as such we denote it by $\tilde{\mathcal{R}}_q$.

The second, and more serious, difficulty is that the exponential term $\mathcal{E}_h$ cannot be interpreted as an element of $U_q \otimes U_q$ (or its completion). The way to overcome this problem was pointed out by Tanisaki (1992). We consider the automorphism Ψ_h of $U_h \otimes U_h$ given by

$$\Psi_h(x) = \mathcal{E}_h^{-1}.x.\mathcal{E}_h.$$

It is easy to show, using 6.4.2, that

$$\Psi_h(K_i \otimes 1) = K_i \otimes 1, \quad \Psi_h(1 \otimes K_i) = 1 \otimes K_i,$$
$$\Psi_h(X_i^\pm \otimes 1) = X_i^\pm \otimes K_i^{\mp 1}, \quad \Psi_h(1 \otimes X_i^\pm) = K_i^{\mp 1} \otimes X_i^\pm,$$

where $K_i = e^{d_i h H_i}$, and that these formulas define an algebra automorphism Ψ_q of $U_q \otimes U_q$ which extends to an automorphism of $\hat{U}_q \hat{\otimes} \hat{U}_q$. The following result was proved by Tanisaki (1992) by direct computation.

LEMMA 10.1.18 We have:
(i) $\tilde{\mathcal{R}}_q$ is an invertible element of $\hat{U}_q \hat{\otimes} \hat{U}_q$;
(ii) $\Psi_q(\Delta^{\mathrm{op}}(x)) = \tilde{\mathcal{R}}_q.\Delta(x).\tilde{\mathcal{R}}_q^{-1}$ for all $x \in \hat{U}_q$;
(iii) $(\Psi_q)_{23}(\tilde{\mathcal{R}}_q^{13}).\tilde{\mathcal{R}}_q^{23} = (\Delta \otimes \mathrm{id})(\tilde{\mathcal{R}})$;
(iv) $(\Psi_q)_{12}(\tilde{\mathcal{R}}_q^{13}).\tilde{\mathcal{R}}_q^{12} = (\mathrm{id} \otimes \Delta)(\tilde{\mathcal{R}})$. ∎

Suppose now that $\rho : U_q \to \mathrm{End}(V)$ and $\rho' : U_q \to \mathrm{End}(V')$ are finite-dimensional irreducible U_q-modules of type **1**. From the results of Subsection 10.1A it is clear that all but finitely-many terms in $\tilde{\mathcal{R}}_q$ act as zero on V and V', and hence that $(\rho \otimes \rho)(\tilde{\mathcal{R}}_q) = \tilde{R}_{\rho,\rho'}$ is a well-defined invertible operator on $V \otimes V'$. Moreover, V and V' are the direct sums of their weight spaces V_λ, V_μ', λ, $\mu \in P$, so we may define an invertible operator $E_{\rho,\rho'}$ on $V \otimes V'$ which acts as the scalar $q^{(\lambda,\mu)}$ on the subspace $V_\lambda \otimes V_\mu'$. Set $R_{\rho,\rho'} = E_{\rho,\rho'}\tilde{R}_{\rho,\rho'}$.

PROPOSITION 10.1.19 With the above notation, $R_{\rho,\rho'}$ is an invertible operator on $V \otimes V'$ such that

$$R_{\rho,\rho'}.(\rho \otimes \rho')(\Delta(x)).R_{\rho,\rho'}^{-1} = (\rho \otimes \rho')(\Delta^{\mathrm{op}}(x))$$

for all $x \in \hat{U}_q$. Moreover, $R_{\rho,\rho'}$ satisfies the quantum Yang–Baxter equation.

OUTLINE OF PROOF The operator $E_{\rho,\rho'}$ is defined so that it 'implements' the automorphism Ψ_q, in the sense that

$$(11) \qquad (\rho \otimes \rho')(\Psi_q(x)) = E_{\rho,\rho'}^{-1}(\rho \otimes \rho')(x)E_{\rho,\rho'}$$

for all $x \in U_q \otimes U_q$. To prove this, it suffices to check that the two sides agree on $V_\lambda \otimes V_\mu'$ for all λ, μ; we may also assume that x is one of the generators $X_i^\pm \otimes 1$, $1 \otimes X_i^\pm$, $K_i^{\pm 1} \otimes 1$ or $1 \otimes K_i^{\pm 1}$ of $U_q \otimes U_q$. For example, taking $x = X_i^+ \otimes 1$, the left-hand side of (11) gives

$$(\rho \otimes \rho')(\Psi_q(X_i^+ \otimes 1))\Big|_{V_\lambda \otimes V_\mu'} = \rho(X_i^+) \otimes \rho'(K_i^{-1})\Big|_{V_\lambda \otimes V_\mu'}$$
$$= q^{-(\mu,\alpha_i)}(\rho(X_i^+) \otimes 1)\Big|_{V_\lambda \otimes V_\mu'},$$

while the right-hand side gives

$$E_{\rho,\rho'}^{-1}(\rho \otimes \rho')(X_i^+ \otimes 1)E_{\rho,\rho'}\Big|_{V_\lambda \otimes V'_\mu} = q^{(\lambda,\mu)}E_{\rho,\rho'}^{-1}(\rho(X_i^+) \otimes 1)\Big|_{V_\lambda \otimes V'_\mu}$$

$$= q^{(\lambda,\mu)-(\lambda+\alpha_i,\mu)}(\rho(X_i^+) \otimes 1)\Big|_{V_\lambda \otimes V'_\mu}.$$

Thus, (11) holds in this case.

The first part of the proposition follows immediately from (11) and 10.1.18 (ii). The second part follows in a similar way using 10.1.18(iii) and (iv). ∎

In category-theoretic language:

COROLLARY 10.1.20 *The category of finite-dimensional U_q-modules of type* **1** *is a quasitensor category, with the commutativity isomorphism $V \otimes V' \to V' \otimes V$ given by $\sigma \circ R_{\rho,\rho'}$.* ∎

Since the matrix of $R_{\rho,\rho'}$ with respect to any $\mathbb{Q}(q)$ bases of V and V' has entries in $\mathbb{Q}(q)$, the preceding results also apply when q is specialized to any transcendental number $\epsilon \in \mathbb{C}^\times$. In fact, we could have replaced U_q by $U_{\mathcal{A}}^{\text{res}}$ in the preceding discussion, noting that Ψ_q takes $U_{\mathcal{A}}^{\text{res}}$ to $U_{\mathcal{A}}^{\text{res}}$ and that each term in $\tilde{\mathcal{R}}_q$ is an element of $U_{\mathcal{A}}^{\text{res}} \otimes U_{\mathcal{A}}^{\text{res}}$, so that the conclusion of 10.1.20 applies to U_ϵ^{res} for arbitrary ϵ (see Lusztig (1993) for further discussion of this point).

E Unitary representations We conclude this section by considering briefly the classification of the unitary representations of U_ϵ. See Subsection 4.1F for the definition of a unitary representation of a Hopf $*$-algebra, and Subsection 9.4B for the classification of the Hopf $*$-structures on U_ϵ. We shall work over $\mathbb{C}$, and accordingly set $U_\epsilon^{\mathbb{C}} = U_\epsilon \underset{\mathbb{Q}(\epsilon)}{\otimes} \mathbb{C}$. The complexification $V_\epsilon^{\mathbb{C}}(\lambda)$ of $V_\epsilon(\lambda)$, for any $\lambda \in P$, is an irreducible $U_\epsilon^{\mathbb{C}}$-module.

We first consider the *standard* Hopf $*$-structure on $U_\epsilon^{\mathbb{C}}$, defined, for $\epsilon \in \mathbb{R}$, by

$$K_i^* = K_i, \quad (X_i^+)^* = X_i^- K_i, \quad (X_i^-)^* = K_i^{-1}X_i^+.$$

PROPOSITION 10.1.21 *Let $\epsilon > 0$ and let $V_\epsilon^{\mathbb{C}}(\lambda)$ be the finite-dimensional irreducible $U_\epsilon^{\mathbb{C}}$-module with highest weight $\lambda \in P^+$. Then, $V_\epsilon^{\mathbb{C}}(\lambda)$ is unitary with respect to the standard $*$-structure on $U_\epsilon^{\mathbb{C}}$.*

OUTLINE OF PROOF One first shows, as in 10.1.8 and 10.1.9, that $V_\epsilon^{\mathbb{C}}(\lambda)$ admits a non-degenerate hermitian form $(\ ,\ )_\epsilon$ such that

$$(x.v_1,\, v_2)_\epsilon = (v_1,\, x^*.v_2)_\epsilon$$

for all $x \in U_\epsilon^{\mathbb{C}}$, v_1, $v_2 \in V_\epsilon^{\mathbb{C}}(\lambda)$. By 10.1.15, one can identify $V_\epsilon^{\mathbb{C}}(\lambda)$ and $V^{\mathbb{C}}(\lambda)$ as vector spaces over $\mathbb{C}$, where $V^{\mathbb{C}}(\lambda)$ denotes the irreducible $U^{\mathbb{C}}$-module with highest weight λ. Thus, we may regard $(\ ,\ )_\epsilon$ as a family of hermitian forms on the fixed vector space $V^{\mathbb{C}}(\lambda)$, parametrized by the set $\mathbb{R}_+$ of positive real numbers. Equivalently, we may consider the family T_ϵ of self-adjoint operators on $V^{\mathbb{C}}(\lambda)$, where

$$(v_1, v_2)_\epsilon = (v_1, T_\epsilon(v_2))_1.$$

It suffices to prove that all the eigenvalues of T_ϵ are positive for all $\epsilon \in \mathbb{R}_+$.

Suppose for a contradiction that T_ϵ is not positive definite for some $\epsilon \in \mathbb{R}_+$. Then, for some $\epsilon' \in \mathbb{R}_+$ between 1 and ϵ, $T_{\epsilon'}$ has a zero eigenvalue. But then $\ker(T_{\epsilon'})$ is a proper $U_{\epsilon'}^{\mathbb{C}}$-submodule of $V_{\epsilon'}^{\mathbb{C}}(\lambda)$, contradicting irreducibility. ∎

For the non-standard ∗-structures (i.e. those which are not equivalent to the standard structure), the question of unitarity of representations seems to be open in general. In the $sl_2(\mathbb{C})$ case, however, we have the following negative result.

PROPOSITION 10.1.22 *There are no non-trivial finite-dimensional unitary representations for any of the non-standard Hopf ∗-structures on $U_\epsilon^{\mathbb{C}}(sl_2(\mathbb{C}))$* ∎

The proof is a straightforward exercise, using complete reducibility and the explicit description of the irreducible $U_\epsilon^{\mathbb{C}}$-modules (cf. 6.4.10) and of the Hopf ∗-structures (cf. 9.4.3).

The question of infinite-dimensional unitary representations is more interesting, and seems still to be open in general. We content ourselves with the following examples, which give quantum analogues of the discrete and continuous series representations of $su_{1,1}$.

Example 10.1.23 Suppose that ϵ is real and > 1 and consider the ∗-structure $K_1^* = K_1$, $(X_1^+)^* = -X_1^- K_1$, $(X_1^-)^* = -K_1^{-1} X_1^+$ (quantum $su_{1,1}$). Then, there are three families of infinite-dimensional irreducible unitary representations:

(i) *Lowest weight discrete series* The Hilbert space with orthonormal basis $\{v_1, v_2, \ldots\}$ and action

$$K_1.v_r = \kappa^2 \epsilon^{2r-2} v_r, \quad X_1^+.v_r = \kappa \epsilon^r \left([r]_\epsilon \frac{\kappa^2 \epsilon^{r-1} - \kappa^{-2} \epsilon^{1-r}}{\epsilon - \epsilon^{-1}}\right)^{1/2} v_{r+1},$$

where κ is a real constant greater than 1. (The formula for the action of X_1^- can, of course, be deduced from those above using unitarity.)

(ii) *Highest weight discrete series* The same Hilbert space as in (i), with action

$$K_1.v_r = \kappa^2 \epsilon^{-2r} v_r, \quad X_1^+.v_r = \kappa \epsilon^{-r} \left([r-1]_\epsilon \frac{\kappa^{-2} \epsilon^r - \kappa^2 \epsilon^{-r}}{\epsilon - \epsilon^{-1}}\right)^{1/2} v_{r-1},$$

where $0 < \kappa < \epsilon$ and $v_0 = 0$.

(iii) *Continuous series*   The Hilbert space with orthonormal basis $\{v_r\}_{r \in \mathbf{Z}}$ and action

$$K_1.v_r = \kappa^2 \epsilon^{-2r} v_r, \quad X_1^+.v_r = \kappa \epsilon^{1-r} \left(c + [r]_\epsilon \frac{\kappa^{-2}\epsilon^{r-1} - \kappa^2 \epsilon^{1-r}}{\epsilon - \epsilon^{-1}} \right)^{1/2} v_{r-1},$$

where $1 \le \kappa < \epsilon$ and $c > 0$. $\Diamond$

Remark There are $*$-algebra structures on U_ϵ which are not Hopf $*$-algebra structures, and the definition of a unitary representation makes sense for them. For example, if $|\epsilon| = 1$ the following formulas define a $*$-algebra structure on U_ϵ:

(12) $$K_1^* = K_1^{-1}, \quad (X_1^+)^* = X_1^-, \quad (X_1^-)^* = X_1^+,$$

although this is not compatible with the coalgebra structure on U_ϵ. The classification of the unitary representations in this case is rather startling.

Suppose that $\epsilon = e^{\sqrt{-1}\pi\theta}$, where $0 < \theta < 1$. Let

$$\theta = \cfrac{1}{t_1 + \cfrac{1}{t_2 + \cfrac{1}{t_3 + \cdots}}}$$

be the continued fraction expansion of θ, where $t_1, t_2, \ldots \in \mathbf{N}$, and let $d_1 = t_1$, $d_2 = t_2 t_1 + 1$, $d_3 = t_3 t_2 t_1 + t_3 + t_1, \ldots$ be the sequence of denominators of the successive convergents

$$\frac{1}{t_1}, \quad \frac{t_2}{t_2 t_1 + 1}, \quad \frac{t_3 t_2 + 1}{t_3 t_2 t_1 + t_3 + t_1}, \ldots$$

Then, the m-dimensional irreducible representation of U_ϵ is unitary for the $*$-structure (12) (with respect to some inner product) if and only if m belongs to the sequence

$$1, \ 2, \ldots, t_1, \ t_1 + 1, \ 2t_1 + 1, \ldots, t_2 t_1 + 1, \ldots$$
$$\ldots, \ d_k + d_{k+1}, \ d_k + 2d_{k+1}, \ldots, \ d_k + t_{k+2} d_{k+1}, \ d_{k+1} + d_{k+2}, \ldots.$$

For example, if $\theta = \frac{1}{2}(\sqrt{5} - 1)$, the possible dimensions form the Fibonacci sequence $1, 2, 3, 5, 8, \ldots$. $\triangle$

10.2 Quantum invariant theory

One of the most beautiful results from the early years of the representation theory of Lie groups is the duality, established by Frobenius and Schur, between the representation theory of general linear groups and symmetric groups. An analogous theory was developed by Brauer for the orthogonal and symplectic groups. In this section, we describe a quantum analogue of these theories.

A Hecke and Birman–Murakami–Wenzl algebras First, we briefly recall the classical theory. Let $V^\natural$ be the natural $(n+1)$-dimensional representation of $sl_{n+1}(\mathbb{C})$, and let $\ell \geq 1$. One is interested in the algebra $\mathrm{Int}((V^\natural)^{\otimes \ell})$ of intertwining operators of the representation $(V^\natural)^{\otimes \ell}$, i.e. the linear maps from $(V^\natural)^{\otimes \ell}$ to itself which commute with the action of $sl_{n+1}(\mathbb{C})$. According to Frobenius and Schur, this algebra is generated by the permutations of adjacent pairs of factors in the tensor product. More precisely, assigning to the transposition $(i, i+1)$ in the symmetric group Σ_ℓ the flip of the ith and $(i+1)$th factors in $(V^\natural)^{\otimes \ell}$ defines a surjective homomorphism of algebras $\mathbb{C}[\Sigma_\ell] \to \mathrm{Int}((V^\natural)^{\otimes \ell})$.

For the orthogonal (resp. symplectic) groups, the natural representation $V^\natural$ is equipped with a non-degenerate symmetric (resp. skew-symmetric) bilinear form. This enables one to identify $V^\natural$ and $(V^\natural)^*$ as representations, and hence also $\mathrm{End}(V^\natural)$ and $\mathrm{End}(V^\natural)^*$. Let $\tau \in \mathrm{End}(V^\natural \otimes V^\natural)$ correspond to the map $\mathrm{End}(V^\natural) \to \mathrm{End}(V^\natural)$ given by $T \mapsto \mathrm{trace}(T).\mathrm{id}_{V^\natural}$. Then, Brauer's assertion is that the algebra of intertwining operators of $(V^\natural)^{\otimes \ell}$ is generated by the flips σ_i and the maps τ_i, $1 \leq i < \ell$, where τ_i is τ acting in the ith and $(i+1)$th factors in $(V^\natural)^{\otimes \ell}$.

We now turn to the quantum situation, beginning with the sl_{n+1} case. The role of the symmetric group in the Frobenius–Schur theory is played in the quantum theory by the algebra defined as follows.

DEFINITION 10.2.1 *The Hecke algebra $\mathcal{H}_\ell(\epsilon)$, where $\epsilon \in \mathbb{C}$ and $\ell \geq 1$, is the associative algebra over $\mathbb{C}$ with generators $\sigma_1, \ldots, \sigma_{\ell-1}$ and defining relations*

$$(13) \qquad \sigma_i \sigma_i^{-1} = \sigma_i^{-1} \sigma_i = 1,$$

$$(14) \qquad \sigma_i \sigma_j = \sigma_j \sigma_i \quad \text{if } |i - j| > 1,$$

$$(15) \qquad \sigma_i \sigma_{i+1} \sigma_i = \sigma_{i+1} \sigma_i \sigma_{i+1},$$

$$(16) \qquad (\sigma_i + 1)(\sigma_i - \epsilon) = 0.$$

(We set $\mathcal{H}_1(\epsilon) = \mathbb{C}$.)

Remarks [1] Putting $\epsilon = 1$ in (13), (14), (15) and (16) gives the defining relations of the symmetric group Σ_ℓ. Further, it is known that $\dim(\mathcal{H}_\ell(\epsilon)) = \ell!$, so $\mathcal{H}_\ell(\epsilon)$ may be regarded as a 'deformation' of the group algebra $\mathbb{C}[\Sigma_\ell]$.

[2] Since $\mathcal{H}_\ell(1) \cong \mathbb{C}[\Sigma_\ell]$ is semisimple, and since semisimplicity is an 'open' condition on an algebra (for it is equivalent to the non-degeneracy of the Killing trace), one expects that $\mathcal{H}_\ell(\epsilon)$ is semisimple for 'generic' values of ϵ. In fact, it is known that $\mathcal{H}_\ell(\epsilon) \cong \mathbb{C}[\Sigma_\ell]$ as algebras, provided $\epsilon \in \mathbb{C}^\times$ is not a root of unity (this is a sufficient, not a necessary, condition for semisimplicity).

[3] Historically, Hecke algebras arose in connection with algebraic groups over finite fields. Let $G = \mathrm{GL}_\ell(k)$, where k is a finite field with r elements, and let B be the Borel subgroup of G consisting of the upper triangular matrices. We apply the construction given at the beginning of Subsection 9.1D with $X = G/B$. Using the Bruhat decomposition

$$G = \coprod_{w \in \Sigma_\ell} BwB,$$

it is easy to see that the orbits of the diagonal action of G on $X \times X$ are parametrized by Σ_ℓ, with representatives (wB, B). The algebra $\boldsymbol{F}$ defined in Subsection 9.1D is generated by the orbits indexed by the transpositions $s_i = (i, i+1)$, $i = 1, \ldots, \ell - 1$, and in fact assigning to σ_i the orbit of $(s_i B, B)$ defines an isomorphism of algebras $\mathcal{H}_\ell(r) \to \boldsymbol{F}$. $\triangle$

Before discussing the quantum analogues of Brauer's algebras, we mention another algebra, closely related to the Hecke algebra, which plays an important role in this circle of ideas. It was introduced by Temperley & Lieb (1971) in their discussion of certain lattice models in statistical mechanics.

DEFINITION 10.2.2 *The Temperley–Lieb algebra* $\mathcal{TL}_\ell(\lambda)$, *where* $\lambda \in \mathbb{C}^\times$, *is the unital associative algebra over* $\mathbb{C}$ *with generators* $t_1, \ldots, t_{\ell-1}$ *and defining relations*

$$t_i^2 = t_i, \quad t_i t_{i \pm 1} t_i = \lambda t_i, \quad t_i t_j = t_j t_i \quad \text{if } |i - j| > 1.$$

PROPOSITION 10.2.3 *If* $\epsilon^2 \neq -1$, *there is a surjective homomorphism of algebras* $\mathcal{H}_\ell(\epsilon^2) \to \mathcal{TL}_\ell((\epsilon + \epsilon^{-1})^{-2})$ *given on generators by*

(17) $$\sigma_i \mapsto \epsilon^2 - (\epsilon^2 + 1)t_i. \quad \blacksquare$$

The proof is a straightforward verification that the assignment (17) respects the defining relations of $\mathcal{H}_\ell(\epsilon^2)$. It is easy to see that the homomorphism in the proposition is an isomorphism if $\ell = 2$, but not if $\ell > 2$.

For the quantum orthogonal and symplectic groups, the Hecke algebras are replaced by the algebras defined as follows.

DEFINITION 10.2.4 *Let ϵ, $r \in \mathbb{C}^\times$. The Birman–Murakami–Wenzl algebra $\mathcal{BMW}_\ell(r,\epsilon)$ is the associative algebra over $\mathbb{C}$ with generators $\sigma_1, \ldots, \sigma_{\ell-1}$ and defining relations (13), (14), (15) and*

(18) $$\tau_i \sigma_i = r^{-1}\tau_i,$$

(19) $$\tau_i \sigma_{i-1}^{\pm 1}\tau_i = r^{\pm 1}\tau_i,$$

where

$$\tau_i = 1 - \frac{\sigma_i - \sigma_i^{-1}}{\epsilon - \epsilon^{-1}}.$$

Remark The Hecke algebra $\mathcal{H}_\ell(\epsilon^2)$ is a quotient of $\mathcal{BMW}_\ell(r,\epsilon)$ for all r. In fact, the map $\sigma_i \mapsto \epsilon^{-1}\sigma_i$, $\tau_i \mapsto 0$ obviously extends to a surjective homomorphism $\mathcal{BMW}_\ell(r,\epsilon) \to \mathcal{H}_\ell(\epsilon^2)$. $\triangle$

B Quantum Brauer–Frobenius–Schur duality Let $\mathfrak{g}$ be of type A_n, B_n, C_n or D_n and assume that $\epsilon \in \mathbb{C}^\times$ is not a root of unity. From Subsection 10.1C, we know that there is a representation $\rho_\epsilon^\flat : U_\epsilon(\mathfrak{g}) \to \mathrm{End}(V_\epsilon^\flat)$ which has the natural representation $\rho^\flat$ of $\mathfrak{g}$ as its classical limit. In fact, as we observed in Subsection 8.3G in the case of $U_h(\mathfrak{g})$, $V_\epsilon^\flat = V^\flat$ as a vector space and, in a suitable basis, $\rho_\epsilon^\flat$ is given by the *same* formulas as $\rho^\flat$. The arguments in Subsection 10.1D show that, although we do not have a universal R-matrix for $U_\epsilon(\mathfrak{g})$, we can associate to $\rho_\epsilon^\flat$ a matrix $R_\epsilon \in \mathrm{End}(V_\epsilon^\flat \otimes V_\epsilon^\flat)$ which satisfies the quantum Yang-Baxter equation and is such that $I_\epsilon = \sigma \circ R_\epsilon$ commutes with the action of $U_\epsilon(\mathfrak{g})$ on $V_\epsilon^\flat \otimes V_\epsilon^\flat$. In fact, up to a scalar multiple, R_ϵ can be obtained from the R-matrices written down in (18) and (20) of Chapter 7 by replacing e^h by ϵ.

THEOREM 10.2.5 *Suppose that ϵ is not a root of unity, let $\mathfrak{g}$ be of type A_n, B_n, C_n or D_n, and let $\rho_\epsilon^\flat : U_\epsilon(\mathfrak{g}) \to \mathrm{End}(V_\epsilon^\flat)$ be the natural representation of $U_\epsilon(\mathfrak{g})$. Let $R_\epsilon \in \mathrm{End}(V_\epsilon^\flat \otimes V_\epsilon^\flat)$ be the result of replacing e^h by ϵ in (18) and (20) of Chapter 7, and let $I_\epsilon = \sigma \circ R_\epsilon$, where σ is the flip operator.*

(i) If $\mathfrak{g}$ is of type A_n, setting $\eta_\ell(\sigma_i) = \epsilon I_\epsilon^{i\,i+1}$ defines a representation η_ℓ of $\mathcal{H}_\ell(\epsilon^2)$ on $(V_\epsilon^\flat)^{\otimes \ell}$. Moreover, each of the operator algebras $\eta_\ell(\mathcal{H}_\ell(\epsilon^2))$ and $(\rho_\epsilon^\flat)^{\otimes \ell}(U_\epsilon^{\otimes \ell})$ is the commutant of the other.

(ii) If $\mathfrak{g}$ is of type B_n, C_n or D_n, setting $\beta_\ell(\sigma_i) = \epsilon I_\epsilon^{i\,i+1}$ defines a representation β_ℓ of $\mathcal{BMW}_\ell(r,\epsilon^2)$ on $(V_\epsilon^\flat)^{\otimes \ell}$, where

$$r = \begin{cases} \epsilon^{2n} & \text{if } \mathfrak{g} \text{ is of type } B_n, \\ -\epsilon^{(2n-1)/2} & \text{if } \mathfrak{g} \text{ is of type } C_n, \\ -\epsilon^{-2n-1} & \text{if } \mathfrak{g} \text{ is of type } D_n. \end{cases}$$

Moreover, each of the operator algebras $\beta_\ell(\mathcal{BMW}_\ell(r, \epsilon^2))$ and $(\rho_\epsilon^\natural)^{\otimes \ell}(U_\epsilon^{\otimes \ell})$ is the commutant of the other. $\blacksquare$

Let us make the sl_2 case a little more explicit. From 6.4.12, we recall that the natural representation $V_\epsilon^\natural$ has a basis $\{v_0, v_1\}$ on which the action of $U_\epsilon(sl_2(\mathbb{C}))$ is given by

$$K_1.v_0 = \epsilon v_0, \quad X_1^+.v_0 = 0, \quad X_1^-.v_0 = v_1,$$
$$K_1.v_1 = \epsilon^{-1} v_1, \quad X_1^+.v_1 = v_0, \quad X_1^-.v_1 = 0.$$

With respect to the basis $\{v_0 \otimes v_0, \ v_1 \otimes v_0, \ v_0 \otimes v_1, \ v_1 \otimes v_1\}$ of $V_\epsilon^\natural \otimes V_\epsilon^\natural$, the R-matrix is given by

$$R_\epsilon = \begin{pmatrix} \epsilon & 0 & 0 & 0 \\ 0 & 1 & 0 & 0 \\ 0 & \epsilon - \epsilon^{-1} & 1 & 0 \\ 0 & 0 & 0 & \epsilon \end{pmatrix}.$$

It is easy to verify directly that the matrices

(20)
$$\eta_\ell(\sigma_i) = \begin{pmatrix} \epsilon^2 & 0 & 0 & 0 \\ 0 & \epsilon^2 - 1 & \epsilon & 0 \\ 0 & \epsilon & 0 & 0 \\ 0 & 0 & 0 & \epsilon^2 \end{pmatrix},$$

acting in the ith and $(i+1)$th places, satisfy the defining relations of $\mathcal{H}_\ell(\epsilon^2)$. In fact, they satisfy a stronger condition:

PROPOSITION 10.2.6 *Let $V_\epsilon^\natural$ be the natural representation of $U_\epsilon(sl_2(\mathbb{C}))$, and let $\ell \geq 2$. Then, the representation η_ℓ of $\mathcal{H}_\ell(\epsilon^2)$ on $(V_\epsilon^\natural)^{\otimes \ell}$ given by 10.2.5 factors through the Temperley–Lieb algebra $TL_\ell((\epsilon + \epsilon^{-1})^{-2})$.*

OUTLINE OF PROOF We send t_i to the matrix

$$\begin{pmatrix} 0 & 0 & 0 & 0 \\ 0 & \frac{1}{\epsilon^2 + 1} & -\frac{\epsilon}{\epsilon^2 + 1} & 0 \\ 0 & -\frac{\epsilon}{\epsilon^2 + 1} & \frac{\epsilon^2}{\epsilon^2 + 1} & 0 \\ 0 & 0 & 0 & 0 \end{pmatrix}$$

acting in the ith and $(i+1)$th places in $(V_\epsilon^\natural)^{\otimes \ell}$. It is easy to check that this is compatible with (20) and the homomorphism (17), and that the images of the t_i satisfy the defining relations of $TL_\ell((\epsilon + \epsilon^{-1})^{-2})$ (it is obviously enough to do the case $\ell = 2$). $\blacksquare$

From this and the comment following 10.2.3, it follows that η_ℓ is not faithful if $\ell > 2$.

C Another realization of Hecke algebras The description of the Hecke algebra given in the third remark following 10.2.1 can obviously be generalized by replacing $GL_\ell(k)$ by an arbitrary reductive algebraic group over the finite field k. In particular, taking G to be the group over k associated to a complex simple Lie algebra $\mathfrak{g}$, we obtain Hecke algebras $\mathcal{H}_\mathfrak{g}(\epsilon)$: then $\mathcal{H}_\ell(\epsilon)$ is the special case $\mathfrak{g} = sl_\ell(\mathbb{C})$. Restricting ourselves to the simply-laced case for simplicity, $\mathcal{H}_\mathfrak{g}(\epsilon)$ is defined by generators $\sigma_1, \ldots, \sigma_n$ and relations (13), (14) (if $a_{ij} = 0$), (15) (if $a_{ij} = -1$) and (16).

It turns out that there is a realization of $\mathcal{H}_\mathfrak{g}(\epsilon^{-2})$, for any (simply-laced) $\mathfrak{g}$, in terms of intertwining operators of $U_\epsilon(\mathfrak{g})$ (although this is quite different from the realization in 10.2.5 when $\mathfrak{g}$ is of type A, B, C or D). We recall the action of the braid group $\mathcal{B}_\mathfrak{g}$ on $U_\epsilon(\mathfrak{g})$ from 9.1.2 (we continue to assume that ϵ is not a root of unity). If V is a finite-dimensional irreducible U_ϵ-module of type **1**, we can twist V by the automorphism T_i to get new representations V^i of U_ϵ on the same underlying vector space. Note that

$$\dim(V_\mu^i) = \dim(V_{s_i(\mu)})$$

for any $\mu \in P$, where s_i is the ith fundamental reflection in the Weyl group of $\mathfrak{g}$. But, by 10.1.15, the dimension on the right-hand side is given by the Weyl character formula. Since, in the classical case, the dimensions of the weight spaces of a finite-dimensional irreducible $\mathfrak{g}$-module are invariant under the action of the the Weyl group on the weights, it follows that

$$\dim(V_\mu^i) = \dim(V_\mu)$$

for all $\mu \in P$. Since V and V^i have the same character, they are isomorphic as U_ϵ-modules (this follows from 10.1.15 and the corresponding classical statement). Hence, there are isomorphisms $\sigma_i \in \mathrm{End}(V)$, unique up to a scalar multiple, such that

$$\sigma_i(x.v) = T_i(x).\sigma_i(v)$$

for all $x \in U_\epsilon$, $v \in V$. From the defining relations of $\mathcal{B}_\mathfrak{g}$, it is obvious that

$$\sigma_i\sigma_j\sigma_i = c_{ij}\sigma_j\sigma_i\sigma_j \quad \text{if } a_{ij} = -1,$$
$$\sigma_i\sigma_j = c_{ij}\sigma_j\sigma_i \quad \text{if } a_{ij} = 0,$$

for some non-zero scalars c_{ij}.

PROPOSITION 10.2.7 *Assume that ϵ is not a root of unity and that $\mathfrak{g}$ is of type A, D or E. Let θ be the highest root of $\mathfrak{g}$. When suitably normalized, the intertwiners σ_i, restricted to the zero weight space $V_\epsilon(\theta)_0$ of the irreducible $U_\epsilon(\mathfrak{g})$-module of highest weight θ, satisfy the defining relations of $\mathcal{H}_\mathfrak{g}(\epsilon^{-2})$.* ∎

Note that, by 10.1.15, the character of $V_\epsilon(\theta)$ is the same as that of the adjoint representation of $\mathfrak{g}$. In particular, $\dim(V_\epsilon(\theta)_0) = \mathrm{rank}(\mathfrak{g})$.

Bibliographical notes

10.1 Most of this subsection is based on Lusztig (1988). One alternative approach, using cohomological methods, can be seen in Andersen, Polo & Wen (1991), and another, using methods from non-commutative ring theory, can be found in Joseph & Letzter (1992).

Complete reducibility in the quantum case was first proved by Rosso (1990b), and it is his proof we have presented, simplified using some remarks from Lusztig (1990c). Different proofs are given by Andersen *et al.* (1991) and by Joseph & Letzter (1992). See also Lusztig (1993).

The quantum analogue of the Harish Chandra homomorphism and the description of the centre of U_ϵ are essentially due to Rosso (1990b). The latter was made more precise by De Concini & Kac (1990), who also obtained the quantum analogue of the determinant formula.

Example 10.1.23 and the remark which follows it are taken from Vaysleb (1992).

10.2 The connection between Hecke algebras and representation theory was first noticed by Iwahori (1964). The BMW algebra was introduced by Murakami (1987) and Birman & Wenzl (1989). A good exposition of the theory of Hecke and BMW algebras can be found in Goodman, de la Harpe & Jones (1989). See Wenzl (1988b) for the theory of Brauer's algebras in the classical situation.

Theorem 10.2.5 is due to Jimbo (1986a) for type A, and to Kirillov & Reshetikhin (1989) in the remaining cases; see also Hayashi (1992b). Jimbo even defines quantum analogues of Young symmetrizers; further steps in this direction are taken by Goodman & Wenzl (1990). It is interesting to note that Jimbo obtained the R-matrix R_ϵ without the benefit of the universal R-matrix.

Representations of Hecke and BMW algebras have been used to construct link invariants. This goes back, in the sl_2 case, to Jones (1987).

Proposition 10.2.7 was noticed independently by Levendorskii & Soibelman (1990) and Lusztig (1990c).

11
Representations of QUE algebras: the root of unity case

In this chapter, we study the representation theory of $U_q(\mathfrak{g})$ when q is specialized to a root of unity ϵ. We have seen in Chapter 9 that there are two ways to carry out this specialization, leading to two different algebras, $U_\epsilon(\mathfrak{g})$ and $U_\epsilon^{\mathrm{res}}(\mathfrak{g})$. Their representation theories turn out to be very different, both from each other and from that of $U_q(\mathfrak{g})$ (or of $\mathfrak{g}$).

The first important difference between the representation theory of the non-restricted specialization $U_\epsilon(\mathfrak{g})$ and that of $\mathfrak{g}$ is that most of the representations of $U_\epsilon(\mathfrak{g})$ are neither highest nor lowest weight. Every irreducible representation has, of course, a well-defined central character, and we shall see that, conversely, there are a finite number of representations (at least one) of $U_\epsilon(\mathfrak{g})$ with any given central character, whereas for $\mathfrak{g}$ only a discrete set of central characters arise from finite-dimensional representations. Using the fact, proved in Chapter 9, that the centre of $U_\epsilon(\mathfrak{g})$ is essentially (but not quite) a polynomial algebra on $\dim(\mathfrak{g})$ generators, it follows that there are a finite number of irreducible representations of $U_\epsilon(\mathfrak{g})$ associated to each point of a complex algebraic variety of dimension equal to that of $\mathfrak{g}$. Almost all of these representations have the same dimension ℓ^N, and all the others have strictly smaller dimension. The methods used to establish these results, which we describe in Section 11.1, come from the theory of finite-dimensional associative algebras, which applies to this situation because $U_\epsilon(\mathfrak{g})$ is finite dimensional over its centre.

Although the representation theory of $U_\epsilon(\mathfrak{g})$ differs so markedly from that of $\mathfrak{g}$, we show that there is a remarkable connection between the quantum and classical situations. In fact, one can associate to every irreducible representation of $U_\epsilon(\mathfrak{g})$ an element of the classical (adjoint) complex Lie group G whose Lie algebra is $\mathfrak{g}$, in such a way that representations which are associated to conjugate elements of G are essentially 'the same', in the sense that they are related by twisting with an automorphism of $U_\epsilon(\mathfrak{g})$. One expects that the structure of a representation of $U_\epsilon(\mathfrak{g})$ will be reflected in the geometric properties of the corresponding conjugacy class in G. In particular, it has been conjectured that the dimension of a representation should be divisible by ℓ^d if the (complex) dimension of the associated conjugacy class is $2d$.

To test this conjecture, it is important to describe the representations of $U_\epsilon(\mathfrak{g})$ as explicitly as possible. This problem is only partially solved, but we review what is currently known.

In Section 11.2, we turn to the representation theory of the restricted specialization $U_\epsilon^{res}(\mathfrak{g})$. This time, the finite-dimensional irreducible representations are parametrized by dominant weights, exactly as in the representation theory of $\mathfrak{g}$. However, the structure of such a representation of $U_\epsilon^{res}(\mathfrak{g})$, including its dimension, is in general different from that of the representation of $\mathfrak{g}$ with the same highest weight. The characters of the irreducible representations are given, rather implicitly, by a formula conjectured by Lusztig and proved by Kazhdan and Lusztig. We discuss the conjecture in this chapter, and its proof in Chapter 16.

The study of the irreducible representations can be reduced to that of two special cases, namely that in which all the components of the highest weight are less than ℓ, and that in which all the components are divisible by ℓ. In fact, every irreducible representation is isomorphic to a tensor product of representations of the two types. This result is analogous to a classical theorem of Steinberg in characteristic p, adding to the evidence seen in Chapter 9 of a close connection between quantum groups at roots of unity and classical groups in characteristic p.

In Section 11.3, we study tensor products of representations of $U_\epsilon^{res}(\mathfrak{g})$. Such tensor products are not, in general, completely reducible, but we show that if one discards the part of the tensor product which has quantum dimension zero, the resulting 'truncated' tensor product is a direct sum of irreducibles. This new tensor product is of interest because it leads to the construction of quasitensor categories whose fusion rings, at least in some cases, are found experimentally to be the same as those which arise from certain quantum and conformal field theories. This coincidence is strong evidence of a deep connection between quantum groups and quantum field theory, a connection which is under very active investigation at the present time.

Note In this chapter, $\mathfrak{g}$ denotes a finite-dimensional complex simple Lie algebra. We use the notation in the Appendix.

11.1 The non-restricted case

In this section, we assume that ϵ is a primitive ℓth root of unity, where ℓ is odd and greater than d_i for all i. We use the notation set up in Section 9.2. All representations are on complex vector spaces.

A Parametrization of the irreducible representations of U_ϵ We begin by observing that every irreducible U_ϵ-module V is finite dimensional. Indeed, let $\mathcal{Z}(V)$ be the subalgebra of the algebra of intertwining operators of V given by the action of the elements of the centre Z_ϵ of U_ϵ. Since, by 9.2.7, U_ϵ is finitely generated as a Z_ϵ-module, V is finitely generated as a $\mathcal{Z}(V)$-module. If

$0 \neq f \in \mathcal{Z}(V)$, then $f(V) = V$, otherwise $f(V)$ would be a proper submodule of V. Hence, by Nakayama's lemma, there exists an endomorphism $g \in \mathcal{Z}(V)$ such that $(1 - fg)(V) = 0$, i.e. f is invertible. Thus, $\mathcal{Z}(V)$ is a field. It follows easily (by using Hilbert's Nullstellensatz, for example) that $\mathcal{Z}(V)$ consists of scalar operators. Thus, V is a finite-dimensional vector space.

Since Z_ϵ acts by scalar operators on V, there is a homomorphism $\chi_V :$ $Z_\epsilon \to \mathbb{C}$, the *central character* of V, such that

$$z.v = \chi_V(z)v$$

for all $z \in Z_\epsilon$, $v \in V$. Isomorphic representations obviously have the same central character, so assigning to a U_ϵ-module its central character gives a well-defined map $\Xi : \mathrm{Rep}(U_\epsilon) \to \mathrm{Spec}(Z_\epsilon)$, where $\mathrm{Rep}(U_\epsilon)$ is the set of isomorphism classes of irreducible U_ϵ-modules, and $\mathrm{Spec}(Z_\epsilon)$ is the set of algebra homomorphisms $Z_\epsilon \to \mathbb{C}$.

To see that Ξ is surjective, let I_ϵ^χ, for $\chi \in \mathrm{Spec}(Z_\epsilon)$, be the ideal in U_ϵ generated by

$$\ker(\chi) = \{z - \chi(z).1 \mid z \in Z_\epsilon\}.$$

To construct $V \in \Xi^{-1}(\chi)$ is the same thing as to construct an irreducible representation of the algebra $U_\epsilon^\chi = U_\epsilon / I_\epsilon^\chi$; note that U_ϵ^χ is finite-dimensional by 9.2.7(vi) and non-zero by Nakayama's lemma. Thus, we may take V, for example, to be any irreducible subrepresentation of the regular representation of U_ϵ^χ.

However, Ξ is not quite injective:

THEOREM 11.1.1 *There is a (non-empty) closed proper subvariety $\mathcal{D}$ of* $\mathrm{Spec}(Z_\epsilon)$ *such that*

(i) if $\chi \in \mathrm{Spec}(Z_\epsilon) \backslash \mathcal{D}$, then $\Xi^{-1}(\chi)$ consists of a single irreducible U_ϵ-module of dimension ℓ^N;

(ii) if $\chi \in \mathcal{D}$, then $\Xi^{-1}(\chi)$ consists of a finite number of irreducible U_ϵ-modules of dimensions strictly less than ℓ^N.

The remainder of this subsection is devoted to an outline of the proof of this theorem. The strategy is to reduce it to the representation theory of certain associative algebras over a field.

Note first that, since U_ϵ is an integral domain (see 9.2.4), so is Z_ϵ, and we may consider the field of fractions $Q(Z_\epsilon)$ of Z_ϵ. Then,

$$Q(U_\epsilon) = U_\epsilon \underset{Z_\epsilon}{\otimes} Q(Z_\epsilon)$$

is a division algebra, finite-dimensional over its centre $Q(Z_\epsilon)$ by 9.2.7 again. Let $\mathbb{F}$ be a maximal commutative subfield of $Q(U_\epsilon)$; then $\mathbb{F}$ is a finite extension of $Q(Z_\epsilon)$ of degree m, say. It follows from standard results in the theory of associative algebras (see, for example, Pierce (1982)) that $Q(U_\epsilon)$ has dimension m^2 over $Q(Z_\epsilon)$, and that $Q(U_\epsilon) \underset{Q(Z_\epsilon)}{\otimes} \mathbb{F}$ is isomorphic to the algebra $M_m(\mathbb{F})$ of all $m \times m$ matrices over $\mathbb{F}$. Thus, if $x \in U_\epsilon$, we may regard x as an $m \times m$ matrix and consider its characteristic polynomial

$$\det(\lambda - x) = \lambda^m - (\operatorname{tr} x)\lambda^{m-1} + \cdots + (-1)^m \det x.$$

It follows from the next result that all the coefficients of the powers of λ on the right-hand side of this equation are in Z_ϵ (see Bourbaki (1985), Chap. 5, §1, no. 6, Corollaire 1).

PROPOSITION 11.1.2 *U_ϵ is integrally closed and (hence) so is its centre Z_ϵ.* ∎

We can now define the subvariety $\mathcal{D} \subset \operatorname{Spec}(Z_\epsilon)$ in the statement of the theorem: it is the set of zeros of the ideal in Z_ϵ generated by the elements $\det\left(\operatorname{tr}(u_i u_j)\right)$, where $u_1, \ldots, u_{m^2}$ are any elements of U_ϵ. By standard associative algebra techniques applied to the algebras U_ϵ^χ, one shows

LEMMA 11.1.3 *(i) If $\chi \in \operatorname{Spec}(Z_\epsilon) \backslash \mathcal{D}$, then U_ϵ^χ is isomorphic to $M_m(\mathbb{C})$, and (hence) there is, up to isomorphism, exactly one irreducible U_ϵ-module V_χ with character χ. One has $\dim(V_\chi) = m$.*

(ii) If $\chi \in \mathcal{D}$, then $\dim(U_\epsilon^\chi) \geq m^2$, but the dimension of every irreducible U_ϵ^χ-module is strictly less than m. ∎

It only remains to prove that $m = \ell^N$; at the same time, we shall prove that the canonical projection $\tau : \operatorname{Spec}(Z_\epsilon) \to \operatorname{Spec}(Z_0)$ has degree ℓ^n (see 9.2.8).

It is enough to show that there is an open subset $\mathcal{O} \subset \operatorname{Spec}(Z_0)$ (in the complex topology) such that, for every $\chi \in \tau^{-1}(\mathcal{O})$, $\Xi^{-1}(\chi)$ consists of irreducible modules of dimension ℓ^N, for $\tau^{-1}(\mathcal{O})$ obviously cannot be contained in $\mathcal{D}$. But $\ell^{2N+n} = m^2 \deg \tau$, since

$$\dim_{Q(Z_0)}(Q(Z_\epsilon)) = \deg \tau,$$
$$\dim_{Q(Z_\epsilon)}(Q(U_\epsilon)) = m^2,$$
$$\dim_{Q(Z_0)}(Q(U_\epsilon)) = \ell^{2N+n}.$$

Here, the first equality is a definition, the second has been pointed out above, and the third follows from 9.2.2.

To construct $\mathcal{O}$, we begin by considering the Verma modules $M_\epsilon(\boldsymbol{\omega})$, $\boldsymbol{\omega} \in (\mathbb{C}^\times)^n$, which are defined exactly as in Subsection 10.1A. Note that, since the elements $x_\beta^- = (X_\beta^-)^\ell$, $\beta \in \Delta^+$, are in Z_ϵ, the vectors $x_\beta^- . v_{\boldsymbol{\omega}} \in M_\epsilon(\boldsymbol{\omega})$ are primitive, where $v_{\boldsymbol{\omega}}$ is a highest weight vector of $M_\epsilon(\boldsymbol{\omega})$. Let $\overline{M_\epsilon(\boldsymbol{\omega})}$ be the quotient of $M_\epsilon(\boldsymbol{\omega})$ by the U_ϵ-submodule generated by all such vectors. Note that the central elements $x_\beta^\pm$ act as zero on $\overline{M_\epsilon(\boldsymbol{\omega})}$ for all $\boldsymbol{\omega}$.

It follows from 9.2.2 that $\{(X^-_{\beta_1})^{t_1}\dots(X^-_{\beta_N})^{t_N}.v_{\boldsymbol{\omega}}\}_{0 \le t_r < \ell, r=1,\dots,N}$ is a basis of $\overline{M_\epsilon(\boldsymbol{\omega})}$; in particular, $\overline{M_\epsilon(\boldsymbol{\omega})}$ has dimension ℓ^N. Moreover, $\overline{M_\epsilon(\boldsymbol{\omega})}$ is irreducible for generic $\boldsymbol{\omega}$. In fact, it follows from 10.1.17 that $\overline{M_\epsilon(\boldsymbol{\omega})}$ is irreducible if and only if $\omega_\beta^2 \epsilon^{2(\rho,\beta)-m(\beta,\beta)} \ne 1$ for all $\beta \in \Delta^+$, $0 \le m < \ell$. This condition obviously holds, in particular, if the central elements $k_\beta^2 \in Z_0$ act as scalars $\ne 1$ on $\overline{M_\epsilon(\boldsymbol{\omega})}$ for all $\beta \in \Delta^+$. To summarize, we have shown that, if

$$\mathcal{H} = \{\chi \in \mathrm{Spec}(Z_0) \mid \chi(x_\beta^\pm) = 0, \text{ and } \chi(k_\beta^2) \ne 1 \text{ for all } \beta \in \Delta^+\},$$

then $(\tau \circ \Xi)^{-1}(\mathcal{H})$ consists of representations of dimension ℓ^N.

The desired open set $\mathcal{O} \subset \mathrm{Spec}(Z_0)$ is manufactured from $\mathcal{H}$ by using the action of the automorphism group $\mathcal{G}$ on $\hat{U}_\epsilon = U_\epsilon \underset{Z_0}{\otimes} \hat{Z}_0$ (see Subsection 9.2C). If $\phi : U_\epsilon \to \mathrm{End}(V)$ is any irreducible U_ϵ-module, the central elements $x_\beta^\pm$, $k_\beta^{\pm 1}$ act as scalars on V, and hence $\hat{Z}_0$, which consists of holomorphic functions of these elements, also acts on V. Thus, V is naturally a $\hat{U}_\epsilon$-module. We can therefore 'twist' V by any element $\gamma \in \mathcal{G}$ to get a new representation $\phi_\gamma = \phi \circ \gamma$ on the same vector space. In particular, every representation in $(\tau \circ \Xi)^{-1}(\mathcal{G}.\mathcal{H})$ has dimension ℓ^N. The proof of 11.1.1 is thus completed by the following lemma, which is proved by using the explicit formulas in 9.2.13 for the action of $\mathcal{G}$ on U_ϵ.

LEMMA 11.1.4 *$\mathcal{G}.\mathcal{H}$ contains an open subset $\mathcal{O}$ of $\mathrm{Spec}(Z_0)$ (in the complex topology).* ∎

Remark The modules $\overline{M_\epsilon(\boldsymbol{\omega})}$ used in the proof of 11.1.1 are called *diagonal*. The reason is clear: under the map $\pi : \mathrm{Spec}(Z_0) \to G^0$ defined in Subsection 9.2C, their Z_0-characters go to elements of the Cartan subgroup H of G (we recall that G is the adjoint group of $\mathfrak{g}$). Similarly, one can define an irreducible U_ϵ-module to be *triangular* if it contains a (non-zero) vector which is annihilated by the X_i^+ and is a common eigenvector for the K_i, $i = 1, \dots, n$. The image of the Z_0-character of such a module under π is an element of the Borel subgroup $B_+ = HN_+$ of G. Corollary 9.2.19, together with the fact that every element of G is conjugate to an element of B_+, implies that if V is any irreducible U_ϵ-module, there is an automorphism $\gamma \in \mathcal{G}$ such that the twist of V by γ is triangular. △

Theorem 11.1.1 gives an adequate parametrization of the irreducible representations of U_ϵ, but gives little idea of their structure. In particular, one does not know the dimensions of the representations in $\Xi^{-1}(\mathcal{D})$. However, de Concini, Kac and Procesi (1992a) make the following interesting conjecture:

CONJECTURE 11.1.5 *Let V be an irreducible representation of U_ϵ, and let g_V be the image of V under the composite map $\pi \circ \tau \circ \Xi : \mathrm{Rep}(U_\epsilon) \to G$. Let*

C_V be the conjugacy class of g_V in G, and let $\dim(C_V)$ be its dimension (as a complex manifold). Then, $\dim(V)$ is divisible by $\ell^{\frac{1}{2}\dim(C_V)}$.

More generally, one expects that the structure of a representation V will be reflected in the geometric properties of C_V.

The conjugacy classes in G of maximal dimension are those of the regular elements, and their dimension is $2N$. For such classes, the conjecture is true:

THEOREM 11.1.6 *With the notation in 11.1.5, suppose that* $\dim(C_V) = 2N$. *Then,* $\chi_V \in \mathrm{Spec}(Z_\epsilon)\backslash\mathcal{D}$. ∎

In the next subsection, we shall test 11.1.5 by giving some explicit constructions of representations of U_ϵ. But first, we discuss the simplest example, namely $U_\epsilon(sl_2(\mathbb{C}))$.

Example 11.1.7 By elementary methods, one can easily show that every irreducible representation of $U_\epsilon(sl_2(\mathbb{C}))$ is isomorphic to exactly one of the following:

(i) the ℓ-dimensional representation $V(\lambda, a, b)$, where $\lambda \in \mathbb{C}^\times$, $a, b \in \mathbb{C}$, with basis $\{v_0, v_1, \ldots, v_{\ell-1}\}$ and action

$$K_1.v_j = \lambda\epsilon^{-2j}v_j, \quad X_1^-.v_j = v_{j+1}, \; j < \ell - 1, \quad X_1^-.v_{\ell-1} = bv_0,$$
$$X_1^+.v_j = \left(\frac{(\lambda\epsilon^{1-j} - \lambda^{-1}\epsilon^{j-1})(\epsilon^j - \epsilon^{-j})}{(\epsilon - \epsilon^{-1})^2} + ab\right)v_{j-1}, \quad X_1^+.v_0 = av_{\ell-1};$$

(ii) the $(r+1)$-dimensional representation V_r^σ, where $0 \le r < \ell - 1$ and $\sigma = \pm$, with basis $\{v_0, v_1, \ldots, v_r\}$ and action

$$K_1.v_j = \sigma\epsilon^{r-2j}v_j, \quad X_1^+.v_j = \sigma[r-j+1]_\epsilon v_{j-1}, \quad X_1^-.v_j = [j+1]_\epsilon v_{j+1}$$

$(v_{-1} = v_{r+1} = 0)$.

The discriminant set $\mathcal{D}$ consists of the $\ell-1$ points $\chi_r^\pm$, $r = 0, 1, \ldots, \frac{1}{2}(\ell-3)$, where

$$\chi_r^\pm(x_1^+) = \chi_r^\pm(x_1^-) = 0, \quad \chi_r^\pm(k_1) = \pm 1, \quad \chi_r^\pm(\Omega) = \pm\frac{\epsilon^{r+1} + \epsilon^{-r-1}}{(\epsilon - \epsilon^{-1})^2}.$$

If $\chi \in \mathrm{Spec}(Z_\epsilon)\backslash\mathcal{D}$, $\Xi^{-1}(\chi)$ consists of a single representation, namely the $V(\lambda, a, b)$ such that

$$\chi(x_1^-) = b, \quad \chi(k_1) = \lambda^\ell,$$
$$\chi(x_1^+) = a\prod_{j=1}^{\ell-1}\left(ab + \frac{(\lambda\epsilon^{1-j} - \lambda^{-1}\epsilon^{-1+j})(\epsilon^j - \epsilon^{-j})}{(\epsilon - \epsilon^{-1})^2}\right).$$

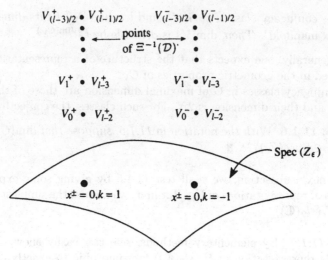

Fig. 18 The space of representations of $U_\epsilon(sl_2(\mathbb{C}))$

On the other hand, there are exactly two representations with any given central character in $\mathcal{D}$. In fact, $\Xi^{-1}(\chi_r^\pm) = \{V_r^\pm, V_{\ell-r-2}^\pm\}$.

We can visualize $\mathrm{Rep}(U_\epsilon)$ and $\mathrm{Spec}(Z_\epsilon)$ as shown in Fig. 18.

Note finally that, since every irreducible representation with central character outside $\mathcal{D}$ has dimension ℓ, and every non-trivial conjugacy class in $PSL_2(\mathbb{C})$ has (complex) dimension 2, 11.1.5 is true for $U_\epsilon(sl_2(\mathbb{C}))$. $\Diamond$

B Some explicit constructions In this subsection, we make the additional assumption that *ℓ is coprime to the determinant of the symmetrized Cartan matrix* $(d_i a_{ij})$ *of* $\mathfrak{g}$.

We should like to have an explicit description of the irreducible representations of $U_\epsilon(\mathfrak{g})$ for arbitrary $\mathfrak{g}$ analogous to that given for $U_\epsilon(sl_2(\mathbb{C}))$ at the end of the preceding subsection. Unfortunately, this problem is only partially solved. We shall restrict ourselves to the construction of representations of the following kind.

DEFINITION 11.1.8 *An irreducible representation* V *of* U_ϵ *is* cyclic *if the elements* $(X_i^\pm)^\ell \in Z_0$ *act as non-zero scalars on* V.

The reason for this terminology is clear: in a cyclic representation of $U_\epsilon(sl_2(\mathbb{C}))$, for example, one can get from any weight vector of V to (a non-zero multiple of) any other by applying some power of X_1^+ or X_1^-, in contrast to the linear structure of the representations of $U(sl_2(\mathbb{C}))$ (see Fig. 19).

We note the following simple observation:

PROPOSITION 11.1.9 *If* V *is a cyclic representation of* U_ϵ, *all the weight spaces of* V *have the same dimension and (hence)* $\dim(V)$ *is divisible by* ℓ^n.

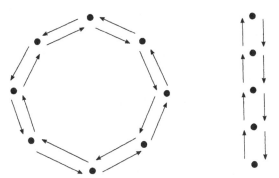

Fig. 19 Cyclic representations of $U_\epsilon(sl_2(\mathbb{C}))$ vs. representations of $sl_2(\mathbb{C})$

PROOF Let $V_\lambda = \{v \in V \mid K_i.v = \epsilon^{(\lambda,\alpha_i)}v,\ \text{for } i = 1,\ldots,n\}$ and V_μ be two non-zero weight spaces of V. Then, $\lambda - \mu = \sum_i k_i\alpha_i$ for some $k_i \in \mathbb{Z}$. Write

$$X_i^{k_i} = \begin{cases} (X_i^+)^{k_i} & \text{if } k_i > 0, \\ (X_i^-)^{-k_i} & \text{if } k_i < 0, \\ 1 & \text{if } k_i = 0. \end{cases}$$

Then, $X_1^{k_1} X_2^{k_2} \ldots X_n^{k_n}$ is an invertible operator $V_\mu \to V_\lambda$. In particular, $\dim(V_\lambda) = \dim(V_\mu)$.

Since V is the sum of its weight spaces, the last part will follow if we can show that $V_\lambda = V_\mu$ iff $\lambda - \mu \in \ell Q$. But the condition that $V_\lambda = V_\mu$ is clearly that

$$\sum_j d_i a_{ij} k_j \equiv 0 \pmod{\ell}$$

for all i. By the assumption on ℓ, this holds if and only if $k_j \equiv 0 \pmod{\ell}$ for all j. ∎

Thus, the dimension of a cyclic representation of U_ϵ is $\geq \ell^n$. The next two results show that the dimension ℓ^n is realized if $\mathfrak{g}$ is of type A_n, B_n or C_n, but not otherwise. To state them, we first introduce an auxiliary quantum algebra.

DEFINITION 11.1.10 $\mathbb{C}_\epsilon[x, z, x^{-1}, z^{-1}]$ *is the associative algebra over* $\mathbb{C}$ *with generators* x, x^{-1}, z, z^{-1} *and defining relations*

$$zx = \epsilon^2 xz, \quad xx^{-1} = x^{-1}x = zz^{-1} = z^{-1}z = 1.$$

We shall also use the following notation: if u is an invertible element of a ring,

$$\{u\} = u + u^{-1}.$$

PROPOSITION 11.1.11 *The following formulas define homomorphisms of algebras:*

(i) ($\mathfrak{g}$ *of type* A_n) $U_{\epsilon^2}(\mathfrak{g}) \to \bigotimes_{j=1}^{n+1} \mathbf{C}_\epsilon[x_j, x_j^{-1}, z_j, z_j^{-1}]$,

$$X_i^+ \mapsto \frac{x_i^{-1} x_{i+1}}{(\epsilon + \epsilon^{-1})^2} \{\epsilon^{-1} z_i\}, \quad X_i^- \mapsto -\frac{x_i x_{i+1}^{-1}}{(\epsilon - \epsilon^{-1})^2} \{\epsilon^{-1} z_{i+1}\}, \quad K_i \mapsto z_i^{-1} z_{i+1}.$$

(ii) ($\mathfrak{g}$ *of type* B_n) $U_{\epsilon^2}(\mathfrak{g}) \to \bigotimes_{j=1}^{n} \mathbf{C}_\epsilon[x_j, x_j^{-1}, z_j, z_j^{-1}]$,

$$X_i^+, \ X_i^-, \ K_i, \ i < n, \ \text{map as in (i)},$$

$$X_n^+ \mapsto -\frac{x_n^{-1}}{(\epsilon - \epsilon^{-1})^2} \{\epsilon^{-1} z_n\}, \quad X_n^- \mapsto x_n, \quad K_n \mapsto z_n^{-1}.$$

(iii) ($\mathfrak{g}$ *of type* C_n) $U_\epsilon(\mathfrak{g}) \to \bigotimes_{j=1}^{n} \mathbf{C}_\epsilon[x_j, x_j^{-1}, z_j, z_j^{-1}]$,

$$X_i^+, \ X_i^-, \ K_i, \ i < n, \ \text{map as in (i)},$$

$$X_n^+ \mapsto -\frac{x_n^{-2}}{(\epsilon^4 - \epsilon^{-4})^2} \{\epsilon^{-1} z_n\} \{\epsilon^{-3} z_n\}, \quad X_n^- \mapsto x_n^2, \quad K_n \mapsto z_n^{-2}. \ \blacksquare$$

The numbering of the simple roots is as in Bourbaki (1972). The proof is a direct verification.

Representations of U_ϵ can now be constructed by pulling back representations of tensor products of copies of $\mathbf{C}_\epsilon[x, x^{-1}, z, z^{-1}]$ by the homomorphisms defined in the proposition. It is an easy exercise to show that every irreducible representation V of $\mathbf{C}_\epsilon[x, x^{-1}, z, z^{-1}]$ is of dimension ℓ, and is given, in a suitable basis $\{v_0, v_1, \ldots, v_{\ell-1}\}$, by

$$x.v_j = \lambda v_{j+1}, \quad z.v_j = \mu \epsilon^{2j} v_j$$

for some $\lambda, \mu \in \mathbf{C}^\times$ (the indices are counted (mod ℓ)). We denote this representation by $V(\lambda, \mu)$.

PROPOSITION 11.1.12 *(i) The pull-back of a representation* $\bigotimes_{j=1}^{n+1} V(\lambda_j, \mu_j)$ *of* $\bigotimes_{j=1}^{n+1} \mathbf{C}_\epsilon[x_j, x_j^{-1}, z_j, z_j^{-1}]$ *by the homomorphism in 11.1.11(i) is the direct sum of* ℓ *irreducible representations of* $U_{\epsilon^2}(\mathfrak{g})$ *of dimension* ℓ^n, *namely the eigenspaces of* $z_1 z_2 \ldots z_{n+1}$.

(ii) The pull-back of a representation $\bigotimes_{j=1}^{n} V(\lambda_j, \mu_j)$ *of* $\bigotimes_{j=1}^{n} \mathbf{C}_\epsilon[x_j, x_j^{-1}, z_j, z_j^{-1}]$ *by the homomorphisms in 11.1.11(ii) and (iii) are irreducible.*

(iii) If $\mathfrak{g}$ *is of type* A_n, B_n *or* C_n, *every cyclic representation of* $U_\epsilon(\mathfrak{g})$ *(or* $U_{\epsilon^2}(\mathfrak{g})$*) of dimension* ℓ^n *arises, up to isomorphism, in this way.*

(iv) If $\mathfrak{g}$ is not of type A_n, B_n or C_n, the minimal dimension of a cyclic representation of $U_\epsilon(\mathfrak{g})$ is strictly greater than ℓ^n. ∎

See Chari and Pressley (1991c) for the proof, and Arnaudon & Chakrabarti (1991a) for an alternative approach.

It is interesting to examine this result in light of 11.1.5. The minimal dimension d of a non-trivial conjugacy class is given below:

A_n	B_n	C_n	D_n	E_6	E_7	E_8	F_4	G_2
$2n$	$2n$	$2n$	$4n-6$	22	34	58	16	6

Except in the B_n case, d is the dimension of the conjugacy class of the unipotent element $\exp(X_\theta^+)$, where X_θ^+ is a root vector corresponding to the highest root θ (it is known that this latter dimension is $4(\rho, \theta)/(\theta, \theta)$); in the B_n case, it is the dimension of the conjugacy class of the exceptional semisimple element $\mathrm{diag}(-1, -1, \ldots, -1, 1)$ in $\mathrm{SO}(2n+1)$. Note that $d = 2n$ when $\mathfrak{g}$ is of type A_n, B_n or C_n, so that 11.1.12 is in agreement with 11.1.5.

Further, it is known that there are at most a finite number of conjugacy classes of dimension d, except for type A_n when there is a one-parameter family of semisimple classes of minimal dimension. The general theory outlined above would therefore lead us to expect a d-parameter family of minimal cyclic representations except in the A_n case, when there should be a $(d+1)$-parameter family. Inspection of 11.1.12 shows that this is indeed the case (there is apparently a $(2n+2)$-parameter family in the A_n case, but the representations are unchanged, up to isomorphism, by replacing λ_i for all i by $\lambda\lambda_i$, for any $\lambda \in \mathbb{C}^\times$).

It is not known how to construct the minimal cyclic representations for arbitrary $\mathfrak{g}$. But, for type D_4, the prototype of the cases not covered by the preceding proposition, we have the following result. It confirms the expectation that the minimal dimension should be ℓ^5, and that the minimal representations should depend on ten parameters.

PROPOSITION 11.1.13 *Suppose that $\mathfrak{g}$ is of type D_4.*

(i) The minimal dimension of a cyclic representation of $U_\epsilon(\mathfrak{g})$ is ℓ^5.

(ii) The following formulas define a homomorphism of algebras $U_{\epsilon^2}(\mathfrak{g}) \to \otimes_{j=1}^4 \mathbb{C}_\epsilon[x_j, x_j^{-1}, z_j, z_j^{-1}] \otimes \mathbb{C}_{\epsilon^2}[x_5, x_5^{-1}, z_5, z_5^{-1}]$:

$$K_i \mapsto z_i^2 z_4, \quad K_4 \mapsto (z_1 z_2 z_3 z_4^2)^{-1}, \quad X_i^+ \mapsto x_i,$$

$$X_4^+ \mapsto -\frac{x_4^{-1}}{(\epsilon^2 - \epsilon^{-2})^2} \left(\{\epsilon^{-2} z_1 z_2 z_3 z_4^{-2}\} + z_1 z_2 z_3 (c_0 + c(z_1^{-2} + z_2^{-2} + z_3^{-2})) \right.$$

$$\left. + (z_1 z_2 z_3)^{-1}(d_0 + d(z_1^2 + z_2^2 + z_3^2))) \right),$$

$$X_i^- \mapsto -\frac{x_i^{-1}}{(\epsilon^2 - \epsilon^{-2})^2} \left(\{\epsilon^{-2} z_i^2 z_4\} + a z_4 + b z_4^{-1} \right), \quad X_4^- \mapsto x_4,$$

where i runs from 1 to 3 and

$$a = \epsilon^2 x_5^{-1} z_5^{-1} + z_5^{-1} + x_5^{-1}, \quad b = x_5, \quad c = z_5, \quad c_0 = \epsilon^2 x_5 z_5,$$

$$d = x_5^{-1}(1 + \epsilon^{-2} z_5), \quad d_0 = x_5^{-1}(x_5^{-1}(\epsilon^{-4} z_5 + \epsilon^2) + 1)(1 + \epsilon^2 z_5^{-1}).$$

(iii) *Pulling back a representation* $\otimes_{j=1}^4 V(\lambda_j, \mu_j) \otimes V(\lambda_5, z_5)$ *of*

$$\otimes_{j=1}^4 \mathbb{C}_\epsilon[x_j, x_j^{-1}, z_j, z_j^{-1}] \otimes \mathbb{C}_{\epsilon^2}[x_5, x_5^{-1}, z_5, z_5^{-1}]$$

under the homomorphism in (ii) gives an irreducible representation of $U_{\epsilon^2}(\mathfrak{g})$, *and every cyclic representation of dimension* ℓ^5 *arises, up to isomorphism, in this way.* ∎

The proof may be found in Chari and Pressley (1992, 1994a).

Let us now turn to representations of maximal dimension ℓ^N. The following result is due to Date *et al.* (1991b). It will be convenient to use the notation

$$[u] = \frac{u - u^{-1}}{\epsilon - \epsilon^{-1}}.$$

PROPOSITION 11.1.14 *Let* $\mathfrak{g}$ *be of type* A_n. *For any* $r_1, \ldots, r_n \in \mathbb{C}^\times$, *there is a homomorphism of algebras* $U_{\epsilon^2}(\mathfrak{g}) \to \otimes_{1 \leq i < j \leq n} \mathbb{C}_\epsilon[x_{ij}, x_{ij}^{-1}, z_{ij}, z_{ij}^{-1}]$ *given by*

$$X_i^+ \mapsto \sum_{j=i}^n [r_i z_{ij} z_{i\,j-1} z_{i-1\,j-1}^{-1} z_{i+1\,j}^{-1}] x_{ij} x_{i\,j+1} \ldots x_{in},$$

$$X_i^- \mapsto \sum_{j=1}^i [z_{i+1-j\,n-j} z_{i-j\,n+1-j} z_{i+1-j\,n+1-j}^{-1} z_{i-j\,n-j}^{-1}]$$
$$\times \, x_{i+1-j\,n+1-j}^{-1} x_{i+2-j\,n+2-j}^{-1} \cdots x_{in}^{-1},$$

$$K_i \mapsto r_i z_{in}^2 z_{i-1\,n}^{-1} z_{i+1\,n}^{-1}.$$

(one sets x_{ij} *and* z_{ij} *equal to one if* i *or* j *is* $> n$).

Pulling back a representation $\otimes_{i,j} V(\lambda_{ij}, \mu_{ij})$ *of* $\otimes_{i,j} \mathbb{C}_\epsilon[x_{ij}, x_{ij}^{-1}, z_{ij}, z_{ij}^{-1}]$ *under these homomorphisms gives (irreducible) cyclic representations of* $U_{\epsilon^2}(\mathfrak{g})$ *of dimension* ℓ^N. *Their* Z_0-*characters fill a Zariski open subset of* Spec(Z_0). ∎

C Intertwiners and the QYBE We recall from 5.2.5 that, if $(A, \mathcal{R})$ is a quasitriangular Hopf algebra and $\rho_1 : A \to \text{End}(V_1)$ and $\rho_2 : A \to \text{End}(V_2)$ are two finite-dimensional representations of A, the tensor products $V_1 \otimes V_2$ and $V_2 \otimes V_1$ are isomorphic as representations. In fact, as we saw in 5.2.2, the map

$$I = \sigma \circ (\rho_1 \otimes \rho_2)(\mathcal{R}) : V_1 \otimes V_2 \to V_2 \otimes V_1$$

is an intertwiner, i.e. it commutes with the action of A. It follows almost immediately that

PROPOSITION 11.1.15 $U_\epsilon(\mathfrak{g})$ *is not quasitriangular.*

PROOF Let V_1 and V_2 be irreducible U_ϵ-modules. From 9.2.11, it is clear that, if $V_1 \otimes V_2 \cong V_2 \otimes V_1$ as U_ϵ-modules, the Z_0-characters of V_1 and V_2 commute in the group $\mathrm{Spec}(Z_0)$. Since $\mathrm{Spec}(Z_0)$ is not commutative, and since the map which assigns to an irreducible U_ϵ-module its Z_0-character is surjective, this completes the proof. ∎

Remark For later purposes, it is useful to make this argument a little more explicit. We note first that the central elements $x_i^\pm = (X_i^\pm)^\ell$ and $k_i = K_i^\ell$ act as scalars on any tensor product $V_1 \otimes V_2$ of irreducible representations of U_ϵ, even though this representation cannot be irreducible in general, for dimensional reasons (see 11.1.1). Taking $a = X_i^+ \otimes K_i$ and $b = 1 \otimes X_i^+$ in 9.2.10, we have

$$(1) \qquad \begin{aligned} (\rho_1 \otimes \rho_2)(x_i^+) &= (\rho_1 \otimes \rho_2)(X_i^+ \otimes K_i + 1 \otimes X_i^+)^\ell \\ &= \rho_1(x_i^+) \otimes \rho_1(k_i) + 1 \otimes \rho_2(x_i^+). \end{aligned}$$

Thus, a necessary condition for $V_1 \otimes V_2$ and $V_2 \otimes V_1$ to be isomorphic as representations of U_ϵ is

$$(2) \qquad \rho_1(x_i^+)\rho_2(k_i) + \rho_2(x_i^+) = \rho_2(x_i^+)\rho_1(k_i) + \rho_1(x_i^+)$$

for all i. Similarly, by considering the action of the X_i^-, we find the necessary conditions

$$(3) \qquad \rho_1(x_i^-) + \rho_1(k_i)^{-1}\rho_2(x_i^-) = \rho_2(x_i^-) + \rho_2(k_i)^{-1}\rho_1(x_i^-). \quad \triangle$$

Despite the lack of a universal R-matrix for U_ϵ, the following general result, which will also be needed in Chapter 12, allows one to obtain matrix-valued solutions of the QYBE by restricting to a set of representations of U_ϵ on which the tensor product operation is commutative (up to isomorphism).

PROPOSITION 11.1.16 *Let* $\{V(v)\}_{v \in \mathcal{V}}$ *be a family of representations of a Hopf algebra A, all with the same underlying vector space V and parametrized by the elements v of some set $\mathcal{V}$, such that:*

(i) for all $v_1, v_2 \in \mathcal{V}$, there is an isomorphism of representations

$$I(v_1, v_2) : V(v_1) \otimes V(v_2) \to V(v_2) \otimes V(v_1);$$

(ii) for all $v_1, v_2, v_3 \in \mathcal{V}$, the only isomorphisms of representations

$$V(v_1) \otimes V(v_2) \otimes V(v_3) \to V(v_1) \otimes V(v_2) \otimes V(v_3)$$

are the scalar multiples of the identity.

Then, if $R = \sigma \circ I$, where σ is the interchange of the factors in the tensor product,

$$(4) \quad R_{12}(v_1, v_2)R_{13}(v_1, v_3)R_{23}(v_2, v_3) = cR_{23}(v_2, v_3)R_{13}(v_1, v_3)R_{12}(v_1, v_2),$$

where c is a scalar (possibly depending on v_1, v_2 and v_3).

PROOF In the diagram

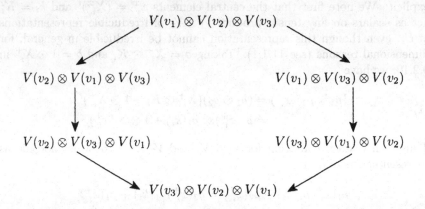

the composite of the maps on the left-hand side gives the intertwiner

$$I_{12}(v_2, v_3)I_{23}(v_1, v_3)I_{12}(v_1, v_2) : V(v_1) \otimes V(v_2) \otimes V(v_3) \to V(v_3) \otimes V(v_2) \otimes V(v_1),$$

while the right-hand side gives the intertwiner $I_{23}(v_1, v_2)I_{12}(v_1, v_3)I_{23}(v_2, v_3)$. By property (ii), these can differ only by a scalar multiple:

$$(5) \quad I_{23}(v_1, v_2)I_{12}(v_1, v_3)I_{23}(v_2, v_3) = cI_{12}(v_2, v_3)I_{23}(v_1, v_3)I_{12}(v_1, v_2).$$

It is easy to see that this is equivalent to (4). ∎

Note that the intertwiners I are only determined up to a scalar multiple, so that it may be possible to normalize them so that $c = 1$. In this case, (4) is the QYBE with spectral parameters.

In practice, the hypotheses of this proposition are a little too restrictive. Typically, $\mathcal{V}$ is an open set in the complex plane, $I(v_1, v_2)$ is a meromorphic function on $\mathcal{V}_1 \times \mathcal{V}_2$, and conditions (i) and (ii) hold for *generic* values of the parameters. But then (5) also holds generically, and hence identically since both sides are meromorphic functions.

Returning to $U_\epsilon(sl_2(\mathbb{C}))$, we note that (2) and (3) are satisfied if $\rho_1(k_1) = \rho_2(k_1) = 1$. This suggests that we consider the family $\mathcal{I}$ of representations of U_ϵ on which $k_1 = 1$ and $x_1^\pm \neq 0$. The latter condition guarantees that all the representations in $\mathcal{I}$ have dimension ℓ, and that they are uniquely determined by their central character. Date *et al.* (1991c) show that the tensor product is indeed commutative on $\mathcal{I}$, compute the intertwiners and show, using 11.1.16, that they lead to solutions of the QYBE.

To state their result precisely, we need to modify our description 11.1.7 of the representations $V(\lambda, a, b)$ in $\mathcal{I}$. For $\alpha, \beta \in \mathbb{C}^{\times}$, set

$$a = \alpha \left(\frac{\epsilon\beta - \epsilon^{-1}\beta^{-1}}{\epsilon - \epsilon^{-1}} \right), \quad b = \alpha^{-1} \left(\frac{\beta - \beta^{-1}}{\epsilon - \epsilon^{-1}} \right),$$

and define a new basis $\{\bar{v}_i\}$ of $V(\lambda, a, b)$ by $\bar{v}_i = \mu_i v_i$, where the constants μ_i are such that

$$\frac{\mu_i}{\mu_{i+1}} = \frac{\alpha^{-1}(\epsilon^i \beta - \epsilon^{-i}\beta^{-1})}{\epsilon - \epsilon^{-1}}.$$

PROPOSITION 11.1.17 *Let* $V_1 = V(1, a_1, b_1)$, $V_2 = V(1, a_2, b_2)$ *be representations in* $\mathcal{I}$. *There is an intertwiner* $I : V_1 \otimes V_2 \to V_2 \otimes V_1$, *unique up to scalar multiples, given, with respect to the basis* $\{\bar{v}_{i_1} \otimes \bar{v}_{i_2}\}$ *of* $V_1 \otimes V_2$ *defined above, by*

$$I_{j_1 j_2}^{i_1 i_2} = W\left(-\frac{\epsilon\alpha_1\beta_1\beta_2}{\alpha_2} \right)_{i_1 - i_2} W\left(-\frac{\alpha_1}{\epsilon\alpha_2\beta_1\beta_2} \right)_{j_1 - j_2}$$
$$\times \widetilde{W}\left(\frac{\alpha_1\beta_2}{\alpha_2\beta_1} \right)_{i_1 + j_1 - i_2 - j_2} \widetilde{W}\left(\frac{\alpha_1\beta_1}{\alpha_2\beta_2} \right)_{-i_1 - j_1 + i_2 + j_2},$$

where

$$\frac{W(\xi)_{r+4}}{W(\xi)_r} = \frac{1 + \epsilon^{r+1}\xi}{\epsilon^{r+1} + \xi}, \quad \frac{\widetilde{W}(\xi)_{r+4}}{\widetilde{W}(\xi)_r} = \frac{\epsilon(\xi - \epsilon^r)}{\epsilon^{r+1}\xi - 1}.$$

The matrix $R = \sigma \circ I$ *satisfies the quantum Yang–Baxter equation.* ∎

The functions $W(\xi)_r$ and $\widetilde{W}(\xi)_r$ are the Boltzmann weights of the model of Fateev & Zamolodchikov (1982) in statistical mechanics.

11.2 The restricted case

In this section we continue to assume that ϵ is a primitive ℓth root of unity, where ℓ is odd and greater than d_i for all i. We use the notation defined in Section 9.3. Except in Subsection 11.2D, all representations are on complex vector spaces.

A Highest weight representations If V is a finite-dimensional representation of U_ϵ^{res}, then $K_1, \ldots, K_n$ act by commuting operators on V, and, by 9.3.5(ii), these operators are simultaneously diagonalizable with eigenvalues in the set $\{\pm \epsilon^r\}_{r=0,1,\ldots,\ell-1}$. As in Subsection 10.1A, by tensoring with a one-dimensional representation, *we may and shall restrict attention to representations on which* $K_i^\ell = 1$ *for all* i; such representations are said to be of *type* **1**.

Following the discussion for U_q, one is tempted to define the weight spaces of V to be

$$V_\lambda = \{v \in V \mid K_i.v = \epsilon^{(\lambda, \alpha_i)} v \text{ for all } i\}, \quad (\lambda \in P).$$

But this definition is unsatisfactory, for V_λ would be equal to V_μ whenever $\lambda - \mu \in \ell P$, so the sum of the weight spaces would not be direct. The 'reason' for this is that the $K_i^{\pm 1}$ do not generate a maximal abelian subalgebra of U_ϵ^{res}. In fact, the elements $\begin{bmatrix} K_i;0 \\ \ell \end{bmatrix}_{\epsilon_i}$ commute with each other and with the K_i, and, by Remark [1] following 9.3.5, these elements together generate $U_\epsilon^{res,0}$.

LEMMA 11.2.1 *If V is a finite-dimensional U_ϵ^{res}-module of type* **1**, *then*

(i) *the $(X_i^\pm)^{(\ell)}$ act nilpotently on V;*

(ii) *the eigenvalues of the $\begin{bmatrix} K_i;0 \\ \ell \end{bmatrix}_{\epsilon_i}$ on V are integers.*

OUTLINE OF PROOF It is enough to do the sl_2 case. Since K_1 and $\begin{bmatrix} K_1;0 \\ \ell \end{bmatrix}_\epsilon$ are commuting operators on V and $K_1^\ell = 1$, we can write $V = \oplus_{m_0, m_1} V_{m_0, m_1}$, where $K_1 = \epsilon^{m_0}.\text{id}$ and $\begin{bmatrix} K_1;0 \\ \ell \end{bmatrix}_\epsilon - m_1$ is nilpotent on V_{m_0, m_1}; here, $m_0 \in \mathbb{Z}$, $0 \le m_0 < \ell$ and $m_1 \in \mathbb{C}$.

Using the relations in 9.3.4, it follows that

$$(X_1^+)^{(\ell)}.V_{m_0, m_1} \subseteq V_{m_0, m_1+2},$$

$$X_1^+.V_{m_0, m_1} \subseteq \begin{cases} V_{m_0+2, m_1} & \text{if } m_0 + 2 < \ell, \\ V_{m_0+2-\ell, m_1+1} & \text{if } m_0 + 2 \ge \ell, \end{cases}$$

with similar results for X_1^- and $(X_1^-)^{(\ell)}$. Since $\dim(V) < \infty$, the possible values of m_1 lie in a bounded set, so the first inclusion implies (i).

Now assume, for a contradiction, that $V_{m_0, m_1} \ne 0$ for some $m_1 \in \mathbb{C} \backslash \mathbb{Z}$, and that $m_0 + \ell \Re(m_1)$ is maximal with this property ($\Re(m_1)$ denotes the real part of m_1). The above inclusions show that X_1^+ and $(X_1^+)^{(\ell)}$ both increase the latter quantity by 2, so must annihilate V_{m_0, m_1}. It follows that V_{m_0, m_1} is annihilated by $(X_1^+)^{(k)}$ for all $k \ge 1$. On the other hand, part (i) implies that V_{m_0, m_1} is annihilated by $((X_1^-)^{(\ell)})^r$ for some r, and hence by $(X_1^-)^{(\ell r)}$. Applying both sides of the relation in (13) of Chapter 9 (with r and s both replaced by ℓr) to V_{m_0,m_1}, it follows that $\begin{bmatrix} K_1 ; 0 \\ \ell r \end{bmatrix}_\epsilon$ also annihilates V_{m_0,m_1}. But, by an inductive argument using the third and fifth relations in 9.3.4, it is not difficult to prove the identity

$$\begin{bmatrix} K_1 ; 0 \\ \ell r \end{bmatrix}_\epsilon = \frac{1}{r!} \prod_{s=0}^{r-1} \left(\begin{bmatrix} K_1 ; 0 \\ \ell \end{bmatrix}_\epsilon - s \right)$$

in $U_\epsilon^{\mathrm{res}}$. Since $\begin{bmatrix} K_1 ; 0 \\ \ell \end{bmatrix}_\epsilon - m_1$ is nilpotent on V_{m_0,m_1}, it follows that m_1 belongs to the set $\{0, 1, \ldots, r-1\}$, contrary to our hypothesis. ∎

In view of this lemma, if v is a simultaneous eigenvector for the elements of $U_\epsilon^{\mathrm{res},0}$, then

$$K_i.v = \epsilon_i^{m_{i,0}} v, \qquad \begin{bmatrix} K_i ; 0 \\ \ell \end{bmatrix}_{\epsilon_i} .v = m_{i,1} v,$$

where $m_{i,0}$, $m_{i,1} \in \mathbf{Z}$, and, without loss of generality, $0 \le m_{i,0} < \ell$. It follows from 9.3.6 that, if $m_i = m_{i,0} + \ell m_{i,1}$, then $\begin{bmatrix} m_i \\ \ell \end{bmatrix}_{\epsilon_i} = m_{i,1}$. This suggests the following definition.

We write, as usual, $\check{\alpha} = 2\alpha/(\alpha,\alpha)$ for any root α; in particular, $\check{\alpha}_i = d_i^{-1} \alpha_i$.

DEFINITION 11.2.2 *The weight spaces V_λ ($\lambda \in P$) of a $U_\epsilon^{\mathrm{res}}$-module V of type $\mathbf{1}$ are defined by*

$$V_\lambda = \left\{ v \in V \mid K_i.v = \epsilon_i^{(\lambda,\check{\alpha}_i)} v, \ \begin{bmatrix} K_i ; 0 \\ \ell \end{bmatrix}_{\epsilon_i} .v = \begin{bmatrix} (\lambda,\check{\alpha}_i) \\ \ell \end{bmatrix}_{\epsilon_i} v, \ i = 1,\ldots,n \right\}.$$

If V is finite-dimensional, its character $\mathrm{ch}(V)$ is the element of the group ring $\mathbf{Z}[P]$ given by

$$\mathrm{ch}(V) = \sum_{\lambda \in P} \dim(V_\lambda) e^\lambda.$$

Remark It follows from the relations in 9.3.4 that $\begin{bmatrix} K_i ; c \\ \ell \end{bmatrix}_{\epsilon_i}$ acts on V_λ as $\begin{bmatrix} (\lambda,\check{\alpha}_i)+c \\ \ell \end{bmatrix}_{\epsilon_i}$ for all $c \in \mathbf{Z}$. △

PROPOSITION 11.2.3 *If V is a finite-dimensional irreducible $U_\epsilon^{\mathrm{res}}$-module of type $\mathbf{1}$, then V is the direct sum of its weight spaces.*

PROOF By 9.3.4,

(6) $$X_i^\pm(V_\lambda) \subset V_{\lambda \pm \alpha_i}, \quad (X_i^\pm)^{(\ell)}(V_\lambda) \subset V_{\lambda \pm \ell \alpha_i}.$$

It follows that $\sum_{\lambda \in P} V_\lambda$ is a submodule of V. To see that the sum is direct, we note that, by 9.3.6, if $m, m' \in \mathbf{Z}$ are such that $\epsilon_i^m = \epsilon_i^{m'}$ and $\begin{bmatrix} m \\ \ell \end{bmatrix}_{\epsilon_i} = \begin{bmatrix} m' \\ \ell \end{bmatrix}_{\epsilon_i}$, then $m = m'$. ∎

Later in the chapter, we shall need to know that, as in the classical case, characters are multiplicative:

PROPOSITION 11.2.4 *If V and W are finite-dimensional $U_\epsilon^{\mathrm{res}}$-modules of type* **1**, *then*

$$\mathrm{ch}(V \otimes W) = \mathrm{ch}(V) \otimes \mathrm{ch}(W).$$

PROOF It suffices to show that, if $v_\lambda \in V$ and $w_\mu \in W$ are vectors of the indicated weights, then $v_\lambda \otimes w_\mu$ has weight $\lambda + \mu$. For this, we might as well assume that $\mathfrak{g} = sl_2(\mathbb{C})$, so that $\lambda, \mu \in \mathbb{Z}$. We must show that $K_1.(v_\lambda \otimes w_\mu) = \epsilon^{\lambda+\mu}(v_\lambda \otimes w_\mu)$ and that $\begin{bmatrix} K_1;0 \\ \ell \end{bmatrix}_\epsilon .(v_\lambda \otimes w_\mu) = \begin{bmatrix} \lambda+\mu \\ \ell \end{bmatrix}_\epsilon (v_\lambda \otimes w_\mu)$. The first equation is obvious since K_1 is group-like. To prove the second, we use the identity

$$\Delta\left(\begin{bmatrix} K_1;0 \\ t \end{bmatrix}_\epsilon\right) = \sum_{s=0}^{t} \begin{bmatrix} K_1;0 \\ t-s \end{bmatrix}_\epsilon K_1^{-s} \otimes \begin{bmatrix} K_1;0 \\ s \end{bmatrix}_\epsilon K_1^{t-s}$$

for $t \in \mathbb{N}$, which is easily proved by induction on t. Then, taking $t = \ell$, what we must prove is

$$\begin{bmatrix} \lambda+\mu \\ \ell \end{bmatrix}_\epsilon = \sum_{s=0}^{\ell} \begin{bmatrix} \lambda \\ \ell-s \end{bmatrix}_\epsilon \begin{bmatrix} \mu \\ s \end{bmatrix}_\epsilon \epsilon^{-(\lambda+\mu)s}.$$

This follows easily from 9.2.10 by expanding both sides of

$$(a+b)^{\lambda+\mu} = (a+b)^\lambda (a+b)^\mu. \quad \blacksquare$$

The definition of highest weight $U_\epsilon^{\mathrm{res}}$-modules now proceeds as in the U_q theory. Namely, a $U_\epsilon^{\mathrm{res}}$-module V of type **1** is a *highest weight module* if it is generated by a *primitive vector*, i.e. a vector $v_\lambda \in V_\lambda$, for some $\lambda \in P$, such that $X_i^+.v_\lambda = (X_i^+)^{(\ell)}.v_\lambda = 0$, $i = 1, \ldots, n$. It is obvious from 9.3.3(iv) and (6) that $V = \oplus_{\mu \leq \lambda} V_\mu$, so that λ is the *highest weight* of V; moreover, $V_\lambda = \mathbb{C}.v_\lambda$. It follows by the usual argument that V has a unique irreducible quotient module.

Let $V_q(\lambda)$ be the irreducible U_q-module with highest weight $\lambda \in P$ and highest weight vector v_λ (see Section 10.1). Let $V_{\mathcal{A}}^{\mathrm{res}}(\lambda)$ be the $U_{\mathcal{A}}^{\mathrm{res}}$-submodule of $V_q(\lambda)$ generated by v_λ, and define the *Weyl module*

$$W_\epsilon^{\mathrm{res}}(\lambda) = V_{\mathcal{A}}^{\mathrm{res}}(\lambda) \underset{\mathcal{A}}{\otimes} \mathbb{C},$$

via the homomorphism $\mathcal{A} \to \mathbb{C}$ given by $q \mapsto \epsilon$. This is clearly a highest weight $U_\epsilon^{\mathrm{res}}$-module with highest weight λ, but it is not necessarily irreducible.

PROPOSITION 11.2.5 *Suppose that* $\lambda \in P^+$. *Then:*

(i) $\dim(W_\epsilon^{\mathrm{res}}(\lambda)) < \infty$;

(ii) the character of $W_\epsilon^{\mathrm{res}}(\lambda)$ *is given by the Weyl character formula;*

(iii) $W_\epsilon^{\mathrm{res}}(\lambda)$ *is irreducible if either*
   (a) $(\lambda + \rho, \check{\alpha}) < \ell$ *for all positive roots* α, *or*
   (b) $\lambda = (\ell - 1)\rho + \ell\mu$ *for some* $\mu \in P^+$.

OUTLINE OF PROOF Parts (i) and (ii) follow immediately from 10.1.5. Part (iii) is proved in Andersen, Polo & Wen (1991). ∎

In general, since $W_\epsilon^{\mathrm{res}}(\lambda)$ is a highest weight module, it has a unique irreducible quotient module $V_\epsilon^{\mathrm{res}}(\lambda)$. From 11.2.5 we obtain immediately

COROLLARY 11.2.6 *If* $\lambda \in P^+$, *then* $\dim(V_\epsilon^{\mathrm{res}}(\lambda)) < \infty$. *If, in addition,* λ *satisfies one of the conditions in 11.2.5(iii), the character of* $V_\epsilon^{\mathrm{res}}(\lambda)$ *is given by the Weyl character formula.* ∎

Example 11.2.7 Suppose that $\mathfrak{g} = sl_2(\mathbb{C})$. For any integer n, write $n = n_0 + \ell n_1$, where $n_0, n_1 \in \mathbb{Z}$ and $0 \leq n_0 < \ell$. Then, if $m \in \mathbb{N}$, the Weyl module $W_\epsilon^{\mathrm{res}}(m)$ with maximal weight m has a basis $\{v_0^{(m)}, \ldots, v_m^{(m)}\}$ on which the action of $U_\epsilon^{\mathrm{res}}$ is given by

$$X_1^+ . v_r^{(m)} = [m - r + 1]_\epsilon v_{r-1}^{(m)}, \quad X_1^- . v_r^{(m)} = [r+1]_\epsilon v_{r+1}^{(m)}, \quad K_1 . v_r^{(m)} = \epsilon^{m-2r} v_r^{(m)},$$

$$(X_1^-)^{(\ell)} . v_r^{(m)} = (r_1 + 1) v_{r+\ell}^{(m)}, \quad (X_1^+)^{(\ell)} . v_r^{(m)} = ((m - r)_1 + 1) v_{r-\ell}^{(m)}.$$

Let V' be the subspace of $W_\epsilon^{\mathrm{res}}(m)$ spanned by the $v_r^{(m)}$ such that $m_0 < r_0 < \ell$ and $r_1 < m_1$. Then, it is not difficult to see that:

(i) V' is the unique proper $U_\epsilon^{\mathrm{res}}$-submodule of $W_\epsilon^{\mathrm{res}}(m)$;

(ii) $V' \cong V_\epsilon^{\mathrm{res}}(m')$, where $m' = \ell - 2 - m_0 + \ell(m_1 - 1)$;

(iii) $W_\epsilon^{\mathrm{res}}(m)/V' \cong V_\epsilon^{\mathrm{res}}(m)$.

It is clear from these remarks that $W_\epsilon^{\mathrm{res}}(m)$ is irreducible if (and only if) either $m < \ell$ or $m_0 = \ell - 1$, confirming 11.2.5(iii) in this case. Note that, by (i), if $W_\epsilon^{\mathrm{res}}(m)$ is reducible, it is not completely reducible.

$(X_1^-)^{(5)}$
$\xrightarrow{\hspace{3cm}}$

$X_1^- \quad X_1^-$
$\xrightarrow{\hspace{1cm}} \xrightarrow{\hspace{1cm}}$

•	•	•	○	○	•	•	•	○	○	•	•	•	○	○	•	•	•
17	15	13	11	9	7	5	3	1	-1	-3	-5	-7	-9	-11	-13	-15	-17

As a specific example, take $\ell = 5$, $m = 17$. The weight structure of $V_\epsilon^{\mathrm{res}}(17)$ is shown above, in which the solid circles denote basis vectors of $V_\epsilon^{\mathrm{res}}(17)$ with the indicated weights, and the open circles denote the basis vectors of the unique maximal submodule of $W_\epsilon^{\mathrm{res}}(17)$, which is isomorphic to $V_\epsilon^{\mathrm{res}}(11)$.

Note that $\dim(V_\epsilon^{\mathrm{res}}(17)) = 12$. In general, it is clear from the above description that $\dim(V') = m_1(\ell - m_0 - 1)$, so that

$$\dim(V_\epsilon^{\mathrm{res}}(m)) = m + 1 - m_1(\ell - m_0 - 1) = (m_0 + 1)(m_1 + 1). \quad \Diamond$$

We now prove the main result of this subsection.

THEOREM 11.2.8 *Every finite-dimensional irreducible $U_\epsilon^{\mathrm{res}}$-module V of type* **1** *is isomorphic to $V_\epsilon^{\mathrm{res}}(\lambda)$ for some (unique) $\lambda \in P^+$.*

PROOF This follows closely the proof of the analogous result in the classical situation. By 11.2.3, V is the direct sum of its weight spaces. Let $\lambda \in P$ be maximal among the weights of V. Then, any non-zero vector $v_\lambda \in V_\lambda$ is primitive, and, since V is irreducible, it follows that v_λ generates V and hence that $V \cong V_\epsilon^{\mathrm{res}}(\lambda)$.

It remains to prove that $\lambda \in P^+$. By 11.2.1(i) and the fact that

$$(X_i^-)^{(m\ell)} = \frac{1}{m!}((X_i^-)^{(\ell)})^m$$

in $U_\epsilon^{\mathrm{res}}$, it follows that there exists $m \geq 1$ such that

$$(X_i^-)^{((m-1)\ell)}.v_\lambda \neq 0, \quad (X_i^-)^{(m\ell)}.v_\lambda = 0.$$

Applying $(X_i^+)^{(\ell)}$ to the second equation and using the relations in 9.3.4, we find that

$$(X_i^-)^{((m-1)\ell)} \begin{bmatrix} K_i \,; \; -(m-1)\ell \\ \ell \end{bmatrix}_{\epsilon_i} .v_\lambda = 0.$$

Hence, $\begin{bmatrix} (\lambda, \check\alpha_i) - (m-1)\ell \\ \ell \end{bmatrix}_{\epsilon_i} = 0$. By 9.3.6, this implies that $(m-1)\ell \leq (\lambda, \check\alpha_i) < m\ell$; in particular, $(\lambda, \check\alpha_i) \geq 0$. ∎

Remark One can also obtain $V_\epsilon^{\mathrm{res}}(\lambda)$ by a Verma module construction. Let $M_\epsilon^{\mathrm{res}}(\lambda)$ be the quotient of $U_\epsilon^{\mathrm{res}}$ by the two-sided ideal generated by the following elements:

$$K_i - \epsilon_i^{(\lambda, \check\alpha_i)}, \quad \begin{bmatrix} K_i ; 0 \\ \ell \end{bmatrix}_{\epsilon_i} - \begin{bmatrix} (\lambda, \check\alpha_i) \\ \ell \end{bmatrix}_{\epsilon_i}, \quad X_i^+, \quad (X_i^+)^{(\ell)}.$$

Then, $M_\epsilon^{\mathrm{res}}(\lambda)$ is a highest weight $U_\epsilon^{\mathrm{res}}$-module, and $V_\epsilon^{\mathrm{res}}(\lambda)$ is its unique irreducible quotient. △

B A tensor product theorem For any $\lambda \in P^+$, we can find unique λ_0, $\lambda_1 \in P^+$ such that $\lambda = \lambda_0 + \ell\lambda_1$ and $0 \leq (\lambda_0, \check{\alpha}_i) < \ell$ for $i = 1, \ldots, n$. The following theorem reduces the study of $V_\epsilon^{res}(\lambda)$ to that of $V_\epsilon^{res}(\lambda_0)$ and $V_\epsilon^{res}(\ell\lambda_1)$.

THEOREM 11.2.9 *Let $\lambda \in P^+$ and write $\lambda = \lambda_0 + \ell\lambda_1$ as above. Then, there is an isomorphism of U_ϵ-modules*

$$V_\epsilon^{res}(\lambda) \cong V_\epsilon^{res}(\lambda_0) \otimes V_\epsilon^{res}(\ell\lambda_1).$$

Before we can sketch the proof of this theorem, we need to gain some understanding of the structure of the modules $V_\epsilon^{res}(\lambda_0)$ and $V_\epsilon^{res}(\ell\lambda_1)$.

The representations of the first kind are essentially the representations of the finite-dimensional Hopf subalgebra U_ϵ^{fin} of U_ϵ^{res} generated by the $X_i^{\pm}$ and $K_i^{\pm 1}$ (see Subsection 9.3B).

PROPOSITION 11.2.10 *Let $\lambda \in P^+$ satisfy $0 \leq (\lambda, \check{\alpha}_i) < \ell$ for $i = 1, \ldots, n$. As a U_ϵ^{fin}-module, $V_\epsilon^{res}(\lambda)$ is irreducible. In fact, restriction to U_ϵ^{fin} sets up a one-to-one correspondence between the finite-dimensional irreducible U_ϵ^{res}-modules of type 1 whose highest weights satisfy the above condition and the irreducible U_ϵ^{fin}-modules on which the K_i^ℓ act as 1. In particular, $\tilde{U}_\epsilon^{fin}$ has, up to isomorphism, exactly ℓ^n irreducible representations.* ∎

Remark It follows by an argument used in the proof of 11.2.8 that, if $V_\epsilon^{res}(\lambda)$ is as in the preceding proposition and v_λ is a highest weight vector in $V_\epsilon^{res}(\lambda)$, then $(X_i^-)^{(\ell)}.v_\lambda = 0$. △

To describe the modules of the second kind, we need the Frobenius homomorphism Fr_ϵ from U_ϵ^{res} to the classical enveloping algebra U defined in 9.3.12 (we extend the base field from $\mathbb{Q}(\epsilon)$ to $\mathbb{C}$).

PROPOSITION 11.2.11 *Let $\lambda \in P^+$ and let $V(\lambda)$ be the irreducible U-module with highest weight λ. Then, $V_\epsilon^{res}(\ell\lambda)$ is isomorphic as a U_ϵ^{res}-module to the pull-back of $V(\lambda)$ by Fr_ϵ.* ∎

Remark It follows from this that the character of $V_\epsilon^{res}(\ell\lambda)$ can be deduced from the Weyl character formula. In fact, recall from 9.3.12 that $\text{Fr}_\epsilon(K_i) = 1$, and note that $\text{Fr}_\epsilon\left(\left[\begin{smallmatrix} K_i\,; 0 \\ \ell \end{smallmatrix}\right]_{\epsilon_i}\right) = \bar{H}_i$ (to see this, compute $\text{Fr}_\epsilon([(X_i^+)^{(\ell)}, (X_i^-)^{(\ell)}])$ using 9.3.12 and (13) from Subsection 9.3A). It follows that the weights of $V_\epsilon^{res}(\ell\lambda)$ are obtained (with the same multiplicities) exactly by multiplying those of $V(\lambda)$ by ℓ.

This implies, in particular, that $X_i^{\pm}$ and $K_i - 1$ are zero on $V_\epsilon^{res}(\ell\lambda)$. △

We can now give an

OUTLINE OF PROOF OF 11.2.9 We first show that $V_\epsilon^{res}(\lambda_0) \otimes V_\epsilon^{res}(\ell\lambda_1)$ $(= V$, say) is a highest weight U_ϵ^{res}-module with highest weight λ. Let v_0, v_1 be highest weight vectors in $V_\epsilon^{res}(\lambda_0)$ and $V_\epsilon^{res}(\ell\lambda_1)$, respectively. By 9.2.10 and the preceding remark, it follows that

$$(X_i^\pm)^{(\ell)}.(v_0 \otimes v_1) = (X_i^\pm)^{(\ell)}.v_0 \otimes v_1 + v_0 \otimes (X_i^\pm)^{(\ell)}.v_1,$$
$$K_i.(v_0 \otimes v_1) = K_i.v_0 \otimes K_i.v_1.$$

(In fact, this is true even if v_0 and v_1 are not highest weight vectors.) From the first of these equations and (13) from Subsection 9.3A, we deduce that

$$\begin{bmatrix} K_i; 0 \\ \ell \end{bmatrix}_{\epsilon_i} .(v_0 \otimes v_1) = \begin{bmatrix} K_i; 0 \\ \ell \end{bmatrix}_{\epsilon_i} .v_0 \otimes v_1 + v_0 \otimes \begin{bmatrix} K_i; 0 \\ \ell \end{bmatrix}_{\epsilon_i} .v_1.$$

These equations imply that $v_0 \otimes v_1$ is a primitive vector in V of weight $\lambda = \lambda_0 + \ell\lambda_1$.

To show that $v_0 \otimes v_1$ generates V, let $v_0' \in V_\epsilon^{res}(\lambda_0)$, $v_1' \in V_\epsilon^{res}(\ell\lambda_1)$. Then, $v_0' = x_0.v_0$, $v_1' = x_1.v_1$, where x_0 (resp. x_1) is a linear combination of products of elements X_i^- (resp. $(X_i^-)^{(\ell)}$), $i = 1, \ldots, n$. By the remark following 11.2.10 and the fact that the $(X_i^-)^{(\ell)}$ behave like primitive elements on V, we have

$$x_0 x_1.(v_0 \otimes v_1) = x_0.(v_0 \otimes x_1.v_1),$$

and, since X_i^- acts trivially on $V_\epsilon^{res}(\ell\lambda_1)$, this equals

$$x_0.v_0 \otimes x_1.v_1 = v_0' \otimes v_1'.$$

It remains to prove that V is irreducible. Let V' be a proper submodule of V, and let λ' be maximal among the weights of V'. Any non-zero vector $v' \in V'_{\lambda'}$ is primitive. Fix a vector space basis $\{w_1, \ldots, w_p\}$ of $V_\epsilon^{res}(\ell\lambda_1)$ and write

$$v' = \sum_{r=1}^{p} u_r \otimes w_r, \qquad (u_r \in V_\epsilon^{res}(\lambda_0)).$$

Arguing as above, we have

$$0 = X_i^+.v' = \sum_{r=1}^{p} X_i^+.u_r \otimes w_r,$$

so $X_i^+.u_r = 0$ for all r and all i. Hence, u_r is a multiple of v_0 and $v' = v_0 \otimes v_1'$ for some $v_1' \in V_\epsilon^{res}(\ell\lambda_1)$. The same argument now gives $(X_i^+)^{(\ell)}.v_1' = 0$, so v_1' is a multiple of v_1. Hence, V' contains $v_0 \otimes v_1$, which generates V. Thus, $V' = V$, a contradiction. ∎

Theorem 11.2.9 is analogous to a theorem of Steinberg in the representation theory of the hyperalgebra $U_{\mathbb{F}_p}$ (see Subsection 9.3C). It is known that the irreducible representations of $U_{\mathbb{F}_p}$ are again parametrized by P^+ (every such representation is finite dimensional); let $V_{\mathbb{F}_p}(\lambda)$ be the irreducible module associated to $\lambda \in P^+$.

STEINBERG'S TENSOR PRODUCT THEOREM *Let $\lambda \in P^+$ and write*

$$\lambda = \lambda_0 + p\lambda_1 + p^2\lambda_2 + \cdots + p^k\lambda_k,$$

where $0 \leq (\lambda_j, \check{\alpha}_i) < p$ for $i = 1, \ldots, n$, $j = 0, 1, \ldots, k$. Then, as $U_{\mathbf{F}_p}$-modules,

$$V_{\mathbf{F}_p}(\lambda) \cong V_{\mathbf{F}_p}(\lambda_0) \otimes V_{\mathbf{F}_p}(\lambda_1)^{\mathrm{Fr}} \otimes V_{\mathbf{F}_p}(\lambda_2)^{\mathrm{Fr}^2} \otimes \cdots \otimes V_{\mathbf{F}_p}(\lambda_k)^{\mathrm{Fr}^k},$$

where $V_{\mathbf{F}_p}(\lambda)^{\mathrm{Fr}^j}$ is the twist of $V_{\mathbf{F}_p}(\lambda)$ by the jth iterate of the Frobenius map Fr. ∎

This theorem reduces the study of the representations $V_{\mathbf{F}_p}(\lambda)$ to the special case in which $0 \leq (\lambda, \check{\alpha}_i) < p$ for all i: the irreducible representations whose highest weights satisfy this condition are called *restricted*.

The most obvious difference between this result and 11.2.9 is that there is no upper bound (independent of λ) on the number of factors in the tensor product in Steinberg's theorem, whereas there are only two factors in 11.2.9. The 'reason' for this is the fact, already noted at the end of Subsection 9.3C, that $U_\epsilon^{\mathrm{res}}$ is finitely generated, but $U_{\mathbf{F}_p}$ is not. Note also that, unlike the characteristic p Frobenius map, Fr_ϵ is a map between two different spaces, and so cannot be iterated.

C Quasitensor structure We remarked at the end of Subsection 10.1D that, although we do not have a universal R-matrix for $U_\epsilon^{\mathrm{res}}$, the conclusion of 10.1.19 holds for $U_\epsilon^{\mathrm{res}}$, even when ϵ is a root of unity. We record this result here.

PROPOSITION 11.2.12 *With the usual operations of direct sum, dual and tensor product, the category of finite-dimensional $U_\epsilon^{\mathrm{res}}$-modules of type $\mathbf{1}$ which are the direct sums of their weight spaces is a rigid, $\mathbb{C}$-linear, quasitensor category.* ∎

D Some conjectures The results of Subsection 11.2B reduce the study of the finite-dimensional irreducible representations of $U_\epsilon^{\mathrm{res}}$ to those of the finite-dimensional Hopf algebra $U_\epsilon^{\mathrm{fin}}$. Recall from 9.3.11 that, if $\ell = p$ is an odd prime, there is a close relationship between the $\mathbb{Z}[\epsilon, \epsilon^{-1}]$-algebra $\tilde{U}_{\epsilon,\mathbf{Z}}^{\mathrm{fin}}$, the quotient of $U_{\epsilon,\mathbf{Z}}^{\mathrm{fin}}$ by the ideal generated by the $K_i^\ell - 1$, and $U_{\mathbf{F}_p}^{\mathrm{fin}}$, the subalgebra of the hyperalgebra $U_{\mathbf{F}_p}$ generated by the $\bar{X}_i^\pm$ (recall also that bars are used to distinguish generators of $U_{\mathbf{F}_p}$ from those of $U_\epsilon^{\mathrm{res}}$). It is known that the irreducible representations of $U_{\mathbf{F}_p}^{\mathrm{fin}}$ are parametrized by $(\mathbb{Z}/p\mathbb{Z})^n$; in fact, these representations are exactly the restricted modules $V_{\mathbf{F}_p}(\lambda)$, whose highest weights are of the form $\lambda = \sum_i r_i \lambda_i$, where $0 \leq r_i < p$. For this reason, $U_{\mathbf{F}_p}^{\mathrm{fin}}$

is called the *restricted enveloping algebra* of $\mathfrak{g}$ over $\mathbb{F}_p$. By 11.2.10, the same set $(\mathbb{Z}/p\mathbb{Z})^n$ parametrizes the finite-dimensional irreducible representations of $\tilde{U}_{\epsilon,\mathbb{Z}}^{\text{fin}}$ (i.e. those which are free and of finite rank as $\mathbb{Z}[\epsilon, \epsilon^{-1}]$-modules). To make this connection on the level of representations, note that if V is a finite-dimensional irreducible $\tilde{U}_{\epsilon,\mathbb{Z}}^{\text{fin}}$-module, and if V has parameter (highest weight) λ, the second isomorphism in 9.3.11 allows one to regard V as a $U_{\mathbb{F}_p}^{\text{fin}}$-module, and as such it has a unique irreducible quotient corresponding to the same parameter λ. Lusztig (1990a) conjectures that V is actually irreducible as a $U_{\mathbb{F}_p}^{\text{fin}}$-module, and in fact

CONJECTURE 11.2.13 *If $\ell = p$ is an odd prime, and p is sufficiently large, then $\tilde{U}_\epsilon^{\text{fin}}$ and $U_{\mathbb{F}_p}^{\text{fin}}$ have identical representation theories.*

This conjecture has largely been proved by Andersen, Jantzen & Soergel (1992). They show that every irreducible $U_{\mathbb{F}_p}^{\text{fin}}$-module arises by reduction (mod p) from an irreducible $U_{\epsilon,\mathbb{Z}}^{\text{fin}}$-module, provided that p is sufficiently large (although no bound is given on how large p should be). We shall discuss their work in a little more detail at the end of this subsection, but, before doing so, we describe a second conjecture of Lusztig (1989), relating the representation theory of $U_\epsilon^{\text{res}}(\mathfrak{g})$ to that of the affine Lie algebra $\tilde{\mathfrak{g}}$ (see the Appendix). This conjecture enables one to determine (in principle) the characters of the modules $V_\epsilon^{\text{res}}(\lambda)$. A proof of the preceding conjecture would then determine (in principle) the characters of the irreducible representations of $U_{\mathbb{F}_p}^{\text{fin}}$ and hence, by Steinberg's tensor product theorem, those of $U_{\mathbb{F}_p}$. It is known that this would completely determine the rational representation theory of the algebraic group (over the algebraic closure of $\mathbb{F}_p$) associated to $\mathfrak{g}$.

To state Lusztig's conjecture, we assume for simplicity that $\mathfrak{g}$ is of type A, D or E. Let

$$\Delta_\ell = \{\lambda \in \mathfrak{h}^* \mid (\lambda + \rho, \theta) \geq -\ell, \ (\lambda, \alpha_i) \leq -1, \ i = 1, \ldots, n\}.$$

Here, $\mathfrak{h}$ is the Cartan subalgebra of $\mathfrak{g}$, ρ is half the sum of the positive roots and θ is the highest root. Then, Δ_ℓ is a simplex bounded by the hyperplanes

$$(\lambda + \rho, \theta) = -\ell, \ (\lambda, \alpha_i) = -1, \ i = 1, \ldots, n.$$

The reflections $s_0, s_1, \ldots, s_n$ in these planes generate the affine Weyl group $\tilde{W}_\ell$ of $\mathfrak{g}$; the length $\ell(w)$ of an element $w \in \tilde{W}_\ell$ is defined in the obvious way. If $\lambda \in \mathfrak{h}^*$, there is a unique $w_\lambda \in \tilde{W}_\ell$ of minimal length such that $w_\lambda^{-1}(\lambda) \in \Delta_\ell$. If w' is any element of $\tilde{W}_\ell$, we set $\lambda_{w'} = w'(w_\lambda^{-1}(\lambda))$.

For any w', $w \in \tilde{W}_\ell$, let $P_{w',w} \in \mathbb{Z}[q]$ be the associated *Kazhdan–Lusztig polynomial* (see Humphreys (1990), Chap. 7). We recall that $P_{w',w} = 0$ unless $w' \leq w$, where $\leq$ is the *Bruhat ordering* on $\tilde{W}_\ell$.

CONJECTURE 11.2.14 *Let* $\lambda \in P^+$. *Then, in the Grothendieck group of finite-dimensional* U_ϵ^{res}-*modules,*

$$V_\epsilon^{\text{res}}(\lambda) = \sum (-1)^{\ell(w'w_\lambda)} P_{w',w_\lambda}(1) \, W_\epsilon^{\text{res}}(\lambda_{w'}),$$

where the sum is over $\{w' \in \tilde{W}_\ell \mid w' \le w_\lambda, \ \lambda_{w'} \in P^+\}$.

Remark The equality in 11.2.14 implies that

$$\text{ch}(V_\epsilon^{\text{res}}(\lambda)) = \sum (-1)^{\ell(w'w_\lambda)} P_{w',w_\lambda}(1) \, \text{ch}(W_\epsilon^{\text{res}}(\lambda_{w'})).$$

By 11.2.5, the characters of the Weyl modules $W_\epsilon^{\text{res}}(\lambda_{w'})$ are given by the Weyl character formula, so a knowledge of the coefficients $P_{w',w_\lambda}(1)$ enables one to compute the characters of the irreducible modules $V_\epsilon^{\text{res}}(\lambda)$. $\triangle$

Conjecture 11.2.14 has been proved by Kazhdan & Lusztig (1991). They show that the category of finite-dimensional $U_\epsilon^{\text{res}}(\mathfrak{g})$-modules is equivalent, as an abelian quasitensor category, to a certain category of representations of $\tilde{\mathfrak{g}}$ of level $-\ell - h$, where h is the Coxeter number of $\mathfrak{g}$ (see the Appendix). The main point is to construct a new kind of tensor product $\overset{\bullet}{\otimes}$ on this category, which has the property that the tensor product of two representations of level $-\ell - h$ is again of level $-\ell - h$ (of course, with the usual tensor product, the levels add). We shall outline their arguments in Chapter 16.

Conjecture 11.2.14 resembles another conjecture of Lusztig (1980), concerning the characters of representations of $\mathfrak{g}$ in characteristic p. It follows from the work of Andersen *et al.* (1992), and the proof of 11.2.14, that this conjecture is equivalent to an analogous conjecture about representations of the (classical) affine Lie algebra $\tilde{\mathfrak{g}}$ of level $-p - h$. A proof of this last conjecture was announced by Casian (1993), and his arguments have been completed by Kashiwara & Tanisaki (1994).

11.3 Tilting modules and the fusion tensor product

In this section, we discuss tensor products of representations of $U_\epsilon^{\text{res}}(\mathfrak{g})$ and show that it leads to a remarkable new algebraic structure. We assume that ϵ is a primitive ℓth root of unity, where, as usual, ℓ is odd and greater than d_i for all i. We also assume that $\ell > h$, the Coxeter number of $\mathfrak{g}$, and that ℓ is not divisible by three if $\mathfrak{g}$ is of type G_2. All representations are on complex vector spaces.

A Tilting modules We begin by introducing an important technical tool.

DEFINITION 11.3.1 *A finite-dimensional $U_\epsilon^{\mathrm{res}}$-module V of type 1 has a Weyl filtration if there exists a sequence of submodules*

$$0 = V_0 \subset V_1 \subset \cdots \subset V_p = V$$

with $V_r/V_{r-1} \cong W_\epsilon^{\mathrm{res}}(\lambda_r)$ for some $\lambda_r \in P^+$, $r = 1, \ldots, p$.

A finite-dimensional $U_\epsilon^{\mathrm{res}}$-module V of type 1 is a tilting module if both V and its dual V^* have Weyl filtrations.

Example 11.3.2 Define the *principal alcove*

$$\mathcal{C}_\ell = \{\lambda \in P^+ \mid (\lambda + \rho, \check\alpha) < \ell \text{ for all } \alpha \in \Delta^+\},$$

where ρ is half the sum of the positive roots of $\mathfrak{g}$. Suppose that $\lambda \in \mathcal{C}_\ell$. Then, by 11.2.5, $W_\epsilon^{\mathrm{res}}(\lambda) = V_\epsilon^{\mathrm{res}}(\lambda)$ is irreducible, and so trivially has a Weyl filtration. But $V_\epsilon^{\mathrm{res}}(\lambda)^* \cong V_\epsilon^{\mathrm{res}}(-w_0(\lambda))$, and it is clear that $-w_0(\lambda) \in \mathcal{C}_\ell$. Hence, $V_\epsilon^{\mathrm{res}}(\lambda)$ is a tilting module in this case. ◊

The next result lists some of the basic properties of tilting modules.

PROPOSITION 11.3.3 *(i) The dual of a tilting module is tilting.*
(ii) Any (finite) direct sum of tilting modules is tilting.
(iii) Any direct summand of a tilting module is tilting.
(iv) Any (finite) tensor product of tilting modules is tilting.

OUTLINE OF PROOF Parts (i) and (ii) are obvious, and part (iii) is nearly so. Part (iv) is proved by Paradowski (1992), using the theory of canonical bases, which we discuss in Chapter 14. ∎

In view of parts (ii) and (iii) of this proposition, we may restrict our attention to indecomposable tilting modules. These are parametrized by P^+:

PROPOSITION 11.3.4 *For any $\lambda \in P^+$, there exists, up to isomorphism, a unique indecomposable tilting module $T_\epsilon(\lambda)$ with the following properties:*
(i) the set of weights of $T_\epsilon(\lambda)$ is contained in the convex hull of the Weyl group orbit of λ;
(ii) λ is the unique maximal weight of $T_\epsilon(\lambda)$;
(iii) the weight space $T_\epsilon(\lambda)_\lambda$ is one-dimensional;
(iv) $T_\epsilon(\lambda)^* \cong T_\epsilon(-w_0(\lambda))$.
Conversely, every indecomposable tilting module is isomorphic to some (unique) $T_\epsilon(\lambda)$.

OUTLINE OF PROOF We indicate briefly how $T_\epsilon(\lambda)$ is constructed. To do so, we shall need the following criterion for the existence of a Weyl filtration.

LEMMA 11.3.5 *A finite-dimensional $U_\epsilon^{\mathrm{res}}$-module V of type $\mathbf{1}$ has a Weyl filtration if and only if, for all $\lambda \in P^+$, every short exact sequence of $U_\epsilon^{\mathrm{res}}$-modules*

$$0 \to W_\epsilon^{\mathrm{res}}(\lambda) \to V' \to V \to 0$$

splits. ∎

Fix $\lambda \in P^+$ and let $V_1 = W_\epsilon^{\mathrm{res}}(\lambda)$. If V_1^* has a Weyl filtration, we set $T_\epsilon(\lambda) = V_1$. Otherwise, by the lemma, there is a non-split extension

$$0 \to V_1 \to V_2 \to W_\epsilon^{\mathrm{res}}(\mu) \to 0$$

for some $\mu \in P^+$. This forces $\lambda > \mu$ (since $W_\epsilon^{\mathrm{res}}(\lambda)$ and $W_\epsilon^{\mathrm{res}}(\mu)$ are deformations of irreducible representations of $\mathfrak{g}$, it is enough to prove the analogous assertion in the classical situation, which is well known). Since $0 \subset V_1 \subset V_2$ is a Weyl filtration of V_2, if V_2^* has a Weyl filtration, we may take $T_\epsilon(\lambda) = V_2$.

Iterating this procedure, one eventually arrives at the desired module $T_\epsilon(\lambda)$; there can be only finitely many iterations, because there are only finitely many dominant weights less than λ. ∎

Propositions 11.3.3 and 11.3.4 have a number of interesting corollaries.

COROLLARY 11.3.6 *Let T be a tilting module for $U_\epsilon^{\mathrm{res}}$. Then,*

$$(7) \qquad\qquad T \cong \bigoplus_{\lambda \in P^+} T_\epsilon(\lambda)^{\oplus n_\lambda(T)},$$

where the multiplicities $n_\lambda(T)$ are uniquely determined by T.

PROOF It is immediate from 11.3.3 and 11.3.4 that T is a direct sum of $T_\epsilon(\lambda)$'s. For the last part, note that, if λ is a maximal weight of T, then $T_\epsilon(\lambda)$ must occur in any decomposition of T of the form (7). The result now follows by an obvious induction on $\dim(T)$. ∎

Essentially the same argument proves

COROLLARY 11.3.7 *Let T_1 and T_2 be tilting modules for $U_\epsilon^{\mathrm{res}}$, and suppose that $\mathrm{ch}(T_1) = \mathrm{ch}(T_2)$. Then, T_1 and T_2 are isomorphic.*

PROOF In fact, the hypothesis implies that T_1 and T_2 have the same weights, hence the same maximal weights. So we can use the same argument by induction on $\dim(T_1) = \dim(T_2)$ as above. ∎

This result and 11.2.4 immediately give

COROLLARY 11.3.8 *If T_1 and T_2 are tilting modules for $U_\epsilon^{\mathrm{res}}$, then $T_1 \otimes T_2$ and $T_2 \otimes T_1$ are isomorphic as $U_\epsilon^{\mathrm{res}}$-modules.* ∎

Of course, this result is also contained in 11.2.12, but the present proof makes no use of R-matrices.

Example 11.3.9 If $\mathfrak{g} = sl_2(\mathbb{C})$, 11.3.2 shows that the tilting modules $T_\epsilon(\lambda)$ with maximal weights in the range $0 \leq \lambda < \ell - 1$ are irreducible (we identify $\lambda \in P$ with the integer (λ, α), α being the positive root of $\mathfrak{g}$). When $\ell - 1 \leq \lambda \leq 2\ell - 2$, $T_\epsilon(\lambda)$ can be described explicitly as follows. First, by 11.2.5, $W_\epsilon^{\mathrm{res}}(\ell - 1)$ is irreducible, so $T_\epsilon(\ell - 1) = V_\epsilon^{\mathrm{res}}(\ell - 1)$. If $\ell \leq \lambda \leq 2\ell - 2$, $T_\epsilon(\lambda)$ is the 2ℓ-dimensional module with basis $\{t_r\}_{r=0,1,\ldots,\lambda} \cup \{t_r'\}_{r=0,1,\ldots,2\ell-2-\lambda}$ and the following action (as usual, we write $r = r_0 + \ell r_1$ with $0 \leq r_0 < \ell$, $r_1 = 0$ or 1):

$$K_1.t_r = \epsilon^{\lambda - 2r}t_r, \quad \begin{bmatrix} K_1;0 \\ \ell \end{bmatrix}_\epsilon .t_r = r_1 t_r,$$

$$X_1^+.t_r = [\lambda - r + 1]_\epsilon t_{r-1}, \quad X_1^-.t_r = [r + 1]_\epsilon t_{r+1},$$

$$(X_1^+)^{(\ell)}.t_r = ((\lambda - r)_1 + 1)t_{r-\ell}, \quad (X_1^-)^{(\ell)}.t_r = (r_1 + 1)t_{r+\ell},$$

$$K_1.t_r' = \epsilon^{2\ell - 2r - 2 - \lambda}t_r', \quad \begin{bmatrix} K_1;0 \\ \ell \end{bmatrix}_\epsilon .t_r' = 0,$$

$$X_1^+.t_r' = [2\ell - 1 - \lambda - r]_\epsilon t_{r-1}' + \begin{bmatrix} \lambda + r - \ell \\ r \end{bmatrix}_\epsilon t_{\lambda + r - \ell}, \quad \text{if } 0 < r \leq 2\ell - 2 - \lambda,$$

$$X_1^-.t_r' = [r + 1]_\epsilon t_{r+1}', \text{ if } 0 \leq r < 2\ell - 2 - \lambda, \quad (X_1^\pm)^{(\ell)}.t_r' = 0,$$

$$X_1^+.t_0' = [\lambda - \ell + 1]_\epsilon t_{\lambda - \ell}, \quad X_1^-.t_{2\ell - 2 - \lambda}' = \begin{bmatrix} \ell - 1 \\ \lambda - \ell + 1 \end{bmatrix}_\epsilon t_\ell.$$

(Vectors t_r and t_r' with r outside the appropriate range are understood to be zero.)

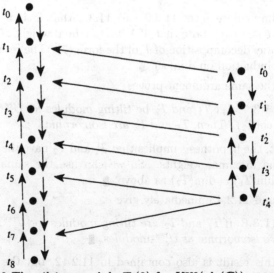

Fig. 20 The tilting module $T_\epsilon(8)$ for $U_\epsilon^{\mathrm{res}}(sl_2(\mathbb{C}))$, $\epsilon = e^{2\pi i/7}$.

The structure of $T_\epsilon(\lambda)$ for $\ell \le \lambda \le 2\ell - 2$ is illustrated in Fig. 20, where dots represent basis vectors, the horizontal level indicates the weight, and upward (resp. downward) sloping arrows indicate the action of X_1^+ (resp. of X_1^-). It is clear that $T_\epsilon(\lambda)$ is indecomposable, that the t_r span a submodule of $T_\epsilon(\lambda)$ isomorphic to $W_\epsilon^{\mathrm{res}}(\lambda)$, and that the quotient, spanned by the t'_r, is isomorphic to $W_\epsilon^{\mathrm{res}}(2\ell - 2 - \lambda)$. Thus, $T_\epsilon(\lambda)$ has a Weyl filtration. The structure of $T_\epsilon(\lambda)^*$ can be understood essentially by turning Fig. 20 upside down, which makes it clear that $T_\epsilon(\lambda)^*$ has a Weyl filtration involving the same Weyl modules as that of $T_\epsilon(\lambda)$. Hence, $T_\epsilon(\lambda)$ is a tilting module.

Note that λ is the unique maximal weight of $T_\epsilon(\lambda)$, but that $T_\epsilon(\lambda)$ is *not* a highest weight module, because it is not generated by the maximal weight vector t_0. Note also that the character of $T_\epsilon(\lambda)$ can be written down without further computation:

$$\mathrm{ch}(T_\epsilon(\lambda)) = \begin{cases} \mathrm{ch}(W_\epsilon^{\mathrm{res}}(\lambda)), & \text{if } \lambda < \ell, \\ \mathrm{ch}(W_\epsilon^{\mathrm{res}}(\lambda)) + \mathrm{ch}(W_\epsilon^{\mathrm{res}}(2\ell - 2 - \lambda)), & \text{if } \ell < \lambda \le 2\ell - 2, \end{cases}$$

the characters of the Weyl modules being given by the Weyl character formula. $\Diamond$

B Quantum dimensions By analogy with the discussion leading to 8.3.15, we define the quantum trace of an endomorphism f of a finite-dimensional $U_\epsilon^{\mathrm{res}}$-module V to be

$$\mathrm{qtr}(f) = \mathrm{trace}(K_{\rho^\bullet} f),$$

where, if $\sum_i r_i \alpha_i$, $r_i \in \mathbb{Z}$, is the sum of the positive roots of $\mathfrak{g}$, then $K_{\rho^\bullet} = \prod_i K_i^{r_i}$. Taking $f = \mathrm{id}_V$, we get the quantum dimension

$$\mathrm{qdim}(V) = \mathrm{trace}(K_{\rho^\bullet}).$$

Note that we do not have a universal R-matrix for $U_\epsilon^{\mathrm{res}}$, much less a ribbon Hopf structure, so this is not, strictly speaking, an application of the theory in Subsection 4.2C.

Example 11.3.10 Since, by 11.2.5, the character of the Weyl module $W_\epsilon^{\mathrm{res}}(\lambda)$ with highest weight $\lambda \in P^+$ is given by the Weyl character formula, it is easy to compute its quantum dimension. One obtains a 'quantum Weyl dimension formula':

$$\mathrm{qdim}(W_\epsilon^{\mathrm{res}}(\lambda)) = \prod_{\beta > 0} \frac{[(\lambda + \rho, \beta)]_\epsilon}{[(\rho, \beta)]_\epsilon},$$

the product being over all the positive roots β of $\mathfrak{g}$. It follows that $W_\epsilon^{\mathrm{res}}(\lambda)$ has non-zero quantum dimension if and only if, for all $\beta > 0$, $(\lambda + \rho, \beta)$ is not divisible by ℓ. $\Diamond$

Example 11.3.11 Let us compute the quantum dimension of the tilting modules $T_\epsilon(\lambda)$ for $sl_2(\mathbb{C})$ described in 11.3.9. If $\lambda < \ell$, we find, using the preceding example, that $\mathrm{qdim}(T_\epsilon(\lambda)) = [\lambda + 1]_\epsilon$; in particular, $\mathrm{qdim}(T_\epsilon(\ell - 1)) = 0$. On the other hand, if $\ell \le \lambda < 2\ell - 2$, we have, using the Weyl filtration of $T_\epsilon(\lambda)$,

$$\mathrm{qdim}(T_\epsilon(\lambda)) = \mathrm{qdim}(W_\epsilon^{\mathrm{res}}(\lambda)) + \mathrm{qdim}(W_\epsilon^{\mathrm{res}}(2\ell - 2 - \lambda))$$
$$= [\lambda + 1]_\epsilon + [2\ell - \lambda - 1]_\epsilon = 0. \quad \diamondsuit$$

It is obvious that, if $\lambda \in \mathcal{C}_\ell$, the condition on λ in the last sentence of 11.3.10 is satisfied, and hence $W_\epsilon^{\mathrm{res}}(\lambda) = V_\epsilon^{\mathrm{res}}(\lambda) = T_\epsilon(\lambda)$ has non-zero quantum dimension. Example 11.3.11 shows that the converse of this assertion is also true if $\mathfrak{g} = sl_2(\mathbb{C})$ and λ is not too large. In fact,

PROPOSITION 11.3.12 *Let $\lambda \in P^+$. Then, $\mathrm{qdim}(T_\epsilon(\lambda)) \ne 0$ if and only if $\lambda \in \mathcal{C}_\ell$.* ∎

See Andersen (1992a) for the proof.

COROLLARY 11.3.13 *Let $\lambda \in P^+ \backslash \mathcal{C}_\ell$ and let $f : T_\epsilon(\lambda) \to T_\epsilon(\lambda)$ be a homomorphism of $U_\epsilon^{\mathrm{res}}$-modules. Then, $\mathrm{qtr}(f) = 0$.*

PROOF Since $T_\epsilon(\lambda)$ is indecomposable, f is the sum of a scalar operator and a nilpotent operator. The quantum trace of the former vanishes by the proposition, and that of the latter vanishes because $K_\rho \cdot f$ is nilpotent if f is nilpotent, $T_\epsilon(\lambda)$ being the direct sum of its weight spaces by 11.2.3. ∎

We shall need one further corollary of 11.3.12 in the next subsection.

COROLLARY 11.3.14 *Let $\lambda \in P^+ \backslash \mathcal{C}_\ell$ and let V be any $U_\epsilon^{\mathrm{res}}$-module of type 1. Then, every direct summand of $T_\epsilon(\lambda) \otimes V$ has quantum dimension zero.*

OUTLINE OF PROOF By considering the projection onto a direct summand, it is enough to prove that $\mathrm{qtr}(f) = 0$ for all $U_\epsilon^{\mathrm{res}}$-module homomorphisms $f : T_\epsilon(\lambda) \otimes V \to T_\epsilon(\lambda) \otimes V$. Consider the following isomorphisms of $U_\epsilon^{\mathrm{res}}$-modules:

$$\mathrm{End}(T_\epsilon(\lambda) \otimes V) \cong (T_\epsilon(\lambda) \otimes V) \otimes (T_\epsilon(\lambda) \otimes V)^*$$
$$\cong T_\epsilon(\lambda) \otimes V \otimes V^* \otimes T_\epsilon(\lambda)^*$$
$$\cong T_\epsilon(\lambda) \otimes \mathrm{End}(V) \otimes T_\epsilon(\lambda)^*.$$

Applying the quantum trace $\mathrm{End}(V) \to \mathbb{C}$, which is a $U_\epsilon^{\mathrm{res}}$-module homomorphism, to the middle factor in the last module, we obtain a $U_\epsilon^{\mathrm{res}}$-module homomorphism $\tilde{f} : T_\epsilon(\lambda) \to T_\epsilon(\lambda)$. It is easy to check that

$$\mathrm{qtr}(f) = \mathrm{qtr}(\tilde{f}).$$

But the right-hand side vanishes by 11.3.13. ∎

C Tensor products We would like to understand how to 'decompose' tensor products $V_\epsilon^{\mathrm{res}}(\lambda) \otimes V_\epsilon^{\mathrm{res}}(\mu)$ for λ, $\mu \in P^+$. Of course, we cannot expect that such a tensor product is a direct sum of irreducible modules, since $U_\epsilon^{\mathrm{res}}$-modules are not completely reducible in general (see 11.2.7). However, it follows from 11.3.2, 11.3.3 and 11.3.6 that, if λ, $\mu \in \mathcal{C}_\ell$, then $V_\epsilon^{\mathrm{res}}(\lambda) \otimes V_\epsilon^{\mathrm{res}}(\mu)$ can be expressed as a direct sum of indecomposable tilting modules. Thus, one could try to define a 'semisimple' tensor product by discarding those tilting components which are not irreducible. Unfortunately, this procedure does not lead to an *associative* tensor product (see the remark following 11.3.17). However, a small modification of it does.

We begin to make this precise with the easy, but important,

PROPOSITION 11.3.15 *Let T_1 and T_2 be tilting modules. Then,*

$$(8) \qquad T_1 \otimes T_2 \cong \left(\bigoplus_{\lambda \in \mathcal{C}_\ell} V_\epsilon^{\mathrm{res}}(\lambda)^{\oplus m_\lambda} \right) \oplus Z,$$

where the multiplicities $m_\lambda \in \mathbf{N}$ and Z is a $U_\epsilon^{\mathrm{res}}$-module with the property that $\mathrm{qtr}(f) = 0$ for all homomorphisms of $U_\epsilon^{\mathrm{res}}$-modules $f : Z \to Z$ (in particular, $\mathrm{qdim}(Z) = 0$).

PROOF By 11.3.3, $T_1 \otimes T_2$ is a tilting module, and hence, by 11.3.6,

$$T_1 \otimes T_2 \cong \bigoplus_{\lambda \in P^+} T_\epsilon(\lambda)^{\oplus n_\lambda}$$

for some $n_\lambda \in \mathbf{N}$. We take $m_\lambda = n_\lambda$ if $\lambda \in \mathcal{C}_\ell$ and set

$$Z = \bigoplus_{\lambda \in P^+ \backslash \mathcal{C}_\ell} T_\epsilon(\lambda)^{\oplus n_\lambda}.$$

The required property of Z follows from 11.3.13. ∎

Example 11.3.16 If $\mathfrak{g} = sl_2(\mathbf{C})$ and $0 \leq \lambda, \mu \leq \ell - 1$, the decomposition of the tensor product $V_\epsilon^{\mathrm{res}}(\lambda) \otimes V_\epsilon^{\mathrm{res}}(\mu)$ into indecomposable tilting modules is as follows:

$$V_\epsilon^{\mathrm{res}}(\lambda) \otimes V_\epsilon^{\mathrm{res}}(\mu) \cong \bigoplus_{\substack{\nu = |\lambda - \mu| \\ \nu \equiv \lambda + \mu \ (\mathrm{mod}\ 2)}}^{\lambda + \mu} V_\epsilon^{\mathrm{res}}(\nu),$$

if $\lambda + \mu \leq \ell - 2$, and

$$V_\epsilon^{\mathrm{res}}(\lambda) \otimes V_\epsilon^{\mathrm{res}}(\mu) \cong \bigoplus_{\substack{\nu = |\lambda - \mu| \\ \nu \equiv \lambda + \mu \ (\mathrm{mod}\ 2)}}^{2\ell - 4 - \lambda - \mu} V_\epsilon^{\mathrm{res}}(\nu) \oplus \bigoplus_{\substack{\nu = \ell - 1 \\ \nu \equiv \lambda + \mu \ (\mathrm{mod}\ 2)}}^{\lambda + \mu} T_\epsilon(\nu),$$

if $\ell - 1 \leq \lambda + \mu \leq 2\ell - 2$. To verify this, it is enough, by 11.3.7, to check that the modules in question have the same character. This is straightforward, since the characters of all modules appearing are already known from 11.3.9. ◊

Returning to the general case, 11.3.15 suggests that we can 'make $T_1 \otimes T_2$ completely reducible' by redefining the tensor product to be the result of discarding Z on the right-hand side of (8). More precisely, if T is any tilting module, let $\overline{T}$ be the sum of the indecomposable summands in a decomposition (7) of T whose maximal weights lie in C_ℓ; then $\overline{T}$ is well defined up to isomorphism. Further, if T_1 and T_2 are tilting modules, define a new kind of tensor product of T_1 and T_2 by

$$T_1 \overline{\otimes} T_2 = \overline{T_1 \otimes T_2}.$$

PROPOSITION 11.3.17 *Let T_1, T_2 and T_3 be tilting modules for $U_\epsilon^{\mathrm{res}}$. Then, as $U_\epsilon^{\mathrm{res}}$-modules:*

(i) $T_1 \overline{\otimes} T_2 \cong T_2 \overline{\otimes} T_1$;

(ii) $T_1 \overline{\otimes} (T_2 \overline{\otimes} T_3) \cong (T_1 \overline{\otimes} T_2) \overline{\otimes} T_3$.

PROOF Part (i) is immediate from 11.3.8. For part (ii), we have

$$T_1 \otimes T_2 = (T_1 \overline{\otimes} T_2) \oplus Z_3, \quad T_2 \otimes T_3 = (T_2 \overline{\otimes} T_3) \oplus Z_1,$$

where $n_\lambda(Z_1) = n_\lambda(Z_3) = 0$ for all $\lambda \in C_\ell$. By 11.3.14, $n_\lambda(T_1 \otimes Z_2) = n_\lambda(Z_3 \otimes T_3) = 0$ if $\lambda \in C_\ell$. Hence,

$$
\begin{aligned}
(T_1 \overline{\otimes} T_2) \overline{\otimes} T_3 &= \overline{(T_1 \otimes T_2) \overline{\otimes} T_3}, \quad \text{since } Z_3 \overline{\otimes} T_3 = 0, \\
&\cong \overline{(T_1 \otimes T_2) \otimes T_3} \\
&\cong \overline{T_1 \otimes (T_2 \otimes T_3)} \\
&\cong T_1 \overline{\otimes} (T_2 \overline{\otimes} T_3) \\
&= T_1 \overline{\otimes} (T_2 \overline{\otimes} T_3), \quad \text{since } T_1 \overline{\otimes} Z_1 = 0. \quad \blacksquare
\end{aligned}
$$

Remark Let $\ell = 5$. From 11.3.16, we find that

$$V_\epsilon^{\mathrm{res}}(2) \otimes V_\epsilon^{\mathrm{res}}(2) \cong V_\epsilon^{\mathrm{res}}(0) \oplus V_\epsilon^{\mathrm{res}}(2) \oplus V_\epsilon^{\mathrm{res}}(4),$$
$$V_\epsilon^{\mathrm{res}}(2) \otimes V_\epsilon^{\mathrm{res}}(4) \cong V_\epsilon^{\mathrm{res}}(4) \oplus T_\epsilon(6),$$
$$V_\epsilon^{\mathrm{res}}(4) \otimes V_\epsilon^{\mathrm{res}}(4) \cong V_\epsilon^{\mathrm{res}}(4) \oplus T_\epsilon(6) \oplus T_\epsilon(8).$$

Thus, if we define a tensor product $\tilde{\otimes}$ by discarding those indecomposable tilting components which are not irreducible, we have

$$(V_\epsilon^{\mathrm{res}}(2) \tilde{\otimes} V_\epsilon^{\mathrm{res}}(2)) \tilde{\otimes} V_\epsilon^{\mathrm{res}}(4) \cong (V_\epsilon^{\mathrm{res}}(0) \oplus V_\epsilon^{\mathrm{res}}(2) \oplus V_\epsilon^{\mathrm{res}}(4)) \tilde{\otimes} V_\epsilon^{\mathrm{res}}(4)$$
$$\cong V_\epsilon^{\mathrm{res}}(4) \oplus V_\epsilon^{\mathrm{res}}(4) \oplus V_\epsilon^{\mathrm{res}}(4),$$

whereas $V_\epsilon^{\mathrm{res}}(2)\tilde{\otimes}(V_\epsilon^{\mathrm{res}}(2)\tilde{\otimes}V_\epsilon^{\mathrm{res}}(4)) \cong V_\epsilon^{\mathrm{res}}(2)\tilde{\otimes}V_\epsilon^{\mathrm{res}}(4) \cong V_\epsilon^{\mathrm{res}}(4)$. Thus, $\tilde{\otimes}$ is not associative, even up to isomorphism. On the other hand,

$$V_\epsilon^{\mathrm{res}}(2)\overline{\otimes}(V_\epsilon^{\mathrm{res}}(2)\overline{\otimes}V_\epsilon^{\mathrm{res}}(4)) = 0 = (V_\epsilon^{\mathrm{res}}(2)\overline{\otimes}V_\epsilon^{\mathrm{res}}(2))\overline{\otimes}V_\epsilon^{\mathrm{res}}(4). \quad \triangle$$

The module $\overline{T}$ associated to a tilting module T is well defined only up to isomorphism, for the decomposition of T into indecomposables is not unique, in general (although the maximal weights of the summands are unique). It is important, especially for the categorical arguments presented below, to make a canonical choice of a module in the isomorphism class of $\overline{T}$. This can be done as follows.

Let $T^{\check{}}$ be the unique maximal submodule of T, all of whose composition factors are of the form $V_\epsilon^{\mathrm{res}}(\lambda)$ for $\lambda \in \mathcal{C}_\ell$. We note that $T^{\check{}}$ is completely reducible. Indeed, we might as well assume that T is indecomposable, and then the highest weights of all the composition factors of T belong to the same orbit of the affine Weyl group on P – this follows from the 'linkage principle' (see Andersen, Polo & Wen Kexin (1991)). In particular, this is so for $T^{\check{}}$, but all its composition factors have highest weights in $\mathcal{C}_\ell$, no two distinct points of which are in the same orbit. Hence, all composition factors of $T^{\check{}}$ are the same, and our assertion now follows from the remarks made in the proof of 11.3.4.

Similarly, let $T^{\hat{}} = T/T'$ be the largest quotient of T which has all its composition factors of the form $V_\epsilon^{\mathrm{res}}(\lambda)$ for $\lambda \in \mathcal{C}_\ell$. The argument in the preceding paragraph shows that $T^{\hat{}}$ is also completely reducible.

The following result was pointed out to us by D. Kazhdan. We include the proof, which was given to us by S. Donkin, since it does not seem to be in the literature.

PROPOSITION 11.3.18 *The image of* $T^{\check{}} \subset T$ *in* $T^{\hat{}} = T/T'$ *is isomorphic to* $\overline{T}$.

PROOF We first show that $T_\epsilon(\lambda)^{\check{}} \subseteq T_\epsilon(\lambda)'$ for any $\lambda \in P^+\backslash\mathcal{C}_\ell$. Indeed, if $T_\epsilon(\lambda)^{\check{}}$ is not contained in $T_\epsilon(\lambda)'$, then, since $T_\epsilon(\lambda)^{\check{}}$ is completely reducible, there is an irreducible submodule V of $T_\epsilon(\lambda)^{\check{}}$ which is not contained in $T_\epsilon(\lambda)'$. Since $T_\epsilon(\lambda)^{\check{}}$ is completely reducible, we have

$$\frac{T_\epsilon(\lambda)}{T_\epsilon(\lambda)'} = \frac{V \oplus T_\epsilon(\lambda)'}{T_\epsilon(\lambda)'} \oplus \frac{W}{T_\epsilon(\lambda)'}$$

for some submodule W of T containing $T_\epsilon(\lambda)'$. Hence, $T_\epsilon(\lambda) = V + W$, and, by counting dimensions, one sees that the sum is direct. Since $T_\epsilon(\lambda)$ is indecomposable, $W = 0$. But then $T(\lambda)' = 0$, which contradicts the fact that $\lambda \notin \mathcal{C}_\ell$.

If T is any tilting module, write $T = T_1 \oplus \cdots \oplus T_r$, where the T_j are indecomposable and their maximal weights λ_j lie in $\mathcal{C}_\ell$ if $1 \leq j \leq s$ (say), but not if $s + 1 \le j \le r$. Now, the highest weights of all the composition factors of $T/\oplus_{j>s} T_j$ are in $\mathcal{C}_\ell$, so that $T' \subseteq \oplus_{j>s} T_j$. Hence, $\oplus_{j \le s} T_j \check{} = \oplus_{j \le s} T_j \check{}$ maps injectively into T/T'. On the other hand, by the result of the previous paragraph we have $T_j \check{} \subseteq T_j' \subseteq T'$ for $j > s$, so that $\oplus_{j>s} T_j \check{}$ maps to zero in T/T'. ∎

From now on, $\overline{T}$ will mean the module described in the statement of 11.3.18.

D The categorical formulation Let $\mathbf{tilt}_\ell$ be the full subcategory of the category of finite-dimensional U_ϵ^{res}-modules of type $\mathbf{1}$ whose objects are the tilting modules for U_ϵ^{res}. Proposition 11.3.3 shows that $\mathbf{tilt}_\ell$ is closed under taking direct sums, duals and tensor products. Thus, from 11.2.12, it follows that

PROPOSITION 11.3.19 $\mathbf{tilt}_\ell$ *is a rigid, $\mathbb{C}$-linear, quasitensor category.* ∎

Now let $\overline{\mathbf{tilt}}_\ell$ be the full subcategory of $\mathbf{tilt}_\ell$ consisting of the finite-dimensional U_ϵ^{res}-modules all of whose weights lie in

$$\{\lambda \in P \mid \langle \lambda + \rho, \check{\alpha} \rangle < \ell \text{ for all roots } \alpha\}.$$

From 11.3.2 and 11.3.4, it follows that $\overline{\mathbf{tilt}}_\ell$ is a semisimple, $\mathbb{C}$-linear category.

PROPOSITION 11.3.20 *Let $\mathbf{tilt}_\ell'$ be the full subcategory of $\mathbf{tilt}_\ell$ consisting of the tilting modules T for which $\overline{T} = 0$. Then:*

(i) $\mathbf{tilt}_\ell = \overline{\mathbf{tilt}}_\ell \oplus \mathbf{tilt}_\ell'$, *i.e. every tilting module is, in exactly one way, the direct sum of objects in $\overline{\mathbf{tilt}}_\ell$ and $\mathbf{tilt}_\ell'$;*

(ii) $\mathbf{tilt}_\ell'$ *is an ideal in $\mathbf{tilt}_\ell$, i.e. if T' is in $\mathbf{tilt}_\ell'$ and T is any tilting module, then $T \otimes T'$ is in $\mathbf{tilt}_\ell'$;*

(iii) if $T_1, T_2 \in \overline{\mathbf{tilt}}_\ell$, $T' \in \mathbf{tilt}_\ell'$, $f \in \text{Hom}_{\mathbf{tilt}_\ell}(T_1, T')$ and $g \in \text{Hom}_{\mathbf{tilt}_\ell}(T', T_2)$, then $g \circ f = 0$.

PROOF Part (i) is immediate from the definitions. For part (ii), note that, by 11.3.13, if T is an object of $\mathbf{tilt}_\ell'$, then $\text{qtr}(f) = 0$ for all U_ϵ^{res}-module endomorphisms of T. But, the proof of 11.3.14 shows that, if f is a U_ϵ^{res}-endomorphism of $T \otimes T'$, then $\text{qtr}(f) = \text{qtr}(\tilde{f})$ for some $\tilde{f} \in \text{End}(T')$. Thus, (ii) follows.

Finally, since $\overline{\mathbf{tilt}}_\ell$ is semisimple, it is enough to prove (iii) when T_1 and T_2 are irreducible; also, we might as well assume that $T_1 = T_2$, since if T_1 and T_2 are not isomorphic the result is trivial. But then if $g \circ f \ne 0$, $T_1 = T_2 = T$ would be a summand of T', which is impossible since T has non-zero quantum dimension. ∎

From this, we deduce that $\overline{\mathbf{tilt}}_\ell$ is a 'quotient category' of $\mathbf{tilt}_\ell$.

THEOREM 11.3.21 *With the usual operations of direct sum and dual, and with the tensor product $\overline{\otimes}$ defined in Subsection 11.3C, $\overline{\text{tilt}}_\ell$ becomes a rigid, semisimple, $\mathbb{C}$-linear, quasitensor category.*

OUTLINE OF PROOF Let $\mathbf{C}$ be the category whose objects are those of $\overline{\text{tilt}}_\ell$, but whose sets of morphisms are given, as $\mathbb{C}$-vector spaces, by

$$\text{Hom}_{\mathbf{C}}(T_1, T_2) = \text{Hom}_{\text{tilt}_\ell}(T_1, T_2)/\text{Hom}'_{\text{tilt}_\ell}(T_1, T_2),$$

where $\text{Hom}'_{\text{tilt}_\ell}(T_1, T_2)$ consists of the morphisms $f : T_1 \to T_2$ which factorize as $f = g \circ h$, where $g \in \text{Hom}_{\text{tilt}_\ell}(T', T_2)$, $h \in \text{Hom}_{\text{tilt}_\ell}(T_1, T')$ and T' is an object of tilt'_ℓ (it is easy to see that $\text{Hom}'_{\text{tilt}_\ell}(T_1, T_2)$ is a vector subspace of $\text{Hom}_{\text{tilt}_\ell}(T_1, T_2)$). By using 11.3.20, it is easy to see that this makes $\mathbf{C}$ into a $\mathbb{C}$-linear category.

Moreover, there is a functor $\mathcal{F} : \text{tilt}_\ell \to \mathbf{C}$ which is the identity on objects, and the natural projection

$$\text{Hom}_{\text{tilt}_\ell}(T_1, T_2) \to \text{Hom}_{\text{tilt}_\ell}(T_1, T_2)/\text{Hom}'_{\text{tilt}_\ell}(T_1, T_2)$$

on morphisms. If one defines commutativity and associativity isomorphisms in $\mathbf{C}$ to be the images of those in tilt_ℓ, then $\mathbf{C}$ becomes a $\mathbb{C}$-linear, quasitensor category. Rigidity is also easily checked.

Restricting $\mathcal{F}$ to $\overline{\text{tilt}}_\ell$ gives an equivalence of categories $\overline{\mathcal{F}} : \overline{\text{tilt}}_\ell \to \mathbf{C}$. Indeed, to prove this, it is enough to show that every object of $\mathbf{C}$ is isomorphic to the image under $\overline{\mathcal{F}}$ of an object of $\overline{\text{tilt}}_\ell$, and that $\overline{\mathcal{F}}$ is a bijection on sets of morphisms (see MacLane (1971), page 91). Both assertions follow easily from 11.3.20. Since $\mathbf{C}$ is a rigid, $\mathbb{C}$-linear, quasitensor category, so is $\overline{\text{tilt}}_\ell$. ∎

It is natural to ask for a description of the fusion ring of the quasitensor category $\overline{\text{tilt}}_\ell$ (see Subsection 5.2D). According to D. Kazhdan (private communication), this fusion ring is isomorphic to the quotient of the Grothendieck ring of the category of finite-dimensional g-modules by the ideal generated by the irreducible modules whose highest weights lie on the upper wall

$$\{\lambda \in P^+ \mid (\lambda + \rho, \check{\theta}) = \ell\}$$

of $\mathcal{C}_\ell$.

Example 11.3.22 We consider the fusion ring associated to $U_\epsilon^{\text{res}}(sl_2(\mathbb{C}))$ for $\epsilon = \sqrt{-1}$. Unfortunately, this example is not covered by the results of this section, since we have always assumed that ℓ is odd. However, it is easy to check directly that there are exactly three irreducible $U_{\sqrt{-1}}^{\text{res}}(sl_2(\mathbb{C}))$-modules of type **1** with non-zero quantum dimension, namely the representations $V_{\sqrt{-1}}^{\text{res}}(0)$, $V_{\sqrt{-1}}^{\text{res}}(1)$ and $V_{\sqrt{-1}}^{\text{res}}(2)$ of the indicated highest weights (and given by the usual formulas). Their dimensions are 1, 2 and 3, and their quantum dimensions are 1, $\sqrt{2}$ and 1, respectively. Defining the truncated tensor product $\overline{\otimes}$

as usual by discarding summands of quantum dimension zero, it is easy to check that

$$V_{\sqrt{-1}}^{\text{res}}(1)\overline{\otimes}\, V_{\sqrt{-1}}^{\text{res}}(1) \cong V_{\sqrt{-1}}^{\text{res}}(0) \oplus V_{\sqrt{-1}}^{\text{res}}(2),$$
$$V_{\sqrt{-1}}^{\text{res}}(1)\overline{\otimes}\, V_{\sqrt{-1}}^{\text{res}}(2) \cong V_{\sqrt{-1}}^{\text{res}}(2)\overline{\otimes}\, V_{\sqrt{-1}}^{\text{res}}(1) \cong V_{\sqrt{-1}}^{\text{res}}(1),$$
$$V_{\sqrt{-1}}^{\text{res}}(2)\overline{\otimes}\, V_{\sqrt{-1}}^{\text{res}}(2) \cong V_{\sqrt{-1}}^{\text{res}}(0),$$
$$V_{\sqrt{-1}}^{\text{res}}(0)\overline{\otimes}\, V_{\sqrt{-1}}^{\text{res}}(\lambda) \cong V_{\sqrt{-1}}^{\text{res}}(\lambda)\overline{\otimes}\, V_{\sqrt{-1}}^{\text{res}}(0) \cong V_{\sqrt{-1}}^{\text{res}}(\lambda), \quad \text{for } \lambda = 0, 1, 2.$$

Comparing with the formulas in 5.2.11, we see that this fusion ring is isomorphic to that of the 'scaling model of the Ising model at the critical point': one should identify the physical representation ρ_s in 5.2.11 with the $U_{\sqrt{-1}}^{\text{res}}(sl_2(\mathbb{C}))$-module $V_{\sqrt{-1}}^{\text{res}}(2s)$.

This coincidence strongly suggests that $U_{\sqrt{-1}}^{\text{res}}(sl_2(\mathbb{C}))$ should act on the Hilbert space of the Ising model as a 'quantum symmetry group'. Mack & Schomerus (1990) show that this is indeed the case. $\diamondsuit$

Bibliographical notes

11.1 The main references for Subsection 11.1A are De Concini & Kac (1990) and De Concini, Kac & Procesi (1992). For the proof of 11.1.6, see De Concini, Kac & Procesi (1993a).

The irreducible representations of $U_\epsilon(sl_2(\mathbb{C}))$ were first described explicitly by Roche & Arnaudon (1989). See Arnaudon (1990), Date, Jimbo, Miki & Miwa (1990, 1991a,b) and Arnaudon & Chakrabarti (1991b) for the sl_n case, Arnaudon & Chakrabarti (1991a,c) and Chari & Pressley (1991c) for minimal cyclic representations, Schnizer (1992) for maximal cyclic representations of the quantum orthogonal and symplectic groups, and Chari & Pressley (1992, 1994a) for the construction of some representations of intermediate dimension.

11.2 Most of the results and conjectures in this section are taken from Lusztig (1989, 1990a–c). We have refrained from entering into the cohomological theory of representations of U_ϵ^{res} (Borel–Weil–Bott theory). For this, see Andersen, Polo & Wen Kexin (1991, 1992), and Andersen & Wen Kexin (1992).

For a sophisticated treatment of the (classical) characteristic p theory, see Jantzen (1987).

11.3 For further details on tilting modules for U_ϵ^{res}, see Andersen (1992a). Similar results, when ℓ is a prime, have been obtained by Gel'fand & Kazhdan (1992). The proof of 11.3.21 was taken from this last reference. The theory of tilting modules for classical algebraic groups is due to Donkin (1993), whose work is based on Ringel (1991). The idea of taking truncated tensor products seems to have appeared first in the physics literature, where the representations of quantum dimension zero are regarded as 'unphysical'; see Pasquier & Saleur (1990). See Fröhlich & Kerler (1993) for further discussion of the relation between U_ϵ^{res} and quantum field theory.

12
Infinite-dimensional quantum groups

All the quantum groups we have considered in detail so far are deformations of the universal enveloping algebra of some finite-dimensional Lie algebra (or of the algebra of functions on the associated group). In this chapter, we look at two families of infinite-dimensional quantum groups, called *Yangians* and *quantum affine algebras*.

Classical affine algebras are the simplest infinite-dimensional Kac–Moody algebras. They are more interesting and tractable than arbitrary Kac–Moody algebras because they have another realization besides that given by the usual generators and relations; namely, they are central extensions of the Lie algebra of maps $\mathbb{C}^\times \to \mathfrak{g}$, where $\mathfrak{g}$ is finite dimensional. One of the crucial results in this chapter gives another realization of quantum affine algebras which, although still in terms of generators and relations, retains many of the features of a space of maps. For example, in some cases, one can 'evaluate' an element of a quantum affine algebra at a point of $\mathbb{C}^\times$, although this is more subtle than in the classical case.

We discuss the structure and representation theory of quantum affine algebras in Section 12.2, but we begin with Yangians, which are deformations of the Lie algebra of maps $\mathbb{C} \to \mathfrak{g}$. They are similar to, but in some ways simpler than, quantum affine algebras. For one thing, they involve no central extension; for another, one can scale away their deformation parameter completely, so that one is dealing with a single algebra (over $\mathbb{C}$) instead of a one-parameter family (in particular, there is no 'root of unity phenomenon' for Yangians). Finally, Yangians may be viewed as 'degenerate' versions of quantum affine algebras. In Section 12.1, we summarize what is known about the structure and representation theory of Yangians, giving a classification of their finite-dimensional irreducible representations analogous to that of finite-dimensional complex simple Lie algebras in terms of highest weights.

In Section 12.3, we describe a duality between quantum affine $sl_{n+1}(\mathbb{C})$ and affine Hecke algebras, analogous to the classical Frobenius–Schur duality between $sl_{n+1}(\mathbb{C})$ and symmetric groups. Affine Hecke algebras arise classically in connection with the representation theory of p-adic groups, so these results give a surprising connection between the quantum and p-adic worlds. There is also a parallel theory relating the Yangian of $sl_{n+1}(\mathbb{C})$ and 'degenerate' affine Hecke algebras.

In Section 12.4, we mention briefly a second connection between Yangians and the representation theory of classical objects, namely the orthogonal groups of infinite-dimensional (real) Hilbert space.

The main reason for the importance of Yangians and quantum affine algebras, at least historically, is that from their finite-dimensional representations one can construct rational and trigonometric solutions of the quantum Yang–Baxter equation, respectively. This is despite the fact that neither Yangians nor quantum affine algebras are quasitriangular in the sense used in this book. However, they both have 'pseudo-universal R-matrices' which, although not well-defined elements of the tensor product of the algebra with itself (even after h-adic completion), give well-defined operators on the tensor product of finite-dimensional representations, after being suitably 'renormalized'. We describe the properties of these pseudo-universal R-matrices, and the resulting solutions of the QYBE, in Section 12.5.

Note In this chapter, $\mathfrak{g}$ denotes a finite-dimensional complex simple Lie algebra.

12.1 Yangians and their representations

A Three realizations If $\mathfrak{g}$ is any Lie algebra over $\mathbb{C}$ and u is an indeterminate, the set $\mathfrak{g}[u] = \mathfrak{g} \underset{\mathbb{C}}{\otimes} \mathbb{C}[u]$ may be identified with the set of polynomial maps $f : \mathbb{C} \to \mathfrak{g}$, and is clearly a Lie algebra under pointwise operations. In case $\mathfrak{g}$ is finite dimensional and simple, a Lie bialgebra structure $\delta : \mathfrak{g}[u] \to \mathfrak{g}[u] \otimes \mathfrak{g}[u] = (\mathfrak{g} \otimes \mathfrak{g})[u, v]$ (where v is a second indeterminate, independent of u) was defined in 1.3.9:

$$\delta(f)(u, v) = (\mathrm{ad}_{f(u)} \otimes \mathrm{id} + \mathrm{id} \otimes \mathrm{ad}_{f(v)}) \left(\frac{t}{u - v} \right),$$

where $t \in \mathrm{Sym}^2(\mathfrak{g})$ is the Casimir element of $\mathfrak{g}$ associated to a fixed invariant bilinear form $(\ ,\ )$ on $\mathfrak{g}$. More explicitly, if $\{x_\lambda\}$ is any orthonormal basis of $\mathfrak{g}$ with respect to $(\ ,\ )$, then $t = \sum_\lambda x_\lambda \otimes x_\lambda$. Note that, if we view $\mathrm{Sym}^2(\mathfrak{g})$ as a subset of $U(\mathfrak{g})$, t is identified with the central (Casimir) element $\sum_\lambda x_\lambda^2 \in U(\mathfrak{g})$.

Now $\mathfrak{g}[u]$ is a graded Lie algebra:

$$\mathfrak{g}[u] = \bigoplus_{r \in \mathbb{N}} \mathfrak{g}u^r, \qquad [\mathfrak{g}u^r, \mathfrak{g}u^s] \subseteq \mathfrak{g}u^{r+s}.$$

Moreover, with the induced grading on $\mathfrak{g}[u] \otimes \mathfrak{g}[u]$, δ lowers degree by one, so it is natural to look for quantizations $U_h(\mathfrak{g}[u])$ of $(\mathfrak{g}[u], \delta)$ which are *homogeneous* in the sense that there is a dense $\mathbb{C}[[h]]$-sub-Hopf algebra $\tilde{U}_h(\mathfrak{g}[u])$ of $U_h(\mathfrak{g}[u])$ such that

(a) $\tilde{U}_h(\mathfrak{g}[u])$ is a graded associative algebra over $\mathbb{C}[[h]]$ (setting $\deg(h) = 1$);

(b) $\tilde{U}_h(\mathfrak{g}[u])/h\tilde{U}_h(\mathfrak{g}[u]) \cong U(\mathfrak{g}[u])$ as graded algebras over $\mathbb{C}$.

The next result shows that there is an essentially unique quantization with these properties.

For any elements z_1, z_2, z_3 of any associative algebra over $\mathbb{C}$, set

$$\{z_1, z_2, z_3\} = \frac{1}{24} \sum_\pi z_{\pi(1)} z_{\pi(2)} z_{\pi(3)},$$

the sum being over all permutations π of $\{1, 2, 3\}$.

THEOREM 12.1.1 *Let $\mathfrak{g}$ be a finite-dimensional complex simple Lie algebra. Fix a (non-zero) invariant bilinear form $(\ ,\ )$ on $\mathfrak{g}$, and let $\{x_\lambda\}$ be an orthonormal basis of $\mathfrak{g}$ with respect to $(\ ,\ )$. There is, up to isomorphism, a unique homogeneous quantization $U_h(\mathfrak{g}[u])$ of $(\mathfrak{g}[u], \delta)$. It is topologically generated by elements x, $J(x)$, for $x \in \mathfrak{g}$, with the following defining relations:*

(1) $\qquad [x, y] \ (in \ U_h(\mathfrak{g}[u])) \ = \ [x, y] \ (in \ \mathfrak{g})$,

(2) $\qquad J(ax + by) = aJ(x) + bJ(y)$,

(3) $\qquad [x, J(y)] = J([x, y])$,

$$[J(x), J([y, z])] + [J(z), J([x, y])] + [J(y), J([z, x])] =$$

(4) $$h^2 \sum_{\lambda, \mu, \nu} ([x, x_\lambda], [[y, x_\mu], [z, x_\nu]]) \{x_\lambda, x_\mu, x_\nu\},$$

$$[[J(x), J(y)], [z, J(w)]] + [[J(z), J(w)], [x, J(y)]] =$$

(5) $$h^2 \sum_{\lambda, \mu, \nu} ([x, x_\lambda], [[y, x_\mu], [[z, w], x_\nu]]) \{x_\lambda, x_\mu, J(x_\nu)\},$$

for all x, y, $z \in \mathfrak{g}$, a, $b \in \mathbb{C}$.
The Hopf structure of $U_h(\mathfrak{g}[u])$ is given by

(6) $$\Delta_h(x) = x \otimes 1 + 1 \otimes x,$$

(7) $$\Delta_h(J(x)) = J(x) \otimes 1 + 1 \otimes J(x) + \frac{1}{2}h[x \otimes 1, t],$$

(8) $$S_h(x) = -x, \quad S_h(J(x)) = -J(x) + \frac{1}{4}cx,$$

(9) $$\epsilon_h(x) = \epsilon_h(J(x)) = 0.$$

Here, c is the eigenvalue of the Casimir element $t \in U(\mathfrak{g})$ in the adjoint representation of $\mathfrak{g}$ (in (7), we regard t as an element of $U(\mathfrak{g})^{\otimes 2}$).
The grading on $U_h(\mathfrak{g}[u])$ is given by

$$\deg(x) = 0, \quad \deg(J(x)) = 1. \quad \blacksquare$$

Remarks [1] Although the defining relations of $U_h(\mathfrak{g}[u])$ depend on the choice of the inner product $(\ ,\ )$, it is easy to see that, up to isomorphism, the Hopf algebra $U_h(\mathfrak{g}[u])$ does not. The right-hand sides of (4) and (5) are independent of the choice of orthonormal basis of $\mathfrak{g}$.

[2] If $\mathfrak{g} = sl_2(\mathbb{C})$, (4) is a consequence of (1)–(3), while if $\mathfrak{g} \not\cong sl_2(\mathbb{C})$, (5) follows from (1)–(4).

[3] The fact that $U_h(\mathfrak{g}[u])$ is a genuine deformation of $U(\mathfrak{g}[u])$, i.e. that $U_h(\mathfrak{g}[u]) \cong U(\mathfrak{g}[u])$ as $\mathbb{C}[[h]]$-modules, is a consequence of 12.1.8. $\triangle$

The relations in 12.1.1 can be 'derived' as follows. Using the fact that $[\mathfrak{g}, \mathfrak{g}] = \mathfrak{g}$, it is easy to see that $\mathfrak{g}[u]$ is generated as a Lie algebra by the elements x, $xu = J_0(x)$ (say) for $x \in \mathfrak{g}$, and that these elements satisfy the relations obtained by putting $h = 0$ in those of 12.1.1. To motivate the definition of Δ_h, note that since $\delta(x) = 0$, it is natural to take $\Delta_h(x)$ to be given by the classical formula. On the other hand,

$$\delta(J_0(x)) = \left[xu \otimes 1 + 1 \otimes xv,\ \frac{t}{u-v}\right]$$

$$= \frac{u}{u-v}[x \otimes 1,\ t] + \frac{v}{u-v}[1 \otimes x,\ t].$$

But, since t lies in the centre of $U(\mathfrak{g})$, $[x \otimes 1, t] = -[1 \otimes x, t]$, so

$$\delta(J_0(x)) = [x \otimes 1,\ t].$$

It is now clear that (7) is the simplest way to satisfy

$$\frac{\Delta_h(J(x)) - \Delta_h^{\mathrm{op}}(J(x))}{h} \equiv \delta(J_0(x)) \pmod h.$$

The right-hand sides of (4) and (5) are designed to make Δ_h extend to a homomorphism of algebras $U_h(\mathfrak{g}[u]) \to U_h(\mathfrak{g}[u]) \otimes U_h(\mathfrak{g}[u])$ (cf. the proof of 6.4.3).

Let $\tilde{U}_h(\mathfrak{g}[u])$ be the $\mathbb{C}[h]$-subalgebra of $U_h(\mathfrak{g}[u])$ generated (in the usual sense, not topologically) by the x, $J(x)$ for $x \in \mathfrak{g}$. Then, $\tilde{U}_h(\mathfrak{g}[u])$ is a graded Hopf algebra over $\mathbb{C}[h]$, and we may now specialize h to any complex number ϵ. However, the resulting Hopf algebra $U_\epsilon(\mathfrak{g}[u])$ (over $\mathbb{C}$) is essentially independent of ϵ, provided that $\epsilon \neq 0$. In fact, if ϵ, $\epsilon' \in \mathbb{C}^\times$, the map $U_\epsilon(\mathfrak{g}[u]) \to U_{\epsilon'}(\mathfrak{g}[u])$ given by $x \mapsto x$, $J(x) \mapsto \frac{\epsilon'}{\epsilon}J(x)$ is clearly an isomorphism of Hopf algebras. Thus, we might as well take $\epsilon = 1$. (On the other hand, $U_0(\mathfrak{g}[u]) = U(\mathfrak{g}[u])$ is certainly *not* isomorphic to $U_\epsilon(\mathfrak{g}[u])$ if $\epsilon \neq 0$, since $U_\epsilon(\mathfrak{g}[u])$ is not cocommutative.)

DEFINITION 12.1.2 *For any finite-dimensional complex simple Lie algebra* $\mathfrak{g}$*, the* Yangian $Y(\mathfrak{g})$ *is the Hopf algebra over* $\mathbb{C}$ *with generators* x*,* $J(x)$ *for*

$x \in \mathfrak{g}$ and defining relations and coalgebra structure given by setting $h = 1$ in (1)–(9) (we drop the subscripts on the coalgebra maps from now on).

Although $U(\mathfrak{g}[u])$ is given by the generators and relations of 12.1.1 with $h = 0$, this description obscures the fact that $\mathfrak{g}[u]$ is a space of maps $\mathbb{C} \to \mathfrak{g}$. It is more natural to take as generators of $U(\mathfrak{g}[u])$ the elements $X_i^{\pm} u^r$, $H_i u^r$ for $i = 1, \ldots, n$, $r \in \mathbb{N}$, where $\{X_i^{\pm}, H_i\}$ are the usual generators of $\mathfrak{g}$ (see the Appendix). Drinfel'd (1988b) gives a second realization of $Y(\mathfrak{g})$ in terms of generators which have the $X_i^{\pm} u^r$, $H_i u^r$ as their classical limit.

From now on, we fix the inner product on $\mathfrak{g}$ to be the standard one, and define root vectors $X_\beta^{\pm} \in \mathfrak{g}$ for all positive roots β as described in the Appendix.

THEOREM 12.1.3 *The Yangian $Y(\mathfrak{g})$ is isomorphic to the associative algebra with generators $X_{i,r}^{\pm}$, $H_{i,r}$, $i = 1, \ldots, n$, $r \in \mathbb{N}$, and the following defining relations:*

(10)
$$[H_{i,r}, H_{j,s}] = 0,$$

(11)
$$[H_{i,0}, X_{j,s}^{\pm}] = \pm d_i a_{ij} X_{j,s}^{\pm},$$

(12)
$$[H_{i,r+1}, X_{j,s}^{\pm}] - [H_{i,r}, X_{j,s+1}^{\pm}] = \pm \frac{1}{2} d_i a_{ij} (H_{i,r} X_{j,s}^{\pm} + X_{j,s}^{\pm} H_{i,r}),$$

(13)
$$[X_{i,r}^{+}, X_{j,s}^{-}] = \delta_{i,j} H_{i,r+s},$$

(14)
$$[X_{i,r+1}^{\pm}, X_{j,s}^{\pm}] - [X_{i,r}^{\pm}, X_{j,s+1}^{\pm}] = \pm \frac{1}{2} d_i a_{ij} (X_{i,r}^{\pm} X_{j,s}^{\pm} + X_{j,s}^{\pm} X_{i,r}^{\pm}),$$

(15)
$$\sum_{\pi} [X_{i,r_{\pi(1)}}^{\pm}, [X_{i,r_{\pi(2)}}^{\pm}, \ldots, [X_{i,r_{\pi(m)}}^{\pm}, X_{j,s}^{\pm}] \cdots]] = 0,$$

for all sequences of non-negative integers $r_1, \ldots, r_m$, where $m = 1 - a_{ij}$ and the sum is over all permutations π of $\{1, \ldots, m\}$.

The isomorphism φ between the two realizations of $Y(\mathfrak{g})$ is given by

(16)
$$\varphi(H_i) = d_i^{-1} H_{i,0}, \quad \varphi(J(H_i)) = d_i^{-1} H_{i,1} + \varphi(v_i),$$
$$\varphi(X_i^{\pm}) = X_{i,0}^{\pm}, \quad \varphi(J(X_i^{\pm})) = X_{i,1}^{\pm} + \varphi(w_i^{\pm}),$$

where

$$v_i = \frac{1}{4} \sum_{\beta \in \Delta^+} \frac{d_\beta}{d_i} (\beta, \alpha_i)(X_\beta^{+} X_\beta^{-} + X_\beta^{-} X_\beta^{+}) - \frac{d_i}{2} H_i^2,$$

$$w_i^{\pm} = \pm \frac{1}{4} \sum_{\beta \in \Delta^+} d_\beta \left([X_i^{\pm}, X_\beta^{\pm}] X_\beta^{\mp} + X_\beta^{\mp} [X_i^{\pm}, X_\beta^{\pm}]\right) - \frac{1}{4} d_i (X_i^{\pm} H_i + H_i X_i^{\pm}). \blacksquare$$

Remarks [1] Multiplying the right-hand sides of (12) and (14) by h, one obtains another presentation of $U_h(\mathfrak{g}[u])$; replacing the right-hand sides of (12) and (14) by zero gives another presentation of $U(\mathfrak{g}[u])$.

[2] The reader will have noticed that we have not given the formulas for the coalgebra structure of $Y(\mathfrak{g})$ in this new realization. While this is determined, in principle, by the isomorphism φ and equations (6)–(9), no *explicit* formula for the action of the comultiplication on the generators $X_{i,r}^{\pm}$, $H_{i,r}$ is known. $\triangle$

There is a third realization of the Yangian, at least when $\mathfrak{g} = sl_{n+1}(\mathbb{C})$, which preceded the other two historically. We take the simple root vectors in $sl_{n+1}(\mathbb{C})$ to be $X_i^+ = E_{i\,i+1}$, $X_i^- = E_{i+1\,i}$, and the basis vectors of the Cartan subalgebra to be $H_i = E_{i\,i} - E_{i+1\,i+1}$. Here, E_{ij} is the $(n+1) \times (n+1)$ matrix with one in the ith row and jth column and zeros elsewhere. The standard inner product in this case is just the trace form $(x, y) = \text{trace}(xy)$.

Let T be the free associative algebra over $\mathbb{C}$ with generators $t_{ij}^{(r)}$, where $r \geq 1$, $1 \leq i, j \leq n + 1$. Introduce the formal power series

$$t_{ij}(u) = \delta_{i,j} + \sum_{r=1}^{\infty} t_{ij}^{(r)} u^{-r} \in T[[u^{-1}]],$$

and define the *quantum determinant* of the matrix $T(u) = (t_{ij}(u))$ by

$$\text{qdet}(T(u)) = \sum_{\pi} \text{sgn}(\pi) t_{1\,\pi(1)}\left(u + \frac{n}{2}\right) t_{2\,\pi(2)}\left(u + \frac{n}{2} - 1\right) \ldots t_{n+1\,\pi(n+1)}\left(u - \frac{n}{2}\right),$$

the sum being over all permutations of $\{1, 2, \ldots, n + 1\}$. The quantum determinant of any square submatrix of $T(u)$ is defined in an analogous way, taking care to preserve the ordering of the rows and columns.

THEOREM 12.1.4 *The Yangian $Y(sl_{n+1}(\mathbb{C}))$ is isomorphic to the quotient of T by the two-sided ideal generated by the following relations:*

(17) $$\check{R}(u - v)(T(u) \odot T(v)) = (T(u) \odot T(v))\check{R}(u - v),$$
(18) $$\text{qdet}(T(u)) = 1.$$

In the first relation, $\check{R}(u) = \sigma - (1 \otimes 1)u^{-1} \in (M_{n+1}(\mathbb{C}) \otimes M_{n+1}(\mathbb{C}))[u^{-1}]$, σ being the flip of the two factors in $\mathbb{C}^{n+1} \otimes \mathbb{C}^{n+1}$.

There is an isomorphism ψ between this realization and that of 12.1.3 given by

$$\psi(X_i^+(u)) = b_i(u)a_i(u)^{-1}, \quad \psi(X_i^-(u)) = a_i(u)^{-1}c_i(u),$$
$$\psi(H_i(u)) = a_i(u)^{-1}a_i(u + 1)^{-1}a_{i-1}\left(u + \frac{1}{2}\right)a_{i+1}\left(u + \frac{1}{2}\right).$$

Here, $X_i^{\pm}(u) = \sum_{r=0}^{\infty} X_{i,r}^{\pm} u^{-r-1}$, $H_i(u) = 1 + \sum_{r=0}^{\infty} H_{i,r} u^{-r-1}$, and $a_i(u)$ (resp. $b_i(u)$, resp. $c_i(u)$) denotes the quantum determinant of the $i \times i$ submatrix of $T(u)$ formed by the entries $t_{jk}(u)$ for which $1 \leq j, k \leq i+1$ and $j, k \neq i+1$ (resp. $j \neq i+1$, $k \neq i$, resp. $j \neq i$, $k \neq i+1$). (We set $a_0 = a_{-1} = 1$.) ∎

Remarks [1] The two relations (17) and (18) are equivalent to the infinitely many relations which result from expanding both sides in powers of u and v. For the meaning of $T(u) \odot T(v)$, see Subsection 7.2A.

[2] Let $Y(gl_{n+1}(\mathbb{C}))$ be the quotient of $\mathcal{T}$ by the relations (17). Then qdet$(T(u))$ lies in the centre of $Y(gl_{n+1}(\mathbb{C}))$. Moreover, $Y(gl_{n+1}(\mathbb{C}))$ is a bialgebra with comultiplication and counit given by

$$\Delta(T(u)) = T(u) \boxdot T(u), \quad \epsilon(T(u)) = 1$$

(see Subsection 7.2A again for the meaning of these formulas), and qdet$(T(u))$ is a group-like element of $Y(gl_{n+1}(\mathbb{C}))$. Finally, the antipode of $Y(gl_{n+1}(\mathbb{C}))$ is given by $S(T(u)) = T(u)^{-1}$, and we have $S(\text{qdet}(T(u))) = \text{qdet}(T(u))^{-1}$.

[3] The matrix $R(u,v) = \sigma \circ \check{R}(u-v) = 1 \otimes 1 - \sigma/(u-v)$ is the simplest (non-constant) rational solution of the *quantum Yang–Baxter equation with spectral parameters*

$$(19) \quad R_{12}(u-v)R_{13}(u-w)R_{23}(v-w) = R_{23}(v-w)R_{13}(u-w)R_{12}(u-v).$$

In Section 12.5, we show how, conversely, given a finite-dimensional representation of a Yangian $Y(\mathfrak{g})$, one can construct a rational solution of the QYBE. Using these solutions, one can extend 12.1.4 to arbitrary $\mathfrak{g}$, although the description is not quite so explicit in the general case. The possibility of constructing rational solutions of the QYBE is one of the main motivations for the study of the representations of Yangians, which we take up later in this section. △

B Basic properties To obtain interesting results about Yangians, one often has to use more than one of the realizations 12.1.2, 12.1.3 and 12.1.4 simultaneously. While 12.1.3 reminds one that the classical limit of $Y(\mathfrak{g})$ is a space of maps, the others have the advantage that the comultiplication is known on all of their generators. As an example of this interplay, we prove the following result, which provides a quantum analogue of the 1-parameter group of translation automorphisms of $\mathfrak{g}[u]$ given by $u \mapsto u + a$, $a \in \mathbb{C}$.

PROPOSITION 12.1.5 *There is a one-parameter group of Hopf algebra automorphisms τ_a of $Y(\mathfrak{g})$, $a \in \mathbb{C}$, given by*

$$(20) \quad \tau_a(H_{i,r}) = \sum_{s=0}^{r} \binom{r}{s} a^{r-s} H_{i,s}, \quad \tau_a(X_{i,r}^{\pm}) = \sum_{s=0}^{r} \binom{r}{s} a^{r-s} X_{i,s}^{\pm}.$$

PROOF The first step is to define τ_a in terms of the presentation 12.1.2:

(21) $$\tau_a(x) = x, \quad \tau_a(J(x)) = J(x) + ax \qquad (x \in \mathfrak{g}).$$

This is the most obvious quantum analogue of the translation $u \mapsto u + a$, since the classical limits of x and $J(x)$ are x and xu, respectively. It is easy to see that this definition is compatible with the relations (1)–(5) and the Hopf structure (6)–(9).

To prove the formulas in (20), we proceed by induction on r, using the isomorphism (16) and the definition (21) for the cases $r = 0$ and 1, and the relations

(22)
$$X_{i,r+1}^{\pm} = \pm\frac{1}{2}[H_{i,1}, X_{i,r}^{\pm}] - \frac{1}{2}d_i(H_{i,0}X_{i,r}^{\pm} + X_{i,r}^{\pm}H_{i,0}),$$
$$H_{i,r+1} = [X_{i,r+1}^{+}, X_{i,0}^{-}]$$

for the inductive step. ∎

Remark The automorphisms τ_a are closely related to the antipode of $Y(\mathfrak{g})$: by checking on the generators x, $J(x)$ in 12.1.2, it is easy to see that $S^2 = \tau_{\frac{1}{2}c}$.

We have not so far made use of the fact that $U_h(\mathfrak{g}[u])$ is a graded algebra. In fact, this grading was lost when we set $h = 1$, since h had positive degree. However,

PROPOSITION 12.1.6 *The Yangian $Y(\mathfrak{g})$ has the structure of a filtered algebra, such that the associated graded algebra is isomorphic to $U(\mathfrak{g}[u])$.*

PROOF Using realization 12.1.3, let $Y(\mathfrak{g})_r$, for $r \in \mathbf{N}$, be the linear span of all monomials in the generators $X_{i,s}^{\pm}$, $H_{i,s}$ for which the sum of the indices s is at most r. It is clear that $Y(\mathfrak{g})_r \subset Y(\mathfrak{g})_{r+1}$ and that $Y(\mathfrak{g})_r.Y(\mathfrak{g})_s \subset Y(\mathfrak{g})_{r+s}$.

The associated graded algebra has generators $X_{i,r}^{\pm}$, $H_{i,r}$ satisfying the same relations as in 12.1.3, except that the right-hand sides of (12) and (14) are replaced by zero. By Remark [1] following 12.1.3, this algebra is just $U(\mathfrak{g}[u])$. ∎

COROLLARY 12.1.7 *The centre of $Y(\mathfrak{g})$ consists of the scalar multiples of the identity element.*

PROOF Let z be an element of the centre of $Y(\mathfrak{g})$, and suppose that $z \in Y(\mathfrak{g})_r \backslash Y(\mathfrak{g})_{r-1}$ (we set $Y(\mathfrak{g})_{-1} = 0$). By the proposition, z corresponds to a non-zero element of the centre of $U(\mathfrak{g}[u])$ of degree r. But it is known that $U(\mathfrak{g}[u])$ has trivial centre (cf. Chari & Ilangovan (1984)), so we must have $r = 0$. ∎

It will be important for the discussion of the representations of Yangians to have available an analogue of the Poincaré–Birkhoff–Witt theorem. For

$U(\mathfrak{g}[u])$, the most obvious PBW basis consists of ordered monomials in the elements $X_\beta^\pm u^r$ and $H_i u^r$, for $i = 1, \ldots, n$, $\beta \in \Delta^+$, $r \in \mathbb{N}$. Thus, to construct a PBW basis for $Y(\mathfrak{g})$, we must define quantum analogues of the $X_\beta^\pm u^r$. Now, the root vectors $X_\beta^\pm$ of $\mathfrak{g}$ are defined in terms of the simple root vectors $X_i^\pm$ by formulas of the form

$$X_\beta^\pm = c[X_{i_1}^\pm, [X_{i_2}^\pm, \ldots, [X_{i_{k-1}}^\pm, X_{i_k}^\pm]\cdots]],$$

where c is a non-zero complex number. For each $r \in \mathbb{N}$ and each $k \geq 1$, choose any partition $r = r_1 + \cdots + r_k$ of r into a sum of k non-negative integers, and define

$$X_{\beta,r}^\pm = c[X_{i_1,r_1}^\pm, [X_{i_2,r_2}^\pm, \ldots, [X_{i_{k-1},r_{k-1}}^\pm, X_{i_k,r_k}^\pm]\cdots]].$$

Although the $X_{\beta,r}^\pm$ depend on the choice of partition of r, this dependence is rather weak, for one can show that, if $\tilde{X}_{\beta,r}^\pm$ are the elements obtained using a different partition, then $\tilde{X}_{\beta,r}^\pm - X_{\beta,r}^\pm \in Y(\mathfrak{g})_{r-1}$.

PROPOSITION 12.1.8 *Fix a total ordering on the set*

$$\Sigma = \{X_{\beta,r}^\pm\}_{\beta \in \Delta^+, r \in \mathbb{N}} \cup \{H_{i,r}\}_{i=1,\ldots,n, r \in \mathbb{N}}.$$

Then, the set of all ordered monomials in the elements of Σ is a vector space basis of $Y(\mathfrak{g})$.

OUTLINE OF PROOF   It is straightfoward to prove that Σ spans $Y(\mathfrak{g})$. To prove that the elements of Σ are linearly independent, one makes use of the automorphisms τ_a. See Levendorskii (1993) for the details. ∎

Remark   A similar argument shows that Σ is also a topological basis of $U_h(\mathfrak{g}[u])$. This proves that $U_h(\mathfrak{g}[u])$ is a genuine quantization of $\mathfrak{g}[u]$, i.e. that $U_h(\mathfrak{g}[u]) \cong U(\mathfrak{g}[u])[[h]]$ as $\mathbb{C}[[h]]$-modules. △

Let $Y^\pm$ (resp. Y^0) be the subalgebra of $Y(\mathfrak{g})$ generated by the $X_{i,r}^\pm$ (resp. by the $H_{i,r}$), $i = 1, \ldots, n$, $r \in \mathbb{N}$. If we choose the total ordering in the proposition to be such that each $X_{i,r}^-$ precedes each $H_{j,s}$, which in turn precedes each $X_{k,t}^+$, we obtain

COROLLARY 12.1.9   (i) *The ordered monomials in the $X_{\beta,r}^\pm$, $\beta \in \Delta^+$, $r \in \mathbb{N}$ (resp. in the $H_{i,r}$, $i = 1, \ldots, n$, $r \in \mathbb{N}$) form a vector space basis of $Y^\pm$ (resp. of Y^0).*
(ii) *Multiplication induces an isomorphism of vector spaces*

$$Y^- \otimes Y^0 \otimes Y^+ \xrightarrow{\cong} Y(\mathfrak{g}). \quad \blacksquare$$

C Classification of the finite-dimensional representations In this sub-section, we describe a classification of the finite-dimensional representations of $Y(\mathfrak{g})$ analogous to the classical theory for $\mathfrak{g}$ in terms of highest weights.

Let $h_{i,r}$, for $i = 1, \ldots, n$, $r \in \mathbb{N}$, be complex numbers, collectively denoted by $\boldsymbol{h}$. If V is a $Y(\mathfrak{g})$-module, its *weight spaces* are the subspaces

$$V_{\boldsymbol{h}} = \{v \in V \mid H_{i,r}.v = h_{i,r}v \text{ for all } i, r\}.$$

A *primitive vector* in V is a non-zero vector $v \in V_{\boldsymbol{h}}$, for some $\boldsymbol{h}$, such that $X_{i,r}^+.v = 0$ for all i, r. One says that V is a *highest weight module* with *highest weight* $\boldsymbol{h}$ if $V = Y(\mathfrak{g}).v$ for some primitive vector $v \in V_{\boldsymbol{h}}$.

Note that there is a homomorphism of Hopf algebras $\iota : U(\mathfrak{g}) \to Y(\mathfrak{g})$ given by $\iota(x) = x$ for all $x \in \mathfrak{g}$. In particular, we may regard any representation of $Y(\mathfrak{g})$ as a representation of $\mathfrak{g}$.

PROPOSITION 12.1.10 *Every finite-dimensional irreducible $Y(\mathfrak{g})$-module V is highest weight.* ∎

The proof of this result is similar to, but simpler than, that of Proposition 12.2.3 below.

Highest weight $Y(\mathfrak{g})$-modules can be constructed by the usual methods. Define the Verma module $M(\boldsymbol{h})$ to be the quotient of $Y(\mathfrak{g})$ by the left ideal generated by the elements $X_{i,r}^+$ and $H_{i,r} - h_{i,r}.1$, for $i = 1, \ldots, n$, $r \in \mathbb{N}$. By 12.1.9(ii), $M(\boldsymbol{h}) = Y^-.1$. It follows that, regarded as a $\mathfrak{g}$-module, the weight space of $M(\boldsymbol{h})$ with weight $\sum_i d_i^{-1} h_{i,0} \lambda_i$ is one-dimensional (the λ_i are the fundamental weights of $\mathfrak{g}$ – see the Appendix). By the standard classical argument, this implies that $M(\boldsymbol{h})$ has a unique irreducible quotient $Y(\mathfrak{g})$-module, say $V(\boldsymbol{h})$.

It only remains to determine which of the $V(\boldsymbol{h})$ are finite dimensional.

THEOREM 12.1.11 *The irreducible $Y(\mathfrak{g})$-module $V(\boldsymbol{h})$ of highest weight $\boldsymbol{h}$ is finite dimensional if and only if there exist polynomials $P_i \in \mathbb{C}[u]$ such that*

$$(23) \qquad \frac{P_i(u + d_i)}{P_i(u)} = 1 + \sum_{r=0}^{\infty} h_{i,r} u^{-r-1},$$

in the sense that the right-hand side is the Laurent expansion of the left-hand side about $u = \infty$. ∎

We sketch the proof of a result very similar to 12.1.11 in the next section (see 12.2.6).

Remarks [1] It is clear that the highest weight $\boldsymbol{h}$ determines the polynomials P_i up to a scalar multiple. Thus, 12.1.10 and 12.1.11 show that the finite-dimensional irreducible representations of $Y(\mathfrak{g})$ are parametrized by the set of n-tuples of monic polynomials.

[2] Let λ be the unique maximal weight of $V(\boldsymbol{h})$ as a $\mathfrak{g}$-module. Using the isomorphism φ in 12.1.3, we find that $h_{i,0} = d_i\lambda(H_i)$. On the other hand, from (23) we obtain $h_{i,0} = d_i\deg(P_i)$. Hence, $\deg(P_i) = \lambda(H_i)$.

[3] If $\rho : Y(\mathfrak{g}) \to \text{End}(V)$ is a finite-dimensional irreducible module with associated polynomials $P_i(u)$, the twist $\rho\circ\tau_a$ by the translation automorphism τ_a defined in 12.1.5, which we denote by $\tau_a^*(V)$, has polynomials $P_i(u - a)$ (this follows easily from the formulas (20)).

[4] If V is any finite-dimensional representation of $Y(\mathfrak{g})$, the remark following 12.1.5 implies that the canonical isomorphism of vector spaces

$$V^{**} \cong \tau_{\frac{1}{2}c}^*(V)$$

is an isomorphism of representations of $Y(\mathfrak{g})$ (the $*$ on the left-hand side indicates the left dual). From the preceding remark, it follows that $V^{**} \not\cong V$ in general (cf. Remark [3] following 5.1.2). $\triangle$

The 'simplest' highest weight $Y(\mathfrak{g})$-modules are those whose associated polynomials are of smallest degree. We shall say that a finite-dimensional irreducible $Y(\mathfrak{g})$-module is *fundamental* if its associated polynomials are given by

$$(24) \qquad P_j(u) = \begin{cases} 1 & \text{if } j \neq i, \\ u - a & \text{if } j = i, \end{cases}$$

for some $i = 1, \ldots, n$; the corresponding representation will be denoted by $V(\lambda_i, a)$. It follows from the preceding remarks that, as a $\mathfrak{g}$-module, $V(\lambda_i, a)$ has λ_i as its unique maximal weight, and so contains the irreducible $\mathfrak{g}$-module $V(\lambda_i)$ with multiplicity one. We shall describe the structure of the $V(\lambda_i, a)$ in more detail later in this section.

The justification for the adjective 'fundamental' is contained in

PROPOSITION 12.1.12 *Let V and W be finite-dimensional $Y(\mathfrak{g})$-modules and let $v \in V$, $w \in W$ be primitive vectors of weights $\boldsymbol{h}_V$ and $\boldsymbol{h}_W$. Then, $v \otimes w$ is a primitive vector in $V \otimes W$ of weight $\boldsymbol{h}_{V\otimes W}$, where*

$$P_{i,V\otimes W} = P_{i,V}P_{i,W},$$

the monic polynomials $P_{i,V\otimes W}$, $P_{i,V}$ and $P_{i,W}$ being related to the weights $\boldsymbol{h}_{V\otimes W}$, $\boldsymbol{h}_V$ and $\boldsymbol{h}_W$ as in (23).

In fact, this result has the almost immediate

COROLLARY 12.1.13 *Every finite-dimensional irreducible $Y(\mathfrak{g})$-module is isomorphic to a subquotient of a tensor product of fundamental representations.*

PROOF Consider a finite-dimensional module $V(\boldsymbol{h})$ with associated polynomials P_i, $i = 1. \ldots, n$. Let $a_{i1}, a_{i2}, \ldots$ be the roots of P_i (counted with multiplicity). By the proposition, the tensor product v of the highest weight vectors in the fundamental representations $V(\lambda_1, a_{11}), V(\lambda_1, a_{12}), \ldots, V(\lambda_n, a_{n1}), \ldots$ (in any order) is a primitive vector of weight $\boldsymbol{h}$. Hence, $V(\boldsymbol{h})$ is a quotient of the submodule $Y(\mathfrak{g}).v$ of $\otimes_{ij} V(\lambda_i, a_{ij})$. ∎

We note also the obvious

COROLLARY 12.1.14 *Let V and W be finite-dimensional irreducible $Y(\mathfrak{g})$-modules and assume that $V \otimes W$ is also irreducible. Then,*

(i) the polynomials $P_{i,V\otimes W}$, $P_{i,V}$ and $P_{i,W}$ associated to $V \otimes W$, V and W are related by $P_{i,V\otimes W} = P_{i,V} P_{i,W}$;
(ii) $V \otimes W$ and $W \otimes V$ are isomorphic as representations of $Y(\mathfrak{g})$. ∎

Remark We shall see later in this section that there are finite-dimensional irreducible representations of $Y(\mathfrak{g})$ on which the tensor product operation is *not* commutative. △

We conclude this subsection with an

OUTLINE OF PROOF OF 12.1.12 This is an easy consequence of the following, still not quite explicit, formulas for the comultiplication of $Y(\mathfrak{g})$ in the realization 12.1.3. Let $N^+ = \sum_{i,r} X_{i,r}^+ Y^+ = \sum_{i,r} Y^+ X_{i,r}^+$. Then, modulo $Y(\mathfrak{g}) \otimes Y(\mathfrak{g})N^+$,

(25)
$$\Delta(X_{i,r}^+) \equiv X_{i,r}^+ \otimes 1 + 1 \otimes X_{i,r}^+ + \sum_{s=1}^{r} H_{i,s-1} \otimes X_{i,r-s}^+,$$
$$\Delta(H_{i,r}) \equiv H_{i,r} \otimes 1 + 1 \otimes H_{i,r} + \sum_{s=1}^{r} H_{i,s-1} \otimes H_{i,r-s}.$$

This is proved by induction on r, using (6)–(9) and the isomorphism φ in 12.1.3 to check the cases $r = 0$, 1, and the relations (22) for the inductive step.

To see that these formulas imply 12.1.12, note that it follows from the first formula that $v \otimes w$ is annihilated by the $X_{i,r}^+$, and from the second that $v \otimes w$ is an eigenvector of the $H_{i,r}$ with eigenvalues

$$h_{i,r,V\otimes W} = h_{i,r,V} + h_{i,r,W} + \sum_{s=1}^{r} h_{i,s-1,V} h_{i,r-s,W}$$

(in an obvious notation). It is easy to see that the scalars on the right-hand side of this equation are the coefficients in the Laurent expansion of

$$\frac{P_{i,V}(u+d_i) P_{i,W}(u+d_i)}{P_{i,V}(u) P_{i,W}(u)} = \left\{ 1 + \sum_{r=0}^{\infty} h_{i,r,V} u^{-r-1} \right\} \left\{ 1 + \sum_{r=0}^{\infty} h_{i,r,W} u^{-r-1} \right\}. ∎$$

D Evaluation representations We now consider the problem of giving 'explicit' descriptions of the finite-dimensional modules $V(\boldsymbol{h})$. Before attempting this, it is instructive to consider what happens in the classical limit.

The most obvious way to construct representations of $\mathfrak{g}[u]$ is to use the homomorphisms $\mathrm{ev}_a : \mathfrak{g}[u] \to \mathfrak{g}$ which evaluate a $\mathfrak{g}$-valued polynomial at a point $a \in \mathbb{C}$. If V is a finite-dimensional irreducible $\mathfrak{g}$-module, its pull-back by ev_a is still irreducible because ev_a is obviously surjective. In fact, it can be shown that every finite-dimensional irreducible representation of $\mathfrak{g}[u]$ is isomorphic to a tensor product of such 'evaluation representations' (cf. Chari & Pressley (1986)).

Thus, it is natural to try to find an analogue of the homomorphisms ev_a for $Y(\mathfrak{g})$. Using the realization 12.1.2 and recalling that the classical limit of $J(x)$ is xu, the most obvious guess is $\mathrm{ev}_a(x) = x$, $\mathrm{ev}_a(J(x)) = ax$. Unfortunately, one finds that this map respects the defining relations of $Y(\mathfrak{g})$ only if $\mathfrak{g} \cong sl_2(\mathbb{C})$. In general, we have

PROPOSITION 12.1.15 *(i) If $\mathfrak{g} = sl_2(\mathbb{C})$, there is a unique homomorphism of algebras $\mathrm{ev}_a : Y(\mathfrak{g}) \to U(\mathfrak{g})$, for all $a \in \mathbb{C}$, such that*

$$\mathrm{ev}_a(x) = x, \quad \mathrm{ev}_a(J(x)) = ax$$

for all $x \in \mathfrak{g}$.
(ii) If $\mathfrak{g} = sl_{n+1}(\mathbb{C})$, where $n > 1$, there is a unique homomorphism of algebras $\mathrm{ev}_a : Y(\mathfrak{g}) \to U(\mathfrak{g})$, for all $a \in \mathbb{C}$, such that

$$(26) \quad \mathrm{ev}_a(x) = x, \quad \mathrm{ev}_a(J(x)) = ax + \frac{1}{4} \sum_{\lambda,\mu} \mathrm{trace}(x(x_\lambda x_\mu + x_\mu x_\lambda)) \, x_\lambda x_\mu$$

for all $x \in \mathfrak{g}$. Here, $\{x_\lambda\}$ is an orthonormal basis of $\mathfrak{g}$, and the trace is computed by multiplying together x, x_λ and x_μ, thought of as $(n+1) \times (n+1)$ matrices.
(iii) If $\mathfrak{g}$ is not of type A_n for any n, there is no homomorphism of algebras $\mathrm{ev} : Y(\mathfrak{g}) \to U(\mathfrak{g})$ such that the composite $U(\mathfrak{g}) \xrightarrow{\iota} Y(\mathfrak{g}) \xrightarrow{\mathrm{ev}} U(\mathfrak{g})$ is the identity map. ∎

The first two parts are proved by direct verification; the last part will become clear later in this section.

Remarks [1] The maps ev_a in (i) and (ii) are *not* homomorphisms of Hopf algebras.
[2] The reason for the special role played by type A_n, $n > 1$, in this result derives from the following strange fact. Consider the tensor product $\mathfrak{g} \otimes \mathfrak{g}$ of two copies of the adjoint representation of $\mathfrak{g}$. This always contains at least one copy of the adjoint representation, since the Lie bracket $\mathfrak{g} \otimes \mathfrak{g} \to \mathfrak{g}$ commutes with the action of $\mathfrak{g}$ (this assertion is equivalent to the Jacobi identity). But

if $\mathfrak{g}$ is of type A_n, $n > 1$, and only in that case, there is a second copy of $\mathfrak{g}$ in $\mathfrak{g} \otimes \mathfrak{g}$. The projection π onto this second copy is given by

$$\pi(x \otimes y) = xy + yx - \frac{2}{n+1}\text{trace}(xy).1.$$

The mysterious extra term on the right-hand side of the formula (26) for $\text{ev}_a(J(x))$ is $\frac{1}{4} \sum_\lambda \pi(x \otimes x_\lambda)x_\lambda$.

[3] If $n > 1$, there is a second family of evaluation homomorphisms ev^a : $Y(sl_{n+1}(\mathbb{C})) \to U(sl_{n+1}(\mathbb{C}))$, which also act as the identity on $U(sl_{n+1}(\mathbb{C}))$ $\subset Y(sl_{n+1}(\mathbb{C}))$, and which are given by

$$\text{ev}_a(x) = x, \quad \text{ev}_a(J(x)) = ax - \frac{1}{4} \sum_{\lambda,\mu} \text{trace}(x(x_\lambda x_\mu + x_\mu x_\lambda)) x_\lambda x_\mu.$$

In fact, $\text{ev}^a = S_U \circ \text{ev}_{a+\frac{1}{2}c} \circ S_Y$, where S_U and S_Y are the antipodes of $U(sl_{n+1}(\mathbb{C}))$ and $Y(sl_{n+1}(\mathbb{C}))$, respectively. If $n = 1$, we set $\text{ev}^a = \text{ev}_a$. $\triangle$

In general, if $\mathfrak{g}$ is of type A and V is a representation of $\mathfrak{g}$, the representation of $Y(\mathfrak{g})$ obtained by pulling back V by an evaluation homomorphism $\text{ev}_a : Y(\mathfrak{g}) \to U(\mathfrak{g})$ (resp. $\text{ev}^a : Y(\mathfrak{g}) \to U(\mathfrak{g})$) is called an *evaluation representation*, and denoted by V_a (resp. V^a). It is clear from the second remark following 12.1.11 that, if $V = V(\lambda_i)$ is the irreducible $\mathfrak{g}$-module with highest weight a fundamental weight λ_i, any associated evaluation representation is a fundamental representation $V(\lambda_i, b)$ of $Y(\mathfrak{g})$ for some b. Hence, from 12.1.13, we obtain

COROLLARY 12.1.16 *If $\mathfrak{g} \cong sl_{n+1}(\mathbb{C})$, every finite-dimensional irreducible $Y(\mathfrak{g})$-module is isomorphic to a subquotient of a tensor product of evaluation representations (which we may assume are all of the form V_a, or all of the form V^a, if we wish).* ∎

Later in this section, we give a much more precise result in the sl_2 case, but first we consider what happens when $\mathfrak{g}$ is not of type A_n.

PROPOSITION 12.1.17 *Let m_i be the multiplicity of the simple root α_i in the highest root θ of $\mathfrak{g}$, and let $d_\theta = d_i$ if θ is conjugate to α_i under the Weyl group of $\mathfrak{g}$. If $m_i = 1$ or d_θ/d_i, the fundamental representation $V(\lambda_i)$ of $\mathfrak{g}$ can be made into a fundamental representation of $Y(\mathfrak{g})$ by letting $J(x)$ act as bx for any $b \in \mathbb{C}$.*

PROOF Let $\rho : Y(\mathfrak{g}) \to \text{End}(V)$ be the fundamental representation of $Y(\mathfrak{g})$ with associated polynomials (24). Then, as a $\mathfrak{g}$-module,

$$V \cong V(\lambda_i) \oplus \bigoplus_{\mu < \lambda_i} V(\mu)^{\oplus r_\mu}$$

for certain multiplicities $r_\mu \geq 0$. Now, the map $\mathfrak{g} \otimes V \to V$ given by $x \otimes v \to \rho(J(x))(v)$ commutes with the action of $\mathfrak{g}$ (this follows from relation (3)). The proposition therefore follows from the following facts:

(i) if $m_i = 0$ or d_θ/d_i, and μ is a dominant weight less than λ_i, there is no non-trivial $\mathfrak{g}$-module map $\mathfrak{g} \otimes V(\lambda_i) \to V(\mu)$;
(ii) for any fixed fundamental weight λ_i of $\mathfrak{g}$, any $\mathfrak{g}$-module map $\mathfrak{g} \otimes V(\lambda_i) \to V(\lambda_i)$ is a scalar multiple of the action of $\mathfrak{g}$ on $V(\lambda_i)$.

The first of these can be checked case-by-case; the second is an easy exercise. ∎

Inspecting the tables in Bourbaki (1968), we see that this result covers all the fundamental representations if $\mathfrak{g}$ is of type A_n (where, of course, we already know the result) or C_n, as well as the natural and spin representations if $\mathfrak{g}$ is of type B_n or D_n. It also covers nodes 1 and 6 for E_6, node 7 for E_7, node 4 for F_4 and node 2 for G_2 (numbering as in Bourbaki (1968)). The structure of the remaining fundamental representations in the B_n and D_n cases is given by

PROPOSITION 12.1.18 *Let $\mathfrak{g}$ be of type B_n or D_{n+1}. Then, for any $a \in \mathbb{C}$,*

$$V(\lambda_i, a) \cong \bigoplus_{j=0}^{[i/2]} V(\lambda_{i-2j}), \quad \text{if } 1 < i < n,$$

as $\mathfrak{g}$-modules. ∎

For the proof of this result, and the description of the fundamental representations of $Y(\mathfrak{g})$ when $\mathfrak{g}$ is of exceptional type, see Chari & Pressley (1991a). Inspecting the results of that work shows that the converse of 12.1.17 is also true, i.e. if m_i does not satisfy the conditions in the proposition, the fundamental representation $V(\lambda_i, a)$ is reducible as a $\mathfrak{g}$-module. Further, in the C_n case, the adjoint representation of $\mathfrak{g}$ (which is not fundamental) does not extend to an action of $Y(\mathfrak{g})$ (on the same space). This result, and those on the Yangians of the exceptional Lie algebras obtained in Chari & Pressley (1991a), prove part (iii) of 12.1.15. See also Drinfel'd (1985).

E The sl_2 case We conclude this section by describing the finite-dimensional irreducible representations of $Y(sl_2(\mathbb{C}))$ in more detail.

Let $V(r)_a$ be the pull-back by ev_a of the $(r + 1)$-dimensional irreducible $sl_2(\mathbb{C})$-module $V(r)$. Let us compute the polynomial associated to $V(r)_a$. There is a basis $\{v_0, \ldots, v_r\}$ of $V(r)$ on which the action of $sl_2(\mathbb{C})$ is given by

$$X_1^+.v_s = (r - s + 1)v_{s-1}, \quad X_1^-.v_s = (s + 1)v_{s+1}, \quad H_1.v_s = (r - 2s)v_s$$

for $s = 0, 1, \ldots, r$ (we set $v_{-1} = v_{r+1} = 0$). Since $J(x)$ acts on $V(r)_a$ as ax acts on $V(r)$, the action of $Y(sl_2(\mathbb{C}))$ is determined. But to compute the associated polynomial, we must describe the action of the generators $X_{1,k}^{\pm}$ and $H_{1,k}$ of the realization 12.1.3. We claim that

(27)
$$X_{1,k}^{+}.v_s = \left(a + \frac{1}{2}r - s + \frac{1}{2}\right)^k (r - s + 1)v_{s-1},$$

$$X_{1,k}^{-}.v_s = \left(a + \frac{1}{2}r - s - \frac{1}{2}\right)^k (s + 1)v_{s+1},$$

$$H_{1,k}.v_s = \left((a + \frac{1}{2}r - s - \frac{1}{2})^k(r - s)(s + 1)\right.$$

$$\left. - (a + \frac{1}{2}r - s + \frac{1}{2})^k(r - s + 1)s\right) v_s.$$

To prove this, one checks first that these formulas do define a representation of $Y(sl_2(\mathbb{C}))$, and then, by using the isomorphism φ in 12.1.3, that they agree with the known action of x and $J(x)$ for $x = X_1^{\pm}$ and H_1 (recall that the relation in (4) does not have to be checked in the sl_2 case because it is a consequence of the other relations).

From (27), we read off the eigenvalues of the $H_{1,k}$ on the highest weight vector v_0 of $V(r)_a$:

$$h_{1,k} = r\left(a + \frac{1}{2}r - \frac{1}{2}\right)^k,$$

from which we find that the monic polynomial satisfying (23) is

$$P_1(u) = \prod_{s=0}^{r-1}\left(u - a - \frac{1}{2}r + \frac{1}{2} + s\right).$$

Note that the roots of P_1 form a 'string' of length r, i.e. a sequence z_0, $z_1, \ldots, z_{r-1}$ of complex numbers such that $z_{s+1} - z_s = 1$ for all s.

We shall need a simple combinatorial property of such strings, whose proof may safely be left to the reader. Let us say that two strings S_1 and S_2 are in *special position* if their union is a string which properly contains S_1 and S_2; otherwise, S_1 and S_2 are in *general position*. For example, the strings $S(r)_a$ and $S(s)_b$ associated to evaluation representations $V(r)_a$ and $V(s)_b$ are in special position if and only if $a - b$ or $b - a = \frac{1}{2}(r + s) - k + 1$ for some $0 < k \leq \min\{r, s\}$.

LEMMA 12.1.19 *Any finite set of complex numbers with multiplicities can be written as a union of strings in general position. Moreover, the resulting set of strings is unique.* ∎

The 'union' here is in the sense of sets with multiplicity: the multiplicity of a complex number z in the union of S_1 and S_2 is the sum of its multiplicities in S_1 and in S_2.

Continuing to imitate the representation theory of $sl_2(\mathbb{C})[u]$, we next consider tensor products of evaluation representations. The crucial result, proved in Chari & Pressley (1990b), is

LEMMA 12.1.20 *A (finite) tensor product $\otimes_j V(r_j)_{a_j}$ of evaluation representations is irreducible if and only if, for all j, k, the strings $S(r_j)_{a_j}$ and $S(r_k)_{a_k}$ are in general position.* ∎

We can now prove

THEOREM 12.1.21 *Every finite-dimensional irreducible $Y(sl_2(\mathbb{C}))$-module is isomorphic to a tensor product of evaluation modules. Moreover, two such tensor products are isomorphic if and only if one is obtained from the other by a permutation of the factors.*

PROOF Let V be a finite-dimensional irreducible $Y(sl_2(\mathbb{C}))$-module with associated polynomial P. The set of roots of P, counted with multiplicity, can be written as a union of strings in general position. By 12.1.20, the tensor product of the evaluation representations corresponding to these strings (in any order) is irreducible, and by 12.1.12 it has associated polynomial P, and is therefore isomorphic to V. The second statement follows immediately from 12.1.12 and the uniqueness part of 12.1.19. ∎

Although we shall omit the proof of 12.1.20, it will be instructive to examine the simpler case of the tensor product of two evaluation modules in more detail.

PROPOSITION 12.1.22 *The representation $V = V(r)_a \otimes V(s)_b$ of $Y(sl_2(\mathbb{C}))$ is reducible if and only if*

$$a - b \text{ or } b - a = \frac{1}{2}(r + s) - j + 1 \quad \text{for some } 0 < j \leq \min\{r, s\}.$$

In this case, V has a unique proper subrepresentation W. Moreover:

(i) if $a - b = \frac{1}{2}(r + s) - j + 1$,

$$W \cong V(r - j)_{a+j/2} \otimes V(s - j)_{b-j/2}$$
$$W/V \cong V(j - 1)_{a-(r-j+1)/2} \otimes V(r + s - j + 1)_{b+(r-j+1)/2};$$

(ii) if $b - a = \frac{1}{2}(r + s) - j + 1$,

$$W \cong V(j - 1)_{a+(r-j+1)/2} \otimes V(r + s - j + 1)_{b-(r-j+1)/2}$$
$$W/V \cong V(r - j)_{a-j/2} \otimes V(s - j)_{b+j/2}.$$

OUTLINE OF PROOF As an $sl_2(\mathbb{C})$-module,

$$V(r)_a \otimes V(s)_b \cong \bigoplus_{j=0}^{\min\{r,s\}} V(r + s - 2j).$$

If w_j is the unique (up to a scalar multiple) $sl_2(\mathbb{C})$-primitive vector of weight $r + s - 2j$ in $V(r)_a \otimes V(s)_b$, one checks by a direct computation that w_j is a $Y(sl_2(\mathbb{C}))$-highest weight vector if and only if $a - b = \frac{1}{2}(r + s) - j + 1$. This gives the necessary and sufficient condition for $V(r)_a \otimes V(s)_b$ to have a $Y(sl_2(\mathbb{C}))$-submodule not containing the tensor product of the highest weight vectors in $V(r)_a$ and $V(s)_b$.

On the other hand, $V(r)_a \otimes V(s)_b$ has a proper submodule containing the tensor product of these highest weight vectors if and only if

$$b - a = \frac{1}{2}(r + s) - j + 1 \quad \text{for some } 0 < j \leq \min\{r, s\}.$$

In fact, $V(r)_a \otimes V(s)_b$ has such a submodule if and only if the (left) dual $(V(r)_a \otimes V(s)_b)^* \cong (V(s)_b)^* \otimes (V(r)_a)^*$ has a submodule which does not contain the tensor product of the highest weight vectors in $(V(s)_b)^*$ and $(V(r)_a)^*$. But $(V(r)_a)^* \cong V(r)_{a-1}$: this follows from the fact that the antipode S of $Y(sl_2(\mathbb{C}))$ satisfies $\mathrm{ev}_a \circ S = S \circ \mathrm{ev}_{a-1}$ (this is easily checked on the generators of the realization 12.1.2). Our assertion thus follows from the first part. ∎

COROLLARY 12.1.23 *Suppose that* $a - b$ *or* $b - a = \frac{1}{2}(r + s) - j + 1$ *for some* $0 < j \leq \min\{r, s\}$. *Then:*

(i) $V(r)_a \otimes V(s)_b$ *is not completely reducible;*
(ii) $V(r)_a \otimes V(s)_b$ *is not isomorphic to* $V(s)_b \otimes V(r)_a$. ∎

Remarks [1] We have already encountered failure of complete reducibility in Chapter 11, but it is worth pointing out that its failure here has nothing to do with any 'root of unity phenomenon'. In fact, the category of finite-dimensional representations of the Lie algebra $\mathfrak{g}[u]$ is not semisimple either. For example, it is easy to see that, if $k \geq 1$ and $a \in \mathbb{C}$, the natural representation of $\mathfrak{g}[u]$ on the space of k-jets

$$J_k(\mathfrak{g}) = \mathfrak{g}[u]/(u - a)^{k+1}\mathfrak{g}[u]$$

is indecomposable but not irreducible (since $(u - a)J_k(\mathfrak{g})$ is a proper submodule). However, tensor products of evaluation representations of $\mathfrak{g}[u]$ are always completely reducible.

[2] Part (ii) implies that $Y(sl_2(\mathbb{C}))$ *is not quasitriangular* (or even almost cocommutative). In fact, it implies that the category of finite-dimensional representations of $Y(sl_2(\mathbb{C}))$ is not quasitensor. Nevertheless, we shall see in Section 12.5 that $Y(\mathfrak{g})$ (for arbitrary $\mathfrak{g}$) is quasitriangular in a certain 'limiting' sense. △

12.2 Quantum affine algebras

Many of the results in this section are analogues of those obtained for Yangians in Section 12.1, or of those obtained for finite-dimensional Lie algebras in Chapters 9 and 10, so we shall keep our discussion brief to avoid too much repetition. The term 'affine Lie algebra' will always mean untwisted affine Lie algebra.

A Another realization: quantum loop algebras The affine Lie algebra $\tilde{\mathfrak{g}}$ associated to a finite-dimensional complex simple Lie algebra $\mathfrak{g}$ is an example of a Kac–Moody algebra $\mathfrak{g}'(A)$, where A is a symmetrizable generalized Cartan matrix (see the Appendix). Thus, 6.1.5 gives a quantization $U_h(\tilde{\mathfrak{g}})$ of $\tilde{\mathfrak{g}}$.

But affine Lie algebras are simpler, and better understood, than arbitrary Kac–Moody algebras because they have another realization. Namely, $\tilde{\mathfrak{g}}$ is a central extension, with one-dimensional centre, of the *loop algebra* $L(\mathfrak{g}) = \mathfrak{g}[u, u^{-1}]$, where u is an indeterminate; note that $L(\mathfrak{g})$ may be thought of as the space of Laurent polynomial maps $\mathbb{C}^\times \to \mathfrak{g}$. As in the case of Yangians, Drinfel'd (1988b) gave another realization of $U_h(\tilde{\mathfrak{g}})$ which, although still in terms of generators and relations, is closer to the description of affine Lie algebras as a space of maps. It will be more convenient for us to work with $U_q(\tilde{\mathfrak{g}})$ than with $U_h(\tilde{\mathfrak{g}})$; we refer to $U_q(\tilde{\mathfrak{g}})$, or any of its specializations, as a *quantum affine algebra*. The following presentation of $U_q(\tilde{\mathfrak{g}})$ is proved in Beck (1993).

THEOREM 12.2.1 *Let $\tilde{\mathfrak{g}}$ be the untwisted affine Lie algebra associated to a finite-dimensional complex simple Lie algebra $\mathfrak{g}$. The standard quantization $U_q(\tilde{\mathfrak{g}})$ of $\tilde{\mathfrak{g}}$ defined in 9.1.1 is isomorphic, as an algebra over $\mathbb{C}(q)$, to the algebra A_q with generators $\mathcal{C}^{\pm 1/2}$, $\mathcal{K}_i^{\pm 1}$, $\mathcal{H}_{i,r}$, $1 \le i \le n$, $r \in \mathbb{Z}\backslash\{0\}$ and $\mathcal{X}_{i,r}^{\pm}$, $1 \le i \le n$, $r \in \mathbb{Z}$, and with the following defining relations:*

$$\mathcal{K}_i\mathcal{K}_i^{-1} = \mathcal{K}_i^{-1}\mathcal{K}_i = 1, \quad \mathcal{C}^{1/2}\mathcal{C}^{-1/2} = 1,$$

$$\mathcal{C}^{\pm 1/2} \text{ are central,}$$

$$[\mathcal{K}_i, \mathcal{K}_j] = [\mathcal{K}_i, \mathcal{H}_{j,r}] = 0,$$

$$[\mathcal{H}_{i,r}, \mathcal{H}_{j,s}] = \delta_{r,-s}\frac{1}{r}[ra_{ij}]_{q_i}\frac{\mathcal{C}^r - \mathcal{C}^{-r}}{q_j - q_j^{-1}},$$

$$\mathcal{K}_i\mathcal{X}_{j,r}^{\pm}\mathcal{K}_i^{-1} = q_i^{\pm a_{ij}}\mathcal{X}_{j,r}^{\pm},$$

$$[\mathcal{H}_{i,r}, \mathcal{X}_{j,s}^{\pm}] = \pm\frac{1}{r}[ra_{ij}]_{q_i}\mathcal{C}^{\mp|r|/2}\mathcal{X}_{j,r+s}^{\pm},$$

$$\mathcal{X}_{i,r+1}^{\pm}\mathcal{X}_{j,s}^{\pm} - q_i^{\pm a_{ij}}\mathcal{X}_{j,s}^{\pm}\mathcal{X}_{i,r+1}^{\pm} = q_i^{\pm a_{ij}}\mathcal{X}_{i,r}^{\pm}\mathcal{X}_{j,s+1}^{\pm} - \mathcal{X}_{j,s+1}^{\pm}\mathcal{X}_{i,r}^{\pm},$$

$$[\mathcal{X}_{i,r}^{+}, \mathcal{X}_{j,s}^{-}] = \delta_{i,j}\frac{\mathcal{C}^{(r-s)/2}\Phi_{i,r+s}^{+} - \mathcal{C}^{-(r-s)/2}\Phi_{i,r+s}^{-}}{q_i - q_i^{-1}},$$

$$\sum_{\pi \in \Sigma_m} \sum_{k=0}^{m} (-1)^k \begin{bmatrix} m \\ k \end{bmatrix}_{q_i} \mathcal{X}^{\pm}_{i,r_{\pi(1)}} \cdots \mathcal{X}^{\pm}_{i,r_{\pi(k)}} \mathcal{X}^{\pm}_{j,s} \mathcal{X}^{\pm}_{i,r_{\pi(k+1)}} \cdots \mathcal{X}^{\pm}_{i,r_{\pi(m)}} = 0, i \neq j,$$

for all sequences of integers $r_1, \ldots, r_m$, where $m = 1 - a_{ij}$ and the elements $\Phi^{\pm}_{i,r}$ are determined by equating coefficients of powers of u in the formal power series

$$\sum_{r=0}^{\infty} \Phi^{\pm}_{i,\pm r} u^{\pm r} = \mathcal{K}^{\pm 1}_i \exp \left(\pm (q_i - q_i^{-1}) \sum_{s=1}^{\infty} \mathcal{H}_{i, \pm s} u^{\pm s} \right).$$

Let $\theta = \sum_{i=1}^{n} m_i \alpha_i$ be the highest root of $\mathfrak{g}$, set $q_\theta = q_i$ if θ is Weyl group conjugate to α_i, and set $\mathcal{K}_\theta = \prod_{i=1}^{n} \mathcal{K}_i^{m_i}$. Suppose that the root vector $\bar{X}_\theta^+$ of $\mathfrak{g}$ is expressed in terms of the simple root vectors as

$$\bar{X}_\theta^+ = \lambda [\bar{X}_{i_1}^+, [\bar{X}_{i_2}^+, \ldots, [\bar{X}_{i_k}^+, \bar{X}_j^+] \cdots]]$$

for some $\lambda \in \mathbb{C}$. Define maps $w_i^{\pm} : U_q(\tilde{\mathfrak{g}}) \to U_q(\tilde{\mathfrak{g}})$ by

$$w_i^{\pm}(a) = \mathcal{X}^{\pm}_{i,0} a - \mathcal{K}^{\pm 1}_i a \mathcal{K}^{\mp 1}_i \mathcal{X}^{\pm}_{i,0}.$$

Then, there exists an isomorphism of $\mathbb{C}(q)$-algebras $f : U_q(\tilde{\mathfrak{g}}) \to A_q$ defined on generators by

$$f(K_0) = C \mathcal{K}_\theta^{-1}, \quad f(K_i) = \mathcal{K}_i, \quad f(X_i^{\pm}) = \mathcal{X}^{\pm}_{i,0}, \ i = 1, \ldots, n,$$
$$f(X_0^+) = \mu w_{i_1}^- \ldots w_{i_k}^-(\mathcal{X}^-_{j,1}) \mathcal{K}_\theta^{-1}, \quad f(X_0^-) = \lambda \mathcal{K}_\theta w_{i_1}^+ \ldots w_{i_k}^+(\mathcal{X}^+_{j,-1}),$$

where $\mu \in \mathbb{C}(q)$ is determined by the condition

$$[X_0^+, X_0^-] = \frac{K_0 - K_0^{-1}}{q_\theta - q_\theta^{-1}}. \quad \blacksquare$$

Remarks [1] To obtain the analogous result for $U_h(\tilde{\mathfrak{g}})$ given in Drinfel'd (1988b), one replaces the generators $\mathcal{K}_i$ and $C^{1/2}$ by new generators $\mathcal{H}_{i,0}$ and $c/2$. The element c is central and the third and fifth sets of relations above are replaced by

$$[\mathcal{H}_{i,0}, \mathcal{H}_{j,r}] = 0, \quad [\mathcal{H}_{i,0}, \mathcal{X}^{\pm}_{j,r}] = \pm d_i a_{ij} \mathcal{X}^{\pm}_{j,r}.$$

The remaining relations are the same as those above, except that q_i is replaced by $e^{d_i h}$ and $C^{1/2}$ by $e^{hc/2}$.

[2] If A is the generalized Cartan matrix of $\tilde{\mathfrak{g}}$, then $U_q(\tilde{\mathfrak{g}}) = U_q(\mathfrak{g}'(A))$. To obtain an analogous presentation of $U_q(\mathfrak{g}(A))$, one introduces additional

generators $\mathcal{D}^{\pm 1}$ and relations

$$\mathcal{D}\mathcal{D}^{-1} = \mathcal{D}^{-1}\mathcal{D} = 1,$$
$$\mathcal{D}\mathcal{H}_{i,r}\mathcal{D}^{-1} = q^r \mathcal{H}_{i,r}, \quad [\mathcal{D}, \mathcal{K}_i] = [\mathcal{D}, \mathcal{C}] = 0,$$
$$\mathcal{D}\mathcal{X}_{i,r}^{\pm}\mathcal{D}^{-1} = q^r \mathcal{X}_{i,r}^{\pm}.$$

[3] By replacing q by any *transcendental* number $\epsilon \in \mathbb{C}^{\times}$, one obtains from 12.2.1 another presentation of the specialization $U_{\epsilon}(\tilde{\mathfrak{g}})$.

[4] Setting

$$\deg(\mathcal{K}_i) = \deg(\mathcal{C}) = 0, \quad \deg(\mathcal{H}_{i,r}) = \deg(\mathcal{X}_{i,r}^{\pm}) = r, \quad \deg(q) = 0$$

gives $U_q(\tilde{\mathfrak{g}})$ the structure of a *graded* algebra.

[5] The elements $\Phi_{i,r}^{\pm}$ are given in terms of the $\mathcal{H}_{j,s}$ by formulas of the form

$$\Phi_{i,r}^{\pm} = \pm \mathcal{K}_i^{\pm 1}((q_i - q_i^{-1})\mathcal{H}_{i,r} + g_{i,r}^{\pm}(\mathcal{H}_{i,\pm 1}, \ldots, \mathcal{H}_{i,\pm(r-1)})),$$

where $g_{i,r}^{\pm}$ are polynomials, with coefficients in $\mathbb{C}(q)$, which are homogeneous of degree r with respect to the above grading. $\triangle$

Beck (1993) also gives an analogue of the Poincaré–Birkhoff–Witt theorem for $U_q(\tilde{\mathfrak{g}})$. We state only the following weak version, which follows easily from 12.2.1.

PROPOSITION 12.2.2 Let $\epsilon \in \mathbb{C}^{\times}$ be transcendental. Let $U_{\epsilon}^{\pm}$ (resp. U_{ϵ}^0) be the subalgebras of $U_{\epsilon}(\tilde{\mathfrak{g}})$ generated by the elements $\mathcal{X}_{i,r}^{\pm}$ (resp. by the elements $\mathcal{C}^{\pm 1/2}$, $\mathcal{K}_i^{\pm 1}$, $\mathcal{H}_{i,r}$) for all i, r. Then,

$$U_{\epsilon}(\tilde{\mathfrak{g}}) = U_{\epsilon}^{-}.U_{\epsilon}^0.U_{\epsilon}^{+}. \quad \blacksquare$$

Note that, by the Remark [5] following 12.2.1, U_{ϵ}^0 is also generated by the $\mathcal{C}^{\pm 1/2}$, $\mathcal{K}_i^{\pm 1}$ and $\Phi_{i,r}^{\pm}$.

B Finite-dimensional representations of quantum loop algebras For the remainder of this section, $\epsilon \in \mathbb{C}^{\times}$ is assumed to be transcendental.

We shall say that a representation V of $U_{\epsilon}(\tilde{\mathfrak{g}})$ is of *type* **1** if the generators $K_0, K_1, \ldots, K_n$ act semisimply on V with eigenvalues which are integer powers of ϵ, and if $\mathcal{C}^{1/2}$ acts as 1 on V. As in 9.1.4, there are 2^{n+1} $\mathbb{C}(q)$-algebra automorphisms of $U_{\epsilon}(\tilde{\mathfrak{g}})$ given on generators by

$$(28) \qquad K_i \mapsto \sigma_i K_i, \quad X_i^{+} \mapsto \sigma_i X_i^{+}, \quad X_i^{-} \mapsto X_i^{-},$$

for any set of signs $\sigma_0, \ldots, \sigma_n \in \{\pm 1\}$. We shall also make use of the automorphism given by

$$(29) \qquad \begin{aligned} \mathcal{C}^{1/2} &\mapsto -\mathcal{C}^{1/2}, \quad \mathcal{X}_{i,r}^{\pm} \mapsto (-1)^r \mathcal{X}_{i,r}^{\pm}, \\ \mathcal{K}_i &\mapsto \mathcal{K}_i, \quad \mathcal{H}_{i,r} \mapsto \mathcal{H}_{i,r}. \end{aligned}$$

PROPOSITION 12.2.3 *(i) Every finite-dimensional irreducible representation V of $U_\epsilon(\tilde{\mathfrak{g}})$ can be obtained from a type* **1** *representation by twisting with a product of some of the automorphisms (28), (29).*
(ii) Every finite-dimensional (non-zero) type **1** *representation V of $U_\epsilon(\tilde{\mathfrak{g}})$ contains a non-zero vector v which is annihilated by the $\mathcal{X}_{i,r}^+$ for all i, r, and is a simultaneous eigenvector for the elements of U_ϵ^0.*

PROOF Let V be a finite-dimensional representation of $U_\epsilon(\tilde{\mathfrak{g}})$, and let

$$V^0 = \{v \in V \mid \mathcal{X}_{i,r}^+.v = 0 \text{ for all } i, r\}.$$

Assume for a contradiction that $V^0 = 0$. Let w be any non-zero joint eigenvector of $K_1, \ldots, K_n$. By the assumption, there is an infinite sequence of pairs (i_p, r_p), $p = 1, 2, \ldots$, such that the vectors w, $\mathcal{X}_{i_1,r_1}^+.w$, $\mathcal{X}_{i_2,r_2}^+ \mathcal{X}_{i_1,r_1}^+.w, \ldots$ are all non-zero. Since these vectors have different weights for the action of $U_\epsilon(\mathfrak{g})$, they are linearly independent, contradicting the finite dimensionality of V.

Let $0 \neq v \in V^0$. By considering the $U_{\epsilon_i}(sl_2(\mathbb{C}))$-subalgebra of $U_\epsilon(\tilde{\mathfrak{g}})$ generated by $X_i^\pm$, $K_i^{\pm 1}$, and using 10.1.6, it follows that

$$K_i.v = \sigma_i q^{s_i} v, \quad i = 0, 1, \ldots, n,$$

for some $\sigma_i = \pm 1$, $s_i \in \mathbb{Z}$, where $s_0 \leq 0$, $s_1, \ldots, s_n \geq 0$. Hence,

$$\mathcal{C}.v = \sigma_0 \prod_{i=1}^n \sigma_i^{m_i} q^{s_0 + \sum_{i=1}^n m_i s_i} v.$$

On the other hand, considering V as a representation of $U_{\epsilon_i}(sl_2(\mathbb{C}))$ via the homomorphism $U_{\epsilon_i}(sl_2(\mathbb{C})) \to U_\epsilon(\tilde{\mathfrak{g}})$ given by

$$X_1^\pm \mapsto \mathcal{X}_{i,\pm r}^\pm, \quad K_1 \mapsto \mathcal{C}^r \mathcal{K}_i, \quad i = 1, \ldots, n, \ r \in \mathbb{Z},$$

one sees that we must have

$$r\left(s_0 + \sum_{i=1}^n m_i s_i\right) + s_i \geq 0$$

for all $r \in \mathbb{Z}$. Thus, $s_0 + \sum_i m_i s_i = 0$ and

$$\mathcal{C}.v = \sigma_0 \prod_{i=1}^n \sigma_i^{m_i} v.$$

By twisting with an automorphism (28), we can assume that $\sigma_i = 1$ for $i = 0, 1, \ldots, n$, and then $\mathcal{C}.v = v$. If V is irreducible, $\mathcal{C}$ acts as 1 on V, and hence either $\mathcal{C}^{1/2} = 1$ on V or $\mathcal{C}^{1/2} = -1$ on V. The second case can be transformed into the first by twisting with the automorphism (29). This proves (i).

If V is of type **1**, it follows from the fourth set of relations in 12.2.1 that the elements of U_ϵ^0 act on V as commuting operators, and from the fifth and sixth relations that these operators preserve V^0. Any simultaneous eigenvector $v \in V^0$ for the action of U_ϵ^0 satisfies the requirements in part (ii). ∎

Thus, studying the finite-dimensional irreducible representations of $U_\epsilon(\tilde{\mathfrak{g}})$ reduces to studying the representations of the *quantum loop algebra* $U_\epsilon(L(\mathfrak{g}))$, the quotient of $U_\epsilon(\tilde{\mathfrak{g}})$ by the ideal generated by $C^{1/2} - 1$. Proposition 12.2.3 also suggests

DEFINITION 12.2.4 *A type* **1** *representation* V *of* $U_\epsilon(L(\mathfrak{g}))$ *is pseudo-highest weight if it is generated by a vector* v_0 *which is annihilated by the* $\mathcal{X}_{i,r}^+$, *and is a simultaneous eigenvector for the* $\mathcal{K}_i$ *and the* $\mathcal{H}_{i,r}$, $i = 1, \ldots, n$, $r \neq 0$. *If* $\Phi_{i,r}^\pm . v_0 = \varphi_{i,r}^\pm v_0$, *the collection of complex numbers* $\varphi_{i,r}^\pm$, *denoted by* φ, *is called the* pseudo-highest weight *of* V.

Remark We use the term 'pseudo-highest weight' because this notion does not reduce in the classical limit to the usual notion of highest weight for representations of Kac–Moody algebras. See Chari (1985) for a discussion of this point in the classical case. △

The proof of 12.2.3 gives

COROLLARY 12.2.5 *Every finite-dimensional irreducible type* **1** *representation* V *of* $U_\epsilon(L(\mathfrak{g}))$ *is pseudo-highest weight.* ∎

Note that $\varphi_{i,0}^+ \varphi_{i,0}^- = 1$, and that $\varphi_{i,r}^+ = 0$ for $r < 0$, $\varphi_{i,r}^- = 0$ for $r > 0$. One proves, as in the case of Yangians, that for any collection of scalars $\varphi = \{\varphi_{i,r}^\pm\}_{i=1,\ldots,n, r \in \mathbf{Z}}$ satisfying these conditions, there is a unique irreducible $U_\epsilon(L(\mathfrak{g}))$-module $V(\varphi)$ with pseudo-highest weight φ.

We can now state the following analogue of 12.1.11.

THEOREM 12.2.6 *Let* V *be a finite-dimensional irreducible* $U_\epsilon(L(\mathfrak{g}))$-*module of type* **1**, *let* v_0 *be a pseudo-highest weight vector of* V, *and define* $\varphi_{i,r}^\pm \in \mathbb{C}$, *for* $i = 1, \ldots, n$, $r \in \mathbf{Z}$, *as in 12.2.4. Then, there exist unique monic polynomials* $P_{i,V} \in \mathbb{C}[u]$, $i = 1, \ldots, n$, *all with non-zero constant term, such that*

$$\sum_{r=0}^{\infty} \varphi_{i,r}^+ u^r = q_i^{\deg(P_{i,V})} \frac{P_{i,V}(q_i^{-2}u)}{P_{i,V}(u)} = \sum_{r=0}^{\infty} \varphi_{i,-r}^- u^{-r},$$

in the sense that the left- and right-hand sides are the Laurent expansions of the middle term about 0 *and* ∞, *respectively. Moreover, every* n-*tuple* $(P_i)_{i=1,\ldots,n}$ *of monic polynomials with non-zero constant term arises from a finite-dimensional irreducible* $U_\epsilon(L(\mathfrak{g}))$-*module of type* **1** *in this way.*

Moreover, if V and W are finite-dimensional type **1** representations of $U_\epsilon(L(\mathfrak{g}))$ such that $V \otimes W$ is irreducible, then

$$P_{i,V \otimes W} = P_{i,V} P_{i,W}.$$

In particular, $V \otimes W \cong W \otimes V$.

OUTLINE OF PROOF This is based on the following lemma. To state it, let $N_\epsilon^+ = \sum_{i,r} U_\epsilon(L(\mathfrak{g})).\mathcal{X}_{i,r}^+$.

LEMMA 12.2.7 Define elements $\mathcal{P}_{i,\pm r}^\pm \in U_\epsilon^0$, for $i = 1,\dots,n$, $r \geq 0$, inductively by setting $\mathcal{P}_{i,0}^\pm = 1$ and, for $r > 0$,

(30)
$$\mathcal{P}_{i,\pm r}^\pm = \frac{\mp q_i^{\pm r}}{q_i^r - q_i^{-r}} \sum_{s=0}^{r-1} \Phi_{i,\pm(s+1)}^\pm \mathcal{P}_{i,\pm(r-s-1)}^\pm \mathcal{K}_i^{\mp 1}.$$

Then,

(31)
$$\mathcal{P}_{i,r}^+ \equiv (-1)^r q_i^{r^2} \frac{(\mathcal{X}_{i,0}^+)^r (\mathcal{X}_{i,1}^-)^r}{([r]_{q_i})^2},$$

$$\mathcal{P}_{i,-r}^- \equiv (-1)^r q_i^{-r^2} \frac{(\mathcal{X}_{i,-1}^+)^r (\mathcal{X}_{i,0}^-)^r}{([r]_{q_i})^2},$$

(32)
$$(-1)^r q_i^{r(r-1)} \frac{(\mathcal{X}_{i,0}^+)^{r-1}(\mathcal{X}_{i,1}^-)^r}{[r-1]_{q_i}[r]_{q_i}} \equiv -\sum_{s=0}^{r-1} \mathcal{X}_{i,s+1}^- \mathcal{P}_{i,r-s-1}^+ \mathcal{K}_i^{r-1},$$

$$(-1)^r q_i^{-r(r-1)} \frac{(\mathcal{X}_{i,-1}^+)^{r-1}(\mathcal{X}_{i,0}^-)^r}{[r-1]_{q_i}[r]_{q_i}} \equiv -\sum_{s=0}^{r-1} \mathcal{X}_{i,-s}^- \mathcal{P}_{i,-r+s+1}^- \mathcal{K}_i^{-r+1},$$

the congruences being (mod N_ϵ^+). ∎

The proof is by induction on r. See Chari & Pressley (1991b) for the details (only the sl_2 case is treated there, but the proof in the general case is the same).

It is convenient to formulate the definition (30) by introducing the formal power series

$$\boldsymbol{\mathcal{P}}_i^\pm(u) = \sum_{r=0}^\infty \mathcal{P}_{i,\pm r}^\pm u^{\pm r}, \quad \boldsymbol{\Phi}_i^\pm(u) = \sum_{r=0}^\infty \Phi_{i,\pm r}^\pm u^{\pm r}$$

in $U_\epsilon^0[[u^{\pm 1}]]$. Then, we have

(33)
$$\boldsymbol{\Phi}_i^\pm(u) = K_i^{\pm 1} \frac{\boldsymbol{\mathcal{P}}_i^\pm(q_i^{\mp 2} u)}{\boldsymbol{\mathcal{P}}_i^\pm(u)}.$$

Suppose then that $V(\varphi)$ is finite dimensional with pseudo-highest weight φ, and let v_0 be a pseudo-highest weight vector. Then,

$$\mathcal{K}_i.v_0 = q_i^{r_i} v_0$$

for some $r_i \in \mathbf{N}$. From the results of Section 10.1, it follows that the $U_\epsilon(\mathfrak{g})$-submodule of $V(\varphi)$ generated by v_0 is isomorphic to the irreducible module $V_\epsilon(\lambda)$, where $\lambda = \sum_i r_i \lambda_i$. In particular, we have $(\mathcal{X}_{i,0}^-)^{r_i+1}.v_0 = 0$. From (31), we obtain

$$\mathcal{P}_i^+(u).v_0 = P_i(u)v_0$$

for some polynomial $P_i(u) = \sum_{r=0}^{r_i} p_{i,r} u^r$ of degree r_i. The first equation in 12.2.6 now follows from (33).

To prove the second equation, apply both sides of (32), with $r = r_i + 1$, to v_0. Since $(\mathcal{X}_{i,1}^-)^{r_i+1}.v_0 = 0$ for weight reasons, we get

$$\sum_{s=0}^{r_i} \mathcal{X}_{i,s+1}^- \mathcal{P}_{i,r_i-s}^+ K_i^{r_i}.v_0 = 0.$$

Applying $\mathcal{X}_{i,-s-1}^+$ for $s \geq 0$ and using the relation

$$[\mathcal{X}_{i,r}^+, \mathcal{X}_{i,s}^-] = \frac{\Phi_{i,r+s}^+ - \Phi_{i,r+s}^-}{q_i - q_i^{-1}},$$

we find that

$$\sum_{t=0}^{s} \varphi_{i,-t}^- p_{i,r_i-s+t} = \sum_{t=0}^{r_i-s} \varphi_{i,t}^+ p_{i,r_i-s-t}$$

for $0 \leq s \leq r_i$, and that

$$\sum_{s-r_i}^{s} \varphi_{i,-t}^+ p_{i,r_i-s+t} = 0$$

for $s > r_i$. Note that, by (30) (with r replaced by $r_i - s$), the right-hand side of the first of these equations is equal to $q_i^{r_i} q_i^{-2(r_i-s)} p_{i,r_i-s}$. Multiplying the sth equation by u^{r_i-s} and summing from $s = 0$ to ∞, we find

$$\left(\sum_{t=0}^{\infty} \varphi_{i,-t}^- u^{-t} \right) P_i(u) = q_i^{r_i} P_i(q_i^{-2}u)$$

as required.

The proof that every n-tuple $(P_i)_{i=1,\ldots,n}$ of monic polynomials with non-zero constant term arises in this way is very similar to the classical proof, given in Humphreys (1972), for example, that the irreducible $\mathfrak{g}$-module $V(\lambda)$

of highest weight $\lambda \in P^+$ is finite dimensional. See Chari & Pressley (1994c) for the details. (A constructive proof is given in the next section in the special case $\mathfrak{g} = sl_{n+1}(\mathbb{C})$.)

The multiplicative property of the polynomials P_i follows by an argument identical to that used for Yangians in 12.1.12.

Remark We have made no use of the $\boldsymbol{P}_i^-$ in this proof. In fact, similar arguments show that $\boldsymbol{P}_i^-.v_0 = P_{i,V}^-(u^{-1})v_0$ for some polynomials $P_{i,V}^-$, and that

$$\sum_{r=0}^{\infty} \varphi_{i,-r}^- u^{-r} = q_i^{-r_i} \frac{P_{i,V}^-(q_i^2 u^{-1})}{P_{i,V}^-(u^{-1})}.$$

It follows that, up to a scalar multiple, $P_{i,V}^-(u) = u^{r_i} P_{i,V}(u^{-1})$. In particular, the action of the $\Phi_{i,-r}^-$ on the highest weight vector v_0 is determined by that of the $\Phi_{i,r}^+$, $i = 1, \ldots, n$, $r \geq 0$. $\triangle$

As for Yangians, the last part of 12.2.6 suggests that we define a finite-dimensional irreducible $U_\epsilon(L(\mathfrak{g}))$-module of type **1** to be *fundamental* if its associated polynomials are given by

$$P_j(u) = \begin{cases} 1 & \text{if } j \neq i, \\ u - a & \text{if } j = i, \end{cases}$$

for some $i = 1, \ldots, n$, $a \in \mathbb{C}$; the corresponding representation will be denoted by $V_\epsilon(\lambda_i, a)$. As a $U_\epsilon(\mathfrak{g})$-module, $V_\epsilon(\lambda_i, a)$ has λ_i as its unique maximal weight, and so contains the irreducible $U_\epsilon(\mathfrak{g})$-module $V_\epsilon(\lambda_i)$ with multiplicity one. In view of 12.2.6, it is clear that

COROLLARY 12.2.8 *Every finite-dimensional irreducible $U_\epsilon(L(\mathfrak{g}))$-module is isomorphic to a subquotient of a tensor product of fundamental representations.* ∎

C Evaluation representations Recalling that the generators $\mathcal{X}_{i,r}^\pm$ have classical limits $\bar{X}_i^\pm u^r$, the bar denoting elements of $\mathfrak{g}$, we might hope that, for any $a \in \mathbb{C}$, there is a homomorphism of algebras $U_\epsilon(L(\mathfrak{g})) \to U_\epsilon(\mathfrak{g})$ such that $\mathcal{X}_{i,r}^\pm \mapsto a^r X_i^\pm$ for all i. In fact, this does not work for any $\mathfrak{g}$. Nevertheless, Jimbo (1986a) defined a quantum analogue of the latter homomorphism when $\mathfrak{g} = sl_{n+1}(\mathbb{C})$. It takes values in an 'enlargement' of $U_\epsilon(sl_{n+1}(\mathbb{C}))$.

Fix a square root $\epsilon^{1/2}$ of ϵ and, for any elements u, v of an algebra over $\mathbb{C}$, set

$$[u, v]_{\epsilon^{1/2}} = \epsilon^{1/2} uv - \epsilon^{-1/2} vu.$$

DEFINITION–PROPOSITION 12.2.9 $U_\epsilon(gl_{n+1}(\mathbb{C}))$ *is the associative algebra over* $\mathbb{C}$ *with generators* $x_i^\pm$, $i = 1, \ldots, n$, $t_r^{\pm 1}$, $r = 0, 1, \ldots, n$, *and the following defining relations:*

$$t_r t_r^{-1} = 1 = t_r^{-1} t_r,$$

$$t_r t_s = t_s t_r,$$

$$t_r x_i^\pm t_r^{-1} = \epsilon^{\delta_{r,i-1} - \delta_{r,i}} x_i^\pm,$$

$$[x_i^\pm, [x_j^\pm, x_i^\pm]_{\epsilon^{1/2}}]_{\epsilon^{1/2}} = 0 \quad \text{if } |i - j| = 1,$$

$$[x_i^\pm, x_j^\pm] = 0 \quad \text{if } |i - j| > 1,$$

$$[x_i^+, x_j^-] = \delta_{i,j} \frac{k_i - k_i^{-1}}{\epsilon - \epsilon^{-1}},$$

where $k_i = t_{i-1} t_i^{-1}$.

There is a Hopf algebra structure on $U_\epsilon(gl_{n+1}(\mathbb{C}))$ *given by the same formulas as in 9.1.1, together with*

$$\Delta(t_r) = t_r \otimes t_r, \quad S(t_r) = t_r^{-1}, \quad \epsilon(t_r) = 1. \quad \blacksquare$$

The proof is straightforward. Note that there is a homomorphism of Hopf algebras $U_\epsilon(sl_{n+1}(\mathbb{C})) \to U_\epsilon(gl_{n+1}(\mathbb{C}))$ which takes $X_i^\pm$ to $x_i^\pm$ and K_i to k_i.

PROPOSITION 12.2.10 *For any* $a \in \mathbb{C}$, *there exist unique homomorphisms of algebras* ev_a *and* ev^a *from* $U_\epsilon(L(sl_{n+1}(\mathbb{C})))$ *to* $U_\epsilon(gl_{n+1}(\mathbb{C}))$ *such that*

$$\text{ev}_a(X_i^\pm) = x_i^\pm = \text{ev}^a(X_i^\pm), \quad i = 1, \ldots, n,$$

$$\text{ev}_a(K_0) = (k_1 \ldots k_n)^{-1} = \text{ev}^a(K_0),$$

$$\text{ev}_a(X_0^\pm) = (\pm 1)^{n-1} \epsilon^{\mp(n+1)/2} a^{\pm 1} [x_n^\mp, [x_{n-1}^\mp, \ldots, [x_2^\mp, x_1^\mp]_{\epsilon^{1/2}} \cdots]_{\epsilon^{1/2}}]_{\epsilon^{1/2}}$$
$$\times (t_0 t_n)^{\pm 1},$$

$$\text{ev}^a(X_0^\pm) = (\pm 1)^{n-1} \epsilon^{\mp(n+1)/2} a^{\pm 1} [x_1^\mp, [x_2^\mp, \ldots, [x_{n-1}^\mp, x_n^\mp]_{\epsilon^{1/2}} \cdots]_{\epsilon^{1/2}}]_{\epsilon^{1/2}}$$
$$\times (t_0 t_n)^{\mp 1}. \quad \blacksquare$$

The proof is a direct verification that ev_a and ev^a respect the defining relations of $U_\epsilon(L(sl_{n+1}(\mathbb{C})))$ (in the form 9.1.1).

Remarks [1] The homomorphism ev_a was given in Jimbo (1986a). The homomorphism $\text{ev}^a = \sigma \circ \text{ev}_a \circ \tilde{\sigma}$, where σ and $\tilde{\sigma}$ are the automorphisms of $U_\epsilon(gl_{n+1}(\mathbb{C}))$ and $U_\epsilon(L(sl_{n+1}(\mathbb{C})))$, respectively, which are given on generators by

$$\sigma(x_i^\pm) = x_{n-i+1}^\pm, \quad \sigma(t_r^{\pm 1}) = t_{n-r}^{\mp 1},$$

and

$$\tilde{\sigma}(\mathcal{X}_{i,r}^\pm) = \mathcal{X}_{n-i+1,r}^\pm, \quad \tilde{\sigma}(\mathcal{H}_{i,r}) = \mathcal{H}_{n-i+1,r}, \quad \tilde{\sigma}(\mathcal{K}_i^{\pm 1}) = \mathcal{K}_{n-i+1}^{\pm 1}.$$

Alternatively, ev^a may be obtained from ev_a essentially by twisting with the antipodes S and $\tilde{S}$ of $U_\epsilon(gl_{n+1}(\mathbb{C}))$ and $U_\epsilon(L(sl_{n+1}(\mathbb{C})))$, respectively: by checking on the generators $X_0^\pm$, one finds that

$$S \circ \text{ev}_a \circ \tilde{S} = \text{ev}^b,$$

where $b = (-1)^{n-1} a \epsilon^{-n+1}$.

[2] Neither ev_a nor ev^a is a homomorphism of Hopf algebras.

[3] If $n = 1$, $\text{ev}_a = \text{ev}^a$ for all $a \in \mathbb{C}^\times$. $\triangle$

Fix an $(n+1)$th root $\epsilon^{1/(n+1)}$ of ϵ. We say that a representation W of $U_\epsilon(gl_{n+1}(\mathbb{C}))$ is of type **1** if the following conditions are satisfied:

(a) W is of type **1** regarded as a representation of $U_\epsilon(sl_{n+1}(\mathbb{C}))$;

(b) the t_r act semisimply on W with eigenvalues which are integer powers of $\epsilon^{1/(n+1)}$;

(c) $t_0 t_1 \ldots t_n$ acts as the identity on W.

It is easy to see that the restriction to $U_\epsilon(sl_{n+1}(\mathbb{C}))$ is an equivalence from the category of finite-dimensional type **1** representations of $U_\epsilon(gl_{n+1}(\mathbb{C}))$ to the category of finite-dimensional type **1** representations of $U_\epsilon(sl_{n+1}(\mathbb{C}))$. Thus, if W is a finite-dimensional type **1** representation of $U_\epsilon(sl_{n+1}(\mathbb{C}))$, we may regard W as a type **1** representation of $U_\epsilon(gl_{n+1}(\mathbb{C}))$. We denote by W_a (resp. W^a) the pull-back of W by ev_a (resp. by ev^a). Both W_a and W^a are called *evaluation representations*.

If W is irreducible as a representation of $U_\epsilon(sl_{n+1}(\mathbb{C}))$, it is obvious that both W_a and W^a are irreducible representations of $U_\epsilon(L(sl_{n+1}(\mathbb{C})))$, for all $a \in \mathbb{C}^\times$. Using the formulas in 12.2.2 and 12.2.9, one can compute the polynomials associated to W_a and W^a, as in 12.2.5.

Example 12.2.11 Let $a \in \mathbb{C}^\times$, $r \in \mathbb{N}$, and let $W = V_\epsilon(r)$ be the $(r+1)$-dimensional irreducible type **1** representation of $U_\epsilon(sl_2(\mathbb{C}))$. In the usual basis $\{v_0, v_1, \ldots, v_n\}$, the action of $U_\epsilon(gl_2(\mathbb{C}))$ on $V_\epsilon(r)$ is given by

$$t_0.v_p = \epsilon^{(r-2p)/2} v_p, \quad t_1.v_p = \epsilon^{-(r-2p)/2} v_p,$$
$$x_1^+.v_p = [r-p+1]_\epsilon v_{p-1}, \quad x_1^-.v_p = [p+1]_\epsilon v_{p+1}.$$

Using these formulas, and the method used in Subsection 12.1E, it is not difficult to compute the polynomial P_1 associated to $V_\epsilon(r)_a$:

$$P_1(u) = \prod_{p=1}^{r} (u - a^{-1}\epsilon^{2p-r-1}). \quad \Diamond$$

This example suggests the following definition.

DEFINITION 12.2.12 *Let* $a \in \mathbb{C}^\times$, $r \in \mathbb{N}$. *The ϵ-segment* $\Sigma(r)_a$ *with centre a and length r is the r-tuple* $(a\epsilon^{-r+1}, a\epsilon^{-r+3}, \ldots, a\epsilon^{r-1})$.

The general result, proved in Chari & Pressley (1993d), is given in

PROPOSITION 12.2.13 *Let W be the finite-dimensional irreducible type* **1** *representation of* $U_\epsilon(sl_{n+1}(\mathbb{C}))$ *with highest weight* $\lambda = \sum_i r_i \lambda_i$, *and let $a \in \mathbb{C}^\times$. Then,*

(i) if $r_i = 0$, we have $P_{i,W_a} = P_{i,W^a} = 1$;

(ii) if $r_i > 0$, the roots of P_{i,W_a} (resp. of P_{i,W^a}) form, when suitably ordered, the ϵ-segment of length r_i and with centre

$$c_i = a^{-1} \epsilon^{i-1-\frac{2}{n+1}\sum_{j=1}^{n}(n-j+1)r_j + 2\sum_{j=1}^{i-1}r_j + r_i}$$

(resp. with centre

$$c^i = a^{-1} \epsilon^{n-i-\frac{2}{n+1}\sum_{j=1}^{n} jr_j + 2\sum_{j=i+1}^{n}r_j + r_i}).$$ ∎

It follows immediately from this computation that every fundamental representation of $U_\epsilon(L(sl_{n+1}(\mathbb{C})))$ is isomorphic to an evaluation representation. By 12.2.7, we have

COROLLARY 12.2.13 *Every finite-dimensional irreducible type* **1** *representation of* $U_\epsilon(L(sl_{n+1}(\mathbb{C})))$ *is isomorphic to a subquotient of a tensor product of evaluation representations.* ∎

This result establishes the converse of the first part of 12.2.5 in the case $\mathfrak{g} = sl_{n+1}(\mathbb{C})$.

Corollary 12.2.14 can be made much more precise in the sl_2 case.

PROPOSITION 12.2.15 *Let the notation be as in 12.2.10 and 12.2.11.*
(i) Every finite-dimensional irreducible type **1** *representation of* $U_\epsilon(L(sl_2(\mathbb{C})))$ *is isomorphic to a tensor product*

$$(34) \qquad\qquad V_\epsilon(r_1)_{a_1} \otimes \cdots \otimes V_\epsilon(r_k)_{a_k}$$

of evaluation representations.
(ii) Two such irreducible tensor products are isomorphic if and only if one is obtained from the other by a permutation of the factors.
(iii) An arbitrary tensor product (34) is reducible if and only at least one pair of ϵ-segments $\Sigma(r_i)_{a_i}$, $\Sigma(r_j)_{a_j}$ is in special position, in the sense that their union is an ϵ-segment which properly contains them both. ∎

The proof of this result is similar to that of 12.1.21; see Chari & Pressley (1990b) for the details.

There is also an analogue of 12.1.22. For example, $V_\epsilon(r)_a \otimes V_\epsilon(s)_b$ is reducible if and only if r/s or $s/r = \epsilon^{r+s-2k+2}$ for some $0 < k \leq \min\{r, s\}$, in which case it has a unique proper submodule. In particular, whenever $V_\epsilon(r)_a \otimes V_\epsilon(s)_b$ is reducible, it is not completely reducible, nor is $V_\epsilon(r)_a \otimes V_\epsilon(s)_b$ $\cong V_\epsilon(s)_b \otimes V_\epsilon(r)_a$. Thus, $U_\epsilon(L(\mathfrak{g}))$ is not quasitriangular, although, as in the case of Yangians, it turns out to be quasitriangular in a certain 'limiting' sense.

Remark The close similarity between the representation theory of Yangians and quantum loop algebras is 'explained' by the following observation of Drinfel'd (1987). Let $U_h(L(\mathfrak{g}))$ be the deformation of $U(L(\mathfrak{g}))$ generated by elements $\mathcal{H}_{i,r}$, $X_{i,r}^\pm$ for $i = 1, \ldots, n$, $r \in \mathbf{Z}$ and with defining relations as in 12.2.1, but with q replaced by e^h, $\mathcal{K}_i$ by $e^{d_i h \mathcal{H}_{i,0}}$ and $\mathcal{C}^{1/2}$ by 1. Let φ be the composite map $U_h(L(\mathfrak{g})) \to U(L(\mathfrak{g})) \to U(\mathfrak{g})$ given by first setting $h = 0$ and then $u = 1$. Finally, let A be the $\mathbf{C}[[h]]$-subalgebra of $U_h(L(\mathfrak{g})) \underset{\mathbf{C}[[h]]}{\otimes} \mathbf{C}((h))$ generated by $U_h(L(\mathfrak{g}))$ and $h^{-1}\mathrm{ker}(\varphi)$. Then, $A/hA \cong Y(\mathfrak{g})$. △

12.3 Frobenius–Schur duality for Yangians and quantum affine algebras

In this section, we describe analogues, for Yangians and quantum affine algebras, of the classical duality between the representation theories of special linear and symmetric groups (another such analogue is discussed in Subsection 10.2B). We first introduce the algebras which play the role of the symmetric groups in the infinite-dimensional (quantum) theory.

Throughout this section, $\epsilon \in \mathbf{C}^\times$ is assumed to be transcendental.

A Affine Hecke algebras and their degenerations We begin with

DEFINITION 12.3.1 *Fix* $\ell \geq 1$, $\epsilon \in \mathbf{C}^\times$. *The affine Hecke algebra* $\tilde{\mathcal{H}}_\ell(\epsilon)$ *is the associative algebra over* $\mathbf{C}$ *with generators* $\sigma_1^{\pm 1}$, $\sigma_2^{\pm 1}, \ldots, \sigma_{\ell-1}^{\pm 1}$, $z_1^{\pm 1}$, $z_2^{\pm 1}, \ldots, z_\ell^{\pm 1}$, *and the following defining relations:*

$$\sigma_i \sigma_i^{-1} = \sigma_i^{-1} \sigma_i = 1,$$
$$\sigma_i \sigma_{i+1} \sigma_i = \sigma_{i+1} \sigma_i \sigma_{i+1},$$
$$\sigma_i \sigma_j = \sigma_j \sigma_i \quad \text{if } |i - j| > 1,$$
$$(\sigma_i + 1)(\sigma_i - \epsilon) = 0,$$
$$z_j z_j^{-1} = z_j^{-1} z_j = 1,$$
$$z_j z_k = z_k z_j,$$
$$z_j \sigma_i = \sigma_i z_j \quad \text{if } j \neq i \text{ or } i + 1,$$
$$\sigma_i z_i \sigma_i = \epsilon z_{i+1}.$$

Note that the first four sets of relations define the Hecke algebra $\mathcal{H}_\ell(\epsilon)$ (see 10.2.1). Hence, there is a natural homomorphism $\mathcal{H}_\ell(\epsilon) \to \tilde{\mathcal{H}}_\ell(\epsilon)$. It follows from 12.3.6 below that this map is injective.

Remark Affine Hecke algebras arise classically in connection with the theory of p-adic groups. Let $G = GL_\ell(\mathbf{Q}_p)$, and let $\mathcal{C}$ be the set of locally constant functions $f : G/B \to \mathbf{C}$ of compact support. Then, $\mathcal{C}$ is a representation of G by left translation, and an algebra under the convolution product

$$(f_1 * f_2)(g) = \int_G f_1(x) f_2(x^{-1}g)\, dx,$$

where dx is left-invariant Haar measure on G (note that, since the integrand is locally constant and compactly supported, the integral is actually a finite sum).

Let B be the compact open subgroup of G given by

$$B = \{g = (g_{ij}) \in GL_\ell(\mathbf{Q}_p) \mid g_{ij} \in \mathbf{Z}_p \text{ for } i > j\}.$$

The *Hecke algebra $\mathcal{H}_\ell(G, B)$ of G with respect to B* is the subalgebra of $\mathcal{C}$ consisting of the B-bi-invariant functions:

$$\mathcal{H}_\ell(G, B) = \{f \in \mathcal{C} \mid f(b_1 g b_2) = f(g) \text{ for all } b_1, b_2 \in B, g \in G\}.$$

It was proved by Bernstein & Zelevinsky (1977) that $\mathcal{H}_\ell(G, B) \cong \tilde{\mathcal{H}}_\ell(p)$.

If W is a (right) $\mathcal{H}_\ell(G, B)$-module, we may define the induced representation

$$\mathcal{C}_W = W \underset{\mathcal{H}_\ell(G,B)}{\otimes} \mathcal{C}$$

of G, regarding $\mathcal{C}$ as a (left) $\mathcal{H}_\ell(G, B)$-module by the convolution product. It is known that $\mathcal{C}_W$ is an admissible representation of G, and that W is irreducible if and only if $\mathcal{C}_W$ is irreducible. We shall obtain an analogous result later in this section in which G is replaced by $U_\epsilon(L(sl_{n+1}(\mathbf{C})))$ (see 12.3.13).

See Cartier (1979) for the notion of an admissible representation and for further background information on p-adic groups. $\triangle$

Let $u_j = (1 - z_j)/(\epsilon - 1)$. Rewriting the defining relations of $\tilde{\mathcal{H}}_\ell(\epsilon)$ in terms of the σ_i and u_j, clearing denominators and dividing both sides of each relation by the highest power of $\epsilon - 1$, and finally replacing ϵ by 1, one obtains the following relations:

$$\sigma_i \sigma_i^{-1} = \sigma_i^{-1} \sigma_i = 1, \quad \sigma_i^2 = 1,$$
$$\sigma_i \sigma_j = \sigma_j \sigma_i \quad \text{if } |i - j| > 1,$$
$$\sigma_i \sigma_{i+1} \sigma_i = \sigma_{i+1} \sigma_i \sigma_{i+1},$$
$$u_i \sigma_i = \sigma_i u_{i+1} + 1,$$
$$u_i u_j = u_j u_i.$$

DEFINITION 12.3.2 *The degenerate affine Hecke algebra* Λ_ℓ *is the asssociative algebra with generators* $\sigma_1^{\pm 1}, \ldots, \sigma_{\ell-1}^{\pm 1}, u_1, \ldots, u_\ell$, *and the above defining relations.*

Note that the first three sets of relations above are the defining relations of the symmetric group Σ_ℓ (take σ_i to be the transposition $(i, i+1)$). Hence, there is a canonical homomorphism $\mathbb{C}[\Sigma_\ell] \to \Lambda_\ell$.

It will be convenient to introduce the elements

$$y_j = u_j + \frac{1}{2} \sum_{\substack{k=1 \\ k \neq j}}^{\ell} \mathrm{sign}(j - k)(j, k), \quad j = 1, \ldots, \ell.$$

In terms of these elements, the defining relations of Λ_ℓ take the following form:

(35) $$\sigma y_j \sigma^{-1} = y_{\sigma(j)}, \quad \sigma \in \Sigma_\ell, \ j = 1, \ldots, \ell,$$

$$[y_j, y_k] = \frac{1}{4} \sum_{i \notin \{j,k\}} ((j, k, i) - (k, j, i)).$$

B Representations of affine Hecke algebras It is known that when ϵ is transcendental (or simply not a root of unity), the Hecke algebra $\mathcal{H}_\ell(\epsilon)$ is isomorphic as an algebra to the group algebra $\mathbb{C}[\Sigma_\ell]$. Its irreducible representations are therefore parametrized by the partitions of ℓ. To describe this correspondence explicitly, we make use of certain elements of $\mathcal{H}_\ell(\epsilon)$ introduced by Kazhdan & Lusztig (1979).

Let $w \in \Sigma_\ell$ and let

(36) $$w = \tau_{i_1} \tau_{i_2} \cdots \tau_{i_{\ell(w)}}$$

be a reduced expression of w as a product of the simple transpositions $\tau_i = (i, i+1)$ (see the Appendix). It follows from the defining relations in $\mathcal{H}_\ell(\epsilon)$ that, if (36) is any reduced expression for w, the element

$$\sigma_w = \sigma_{i_1} \sigma_{i_2} \cdots \sigma_{i_{\ell(w)}}$$

depends only on w, and not on the choice of reduced expression (see 8.1.3: in fact, this statement is a consequence of the corresponding assertion for the braid group of $sl_{n+1}(\mathbb{C})$, since $\mathcal{H}_\ell(\epsilon)$ is the quotient of the group algebra of the braid group by the two-sided ideal generated by $(\sigma_i + 1)(\sigma_i - \epsilon)$ for $i = 1, \ldots, \ell-1$). Let $\leq$ be the Bruhat ordering on Σ_ℓ determined by the choice of generators $\tau_1, \ldots, \tau_{\ell-1}$, and for $w' \leq w$ let $P_{w',w}$ be the Kazhdan–Lusztig polynomial (see Humphreys (1990), Chap. 7, for background information). Define elements $C_w \in \mathcal{H}_\ell(\epsilon)$, for all $w \in \Sigma_\ell$, by

$$C_w = \epsilon^{\ell(w)/2} \sum_{\{w' | w' \le w\}} (-1)^{\ell(w)-\ell(w')} \epsilon^{-\ell(w')} P_{w',w}(\epsilon^{-1}) \sigma_w.$$

Let $\ell = \ell_1 + \ell_2 + \cdots + \ell_p$ be a partition π of ℓ, with each $\ell_r > 0$, and let Σ_ℓ^π be the subgroup $\Sigma_{\ell_1} \times \cdots \times \Sigma_{\ell_p}$ of Σ_ℓ which fixes π. Let w_r be the longest element of the subgroup Σ_{ℓ_r}, i.e. the permutation which reverses the order of

$$(\ell_1 + \cdots + \ell_{r-1} + 1, \ldots, \ell_1 + \cdots + \ell_r).$$

Set $w_\pi = w_1 w_2 \ldots w_p$ and let I_π be the right ideal in $\mathcal{H}_\ell(\epsilon)$ generated by C_{w_π}.

PROPOSITION 12.3.3 *For every partition π of ℓ, I_π has a unique irreducible quotient J_π in which C_{w_π} has non-zero image. Every finite-dimensional irreducible representation of $\mathcal{H}_\ell(\epsilon)$ is isomorphic to exactly one J_π.* ∎

See Rogawski (1985) for a proof.

Turning now to affine Hecke algebras, we first introduce a family of 'universal' representations, analogous to the Verma representations of quantum groups (see Chapter 9). Let $\mathbf{a} = (a_1, \ldots, a_\ell) \in (\mathbb{C}^\times)^\ell$, and let

$$M_{\mathbf{a}} = \mathcal{H}_{\mathbf{a}} \backslash \tilde{\mathcal{H}}_\ell(\epsilon),$$

the quotient of $\tilde{\mathcal{H}}_\ell(\epsilon)$ by the right ideal $\mathcal{H}_{\mathbf{a}}$ generated by $z_j - a_j.1$, $j = 1, \ldots, \ell$. The following result is proved in Rogawski (1985).

PROPOSITION 12.3.4 *(i) Every finite-dimensional irreducible (right) $\tilde{\mathcal{H}}_\ell(\epsilon)$-module is isomorphic to a quotient of some $M_{\mathbf{a}}$.*
(ii) For all $\mathbf{a} \in (\mathbb{C}^\times)^\ell$, the natural map $\mathcal{H}_\ell(\epsilon) \to M_{\mathbf{a}}$ is an isomorphism of $\mathcal{H}_\ell(\epsilon)$-modules (where $\mathcal{H}_\ell(\epsilon)$ acts on itself in the right regular representation).
(iii) $M_{\mathbf{a}}$ is reducible as an $\tilde{\mathcal{H}}_\ell(\epsilon)$-module if and only if $a_j = \epsilon a_k$ for some j, k. ∎

Zelevinsky (1980) and Rogawski (1985) identified certain distinguished quotients of the $M_{\mathbf{a}}$, for certain $\mathbf{a}$, which exhaust the irreducible representations of $\tilde{\mathcal{H}}_\ell(\epsilon)$:

THEOREM 12.3.5 *Let $\mathbf{s} = \{s_1, \ldots, s_p\}$ be any (unordered) collection of $\epsilon^{1/2}$-segments, the sum of whose lengths is ℓ. Let ℓ_r be the length of s_r, and denote by $\pi(\mathbf{s})$ the partition $\ell = \ell_1 + \cdots + \ell_p$. Let $\mathbf{a} = s_1 s_2 \ldots s_p \in (\mathbb{C}^\times)^\ell$ be the result of juxtaposing the segments in $\mathbf{s}$. Then,*
(i) $I_{\pi(\mathbf{s})}$ is an $\tilde{\mathcal{H}}_\ell(\epsilon)$-submodule of $M_{\mathbf{a}}$ (this makes sense in view of 12.3.4(ii));
(ii) with the $\tilde{\mathcal{H}}_\ell(\epsilon)$-module structure inherited from $M_{\mathbf{a}}$, $I_{\pi(\mathbf{s})}$ has a unique irreducible subquotient $V_{\mathbf{a}}$ in which $C_{w_{\pi(\mathbf{s})}}$ has non-zero image.

Moreover, every finite-dimensional irreducible right $\tilde{\mathcal{H}}_\ell(\epsilon)$-module is isomorphic to some $V_{\mathbf{a}}$. ∎

The next proposition allows one to give a simple construction of some, but not all, of the irreducible representations of $\tilde{\mathcal{H}}_\ell(\epsilon)$. We shall see in 12.3.15 that the similarity between this result and 12.2.10 is not accidental.

PROPOSITION 12.3.6 *For any $a \in \mathbb{C}^\times$, there exist unique homomorphisms $\tilde{\mathrm{ev}}_a$ and $\tilde{\mathrm{ev}}^a$ from $\tilde{\mathcal{H}}_\ell(\epsilon)$ to $\mathcal{H}_\ell(\epsilon)$ such that*

$$\tilde{\mathrm{ev}}_a(\sigma_i) = \sigma_i = \tilde{\mathrm{ev}}^a(\sigma_i),$$
$$\tilde{\mathrm{ev}}_a(z_j) = a\epsilon^{-j+1}\sigma_{j-1}\sigma_{j-2}\ldots\sigma_2\sigma_1^2\sigma_2\ldots\sigma_{j-2}\sigma_{j-1},$$
$$\tilde{\mathrm{ev}}^a(z_j) = a\epsilon^{\ell-j}\sigma_j^{-1}\sigma_{j+1}^{-1}\ldots\sigma_{\ell-2}^{-1}\sigma_{\ell-1}^{-2}\sigma_{\ell-2}^{-1}\ldots\sigma_{j+1}^{-1}\sigma_j^{-1}. \ \blacksquare$$

The proof is straightforward.

If M is any $\mathcal{H}_\ell(\epsilon)$-module, pulling back M by $\tilde{\mathrm{ev}}_a$ (resp. by $\tilde{\mathrm{ev}}^a$) gives an $\tilde{\mathcal{H}}_\ell(\epsilon)$-module M_a (resp. M^a) which is isomorphic to M as an $\mathcal{H}_\ell(\epsilon)$-module. In particular, M_a and M^a are irreducible if M is irreducible. If $\ell > 1$, a module M^a is not in general isomorphic to any module of the form $M'_{a'}$. We call M_a and M^a *evaluation representations* of $\tilde{\mathcal{H}}_\ell(\epsilon)$ (the justification for this terminology appears in 12.3.15).

One can extend the class of 'constructible' $\tilde{\mathcal{H}}_\ell(\epsilon)$-modules by using a kind of tensor product, the definition of which is based on the following easy observation:

PROPOSITION 12.3.7 *For any ℓ_1, $\ell_2 \geq 1$, there exists a unique homomorphism of algebras*

$$\tilde{\iota}_{\ell_1,\ell_2} : \tilde{\mathcal{H}}_{\ell_1}(\epsilon) \otimes \tilde{\mathcal{H}}_{\ell_2}(\epsilon) \to \tilde{\mathcal{H}}_{\ell_1+\ell_2}(\epsilon)$$

such that

$$\tilde{\iota}_{\ell_1,\ell_2}(\sigma_i \otimes 1) = \sigma_i, \quad \tilde{\iota}_{\ell_1,\ell_2}(z_j \otimes 1) = z_j,$$

for $i = 1, \ldots, \ell_1 - 1$, $j = 1, \ldots, \ell_1$, and

$$\tilde{\iota}_{\ell_1,\ell_2}(1 \otimes \sigma_i) = \sigma_{i+\ell_1}, \quad \tilde{\iota}_{\ell_1,\ell_2}(1 \otimes z_j) = z_{j+\ell_1},$$

for $i = 1, \ldots, \ell_2 - 1$, $j = 1, \ldots, \ell_2$. ∎

Clearly, $\tilde{\iota}_{\ell_1,\ell_2}$ restricts to give a homomorphism $\iota_{\ell_1,\ell_2} : \mathcal{H}_{\ell_1}(\epsilon) \otimes \mathcal{H}_{\ell_2}(\epsilon) \to \mathcal{H}_{\ell_1+\ell_2}(\epsilon)$.

Suppose now that M_1 (resp. M_2) is an $\tilde{\mathcal{H}}_{\ell_1}(\epsilon)$-module (resp. an $\tilde{\mathcal{H}}_{\ell_2}(\epsilon)$-module), and let $M_1 \otimes M_2$ be their outer tensor product (an $\tilde{\mathcal{H}}_{\ell_1}(\epsilon) \otimes \tilde{\mathcal{H}}_{\ell_2}(\epsilon)$-module). Then, we define an $\tilde{\mathcal{H}}_{\ell_1+\ell_2}(\epsilon)$-module $M_1 \boxtimes M_2$, sometimes called the *Zelevinsky tensor product* of M_1 and M_2, to be the induced module

$$M_1 \boxtimes M_2 = (M_1 \otimes M_2) \underset{\tilde{\mathcal{H}}_{\ell_1}(\epsilon) \otimes \tilde{\mathcal{H}}_{\ell_2}(\epsilon)}{\otimes} \tilde{\mathcal{H}}_{\ell_1+\ell_2}(\epsilon).$$

The Zelevinsky tensor product $\boxtimes$ of representations of Hecke algebras is defined in the same way. Standard properties of induced modules show that the Zelevinsky tensor products are associative up to isomorphism (but we shall see that $\boxtimes$ is not commutative, in general).

Remark We have already noted that $\mathcal{H}_\ell(\epsilon)$ is isomorphic to $\mathbb{C}[\Sigma_\ell]$ as an algebra. Thus, $\mathcal{H}_\ell(\epsilon)$ has a Hopf algebra structure. This gives another way to take tensor products of Hecke algebra modules, quite different from the Zelevinsky tensor product. $\triangle$

To understand these constructions, the reader may find it useful to establish the result of the following example.

Example 12.3.8 Let $\mathbb{C}$ be the one-dimensional trivial representation of $\mathcal{H}_1(\epsilon) \cong \mathbb{C}$ (on which $1 \in \mathcal{H}_1(\epsilon)$ acts as 1). Then, for any $\mathbf{a} = (a_1, \ldots, a_\ell) \in (\mathbb{C}^\times)^\ell$,

$$M_{\mathbf{a}} \cong \mathbb{C}_{a_1} \tilde{\boxtimes} \mathbb{C}_{a_2} \tilde{\boxtimes} \cdots \tilde{\boxtimes} \mathbb{C}_{a_\ell}$$

as $\tilde{\mathcal{H}}_\ell(\epsilon)$-modules. $\Diamond$

All of the above constructions have analogues for degenerate affine Hecke algebras. For example, using the limiting procedure described in Subsection 12.3A by which Λ_ℓ is obtained from $\tilde{\mathcal{H}}_\ell(\epsilon)$, it is easy to deduce the existence of a homomorphism $\Lambda_{\ell_1} \otimes \Lambda_{\ell_2} \to \Lambda_{\ell_1+\ell_2}$ such that

$$y_j \otimes 1 \mapsto y_j + \frac{1}{2} \sum_{k=1}^{\ell_2} (j, k+\ell_1), \quad 1 \otimes y_j \mapsto y_{j+\ell_1} - \frac{1}{2} \sum_{k=1}^{\ell_1} (k, j+\ell_1).$$

Using this, one can define a tensor product of degenerate affine Hecke algebra modules analogous to $\tilde{\boxtimes}$.

We leave it to the reader to find in a similar way the homomorphisms $\Lambda_\ell \to \mathbb{C}[\Sigma_\ell]$ which are the $\epsilon = 1$ limits of ev_a and ev^a (see Drinfel'd (1986b) for the answer).

However, the classification of the finite-dimensional irreducible Λ_ℓ-modules cannot be deduced so easily from that for $\tilde{\mathcal{H}}_\ell(\epsilon)$. See Cherednik (1988b).

C Duality for $U_\epsilon(sl_{n+1}(\mathbb{C}))$ – revisited We begin our discussion of quantum Frobenius–Schur duality with the case of $U_\epsilon(sl_{n+1}(\mathbb{C}))$. Recall from Section 10.2 that $U_\epsilon(sl_{n+1}(\mathbb{C}))$ has a natural $(n+1)$-dimensional irreducible type **1** representation $V_\epsilon^\natural$ with basis $\{v_0, \ldots, v_n\}$ and action given by

$$K_i.v_r = \epsilon^{\delta_{r,i-1} - \delta_{r,i}} v_r, \quad X_i^+.v_r = \delta_{r,i} v_{r-1}, \quad X_i^-.v_r = \delta_{r,i-1} v_{r+1}$$

(we set $v_{-1} = v_{n+1} = 0$). A finite-dimensional representation of $U_\epsilon(sl_{n+1}(\mathbb{C}))$ is said to be of *level* ℓ if each of its irreducible components is isomorphic to some irreducible component of $(V_\epsilon^\natural)^{\otimes \ell}$ (recall from 10.1.7 that every finite-dimensional representation of $U_\epsilon(sl_{n+1}(\mathbb{C}))$ is completely reducible). The level of a representation is not, in general, uniquely defined, but the following result shows that the level is unique if it is $\leq n$.

PROPOSITION 12.3.9 *The finite-dimensional irreducible $U_\epsilon(sl_{n+1}(\mathbb{C}))$-module $V_\epsilon(\lambda)$ is of level $\ell \leq n$ if and only if its highest weight $\lambda = \sum_{i=1}^n r_i \lambda_i$, where $\sum_{i=1}^n i r_i = \ell$.*

OUTLINE OF PROOF By the results of Chapter 10, it suffices to prove the analogous classical result, which is not difficult. ∎

Let $I_\epsilon : (V_\epsilon^\natural)^{\otimes 2} \to (V_\epsilon^\natural)^{\otimes 2}$ be the linear map defined by

$$(37) \qquad I_\epsilon(v_r \otimes v_s) = \begin{cases} \epsilon v_r \otimes v_s & \text{if } r = s, \\ v_s \otimes v_r & \text{if } s > r, \\ v_s \otimes v_r + (\epsilon - \epsilon^{-1}) v_r \otimes v_s & \text{if } r > s. \end{cases}$$

We saw in 10.2.5(i) that there is an action of $\mathcal{H}_\ell(\epsilon^2)$ on $(V_\epsilon^\natural)^{\otimes \ell}$ such that σ_i acts as ϵI_ϵ on the ith and $(i+1)$th factors and as the identity on the other factors. Thus, if M is any right $\mathcal{H}_\ell(\epsilon^2)$-module, the natural action of $U_\epsilon(sl_{n+1}(\mathbb{C}))$ on $(V_\epsilon^\natural)^{\otimes \ell}$ induces an action of $U_\epsilon(sl_{n+1}(\mathbb{C}))$ on the vector space $M \underset{\mathcal{H}_\ell(\epsilon^2)}{\otimes} (V_\epsilon^\natural)^{\otimes \ell}$.

We have the following more precise result, due to Jimbo (1986a).

THEOREM 12.3.10 *Fix $\ell, n \geq 1$. There is a functor $\mathcal{J}$ from the category of finite-dimensional right $\mathcal{H}_\ell(\epsilon^2)$-modules to the category of finite-dimensional left $U_\epsilon(sl_{n+1}(\mathbb{C}))$-modules of level ℓ such that, if M is an $\mathcal{H}_\ell(\epsilon^2)$-module,*

$$\mathcal{J}(M) = M \underset{\mathcal{H}_\ell(\epsilon^2)}{\otimes} (V_\epsilon^\natural)^{\otimes \ell},$$

equipped with the natural $U_\epsilon(sl_{n+1}(\mathbb{C}))$-module structure induced by that on $(V_\epsilon^\natural)^{\otimes \ell}$.

If $\ell \leq n$, $\mathcal{J}$ is an equivalence of categories. ∎

In view of complete reducibility, $\mathcal{J}$ is determined by its effect on the J_π (see 12.3.3).

PROPOSITION 12.3.11 *Let $1 \leq \ell \leq n$ and let $\ell_1 + \ell_2 + \cdots + \ell_p$ be a partition π of ℓ. Then,*

$$\mathcal{J}(J_\pi) \cong V_\epsilon(\lambda_{\ell_1} + \lambda_{\ell_2} + \cdots + \lambda_{\ell_p})$$

as representations of $U_\epsilon(sl_{n+1}(\mathbb{C}))$.

OUTLINE OF PROOF It is proved in Kazhdan & Lusztig (1979) that

$$C_w \sigma_i = -C_w \quad \text{if } w\tau_i < w.$$

Using the formulas in (37), this implies that, if $\mathbf{v} \in (V_\epsilon^\natural)^{\otimes \ell}$ has $v_r \otimes v_r$ in the ith and $(i+1)$th positions, for some $r = 0, 1, \ldots, n$, then in $\mathcal{J}(J_\pi)$ we have

$$(38) \qquad\qquad C_{w_\pi} \otimes \mathbf{v} = 0.$$

By 12.3.10, there exists a partition π' of ℓ, say $\ell = \ell_1' + \cdots + \ell_m'$, such that

$$\mathcal{J}(J_{\pi'}) \cong V_\epsilon(\lambda_{\ell_1} + \cdots + \lambda_{\ell_p}).$$

It is easy to see that the weight space of $(V_\epsilon^\natural)^{\otimes \ell}$ of weight $\lambda_{\ell_1} + \cdots + \lambda_{\ell_p}$ is spanned by the permutations of the vector

$$v_0 \otimes v_1 \otimes \cdots \otimes v_{\ell_1 - 1} \otimes v_0 \otimes \cdots \otimes v_{\ell_2 - 1} \otimes v_0 \otimes \cdots \otimes v_{\ell_p - 1}.$$

If $\mathbf{v}$ is such a vector, it follows from (38) that $C_{w_{\pi'}} \otimes \mathbf{v} = 0$ unless the first ℓ_1' components of $\mathbf{v}$ are distinct, as are the next ℓ_2', $\ldots$, and the last ℓ_m'. Since $C_{w_{\pi'}} \otimes \mathbf{v}$ must be non-zero for some such $\mathbf{v}$, it follows that $\pi' = \pi$. ∎

The functor $\mathcal{J}$ is clearly one of $\mathbb{C}$-linear categories, but it also captures part of the tensor structure of the category of $U_\epsilon(sl_{n+1}(\mathbb{C}))$-modules.

PROPOSITION 12.3.12 Let n, ℓ_1, $\ell_2 \geq 1$, and let M_1 (resp. M_2) be a finite-dimensional $\mathcal{H}_{\ell_1}(\epsilon^2)$-module (resp. $\mathcal{H}_{\ell_2}(\epsilon^2)$-module). Then, there is a canonical isomorphism of $U_\epsilon(sl_{n+1}(\mathbb{C}))$-modules

$$\mathcal{J}(M_1 \boxtimes M_2) \cong \mathcal{J}(M_1) \otimes \mathcal{J}(M_2). \quad \blacksquare$$

The proof is straightforward.

D Quantum affine algebras and affine Hecke algebras The following result, due to Chari & Pressley (1993b), is an analogue of 12.3.8 for $U_\epsilon(L(sl_{n+1}(\mathbb{C})))$. Let $K_\theta = K_1 K_2 \ldots K_n \in U_\epsilon(sl_{n+1}(\mathbb{C}))$, and let $X_\theta^\pm$ be the operators on $V_\epsilon^\natural$ defined by

$$X_\theta^+ . v_r = \delta_{r,n} v_0, \quad X_\theta^- . v_r = \delta_{r,0} v_n.$$

THEOREM 12.3.13 Fix ℓ, $n \geq 1$. There is a functor $\mathcal{F}$ from the category of finite-dimensional right $\tilde{\mathcal{H}}_\ell(\epsilon^2)$-modules to the category of finite-dimensional left $U_\epsilon(L(sl_{n+1}(\mathbb{C})))$-modules which are of level ℓ as $U_\epsilon(sl_{n+1}(\mathbb{C}))$-modules, defined as follows. If M is an $\tilde{\mathcal{H}}_\ell(\epsilon^2)$-module, $\mathcal{F}(M) = \mathcal{J}(M)$ as a $U_\epsilon(sl_{n+1}(\mathbb{C}))$-module, and the action of the remaining generators of $U_\epsilon(L(sl_{n+1}(\mathbb{C})))$ (in the presentation 9.1.1) is given by

$$(39) \qquad \begin{aligned} X_0^\pm.(m \otimes \mathbf{v}) &= \sum_{j=1}^{\ell} m.z_j^{\pm 1} \otimes Z_j^\pm . \mathbf{v}, \\ K_0.(m \otimes \mathbf{v}) &= m \otimes (K_\theta^{-1})^{\otimes \ell} . \mathbf{v}, \end{aligned}$$

for all $m \in M$, $\mathbf{v} \in (V_\epsilon^\natural)^{\otimes \ell}$, where the operators $Z_j^\pm$ on $(V_\epsilon^\natural)^{\otimes \ell}$ are defined by

$$Z_j^+ = \mathrm{id}^{\otimes j-1} \otimes X_\theta^- \otimes (K_\theta^{-1})^{\otimes \ell-j},$$
$$Z_j^- = K_\theta^{\otimes j-1} \otimes X_\theta^+ \otimes \mathrm{id}^{\otimes \ell-j}.$$

If $f : M \to M'$ is a homomorphism of $\tilde{\mathcal{H}}_\ell(\epsilon^2)$-modules, then $\mathcal{F}(f) : \mathcal{F}(M) \to \mathcal{F}(M')$ is defined by

$$\mathcal{F}(f)(m \otimes \mathbf{v}) = f(m) \otimes \mathbf{v}.$$

The functor $\mathcal{F}$ is an equivalence of categories if $\ell \le n$.

OUTLINE OF PROOF To see that the action of X_0^+ (say) is well defined, we must show that, as operators on $\mathcal{J}(M)$, we have

(40) $$\sum_{j=1}^{\ell} \sigma_i z_j \otimes Z_j^+ = \sum_{j=1}^{\ell} z_j \otimes Z_j^+ \sigma_i.$$

If $j \ne i$ or $i+1$, the jth terms on the left- and right-hand sides of (40) are equal, since $\sigma_i z_j = z_j \sigma_i$ and $\sigma_i Z_j^+ = Z_j^+ \sigma_i$. Hence, we must show that

$$\sigma_i z_i \otimes Z_i^+ + \sigma_i z_{i+1} \otimes Z_{i+1}^+ = z_i \otimes Z_i^+ \sigma_i + z_{i+1} \otimes Z_{i+1}^+ \sigma_i.$$

Using the relation $\epsilon^2 \sigma_i^{-1} = \sigma_i - \epsilon^2 + 1$, this reduces to

$$\epsilon^2 z_{i+1} \otimes (\sigma_i^{-1} Z_i^+ - Z_{i+1}^+ \sigma_i^{-1}) + z_i \otimes (\sigma_i Z_{i+1}^+ - Z_i^+ \sigma_i) = 0.$$

Thus, it suffices to prove that $\sigma_i Z_{i+1}^+ = Z_i^+ \sigma_i$, i.e. that

$$I_\epsilon(\mathrm{id} \otimes X_\theta^-) = (X_\theta^- \otimes K_\theta^{-1})I_\epsilon.$$

This equation is easily checked.

The fact that the formulas in the theorem define a $U_\epsilon(L(sl_{n+1}(\mathbb{C})))$-module is proved by directly verifying those defining relations of $U_\epsilon(L(sl_{n+1}(\mathbb{C})))$ which involve X_0^+, X_0^- or K_0.

Suppose now that $\ell \le n$, and let W be a finite-dimensional representation of $U_\epsilon(L(sl_{n+1}(\mathbb{C})))$ of level ℓ. By 12.3.8, we may assume that $W = \mathcal{J}(M)$ as a representation of $U_\epsilon(sl_{n+1}(\mathbb{C}))$, for some $\mathcal{H}_\ell(\epsilon^2)$-module M. Since the permutations of the vector

$$v_1 \otimes v_2 \otimes \cdots \otimes v_{j-1} \otimes v_0 \otimes v_j \otimes v_{j+1} \otimes \cdots \otimes v_{\ell-1}$$

clearly span the subspace of $(V_\epsilon^\natural)^{\otimes \ell}$ of weight λ_ℓ, it follows that there exists $\zeta_j^- \in \mathrm{End}_{\mathbb{C}}(M)$ such that

$$X_0^-.(m \otimes v_1 \otimes v_2 \otimes \cdots \otimes v_{j-1} \otimes v_n \otimes v_j \otimes v_{j+1} \otimes \cdots \otimes v_{\ell-1})$$
$$= \zeta_j^-(m) \otimes v_1 \otimes v_2 \otimes \cdots \otimes v_{j-1} \otimes v_0 \otimes v_j \otimes v_{j+1} \otimes \cdots \otimes v_{\ell-1}$$

for all $m \in M$. Similarly, there exists $\zeta_j^+ \in \text{End}_{\mathbb{C}}(M)$ such that

$$X_0^+ . (m \otimes v_{n-\ell+1} \otimes \cdots \otimes v_{n-\ell+j-1} \otimes v_0 \otimes v_{n-\ell+j} \otimes \cdots \otimes v_{n-1})$$
$$= \zeta_j^+(m) \otimes v_{n-\ell+1} \otimes \cdots \otimes v_{n-\ell+j-1} \otimes v_n \otimes v_{n-\ell+j} \otimes \cdots \otimes v_{n-1}.$$

One shows that letting $z_j^{\pm 1}$ act as $\zeta_j^{\pm}$ for $j = 1, \ldots, \ell$ defines an $\tilde{\mathcal{H}}_\ell(\epsilon^2)$-module structure on M, and that one then has $W = \mathcal{F}(M)$. ∎

The following result determines the action of $\mathcal{F}$ on the irreducible representations of $\tilde{\mathcal{H}}_\ell(\epsilon^2)$. See Chari & Pressley (1993b) for the proof.

THEOREM 12.3.14 *Let* $\mathbf{s} = \{s_1, s_2, \ldots, s_p\}$ *be a collection of ϵ-segments, the sum of whose lengths is* ℓ. *Let* a_r *be the centre of* s_r *and* ℓ_r *its length. Let* $\mathbf{a} = s_1 s_2 \ldots s_p \in (\mathbb{C}^\times)^\ell$ *be the result of juxtaposing* $s_1, \ldots, s_p$, *and let* $V_{\mathbf{a}}$ *be the irreducible* $\tilde{\mathcal{H}}_\ell(\epsilon^2)$-*module defined in 12.3.5. Then, if* $\ell \leq n$, $\mathcal{F}(V_{\mathbf{a}})$ *is the irreducible* $U_\epsilon(L(sl_{n+1}(\mathbb{C})))$-*module whose associated polynomials* P_i *(as in 12.2.6) are*

$$P_i(u) = \prod_{\{r | \ell_r = i\}} (u - a_r^{-1}), \quad i = 1, \ldots, n. \quad ∎$$

The simplest representations of $\tilde{\mathcal{H}}_\ell(\epsilon^2)$ are the 'evaluation representations' defined in 12.3.6 and the remarks following this proposition. The following result justifies this terminology.

PROPOSITION 12.3.15 *Let* $1 \leq \ell \leq n$ *and let* M *be a finite-dimensional* $\mathcal{H}_\ell(\epsilon^2)$-*module. Then, there are canonical isomorphisms of* $U_\epsilon(L(sl_{n+1}(\mathbb{C})))$-*modules*

$$\mathcal{F}(M_{a\epsilon^{-2\ell/(n+1)}}) \cong \mathcal{J}(M)_a, \quad \mathcal{F}(M^{a\epsilon^{-2\ell/(n+1)}}) \cong \mathcal{J}(M)^a.$$

OUTLINE OF PROOF In the first case, we know by 12.3.13 that $\mathcal{J}(M)_a \cong \mathcal{F}(N)$ for some $\tilde{\mathcal{H}}_\ell(\epsilon^2)$-module N which is isomorphic to M as an $\mathcal{H}_\ell(\epsilon^2)$-module. It suffices to prove that $z_1 \in \tilde{\mathcal{H}}_\ell(\epsilon^2)$ acts as the scalar $a\epsilon^{-2\ell/(n+1)}$ on N. To see this, one computes the action of X_0^+ on $m \otimes v_0 \otimes v_{n-\ell+1} \otimes v_{n-\ell+2} \otimes \cdots \otimes v_{n-1}$, for all $m \in M$, in two different ways. First, by the definition of $\mathcal{F}$, we have

$$X_0^+ . (m \otimes v_0 \otimes v_{n-\ell+1} \otimes v_{n-\ell+2} \otimes \cdots \otimes v_{n-1})$$
$$= m.z_1 \otimes v_n \otimes v_{n-\ell+1} \otimes v_{n-\ell+2} \otimes \cdots \otimes v_{n-1}.$$

On the other hand, let

$$f_n = [x_n^-, [x_{n-1}^-, \ldots, [x_2^-, x_1^-]_{\epsilon^{1/2}} \cdots]_{\epsilon^{1/2}}]_{\epsilon^{1/2}} \in U_\epsilon(gl_{n+1}(\mathbb{C})).$$

Then, by 12.2.10,

$$X_0^+.(m \otimes v_0 \otimes v_{n-\ell+1} \otimes v_{n-\ell+2} \otimes \cdots \otimes v_{n-1})$$
$$= m \otimes \mathrm{ev}_a(X_0^+).(v_0 \otimes v_{n-\ell+1} \otimes v_{n-\ell+2} \otimes \cdots \otimes v_{n-1})$$
$$= a\epsilon^{-(n-1)/2-2\ell/(n+1)} m \otimes f_n.(v_0 \otimes v_{n-\ell+1} \otimes v_{n-\ell+2} \otimes \cdots \otimes v_{n-1})$$
$$= a\epsilon^{-2\ell/(n+1)} m \otimes v_n \otimes v_{n-\ell+1} \otimes v_{n-\ell+2} \otimes \cdots \otimes v_{n-1}. \quad \blacksquare$$

Finally, as in 12.3.12, $\mathcal{F}$ respects the tensor product operations on the categories involved: it is straightforward to prove

PROPOSITION 12.3.16 *Let* ℓ_1, $\ell_2 \geq 1$, *and let* M_1 *(resp.* M_2*) be a finite-dimensional* $\tilde{\mathcal{H}}_{\ell_1}(\epsilon^2)$*-module (resp.* $\tilde{\mathcal{H}}_{\ell_2}(\epsilon^2)$*-module). Then, there is a canonical isomorphism of* $U_\epsilon(L(sl_{n+1}(\mathbb{C})))$*-modules*

$$\mathcal{F}(M_1 \tilde{\boxtimes} M_2) \cong \mathcal{F}(M_1) \otimes \mathcal{F}(M_2). \quad \blacksquare$$

We conclude this subsection with

Example 12.3.17 Let $\mathbb{C}$ be the one-dimensional trivial representation of $\mathcal{H}_1(\epsilon) \cong \mathbb{C}$. It is easy to see that $\mathcal{J}(\mathbb{C}) \cong V_\epsilon^\natural$. By 12.3.15, for any $a \in \mathbb{C}^\times$,

$$\mathcal{F}(\mathbb{C}_a) \cong (V_\epsilon^\natural)_{a\epsilon^{2/(n+1)}}.$$

By 12.3.8 and 12.3.16, we see that, for any $\mathbf{a} = (a_1, \ldots, a_\ell) \in (\mathbb{C}^\times)^\ell$,

$$\mathcal{F}(M_\mathbf{a}) \cong (V_\epsilon^\natural)_{a_1\epsilon^{2/(n+1)}} \otimes \cdots \otimes (V_\epsilon^\natural)_{a_\ell\epsilon^{2/(n+1)}}.$$

By 12.3.4(i) and 12.3.13, it follows that every finite-dimensional irreducible representation of $U_\epsilon(L(sl_{n+1}(\mathbb{C})))$ of level $\ell \leq n$ is isomorphic to a quotient of a tensor product of representations of the form $(V_\epsilon^\natural)_a$. $\Diamond$

E Yangians and degenerate affine Hecke algebras In view of the remark at the end of Section 12.2, and the discussion preceding 12.3.2, it is not surprising that there is an analogue of 12.3.13 relating the Yangian $Y(sl_{n+1}(\mathbb{C}))$ to the degenerate affine Hecke algebra Λ_ℓ.

Let $V^\natural$ be the $(n+1)$-dimensional irreducible representation of $sl_{n+1}(\mathbb{C})$. We say that a finite-dimensional representation W of $sl_{n+1}(\mathbb{C})$ is of level ℓ if every irreducible component of W occurs in $(V^\natural)^{\otimes \ell}$. Recall that there is a canonical homomorphism $sl_{n+1}(\mathbb{C}) \to Y(sl_{n+1}(\mathbb{C}))$, so that any representation of $Y(sl_{n+1}(\mathbb{C}))$ may be regarded as one of $sl_{n+1}(\mathbb{C})$.

THEOREM 12.3.18 *Let ℓ, $n \geq 1$. There is a functor $\mathcal{D}$ from the category of finite-dimensional right Λ_ℓ-modules to the category of finite-dimensional $Y(sl_{n+1}(\mathbb{C}))$-modules which are of level ℓ as $sl_{n+1}(\mathbb{C})$-modules, defined as follows. If M is a Λ_ℓ-module, then, as a vector space,*

$$\mathcal{D}(M) = M \underset{\mathbb{C}[\Sigma_\ell]}{\otimes} (V^\natural)^{\otimes \ell},$$

where Σ_ℓ acts on $(V^\natural)^{\otimes \ell}$ by permuting the factors. The action of $sl_{n+1}(\mathbb{C})$ on $\mathcal{D}(M)$ is induced by that on $(V^\natural)^{\otimes \ell}$, while the action of the generators $J(x)$ $(x \in sl_{n+1}(\mathbb{C}))$ in 12.1.2 is given by

$$J(x).(m \otimes w_1 \otimes w_2 \otimes \cdots \otimes w_\ell) = \sum_{j=1}^{\ell} m.y_j \otimes w_1 \otimes \cdots \otimes w_{i-1} \otimes x.w_i \otimes w_{i+1} \otimes \cdots \otimes w_\ell$$

for all $m \in M$, $w_1, \ldots, w_\ell \in V^\natural$.

If $\ell \leq n$, $\mathcal{D}$ is an equivalence of categories. ∎

Note that it follows from (35) that the action of $J(x)$ is well-defined.

This result is due to Drinfel'd (1986b). The same work also contains an analogue of 12.3.16 for Yangians, whose formulation we leave to the reader. The finite-dimensional irreducible Λ_ℓ-modules have been classified in Cherednik (1988b), so 12.3.18 gives a description of all irreducible $Y(sl_{n+1}(\mathbb{C}))$-modules which are of level $\leq n$ for $sl_{n+1}(\mathbb{C})$.

Nazarov (1993b) gives an analogue of 12.3.18 for the orthogonal Yangians $Y(so_n(\mathbb{C}))$.

12.4 Yangians and infinite-dimensional classical groups

The second instance of a relation between Yangians and classical representation theory concerns the infinite-dimensional classical group O_∞ consisting of the $\mathbf{N} \times \mathbf{N}$ matrices of the form

(41)
$$\begin{pmatrix} A & 0 & 0 & 0 \\ 0 & 1 & 0 & 0 \\ 0 & 0 & 1 & 0 \\ 0 & 0 & 0 & \ddots \end{pmatrix}$$

where $A \in O_n$ for some $n \geq 1$; we identify O_n with the subgroup of O_∞ consisting of the matrices (41). We define the Lie algebra o_∞ of O_∞ to be the union of the Lie algebras o_n of O_n under the same embeddings.

A Tame representations A representation ρ of O_∞ on a complex Hilbert space V is said to be *tame* if every vector $v \in V$ can be approximated arbitrarily closely by a vector fixed by O_n for some sufficiently large n. If $x \in o_\infty$, let $\rho'(x)$ be the skew-symmetric operator such that $\exp{(t\rho'(x))} = \rho(\exp{(tx)})$ for all $t \in \mathbb{R}$. This gives an action of o_∞, and hence of its universal enveloping algebra $U(o_\infty)$, on V.

It is not difficult to construct tame representations. The most obvious example is the complexification of the natural representation of O_∞ on the real Hilbert space of all square-summable sequences $(t_1, t_2, \dots)$ (thought of as $\mathbb{N} \times 1$ column vectors). Let us denote this representation by V^{nat}: it is clearly tame. More generally, let $(V^{\mathrm{nat}})^{\otimes k}$ be the k-fold *Hilbert space tensor product* of V^{nat}, i.e. the completion of the k-fold algebraic tensor product equipped with the inner product

$$\langle v_1 \otimes \cdots \otimes v_k, v_1' \otimes \cdots \otimes v_k' \rangle = \langle v_1, v_1' \rangle \dots \langle v_k, v_k' \rangle.$$

The symmetric group Σ_k acts on $(V^{\mathrm{nat}})^{\otimes k}$ by permuting the factors, and this action obviously commutes with that of O_∞. It is known that, if π is any irreducible representation of Σ_k, the isotypical component of $(V^{\mathrm{nat}})^{\otimes k}$ of type π is a direct sum of $\dim(\pi)$ copies of an irreducible representation V^π of O_∞. Further, V^π is tame, and every tame irreducible representation of O_∞ is isomorphic to some V^π.

Remarks [1] This result obviously resembles the Frobenius–Schur duality between the representations of symmetric groups and those of finite-dimensional general linear groups. We shall see an explanation for this at the end of this section.

[2] It is perhaps worth emphasizing that the tensor products $(V^{\mathrm{nat}})^{\otimes k}$ do not decompose into irreducibles in the 'same' way as in the case of finite-dimensional orthogonal groups. For example, the inner product

$$\langle \, , \, \rangle : V^{\mathrm{nat}} \otimes V^{\mathrm{nat}} \to \mathbb{C}$$

clearly does not extend to a continuous linear form $(V^{\mathrm{nat}})^{\otimes 2} \to \mathbb{C}$. This implies that the representation $(V^{\mathrm{nat}})^{\otimes 2}$ of O_∞ does not contain a copy of the trivial representation. $\triangle$

One would like to study tame representations by infinitesimal methods. For example, one would like to classify them by their central characters. Unfortunately, it is known that the centre of $U(o_\infty)$ contains only the scalars. However, there is a way to enlarge the enveloping algebra in such a way that it still acts on tame representations.

DEFINITION 12.4.1 *Fix* $m \in \mathbb{N}$ *and let* X_m *be the set of all sequences* x_1, x_2, x_3, \dots *such that*

(i) $x_n \in U(o_n)$ is invariant under the adjoint action of the subgroup

$$O_{m,n} = \begin{pmatrix} 1 & 0 \\ 0 & O_{n-m} \end{pmatrix}$$

of O_n for all $n > m$;

(ii) the degrees of the x_n are bounded (the degree of an element x of an enveloping algebra $U(\mathfrak{g})$ is the minimal $k \in \mathbb{N}$ such that x is in the image of the canonical embedding $\mathfrak{g}^{\otimes k} \to U(\mathfrak{g})$ – see 4.1.8);

(iii) if ρ is any tame representation of O_∞, the strong limit $\lim_{n\to\infty} \rho'(x_n)$ exists (and is then necessarily a bounded operator).

Two sequences $(x_1, x_2, x_3, \dots)$, $(y_1, y_2, y_3, \dots) \in X_m$ are said to be equivalent if

$$\lim_{n\to\infty} \rho'(x_n) = \lim_{n\to\infty} \rho'(y_n)$$

for all tame representations ρ. The set of equivalence classes is denoted by $\hat{U}_m(o_\infty)$.

Since $O_{m+1,n} \subset O_{m,n}$, there are natural enbeddings $\hat{U}_m(o_\infty) \hookrightarrow \hat{U}_{m+1}(o_\infty)$ and we define

$$\hat{U}(o_\infty) = \varinjlim \hat{U}_m(o_\infty).$$

Note that $\hat{U}_m(o_\infty)$ has an obvious algebra structure, corresponding to componentwise multiplication of sequences in X_m, and hence so does $\hat{U}(o_\infty)$. Note also that there is an embedding $U(o_\infty) \hookrightarrow \hat{U}(o_\infty)$ which associates the sequence $(x, x, x, \dots)$ to any element $x \in U(o_\infty)$. It is tautologous that $\hat{U}(o_\infty)$ acts by bounded operators on any tame representation of O_∞, extending the action of $U(o_\infty)$.

B The relation with Yangians With these preparations, we can state the main result of this section, which shows that the extended enveloping algebra $\hat{U}(o_\infty)$ is closely related to Yangians.

THEOREM 12.4.2 Let Z be the centre of $\hat{U}(o_\infty)$. Then, for all $m \geq 1$, there are isomorphisms of algebras

$$\hat{U}_m(o_\infty) \cong Z \otimes Y(gl_m(\mathbb{C})).$$

These isomorphisms are compatible with the natural embeddings

$$\hat{U}_m(o_\infty) \hookrightarrow \hat{U}_{m+1}(o_\infty) \quad \text{and} \quad Y(gl_m(\mathbb{C})) \hookrightarrow Y(gl_{m+1}(\mathbb{C})). \ \blacksquare$$

Thus, the infinite Yangian

$$Y(gl_\infty(\mathbb{C})) = \varinjlim Y(gl_m(\mathbb{C}))$$

acts on any tame representation of O_∞.

Remarks [1] The last sentence of the theorem implies that

$$\hat{U}(o_\infty) \cong Z \otimes Y(gl_\infty(\mathbb{C})),$$

and hence that the centre of $Y(gl_\infty(\mathbb{C}))$ is trivial (cf. 12.1.7).

[2] The structure of the centre Z of $\hat{U}(o_\infty)$ is known. Let Z_m be the sub-algebra of the polynomial algebra $\mathbb{C}[z_1, \ldots, z_m]$ consisting of the polynomials which are symmetric in the variables $z_1 - 1, z_2 - 2, \ldots, z_m - m$. Restricting a polynomial in Z_m to the hyperplane $z_m = 0$ in $\mathbb{C}^m$ defines a homomorphism $Z_m \to Z_{m-1}$, and

$$Z \cong \varprojlim Z_m.$$

[3] The resemblance between the theory of tame representations of O_∞ and that of the representations of finite-dimensional general linear groups is explained to some extent by the following alternative description of $\hat{U}(o_\infty)$. Let $U_n = U(gl_n(\mathbb{C}))$ and regard U_n as a subalgebra of U_{n+1} in the usual way. Let U_n^m, for each $m \leq n$, be the centralizer in U_n of the elementary matrices E_{ij} with $m < i, j \leq n$. Then, $U_n^{m-1} \subset U_n^m$ for all $m < n$. It is easy to see that, if I_n is the left ideal in U_n generated by $E_{1n}, E_{2n}, \ldots, E_{nn}$, then $U_n^{n-1} = U_{n-1} \oplus I_n$. If we restrict the isomorphism $U_n^{n-1}/I_n \to U_{n-1}$ to U_n^m, we obtain a homomorphism $U_n^m \to U_{n-1}$ whose image clearly lies inside U_{n-1}^m. We define

$$U^m = \varprojlim U_n^m$$

in the category of filtered algebras. Thus, an element of U^m is a compatible sequence $(x_m, x_{m+1}, \ldots)$ with $x_n \in U_n^m$, such that the degrees of the x_n are bounded. There are obvious embeddings $U^m \hookrightarrow U^{m+1}$ and one shows that

$$\hat{U}(o_\infty) \cong \varinjlim U^m$$

in the category of filtered algebras. $\triangle$

12.5 Rational and trigonometric solutions of the QYBE

We have seen that Yangians and quantum affine algebras are not quasitrian-gular. However, we show in this section that these Hopf algebras are 'pseu-dotriangular' (this terminology is due to Drinfel'd (1987)), and that, as for quasitriangular Hopf algebras, this property is sufficient to guarantee that there is a solution of the quantum Yang–Baxter equation associated to the tensor product of any pair of finite-dimensional representations.

A Yangians and rational solutions We recall from Subsection 12.1A (see also 1.3.9) that the Lie bialgebra structure on $\mathfrak{g}[u]$ of which $U_h(\mathfrak{g}[u])$ is the quantization is given by

$$(42) \qquad \delta(f)(u,v) = (\mathrm{ad}_{f(u)} \otimes \mathrm{id} + \mathrm{id} \otimes \mathrm{ad}_{f(v)}) \left(\frac{t}{u-v} \right),$$

where t is the Casimir element of $\mathfrak{g} \otimes \mathfrak{g}$. Although this formula defines a Lie bialgebra $(\mathfrak{g}[u], \delta)$, it does not imply that it is quasitriangular because $r = t/(u-v)$ is not an element of the algebraic tensor product $\mathfrak{g}[u] \otimes \mathfrak{g}[u] \cong (\mathfrak{g} \otimes \mathfrak{g})[u,v]$.

However, let λ be an indeterminate and let $\mathfrak{g}[u]((\lambda^{-1}))$ be the extension of $\mathfrak{g}[u]$ to a Lie algebra over the field of formal power series in λ^{-1}. The translation $f(u) \mapsto f(u+\lambda)$ is a well-defined automorphism of $\mathfrak{g}[u]((\lambda^{-1}))$, which we denote by τ_λ, and the coefficients of powers of λ in the formal power series

$$r(\lambda) = (\tau_\lambda \otimes \mathrm{id}) \left(\frac{t}{u-v} \right) = \frac{t}{u-v+\lambda} = \sum_{r=0}^{\infty} (v-u)^r \lambda^{-r-1} t$$

do belong to $(\mathfrak{g} \otimes \mathfrak{g})[u,v]$. Thus, replacing $t/(u-v)$ on the right-hand side of (42) by $r(\lambda)$ gives a genuinely coboundary Lie bialgebra structure on $\mathfrak{g}[u]((\lambda^{-1}))$. Note that $r_{21}(\lambda) = -r_{12}(-\lambda)$.

The translation automorphisms of $\mathfrak{g}[u]$ have been 'quantized' in 12.1.5. If we replace the complex number a in (20) by the indeterminate λ, we obtain a well-defined algebra automorphism τ_λ of $Y(\mathfrak{g})((\lambda^{-1}))$. From the preceding remarks, it is reasonable to expect that $Y(\mathfrak{g})((\lambda^{-1}))$ has a universal R-matrix $\mathcal{R}(\lambda)$ satisfying $\mathcal{R}_{21}(\lambda) = \mathcal{R}_{12}(-\lambda)^{-1}$. This expectation is fulfilled:

THEOREM 12.5.1 *There is a unique formal power series*

$$(43) \qquad \mathcal{R}(\lambda) = 1 \otimes 1 + \frac{t}{\lambda} + \sum_{r=1}^{\infty} \mathcal{R}_r \lambda^{-r-1} \in (Y(\mathfrak{g}) \otimes Y(\mathfrak{g}))[[\lambda^{-1}]]$$

with the following properties:

$$(\Delta \otimes \mathrm{id})(\mathcal{R}(\lambda)) = \mathcal{R}_{13}(\lambda)\mathcal{R}_{23}(\lambda), \quad (\mathrm{id} \otimes \Delta)(\mathcal{R}(\lambda)) = \mathcal{R}_{13}(\lambda)\mathcal{R}_{12}(\lambda),$$
$$(\tau_\lambda \otimes \mathrm{id})(\Delta^{\mathrm{op}}(a)) = \mathcal{R}(\lambda).((\tau_\lambda \otimes \mathrm{id})(\Delta(a))).\mathcal{R}(\lambda)^{-1}$$

for all $a \in Y(\mathfrak{g})$. Moreover, $\mathcal{R}(\lambda)$ is a triangular solution of the QYBE:

$$(44) \qquad \begin{aligned} \mathcal{R}_{12}(\lambda_1 - \lambda_2)\mathcal{R}_{13}(\lambda_1 - \lambda_3)\mathcal{R}_{23}(\lambda_2 - \lambda_3) = \\ \mathcal{R}_{23}(\lambda_2 - \lambda_3)\mathcal{R}_{13}(\lambda_1 - \lambda_3)\mathcal{R}_{12}(\lambda_1 - \lambda_2), \\ \mathcal{R}_{21}(-\lambda) = \mathcal{R}_{12}(\lambda)^{-1}. \quad \blacksquare \end{aligned}$$

We shall call $\mathcal{R}(\lambda)$ the *pseudo-universal R-matrix* of $Y(\mathfrak{g})$.

Remark In principle, it should be possible to compute the pseudo-universal R-matrix $\mathcal{R}(\lambda)$ by means of the quantum double method, which we used in Chapter 8 to compute the universal R-matrix of $U_h(\mathfrak{g})$. As in that case, to carry out the computation, it would be necessary to calculate the action of the comultiplication on all of the basis elements of $Y(\mathfrak{g})$ given in 12.1.9. As we noted in Remark [1] following 12.1.1, this has not been achieved, even for the generators of $Y(\mathfrak{g})$ given in 12.1.3. $\triangle$

In view of (43) and (44), if $\rho : Y(\mathfrak{g}) \to \mathrm{End}(V)$ is a representation of $Y(\mathfrak{g})$, then $R^\rho(\lambda) = (\rho \otimes \rho)(\mathcal{R}(\lambda)) \in \mathrm{End}(V \otimes V)[[\lambda^{-1}]]$ will be a matrix solution of the QYBE, with λ_1, λ_2, λ_3 playing the role of spectral parameters, of the form

$$(45) \qquad R^\rho(\lambda) = 1 \otimes 1 + (\rho \otimes \rho)(t)\lambda^{-1} + \sum_{r=1}^{\infty} R_r \lambda^{-r-1}.$$

Unfortunately, since no explicit expression for $\mathcal{R}(\lambda)$ is known, this result does not allow one to actually compute $R^\rho(\lambda)$.

However, 11.1.16 gives another method of constructing solutions of the QYBE from representations of Hopf algebras. To apply this to Yangians, we need a one-parameter family of finite-dimensional irreducible representations of $Y(\mathfrak{g})$ on a fixed underlying vector space. If $\rho : Y(\mathfrak{g}) \to \mathrm{End}(V)$ is any finite-dimensional representation, we can take the family $\rho \circ \tau_a$, which we denote by $\tau_a^*(V)$. Then, we expect that, if $I(a,b) : \tau_a^*(V) \otimes \tau_b^*(V) \to \tau_b^*(V) \otimes \tau_a^*(V)$ is a suitably normalized intertwining operator, then $R = \sigma \circ I$, where σ is the map which interchanges the factors in the tensor product, is a solution of the QYBE, the spectral parameters now being complex numbers rather than indeterminates. The following result shows that this expectation is fulfilled, and that the resulting solutions of the QYBE are the same as those obtained by applying ρ to the universal R-matrix $\mathcal{R}(\lambda)$.

THEOREM 12.5.2 *Let $\rho : Y(\mathfrak{g}) \to \mathrm{End}(V)$ be a finite-dimensional irreducible representation of $Y(\mathfrak{g})$. Then, up to a scalar multiple, $R^\rho(\lambda) = (\rho \otimes \rho)(\mathcal{R}(\lambda))$ is the Laurent expansion about ∞ of a rational function of λ, and thus $R^\rho(a)$ is a well-defined element of $\mathrm{End}(V \otimes V)$ for all except finitely many values of $a \in \mathbb{C}$. Moreover, except for finitely many values of $a - b$, $I^\rho(a,b) = \sigma \circ R^\rho(a-b)$ is the intertwining operator $\tau_a^*(V) \otimes \tau_b^*(V) \to \tau_b^*(V) \otimes \tau_a^*(V)$ (up to a scalar multiple).* ∎

Remark It is clear *a priori* that, up to a scalar multiple, every intertwining operator $I(a,b) : \tau_a^*(V) \otimes \tau_b^*(V) \to \tau_b^*(V) \otimes \tau_a^*(V)$ is a rational function of $a - b$. For, using the description (21) of τ_a, the condition that $I(a,b)$ is an intertwiner is that

$$(\rho \otimes \rho)(x \otimes 1 + 1 \otimes x)I(a,b) = I(a,b)(\rho \otimes \rho)(x \otimes 1 + 1 \otimes x),$$

(46) $$(\rho \otimes \rho)((J(x) + bx) \otimes 1 + 1 \otimes (J(x) + ax) + \frac{1}{2}[x \otimes 1, t])I(a,b) =$$

$$I(a,b)(\rho \otimes \rho)((J(x) + ax) \otimes 1 + 1 \otimes (J(x) + bx) + \frac{1}{2}[x \otimes 1, t]),$$

for all $x \in \mathfrak{g}$. Using the first equation in (46), the second is equivalent to

$$(\rho \otimes \rho)(J(x) \otimes 1 + 1 \otimes (J(x) + (a-b)x) + \frac{1}{2}[x \otimes 1, t])I(a,b) =$$

$$I(a,b)(\rho \otimes \rho)((J(x) + (a-b)x) \otimes 1 + 1 \otimes J(x) + \frac{1}{2}[x \otimes 1, t]).$$

Letting x run through a basis of $\mathfrak{g}$, we see that the equations (46) are equivalent to a finite system of linear equations for the entries of the matrix $I(a,b)$, the coefficients of which are polynomials in $a - b$. For elementary reasons, the solution $I(a,b)$ must therefore be a rational function of $a - b$ (up to a scalar multiple). $\triangle$

It is natural to ask which rational solutions $R(\lambda) \in \mathrm{End}(V \otimes V)$ of the QYBE arise from representations of Yangians in this way. An obvious necessary condition is that, up to a scalar multiple, the Laurent expansion of $R(\lambda)$ about $\lambda = \infty$ is of the form (45) for some representation ρ of $\mathfrak{g}$ on V. The next result shows that this condition is also sufficient.

THEOREM 12.5.3 *Let $R(\lambda)$ be a rational solution of the QYBE of the form (45) for some finite-dimensional representation ρ of $\mathfrak{g}$. Then, there is an extension of ρ to a representation $\tilde{\rho} : Y(\mathfrak{g}) \to \mathrm{End}(V)$ on the same vector space such that $R(a)$ is a scalar multiple of $R^{\tilde{\rho}}(a)$ for all but finitely many $a \in \mathbb{C}$. Moreover, $\tilde{\rho}$ is uniquely determined by $R(\lambda)$ and ρ up to twisting with one of the automorphisms τ_b.* ∎

Remark The condition (45) is natural from another point of view, for the most obvious way to construct rational solutions of the CYBE is to apply representations ρ of $\mathfrak{g}$ to the classical r-matrix $t/(u - v)$ of $\mathfrak{g}[u]$. Then, (45) gives a 'quantization' of this classical r-matrix (see 6.2.11). To see this, it is necessary to work with $U_h(\mathfrak{g}[u])$ rather than $Y(\mathfrak{g})$, i.e. to avoid specializing h to 1. If one does that, one finds that the analogue of (45) is

$$R^{\rho}(\lambda) = 1 \otimes 1 + h(\rho \otimes \rho)(t)\lambda^{-1} + \sum_{r=1}^{\infty} h^{r+1} R_r \lambda^{-r-1},$$

so that $(\rho \otimes \rho)(t/\lambda) \equiv h^{-1}(R^{\rho}(\lambda) - 1 \otimes 1) \pmod{h}$. $\triangle$

We now turn to the problem of actually computing the R-matrix associated to a given representation of $Y(\mathfrak{g})$. In general, this is far from solved. If $\mathfrak{g} = sl_2(\mathbb{C})$, however, we can give a complete answer:

PROPOSITION 12.5.4 *Let $a_1, \ldots, a_k \in \mathbb{C}$, $r_1, \ldots, r_k \in \mathbb{N}$, $k \geq 1$, satisfy the condition in 12.2.15(iii), so that*

$$V = V(r_1)_{a_1} \otimes \cdots \otimes V(r_k)_{a_k}$$

is an irreducible representation of $Y(sl_2(\mathbb{C}))$. The R-matrix associated to V can be described as follows. Choose $sl_2(\mathbb{C})$-primitive vectors v_j in $V(r) \otimes V(s)$ of weight $r+s-2j$, whenever $0 \leq j \leq \max\{r, s\}$, such that $(X_1^+ \otimes 1).v_j = v_{j-1}$ for all $j > 0$, and let $P_j^{r,s} : V(r) \otimes V(s) \rightarrow V(s) \otimes V(r)$ be the unique homomorphism of $sl_2(\mathbb{C})$-modules such that $P_j^{r,s}(v_j) = \sigma(v_j)$ and $P_j^{r,s}(v_k) = 0$ if $k \neq j$ (σ is the flip operator). Define

$$I(r, a\,;\, s, b) : V(r)_a \otimes V(s)_b \rightarrow V(s)_b \otimes V(r)_a$$

by

$$I(r, a\,;\, s, b) = \sum_{j=0}^{\min\{r,s\}} \left(\prod_{i=0}^{j-1} \frac{b - a + \frac{1}{2}(r+s) - j}{b - a - \frac{1}{2}(r+s) + j} \right) P_j^{r,s}.$$

Then, the R-matrix associated to V is

(47)
$$R(a - b) = \sigma \circ \left(\prod_{i,j=1}^{k} I(r_i, a + a_i\,;\, r_j, b + a_j) \right),$$

where the operator I in the (i, j)th term acts in the $(i+j-1)$th and $(i+j)$th factors in the $2k$-fold tensor product, and the (i, j)th term appears to the right of the (i', j')th term if and only if $i > i'$ or $i = i'$ and $j < j'$.

The possibility of choosing primitive vectors v_j satisfying the conditions in the proposition will be established in the course of the proof.

PROOF Let $I(r, a\,;\, s, b) : V(r)_a \otimes V(s)_b \rightarrow V(s)_b \otimes V(r)_a$ be the unique intertwining operator of $Y(sl_2(\mathbb{C}))$-modules which flips the tensor product of highest weight vectors in $V(r)_a$ and $V(s)_b$. The unique intertwining operator $\tau_a^*(V) \otimes \tau_b^*(V) \rightarrow \tau_b^*(V) \otimes \tau_a^*(V)$, between the tensor products of the twists of V by τ_a and τ_b, which permutes the $2k$-fold tensor product of highest weight vectors, can be computed as the product of the intertwiners which successively move each of the k factors in $\tau_b^*(V)$ to the left of each of the k factors in $\tau_a^*(V)$. The result is the product in (47).

Thus, it suffices to compute $I(r, a\,;\, s, b)$. Since this operator commutes with the action of $sl_2(\mathbb{C})$, we must have

$$I(r, a\,;\, s, b) = \sum_{j=0}^{\min\{r,s\}} c_j P_j^{r,s}$$

for some scalars c_j; by the chosen normalization, $c_0 = 1$. Let v_j be an $sl_2(\mathbb{C})$-primitive vector in $V(r) \otimes V(s)$ of weight $r + s - 2j$. If $j > 0$, it is easy to see that $(X_1^+ \otimes 1).v_j$ is an $sl_2(\mathbb{C})$-highest weight vector of weight $r + s - 2j + 2$; it is non-zero, for otherwise v_j would be annihilated by $X_1^+ \otimes 1$ and $1 \otimes X_1^+$, and hence would have to be a tensor product of highest weight vectors in $V(r)$ and $V(s)$, contradicting the assumption that $j > 0$. Hence, if we normalize the v_j suitably, we may assume that

$$(48) \qquad\qquad (X_1^+ \otimes 1).v_j = v_{j-1}$$

for $j > 0$. From the proof of 12.1.22, v_j is a $Y(sl_2(\mathbb{C}))$-highest weight vector in $V(r)_a \otimes V(s)_b$ if $a - b = \frac{1}{2}(r + s) - j + 1$. It follows from (7) that, in $V(r)_a \otimes V(s)_b$,

$$J(X_1^+).v_j = \left(a - b - \frac{1}{2}(r + s) + j - 1 \right) (X_1^+ \otimes 1).v_j.$$

If $v_j' = \sigma(v_j) \in V(s)_b \otimes V(r)_a$, we find by the same argument that

$$J(X_1^+).v_j' = \left(b - a - \frac{1}{2}(r + s) + j - 1 \right) (X_1^+ \otimes 1).v_j',$$

and from (48) that $(X_1^+ \otimes 1).v_j' = -v_{j-1}'$. The equation

$$I(r, a\,;\, s, b)(J(X_1^+).v_j) = J(X_1^+).(I(r, a\,;\, s, b)(v_j))$$

now gives

$$\frac{c_j}{c_{j-1}} = \frac{b - a + \frac{1}{2}(r + s) - j + 1}{b - a - \frac{1}{2}(r + s) + j - 1},$$

and hence the stated formula for $I(r, a\,;\, s, b)$. ∎

Remarks [1] The expression for $R(a - b)$ in 12.5.4 was first obtained by Kulish, Reshetikhin & Sklyanin (1981), using methods of inverse scattering theory. Even when the evaluation representations $V(r)_a$ are all two-dimensional (i.e. $r = 1$), the resulting solutions of the QYBE are related to interesting physical models (the quantum non-linear Schrödinger equation, the Heisenberg ferromagnet (XXX model), Toda chains, ...).

Kulish *et al.* apparently proved that $R(a - b)$, as defined above, satisfies the QYBE by a direct verification (which is not entirely trivial). For us, however, this follows from 11.1.16 without further computation. The hypotheses (i) and (ii) in that proposition (for generic values of the parameters) follow from 12.1.20 and 12.1.21, and the scalar c in equation (4) in the same proposition is equal to 1 because the products of the R-matrices on both sides of (4) preserve the tensor products of highest weight vectors.

[2] The kind of direct computation used to prove 12.5.4 is impractical for general $\mathfrak{g}$, except for the simplest representations. However, the R-matrices associated to most of the fundamental representations of $Y(\mathfrak{g})$ are known. See Ogievetsky, Reshetikhin & Wiegmann (1987), Ogievetsky & Wiegmann (1986), Reshetikhin & Wiegmann (1987) and Chari & Pressley (1991a). The arguments used in the first three references depend on the assumption of 'the physical consistency of certain quantum field theories', but most of their results are proved in the last reference by Lie-theoretic methods. $\triangle$

B Quantum affine algebras and trigonometric solutions Like Yangians, (untwisted) quantum affine algebras also turn out to be pseudotriangular. In principle, their pseudo-universal R-matrices can be constructed by the quantum double method used in Chapter 8, but there are two difficulties caused by differences between the structure of the root system of an affine Lie algebra $\tilde{\mathfrak{g}}$ and that of its underlying finite-dimensional Lie algebra $\mathfrak{g}$.

The first important difference is that, unlike the situation for $\mathfrak{g}$, not every root of $\tilde{\mathfrak{g}}$ can be obtained by applying an element of the Weyl group of $\tilde{\mathfrak{g}}$ to a simple root. This means that the procedure used in Section 8.1 to define non-simple root vectors in $U_h(\mathfrak{g})$ cannot be used for $U_h(\tilde{\mathfrak{g}})$. The other difference is more obvious: $\tilde{\mathfrak{g}}$ has infinitely many positive roots. Thus, if one is to prove a formula like that in 8.3.9 for the universal R-matrix $\tilde{\mathcal{R}}$ of $U_h(\tilde{\mathfrak{g}})$, the coefficients in its h-adic expansion will be formal infinite linear combinations of products of root vectors. There would then be no guarantee that $\tilde{\mathcal{R}}$ would act on a tensor product of representations of $U_h(\tilde{\mathfrak{g}})$ (and is why we obtain only a 'pseudo-universal' R-matrix).

Despite these objections, we have the following result of Drinfel'd (1987). Let λ be an indeterminate and let τ_λ be the automorphism of $U_h(\tilde{\mathfrak{g}})((\lambda))$ given by

$$\tau_\lambda(X_i^\pm) = \lambda^{\pm 1} X_i^\pm, \quad \tau_\lambda(H_i) = H_i, \quad i = 0, 1, \ldots, n.$$

THEOREM 12.5.5 *There exists an element $\tilde{\mathcal{R}}(\lambda) \in (U_h(\tilde{\mathfrak{g}}) \otimes U_h(\tilde{\mathfrak{g}}))((\lambda))$ such that, if $\rho : U_h(\tilde{\mathfrak{g}}) \to \mathrm{End}(V)$ is an indecomposable representation of $U_h(\tilde{\mathfrak{g}})$ on a free $\mathbb{C}[[h]]$-module V of finite rank, then $\tilde{R}^\rho(\lambda) = (\rho \otimes \rho)((\tau_\lambda \otimes \mathrm{id})(\tilde{\mathcal{R}}))$ is a well-defined element of $\mathrm{End}(V \otimes V)((\lambda))$ which satisfies the QYBE in the form*

$$(49) \qquad \tilde{R}_{12}^\rho(\lambda/\mu)\tilde{R}_{13}^\rho(\lambda/\nu)\tilde{R}_{23}^\rho(\mu/\nu) = \tilde{R}_{23}^\rho(\mu/\nu)\tilde{R}_{13}^\rho(\lambda/\nu)\tilde{R}_{12}^\rho(\lambda/\mu).$$

Up to a scalar factor, $\sigma \circ \tilde{R}^\rho(\lambda)$ is a rational matrix-valued function of λ and is an intertwining operator between the representations $(\rho \otimes \rho)(\tau_\lambda \otimes \mathrm{id})$ and $(\rho \otimes \rho)(\mathrm{id} \otimes \tau_\lambda)$.

Remarks [1] As in the Yangian case, it is clear a priori that the intertwining operator referred to in the theorem is, up to a scalar multiple, a rational function of λ.

[2] The form (49) of the QYBE can be brought to the more usual form (19) simply by making the change of variable $\lambda = e^u$, $\mu = e^v$, $\nu = e^w$. Then, $\tilde{R}^\rho$ becomes a rational function of e^u: this is what is meant by a 'trigonometric' function.

[3] Khoroshkin & Tolstoy (1992a) give a formula for $\tilde{\mathcal{R}}(\lambda)$ for arbitrary (untwisted) quantum affine algebras. They define the root vectors of $U_h(\tilde{\mathfrak{g}})$ without using the braid group action, using instead the notion of a *normal ordering* of the positive roots of $\tilde{\mathfrak{g}}$. Levendorskii, Soibelman and Stukopin (1993) obtained the formula in the sl_2 case essentially by using the braid group action to define the root vectors, but replacing the non-existent longest element of the Weyl group of $\tilde{\mathfrak{g}}$ by the pair of infinite expressions

$$T_0 T_1 T_0 T_1 T_0 \ldots, \quad T_1 T_0 T_1 T_0 T_1 \ldots . \triangle$$

Even though an 'explicit' formula for the pseudo-universal R-matrix is available, it is not easy to compute its action in any but the simplest representations. Thus, to determine $\tilde{R}^\rho(\lambda)$ (up to a scalar factor), one computes the intertwiner directly.

The proof of the following result is similar to that of 12.5.4. Let $V_h(r)$ be the indecomposable representation of $U_h(sl_2(\mathbb{C}))$ of dimension $r + 1$.

PROPOSITION 12.5.6 *Let $a_1, \ldots, a_k \in \mathbb{C}[[h]]^\times$, $r_1, \ldots, r_k \in \mathbf{N}$, $k \geq 1$, satisfy the obvious analogue of 12.2.15(iii), and let ρ be the indecomposable representation*

$$V = V_h(r_1)_{a_1} \otimes \cdots \otimes V_h(r_k)_{a_k}$$

of $U_h(\tilde{sl}_2(\mathbb{C}))$. Choose $U_h(sl_2(\mathbb{C}))$-primitive vectors $v_j \in V_h(r) \otimes V_h(s)$ (resp. $v'_j \in V_h(s) \otimes V_h(r)$) of weight $r + s - 2j$, whenever $0 \leq j \leq \max\{r, s\}$, such that $(X_1^+ \otimes 1).v_j = v_{j-1}$ (resp. $(X_1^+ \otimes 1).v'_j = v'_{j-1}$) for all $j > 0$. Let $P_j^{r,s} : V_h(r) \otimes V_h(s) \to V_h(s) \otimes V_h(r)$ be the unique homomorphism of $U_h(sl_2(\mathbb{C}))$-modules such that $P_j^{r,s}(v_j) = v'_j$ and $P_j^{r,s}(v_k) = 0$ if $j \neq k$. Define

$$I(r, a\,;\, s, b) : V_h(r)_a \otimes V_h(s)_b \to V_h(s)_b \otimes V_h(r)_a$$

by

$$I(r, a\,;\, s, b) = \sum_{j=0}^{\min\{r,s\}} \left(\prod_{i=0}^{j-1} \frac{b - ae^{(r+s-2i)h}}{b - ae^{-(r+s-2i)h}} \right) P_j^{r,s}.$$

Then,

$$\tilde{R}^\rho(\lambda) = A\sigma \circ \left(\prod_{i,j=1}^{k} I(r_i, \lambda a_i\,;\, r_j, a_j) \right),$$

where A is a scalar and the operator I in the (i,j)th term acts in the $(i+j-1)$th and $(i+j)$th factors in the $2k$-fold tensor product, and the (i,j)th term appears to the right of the (i', j')th term if and only if $i > i'$ or $i = i'$ and $j < j'$. ∎

Example 12.5.7 Let us compute the intertwiner

$$V_h(1)_a \otimes V_h(1)_b \to V_h(1)_b \otimes V_h(1)_a.$$

If $\{v_0, v_1\}$ is the usual basis of $V_h(1)$ (see 6.4.12), we find that the maps $P_j^{1,1}$, $j = 0, 1$, are given by

$$P_0^{1,1}(v_0 \otimes v_0) = v_0 \otimes v_0, \quad P_0^{1,1}(v_1 \otimes v_1) = v_1 \otimes v_1,$$
$$P_0^{1,1}(v_0 \otimes v_1) = \frac{e^{-h} v_0 \otimes v_1 + v_1 \otimes v_0}{e^h + e^{-h}}, \quad P_0^{1,1}(v_1 \otimes v_0) = \frac{v_0 \otimes v_1 + e^h v_1 \otimes v_0}{e^h + e^{-h}},$$

and

$$P_1^{1,1}(v_0 \otimes v_0) = P_1^{1,1}(v_1 \otimes v_1) = 0,$$
$$P_1^{1,1}(v_0 \otimes v_1) = \frac{e^h v_0 \otimes v_1 - v_1 \otimes v_0}{e^h + e^{-h}}, \quad P_1^{1,1}(v_1 \otimes v_0) = \frac{e^{-h} v_1 \otimes v_0 - v_0 \otimes v_1}{e^h + e^{-h}},$$

respectively. The formula in 12.5.7 gives the intertwiner as

$$\tilde{R}_h = \begin{pmatrix} 1 & 0 & 0 & 0 \\ 0 & \frac{b-a}{b-ae^{-2h}} & \frac{a(e^h - e^{-h})}{b - ae^{-2h}} & 0 \\ 0 & \frac{a(e^h - e^{-h})}{b - ae^{-2h}} & \frac{b-a(e^{2h} - 1 + e^{-2h})}{b - ae^{-2h}} & 0 \\ 0 & 0 & 0 & 1 \end{pmatrix},$$

with respect to the basis $\{v_0 \otimes v_0, v_1 \otimes v_0, v_0 \otimes v_1, v_1 \otimes v_1\}$. Note that $\tilde{R}_h$ has a 'pole' when $b/a = e^{-2h}$; further, one checks that $\tilde{R}_h$ fails to be invertible when $b/a = e^{2h}$. On the other hand, from the analogue of 12.2.14 (or by a direct computation), one finds that $V_h(1)_a \otimes V_h(1)_b$ fails to be isomorphic to $V_h(1)_b \otimes V_h(1)_a$ precisely when $b/a = e^{\pm 2h}$, i.e. precisely when $\tilde{R}_h$ has 'singularities'.

The R-matrix $\tilde{R}^\rho(\lambda)$ found by Levendorskii, Soibelman & Stukopin (1993) for this example can be obtained, up to a scalar multiple, by putting $a = \lambda^2$, $b = 1$ in the formula for $\tilde{R}_h$. ◇

Bibliographical notes

12.1 The two realizations 12.1.1 and 12.1.3 are due to Drinfel'd (1987, 1988b). For the R-matrix realization 12.1.4, see Kirillov & Reshetikhin (1986).

Theorem 12.1.11 appears in Drinfel'd (1988b). Fundamental representations of $Y(\mathfrak{g})$ were introduced, and their structure investigated, in Chari & Pressley (1991a). Evaluation representations of $Y(sl_2(\mathbb{C}))$ were implicit in Kulish, Reshetikhin & Sklyanin (1981), although they did not prove that every irreducible representation is a tensor product of evaluation representations: this was done in Chari & Pressley (1990b). Proposition 12.1.17 was stated in Drinfel'd (1985).

The results 12.1.8 and 12.1.9 of Poincaré–Birkhoff–Witt type are due to Levendorskii (1993).

12.2 The classification of pseudo-highest weight representations 12.2.6 is due to Chari & Pressley (1991b). Evaluation homomorphisms and representations for quantum affine algebras were introduced by Jimbo (1986a). Little seems to be known about representations of quantum affine algebras at roots of unity, except for the representations obtained from evaluation representations of $U_\epsilon(\mathfrak{g})$ when $\mathfrak{g}$ is of type A. See, however, Tarasov (1992).

For results on 'standard' highest weight representations of quantum affine algebras, see Lusztig (1993). For the construction of such representations using quantum analogues of vertex operators, see Frenkel & Jing (1988) and Jing (1990).

For another realization of quantum affine algebras, analogous to the realization of Yangians in 12.1.4, see Reshetikhin & Semenov-Tian-Shansky (1990).

Readers unfamiliar with classical affine Lie algebras should see Kac (1983) for background information.

12.3 For proofs of the results in Subsections 12.3A and 12.3B, and for further information about affine Hecke algebras, see Zelevinsky (1980) and Rogawski (1985). The relation between $U_\epsilon(sl_{n+1}(\mathbb{C}))$ and $\mathcal{H}_\ell(\epsilon^2)$ is due to Jimbo (1986a). The results in Subsection 12.3D are due to Chari & Pressley (1993b, 1994b). See Cherednik (1987a) for related results. The relation between affine Hecke algebras and Yangians is due to Drinfel'd (1986b).

12.4 The relation between Yangians and infinite-dimensional classical groups is due to Olshanskii (1988, 1989, 1991, 1992b). For other connections between Yangians and classical representation theory, see Nazarov (1991), which gives a Yangian-theoretic interpretation of the classical Capelli identities (see Weyl (1939), Chap 2, §4), and Nazarov (1993b), which relates Yangians and projective representations of symmetric groups.

12.5 Theorems 12.5.1–12.5.3 are due to Drinfel'd (1985, 1987). The computation 12.5.4 is taken from Chari & Pressley (1990b). Theorem 12.5.5 is due to Drinfel'd (1987).

See Levendorskii, Soibelman & Stukopin (1993) for the computation of the pseudo-universal R-matrix for $U_h(\widetilde{sl}_2(\mathbb{C}))$, and Khoroshkin & Tolstoy (1991a, 1992a) for an alternative, and more general, approach. See Tolstoy (1990) for more on normal orderings.

For a list of some solutions of the QYBE, of rational, trigonometric and elliptic type, see Kulish & Sklyanin (1982a). The connection between elliptic solutions of the QYBE and quantum groups is still something of a mystery (see, however, Frenkel & Reshetikhin (1992a)). More recently, solutions $R(u, v)$ of the QYBE have been found for which the spectral parameters are points of an algebraic curve of genus > 1, in contrast to 3.2.4, which allows only rational, trigonometric and elliptic solutions of the CYBE. See, for example, Au-Yang *et al.* (1987). These solutions are related to representations of $U_\epsilon(\mathfrak{g})$ for ϵ a root of unity (see Subsection 10.1C); see Date *et al.* (1990, 1991a,c).

13
Quantum harmonic analysis

In this chapter, we return to the study of quantized algebras of functions begun in Chapter 7. Classically, the assumption that groups are compact leads to a more complete theory than for topological groups in general, one of the main reasons for this being the existence on such groups of a normalized Haar measure. For quantum groups, one must formulate the notions of compactness and Haar measure algebraically, for one has no space on which to put a topology or measure. We saw how to deal with the question of compactness in Subsection 4.1F: the algebra of functions on a real form of a complex simple Lie group G, and a compact real form K in particular, can be characterized as the set of fixed points of a conjugate-linear involution on the algebra $\mathcal{F}(G)$ of representative functions on G. Thus, one may define a compact quantum group to be a quantized algebra of functions equipped with a conjugate-linear involution such that the involution on $\mathcal{F}(G)$ obtained by taking the classical limit corresponds to a compact real form of G. A Haar integral may then be described as a linear functional having certain invariance and positivity properties.

As in previous chapters, we find it more convenient to work with a specialization $\mathcal{F}_\epsilon(G)$, rather than the formal deformation $\mathcal{F}_h(G)$. It follows from the results of Subsection 9.4B that, to be able to define an involution with classical limit a compact real form, one must assume that ϵ is real. Thus, with the possible exception of $\epsilon = -1$, which we shall ignore, there are no 'root of unity phenomena'. We shall prove the existence and uniqueness of Haar functionals on $\mathcal{F}_\epsilon(G)$, and deduce quantum analogues of a number of well-known classical applications of the Haar measure, such as the Schur orthogonality relations for the matrix elements of finite-dimensional irreducible representations of G.

The *-structure also allows one to discuss unitary representations. In the classical theory, there is a (one-dimensional) unitary representation of $\mathcal{F}(G)$ associated to every point of the real form of G corresponding to the involution, given simply by evaluating a function at the point. In the quantum case, one has no 'points', so the study of the unitary representations replaces, to some extent, the study of the space itself. In the case of compact quantum groups, it turns out that the unitary representations of $\mathcal{F}_\epsilon(G)$ are parametrized by pairs (w, t), where w is an element of the Weyl group of G and t is an element of the maximal torus of the compact real form K of G; these representations are one-dimensional when w is the identity element, but infinite-dimensional otherwise. Comparing this with the results of Subsection 1.5C, one sees that there is a one-to-one correspondence between the irreducible unitary representations of $\mathcal{F}_\epsilon(G)$ and the symplectic leaves of the Poisson–Lie group which is the classical limit of $\mathcal{F}_\epsilon(G)$. This provides support for the tentative generalization of the 'orbit principle' of Kirillov and Kostant which we discussed in 6.2.5. However, we shall see that this principle fails when we discuss the 'twisted' Poisson–Lie group structures on G.

Section 13.2 gives a brief discussion of the notion of a quantum G-space. Classically, an action of a group G on a space M is a map $G \times M \to M$ satisfying certain properties; it therefore gives rise to a homomorphism of algebras $\mathcal{F}(M) \to \mathcal{F}(G) \otimes \mathcal{F}(M)$. We thus define a quantum G-space to be an algebra A together with a homomorphism $\alpha : A \to \mathcal{F}_\epsilon(G) \otimes A$; the defining properties of the classical action translate into the requirement that α is an $\mathcal{F}_\epsilon(G)$-comodule algebra structure on A (see Section 4.1). The examples which have been studied most extensively, apart from quantum groups themselves (in which case α is given by the comultiplication of $\mathcal{F}_\epsilon(G)$) are the quantum analogues of odd-dimensional spheres. Classically, such spheres can be realized as the homogeneous spaces SU_{n+1}/SU_n; in the quantum case, they are $*$-subalgebras of $\mathcal{F}_\epsilon(SL_{n+1}(\mathbb{C}))$, the $*$-structure being the one which defines the compact quantum group. The case of even-dimensional spheres is more difficult: classically, they arise as homogeneous spaces SO_{2n+1}/SO_{2n}, but, as we have mentioned earlier in the book, $U_\epsilon(so_{2n}(\mathbb{C}))$ is not in any obvious way a subalgebra of $U_\epsilon(so_{2n+1}(\mathbb{C}))$, so it is not clear how to define the corresponding quantum homogeneous space. However, one can define quantum 2-spheres, essentially because they can be realized as quantum (complex) projective lines.

The elements of $\mathcal{F}_\epsilon(G)$ all act by bounded operators in any unitary representation. This fact allows one to introduce a C*-norm on $\mathcal{F}_\epsilon(G)$. In the classical theory, the Peter–Weyl theorem tells us that the completion of the algebra of representative functions in the corresponding norm is the space of continous functions on G. Thus, we define the quantized algebra of continuous functions to be the completion of $\mathcal{F}_\epsilon(G)$ in its norm. The resulting C*-algebras are examples of the compact matrix quantum groups first studied by Woronowicz, quite independently of the theory of quantized universal enveloping algebras. We present some of his results in Section 13.3.

In Section 13.4, we discuss briefly an example of a non-compact quantum group, i.e. a $*$-structure on $\mathcal{F}_\epsilon(G)$ whose classical limit defines a non-compact real form of G. As one would expect, the definition of the Haar integral on such a quantum group is more delicate than in the compact case, necessitating the definition of a space of quantized functions which vanish sufficiently rapidly at infinity.

The chapter concludes with a discussion of the relation between quantum groups and the theory of q-special functions. The classical theory of special functions originated in the work of Euler and the Bernoullis in the early 18th century, but its connection with representation theory was not noticed until the 1930s, by Elie Cartan. The theory of q-special functions is more recent, dating back to the introduction of q-hypergeometric series by Erich Heine in 1847. Although some isolated results had been obtained a little earlier, the proper representation-theoretic setting for q-special functions was found only in 1986, when L. Vaksman and Ya. Soibelman observed that the matrix elements of finite-dimensional *-representations of $U_q(sl_2)$ could be expressed in terms of little q-Jacobi polynomials. (So one could say that, after more than 100 years of work on q-special functions, it has finally been discovered what the 'q' stands for!). More recently, several other families of q-special functions have found interpretations in terms of the representation theory of quantum groups, and we outline a few of these in Section 13.5.

13.1 Compact quantum groups and their representations

In this section, $\mathfrak{g}$ denotes a finite-dimensional complex simple Lie algebra and, except in Subsection 13.1F, ϵ is a positive real number. All Hopf algebras and their representations are defined over $\mathbb{C}$. We shall often denote $U_\epsilon(\mathfrak{g})$ by U_ϵ provided this is unambiguous.

A Definitions The quantized algebra of functions $\mathcal{F}_h(G)$ on the connected, simply-connected, complex Lie group G with Lie algebra $\mathfrak{g}$ was defined in 7.3.1 as the dual of the formal deformation $U_h(\mathfrak{g})$ of the universal enveloping algebra $U(\mathfrak{g})$. By analogy, we make

DEFINITION 13.1.1 *The quantized function algebra $\mathcal{F}_\epsilon(G)$ is the subalgebra of the Hopf algebra dual of $U_\epsilon(\mathfrak{g})$ generated by the matrix elements of the finite-dimensional $U_\epsilon(\mathfrak{g})$-modules of type* **1**.

We shall denote $\mathcal{F}_\epsilon(G)$ simply by $\mathcal{F}_\epsilon$ when the group G is understood.

Remarks [1] If V is a finite-dimensional $U_\epsilon(\mathfrak{g})$-module, and $\{v_0, v_1, \ldots, v_k\}$ is a basis of V, the corresponding matrix elements are the functions C^V_{rs}, $0 \leq r, s \leq k$, where

$$x.v_s = \sum_{r=0}^{k} C^V_{rs}(x) v_r \qquad (x \in U_\epsilon(\mathfrak{g})).$$

The C^V_{rs}, $0 \leq r, s \leq k$, depend on the choice of basis, but the subspace they span does not.

[2] Since the trivial representation is of type **1**, and tensor products and duals of representations of type **1** are of type **1**, it follows that $\mathcal{F}_\epsilon$ is actually a Hopf subalgebra of the Hopf dual U_ϵ°.

[3] If we dropped the restriction to type **1** representations, the resulting dual Hopf algebra would be generated by $\mathcal{F}_\epsilon$ together with n group-like elements corresponding to the one-dimensional representations $X_i^\pm \mapsto 0$, $K_j \mapsto 1$ if $j \neq i$, $K_i \mapsto -1$, where $i = 1, \ldots, n$ (cf. Takeuchi (1992a)).

[4] There is an isomorphism of Hopf algebras $\mathcal{F}_\epsilon \cong \mathcal{F}_{\epsilon^{-1}}$. This is because there is an isomorphism of Hopf algebras $U_\epsilon \cong U_{\epsilon^{-1}}$, namely

$$K_i \mapsto K_i, \quad X_i^+ \mapsto K_i X_i^-, \quad X_i^- \mapsto X_i^+ K_i^{-1}.$$

This is easily checked directly using the relations in 9.1.1. Thus, we shall, without loss of generality, *assume throughout this section that $\epsilon > 1$*, except in Subsection 13.1F. $\triangle$

As we saw in Subsection 4.1F, the appropriate language for discussing the algebra of (complex-valued) functions on a real form of G, or the enveloping algebra of a real form of $\mathfrak{g}$, is that of Hopf $*$-structures. Throughout this section, we shall use the standard structure

$$(1) \qquad K_i^* = K_i, \quad (X_i^+)^* = \epsilon_i^{-1} X_i^- K_i, \quad (X_i^-)^* = \epsilon_i K_i^{-1} X_i^+$$

on U_ϵ. (This is slightly different from, although equivalent to, the standard structure used in Subsection 10.1E. The normalization in (1) is convenient in this chapter.) The classical limit of $*$ is given by

$$\bar{H}_i^* = \bar{H}_i, \quad (\bar{X}_i^\pm)^* = \bar{X}_i^\mp,$$

the bars denoting elements of $\mathfrak{g}$. The corresponding real form

$$\{x \in \mathfrak{g} \mid x^* = -x\}$$

is the compact real form $\mathfrak{k}$ of $\mathfrak{g}$, so-called because the Lie subgroup K of G with Lie algebra $\mathfrak{k}$ is compact.

By 13.1.1 and Subsection 4.1F, (1) induces a Hopf $*$-structure, also denoted by $*$, on $\mathcal{F}_\epsilon$.

DEFINITION 13.1.2 *The pair $(\mathcal{F}_\epsilon(G), *)$ is the* quantized algebra of functions on the compact real form K of G.

Remark In the literature, the pair $(\mathcal{F}_\epsilon(G), *)$ is often denoted by $\mathcal{F}_\epsilon(K)$, but we find this abuse of notation confusing, and we shall avoid it. $\triangle$

As usual, for any $\lambda \in P^+$, we denote by $V_\epsilon(\lambda)$ the irreducible U_ϵ-module of type **1** with highest weight λ. Recall from 10.1.21 that $V_\epsilon(\lambda)$ is unitarizable, i.e. it admits a positive-definite hermitian form (,) such that

$$(x.v_1, v_2) = (v_1, x^*.v_2)$$

for all v_1, $v_2 \in V_\epsilon(\lambda)$, $x \in U_\epsilon$ (the algebra denoted by $U_\epsilon^{\mathbb{C}}$ there is denoted by U_ϵ here). Fix an orthonormal basis $\{v_\mu^r\}$ of each weight space $V_\epsilon(\lambda)_\mu$, $\mu \in P$; their union is then an orthonormal basis of $V_\epsilon(\lambda)$. Let

$$C_{\nu,s;\mu,r}^\lambda(x) = (x.v_\mu^r, v_\nu^s)$$

be the associated matrix elements of $V_\epsilon(\lambda)$. To simplify the notation, we omit the subscript r (resp. s) if $\dim(V_\epsilon(\lambda)_\mu) = 1$ (resp. if $\dim(V_\epsilon(\lambda)_\nu) = 1$).

PROPOSITION 13.1.3 *(i) The matrix elements* $C_{\nu,s;\mu,r}^\lambda$, *where* λ *runs through* P^+ *and* (μ, r) *and* (ν, s) *run independently through the index set of a basis of* $V_\epsilon(\lambda)$, *form a vector space basis of* $\mathcal{F}_\epsilon(G)$.

(ii) The action of the involution $*$ *of* $\mathcal{F}_\epsilon(G)$ *in this basis is given by*

$$(C_{\nu,s;\mu,r}^\lambda)^* = \epsilon^{(\mu-\nu,\rho)} C_{-\nu,s;-\mu,r}^{-w_0\lambda}.$$

(iii) The comultiplication and counit of $\mathcal{F}_\epsilon(G)$ *are given by*

$$\Delta_{\mathcal{F}_\epsilon}(C_{\nu,s;\mu,r}^\lambda) = \sum_{\pi,p} C_{\nu,s;\pi,p}^\lambda \otimes C_{\pi,p;\mu,r}^\lambda,$$

$$\epsilon_{\mathcal{F}_\epsilon}(C_{\nu,s;\mu,r}^\lambda) = \delta_{\mu,\nu}\delta_{r,s},$$

respectively.

OUTLINE OF PROOF It is clear that $\mathcal{F}_\epsilon(G)$ is spanned as a vector space by the $C_{\nu,s;\mu,r}^\lambda$. Indeed, this follows from the complete reducibility of finite-dimensional $U_\epsilon(\mathfrak{g})$-modules proved in 10.1.14, and the fact that the product of two such matrix elements is a matrix element of a tensor product of two $U_\epsilon(\mathfrak{g})$-modules.

Now let $\lambda^1, \ldots, \lambda^k$ be distinct elements of P^+. By Wedderburn's theorem, the homomorphism

$$U_\epsilon \to \mathrm{End}(V_\epsilon(\lambda^1)) \times \cdots \times \mathrm{End}(V_\epsilon(\lambda^k)),$$

given by the action of U_ϵ, is surjective. This means that the space spanned by the matrix elements $C_{\nu,s;\mu,r}^{\lambda^j}$ of $V_\epsilon(\lambda^j)$, for $j = 1, \ldots, k$, has dimension $\prod_{j=1}^k \dim(V_\epsilon(\lambda^j))^2$. Since this is the number of matrix elements in question, they must be linearly independent.

Part (ii) follows by a straightforward calculation, using the isomorphisms of U_ϵ-modules $V_\epsilon(\lambda)^* \cong V_\epsilon(-w_0\lambda)$, and the fact that the square of the antipode of U_ϵ is given by conjugation by K_{ρ}. (see Subsections 8.3F and 11.3B).

Part (iii) is obvious. ∎

Remark The antipode $S_{\mathcal{F}_\epsilon}$ of $\mathcal{F}_\epsilon$ is computed in the proof of 13.3.1. △

A representation of $\mathcal{F}_\epsilon$ on a complex vector space V is said to be *unitarizable* if V admits a positive-definite hermitian form $(\ ,\ )$ such that

$$(f.v_1\,,\,v_2) = (v_1\,,\,f^*.v_2)$$

for all $f \in \mathcal{F}_\epsilon$, v_1, $v_2 \in V$. If V is complete in the inner product $(\ ,\ )$ and $\mathcal{F}_\epsilon$ acts by bounded operators, the representation is said to be *unitary*, or a $*$-*representation*. Two unitarizable (or unitary) representations V and V' are *equivalent* if there is an invertible operator $T : V \to V'$ which commutes with the action of $\mathcal{F}_\epsilon$ and is unitary:

$$(T(v_1)\,,\,T(v_2))_{V'} = (v_1\,,\,v_2)_V, \qquad (v_1, v_2 \in V).$$

The following result follows from the classification of the unitarizable representations of $\mathcal{F}_\epsilon$ described below.

PROPOSITION 13.1.4 *Every irreducible unitarizable representation V of $\mathcal{F}_\epsilon(G)$ extends to a unitary representation on the completion of V. Moreover, this sets up a one-to-one correspondence between the irreducible unitary representations of $\mathcal{F}_\epsilon(G)$ on separable Hilbert spaces, and its irreducible unitarizable representations.* ∎

B Highest weight representations We are going to describe a parametrization of the irreducible unitarizable representations of $\mathcal{F}_\epsilon(G)$ by highest weights. An analogous description of the representations of $U_\epsilon(\mathfrak{g})$ was given in Chapters 10 and 11, but the $\mathcal{F}_\epsilon$-theory presents a number of interesting differences. For example, we shall see that there are several subalgebras of $\mathcal{F}_\epsilon$ which have the right to be called Cartan subalgebras, but they are not equivalent, and different representations must in general be described by characters of different Cartan subalgebras.

Let $\mathcal{F}_\epsilon^+$ be the subalgebra of $\mathcal{F}_\epsilon$ generated by the matrix elements $C_{\mu,s;\lambda}^\lambda$ for all λ, μ, s, and let $\mathcal{F}_\epsilon^- = (\mathcal{F}_\epsilon^+)^*$. We shall also need the subalgebra $\mathcal{F}_\epsilon^{++} \subset \mathcal{F}_\epsilon^+$ obtained by restricting λ to the set P^{++} of regular dominant weights (i.e. those $\lambda \in P^+$ such that $\lambda(H_i) > 0$ for all $i = 1, \ldots, n$); we set $\mathcal{F}_\epsilon^{--} = (\mathcal{F}_\epsilon^{++})^* \subset \mathcal{F}_\epsilon^-$. The following result is a kind of Poincaré–Birkhoff–Witt factorization.

LEMMA 13.1.5 *We have $\mathcal{F}_\epsilon = \mathcal{F}_\epsilon^{--}.\mathcal{F}_\epsilon^{++}$. A fortiori, the same result holds if $\mathcal{F}_\epsilon^{++}$ and $\mathcal{F}_\epsilon^{--}$ are replaced by $\mathcal{F}_\epsilon^+$ and $\mathcal{F}_\epsilon^-$.* ∎

We shall see that $\mathcal{F}_\epsilon^+$ plays a role in the representation theory of $\mathcal{F}_\epsilon$ analogous to that played by the 'Borel subalgebra' $U_\epsilon^{\geq 0}$ of U_ϵ (cf. Subsection 9.1B). In particular, we make

DEFINITION 13.1.6 *A representation V of $\mathcal{F}_\epsilon(G)$ is* highest weight *if there exists a vector $v \in V$ such that*

(i) $V = \mathcal{F}_\epsilon(G).v$;
(ii) $\mathcal{F}_\epsilon^+(G)$ preserves the line $\mathbb{C}.v$ in V.

The homomorphism $\chi : \mathcal{F}_\epsilon^+(G) \to \mathbb{C}$ such that $f.v = \chi(f)v$ (which exists by (ii)) is called the highest weight *of V.*

The main result of this subsection is

THEOREM 13.1.7 *Every irreducible unitary representation of $\mathcal{F}_\epsilon(G)$ on a separable Hilbert space is the completion of a unitarizable highest weight representation. Moreover, two such representations are equivalent if and only if they have the same highest weight.*

OUTLINE OF PROOF Let $\rho : \mathcal{F}_\epsilon(G) \to \mathrm{End}(V)$ be an irreducible $*$-representation (assumed to be non-trivial). By 13.1.5, there exists $\lambda \in P^{++}$ such that $\rho(C_{\mu,s;\lambda}^\lambda) \neq 0$ for some μ, s. Choose a μ with this property which is minimal for the usual partial ordering on P (it will turn out that this μ is unique). The first step is to prove the following relation in $\mathcal{F}_\epsilon$:

$$(C_{\mu,s;\lambda}^\lambda)^* C_{\mu,s;\lambda}^\lambda = \epsilon^{-(\lambda,\lambda)+(\mu,\mu)} C_{\mu,s;\lambda}^\lambda (C_{\mu,s;\lambda}^\lambda)^*.$$

This is done by analysing the isomorphism from $V_\epsilon(\lambda) \otimes V_\epsilon(\lambda)$ to itself given by the universal R-matrix of $U_h(\mathfrak{g})$. (One has to be a little careful here, since this universal R-matrix cannot be specialized to give a well-defined element of $U_\epsilon(\mathfrak{g}) \otimes U_\epsilon(\mathfrak{g})$, but this problem can be overcome by using the results in Subsection 10.1D.)

Since $\epsilon > 1$ and $\rho(C_{\mu,s;\lambda}^\lambda)$ is bounded, it follows that $(\lambda, \lambda) = (\mu, \mu)$ and that $\rho(C_{\mu,s;\lambda}^\lambda)$ is a normal operator. Since $\lambda \in P^{++}$, the first equality implies that $\mu = w\lambda$ for some $w \in W$, the Weyl group of $\mathfrak{g}$ (in particular, μ is a weight of multiplicity one).

The next step is to study the spectrum Σ of the operator $\rho(C_{w\lambda;\lambda}^\lambda)$. One shows that it is contained in the unit disc $\{z \in \mathbb{C} \mid |z| \leq 1\}$, that it intersects the unit circle $|z| = 1$ in a unique point, say z_0, and that 0 is its only possible limit point. Thus, $\Sigma \backslash \{0\}$ is contained in the discrete spectrum, and in fact is equal to it since one can show that $\ker \rho(C_{w\lambda;\lambda}^\lambda) = 0$. Thus, we have

$$V = \bigoplus_{z \in \Sigma \backslash \{0\}} V_z,$$

where $V_z = \{v \in V \mid \rho(C_{w\lambda;\lambda}^\lambda)(v) = zv\}$. Although this decomposition depends on λ and w, it turns out that V_{z_0} is actually independent of λ. Moreover, it is one-dimensional and is preserved by $\mathcal{F}_\epsilon^+$.

This completes our sketch proof of the first part of the theorem. The uniqueness statement is proved by constructing, for each character $\chi : \mathcal{F}_\epsilon^+ \to \mathbb{C}$, a 'Verma module' $M_\epsilon(\chi)$ with highest weight χ, and proving as usual that $M_\epsilon(\chi)$ has a unique irreducible quotient. ∎

The complete details of the proof of 13.1.7 may be found in Vaksman & Soibelman (1988) in the SU_2 case, Vaksman & Soibelman (1991) in the SU_{n+1} case, and Levendorskii & Soibelman (1991a) and Soibelman (1991a) in the general case.

In the course of a complete proof, one obtains a more precise result. For any $w \in W$, $\lambda \in P^+$, let $\mathcal{J}_\epsilon(w, \lambda)$ be the two-sided ∗-ideal in $\mathcal{F}_\epsilon$ generated by the matrix elements $C^\lambda_{\mu,s;\lambda}$ such that $v^s_\mu \notin U^{\geq 0}_\epsilon . V_\epsilon(\lambda)_{w\lambda}$. Then, for any irreducible ∗-representation $\rho : \mathcal{F}_\epsilon \to \text{End}(V)$, there exists $w \in W$ (possibly depending on ρ) such that, for all $\lambda \in P^+$:

(i) $\rho(\mathcal{J}_\epsilon(w, \lambda)) = 0$;
(ii) $\rho(C^\lambda_{w\lambda;\lambda}) \neq 0$.

The subalgebra $\mathcal{F}_{\epsilon,w}$ of $\mathcal{F}_\epsilon$ generated by the $C^\lambda_{w\lambda;\lambda}$ for $\lambda \in P^+$ plays the role of a Cartan subalgebra of $\mathcal{F}_\epsilon$, for it can be shown that $\rho(\mathcal{F}_{\epsilon,w})$ is a commutative subalgebra of $\text{End}(V)$. As we mentioned above, different Cartan subalgebras may thus be required to describe different representations.

When (i) and (ii) hold for the Weyl group element w, Levendorskii, Soibelman and Vaksman say that ρ *corresponds to the Schubert cell* $S_w = B_+wB_+/B_+$ in G/B_+ (cf. Subsection 1.5C). The reason for this terminology is discussed in Subsection 13.2B.

In view of the last part of the proof, it seems natural to try to parametrize the irreducible ∗-representations of $\mathcal{F}_\epsilon$ by determining when the irreducible quotient of the Verma module $M_\epsilon(\chi)$ is unitarizable. This seems to be rather difficult, however, and we shall take a different approach, beginning with the sl_2 case.

C The sl_2 case The algebra $\mathcal{F}_\epsilon(SL_2(\mathbb{C}))$ is generated by the matrix elements

$$\begin{pmatrix} a & b \\ c & d \end{pmatrix}$$

of the two-dimensional type **1** representation $V^\natural_\epsilon$ of $U_\epsilon(sl_2(\mathbb{C}))$ with basis $\{v_0, v_1\}$ given by

$$K_1 \mapsto \begin{pmatrix} \epsilon & 0 \\ 0 & \epsilon^{-1} \end{pmatrix}, \quad X^+_1 \mapsto \begin{pmatrix} 0 & 1 \\ 0 & 0 \end{pmatrix}, \quad X^-_1 \mapsto \begin{pmatrix} 0 & 0 \\ 1 & 0 \end{pmatrix}.$$

As in 7.1.4, one shows that the defining relations of $\mathcal{F}_\epsilon(SL_2(\mathbb{C}))$ are

$$ab = \epsilon^{-1}ba, \quad bd = \epsilon^{-1}db, \quad ac = \epsilon^{-1}ca, \quad cd = \epsilon^{-1}dc,$$
(2)
$$bc = cb, \quad ad - da + (\epsilon - \epsilon^{-1})bc = 0,$$
$$ad - \epsilon^{-1}bc = 1,$$

and its coalgebra structure

$$\Delta_{\mathcal{F}_\epsilon}\begin{pmatrix} a & b \\ c & d \end{pmatrix} = \begin{pmatrix} a & b \\ c & d \end{pmatrix} \square \begin{pmatrix} a & b \\ c & d \end{pmatrix},$$

(3)

$$S_{\mathcal{F}_\epsilon}\begin{pmatrix} a & b \\ c & d \end{pmatrix} = \begin{pmatrix} d & -\epsilon b \\ -\epsilon^{-1}c & a \end{pmatrix}, \quad \epsilon_{\mathcal{F}_\epsilon}\begin{pmatrix} a & b \\ c & d \end{pmatrix} = \begin{pmatrix} 1 & 0 \\ 0 & 1 \end{pmatrix}.$$

The $*$-structure on $\mathcal{F}_\epsilon$ dual to the standard structure on U_ϵ is:

(4) $a^* = d, \quad b^* = -\epsilon^{-1}c.$

To prove the second formula, for example, one checks, as in Subsection 7.1D, that

$$\langle b, (X_1^-)^r K_1^s (X_1^+)^t \rangle = \delta_{r,0} \delta_{t,1} \epsilon^s.$$

Hence,

$$\langle b^*, (X_1^-)^r K_1^s (X_1^+)^t \rangle = \overline{\langle b, (S_{\mathcal{F}_\epsilon}((X_1^-)^r K_1^s (X_1^+)^t))^* \rangle}$$
$$= \overline{\langle b, ((-X_1^+ K_1^{-1})^t K_1^{-s}(-K_1 X_1^-)^r))^* \rangle}$$
$$= (-1)^{r+t} \langle b, (\epsilon^{-1} X_1^+)^r K_1^{-s} (\epsilon X_1^-)^t \rangle.$$

By the method used in Subsection 7.1D, this is equal to

$$(-1)^{r+t} \epsilon^{t-r} . \epsilon^s \delta_{r,1} \delta_{t,0} = -\delta_{r,1} \delta_{t,0} \epsilon^{s-1}.$$

On the other hand,

$$\langle c, (X_1^-)^r K_1^s (X_1^+)^t \rangle = \delta_{r,1} \delta_{t,0} \epsilon^s.$$

The equality $b^* = -\epsilon^{-1}c$ follows.

It follows immediately from (4) and the last relation in (2) that

(5) $aa^* + bb^* = 1.$

Thus, it is clear in this case that every unitarizable representation ρ of $\mathcal{F}_\epsilon(SL_2(\mathbb{C}))$ extends to a representation by *bounded* operators on the completion, for (5) obviously implies that $\| \rho(a) \| \leq 1$ and $\| \rho(b) \| \leq 1$, and then (4) shows that $\rho(c)$ and $\rho(d)$ are bounded too. Equations (4) and (5) also imply that the spectra of $\rho(a)$ and $\rho(b)$ lie inside the unit disc (cf. the proof of 13.1.7).

If λ denotes the highest weight of $V_\epsilon^\natural$, it is easy to check that the matrix elements $C_{\lambda;\lambda}^\lambda = a$ and $C_{-\lambda;\lambda}^\lambda = c$. Further, since every finite-dimensional irreducible $U_\epsilon(sl_2(\mathbb{C}))$-module of type $\mathbf{1}$ is a direct summand of a tensor product of copies of $V_\epsilon^\natural$ (see 10.1.16), $\mathcal{F}_\epsilon^+$ is the subalgebra of $\mathcal{F}_\epsilon$ generated by a and c. Note that, by (2) and (4),

(6) $ac = \epsilon^{-1}ca, \quad a^*a + c^*c = 1.$

It is now straightforward to prove

PROPOSITION 13.1.8 *Every irreducible $*$-representation of $\mathcal{F}_\epsilon(SL_2(\mathbb{C}))$ is equivalent to a representation belonging to one of the following families, each of which is parametrized by $S^1 = \{t \in \mathbb{C} \mid |t| = 1\}$:*

(i) the family of one-dimensional representations τ_t given by

$$\tau_t(a) = t, \quad \tau_t(b) = \tau_t(c) = 0, \quad \tau_t(d) = t^{-1};$$

(ii) the family π_t of representations on $\ell^2(\mathbb{N})$ given by

$$\pi_t(a)(e_k) = (1 - \epsilon^{-2k})^{1/2}e_{k-1}, \quad \pi_t(b)(e_k) = -\epsilon^{-k-1}t^{-1}e_k,$$
$$\pi_t(c)(e_k) = \epsilon^{-k}te_k, \quad \pi_t(d)(e_k) = (1 - \epsilon^{-2k-2})^{1/2}e_{k+1}.$$

Here, $\{e_k\}_{k\in\mathbb{N}}$ is the natural orthonormal basis of $\ell^2(\mathbb{N})$, in which e_k is the vector with one in the kth position and zeros elsewhere (and $e_{-1} = 0$).

These representations are all mutually inequivalent.

PROOF It is easy to check that these formulas do define $*$-representations of $\mathcal{F}_\epsilon(SL_2(\mathbb{C}))$. The highest weight of τ_t is obviously given by

$$\chi(a) = t, \quad \chi(c) = 0.$$

For π_t, $\mathbb{C}.e_0$ is the unique line preserved by $\mathcal{F}_\epsilon^+$, so its highest weight is given by

$$\chi(a) = 0, \quad \chi(c) = t.$$

In view of 13.1.7, it is enough to prove that these are the only characters χ of $\mathcal{F}_\epsilon^+$ which arise from $*$-representations.

From the first equation in (6), it is clear that $\chi(a) = 0$ or $\chi(c) = 0$. If $\chi(c) = 0$ and $\chi(a) = \alpha \neq 0$, say, there exists a vector v such that $a.v = \alpha v$, $c.v = 0$. The second equation in (6) now implies that $|\alpha| = 1$. Similarly, if $\chi(a) = 0$, then $|\chi(c)| = 1$. ∎

This result should be compared with the description in Subsection 1.5C of the symplectic leaves of the Poisson–Lie group structure on SU_2 which is the classical limit of $U_\epsilon(sl_2(\mathbb{C}))$. Both the symplectic leaves of SU_2 and the irreducible $*$-representations of $\mathcal{F}_\epsilon(SL_2(\mathbb{C}))$ are parametrized by the disjoint union of two circles. We shall now extend this correspondence to $\mathcal{F}_\epsilon(G)$ for arbitrary G.

D The general case: tensor products The embedding $U_{\epsilon_i}(sl_2(\mathbb{C})) \rightarrow U_\epsilon(\mathfrak{g})$ given by $K_1 \mapsto K_i$, $X^\pm \rightarrow X_i^\pm$ induces, by passing to the dual, a surjective homomorphism $\mathcal{F}_\epsilon(G) \rightarrow \mathcal{F}_{\epsilon_i}(SL_2(\mathbb{C}))$. Composing the representation π_{-1} of $\mathcal{F}_{\epsilon_i}(SL_2(\mathbb{C}))$ with this homomorphism gives a representation of $\mathcal{F}_\epsilon(G)$ on $\ell^2(\mathbb{N})$, which, for reasons which will soon become clear, we shall denote by π_{s_i}, where s_i is the ith fundamental reflection in the Weyl group W of G.

On the other hand, there is a family of one-dimensional $*$-representations τ_t of $\mathcal{F}_\epsilon(G)$ parametrized by the elements t of the maximal torus T of K (we may assume that the complexification of the Lie algebra $\mathfrak{t}$ of T is our chosen Cartan subalgebra of $\mathfrak{g}$). In fact, if $t = \exp(2\pi\sqrt{-1}x)$ for some $x \in \mathfrak{t}$,

$$\tau_t(C^\lambda_{\nu,s;\mu,r}) = \delta_{r,s}\delta_{\mu,\nu}\exp(2\pi\sqrt{-1}\mu(x)).$$

It is clear that the representations τ_t are associated to the zero-dimensional Schubert cell S_e, where e is the identity element of the Weyl group W of $\mathfrak{g}$.

THEOREM 13.1.9 *Let* $w = s_{i_1}s_{i_2}\dots s_{i_k}$ *be a reduced decomposition of an element* w *of the Weyl group* W *of* G. *Then*,

(i) the Hilbert space tensor product

$$\rho_{w,t} = \pi_{s_{i_1}} \otimes \pi_{s_{i_2}} \otimes \cdots \otimes \pi_{s_{i_k}} \otimes \tau_t$$

is an irreducible $*$-representation of $\mathcal{F}_\epsilon(G)$ *which is associated to the Schubert cell* S_w;
(ii) up to equivalence, the representation $\rho_{w,t}$ *does not depend on the choice of the reduced decomposition of* w;
(iii) every irreducible $*$-representation of $\mathcal{F}_\epsilon(G)$ *is equivalent to some* $\rho_{w,t}$. ∎

Although we shall not prove this result, we comment briefly on part (ii). In view of a classical result which we have already used in Subsection 8.1A (see Bourbaki (1968), Chap. 4, § 1.5, Proposition 5), it is enough to prove

PROPOSITION 13.1.10 *Let* (a_{ij}) *be the Cartan matrix of* $\mathfrak{g}$. *Then:*

$$\pi_{s_i} \otimes \pi_{s_j} \cong \pi_{s_j} \otimes \pi_{s_i} \quad \text{if } a_{ij} = 0,$$
$$\pi_{s_i} \otimes \pi_{s_j} \otimes \pi_{s_i} \cong \pi_{s_j} \otimes \pi_{s_i} \otimes \pi_{s_j} \quad \text{if } a_{ij}a_{ji} = 1,$$
$$(\pi_{s_i} \otimes \pi_{s_j})^{\otimes 2} \cong (\pi_{s_j} \otimes \pi_{s_i})^{\otimes 2} \quad \text{if } a_{ij}a_{ji} = 2,$$
$$(\pi_{s_i} \otimes \pi_{s_j})^{\otimes 3} \cong (\pi_{s_j} \otimes \pi_{s_i})^{\otimes 3} \quad \text{if } a_{ij}a_{ji} = 3.$$

(All tensor products are Hilbert space tensor products.) ∎

This can be proved by directly computing the highest weights of the representations involved and appealing to 13.1.7.

Remarks [1] Using 13.1.3, 13.1.8 and 13.1.9, one can show that the intersection of the kernels of the representations $\rho_{w,t}$ is trivial, which implies that their direct integral

$$\bigoplus_{w \in W} \int_T^\oplus \rho_{w,t}\, dt$$

is a faithful representation.

[2] Theorem 13.1.9 is the starting point for another approach to the quantum Weyl group, which we discussed in Section 8.2. In fact, let $\tilde{w}_i$ be the Gel'fand–Naimark–Segal state

$$\tilde{w}_i(f) = (\pi_{s_i}(f)(v_{s_i}), v_{s_i}),$$

where $f \in \mathcal{F}_\epsilon$, v_{s_i} is a normalized highest weight vector of π_{s_i}, and $(\ ,\ )$ is the inner product on π_{s_i}. Then, 13.1.9 immediately implies that the $\tilde{w}_i$ satisfy the braid relations (i)–(iv) in 8.1.1. As in Subsection 8.2C, one proves that conjugation by the $\tilde{w}_i$ agrees, at least up to a scalar multiple, with Lusztig's automorphisms T_i; this gives another proof of 8.1.2.

[3] We saw in 1.5.7 that the set $W \times T$ which, by 13.1.9, parametrizes the irreducible $*$-representations of $\mathcal{F}_\epsilon(G)$, also parametrizes the set of symplectic leaves of the Poisson–Lie group structure on the maximal compact subgroup K of G. In fact, comparing 13.1.9(i) with 1.5.8, it appears that multiplication of symplectic leaves 'corresponds' to taking tensor products of representations. One might view this as support for a general principle, according to which the symplectic leaves of a Poisson–Lie group K (not necessarily compact) should 'correspond' to the irreducible representations of a quantization of the algebra of functions on K. We explained in 6.2.5 that this would be a generalization of the 'orbit principle' of Kirillov and Kostant, according to which the unitary representations of a group should 'correspond' to its coadjoint orbits. Just as the Kirillov–Kostant principle, while very successful in some situations, fails in others, so we shall now see a situation where its putative generalization also fails. $\triangle$

E The twisted case and quantum tori We saw in 1.5.10 that the most general Lie bialgebra structure on the Lie algebra $\mathfrak{k}$ of a compact simple Lie group K is of the form

$$\delta_{a,u}(x) = a\delta_{\mathbf{R}}(x) + (\mathrm{ad}_x \otimes \mathrm{id} + \mathrm{id} \otimes \mathrm{ad}_x)(u),$$

where $a \in \mathbf{R}$, $u \in \wedge^2(\mathfrak{t})$, and $\delta_{\mathbf{R}}$ is the standard Lie bialgebra structure on $\mathfrak{k}$, a quantization of which we have used above. We now consider briefly the corresponding theory for this most general Lie bialgebra structure.

We suppose first that $a \neq 0$ and $u \neq 0$; in this case, we may as well assume that $a = 1$. A quantization of the Lie bialgebra $(\mathfrak{g}, \delta_{1,u})$, where $\delta_{1,u}$ is extended complex-linearly to $\mathfrak{g}$, can be obtained by the twisting construction of Subsection 4.2E. In fact, the desired quantization $U_h^{1,u}(\mathfrak{g})$ is identical to the standard quantization $U_h(\mathfrak{g})$ as an algebra over $\mathbf{C}[[h]]$, but its comultiplication is given in terms of that of $U_h(\mathfrak{g})$ by

$$(7) \qquad \Delta_h^{1,u}(a) = e^{-hu/2}\Delta_h(a)e^{hu/2}.$$

We shall not need the formula for the antipode, but it can be found in 4.2.13. The usual formulas (1) define a $*$-structure on $U_h^{1,u}(\mathfrak{g})$, if one uses the conjugation on $\mathbb{C}[[h]]$ given by $ch^k \mapsto \bar{c}h^k$ for $c \in \mathbb{C}$, $k \in \mathbb{N}$. One is then forced to take $H_i^* = H_i$, and hence $u^* = u$, so that $\Delta_h^{1,u}$ is a $*$-homomorphism.

Unfortunately, it is not immediately clear how to specialize this twisted quantization, since the expression $\epsilon^{u/2}$ is meaningless. However, as we have seen in Section 10.1, the whole of the representation theory of $U_\epsilon(\mathfrak{g})$ can be carried through for $U_h(\mathfrak{g})$ with only notational changes. For example, as we have already seen in the sl_2 case (see Subsection 6.4E), there is a one-to-one correspondence between irreducible type $\mathbf{1}$ representations of U_ϵ and the indecomposable representation of U_h. Further, the indecomposable U_h-module $V_h(\lambda)$ with highest weight $\lambda \in P^+$, which is free and of finite rank as a $\mathbb{C}[[h]]$-module, has an invariant hermitian form (with values in $\mathbb{C}[[h]]$), which induces an isomorphism of $\mathbb{C}[[h]]$-modules $V_h(\lambda) \to \bar{V}_h(\lambda)^*$, where $\bar{V}_h(\lambda)^*$ denotes the dual of V_h with the conjugate $\mathbb{C}[[h]]$-module structure. Thus, if $\{v_\mu^r\}$ is a basis of $V_h(\lambda)$, and $\{\xi_\mu^r\}$ is the dual basis of $V_h(\lambda)^*$, we can define the matrix elements

$$C_{\nu,s;\mu,r}^\lambda(x) = \xi_\nu^s(x.v_\mu^r)$$

as before.

Since $U_h^{1,u}(\mathfrak{g})$ is the same as $U_h(\mathfrak{g})$ as an algebra, it has the same set of indecomposable representations, and hence one can define matrix elements for it with the same names as those of $U_h(\mathfrak{g})$. However, because the coalgebra structure of $U_h^{1,u}$ is different from that of U_h, the algebra structure on the space of such matrix elements is different in the twisted case: from (7), it follows that

$$C_{\nu,s;\mu,r}^\lambda \bullet_u C_{\nu',s';\mu',r'}^{\lambda'} = \exp\left(\frac{h}{2}\langle u, \nu \otimes \nu' - \mu \otimes \mu'\rangle\right) C_{\nu,s;\mu,r}^\lambda \bullet C_{\nu',s';\mu',r'}^{\lambda'},$$

where the $\bullet$ on the right-hand side denotes the multiplication in the untwisted case. It is now clear how to define the twisted function algebra $\mathcal{F}_\epsilon^{1,u}(G)$ (over $\mathbb{C}$):

DEFINITION 13.1.11 Let $u \in \wedge^2(\mathfrak{t})$. The twisted function algebra $\mathcal{F}_\epsilon^{1,u}(G)$ is the algebra over $\mathbb{C}$ spanned by the matrix elements $C_{\nu,s;\mu,r}^\lambda$, with multiplication given by

$$(8) \qquad C_{\nu,s;\mu,r}^\lambda \bullet_u C_{\nu',s';\mu',r'}^{\lambda'} = \epsilon^{\frac{1}{2}\langle u, \nu\otimes\nu' - \mu\otimes\mu'\rangle} C_{\nu,s;\mu,r}^\lambda \bullet C_{\nu',s';\mu',r'}^{\lambda'},$$

where the $\bullet$ on the right-hand side is the multiplication in the untwisted algebra $\mathcal{F}_\epsilon(G)$.

It is clear that $\mathcal{F}_\epsilon^{1,u}$ inherits a Hopf $*$-structure from that on $U_h^{1,u}(\mathfrak{g})$. Note that, since $\epsilon \geq 1$, the scalar on the right-hand side of (8) makes sense even if the exponent is complex.

One can now define the notion of a representation of $\mathcal{F}_\epsilon^{1,u}$ corresponding to a Schubert cell S_w exactly as in the untwisted case, and, as in that case, we have

PROPOSITION 13.1.12 *Every irreducible $*$-representation of $\mathcal{F}_\epsilon^{1,u}$ corresponds to a unique Schubert cell.* ∎

The 'Cartan subalgebras' $\mathcal{F}_{\epsilon,w}^{1,u}$ of $\mathcal{F}_\epsilon^{1,u}$ are defined in exactly the same way as the $\mathcal{F}_{\epsilon,w}$, but in the twisted case their images under a $*$-representation are no longer commutative in general.

LEMMA 13.1.13 *The following relations hold in $\mathcal{F}_{\epsilon,w}^{1,u}$:*

$$C_{w\lambda;\lambda}^\lambda \bullet_u C_{w\mu;\mu}^\mu = \sigma^u(\lambda,\mu) C_{w\lambda+w\mu;\lambda+\mu}^{\lambda+\mu},$$

$$C_{w\lambda;\lambda}^\lambda \bullet_u C_{w\mu;\mu}^\mu = \sigma^u(\lambda,\mu)^2 \, C_{w\mu;\mu}^\mu \bullet_u C_{w\lambda;\lambda}^\lambda,$$

$$C_{w\lambda;\lambda}^\lambda \bullet_u (C_{w\mu;\mu}^\mu)^* = \sigma^u(\lambda,\mu)^2 \, (C_{w\mu;\mu}^\mu)^* \bullet_u C_{w\lambda;\lambda}^\lambda,$$

where

$$\sigma^u(\lambda,\mu) = \epsilon^{\langle u,\, \lambda\otimes\mu - w\lambda\otimes w\mu\rangle}.$$

Moreover, the elements $Z_w^\lambda = C_{w\lambda;\lambda}^\lambda \bullet_u (C_{w\lambda;\lambda}^\lambda)^* = (C_{w\lambda;\lambda}^\lambda)^* \bullet_u C_{w\lambda;\lambda}^\lambda$ are in the centre of $\mathcal{F}_{\epsilon,w}^{1,u}$. ∎

It follows from this lemma that the central subalgebra of $\mathcal{F}_{\epsilon,w}^{1,u}$ generated by the Z_w^λ, $\lambda \in P^+$, is already generated by $Z_w^{\lambda_1},\ldots,Z_w^{\lambda_n}$, where the λ_i are the fundamental weights of $\mathfrak{g}$. As in the proof of 13.1.7, one shows that, if ρ is any irreducible $*$-representation of $\mathcal{F}_\epsilon^{1,u}$ on a separable Hilbert space V, the joint spectrum of the commuting self-adjoint operators $\rho(Z_w^{\lambda_i})$ is of the form $\Sigma_i \cup \{0\}$, where Σ_i is a discrete subset of $\mathbb{R}_+$ with 0 as its only possible limit point. We partially order the joint eigenvalues $\boldsymbol{z} = (z_1,\ldots,z_n)$ of $Z_w^{\lambda_1},\ldots,Z_w^{\lambda_n}$ by $\boldsymbol{z} \geq \boldsymbol{z'}$ if and only if $z_i \geq z_i'$ for all i.

LEMMA 13.1.14 *The n-tuple $(1,1,\ldots,1)$ is the unique maximal joint eigenvalue of the $Z_w^{\lambda_i}$ on V.* ∎

In view of this lemma, the action of $\mathcal{F}_{\epsilon,w}^{1,u}$ on the maximal joint eigenspace V_0 of the $Z_w^{\lambda_i}$ factors through its quotient $\mathcal{T}_w^{1,u}$ by the two-sided $*$-ideal generated by the $Z_w^{\lambda_i} - 1$. The resulting irreducible $*$-representation of $\mathcal{T}_w^{1,u}$ on V_0 is called the *highest weight* of (ρ, V) (note that V_0 is not necessarily one-dimensional).

THEOREM 13.1.15 *Two irreducible $*$-representations of $\mathcal{F}_\epsilon^{1,u}(G)$ corresponding to a Schubert cell S_w are equivalent if and only if their highest weights are equivalent. Conversely, given $w \in W$ and an irreducible unitary representation ρ_0 of $\mathcal{T}_w^{1,u}$, there exists an irreducible $*$-representation of $\mathcal{F}_\epsilon^{1,u}(G)$ which corresponds to S_w and has highest weight ρ_0.* ∎

If ρ_0 is the highest weight of an irreducible $*$-representation corresponding to S_w, the operators $t_\lambda = \rho_0(C^\lambda_{w\lambda;\lambda})$ generate $\rho_0(\mathcal{T}^{1,u}_w)$, and, by 13.1.13 and 13.1.14, they obviously satisfy

$$(9) \qquad t_\lambda t^*_\lambda = t^*_\lambda t_\lambda = 1, \quad t_\lambda t_\mu = \sigma(\lambda, \mu) t_{\lambda+\mu}.$$

A $*$-algebra with generators t_λ for λ in some lattice P, and defining relations (9) for some $\sigma(\lambda, \mu) = e^{\kappa(\lambda, \mu)}$, where $\kappa : P \times P \to \sqrt{-1}\mathbb{R}$ is a skew-symmetric biadditive form, is called a *quantum torus*. See Rieffel (1990a).

The correspondence between irreducible $*$-representations and symplectic leaves fails, in general, in the twisted case. Recall from 1.5.11 that, for generic $u \in \wedge^2(\mathfrak{t})$, if $K^{1,u}$ is the Poisson–Lie group with tangent Lie bialgebra $(\mathfrak{k}, \delta_{1,u})$, the symplectic leaves of $K^{1,u}$ (resp. their closures) are parametrized by W if n is even (resp. odd). One finds that, again for generic u, there is a one-to-one correspondence between $W\backslash\{e\}$ and the set of kernels of irreducible $*$-representations of $\mathcal{F}^{1,u}_\epsilon$ of dimension greater than one. However, an irreducible unitary representation is *not* uniquely determined by its kernel, in general. This is related to the *type* of $\mathcal{F}^{1,u}_\epsilon$, or rather that of a suitable completion of it, as a C*-algebra. We shall discuss this question in Section 13.3.

We conclude this subsection by considering briefly the family of Lie bialgebra structures $\delta_{0,u}$. One can define a quantization $U^{0,u}_h(\mathfrak{g})$ of $\delta_{0,u}$ by twisting the null deformation $U(\mathfrak{g})[[h]]$ by $e^{hu/2}$ as in (7). This leads to a Hopf $*$-algebra $\mathcal{F}^{0,u}_\epsilon(G)$ as before. However, there is now no relation between the representations of $\mathcal{F}^{0,u}_\epsilon$ and Schubert cells. This is not unexpected, since we saw in 1.5.12 that there is no relation between the Bruhat decomposition and the symplectic leaves of the Poisson–Lie group $K^{0,u}$ with tangent Lie bialgebra $(\mathfrak{k}, \delta_{0,u})$. One finds that there is an irreducible $*$-representation of $\mathcal{F}^{0,u}_\epsilon$ associated to each pair (π, χ), where π is an irreducible unitary representation of the classical algebra of functions $\mathcal{F}(G)$ and χ is an irreducible unitary representation of a certain quantum torus. Moreover, every such representation of $\mathcal{F}^{0,u}_\epsilon$ arises in this way. For a precise statement, and proofs of this and of the other results in this subsection, see Levendorskii (1990, 1992) and Levendorskii & Soibelman (1991a).

F Representations at roots of unity Although our main interest in this chapter is in unitary representations, for which ϵ must be real, in this subsection we consider briefly the structure and representation theory of $\mathcal{F}_\epsilon$ when ϵ is a root of unity.

We assume, then, that ϵ is a primitive ℓth root of unity, where ℓ is odd, greater than all the d_i, and not divisible by three in case $\mathfrak{g}$ is of type G_2. Unlike the case when $\epsilon > 1$, we do not define the quantized function algebra to be the Hopf dual of $U_\epsilon(\mathfrak{g})$ – this would be too big in the root of unity case. To define it, we first introduce a new rational form of U_q, namely the $\mathcal{A}$-subalgebra $\tilde{U}_\mathcal{A}^{\mathrm{res}}$ of U_q generated by $(X_i^\pm)^{(t)}$, $K_i^{\pm 1}$, and

$$\binom{K_i; 0}{t}_q = \prod_{s=1}^t \frac{K_i q_i^{-s+1} - 1}{q_i^s - 1},$$

where $i = 1, \ldots, n$, $t \in \mathbb{N}$. (Although we do not need it, we remark that $\tilde{U}_\mathcal{A}^{\mathrm{res}}$ properly contains $U_\mathcal{A}^{\mathrm{res}}$.) One then defines $\tilde{\mathcal{F}}_\mathcal{A}$ to be the set of $\mathcal{A}$-linear maps $\varphi : \tilde{U}_\mathcal{A}^{\mathrm{res}} \to \mathcal{A}$ such that $\ker(\varphi)$ contains an ideal I such that

(a) there is a free $\mathcal{A}$-module M of finite rank such that $\tilde{U}_\mathcal{A}^{\mathrm{res}} = I \oplus M$ as $\mathcal{A}$-modules,

(b) there exists $k \in \mathbb{N}$ such that $\prod_{r=-k}^k (K_i - q_i^r) \in I$ for $i = 1, \ldots, n$.

Note that $\tilde{\mathcal{F}}_\mathcal{A}$ is contained in the Hopf dual $(\tilde{U}_\mathcal{A}^{\mathrm{res}})^\circ$ (see Subsection 4.1D). One can show that it is actually a Hopf subalgebra of $(\tilde{U}_\mathcal{A}^{\mathrm{res}})^\circ$.

The specializations are now defined in the obvious way:

$$\tilde{\mathcal{F}}_\epsilon = \tilde{\mathcal{F}}_\mathcal{A} \underset{\mathcal{A}}{\otimes} \mathbb{C}, \qquad \tilde{U}_\epsilon^{\mathrm{res}} = \tilde{U}_\mathcal{A}^{\mathrm{res}} \underset{\mathcal{A}}{\otimes} \mathbb{C},$$

via the homomorphism $\mathcal{A} \to \mathbb{C}$ which takes q to ϵ.

LEMMA 13.1.16 *There is a non-degenerate pairing of Hopf algebras*

$$\tilde{\mathcal{F}}_\epsilon \otimes \tilde{U}_\epsilon^{\mathrm{res}} \to \mathbb{C}. \quad \blacksquare$$

The following result is analogous to 9.3.11.

PROPOSITION 13.1.17 *There is a surjective homomorphism of Hopf algebras from $\tilde{U}_\epsilon^{\mathrm{res}}$ to the classical universal enveloping algebra U such that*

$$\binom{K_i; 0}{\ell} \mapsto \bar{H}_i, \quad K_i^{\pm 1} \mapsto 1,$$

$$(X_i^\pm)^{(t)} \mapsto \begin{cases} \frac{(\bar{X}_i^\pm)^{t/p}}{(t/p)!} & \text{if } p \text{ divides } t, \\ 0 & \text{otherwise.} \quad \blacksquare \end{cases}$$

As usual, bars denote generators of U.

Passing to the dual, using 13.1.16, and recalling from 4.1.17 that the Hopf dual of the classical enveloping algebra U is the classical function algebra $\mathcal{F}$, we obtain an injective homomorphism $\mu : \mathcal{F} \to (\tilde{U}_\epsilon^{\mathrm{res}})^\circ$; let $\mathcal{F}_0$ be its image.

THEOREM 13.1.18 *The Hopf algebra $\mathcal{F}_0$ over $\mathbb{C}$ defined above is contained in the centre of $\tilde{\mathcal{F}}_\epsilon$. Moreover, $\tilde{\mathcal{F}}_\epsilon$ is a finitely generated algebra over $\mathcal{F}_0$ of degree ℓ^N, where N is the number of positive roots of $\mathfrak{g}$, and a projective $\mathcal{F}_0$-module of rank $\ell^{\dim(\mathfrak{g})}$.* ∎

Remark This result is 'dual' to those in Section 9.2. We found there that the centre of the non-restricted specialization U_ϵ contains a subalgebra Z_0 for which $\mathrm{Spec}(Z_0)$ has the structure of a complex Poisson–Lie group. Moreover, this group is the dual, in the sense of Poisson–Lie groups, of the standard Poisson–Lie group structure on G, which underlies the quantization $U_q(\mathfrak{g})$. △

Example 13.1.19 If $G = SL_2(\mathbb{C})$, $\mathcal{F}_0$ is the subalgebra of $\tilde{\mathcal{F}}_\epsilon$ generated by the ℓth powers of the generators a, b, c and d (cf. Subsection 13.1C). The isomorphism $\mathcal{F}_0(sl_2(\mathbb{C})) \to \mathcal{F}(SL_2(\mathbb{C}))$ is given by $a^\ell \mapsto a$, $b^\ell \mapsto b$, $c^\ell \mapsto c$, $d^\ell \mapsto d$. ◊

It follows from 13.1.18 that, if V is an irreducible (complex) representation of $\tilde{\mathcal{F}}_\epsilon$, then $\dim(V) \leq \ell^N$. In order to state a more precise result, note that $\mathcal{F}_0$ acts by scalars on V, so to V there corresponds an element $g_V \in \mathrm{Spec}(\mathcal{F}_0) \cong G$.

PROPOSITION 13.1.20 *Assume, in addition to the usual restrictions, that ℓ is coprime to the coefficients in the expression of the highest root of $\mathfrak{g}$ in terms of the simple roots. Let V be an irreducible representation of $\mathcal{F}_\epsilon$ (over $\mathbb{C}$), and let g_V be the associated element of G. Then:*

(i) there is a dense open subset G_0 of G such that, if $g_V \in G_0$, then $\dim(V) = \ell^N$;

(ii) for any V, $\dim(V)$ is divisible by $\ell^{\frac{1}{2}\dim(\mathcal{O})}$, where $\mathcal{O}$ is the symplectic leaf of G passing through g_V. ∎

Remark The symplectic leaves of G have been described by Hodges & Levasseur (1993). They are parametrized by triples (t, w_1, w_2), where t is an element of the Cartan subgroup of G and w_1, w_2 are elements of the Weyl group of G. This should not be confused with the study of the symplectic leaves of the compact real form of G in Subsection 1.5C, although the two theories are, of course, closely related. △

To actually construct some representations of $\tilde{\mathcal{F}}_\epsilon$, we can proceed as follows. If $G = SL_2(\mathbb{C})$, the formulas in 13.1.8(ii) define an infinite-dimensional representation of $\tilde{\mathcal{F}}_\epsilon(SL_2(\mathbb{C}))$ (we take $t = 1$ and the branch of the square root which is continuous on the complement of the negative real axis, and such that $\sqrt{1} = 1$). Since $\pi_1(a)(e_\ell) = 0$, the vectors $\{e_\ell, e_{\ell+1}, \dots\}$ span a subrepresentation; let V be the ℓ-dimensional quotient. It is clear that V is irreducible.

One passes to the general case as in Subsection 13.1D. Namely, if G is arbitrary, there is a canonical homomorphism $\tilde{\mathcal{F}}_\epsilon(G) \rightarrow \tilde{\mathcal{F}}_{\epsilon_i}(SL_2(\mathbb{C}))$, for each $i = 1, \ldots, n$, induced by the natural inclusions $U_{q_i}(sl_2(\mathbb{C})) \rightarrow U_q(\mathfrak{g})$. Pulling back V by these homomorphisms gives representations V_{s_i} of $\mathcal{F}_\epsilon(G)$. If $w = s_{i_1} \ldots s_{i_{\ell(w)}}$ is a reduced expression of an element w of the Weyl group W of $\mathfrak{g}$, we can form

$$V_w = V_{s_{i_1}} \otimes \cdots \otimes V_{s_{i_{\ell(w)}}}.$$

We have the following analogue of (part of) 13.1.9.

THEOREM 13.1.21 *The representation V_w defined above depends, up to isomorphism, only on the element $w \in W$, and not on its reduced decomposition $w = s_{i_1} \ldots s_{i_k}$. The representations V_w, for $w \in W$, are irreducible and mutually non-isomorphic.* ∎

In contrast to 13.1.9, however, this theorem does not describe all the irreducible representations of $\mathcal{F}_\epsilon$. Other examples are constructed in de Concini & Lyubashenko (1993), from which the results of this subsection were taken, and to which the reader should refer for proofs and further details.

13.2 Quantum homogeneous spaces

In this section, we assume that ϵ is a real number greater than 1.

A Quantum G-spaces Classically, if a Lie group G acts on a smooth manifold M, then the Lie algebra $\mathfrak{g}$ of G, thought of as the left-invariant vector fields on G, acts by derivations on $C^\infty(M)$:

$$Z.(f_1 f_2) = (Z.f_1)f_2 + f_1(Z.f_2), \quad (Z \in \mathfrak{g}, \ f_1, f_2 \in C^\infty(M)).$$

The action of $\mathfrak{g}$ extends uniquely to one of $U(\mathfrak{g})$, and the preceding equation generalizes to

$$Z.(f_1 f_2) = \mu(\Delta(Z).(f_1 \otimes f_2)),$$

where Δ is the comultiplication of $U(\mathfrak{g})$ and μ is the multiplication in $C^\infty(M)$. This equation says that $C^\infty(M)$ is a $U_\epsilon(\mathfrak{g})$-module algebra (see Subsection 4.1C).

We are now in a position to extend the notion of an action of a group on a space to the action of a quantum group on a 'quantum space'.

DEFINITION 13.2.1 *Let $U_\epsilon(\mathfrak{g})$ be the standard quantization of the Lie algebra $\mathfrak{g}$ of a complex simple Lie group G. A quantum G-space is an associative* algebra A equipped with a $U_\epsilon(\mathfrak{g})$-module algebra structure. A morphism $\varphi : A_1 \to A_2$ of quantum G-spaces is a map which is a homomorphism of algebras and of $U_\epsilon(\mathfrak{g})$-modules.

Remark Classically, an action of G on M can also be defined as a map $\alpha : G \times M \to M$ such that $\alpha(g_1 g_2, m) = \alpha(g_1, \alpha(g_2, m))$ for all g_1, $g_2 \in G$, $m \in M$. This translates into the commutativity of the diagram

$$
\begin{array}{ccc}
\mathcal{F}(M) & \xrightarrow{\;\;\alpha^*\;\;} & \mathcal{F}(G) \otimes \mathcal{F}(M) \\
\alpha^* \downarrow & & \downarrow \Delta_{\mathcal{F}} \otimes \mathrm{id} \\
\mathcal{F}(G) \otimes \mathcal{F}(M) & \xrightarrow[\mathrm{id} \otimes \alpha^*]{} & \mathcal{F}(G) \otimes \mathcal{F}(G) \otimes \mathcal{F}(M)
\end{array}
$$

(10)

where $\alpha^* : \mathcal{F}(M) \to \mathcal{F}(G) \otimes \mathcal{F}(M)$ is the homomorphism induced by α (we ignore the question of the precise definition of $\otimes$).

Replacing $\mathcal{F}(M)$ by A and $\mathcal{F}(G)$ by $\mathcal{F}_\epsilon(G)$ thus gives another reasonable definition of a 'quantum G-space'. However, this is simply the 'transpose' of 13.2.1: modulo questions about duality of infinite-dimensional Hopf algebras, the two definitions are equivalent. $\triangle$

Of course, we expect that a quantum group will act on itself by translations. The most obvious such action is

$$
Z.f = \sum_k \langle Z, f'_k \rangle f''_k,
$$

where $Z \in U_\epsilon(\mathfrak{g})$, $f \in \mathcal{F}_\epsilon(G)$ and $\Delta_{\mathcal{F}_\epsilon}(f) = \sum_k f'_k \otimes f''_k$. We leave it to the reader to check that this is a U_ϵ-module algebra structure. Unfortunately, however, it is a right action of U_ϵ. For psychological reasons, we prefer to work with the left action

(11) $$ Z *_L f = \sum_k \langle S(Z), f'_k \rangle f''_k. $$

This is a U_ϵ^{op}-module algebra structure. We could also have worked with the left action

(11a) $$ Z *_R f = \sum_k f'_k \langle Z, f''_k \rangle. $$

Whether the reader prefers (11a) or (11) will depend on whether (s)he prefers to define a left action of G on $C^\infty(G)$ by $g.f(h) = f(hg)$ or by $g.f(h) = f(g^{-1}h)$, to which they reduce in the classical limit.

Example 13.2.2 As in Subsection 7.1B, it is easy to see that $\mathcal{F}_\epsilon(SL_2(\mathbb{C}))$ is spanned by the products $a^\alpha d^\delta b^\beta c^\gamma$ for α, β, γ, $\delta \in \mathbb{N}$ (in fact, a basis is given by the subset described in the statement of 7.1.2, but we shall not need this here). A rather tedious computation shows that the action of $U_\epsilon(sl_2(\mathbb{C}))$ on $\mathcal{F}_\epsilon(SL_2(\mathbb{C}))$ is given by the following formulas:

$$K_1 *_L c^\gamma b^\beta a^\alpha d^\delta = \epsilon^{-\alpha-\beta+\gamma+\delta} c^\gamma b^\beta a^\alpha d^\delta,$$

$$X_1^+ *_L c^\gamma b^\beta a^\alpha d^\delta = -\epsilon^{\gamma+1}[\beta]_\epsilon c^\gamma b^{\beta-1} a^\alpha d^{\delta+1} - \epsilon^{-\alpha+\beta+\gamma+2}[\alpha]_\epsilon c^{\gamma+1} b^\beta a^{\alpha-1} d^\delta,$$

$$X_1^- *_L c^\gamma b^\beta a^\alpha d^\delta = -\epsilon^{-\alpha-1}[\delta]_\epsilon c^\gamma b^{\beta+1} a^\alpha d^{\delta-1} - \epsilon^{\alpha-\gamma-\delta}[\gamma]_\epsilon c^{\gamma-1} b^\beta a^{\alpha+1} d^\delta. \quad \diamond$$

In the next two subsections, we give a number of less trivial examples of quantum G-spaces.

B Quantum flag manifolds and Schubert varieties A classical (generalized) *flag manifold* F_λ is the orbit of a highest weight vector in the projective space $\mathbb{P}(V(\lambda))$ of the irreducible representation $V(\lambda)$ of a complex simple Lie group G of highest weight $\lambda \in P^+$. The group G obviously acts transitively on F_λ, so that $F_\lambda \cong G/P_\lambda$, where P_λ is the stabilizer of the highest weight vector in $\mathbb{P}(V(\lambda))$; note that P_λ obviously contains B_+. The terminology derives from the case $G = SL_{n+1}(\mathbb{C})$, when such an orbit can be identified with the set of sequences of subspaces

$$0 = V_0 \subseteq V_1 \subseteq V_2 \subseteq \cdots \subseteq V_{n+1} = \mathbb{C}^{n+1},$$

the dimensions of the V_j depending only on j and λ.

We shall define the quantum analogue of the 'affine' version of F_λ, i.e. the orbit of the highest weight vector in the vector space $V(\lambda)$.

DEFINITION–PROPOSITION 13.2.3 *For any $\lambda \in P^+$, the subalgebra of $\mathcal{F}_\epsilon(G)$ generated by the matrix elements $C^\lambda_{\nu,s;\lambda}$ for all ν, s is called the* quantized algebra of functions on the flag manifold F_λ, *and is denoted by $\mathcal{F}_\epsilon(F_\lambda)$. It is a quantum G-space, and the inclusion $\mathcal{F}_\epsilon(F_\lambda) \hookrightarrow \mathcal{F}_\epsilon(G)$ is a morphism of quantum G-spaces.*

PROOF We have to check that $\mathcal{F}_\epsilon(F_\lambda)$ is a $U_\epsilon(\mathfrak{g})$-submodule of $\mathcal{F}_\epsilon(G)$. But this is immediate: the formula for the comultiplication of $\mathcal{F}_\epsilon(G)$ given in 13.1.3 implies that

$$X *_L C^\lambda_{\nu,s;\lambda} = \sum_{\mu,r} \langle S(X), C^\lambda_{\nu,s;\mu,r} \rangle C^\lambda_{\mu,r;\lambda}$$

is an element of $\mathcal{F}_\epsilon(F_\lambda)$ for all $X \in U_\epsilon(\mathfrak{g})$. ∎

Remark $\mathcal{F}_\epsilon(F_\lambda)$ is not a $*$-subalgebra of $\mathcal{F}_\epsilon(G)$, as one sees immediately from the description of the $*$-structure on $\mathcal{F}_\epsilon(G)$ in 13.1.3. $\triangle$

It is possible to describe the algebra structure of $\mathcal{F}_\epsilon(F_\lambda)$ quite explicitly. Let

$$\mathcal{V}_\epsilon^\lambda = \bigoplus_{k=0}^\infty V_\epsilon(k\lambda)^*,$$

(the $*$ indicating the dual representation!), and define an algebra structure on $\mathcal{V}_\epsilon^\lambda$ by using the maps

$$V_\epsilon(j\lambda)^* \otimes V_\epsilon(k\lambda)^* \to V_\epsilon((j+k)\lambda)^*$$

which project onto the (unique) highest component of the tensor product. Since these projections are surjective, $\mathcal{V}_\epsilon^\lambda$ is generated as an algebra by the basis elements ξ_μ^r of $V_\epsilon(\lambda)^*$ defined by $\xi_\mu^r(v) = (v, v_\mu^r)$, where $(\,,\,)$ is the invariant hermitian form on $V_\epsilon(\lambda)$.

PROPOSITION 13.2.4 *The map* $\xi_\mu^r \mapsto C_{\mu,r;\lambda}^\lambda$ *extends to an isomorphism of algebras and of* U_ϵ-*modules* $\mathcal{V}_\epsilon^\lambda \to \mathcal{F}_\epsilon(F_\lambda)$. *Moreover, the ideal of relations of* $\mathcal{V}_\epsilon^\lambda$ *is generated by elements which are homogeneous quadratic functions of the* ξ_μ^r. ∎

The ideal of relations in $\mathcal{V}_\epsilon^\lambda$ (or, equivalently, in $\mathcal{F}_\epsilon(F_\lambda)$) can also be described explicitly – apart from the relations which define $\mathcal{V}_\epsilon^\lambda$ as a 'quantum vector space' (cf. Subsection 7.1A), there are quantum analogues of the *Plücker relations* in the classical theory. See Soibelman (1992c).

We now turn to the quantum analogue of Schubert varieties. The varieties $\bar{S}_w^\lambda$ we have in mind are the closures in the projective space of $V(\lambda)$ of the B_+-orbits of the vector of weight $w\lambda$ (which is unique, up to a scalar multiple). If $\lambda \in P^{++}$, then $\bar{S}_w^\lambda \cong \bar{S}_w$ is the closure of a Schubert cell.

The following definition is motivated by the description of the classical Schubert varieties $\bar{S}_w^\lambda$ using 'standard monomial theory' (see Lakshmibai & Seshadri (1991) for a recent survey).

DEFINITION–PROPOSITION 13.2.5 *For* $\lambda \in P^+$, $w \in W$, *let* $\tilde{\mathcal{J}}_\epsilon(w, \lambda)$ *be the two-sided ideal in* $\mathcal{F}_\epsilon(F_\lambda)$ *generated by the matrix elements* $C_{\nu,s;\lambda}^\lambda$ *such that* $\nu \not\geq w\lambda$. *Then, the quotient algebra* $\mathcal{F}_\epsilon(F_\lambda)/\tilde{\mathcal{J}}_\epsilon(w, \lambda)$ *is called the* quantized algebra of functions on the Schubert variety $\bar{S}_w^\lambda$, *and is denoted by* $\mathcal{F}_\epsilon(\bar{S}_w^\lambda)$. *Restricting the* $U_\epsilon(\mathfrak{g})$-*structure of* $\mathcal{F}_\epsilon(F_\lambda)$ *to* $U_\epsilon(\mathfrak{b}_+)$ *induces the structure of a quantum* B_+-*space on* $\mathcal{F}_\epsilon(\bar{S}_w^\lambda)$ *such that the natural projection* $\mathcal{F}_\epsilon(F_\lambda) \to \mathcal{F}_\epsilon(\bar{S}_w^\lambda)$ *is a morphism of quantum* B_+-*spaces.* ∎

Remark Note the similarity of the ideals $\tilde{\mathcal{J}}_\epsilon(w, \lambda)$ to the ideals $\mathcal{J}_\epsilon(w, \lambda)$ defined at the end of Subsection 13.1B. In fact, it is clear that $\tilde{\mathcal{J}}_\epsilon(w, \lambda) \subset \mathcal{J}_\epsilon(w, \lambda)$ and that the two ideals coincide in the sl_2 case. This is the motivation for the notion of a $*$-representation of $\mathcal{F}_\epsilon(G)$ 'corresponding to a Schubert cell S_w'. In fact, such a representation induces a representation of $\mathcal{F}_\epsilon(\bar{S}_w^\lambda)$ for all $\lambda \in P^+$, and so is 'supported' on these quantum Schubert varieties. △

C Quantum spheres The classical odd-dimensional sphere S^{2n+1} can be realized as the orbit under the action of the compact group SU_{n+1} of the highest weight vector v_0 in its natural $(n+1)$-dimensional representation $V^{\natural}$. If t_{rs}, $0 \leq r, s \leq n$, are the matrix entries of $V^{\natural}$, the algebra of functions on the orbit is generated by the entries in the 'first column' t_{s0} and their complex conjugates. In fact, $\mathcal{F}(S^{2n+1})$ is the quotient of the polynomial algebra $\mathbb{C}[t_{00}, \ldots, t_{n0}, \bar{t}_{00}, \ldots, \bar{t}_{n0}]$ by the relation

$$\sum_{s=0}^{n} t_{s0} \bar{t}_{s0} = 1.$$

To define a quantum analogue of $\mathcal{F}(S^{2n+1})$, we observe that $\mathcal{F}_{\epsilon}(SL_{n+1}(\mathbb{C}))$ is generated by the matrix entries in its natural representation $V_{\epsilon}^{\natural}$, since every irreducible $U_{\epsilon}(sl_{n+1}(\mathbb{C}))$-module occurs in a tensor product of copies of $V_{\epsilon}^{\natural}$. In fact, we have

PROPOSITION 13.2.6 *The Hopf algebra* $\mathcal{F}_{\epsilon}(SL_{n+1}(\mathbb{C}))$ *is given by the same generators and relations as in Subsection 7.3B, with* $e^{\pm h}$ *replaced by* $\epsilon^{\pm 1}$. *The* *-structure is given by*

$$t_{rs}^* = (-\epsilon)^{r-s} \mathrm{qdet}(\hat{T}_{rs}),$$

where $\hat{T}_{rs}$ *is the matrix obtained by removing the rth row and sth column from T.* ∎

Following the above discussion, we make

DEFINITION–PROPOSITION 13.2.7 *The *-subalgebra of* $\mathcal{F}_{\epsilon}(SL_{n+1}(\mathbb{C}))$ *generated by the elements* t_{s0} *and* t_{s0}^*, *for* $s = 0, \ldots, n$, *is called the* quantized algebra of functions on the sphere S^{2n+1}, *and is denoted by* $\mathcal{F}_{\epsilon}(S^{2n+1})$. *It is a quantum* $SL_{n+1}(\mathbb{C})$-*space.*

OUTLINE OF PROOF The last part follows from the comultiplication formulas

$$\Delta_{\mathcal{F}_{\epsilon}}(t_{s0}) = \sum_{r} t_{sr} \otimes t_{r0}, \quad \Delta_{\mathcal{F}_{\epsilon}}(t_{s0}^*) = \sum_{r} t_{sr}^* \otimes t_{r0}^*,$$

as in the proof of 13.2.3. ∎

We set $z_s = t_{s0}$ from now on. Using 13.2.6, it is easy to see that the following relations hold in $\mathcal{F}_{\epsilon}(S^{2n+1})$:

(12)
$$z_r z_s = \epsilon^{-1} z_s z_r, \ \text{if } r < s, \quad z_r z_s^* = \epsilon^{-1} z_s^* z_r, \ \text{if } r \neq s,$$
$$z_r z_r^* - z_r^* z_r + (\epsilon^{-2} - 1) \sum_{s>r} z_s z_s^* = 0, \quad \sum_{s=0}^{n} z_s z_s^* = 1.$$

PROPOSITION 13.2.8 $\mathcal{F}_{\epsilon}(S^{2n+1})$ *has (12) as its defining relations.*

This follows from the representation theory of $\mathcal{F}_\epsilon(S^{2n+1})$, which we now discuss.

The most obvious way to construct representations of $\mathcal{F}_\epsilon(S^{2n+1})$ is, of course, to restrict representations of $\mathcal{F}_\epsilon(SL_{n+1}(\mathbb{C}))$. The following result shows that every irreducible *-representation of $\mathcal{F}_\epsilon(S^{2n+1})$ arises in this way.

THEOREM 13.2.9 *Every irreducible *-representation of $\mathcal{F}_\epsilon(S^{2n+1})$ is equivalent to exactly one of the following:*

(i) *the one-dimensional representations* $\rho_{0,t}$, $t \in S^1$, *given by*

$$\rho_{0,t}(z_0^*) = t^{-1}, \quad \rho_{0,t}(z_r^*) = 0, \quad \text{if } r > 0;$$

(ii) *the representations* $\rho_{r,t}$, $1 \le r \le n$, $t \in S^1$, *on the Hilbert space tensor product* $\ell^2(\mathbb{N})^{\otimes r}$, *given by*

$\rho_{r,t}(z_s^*)(e_{k_1} \otimes \cdots \otimes e_{k_r}) =$

$$\begin{cases} \epsilon^{-(k_1 + \cdots + k_s + s)}(1 - \epsilon^{-2(k_s+1+1)})^{1/2} e_{k_1} \otimes \cdots \otimes e_{k_s} \otimes e_{k_{s+1}+1} \otimes e_{k_s+2} \otimes \cdots \otimes e_{k_r} & \text{if } s < r, \\ t^{-1}\epsilon^{-(k_1 + \cdots + k_r + r)} e_{k_1} \otimes \cdots \otimes e_{k_r} & \text{if } s = r, \\ 0 & \text{if } s > r. \end{cases}$$

The representation $\rho_{0,t}$ *is equivalent to the restriction of the representation* τ_t *of* $\mathcal{F}_\epsilon(SL_{n+1}(\mathbb{C}))$ *(cf. 13.1.9); and, for* $r > 0$, $\rho_{r,t}$ *is equivalent to the restriction of* $\pi_{s_1} \otimes \pi_{s_2} \otimes \cdots \otimes \pi_{s_r} \otimes \tau_t$.

OUTLINE OF PROOF OF 13.2.8 Let $\tilde{\mathcal{F}}$ be the quotient of the free associative algebra over $\mathbb{C}$ with generators z_r, z_r^*, $r = 0, \ldots, n$, by the two-sided ideal generated by the relations in (12). We have to show that the canonical surjective homomorphism of algebras $\varphi : \tilde{\mathcal{F}} \to \mathcal{F}_\epsilon(S^{2n+1})$ given by the comments following 13.2.7 is injective. One shows first that the subalgebra $\tilde{\mathcal{H}}$ of $\tilde{\mathcal{F}}$ generated by the elements $z_r z_r^*$, $r = 0, \ldots, n$, is commutative, and that the monomials $z_0^{p_0} \ldots z_n^{p_n}$, where the exponents p_r are either zero, one or $*$, form an $\tilde{\mathcal{H}}$-basis of $\tilde{\mathcal{F}}$. Next, one checks that the formulas in 13.2.9 define representations of $\tilde{\mathcal{F}}$ in which $\tilde{\mathcal{H}}$ acts diagonally. An explicit computation finally shows that no non-trivial linear combination of the $z_0^{p_0} \ldots z_n^{p_n}$ with coefficients in $\tilde{\mathcal{H}}$ can be annihilated by all the $\rho_{r,t}$. ∎

OUTLINE OF PROOF OF 13.2.9 One proves by induction on n that the representations in (i) and (ii) exhaust the irreducible *-representations of $\mathcal{F}_\epsilon(S^{2n+1})$. The case $n = 1$ is 13.1.8, for it is easy to see that $\mathcal{F}_\epsilon(S^3)$ is the whole of $\mathcal{F}_\epsilon(SL_2(\mathbb{C}))$ ($SU_2 \cong S^3$!), and the inductive step follows from the observation that $\mathcal{F}_\epsilon(S^{2n-1})$ is isomorphic to the quotient of $\mathcal{F}_\epsilon(S^{2n+1})$ by the two-sided ideal generated by z_n and z_n^* (this is immediate from 13.2.8). ∎

In the classical case, the orbit of the highest weight vector of SU_{n+1} in the projective space of its natural representation $V^\natural$ is the entire complex projective space $P(V^\natural) \cong \mathbb{C}P^n$. Thus, it is natural to define $\mathcal{F}_\epsilon(\mathbb{C}P^n)$, the *quantized algebra of functions on the projective space* $\mathbb{C}P^n$, to be the 'homogeneous' version of $\mathcal{F}_\epsilon(S^{2n+1})$, i.e. the $*$-subalgebra of $\mathcal{F}_\epsilon(S^{2n+1})$ consisting of the fixed points of the $*$-automorphisms $z_r \mapsto \zeta z_r$, $\zeta \in S^1$. Taking $n = 1$ gives our first example of the algebra of functions on an even-dimensional quantum sphere $\mathcal{F}_\epsilon(S^2)$. We study it in detail in Subsection 13.5B.

13.3 Compact matrix quantum groups

The Hopf $*$-algebra $\mathcal{F}_\epsilon(G)$ is a quantum analogue of the algebra $\mathcal{F}(K)$ of representative functions on the compact real form K of G. If K is realized as a closed subgroup of a unitary group U_m, an element of $\mathcal{F}(K)$ is a polynomial in the matrix entries and their complex conjugates. There is another approach to quantized algebras of functions, originally developed by Woronowicz (1987a) independently of the study of quantized enveloping algebras, which studies the properties of a quantum analogue of the algebra of *continuous* functions on K. We do not discuss this approach in detail in this book, but we shall now outline its relation to quantized enveloping algebras.

A C* completions and compact matrix quantum groups Since one does not have any 'geometric' realization of quantum groups, one cannot define quantized continuous functions in a topological way. Classically, however, the Peter–Weyl theorem gives an essentially algebraic characterization of the algebra of continuous functions on K as a completion of the algebra generated by the matrix elements of the unitary representations of K. This provides the motivation for

DEFINITION–PROPOSITION 13.3.1 *The quantized algebra of continuous functions on K is the C*-completion $\mathcal{C}_\epsilon(G)$ of the $*$-algebra $\mathcal{F}_\epsilon(G)$ with respect to the norm*

$$\| f \| = \sup_\rho \| \rho(f) \|, \qquad (f \in \mathcal{F}_\epsilon(G)),$$

where ρ runs through the $*$-representations of $\mathcal{F}_\epsilon(G)$ and the norm on the right-hand side is the operator norm.

PROOF It suffices to show that $\| f \|$ is finite for all $f \in \mathcal{F}_\epsilon(G)$, for it is clear that $\| \; \|$ is a C*-norm, i.e. $\| ff^* \| = \| f \|^2$.

We have already proved this finiteness property in the sl_2 case: it was a consequence of the relation in (5). Although we do not know all the defining relations of $\mathcal{F}_\epsilon$ in the general case, we shall prove that the following relations hold for all $\lambda \in P^+$:

(13)
$$\sum_{\nu,s} C^\lambda_{\nu,s;\mu,r}(C^\lambda_{\nu,s;\kappa,t})^* = \delta_{\mu,\kappa}\delta_{r,t}.$$

Taking $\mu = \kappa$ and $r = t$, and then applying any $*$-representation ρ of $\mathcal{F}_\epsilon$, we find that $\| \rho(C^\lambda_{\nu,s;\mu,r}) \| \leq 1$ for all λ, μ, ν, r and s, proving our assertion.

To prove (13), fix an invariant hermitian form $(\ ,\ )$ on $V_\epsilon(\lambda)$ and let T_λ be the matrix of the representation $V_\epsilon(\lambda)$ with respect to an orthonormal basis $\{v^r_\mu\}$. Then, (13) is equivalent to

(14)
$$T_\lambda(T^t_\lambda)^* = 1,$$

where $(T^t_\lambda)^*$ is the result of applying the $*$-structure of $\mathcal{F}_\epsilon$ to each entry of the transpose of the matrix T_λ. Now, as in the sl_2 case considered at the end of the proof of 7.1.4, it is tautologous that

$$\Delta_{\mathcal{F}_\epsilon}(T_\lambda) = T_\lambda \mathbin{\square} T_\lambda.$$

Using the relation $\mu_{\mathcal{F}_\epsilon}(\mathrm{id} \otimes S_{\mathcal{F}_\epsilon})\Delta_{\mathcal{F}_\epsilon} = \epsilon_{\mathcal{F}_\epsilon}$, we obtain $T_\lambda S_{\mathcal{F}_\epsilon}(T_\lambda) = \epsilon_{\mathcal{F}_\epsilon}(T_\lambda)$. But,

$$\epsilon_{\mathcal{F}_\epsilon}(C^\lambda_{\nu,s;\mu,r}) = C^\lambda_{\nu,s;\mu,r}(1) = \delta_{\mu,\nu}\delta_{r,s}$$

since the basis $\{v^r_\mu\}$ is orthonormal. Hence, $T_\lambda S_{\mathcal{F}_\epsilon}(T_\lambda) = 1$.

To prove (14), we are thus reduced to proving that

$$S_{\mathcal{F}_\epsilon}(T_\lambda) = (T^t_\lambda)^*.$$

The $(\mu, r; \nu, s)$ entry of the right-hand side, when applied to an element $x \in U_\epsilon$, is equal to

$$\overline{C^\lambda_{\nu,s;\mu,r}(S_{U_\epsilon}(x)^*)} = \overline{(S_{U_\epsilon}(x)^* . v^r_\mu, v^s_\nu)} = \overline{(v^r_\mu, S_{U_\epsilon}(x).v^s_\nu)}$$
$$= (S_{U_\epsilon}(x).v^s_\nu, v^r_\mu) = C^\lambda_{\nu,s;\mu,r}(S_{U_\epsilon}(x)),$$

which is equal to the $(\mu, r; \nu, s)$ entry of $S_{\mathcal{F}_\epsilon}(T_\lambda)$. ∎

Remarks [1] One can define the twisted C*-algebra $\mathcal{C}^{1,u}_\epsilon(G)$ in the same way.

[2] From the definition of the norm on $\mathcal{C}_\epsilon(G)$, it is obvious that every $*$-representation of $\mathcal{F}_\epsilon(G)$ extends to a continuous $*$-representation of $\mathcal{C}_\epsilon(G)$.

[3] One can define $\mathcal{C}_\epsilon(S^{2n+1})$, the *quantized algebra of continuous functions on* S^{2n+1}, by completing $\mathcal{F}_\epsilon(S^{2n+1})$ in the same way as we defined $\mathcal{C}_\epsilon(SL_{n+1}(\mathbb{C}))$ by completing $\mathcal{F}_\epsilon(SL_{n+1}(\mathbb{C}))$. It is obvious that every $*$-representation of $\mathcal{F}_\epsilon(S^{2n+1})$ extends to $\mathcal{C}_\epsilon(S^{2n+1})$. △

We have now made contact with the theory of compact matrix quantum groups in the sense of Woronowicz (1987a) (he originally used the term 'compact matrix pseudogroup').

DEFINITION 13.3.2 *A compact matrix quantum group is a unital C*-algebra A together with an $m \times m$ matrix $T = (T_{rs})$ of elements of A such that*

(i) the *-subalgebra of A generated by the T_{rs} is dense in A;
(ii) there is a unital C*-homomorphism $\Delta : A \to A \hat{\otimes} A$ for which $\Delta(T) = T \otimes T$;
(iii) there is a linear anti-homomorphism $S : A \to A$ such that $S(S(a^*)^*) = a$ for all $a \in A$ and $S(T)T = TS(T) = 1$.

The tensor product $A \hat{\otimes} A$ in (ii) is defined by completing the algebraic tensor product $A \otimes A$ with respect to the norm defined by

$$\| x \| = \sup_{\rho_1, \rho_2} \| (\rho_1 \otimes \rho_2)(x) \|,$$

where $x \in A \otimes A$ and ρ_1 and ρ_2 run through the *-representations of A.

PROPOSITION 13.3.3 *The C*-algebra $\mathcal{C}_\epsilon(G)$ is a compact matrix quantum group.*

OUTLINE OF PROOF One may take the matrix T in 13.3.2 to be the matrix of the representation $V_\epsilon = \oplus_i V_\epsilon(\lambda_i)$ with respect to the orthonormal basis obtained by taking the union of the usual bases of the $V_\epsilon(\lambda_i)$ (here, $\lambda_1, \ldots, \lambda_n$ are the fundamental weights of $\mathfrak{g}$). Now, every finite-dimensional irreducible U_ϵ-module of type **1** is a direct summand of a tensor product of a sufficiently large number of copies of V_ϵ (this follows from 10.1.16 and the classical fact that every finite-dimensional irreducible $\mathfrak{g}$-module is a direct summand of a tensor product of fundamental representations). It follows that the entries of T generate $\mathcal{F}_\epsilon$.

To show that $\mathcal{F}_\epsilon$ is dense in $\mathcal{C}_\epsilon$, we must show that $\| f \| = 0$ only if $f = 0$. This follows from the existence of a faithful *-representation (see the first remark following 13.1.10).

Finally, the comultiplication $\Delta_{\mathcal{F}_\epsilon}$ of $\mathcal{F}_\epsilon$ obviously extends continuously to $\mathcal{C}_\epsilon$. For, if $f \in \mathcal{F}_\epsilon$, then, by the definition of the norm on $\mathcal{F}_\epsilon \otimes \mathcal{F}_\epsilon$,

$$\| \Delta_{\mathcal{F}_\epsilon}(f) \| = \sup_{\rho_1, \rho_2} \| (\rho_1 \otimes \rho_2) \circ \Delta_{\mathcal{F}_\epsilon}(f) \|,$$

and this is $\leq \| f \|$ because $(\rho_1 \otimes \rho_2) \circ \Delta_{\mathcal{F}_\epsilon}$ is a *-representation of $\mathcal{F}_\epsilon$. ∎

Remark Note that S is not assumed to be continuous. In fact, it is already clear in the SL_2 case that the antipode $S_{\mathcal{F}_\epsilon}$ of $\mathcal{F}_\epsilon$ does not extend continuously to $\mathcal{C}_\epsilon$. For, $S_{\mathcal{F}_\epsilon}$ is given by (see Subsection 13.1C)

$$S_{\mathcal{F}_\epsilon}(a) = d, \quad S_{\mathcal{F}_\epsilon}(b) = -\epsilon b, \quad S_{\mathcal{F}_\epsilon}(c) = -\epsilon^{-1} c, \quad S_{\mathcal{F}_\epsilon}(d) = a,$$

and the relations $\| S_{\mathcal{F}_\epsilon}(b^k) \| = \epsilon^k \| b^k \|$, $\| S_{\mathcal{F}_\epsilon}(c^k) \| = \epsilon^{-k} \| c^k \|$ imply that $S_{\mathcal{F}_\epsilon}$ is unbounded. △

We conclude this subsection by discussing some of the properties of C_ϵ as a C*-algebra.

In the classical case, the C*-algebra $C(G)$ is obviously of type I, since it is abelian. The following result, however, is much less trivial:

PROPOSITION 13.3.4 *(i) The C*-algebra $C_\epsilon(G)$ is of type I.*
(ii) The C-algebra $C_\epsilon^{1,u}(G)$ is of type I if and only if, for all λ, $\mu \in P$,*

$$\langle u, \lambda \otimes \mu - w\lambda \otimes w\mu \rangle \in \frac{\sqrt{-1}\pi}{\ln\epsilon}\mathbb{Q}. \blacksquare$$

In the untwisted case, we have seen that there is a one-to-one correspondence between the symplectic leaves of K and the irreducible *-representations of $\mathcal{F}_\epsilon$. On the space of symplectic leaves, there is a natural quotient topology, which one might hope is reflected in the topology of C_ϵ. In fact, there is a natural topology, the *Jacobson topology*, on the set $\mathrm{prim}(C_\epsilon)$ of *primitive ideals* of C_ϵ, i.e. the kernels of the irreducible *-representations of C_ϵ. The closed sets of this topology are those sets of primitive ideals which contain some fixed, but arbitrary, subset of C_ϵ.

Example 13.3.5 We consider the case $G = SL_2(\mathbb{C})$. Using the notation in 13.1.8, it is easy to see that the kernels of the representations τ_t and π_t of $\mathcal{F}_\epsilon(SL_2(\mathbb{C}))$ are all distinct. Thus, there is a one-to-one correspondence between irreducible *-representations and primitive ideals: in an obvious notation, $\mathrm{prim}(C_\epsilon(SL_2(\mathbb{C}))) = S_\tau^1 \coprod S_\pi^1$ is the disjoint union of two circles.

The Jacobson topology on $\mathrm{prim}(C_\epsilon(SL_2(\mathbb{C})))$ can be computed directly (see Meister (1992) for the details). The most general closed set is either a closed set of S_τ^1 in its usual topology as a subset of $\mathbb{C}$, or the disjoint union of S_τ^1 with a closed set in the usual topology of S_π^1. Comparing with Fig. 2, and recalling that the representations τ_t (resp. π_t) are associated with the zero-dimensional (resp. two-dimensional) symplectic leaves, we see that the topology on the space of symplectic leaves is identical with that on $\mathrm{prim}(C_\epsilon)$. $\Diamond$

Remarks [1] Note that $\mathrm{prim}(C_\epsilon(SL_2(\mathbb{C})))$ and its topology are independent of ϵ. In fact, Woronowicz (1987b) proves that the C*-algebra $C_\epsilon(SL_2(\mathbb{C}))$ is itself independent of ϵ, up to isomorphism, provided that $\epsilon \in \mathbb{R}_+\backslash\{1\}$.

[2] See Hodges & Levasseur (1993) for a description of the primitive ideals of $C_\epsilon(SL_3(\mathbb{C}))$. $\triangle$

B The Haar integral on compact quantum groups Following the discussion in Subsection 4.1E, we say that a linear functional $H : \mathcal{F}_\epsilon(G) \to \mathbb{C}$ is a *normalized bi-invariant integral*, or simply a *Haar integral*, if $H(1) = 1$ and

$$(15) \qquad \langle H \otimes Z, \Delta_{\mathcal{F}_\epsilon}(f)\rangle = \langle Z \otimes H, \Delta_{\mathcal{F}_\epsilon}(f)\rangle = H(f)\epsilon_{U_\epsilon}(Z)$$

for all $f \in \mathcal{F}_\epsilon(G)$, $Z \in U_\epsilon(\mathfrak{g})$.

PROPOSITION 13.3.6 *The quantized algebra of functions $\mathcal{F}_\epsilon(G)$ admits a unique Haar integral.*

PROOF We first establish uniqueness; in the process, we shall obtain an explicit description of the Haar integral.

Suppose, then, that H is a Haar integral. We prove first that H satisfies the 'left-invariance' condition

$$H(Z *_L f) = \epsilon_{U_\epsilon}(Z)H(f)$$

for all $Z \in U_\epsilon(\mathfrak{g})$, $f \in \mathcal{F}_\epsilon(G)$, where $*_L$ denotes the action of U_ϵ on $\mathcal{F}_\epsilon$ defined in (11). We shall need the following identity, which is easily checked:

$$(16) \qquad \Delta_{\mathcal{F}_\epsilon}(Z *_L f) = (Z \otimes 1) *_L \Delta_{\mathcal{F}_\epsilon}(f).$$

Then,

$$
\begin{aligned}
1.H(Z *_L f) &= (\mathrm{id} \otimes H)((Z \otimes 1) *_L \Delta_{\mathcal{F}_\epsilon}(f)) \\
&= Z *_L ((\mathrm{id} \otimes H)\Delta_{\mathcal{F}_\epsilon}(f)) \\
&= Z *_L (H(f)1) = \epsilon_{U_\epsilon}(Z)H(f),
\end{aligned}
$$

proving our assertion.

We now use (16) to show that $H(C^\lambda_{\nu,s;\mu,r}) = 0$ unless $\lambda = 0$. In fact, using the identity $H(K_i *_L f) = H(f)$, we find that

$$H(C^\lambda_{\nu,s;\mu,r}) = \epsilon^{-(\nu,\alpha_i)} H(C^\lambda_{\nu,s;\mu,r}),$$

which implies that $H(C^\lambda_{\nu,s;\mu,r}) = 0$ unless $\nu = 0$. Similarly, from the identity $H(X_i^+ *_L f) = 0$, we obtain

$$0 = H(X_i^+ *_L C^\lambda_{\nu,s;\mu,r}) = -\sum_t (X_i^+ K_i^{-1}.v_0^t, v_\nu^s) H(C^\lambda_{0,t;\mu,r}),$$

where $(\ ,\ )$ is the inner product on $V_\epsilon(\lambda)$. But, if $\lambda \neq 0$, no non-zero vector in the 0-weight space $V_\epsilon(\lambda)_0$ can be annihilated by all the X_i^+, otherwise $V_\epsilon(\lambda)$ would have a primitive vector of weight 0, contradicting its irreducibility. Hence, the preceding equation implies that $H(C^\lambda_{0,t;\mu,r}) = 0$ for all μ, r, t. Combining these two observations, it follows that H must be given by

$$(17) \qquad H(C^\lambda_{\nu,s;\mu,r}) = \begin{cases} 1 & \text{if } \lambda = 0, \\ 0 & \text{otherwise.} \end{cases}$$

Thus, H is uniquely determined.

Conversely, it is easy to check that (17) does define a Haar integral on $\mathcal{F}_\epsilon$. ∎

Remarks [1] Note that we have actually shown that H is uniquely determined by the normalization and left-invariance conditions. Thus, every normalized left-invariant integral on $\mathcal{F}_\epsilon(G)$ is right-invariant, and conversely.

[2] In the quantum case, the Haar integral is *non-central*, i.e. $H(f_1 f_2) \neq H(f_2 f_1)$ in general. $\triangle$

Woronowicz (1987a) proves a result more general than 13.3.6 in the context of compact matrix quantum groups:

THEOREM 13.3.7 *Let* (A, T) *be a compact matrix quantum group. Then, there exists a unique normalized, bi-invariant integral H on A. Moreover, $H(a) > 0$ if $a > 0$, and $H(S(a)) = H(a)$ for every a in the dense subalgebra of A generated by the matrix entries of T.* ∎

By a bi-invariant integral on A, we mean a linear functional $H : A \to \mathbb{C}$ satisfying (15) for all continuous linear functionals Z on A and all $f \in A$. An element $a \in A$ is positive if $a = bb^*$ for some $b \in A$, and $a > 0$ if a is non-zero and positive.

In view of 13.3.3, this last result has the following corollary, which is also easy to prove directly.

COROLLARY 13.3.8 *The Haar integral on $\mathcal{F}_\epsilon(G)$ is positive, in the sense that, for all $f \in \mathcal{F}_\epsilon(G)$, $H(f) > 0$ if $f > 0$. Moreover, $H(S(f)) = H(f)$ for all f and H extends to a continuous linear functional on $\mathcal{C}_\epsilon(G)$.* ∎

Although the description of H given by (17) is, in a sense, 'explicit', one would like a formula for H in terms of a more economical set of generators of $\mathcal{F}_\epsilon$, such as those used in the proof of 13.3.3. Unfortunately, no such description seems to be known except in the sl_2 case.

Example 13.3.9 From the proof of 7.1.2, it follows that $\mathcal{F}_\epsilon(SL_2(\mathbb{C}))$ is spanned by the monomials $a^\alpha d^\delta b^\beta c^\gamma$, where $\alpha, \beta, \gamma, \delta \in \mathbb{N}$. The Haar integral is given by

$$H(a^\alpha d^\delta b^\beta c^\gamma) = \begin{cases} (-\epsilon)^\beta \epsilon^{2\alpha(\beta+1)}(1 - \epsilon^2)\frac{(\epsilon^2;\epsilon^2)_\alpha(\epsilon^2;\epsilon^2)_\beta}{(\epsilon^2;\epsilon^2)_{\alpha+\beta+1}} & \text{if } \alpha = \delta \text{ and } \beta = \gamma, \\ 0 & \text{otherwise.} \end{cases}$$

Here, we have used the symbol

$$(a\,;q)_k = \begin{cases} \prod_{j=0}^{k-1}(1 - aq^j) & \text{if } k > 0, \\ 1 & \text{if } k = 0. \end{cases}$$

To prove this formula, one first observes, using the invariance of H under the left and right actions of $K_1 \in U_\epsilon$, that $H(a^\alpha d^\delta b^\beta c^\gamma) = 0$ unless $\alpha = \delta$ and $\beta = \gamma$. Now, any monomial $a^\alpha d^\alpha b^\beta c^\beta$ can be written as a linear combination of powers of bc (cf. the proof of 7.1.2). Thus, it is enough to compute

$H((bc)^\beta)$. This is done by induction on β. For example, using the right-invariance condition, we have

$$
\begin{aligned}
H(bc).1 &= (H \otimes \mathrm{id})(\Delta_{\mathcal{F}_\epsilon}(bc)) \\
&= (H \otimes \mathrm{id})(a \otimes b + b \otimes d)(c \otimes a + d \otimes c)) \\
&= H(ad)bc + H(bc)da \\
&= H(\epsilon^{-1}bc + 1)bc + H(bc)(\epsilon bc + 1).
\end{aligned}
$$

Since $H(1) = 1$, we obtain $H(bc) = -1/(\epsilon + \epsilon^{-1})$. In general, this argument gives a recurrence relation between $H((bc)^{\beta+1})$ and $H(bc)^\beta$, from which we find that

$$
H((-\epsilon bc)^\beta) = \frac{(1 - \epsilon^2)\epsilon^{2\beta}}{1 - \epsilon^{2\beta+2}}.
$$

The general formula can easily be deduced from this.

The reason for writing the previous fromula as we have is that, if $z = \epsilon bc$ and f is a polynomial in z, it implies that

$$
H(f) = \int_0^1 f(z)\, d_{q^2}z,
$$

where $q = \epsilon^{-1}$ and

$$
\int_0^1 f(z)\, d_q z = (1 - q)\sum_{k=0}^{\infty} q^k f(q^k)
$$

is the *Jackson integral* from q-analysis. We shall see further connections between quantum groups and q-analysis in Section 13.5. $\Diamond$

Perhaps surprisingly, although no explicit formula for the Haar integral is known for general $\mathfrak{g}$, the quantum analogues of the classical Schur orthogonality relations can be obtained rather easily.

PROPOSITION 13.3.10 *We have*

$$
H((C^\lambda_{\nu,s;\mu,r})^* C^{\tilde\lambda}_{\tilde\nu,\tilde s;\tilde\mu,\tilde r}) = \frac{\delta_{\lambda,\tilde\lambda}\delta_{\nu,\tilde\nu}\delta_{s,\tilde s}\delta_{\mu,\tilde\mu}\delta_{r,\tilde r}\epsilon^{-2(\nu,\rho)}}{\sum_\Lambda \dim(V_\epsilon(\lambda)_\Lambda)\epsilon^{2(\Lambda,\rho)}},
$$

where the sum is over the set of weights Λ of $V_\epsilon(\lambda)$.

PROOF Let $\{\xi^r_\mu\}$ be the basis of $V_\epsilon(\lambda)^*$ dual to the orthonormal weight basis $\{v^r_\mu\}$ of $V_\epsilon(\lambda)$, and define a basis $\{\tilde\xi^{\tilde r}_{\tilde\mu}\}$ of $V_\epsilon(\tilde\lambda)$ similarly. Then, if $Z \in U_\epsilon$ and

$$\Delta_{U_\epsilon}(Z) = \sum_k Z'_k \otimes Z''_k,$$

$$
\begin{aligned}
((C^\lambda_{\nu,s;\mu,r})^* C^{\tilde\lambda}_{\tilde\nu,\tilde s;\tilde\mu,\tilde r})(Z) &= \sum_k (C^\lambda_{\nu,s;\mu,r})^*(Z'_k) C^{\tilde\lambda}_{\tilde\nu,\tilde s;\tilde\mu,\tilde r}(Z''_k) \\
&= \sum_k C^\lambda_{\nu,s;\mu,r}(S_{U_\epsilon}(Z'_k)) C^{\tilde\lambda}_{\tilde\nu,\tilde s;\tilde\mu,\tilde r}(Z''_k) \\
&= \sum_k \langle S_{U_\epsilon}(Z'_k).v^s_\nu , \xi^r_\mu \rangle \langle Z''_k.v^{\tilde r}_{\tilde\mu} , \xi^{\tilde s}_{\tilde\nu} \rangle \\
&= \sum_k \langle v^s_\nu , Z'_k.\xi^r_\mu \rangle \langle Z''_k.v^{\tilde r}_{\tilde\mu} , \xi^{\tilde s}_{\tilde\nu} \rangle \\
&= \langle Z.(\xi^r_\mu \otimes v^{\tilde r}_{\tilde\mu}) , v^s_\nu \otimes \xi^{\tilde s}_{\tilde\nu} \rangle.
\end{aligned}
$$

This is a matrix element of the representation $V_\epsilon(\lambda)^* \otimes V_\epsilon(\tilde\lambda)$. By (17), $H((C^\lambda_{\nu,s;\mu,r})^* C^{\tilde\lambda}_{\tilde\nu,\tilde s;\tilde\mu,\tilde r})$ will vanish unless this representation contains the trivial representation, that is, unless $\lambda = \tilde\lambda$. Assuming this, the trivial component is spanned by a linear combination of the vectors $\xi^r_\mu \otimes v^r_\mu$, and it follows that

$$H((C^\lambda_{\nu,s;\mu,r})^* C^{\tilde\lambda}_{\tilde\nu,\tilde s;\tilde\mu,\tilde r}) = 0 \quad \text{unless } \mu = \tilde\mu,\ \nu = \tilde\nu,\ r = \tilde r \text{ and } s = \tilde s.$$

Finally, since $V_\epsilon(\lambda)^* \otimes V_\epsilon(\lambda) \cong \mathrm{Hom}_{\mathbf{C}}(V_\epsilon(\lambda)', V_\epsilon(\lambda)^*)$ as U_ϵ-modules (see Subsection 4.1C), we can interpret $H((C^\lambda_{\nu,s;\mu,r})^* C^\lambda_{\nu,s;\mu,r})$ as a matrix element of an intertwining operator $\Phi : V_\epsilon(\lambda)' \to V_\epsilon(\lambda)^*$. Now, by the discussion preceding 8.3.15, Φ is a multiple of the map $\xi \mapsto K_{\rho^*}.\xi$, where the dot denotes the action in $V(\lambda)^*$ and $K_{\rho^*} = \prod_i K_i^{r_i}$ if $2\rho = \sum_i r_i \alpha_i$. It follows that

$$H((C^\lambda_{\nu,s;\mu,r})^* C^\lambda_{\nu,s;\mu,r}) = c_\lambda \epsilon^{-2(\nu,\rho)}$$

for some constant c_λ, the value of which is fixed by relation (13) to be as stated in the proposition. ∎

Example 13.3.11 If $G = SL_2(\mathbf{C})$, 13.3.10 gives

$$H((C^\lambda_{\nu;\mu})^* C^{\tilde\lambda}_{\tilde\nu;\tilde\mu}) = \delta_{\lambda,\tilde\lambda}\delta_{\mu,\tilde\mu}\delta_{\nu,\tilde\nu} \frac{\epsilon^{-\nu}}{[\lambda+1]_\epsilon}$$

(as usual, we identify a weight $\lambda \in P$ with the integer (λ,α), where α is the positive root of $sl_2(\mathbf{C})$). Noting that $C^1_{1;1} = a$, $C^1_{1;-1} = b$, $C^1_{-1;1} = c$ and $C^1_{-1;-1} = d$, it is easy to check that this formula is in agreement with that in 13.3.9. $\diamond$

13.4 A non-compact quantum group

The Hopf $*$-algebra $\mathcal{F}_\epsilon(G)$ discussed in the preceding section may be called a 'compact' quantum group, because the classical limit of its $*$-structure defines the compact real form of G. In this section, we look at a non-compact quantum group.

We continue to assume that ϵ is a real number greater than one.

A The quantum euclidean group The group E_2 of orientation-preserving isometries of the plane $\mathbb{R}^2$, equipped with its usual euclidean metric, is the semidirect product $SO_2 \ltimes \mathbb{R}^2$ of the group SO_2 of rigid rotations with the group $\mathbb{R}^2$ of translations, with SO_2 acting on $\mathbb{R}^2$ in its natural representation. The group E_2 is solvable, and is not among the groups for which we have given deformations, either on the enveloping algebra or function algebra level. However, it is possible to write down a deformation of E_2 by imitating the following classical construction whereby E_2 is obtained as a 'limit' of the simple group SO_3 (or, equivalently, SO_3 is realized as a deformation of E_2).

We regard SO_3 as the group of rotations of the sphere in $\mathbb{R}^3$ with centre $(0, 0, R)$ and radius R. In the limit as $R \to \infty$, rotations about an axis passing through $(0, 0, R)$ and parallel to the xy-plane induce translations in the tangent plane $\mathbb{R}^2$ to the sphere at the origin, while rotations about the z-axis induce rotations in $\mathbb{R}^2$ (see Fig. 21). More explicitly, recall that every rotation is a product of a rotation through an angle ψ about the z-axis, followed by a rotation θ about the y-axis, followed by a rotation φ about the z-axis, where (ψ, θ, φ) are the Euler angles. The rotation about the y-axis is given by

$$(x, y, z) \mapsto (x \cos\theta - (z - R)\sin\theta, \, y, \, x\sin\theta + (z - R)\cos\theta + R).$$

Writing $\theta = r/R$, where r is fixed, and letting $R \to \infty$, we obtain the translation $(x, y, 0) \mapsto (x + r, y, 0)$.

Since SU_2 is a double covering of SO_3, we obtain a double covering $\tilde{E}_2$ of E_2 as a limit of SU_2 ($\tilde{E}_2$ is obtained simply by replacing SO_2 by its double cover).

At the level of (complexified) Lie algebras, this geometric construction amounts to replacing the generators $X^\pm$ of $sl_2(\mathbb{C})$ by $RX^\pm$. Letting $R \to \infty$ gives the Lie algebra with defining relations

$$[H, X^+] = 2X^+, \quad [H, X^-] = -2X^-, \quad [X^+, X^-] = 0.$$

This is clearly isomorphic to the (complexified) Lie algebra $\mathfrak{e}_2$ of E_2: if τ_+, τ_- and ρ are, respectively, the generators of translations along the x- and y-axes and of rotations, set $H = 2\sqrt{-1}\rho$, $X^\pm = \frac{1}{2}(\tau_+ \pm \sqrt{-1}\tau_-)$.

Turning now to the quantum case, the preceding discussion suggests

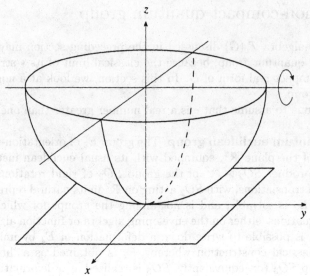

Fig. 21 E_2 as a limiting version of SO_3

DEFINITION–PROPOSITION 13.4.1 *(i) The* quantized universal enveloping al-
gebra *of the euclidean group* E_2 *is the associative algebra* $U_\epsilon(\mathfrak{e}_2)$ *over* $\mathbb{C}$ *with
generators* $L^{\pm 1}$, $X^\pm$ *and the following defining relations:*

(18) $$LX^\pm L^{-1} = \epsilon^{\pm 2}X^\pm, \quad [X^+, X^-] = 0.$$

*It is a Hopf *-algebra with structure maps given by*

$$\Delta_{U_\epsilon}(L) = L \otimes L,$$
(19) $$\Delta_{U_\epsilon}(X^+) = X^+ \otimes L^2 + 1 \otimes X^+, \quad \Delta_{U_\epsilon}(X^-) = X^- \otimes 1 + L^{-2} \otimes X^-,$$
$$S_{U_\epsilon}(L) = L^{-1}, \quad S_{U_\epsilon}(X^+) = -X^+L^{-2}, \quad S_{U_\epsilon}(X^-) = -L^2X^-,$$
$$\epsilon_{U_\epsilon}(L) = 1, \quad \epsilon_{U_\epsilon}(X^\pm) = 0,$$

*and *-structure*

(20) $$L^* = L, \quad (X^+)^* = \epsilon^{-1}X^-L^2, \quad (X^-)^* = \epsilon L^{-2}X^+.$$

(ii) The quantized algebra of functions *on the euclidean group* $\bar{E}_2$ *is the
associative algebra* $\mathcal{F}_\epsilon(\bar{E}_2)$ *over* $\mathbb{C}$ *with generators* a, b, c, d *and the following
defining relations:*

(21) $$ab = \epsilon^{-1}ba, \quad ac = \epsilon^{-1}ca, \quad bd = \epsilon^{-1}db, \quad cd = \epsilon^{-1}dc,$$
$$bc = cb, \quad ad = da = 1.$$

It is a Hopf *-algebra with coalgebra structure

(22)
$$\Delta_{\mathcal{F}_\epsilon} \begin{pmatrix} a & b \\ c & d \end{pmatrix} = \begin{pmatrix} a \otimes a & a \otimes b + b \otimes d \\ c \otimes a + d \otimes c & d \otimes d \end{pmatrix},$$
$$S_{\mathcal{F}_\epsilon} \begin{pmatrix} a & b \\ c & d \end{pmatrix} = \begin{pmatrix} d & -\epsilon b \\ -\epsilon^{-1} c & a \end{pmatrix}, \quad \epsilon_{\mathcal{F}_\epsilon} \begin{pmatrix} a & b \\ c & d \end{pmatrix} = \begin{pmatrix} 1 & 0 \\ 0 & 1 \end{pmatrix},$$

and *-structure

(23)
$$a^* = d, \quad b^* = -\epsilon^{-1} c.$$

The quantized algebra of functions on the euclidean group E_2 is the sub-algebra $\mathcal{F}_\epsilon(E_2)$ of $\mathcal{F}_\epsilon(\tilde{E}_2)$ spanned by the products of an even number of generators. It is a sub-Hopf *-algebra of $\mathcal{F}_\epsilon(\tilde{E}_2)$. ∎

Remark The formulas in (18) and (19) are obtained from those defining the 'simply-connected' form of $U_q(sl_2(\mathbb{C}))$ by replacing L_1 by L and $X_1^\pm$ by $RX^\pm$ and letting $R \to \infty$ (see Remark [4] following 9.1.1). Similarly, the formulas in (21)–(23) are obtained from those in (2)–(4) by replacing a, b, c and d by a, $R^{-1}b$, $R^{-1}c$ and d, respectively, and letting $R \to \infty$. △

Of course, we expect that $\mathcal{F}_\epsilon(\tilde{E}_2)$ will be isomorphic to an appropriately defined dual of $U_\epsilon(\mathfrak{e}_2)$. We shall be content with the following slightly weaker result. Let $E^{1/2}$ be the positive square root of E.

PROPOSITION 13.4.2 *There is a unique non-degenerate pairing of Hopf algebras $\langle\,,\,\rangle : U_\epsilon(\mathfrak{e}_2) \times \mathcal{F}_\epsilon(\tilde{E}_2) \to \mathbb{C}$ which is zero on all pairs of generators except*

$$\langle L, a \rangle = \epsilon^{1/2}, \quad \langle L, d \rangle = \epsilon^{-1/2}, \quad \langle X^+, b \rangle = \epsilon^{1/2}, \quad \langle X^-, c \rangle = \epsilon^{1/2}.$$

The notion of a non-degenerate pairing of Hopf algebras is defined in Subsection 4.1D.

OUTLINE OF PROOF It is not difficult to show that $\{L^p(X^+)^r(X^-)^s\}_{p \in \mathbb{Z}, r, s \in \mathbb{N}}$ is a basis of $U_\epsilon(\mathfrak{e}_2)$, and that the elements $a^l b^m c^n$, for $l \in \mathbb{Z}$, $m, n \in \mathbb{N}$, form a basis of $\mathcal{F}_\epsilon(\tilde{E}_2)$ (if $l < 0$, a^l is interpreted as d^{-l}). The pairing is defined on basis elements by

(24)
$$\langle L^p(X^+)^r(X^-)^s, a^l b^m c^n \rangle$$
$$= \delta_{r,m} \delta_{s,n} \epsilon^{\frac{1}{2}p(n+l-m)+m+n-nl-\frac{1}{2}(m^2+n^2)} \frac{(\epsilon^2; \epsilon^2)_m (\epsilon^2; \epsilon^2)_n}{(1-\epsilon^2)^{m+n}}.$$

One can check that the pairing (24) is invariant and non-degenerate∎

Remark One can define a quantized algebra $C_\epsilon(\tilde{E}_2)$ of continuous functions on $\tilde{E}_2$ which 'vanish at infinity'. However, this is technically more difficult to work with than in the compact case because it is a non-unital C*-algebra. In addition, we shall see shortly that the matrix elements of *-representations of $U_\epsilon(\mathfrak{e}_2)$ do not usually belong to $C_\epsilon(\tilde{E}_2)$ since they are 'unbounded'. This difficulty can be overcome by working with elements 'affiliated' with $C_\epsilon(\tilde{E}_2)$. See Podleś & Woronowicz (1990) and Woronowicz (1991b). $\triangle$

B Representation theory We now turn to the representation theory of $U_\epsilon(\mathfrak{e}_2)$ and $\mathcal{F}_\epsilon(\tilde{E}_2)$. As one would expect, this is a kind of 'limit' of the representation theory of $U_\epsilon(sl_2(\mathbb{C}))$ and $\mathcal{F}_\epsilon(SL_2(\mathbb{C}))$.

THEOREM 13.4.3 *(i) Every irreducible unitarizable representation of $U_\epsilon(\mathfrak{e}_2)$ on which L acts semisimply with finite-dimensional eigenspaces is equivalent to exactly one of the following:*

(a) for $\nu \in \mathbb{R}\backslash\{0\}$, the one-dimensional representation σ^ν given by

$$\sigma^\nu(L) = \nu, \quad \sigma^\nu(X^\pm) = 0;$$

(b) for λ, $\mu \in \mathbb{R}$ with $\lambda > 0$ and $1 \le |\mu| < \epsilon$, the representations $\rho_\pm^{\lambda,\mu}$ on the Hilbert space $\ell^2(\mathbb{Z})$ given by

$$\rho_\pm^{\lambda,\mu}(L)(e_k) = \pm \mu\epsilon^k e_k,$$
$$\rho_\pm^{\lambda,\mu}(X^+)(e_k) = \lambda\mu\epsilon^{k+1} e_{k+1},$$
$$\rho_\pm^{\lambda,\mu}(X^-)(e_k) = \pm \bar{\lambda}\mu^{-1}\epsilon^{1-k} e_{k-1},$$

in terms of the standard orthonormal basis $\{e_k\}_{k\in\mathbb{Z}}$ of $\ell^2(\mathbb{Z})$ (the reason for writing $\bar{\lambda}$ in the last formula is explained below).

(ii) Every irreducible unitarizable representation of $\mathcal{F}_\epsilon(\tilde{E}_2)$ is equivalent to exactly one of the following:

(a) for $t \in S^1$, the one-dimensional representation τ_t given by

$$\tau_t(a) = t, \quad \tau_t(b) = \tau_t(c) = 0, \quad \tau_t(d) = t^{-1};$$

(b) for $t \in \mathbb{C}$, the representation π_t on $\ell^2(\mathbb{Z})$ given by

$$\pi_t(a)(e_k) = e_{k-1}, \quad \pi_t(b)(e_k) = \bar{t}\epsilon^{-k-1} e_k,$$
$$\pi_t(c)(e_k) = t\epsilon^{-k} e_k, \quad \pi_t(d)(e_k) = e_{k+1}. \blacksquare$$

Remarks [1] The formulas in (i)(a) define an irreducible unitarizable representation $\rho_\pm^{\lambda,\mu}$ whenever $\lambda \in \mathbb{C}^\times$ and $\mu \in \mathbb{R}^\times$. However, it is easy to see that one has the following equivalences of representations:

(i) $\rho_{\pm}^{\lambda,\mu} \cong \rho_{\mp}^{\lambda,-\mu}$;

(ii) $\rho_{\pm}^{\lambda,\mu} \cong \rho_{\pm}^{\nu\lambda,\mu}$ for any $\nu \in \mathbb{C}$ with $|\nu| = 1$;

(iii) $\rho_{\pm}^{\lambda,\mu} \cong \rho_{\pm}^{\lambda,\mu\epsilon^r}$ for all $r \in \mathbb{Z}$.

Using these observations, one sees that every $\rho_{\pm}^{\lambda,\mu}$ is equivalent to one for which the parameters λ and μ satisfy the stated restrictions. Moreover, with these restrictions, the representations $\rho_{\pm}^{\lambda,\mu}$ are mutually inequivalent, as one easily sees by looking at the eigenvalues of the operators $\rho_{\pm}^{\lambda,\mu}(L)$ and $\rho_{\pm}^{\lambda,\mu}(X^+X^-)$.

[2] The representation $\rho_{+}^{\lambda,1}$ is a limit of the finite-dimensional representations of $U_\epsilon(sl_2(\mathbb{C}))$. In fact, one writes the basis of the $(2k+1)$-dimensional irreducible representation as $e_k, e_{k-1}, \ldots, e_{-k}$ (instead of the more customary $v_0, v_1, \ldots, v_{2k}$) and takes the radius of the sphere to be

$$R = \frac{\epsilon^k}{\lambda(\epsilon - \epsilon^{-1})},$$

so that $R \to \infty$ as $k \to \infty$. Similarly, the representations $\rho_{+}^{\lambda,\epsilon^{1/2}}$ can be obtained as a limit of the even-dimensional irreducible representations of $U_\epsilon(sl_2(\mathbb{C}))$. Finally, the representations $\rho_{-}^{\lambda,1}$ and $\rho_{-}^{\lambda,\epsilon^{1/2}}$ can be obtained by starting with representations of $U_\epsilon(sl_2(\mathbb{C}))$ which are not of type **1**.

[3] The operators $\rho_{\pm}^{\lambda,\mu}(L)$, $\rho_{\pm}^{\lambda,\mu}(X^+)$ and $\rho_{\pm}^{\lambda,\mu}(X^-)$ are clearly unbounded. However, all the operators in $\rho_{\pm}^{\lambda,\mu}(U_\epsilon(\mathfrak{e}_2))$ obviously have the set of finite linear combinations of the e_k as a common domain of definition.

[4] The algebra $U_\epsilon(\mathfrak{e}_2)$ has a three-parameter family of automorphisms given by arbitrary rescalings of the generators L, $X^{\pm}$. It is clear that the representations $\rho_{\pm}^{\lambda,\mu}$ can all be obtained from $\rho_{+}^{1,1}$ (say) by twisting with an automorphism of this type. $\triangle$

We shall study the representations of $U_\epsilon(\mathfrak{e}_2)$ and $\mathcal{F}_\epsilon(\tilde{E}_2)$ further in the next subsection.

C Invariant integration on the quantum euclidean group The classical euclidean group E_2 is unimodular, i.e. it admits a bi-invariant positive measure. Of course, since E_2 is non-compact, one can only integrate continuous functions which 'vanish sufficiently rapidly at infinity'; in particular, the measure cannot be normalized. We will now show how to obtain analogous results in the quantum case.

We recall that $\mathcal{F}_\epsilon(\tilde{E}_2)$ has a basis consisting of the monomials $a^l b^m c^n$ for $l \in \mathbb{Z}$, $m, n \in \mathbb{N}$. Equivalently, we may take the monomials $a^l c^m (c^* c)^n$ and $a^l (c^*)^m (c^* c)^n$. To define the appropriate class of quantized functions 'vanishing at infinity', we introduce the ϵ-*derivative* of a power series $\varphi \in \mathbb{C}[[z]]$:

$$(D_\epsilon \varphi)(z) = \frac{\varphi(z) - \varphi(\epsilon^{-1}z)}{(1 - \epsilon^{-1})z}, \quad (z \neq 0).$$

For $r > 0$, let $\mathcal{F}_\epsilon^{(r)}$ consist of all formal power series in the generators of $\mathcal{F}_\epsilon(\tilde{E}_2)$ of the form $a^l c^m \varphi(c^*c)$ or $a^l(c^*)^m \varphi(c^*c)$, where φ is a holomorphic function on $\mathbb{C}$ satisfying the following condition: for all $j \in \mathbb{N}$, $s > 0$, there exists a constant $C(j, s)$ for which

$$|(D_{\epsilon^2}^j \varphi)(r\epsilon^{2k})| < C(j, s)(1 + \epsilon^{2k})^{-s}, \quad \text{for all } k \in \mathbb{Z}.$$

LEMMA 13.4.4 For all $r > 0$, $\mathcal{F}_\epsilon^{(r)}$ is a $*$-algebra (with multiplication and $*$-structure obtained by extending those of $\mathcal{F}_\epsilon(\tilde{E}_2)$ in the obvious way). ∎

Now define linear functionals $H_r : \mathcal{F}_\epsilon^{(r)} \to \mathbb{C}$ by

$$H_r(a^l(c \text{ or } c^*)^m \varphi(c^*c)) = \delta_{l,0}\delta_{m,0} \sum_{k=-\infty}^{\infty} \varphi(r\epsilon^{2k})\epsilon^{2k}$$

(cf. 13.3.9).

PROPOSITION 13.4.5 For all $r > 0$, H_r is a Haar integral on $\mathcal{F}_\epsilon^{(r)}$, i.e.

$$H_r(Z *_L f) = \epsilon_{U_\epsilon}(Z)H_r(f) = H_r(Z *_R f)$$

for all $Z \in U_\epsilon(\mathfrak{e}_2)$, $f \in \mathcal{F}_\epsilon^{(r)}$. ∎

As in the case of compact quantum groups, one can use these Haar functionals to prove analogues of the Schur orthogonality relations.

PROPOSITION 13.4.6 Let σ and τ be irreducible $*$-representations of $U_\epsilon(\mathfrak{e}_2)$ on which K acts semisimply with finite-dimensional eigenspaces, and denote their matrix elements by σ_{jk}, τ_{st}. Assume that the products $\sigma_{jk}\tau_{st}^*$ and $\tau_{st}^*\sigma_{jk}$ all lie in $\mathcal{F}_\epsilon^{(r)}$ for some r.

(i) If σ and τ are inequivalent, then

$$H_r(\sigma_{jk}\tau_{st}^*) = H_r(\tau_{st}^*\sigma_{jk}) = 0$$

for all j, k, s, t.

(ii) We have

$$H_r(\tau_{st}\tau_{uv}^*) = c\langle K^{-1}, \tau_{tv}\rangle\delta_{s,u}, \quad H_r(\tau_{st}^*\tau_{uv}) = c\langle K^{-1}, \tau_{us}\rangle\delta_{t,v}$$

for some constant c independent of s, t, u and v. ∎

13.5 q-special functions

In this section, we describe the relation between three types of q-special functions and the representation theory of quantum groups and their quantum homogeneous spaces, with one example drawn from each of Sections 13.1, 13.2 and 13.4. We continue to assume that $\epsilon > 1$, and set $q = \epsilon^{-1}$.

All the q-special functions we shall be concerned with are special cases of the q-hypergeometric series, which are defined as follows. We have already defined the symbol

$$(a\,;q)_k = \prod_{j=0}^{\infty}(1 - aq^j), \quad k \in \mathbf{N}.$$

Note that, since $q < 1$, the limit

$$(a\,;q)_\infty = \lim_{k\to\infty}(a\,;q)_k$$

exists. Set

$$(a_1, a_2, \ldots, a_r\,;q)_k = (a_1\,;q)_k(a_2\,;q)_k\ldots(a_r\,;q)_k.$$

Then, the general q-hypergeometric series are the power series in a complex variable z given by
(25)

$$_r\varphi_s\left(\begin{array}{c}a_1,\ldots,a_r\\b_1,\ldots,b_s\end{array}; q, z\right) = \sum_{k=0}^{\infty}\frac{(a_1,\ldots,a_r\,;q)_k}{(q,b_1,\ldots,b_s\,;q)_k}\left((-1)^k q^{\frac{1}{2}k(k-1)}\right)^{s-r+1}z^k.$$

The terminology is justified by the observation that, if we replace a_i, b_i and z in the right-hand side of (25) by q^{a_i}, q^{b_i} and $(1-q)^{1+s-r}z$, respectively, and then let $q \uparrow 1$ in each term of the sum, we obtain the power series expansion of a classical hypergeometric function.

For generic values of the upper parameters $a_1,\ldots,a_r$ and lower parameters $b_1,\ldots,b_s$, the radius of convergence of this series is 0, 1 or ∞ according to whether $r - s > 1$, $r - s = 1$ or $r - s < 1$, respectively. However, if one of the upper parameters is equal to q^{-n} for some $n \in \mathbf{N}$, the series becomes a polynomial because $(q^{-n}\,;q)_k = 0$ if $k > n$. Further, if one of the lower parameters is equal to q^{-n} for some $n \in \mathbf{N}$, then the series is well defined only if one of the upper parameters is equal to q^{-m} for some $0 \le m \le n$.

A Little *q*-Jacobi polynomials and quantum SU_2 In our first example, we compute the matrix elements $C^\lambda_{\nu;\mu}$ of the irreducible $U_\epsilon(sl_2(\mathbb{C}))$-module $V_\epsilon(\lambda)$ with highest weight $\lambda \in \mathbb{N}$, with respect to an orthonormal basis (see Subsection 13.1A for the relevant definitions). It turns out that they are related to the *little q-Jacobi polynomials*, defined by

$$p_n(z\,;\,\alpha,\,\beta\,;q) = {}_2\varphi_1 \left(\begin{matrix} q^{-n}, q^{n+1}\alpha\beta \\ \alpha q \end{matrix}; q, qz \right).$$

PROPOSITION 13.5.1 *The matrix elements $C^\lambda_{\nu;\mu}$ of the irreducible $U_\epsilon(sl_2(\mathbb{C}))$-module of highest weight λ are given as follows:*

(i) if $\mu + \nu \geq 0$ and $\mu \geq \nu$,

$$C^\lambda_{\nu;\mu} = a^{(\mu+\nu)/2}c^{(\mu-\nu)/2}q^{(\lambda-\mu)(\nu-\mu)/4} \begin{bmatrix} \frac{1}{2}(\lambda - \nu) \\ \frac{1}{2}(\mu - \nu) \end{bmatrix}^{\frac{1}{2}}_{q^2} \begin{bmatrix} \frac{1}{2}(\lambda + \mu) \\ \frac{1}{2}(\mu - \nu) \end{bmatrix}^{\frac{1}{2}}_{q^2}$$
$$\times\; p_{\frac{1}{2}(\lambda-\mu)}(-q^{-1}bc; q^{\mu-\nu}, q^{\mu+\nu}; q^2);$$

(ii) if $\mu + \nu \geq 0$ and $\mu \leq \nu$,

$$C^\lambda_{\nu;\mu} = a^{(\mu+\nu)/2}b^{(\nu-\mu)/2}q^{(\lambda-\nu)(\mu-\nu)/4} \begin{bmatrix} \frac{1}{2}(\lambda + \nu) \\ \frac{1}{2}(\nu - \mu) \end{bmatrix}^{\frac{1}{2}}_{q^2} \begin{bmatrix} \frac{1}{2}(\lambda - \mu) \\ \frac{1}{2}(\nu - \mu) \end{bmatrix}^{\frac{1}{2}}_{q^2}$$
$$\times\; p_{\frac{1}{2}(\lambda-\nu)}(-q^{-1}bc; q^{\nu-\mu}, q^{\mu+\nu}; q^2);$$

(iii) if $\mu + \nu \leq 0$ and $\mu \leq \nu$,

$$C^\lambda_{\nu;\mu} = q^{-(\lambda+\mu)(\nu-\mu)/4} \begin{bmatrix} \frac{1}{2}(\lambda + \nu) \\ \frac{1}{2}(\nu - \mu) \end{bmatrix}^{\frac{1}{2}}_{q^2} \begin{bmatrix} \frac{1}{2}(\lambda - \mu) \\ \frac{1}{2}(\nu - \mu) \end{bmatrix}^{\frac{1}{2}}_{q^2}$$
$$\times\; p_{\frac{1}{2}(\lambda+\nu)}(-q^{-1}bc; q^{\mu-\nu}, q^{-\mu-\nu}; q^2)b^{(\nu-\mu)/2}d^{-(\mu+\nu)/2};$$

(iv) if $\mu + \nu \leq 0$ and $\mu \geq \nu$,

$$C^\lambda_{\nu;\mu} = q^{-(\lambda+\nu)(\mu-\nu)/4} \begin{bmatrix} \frac{1}{2}(\lambda - \nu) \\ \frac{1}{2}(\mu - \nu) \end{bmatrix}^{\frac{1}{2}}_{q^2} \begin{bmatrix} \frac{1}{2}(\lambda + \mu) \\ \frac{1}{2}(\mu - \nu) \end{bmatrix}^{\frac{1}{2}}_{q^2}$$
$$\times\; p_{\frac{1}{2}(\lambda+\mu)}(-q^{-1}bc; q^{\mu-\nu}, q^{-\mu-\nu}; q^2)c^{(\mu-\nu)/2}d^{-(\mu+\nu)/2}. \;\blacksquare$$

Proofs of these formulas may be found in Vaksman & Soibelman (1988), Masuda, Mimachi, Nakagami, Noumi & Ueno (1988, 1991), and Koornwinder (1989a), the first of which is closest in spirit to the methods used in this book.

Remark One interesting special case is that in which λ is even and $\mu = \nu = 0$. Then, one obtains
$$C_{00}^{\lambda} = p_{\lambda/2}(-q^{-1}bc; 1, 1; q^2),$$
which is a *little q-Legendre polynomial* with argument $-q^{-1}bc$. $\triangle$

Substituting the formulas in this proposition into the quantum Schur orthogonality relations in 13.3.11, one obtains the *orthogonality relations for the little q-Jacobi polynomials*:

COROLLARY 13.5.2 *For k, l, m, $n \in \mathbf{N}$ and $0 < q < 1$, we have*

$$\int_0^1 p_m(z; q^k, q^l; q)\, p_n(z; q^k, q^l; q) z^k (qz; q)_l d_q z =$$

$$\delta_{m,n} q^{n(k+1)} \frac{(1-q)(q; q)_k^2 (q; q)_{l+n}(q; q)_n}{(1 - q^{k+l+2n+1})(q; q)_{k+n}(q; q)_{k+l+n}},$$

where $\int_0^1 d_q z$ denotes the Jackson integral defined in 13.3.9. ∎

There are other connections between quantum SU_2 and q-special functions, but we do not have the space to discuss them here. We mention only the relation between quantum Clebsch–Gordan coefficients and *q-Hahn polynomials* obtained by Kirillov & Reshetikhin (1989), Koelink & Koornwinder (1989) and Vaksman (1989).

B Big q-Jacobi polynomials and quantum spheres Our second instance of the connection between the quantum and q-worlds concerns the quantized algebra $\mathcal{F}_\epsilon(S^2)$ of functions on the 2-sphere, introduced at the end of Section 13.2. We recall that $\mathcal{F}_\epsilon(S^2)$ is the subalgebra of $\mathcal{F}_\epsilon(SL_2(\mathbf{C}))$ consisting of the fixed points of the *-automorphisms given by $a \mapsto \zeta a$, $b \mapsto \zeta^{-1}b$, $c \mapsto \zeta c$, $d \mapsto \zeta^{-1}d$, $\zeta \in S^1$. The reader may check that an equivalent description is

(26) $$\mathcal{F}_\epsilon(S^2) = \{x \in \mathcal{F}_\epsilon(SL_2(\mathbf{C})) \mid K_1 *_R x = x\}.$$

(This is analogous to the description of S^2 as the homogeneous space SU_2/S^1, where S^1 is the diagonal matrices in SU_2.) It is easy to see that $\mathcal{F}_\epsilon(S^2)$ is generated, as an algebra over $\mathbf{C}$, by the elements $x = \epsilon ab$, $y = \epsilon cd$ and $z = -\epsilon bc$. Further, by using (2), one checks that these generators satisfy the following relations:

(27) $$xz = \epsilon^{-2}zx, \quad yz = \epsilon^2 zy,$$
$$xy = -\epsilon^{-1}z(1 - \epsilon^{-2}z), \quad yx = -\epsilon z(1 - z).$$

By (4), the *-structure on $\mathcal{F}_\epsilon(S^2)$ is given by

(28) $$x^* = -\epsilon^{-1}y, \quad y^* = -\epsilon x, \quad z^* = z.$$

In fact, we have

PROPOSITION 13.5.3 $\mathcal{F}_\epsilon(S^2)$ is isomorphic to the associative algebra over $\mathbb{C}$ with generators x, y, z, defining relations (27) and $*$-structure (28). The monomials $y^j z^k x^l$, with j, k, $l \in \mathbb{N}$, are a basis of $\mathcal{F}_\epsilon(S^2)$ over $\mathbb{C}$. Moreover, $\mathcal{F}_\epsilon(S^2)$ is a quantum $SL_2(\mathbb{C})$-space.

OUTLINE OF PROOF The last statement follows from the fact that the co-multiplication of $\mathcal{F}_\epsilon(SL_2(\mathbb{C}))$ maps $\mathcal{F}_\epsilon(S^2)$ into $\mathcal{F}_\epsilon(SL_2(\mathbb{C})) \otimes \mathcal{F}_\epsilon(S^2)$. This assertion in turn follows from the description of $\mathcal{F}_\epsilon(S^2)$ given in (26), together with the analogue of (16) for $*_R$ instead of $*_L$:

$$\Delta_{\mathcal{F}_\epsilon}(Z *_R f) = (1 \otimes Z) *_R \Delta_{\mathcal{F}_\epsilon}(f). \quad \blacksquare$$

It is well known that the algebra of polynomial functions on the classical 2-sphere, as a representation of SO_3, contains each irreducible representation exactly once. Precisely the same result holds in the quantum case:

PROPOSITION 13.5.4 As an $\mathcal{F}_\epsilon(SL_2(\mathbb{C}))$-comodule, $\mathcal{F}_\epsilon(S^2)$ decomposes into irreducibles as follows:

$$\mathcal{F}_\epsilon(S^2) \cong \bigoplus_{\lambda \in 2\mathbb{N}} V_\epsilon(\lambda),$$

where $V_\epsilon(\lambda)$ is the irreducible $\mathcal{F}_\epsilon(S^2)$-comodule of dimension $\lambda + 1$.

OUTLINE OF PROOF Since $\mathcal{F}_\epsilon(S^2)$ is a quantum $SL_2(\mathbb{C})$-space, it is a $U_\epsilon(sl_2(\mathbb{C}))$-module. Computing the action of U_ϵ, one checks that the only elements of $\mathcal{F}_\epsilon(S^2)$ which are killed by X_1^+ and are eigenvectors of K_1 are the scalar multiples of powers of x, and that $K_1 *_L x^r = \epsilon^{2r} x^r$. If we knew that $\mathcal{F}_\epsilon(S^2)$ is completely reducible, it would follow immediately that $\mathcal{F}_\epsilon(S^2)$ decomposes as stated as a $U_\epsilon(sl_2(\mathbb{C}))$-module, or equivalently as an $\mathcal{F}_\epsilon(SL_2(\mathbb{C}))$-comodule. But since $\mathcal{F}_\epsilon(S^2)$ is infinite dimensional, we cannot immediately deduce complete reducibility from 10.1.14. However, one can check that the finite-dimensional subspace $\mathcal{F}_\epsilon^d$ of $\mathcal{F}_\epsilon(S^2)$ spanned by the monomials $y^j z^k x^l$ with $j + k + l \leq d$ is a $U_\epsilon(sl_2(\mathbb{C}))$-submodule of $\mathcal{F}_\epsilon(S^2)$, and so is completely reducible by 10.1.14. It follows that the same is true of $\mathcal{F}_\epsilon(S^2)$, since

$$\mathcal{F}_\epsilon(S^2) = \sum_d \mathcal{F}_\epsilon^d. \quad \blacksquare$$

It follows from this proposition that there is a unique basis $\{S_\mu^\lambda\}$ of $\mathcal{F}_\epsilon(S^2)$, where $\mu = \lambda, \lambda - 2, \ldots, -\lambda$, $\lambda \in 2\mathbb{N}$, such that $S_0^0 = 1$ and

(29) $$\Delta_{\mathcal{F}_\epsilon}(S_\mu^\lambda) = \sum_\nu C_{\mu;\nu}^\lambda \otimes S_\nu^\lambda.$$

The S_μ^λ are q-analogues of spherical functions.

The main result of this subsection gives a formula for the S_μ^λ in terms of big q-Jacobi polynomials, which are defined as follows:

$$P_n(z; \alpha, \beta; q) = {}_3\varphi_2 \left(\begin{matrix} q^{-n}, q^{n+1}\alpha\beta, q\alpha z \\ q\alpha, 0 \end{matrix} ; q, q \right).$$

PROPOSITION 13.5.5 *The q-spherical functions S_μ^λ are given by the following formulas:*

(i) if $\mu \geq 0$,

$$S_\mu^\lambda = (-1)^{(\lambda-\mu)/2} q^{-(\lambda-\mu)(\lambda+3\mu+6)/8} \begin{bmatrix} \lambda \\ \frac{1}{2}(\lambda-\mu) \end{bmatrix}_{q^2}^{-1/2} \begin{bmatrix} \frac{1}{2}\lambda \\ \frac{1}{2}(\lambda-\mu) \end{bmatrix}_{q^2}$$
$$\times \; y^{\mu/2} P_{\frac{1}{2}(\lambda-\mu)}(z\,;\,q^\mu, q^\mu\,;\,q^2)\,;$$

(ii) if $\mu \leq 0$,

$$S_\mu^\lambda = (-1)^{(\lambda+\mu)/2} q^{-(\lambda+\mu)(\lambda-3\mu+6)/8} \begin{bmatrix} \lambda \\ \frac{1}{2}(\lambda+\mu) \end{bmatrix}_{q^2}^{-1/2} \begin{bmatrix} \frac{1}{2}\lambda \\ \frac{1}{2}(\lambda+\mu) \end{bmatrix}_{q^2}$$
$$\times \; P_{\frac{1}{2}(\lambda+\mu)}(z\,;\,q^{-\mu}, q^{-\mu}\,;\,q^2) x^{-\mu/2}. \; \blacksquare$$

The proof is a direct verification that the elements $S_\mu^\lambda \in \mathcal{F}_\epsilon(S^2)$ defined by these formulas satisfy (29). This uses certain properties of the big q-Jacobi polynomials which we shall not write down here.

Remark The quantum 2-sphere which we have studied is a special case of a two-parameter family introduced by Podleś (1987). In the form given by Noumi & Mimachi (1990c), the defining relations for the most general quantum 2-sphere are

$$xz = \epsilon^{-2} zx, \quad yz = \epsilon^2 zy,$$
$$xy = -\epsilon(\alpha + \epsilon^{-2}z)(\beta - \epsilon^{-2}z), \quad yx = -\epsilon(\alpha + z)(\beta - z),$$

where α and β are real constants (the $*$-structure is still given by (28)). Our quantum 2-sphere is the special case $\alpha = 0$, $\beta = 1$. $\triangle$

C q-Bessel functions and the quantum euclidean group We first recall the connection between the representations of the classical euclidean group E_2 and the Bessel functions

$$J_n(z) = \frac{1}{2\pi} \int_0^{2\pi} \exp(\sqrt{-1}(z \sin\theta - n\theta))\, d\theta \qquad (n \in \mathbb{Z},\ z \in \mathbb{C}).$$

We define, for any $s \in \mathbb{C}$, a representation ρ^s of E_2 on the Hilbert space $L^2(S^1)$ of square-integrable functions on the unit circle, as follows. The most general element of E_2 is a product of a rotation through an angle α about the origin, followed by a translation of $\mathbb{R}^2$ by a vector $(r \cos\phi, r \sin\phi)$; we denote this element by $g(r, \phi, \alpha)$. Then, if $f \in L^2(S^1)$,

$$\rho^s(g(r, \phi, \alpha))(f)(\theta) = \exp(rs \cos(\theta - \phi)) f(\theta - \alpha).$$

This representation is irreducible if $s \neq 0$, and unitary if s is purely imaginary. Moreover, every irreducible unitary representation of E_2 is equivalent to ρ^s for some $s \neq 0$, or to a (one-dimensional) irreducible component of ρ^0.

Bessel functions enter as matrix elements of ρ^s for $s \neq 0$. In fact, if

$$\rho^s_{mn}(g) = (\rho^s(g)(e^{\sqrt{-1}n\theta}),\, e^{\sqrt{-1}m\theta}),$$

it is easy to see that

$$\rho^s_{mn}(g(r,\phi,\alpha)) = (\sqrt{-1})^{n-m}\exp[-\sqrt{-1}(n\alpha + (n-m)\phi)]J_{m-n}(-\sqrt{-1}rs).$$

There are several q-analogues of Bessel functions in the literature, but we shall only be concerned with the *Hahn–Exton q-Bessel functions*

$$J_n(z\,;\,q) = z^n \frac{(q^{n+1};q)_\infty}{(q;q)_\infty}\,{}_1\varphi_1\left(\begin{matrix} 0 \\ q^{n+1} \end{matrix}\,;\,q, qz^2\right).$$

It can be shown that

$$\lim_{q\uparrow 1} J_n((1-q)z\,;\,q) = J_n(2z).$$

The following result gives the first connection between these q-Bessel functions and the representation theory of $U_\epsilon(\mathfrak{e}_2)$. It is easily proved using the formulas in 13.4.3(ii) and the pairing (24).

PROPOSITION 13.5.6 *The following formulas give the matrix elements of the representations* $\rho^{\lambda,1}_+$ *and* $\rho^{\lambda,\epsilon^{1/2}}_+$:

(i) *for* $j \geq k$,

$$(\rho^{\lambda,1}_+)_{jk} = \left(\frac{\lambda(1-q^2)}{q^{k+1}}\right)^{j-k}\frac{a^{j+k}b^{j-k}}{(q^2;q^2)_{j-k}}\,{}_1\varphi_1\left(\begin{matrix}0\\q^{2(j-k+1)}\end{matrix}\,;\,q^2,-(1-q^2)^2\lambda^2 q^{-2k-1}bc\right),$$

$$(\rho^{\lambda,\epsilon^{1/2}}_+)_{jk} = \left(\frac{\lambda(1-q^2)}{q^{k+\frac{3}{2}}}\right)^{j-k}\frac{a^{j+k+1}b^{j-k}}{(q^2;q^2)_{j-k}}\,{}_1\varphi_1\left(\begin{matrix}0\\q^{2(j-k+1)}\end{matrix}\,;\,q^2,-(1-q^2)^2\lambda^2 q^{-2k-2}bc\right);$$

(ii) *for* $j \leq k$,

$$(\rho^{\lambda,1}_+)_{jk} = \left(\frac{\lambda(1-q^2)}{q^{j+1}}\right)^{k-j}\frac{a^{j+k}c^{k-j}}{(q^2;q^2)_{k-j}}\,{}_1\varphi_1\left(\begin{matrix}0\\q^{2(k-j+1)}\end{matrix}\,;\,q^2,-(1-q^2)^2\lambda^2 q^{-2j-1}bc\right),$$

$$(\rho^{\lambda,\epsilon^{1/2}}_+)_{jk} = \left(\frac{\lambda(1-q^2)}{q^{j+\frac{3}{2}}}\right)^{k-j}\frac{a^{j+k+1}c^{k-j}}{(q^2;q^2)_{k-j}}\,{}_1\varphi_1\left(\begin{matrix}0\\q^{2(k-j+1)}\end{matrix}\,;\,q^2,-(1-q^2)^2\lambda^2 q^{-2j-2}bc\right). \quad\blacksquare$$

Remark Note that these matrix elements do not belong to $\mathcal{F}_\epsilon(\tilde{E}_2)$ because they are not given by polynomials in the generators a, b, c and d. However, they are all contained in the space $\mathcal{F}_\epsilon^{\mathrm{ext}}(\tilde{E}_2)$ consisting of the formal linear combinations

$$\sum_{l,m,n} \gamma_{l,m,n} a^l b^m c^n,$$

where $l \in \mathbb{Z}$, m, $n \in \mathbb{N}$, and where the constants $\gamma_{l,m,n}$ vanish for all m, n except for finitely many values of l. One can check that the multiplication in $\mathcal{F}_\epsilon(\tilde{E}_2)$, as well as the left and right actions of $U_\epsilon(\mathfrak{e}_2)$, extend to $\mathcal{F}_\epsilon^{\mathrm{ext}}(\tilde{E}_2)$. Moreover, $\mathcal{F}_\epsilon^{\mathrm{ext}}(\tilde{E}_2)$ contains the domains $\mathcal{F}_\epsilon^{(r)}$ of all the Haar functionals H_r. $\triangle$

Using 13.5.6 and the unitarity property $\sum_i \rho_{k,i} \rho_{j,i}^* = \delta_{i,j}$ for $\rho = \rho_+^{\lambda,1}$ and $\rho_+^{\lambda,\epsilon^{1/2}}$, one obtains the *Hansel–Lommel orthogonality relations*:

COROLLARY 13.5.7 *The Hahn–Exton q-Bessel functions $J_n(z\,;q)$ satisfy the following identities:*

$$\sum_{k\in\mathbb{Z}} q^k J_{n+k}(z\,;q) J_{m+k}(z\,;q) = \delta_{m,n} q^{-n},$$

for $|z| < q^{-1/2}$, m, $n \in \mathbb{Z}$. ∎

Although these orthogonality relations were known before the advent of quantum groups, we conclude this section by giving an 'addition formula' for the Hahn–Exton q-Bessel functions, which was discovered by Koelink (1991b) using quantum group methods.

THEOREM 13.5.8 *The Hahn–Exton q-Bessel functions satisfy the following identity:*

$$(-q)^p J_{l-p}(q^{n-m}\,;q^2) J_p(\lambda q^{n+p}\,;q^2)$$
$$= \sum_{k\in\mathbb{Z}} q^{2k} J_k(\lambda q^{l+m}\,;q^2) J_{k-p}(\lambda q^m\,;q^2) J_l(q^{n-m+k}\,;q^2)$$

for all $\lambda > 0$, l, m, n, $p \in \mathbb{Z}$.

OUTLINE OF PROOF The idea is to start with the identity

$$\Delta_{\mathcal{F}_\epsilon}(\rho_{0n}^{\lambda,1}) = \sum_{k\in\mathbb{Z}} \rho_{0k}^{\lambda,1} \otimes \rho_{kn}^{\lambda,1}$$

in $\mathcal{F}_\epsilon^{\mathrm{ext}}(\tilde{E}_2)$. Applying the representation $\pi_1 \otimes \pi_1$ to both sides gives an equation between operators on the Hilbert space tensor product $\ell^2(\mathbb{Z}) \bar{\otimes} \ell^2(\mathbb{Z})$, from which we extract a scalar identity by applying both sides to two cleverly chosen vectors and then taking inner products.

In a little more detail, we define the vector

$$f_l^m = \sum_{n \in \mathbf{Z}} (-1)^n q^{m-n} J_l(q^{m-n} ; q^2)\, e_{l+n} \otimes e_n.$$

A direct computation shows that

(30) $(\pi_1 \otimes \pi_1)\Delta_{\mathcal{F}_\epsilon}(c^*c)(f_l^m) = q^{2m} f_l^m.$

Moreover, f_l^m is a unit vector: the proof of this uses the *q-Hankel orthogonality relations*

$$\sum_{k \in \mathbf{Z}} q^k J_r(q^{\frac{1}{2}(n+k)} ; q) J_r(q^{\frac{1}{2}(m+k)} ; q) = \delta_{m,n} q^{-n}$$

for $r > -1$, $m, n \in \mathbf{Z}$, which follow from the Schur orthogonality relations in 13.4.6.

Now we compute each side of the inner product:

(31)
$$((\pi_1 \otimes \pi_1)(\Delta_{\mathcal{F}_\epsilon}(\rho_{0,n}^{\lambda,1}))(e_u \otimes e_v),\ f_l^m) = \sum_{k \in \mathbf{Z}}(((\pi_1 \otimes \pi_1)(\rho_{0,k}^{\lambda,1} \otimes \rho_{k,n}^{\lambda,1}))(e_u \otimes e_v),\ f_l^m).$$

On the one hand, the right-hand side can be computed directly using the formulas in 13.5.6; one finds that it is equal to

$$\sum_{k,p \in \mathbf{Z}} (-1)^v q^{m-v+2k} J_k(\lambda(1-q^2)q^{u-1} ; q^2) J_{k-n}(\lambda(1-q^2)q^{v-n-1} ; q^2)$$

$$\times\ J_l(q^{m-v+k+n} ; q^2)\, \delta_{u-v,l-n}.$$

On the other hand, the property (30) of f_l^m enables one to compute the action of the hypergeometric series in 13.5.6 on f_l^m. Using this, and the fact that π_1 is a *-representation, one finds that, if $n > 0$, the left-hand side of (31) is equal to

$$q^{-n(n+m)} J_n(\lambda(1-q^2)q^{n+m-1} ; q^2)$$
$$\times \sum_{k \in \mathbf{Z}} (-1)^{v-k} \begin{bmatrix} n \\ k \end{bmatrix}_{q^2} q^{(u-k)(n-k)+vk+m-v+2n-k} J_l(q^{m-v+2n-k} ; q^2)\, \delta_{u-v,l-n}.$$

Taking $u - v = l - n$, replacing λ by $\lambda q^{1/2}$, and renaming the indices suitably, the equality of the last two expressions can be seen to be equivalent to the identity in the statement of the theorem. ∎

Remark Because the matrix elements involved are given by formal power series, the argument leading to the formula in 13.5.8 must be supplemented by an analytic proof. This has been supplied by Kalnins, Miller & Mukherjee (1994) and Koelink & Swarttouw (1994). Different addition formulas are proved in Koornwinder & Swarttouw (1992). △

Bibliographical notes

13.1 The representation theory of compact quantum groups began with the work of Vaksman & Soibelman (1988) on the sl_2 case. For the sl_{n+1} case, see Soibelman (1990a) and Vaksman & Soibelman (1991), and, for the general case, see Levendorskii & Soibelman (1991a). For the twisted case, see Levendorskii (1990, 1992) and Levendorskii & Soibelman (1991a).

The sl_2 case has been developed by a number of authors by starting with the definition of $\mathcal{F}_\epsilon(SL_2(\mathbb{C}))$ in terms of generators and relations. See, for example, Masuda, Mimachi, Nakagami, Noumi & Ueno (1988, 1991) and Koelink & Koornwinder (1989). The sl_{n+1} case is treated by the same methods by Koelink (1991a).

The *-representations π_t of $\mathcal{F}_\epsilon(SL_2(\mathbb{C}))$ were studied by Arik & Coon (1976) and Feinsilver (1987) in connection with a q-analogue of the Schrödinger representation, independently of quantum groups.

The results on quantized function algebras at roots of unity are taken from de Concini & Lyubashenko (1993).

13.2 Quantum spheres were first introduced by Podleś (1987); see also Podleś (1989) and Noumi & Mimachi (1990c); we have followed the latter reference. For quantum flag manifolds and Schubert varieties, see Lakshmibai & Reshetikhin (1991, 1992) and Soibelman (1992c).

13.3 The theory of compact matrix quantum groups is due to Woronowicz, and the reader should refer to his papers in the references, beginning with Woronowicz (1987a,b).

The fact that $\mathcal{C}_\epsilon(G)$ is of type I was proved by Vaksman & Soibelman (1988) in the sl_2 case, by Bragiel (1989) in the sl_3 case, and by Koelink (1991b) in the sl_{n+1} case; it is stated in Levendorskii & Soibelman (1991a) in the general case. For the twisted case, see Levendorskii (1990, 1992) and Levendorskii & Soibelman (1991a).

The Haar integral on $\mathcal{F}_\epsilon(SL_2(\mathbb{C}))$ has been computed by a number of authors: see, for example, Woronowicz (1987a) and Masuda *et al.* (1988, 1991). Proposition 13.3.10 is due to Soibelman & Vaksman (1992).

13.4 The results of this section have been taken from Koelink (1991b); see also Koelink (1994a,b) and Vaksman & Korogodskii (1989). A number of other non-compact quantum groups have been treated in the literature. See Masuda *et al.* (1990a, b) and Soibelman & Vaksman (1992) for quantum $SU_{1,1}$, and Podleś & Woronowicz (1990) and Takeuchi (1992b) for the quantum Lorentz group. A discussion of the difficulties in dealing with quantized algebras of continuous functions on non-compact quantum groups is given in Woronowicz (1991b).

13.5 The connection between quantum sl_2 and little q-Jacobi polynomials was first noticed by Vaksman & Soibelman (1988); see also Masuda *et al.* (1988, 1991) and Koornwinder (1989a). For the connection between quantum Clebsch–Gordan coefficients and q-Hahn polynomials, see Koelink & Koornwinder (1989) and Vaksman (1989). Our discussion of quantum 2-spheres and big q-Jacobi polynomials is taken from Noumi & Mimachi (1990c). The relation between q-Bessel functions and the quantum euclidean group is due to Vaksman & Korogodskii (1989) and Koelink (1991b); we have followed the latter.

For an excellent general survey of the connections between quantum groups and q-analysis, see Koornwinder (1990). For general background in the classical theory of q-special functions, see Exton (1983) and Gasper & Rahman (1990).

14
Canonical bases

This is the first of three chapters which describe some applications of quantum groups to problems in 'classical' mathematics. The problem addressed in this chapter is that of giving a 'canonical' basis of any (finite-dimensional) representation of a complex simple Lie algebra $\mathfrak{g}$. Roughly speaking, this means bases which are independent of any arbitrary choices, at least up to an overall scalar multiple. Such bases should also have good properties: for example, the action of the generators of $\mathfrak{g}$ in the basis should be 'simple', the bases should behave well under tensor products, etc.

The construction of canonical bases goes back, at least, to the work of W. V. D. Hodge in the sl_m case in the 1940s, subsequently elaborated into 'standard monomial theory' by V. Lakshmibai, C. Musili and C. Seshadri, which applies to any $\mathfrak{g}$ of classical type. Another canonical basis, and perhaps the simplest, is that of I. M. Gel'fand & M. L. Tsetlin; it applies when $\mathfrak{g} = sl_m$ or so_m. There are several other classical approaches to this problem, but all of them seem to be restricted to the case when $\mathfrak{g}$ is of classical type, and often to the sl_m case. Thus, it is remarkable that the approaches using quantum groups we shall discuss apply when $\mathfrak{g}$ is an arbitrary symmetrizable Kac–Moody algebra.

In Section 14.1, we discuss the construction of *crystal bases*, due to M. Kashiwara. These are bases of any finite-dimensional $U_q(\mathfrak{g})$-module V with the surprising property that the action of the generators $X_i^{\pm}$ of $U_q(\mathfrak{g})$ on the crystal basis of V, when suitably 'renormalized', makes sense at $q = 0$ (we continue to assume that $\mathfrak{g}$ is finite dimensional, although, as we have said, Kashiwara's construction extends to the infinite-dimensional case). In fact, it was an observation, by E. Date, M. Jimbo and T. Miwa, that one can make sense of the action of $U_q(sl_m(\mathbb{C}))$ on its finite-dimensional representations at $q = 0$, and that the quantum analogue of the Gel'fand–Tsetlin basis then becomes drastically simpler, that provided the starting point for Kashiwara's theory. In the application of quantum groups to vertex models in statistical mechanics (see Section 7.5), the parameter q essentially plays the role of temperature, so crystal bases are related to the behaviour of the model at 'absolute zero', whence their name.

The construction of crystal bases is completely algebraic, making use only of the structural properties of $U_q(\mathfrak{g})$-modules developed in Chapter 10. One of the most important properties of crystal bases is that they behave very simply under tensor products: this leads to a simple graphical algorithm for decomposing tensor products of irreducible representations of $U_q(\mathfrak{g})$. Specializing q to 1, we obtain a solution of the corresponding classical problem which is completely different in nature from that provided by the well-known formula of R. Steinberg.

G. Lusztig gave a quite different construction of canonical bases, which we outline in Section 14.2. It is based on the relation between quivers and quantum groups discussed in Subsection 9.3D. The set of isomorphism classes of representations of a given dimension of a quiver can be naturally identified with the set of orbits of an algebraic group on a finite-dimensional vector space. The coefficients of the canonical basis elements in terms of the elements of a Poincaré–Birkhoff–Witt basis (see 9.1.3) turn out to be essentially the Euler characteristic of the local intersection cohomology of the closures of these orbits. Note that the PBW bases themselves are not canonical, for they depend on the choice of a reduced decomposition of the longest element of the Weyl group of $\mathfrak{g}$.

We begin our discussion of Lusztig's canonical bases by constructing them using an elementary algebraic method, although the topological approach seems to be necessary to derive many of their most important properties. In fact, we construct a basis of the subalgebra U_q^+ of $U_q(\mathfrak{g})$ which has the property that, when this basis is applied to the lowest weight vector of any finite-dimensional irreducible $U_q(\mathfrak{g})$-module V, the result is precisely the canonical basis of V (together with zero). It turns out that this is the same as the basis produced by Kashiwara's theory.

It is perhaps worth emphasizing that, while these bases specialize to give canonical bases of all the finite-dimensional irreducible $\mathfrak{g}$-modules, the construction of crystal bases, as well as the algebraic construction of Lusztig's bases, can only be carried out at the quantum level. Thus, quantum groups appear to be essential for the solution of the classical problem with which we began this introduction.

We also remark that Lusztig's construction is very reminiscent of the well-known construction of special bases for Hecke algebras, due to D. Kazhdan and G. Lusztig. This supports the idea, already suggested by the results of Section 10.2, that Hecke algebras should be thought of as 'quantum' objects.

14.1 Crystal bases

A Gel'fand–Tsetlin bases Perhaps the most well-known classical construction of canonical bases is that due to Gel'fand & Tsetlin (1959a,b). We begin with a brief description of the quantum analogue of this construction, as it provides motivation for the introduction of crystal bases later in this section. It is based on the following result (see Jimbo (1986a)). Note that there are embeddings $U_q(sl_n(\mathbb{C})) \to U_q(sl_{n+1}(\mathbb{C}))$ given by $K_i \mapsto K_i$, $X_i^{\pm} \mapsto X_i^{\pm}$ for $i = 1, \ldots, n-1$ (see Section 9.1 for the definition of the $\mathbb{Q}(q)$-Hopf algebra $U_q(\mathfrak{g})$).

LEMMA 14.1.1 *Let V be a finite-dimensional irreducible type $\mathbf{1}$ $U_q(sl_{n+1}(\mathbb{C}))$-module, and regard V as a $U_q(sl_n(\mathbb{C}))$-module via the above embedding. Then, each irreducible $U_q(sl_n(\mathbb{C}))$-module which occurs in V does so with multiplicity one.* ∎

This result implies that the decomposition of V into irreducible $U_q(sl_n(\mathbb{C}))$-modules

$$V = \bigoplus_r V^{(r)}$$

is unique. Hence, GT bases may be constructed by induction on n: one simply takes the GT basis of V to be the union of the GT bases of the $V^{(r)}$. The induction starts with $U_q(sl_2(\mathbb{C}))$. If we choose the bases of the irreducible representations of $U_q(sl_2(\mathbb{C}))$ as in 14.1.6 below, the resulting GT basis is uniquely determined up to scalar multiples.

Remark In the classical situation, GT bases also exist for representations of $so_n(\mathbb{C})$, since the classical analogue of 14.1.1 holds for the natural inclusions $so_n(\mathbb{C}) \hookrightarrow so_{n+1}(\mathbb{C})$. However, there is no obvious homomorphism $U_q(so_n(\mathbb{C})) \to U_q(so_{n+1}(\mathbb{C}))$ which might have the property 14.1.1, essentially because the Dynkin diagram of $so_n(\mathbb{C})$ is not a subdiagram of that of $so_{n+1}(\mathbb{C})$ (of course, this is true classically, but there we have a 'geometric' description of $so_n(\mathbb{C})$). Thus, it is unclear how to construct bases of GT type for $U_q(so_n(\mathbb{C}))$-modules (see, however, Takeuchi (1989)). And for the algebras of exceptional type, GT bases do not even exist in the classical case. △

Let $V_q^{\natural}$ be the natural irreducible $(n+1)$-dimensional $U_q(sl_{n+1}(\mathbb{C}))$-module with basis $\{v_0^{\natural}, \ldots, v_n^{\natural}\}$ and action:

(1) $\quad K_i.v_j^{\natural} = q^{\delta_{i,j-1} - \delta_{i,j}} v_j^{\natural}, \quad X_i^+.v_j^{\natural} = \delta_{i,j} v_{j-1}^{\natural}, \quad X_i^-.v_j^{\natural} = \delta_{i,j+1} v_{j+1}^{\natural}.$

LEMMA 14.1.2 *If V is a finite-dimensional irreducible $U_q(sl_{n+1}(\mathbb{C}))$-module of type $\mathbf{1}$, then every irreducible $U_q(sl_{n+1}(\mathbb{C}))$-module which occurs in $V \otimes V_q^{\natural}$ does so exactly once.* ∎

This follows, as in the classical case, from the fact that the non-zero weight spaces of $V_q^{\natural}$ all have dimension one. (Alternatively, one may deduce the quantum statement from the classical one using the fact that the decomposition of tensor products of $U_q(sl_{n+1}(\mathbb{C}))$-modules is the 'same' as for $sl_{n+1}(\mathbb{C})$-modules (see 10.1.16).)

With V as in the lemma, let V' be any irreducible component of $V \otimes V_q^\natural$, and let $\{v'_\alpha\}$, $\{v_\beta\}$ and $\{v_\gamma^\natural\}$ be the GT bases of V', V and $V_q^\natural$, respectively. Then,

$$v'_\alpha = \sum_{\beta,\gamma} c_\alpha^{\beta\gamma} v_\beta \otimes v_\gamma^\natural$$

for some constants $c_\alpha^{\beta\gamma} \in \mathbb{Q}(q)$. By computing the coefficients $c_\alpha^{\beta\gamma}$ explicitly, Date, Jimbo & Miwa (1990) made the following striking observation. Let $\mathcal{A}^0$ be the subring of $\mathbb{Q}(q)$ consisting of the rational functions which are regular at $q = 0$.

PROPOSITION 14.1.3 *With the above notation, the constants* $c_\alpha^{\beta\gamma} \in \mathcal{A}^0$ *for all* α, β, γ. *In fact, for each* α, $c_\alpha^{\beta\gamma} \in q\mathcal{A}^0$ *except for precisely one pair* (β,γ) *(depending on* α). ∎

COROLLARY 14.1.4 *Let* V *be any irreducible component of an iterated tensor product* $(V_q^\natural)^{\otimes k}$ *for some* $k \geq 2$, *and let* $\{v_\beta\}$ *and* $\{v_\gamma^\natural\}$ *be the Gel'fand–Tsetlin bases of* V *and* $V_q^\natural$, *respectively. Let*

$$v_\beta = \sum_{\gamma_1,\ldots,\gamma_k} c_\beta^{\gamma_1 \ldots \gamma_k} v_{\gamma_1}^\natural \otimes \cdots \otimes v_{\gamma_k}^\natural.$$

Then, $c_\beta^{\gamma_1 \ldots \gamma_k} \in \mathcal{A}^0$ *and, for each* β, $c_\beta^{\gamma_1 \ldots \gamma_k} \in q\mathcal{A}^0$, *except for precisely one* k-*tuple* $(\gamma_1,\ldots,\gamma_k)$ *(depending on* β). ∎

In other words, 'at $q = 0$', each GT basis vector of V becomes, up to a scalar multiple, a 'monomial' $v_{\gamma_1}^\natural \otimes \cdots \otimes v_{\gamma_k}^\natural$ in the GT basis vectors of $V_q^\natural$. This is the first suggestion that the representations of $U_q(sl_{n+1}(\mathbb{C}))$ are meaningful, and drastically simpler, when '$q = 0$'. The crystal bases described in the remainder of this section are the natural development of this observation.

Remark By using the *Robinson–Schensted correspondence*, it is possible to describe exactly which monomial $v_{\gamma_1}^\natural \otimes \cdots \otimes v_{\gamma_k}^\natural$ is equal to v_β at $q = 0$. See Date, Jimbo & Miwa (1990). △

B Crystal bases Throughout the rest of this section, $\mathfrak{g}$ will denote a finite-dimensional complex simple Lie algebra. We denote the $\mathbb{Q}(q)$-Hopf algebra $U_q(\mathfrak{g})$ simply by U_q from now on if there is no risk of ambiguity.

If V is a finite-dimensional $U_q(\mathfrak{g})$-module of type **1**, we may regard V as a module for the $U_{q_i}(sl_2(\mathbb{C}))$-subalgebra of $U_q(\mathfrak{g})$ generated by $X_i^\pm$ and $K_i^{\pm 1}$. By the results of Chapter 10, any such module is generated by its primitive vectors, i.e. vectors lying in some weight space V_λ of V which are annihilated by X_i^+. By the Poincaré–Birkhoff–Witt decomposition 9.1.3, we thus have

$$V = \bigoplus (X_i^-)^{(r)}.(V_\lambda \cap \ker X_i^+),$$

where the sum runs over the pairs $(\lambda, r) \in P \times \mathbb{Z}$ such that $(\lambda, \check{\alpha}_i) \geq r \geq 0$ (recall that $\check{\alpha}_i = d_i^{-1}\alpha_i$). We may therefore define operators $\tilde{x}_i^{\pm} \in \text{End}(V)$ by

$$\tilde{x}_i^{\pm}((X_i^-)^{(r)}.v) = (X_i^-)^{(r \mp 1)}.v$$

for any $v \in V_\lambda \cap \ker X_i^+$ and any (λ, r) as above.

DEFINITION 14.1.5 *A crystal basis of a finite-dimensional $U_q(\mathfrak{g})$-module V is a pair (V^0, B_V) consisting of a free $\mathcal{A}^0$-module V^0 and a basis B_V of the $\mathbb{Q}$-vector space V^0/qV^0 with the following properties:*

(i) V^0 is an $\mathcal{A}^0$-form of V, i.e. $V \cong V^0 \underset{\mathcal{A}^0}{\otimes} \mathbb{Q}(q)$;

(ii) $V^0 = \oplus_\lambda V_\lambda^0$ and $B_V = \coprod_\lambda B_\lambda$, where $B_\lambda = B_V \cap (V_\lambda^0/qV_\lambda^0)$ and $V_\lambda^0 = V^0 \cap V_\lambda$;

(iii) $\tilde{x}_i^{\pm}(V^0) \subseteq V^0$ and $\tilde{x}_i^{\pm}(B_V) \subseteq B_V \cup \{0\}$;

(iv) for any $i = 1, \ldots, n$ and any $u, v \in B_V$, one has $v = \tilde{x}_i^-(u)$ if and only if $u = \tilde{x}_i^+(v)$.

Before we address the question of the existence and uniqueness of crystal bases in general, let us consider the sl_2 case.

Example 14.1.6 Let $V_q(m)$ be the $(m+1)$-dimensional irreducible $U_q(sl_2(\mathbb{C}))$-module of type **1**. Thus, $V_q(m)$ has a basis $\{v_0, \ldots, v_m\}$ such that

$$K_1.v_r = q^{m-2r}v_r, \quad X_1^+.v_r = [m-r+1]_q v_{r-1}, \quad X_1^-.v_r = [r+1]_q v_{r+1}$$

(we set $v_{-1} = v_{m+1} = 0$). The operators $\tilde{x}_1^{\pm}$ in this case are given by $\tilde{x}_1^{\pm}(v_r) = v_{r \mp 1}$. We may therefore take $V_q(m)^0$ to be the free $\mathcal{A}^0$-module with basis $\{v_0, \ldots, v_m\}$, and the images of these vectors in $V_q(m)^0/qV_q(m)^0$ as the basis $B_{V_q(m)}$. In fact, it is clear that, up to an overall scalar multiple, this is the only possible choice for $B_{V_q(m)}$. $\Diamond$

Returning to the general case, denote as usual by $V_q(\lambda)$ the irreducible $U_q(\mathfrak{g})$-module of type **1** with highest weight $\lambda \in P^+$ (see Chapter 10). Let $V_q(\lambda)^0$ be the $\mathcal{A}^0$-submodule of $V_q(\lambda)$ generated by the vectors of the form $\tilde{x}_{i_1}^- \circ \cdots \circ \tilde{x}_{i_k}^-(v_\lambda)$, where $k \geq 0$ and v_λ is the highest weight vector of $V_q(\lambda)$, and let $B_{V_q(\lambda)} = B(\lambda)$ consist of the images of these vectors in $V_q(\lambda)^0/qV_q(\lambda)^0$, deleting those which are zero. We can now state the main result of this section.

THEOREM 14.1.7 *With the above notation, the pair $(V_q(\lambda)^0, B(\lambda))$ is a crystal basis of $V_q(\lambda)$. More generally, if $V = \oplus_\lambda V_q(\lambda)$ is a finite direct sum of irreducible modules, and we define $V^0 = \oplus_\lambda V_q(\lambda)^0$, $B_V = \coprod_\lambda B(\lambda)$, then (V^0, B_V) is a crystal basis of V.*

*Conversely, let V be a finite-dimensional $U_q(\mathfrak{g})$-module of type **1**, and let $V \cong \oplus_\lambda V_q(\lambda)$ be a decomposition of V into irreducible modules. If (V^0, B_V) is* any crystal basis of V, there is an isomorphism $V^0 \cong \oplus_\lambda V_q(\lambda)^0$ of $\mathcal{A}^0$-modules which carries B_V bijectively onto $\coprod_\lambda B(\lambda)$. $\blacksquare$

We shall not discuss the proof of this theorem, which is given in complete detail in Kashiwara (1991).

C Globalization We now show how to 'melt' a crystal basis (V^0, B_V): we shall obtain, from the basis B_V of V^0/qV^0, a basis of V itself. This is done by first constructing a basis of U_q^- (cf. Section 9.1) which has some of the properties of a crystal basis, and then using the fact that V is generated as a U_q^--module by its primitive vectors (this follows from 9.1.3).

We begin with the following simple lemma.

LEMMA 14.1.8 _For any $X \in U_q^-$ and any $i = 1, \ldots, n$, there exist unique elements $Y_i, Z_i \in U_q^-$ such that_

$$(2) \qquad\qquad [X_i^+, X] = \frac{K_i Y_i - K_i^{-1} Z_i}{q_i - q_i^{-1}}.$$

OUTLINE OF PROOF Uniqueness is a consequence of 9.1.3. As for existence, since any element of U_q^- is a linear combination of products of X_i^-'s, and since we can take $Y_i = Z_i = 0$ if $X = 1$, it is enough to show that, if the result holds for X, it also holds for any product $X_j^- X$. This is proved by a simple computation. $\blacksquare$

We may therefore define endomorphisms y_i and z_i of U_q^- by $y_i(X) = Y_i$, $z_i(X) = Z_i$ for all $X \in U_q^-$, with Y_i and Z_i as in (2).

LEMMA 14.1.9 _For all $i = 1, \ldots, n$, we have_

$$U_q^- = \bigoplus_{r=0}^{\infty} (X_i^-)^{(r)}.\ker y_i. \qquad \blacksquare$$

This is analogous to the statement that every finite-dimensional U_q-module V is generated by its primitive vectors. Thus, we can imitate the construction of crystal bases of V by defining endomorphisms $\tilde{x}_i^\pm$ of U_q^- by

$$\tilde{x}_i^\pm((X_i^-)^{(r)}.X) = (X_i^-)^{(r\mp 1)} X$$

for all $r \geq 0$ and $X \in \ker y_i$. Note that, by 14.1.9, $\tilde{x}_i^+ \circ \tilde{x}_i^- = \mathrm{id}$. Let U_0^- be the $\mathcal{A}^0$-submodule of U_q^- generated by the elements of the form $x_{i_1}^- \circ \cdots \circ x_{i_k}^-(1)$ for all $k \geq 0$, and let B_0 be the set of images of these elements in U_0^-/qU_0^-, discarding those which are zero. Then, we have the following analogue of the properties in 14.1.5:

PROPOSITION 14.1.10 *With the above notation,*

(i) B_0 *is a basis of the $\mathbb{Q}$-vector space U_0^-/qU_0^- ;*
(ii) $\tilde{x}_i^\pm(U_0^-) \subseteq U_0^-$ *and* $\tilde{x}_i^\pm(B_0) \subseteq B_0 \cup \{0\}$ *(in fact,* $\tilde{x}_i^-(B_0) \subseteq B_0$*);*
(iii) if $a, b \in B_0$, then $a = \tilde{x}_i^+(b)$ if and only if $b = \tilde{x}_i^-(a)$. ∎

Kashiwara (1991) calls (U_0^-, B_0) the *crystal basis of U_q^-.*

The next proposition is the crucial result which allows one to 'melt' a crystal basis of a finite-dimensional U_q-module V to obtain a genuine basis of V over $\mathbb{Q}(q)$. To state it, let $^-$ be the $\mathbb{Q}$-algebra automorphism of U_q defined by

$$\bar{q} = q^{-1}, \quad \overline{X_i^\pm} = X_i^\pm, \quad \overline{K_i} = K_i^{-1},$$

let $\mathcal{A} = \mathbb{Z}[q, q^{-1}]$, and let $U_{\mathcal{A}}^{\text{res}\,-}$ be the $\mathcal{A}$-subalgebra of U_q generated by the $(X_i^-)^{(r)}$, $i = 1, \ldots, n$, $r \geq 1$ (see Section 9.3).

PROPOSITION 14.1.11 *The natural map*

$$(U_{\mathcal{A}}^{\text{res}\,-} \underset{\mathcal{A}}{\otimes} \mathbb{Q}[q, q^{-1}]) \cap U_0^- \cap \overline{U_0^-} \hookrightarrow U_0^- \to U_0^-/qU_0^-$$

is an isomorphism of $\mathbb{Q}$-vector spaces. ∎

Let $\pi : U_0^-/qU_0^- \to (U_{\mathcal{A}}^{\text{res}\,-} \underset{\mathcal{A}}{\otimes} \mathbb{Q}[q, q^{-1}]) \cap U_0^- \cap \overline{U_0^-}$ be the inverse of the isomorphism in 14.1.11, and let $\mathcal{B}_0 = \pi(B_0)$.

THEOREM 14.1.12 *With the above notation, $\mathcal{B}_0$ is an $\mathcal{A}$-basis of $U_{\mathcal{A}}^{\text{res}\,-}$, and a $\mathbb{Q}(q)$-basis of U_q^-. Moreover, for every element $b \in \mathcal{B}_0$, we have $\bar{b} = b$.* ∎

We can now describe a canonical basis, over $\mathbb{Q}(q)$, of any finite-dimensional irreducible U_q-module V of type **1**. In fact, we obtain an $\mathcal{A}$-basis of the $\mathcal{A}$-form $V_{\mathcal{A}}^{\text{res}} = U_{\mathcal{A}}^{\text{res}\,-}.v_0$ of V, where v_0 is a highest weight vector of V.

THEOREM 14.1.13 *Let V be a finite-dimensional irreducible U_q-module, and let v_0 be a highest weight vector of V. Let $\tilde{B}_V$ be the set of non-zero vectors in V of the form $\pi(b).v_0$, where b runs through the basis $\mathcal{B}_0$ of $U_{\mathcal{A}}^{\text{res}\,-}$ described before 14.1.12. Then, $\tilde{B}_V$ is an $\mathcal{A}$-basis of $V_{\mathcal{A}}^{\text{res}}$ and (hence) a $\mathbb{Q}(q)$-basis of V. Moreover, if V^0 is the $\mathcal{A}^0$-submodule of V with basis $\mathcal{B}_0$, and if B_V is the image of $\tilde{B}_V$ in V^0/qV^0, then (V^0, B_V) is a crystal basis of V.* ∎

Kashiwara (1991) calls B_V the *global crystal basis* of V.

D Crystal graphs and tensor products We have not yet considered the behaviour of crystal bases under tensor products. This is the subject of the next result. As a corollary, we shall give a new combinatorial algorithm for decomposing into irreducibles any tensor product of finite-dimensional representations of $\mathfrak{g}$.

PROPOSITION 14.1.14 *Let* (V_1^0, B_{V_1}) *and* (V_2^0, B_{V_2}) *be crystal bases of two finite-dimensional* U_q*-modules* V_1 *and* V_2 *of type* **1**. *Let* $V^0 = V_1^0 \underset{A^0}{\otimes} V_2^0$ *and* $B_V = B_{V_1} \otimes B_{V_2} = \{b_1 \otimes b_2 \mid b_1 \in B_{V_1}, b_2 \in B_{V_2}\}$. *Then,* (V^0, B_V) *is a crystal basis of* $V_1 \underset{\mathbb{Q}(q)}{\otimes} V_2$. *Moreover, we have*

$$\tilde{x}_i^-(b_1 \otimes b_2) = \begin{cases} \tilde{x}_i^-(b_1) \otimes b_2 & \text{if } \exists\, k \geq 1 \text{ with } (\tilde{x}_i^-)^k(b_1) \neq 0,\ (\tilde{x}_i^+)^k(b_2) = 0, \\ b_1 \otimes \tilde{x}_i^-(b_2) & \text{otherwise}, \end{cases}$$

$$\tilde{x}_i^+(b_1 \otimes b_2) = \begin{cases} b_1 \otimes \tilde{x}_i^+(b_2) & \text{if } \exists\, k \geq 1 \text{ with } (\tilde{x}_i^+)^k(b_2) \neq 0,\ (\tilde{x}_i^-)^k(b_1) = 0, \\ \tilde{x}_i^+(b_1) \otimes b_2 & \text{otherwise}. \quad \blacksquare \end{cases}$$

We shall now reformulate this result graphically. By a *coloured oriented graph*, we mean a set $\mathcal{V}$ of *vertices*, a set $\mathcal{E}$ of ordered pairs of elements of $\mathcal{V}$, called *arrows*, and a set $\mathcal{C}$ of *colours*, together with a map $\kappa : \mathcal{E} \to \mathcal{C}$. If $a = (t(a), h(a))$ is an arrow of the graph, we call $t(a)$ the *tail* of a, and $h(a)$ its *head*. Finally, a said to have colour $c \in \mathcal{C}$ if $\kappa(a) = c$. We leave it to the reader to formulate the notion of an isomorphism of coloured oriented graphs.

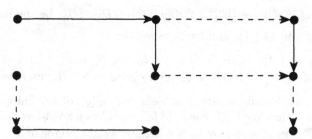

Fig. 22 An admissible coloured oriented graph

The coloured oriented graphs we shall encounter have one further property. To state it, let us say that a (non-coloured) oriented graph is a *string* if it is of the following type:

The *length* of a string is the number of arrows it contains.

A coloured oriented graph $\mathcal{G}$ is *admissible* if, for each colour c, the oriented sub-graph $\mathcal{G}^c$ consisting of all the vertices, but only the arrows of colour c, is a disjoint union of strings.

Example 14.1.15 Fig. 22 shows an admissible coloured oriented graph, in which the dots represent the vertices, and there are two colours, represented by the solid and dashed arrows. ◊

There is a tensor product operation on admissible coloured oriented graphs, which may be described as follows. First, we define the tensor product of strings as indicated in the diagram which follows; the result is a disjoint union of strings.

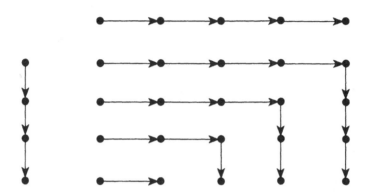

More generally, if (non-coloured) oriented graphs S_1 and S_2 are disjoint unions of strings, their tensor product is the union of the graphs obtained by taking the tensor product of each component of S_1 with each component of S_2.

Suppose now that $\mathcal{G}_1$ and $\mathcal{G}_2$ are admissible coloured oriented graphs. For each colour c, we form the tensor product $\mathcal{G}_1^c \otimes \mathcal{G}_2^c$ as indicated above. Then, the tensor product $\mathcal{G}_1 \otimes \mathcal{G}_2$ is the union of the $\mathcal{G}_1^c \otimes \mathcal{G}_2^c$, with the colour c attached.

Example 14.1.16 Let $\mathcal{G}_1 = \mathcal{G}_2$ be the coloured oriented graph shown below.

The sub-graph corresponding to the solid arrows is

and the tensor product with itself is

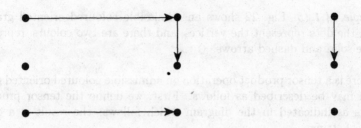

Similarly, the sub-graph corresponding to the dashed arrows gives

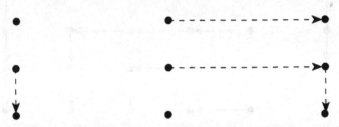

Taking the union, we see that $\mathcal{G}_1 \otimes \mathcal{G}_2$ is precisely the coloured oriented graph shown in Fig. 22. $\Diamond$

Returning now to representation theory, let V be a finite-dimensional $U_q(\mathfrak{g})$-module of type **1**, and let (V^0, B_V) be a crystal basis of V. The *crystal graph* $\mathcal{G}_V$ of V is the coloured oriented graph with set of vertices B_V, set of colours $\mathcal{C} = \{1, \ldots, n\}$, and set of arrows of colour i the pairs (b, b') such that $b' = \tilde{x}_i^-(b)$ (or, equivalently, $b = \tilde{x}_i^+(b')$).

The following result is fairly clear from what we have said above.

PROPOSITION 14.1.17   *The crystal graph* $\mathcal{G}_{V_q(\lambda)}$ *of the finite-dimensional* U_q-*module* $V_q(\lambda)$ *of highest weight* λ *has the following properties:*

(i) it is admissible;

(ii) it is connected (regarded as a non-coloured, non-oriented graph);

(iii) the vector $v_\lambda \in B(\lambda)$ *of weight* λ *is the unique source of* $\mathcal{G}_{V_q(\lambda)}$ *(by a source, we mean a vertex which is the tail of any arrow of which it is a component);*

(iv) the component containing v_λ *of the sub-graph* $\mathcal{G}^i_{V_q(\lambda)}$ *of colour* i *is a string of length* $(\lambda, \check{\alpha}_i)$.

Moreover, the crystal graph $\mathcal{G}_{V_q(\lambda) \otimes V_q(\mu)}$ *of the tensor product of the irreducible finite-dimensional* U_q-*modules of highest weights* λ *and* μ *is isomorphic to the tensor product* $\mathcal{G}_{V_q(\lambda)} \otimes \mathcal{G}_{V_q(\mu)}$ *of the crystal graphs of* $V_q(\lambda)$ *and* $V_q(\mu)$. ∎

This result gives the following algorithm for decomposing $V_q(\lambda) \otimes V_q(\mu)$. Note that, by 10.1.16, the decomposition of this tensor product into irreducibles is the 'same' as in the classical case (the proof of 10.1.16 is given for U_ϵ when $\epsilon \in \mathbb{C}^\times$ is not a root of unity, but the same result and proof are valid for U_q).

KASHIWARA'S TENSOR PRODUCT ALGORITHM 14.1.18 *Let $V_q(\lambda)$ and $V_q(\mu)$ be the irreducible finite-dimensional U_q-modules of the indicated highest weights. For each connected component $\mathcal{G}$ of $\mathcal{G}_{V_q(\lambda)} \otimes \mathcal{G}_{V_q(\mu)}$, locate the unique source s in $\mathcal{G}$, and, for each colour i, let ν_i be the length of the string in $\mathcal{G}$ containing s. Define a weight $\nu \in P^+$ by $(\nu, \check{\alpha}_i) = \nu_i$. Then, $V_q(\lambda) \otimes V_q(\mu)$ is isomorphic to the direct sum of the $V_q(\nu)$, as $\mathcal{G}$ runs through the connected components of $\mathcal{G}_{V_q(\lambda)} \otimes \mathcal{G}_{V_q(\mu)}$.* ∎

Example 14.1.19 We consider the natural 3-dimensional irreducible representation $V_q(\lambda_1)$ of $U_q(sl_3(\mathbb{C}))$. Then, $V_q(\lambda_1)$ has a basis $\{v_0, v_1, v_2\}$ on which the action of the $X_i^\pm$ is given by (1) (with $n = 2$). The crystal graph of $V_q(\lambda_1)$ is thus

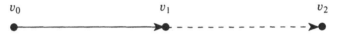

where solid arrows are coloured 1 and dashed arrows are coloured 2. Hence, $\mathcal{G}_{V_q(\lambda_1)} \otimes \mathcal{G}_{V_q(\lambda_1)}$ is the graph shown in Fig. 22. It has two connected components. In the component

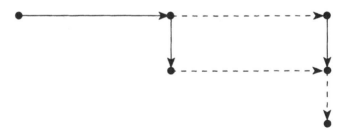

the unique source is the top left vertex, and the strings of colours 1 and 2 originating at this vertex are of lengths 2 and 0, respectively; so this subgraph is the crystal graph of the U_q-module $V_q(2\lambda_1)$. The source of the other component

is the upper vertex, and the strings of colours 1 and 2 originating at this vertex have lengths 0 and 1, respectively; so this component is the crystal graph of $V_q(\lambda_2)$. Kashiwara's algorithm thus gives the isomorphism

$$V_q(\lambda_1) \otimes V_q(\lambda_1) \cong V_q(2\lambda_1) \oplus V_q(\lambda_2)$$

(λ_1 and λ_2 are the fundamental weights of $\mathfrak{g}$ – see the Appendix.) $\Diamond$

14.2 Lusztig's canonical bases

In this section, we describe another construction of canonical bases using a method which looks quite different from that used to produce crystal bases. Eventually, however, we shall see that the two bases are the same.

We assume throughout this section that $\mathfrak{g}$ is a complex simple Lie algebra of type A_n, D_n or E_n.

A The algebraic construction We have seen in 9.3.3 that the set B_σ of products

$$X_{\sigma,t}^+ = (X_{\beta_N}^+)^{(t_N)}(X_{\beta_{N-1}}^+)^{(t_{N-1})} \ldots (X_{\beta_1}^+)^{(t_1)}$$

forms an $\mathcal{A}$-basis of $U_{\mathcal{A}}^{\mathrm{res}\,+}$; here, $t = (t_1, \ldots, t_N) \in \mathbf{N}^N$, and N is the number of positive roots of $\mathfrak{g}$. This basis is certainly not canonical: as our notation suggests, it depends on the choice of a reduced decomposition σ of the longest element w_0 of the Weyl group of $\mathfrak{g}$, for it is only by using such a decomposition that one can define the root vectors X_β^+ (see Subsection 9.1B).

Our aim is to modify the basis B_σ in order to obtain a new basis which is independent of σ. Let $\mathcal{U}_\sigma^+$ be the $\mathbf{Z}[q^{-1}]$-submodule of U_q^+ generated by B_σ.

PROPOSITION 14.2.1 *The $\mathbf{Z}[q^{-1}]$-submodule $\mathcal{U}_\sigma^+$ of U_q^+ is independent of the reduced decomposition σ, and may therefore be denoted by $\mathcal{U}^+$.*

Moreover, if $\tilde{\pi} : \mathcal{U}^+ \to \mathcal{U}^+/q^{-1}\mathcal{U}^+$ is the canonical projection, then $\tilde{\pi}(B_\sigma)$ is a $\mathbf{Z}$-basis $\tilde{\mathcal{B}}$ of $\mathcal{U}^+/q^{-1}\mathcal{U}^+$ which is independent of σ. ∎

Remark It is only the *set* $\tilde{\pi}(B_\sigma)$ which is independent of σ, not the individual elements $\tilde{\pi}(X_{\sigma,t}^+)$. If σ' is another reduced decomposition of w_0, there is a (generally non-trivial) bijection $R_\sigma^{\sigma'} : \mathbf{N}^N \to \mathbf{N}^N$ such that $R_\sigma^{\sigma'}(t) = t'$ if and only if $\tilde{\pi}(X_{\sigma,t}^+) = \tilde{\pi}(X_{\sigma',t'}^+)$. These permutations will play an important role in Subsection 14.2C. $\triangle$

We can now state the main result of this section. We recall the ring automorphism $^-$ of U_q defined before 14.1.11.

THEOREM 14.2.2 *The restriction of $\tilde{\pi}$ to $\mathcal{U}^+ \cap \overline{\mathcal{U}^+}$ induces an isomorphism of $\mathbb{Z}$-modules $\mathcal{U}^+ \cap \overline{\mathcal{U}^+} \to \mathcal{U}^+/q^{-1}\mathcal{U}^+$. Moreover, if X_t^+ is the unique element of $\mathcal{U}^+ \cap \overline{\mathcal{U}^+}$ such that $\tilde{\pi}(X_t^+) = \tilde{\pi}(X_{\sigma,t}^+)$, then $X_t^+ = \overline{X_t^+}$ and $\{X_t^+\}_{t \in \mathbb{N}^N}$ is*
(i) *a $\mathbb{Z}[q^{-1}]$-basis of $\mathcal{U}^+$;*
(ii) *an $\mathcal{A}$-basis of $U_{\mathcal{A}}^{\mathrm{res}\,+}$;*
(iii) *a $\mathbb{Q}(q)$-basis of U_q^+.* ∎

The proof, given in Lusztig (1990d), uses the relation between quantized universal enveloping algebras and quivers described in 9.3.14.

We call the basis $\mathcal{B} = \{X_t^+\}_{t \in \mathbb{N}^N}$ given by this theorem *Lusztig's canonical basis*. From Chapter 9, we know that the specialization of U_q^+ to $q = 1$ gives the classical enveloping algebra U^+. Theorem 14.2.2 thus gives a canonical basis of U^+ over $\mathbb{Q}$.

Examples 14.2.3 If $\mathfrak{g}$ is of type A_1, $\mathcal{B}$ consists of the q-divided powers $(X_1^+)^{(t)}$, for all $t \in \mathbb{N}$.

If $\mathfrak{g}$ is of type A_2, $\mathcal{B}$ consists of the monomials of the form

$$(X_1^+)^{(t)}(X_2^+)^{(t')}(X_1^+)^{(t'')} \quad \text{and} \quad (X_2^+)^{(t)}(X_1^+)^{(t')}(X_2^+)^{(t'')},$$

where $t, t', t'' \in \mathbb{N}$ satisfy $t' \geq t + t''$. These elements are, in fact, all distinct, except when $t' = t + t''$, when one has the relation

$$(X_1^+)^{(t)}(X_2^+)^{(t+t'')}(X_1^+)^{(t'')} = (X_2^+)^{(t'')}(X_1^+)^{(t+t'')}(X_2^+)^{(t)}$$

(this is a q-analogue of a classical identity due to D. N. Verma).

If $\mathfrak{g}$ is of type A_3, $\mathcal{B}$ contains some elements which are not monomials, such as

$$(X_2^+)^{(2)}X_1^+X_3^+X_2^+ - (X_2^+)^{(3)}X_1^+X_3^+ = X_2^+X_1^+X_3^+(X_2^+)^{(2)} - X_1^+X_3^+(X_2^+)^{(3)}$$

(node 2 is the middle node of the Dynkin diagram of $\mathfrak{g}$). ◊

Lusztig's canonical basis has many remarkable properties, some of which we shall now describe. The first result shows that the structure constants of the algebra $U_{\mathcal{A}}^{\mathrm{res}\,+}$ with respect to $\mathcal{B}$ are 'positive'.

PROPOSITION 14.2.4 *The product of any two elements of Lusztig's canonical basis $\mathcal{B}$ is a linear combination of elements of $\mathcal{B}$ with coefficients in $\mathbb{N}[q, q^{-1}]$.* ∎

Specializing to $q = 1$, we see that the structure constants of the classical enveloping algebra U^+, with respect to its canonical basis, are non-negative integers.

The next result shows that $\mathcal{B}$ is beautifully adapted to the study of representations of U_q (cf. 14.1.13). Let $\rho : U_q \to \mathrm{End}(V)$ be a finite-dimensional irreducible U_q-module of type **1**. Then, V is generated as a U_q^+-module by a lowest weight vector, i.e. a weight vector $v_\infty \in V$ which is annihilated by X_i^- for all $i = 1, \ldots, n$. This follows from the corresponding results in Chapter 10 for highest weight vectors, using the involution of U_q, as a $\mathbb{Q}(q)$-algebra, given by $X_i^\pm \mapsto X_i^\mp$, $K_i \mapsto K_i^{-1}$.

THEOREM 14.2.5 *With the above notation, the non-zero vectors in $\rho(\mathcal{B})(v_\infty)$ are a $\mathbb{Q}(q)$-basis of V.* ∎

Specializing to $q = 1$ gives, of course, a canonical basis of any finite-dimensional irreducible g-module.

The proofs of the last two results make use of another construction of $\mathcal{B}$, using topological methods, which we outline in the next subsection. This also yields an 'explicit' formula for the elements of Lusztig's canonical basis.

Before doing this, however, we shall establish the connection between $\mathcal{B}$ and the basis $\mathcal{B}_0$ of U_q^- constructed in 14.1.10. The existence of such a connection is strongly suggested by similarities between the construction of the two bases (cf. 14.1.12 and 14.2.2), and between their properties (cf. 14.1.13 and 14.2.5).

We must first make sure that we are comparing two bases of the same space. By applying to $\mathcal{B}_0$ the $\mathbb{Q}$-algebra automorphism of U_q given by

$$K_i \mapsto K_i, \quad X_i^\pm \mapsto X_i^\mp, \quad q \mapsto q^{-1},$$

we obtain an $\mathcal{A}$-basis of $U_\mathcal{A}^+$, and a $\mathbb{Q}(q)$-basis of U_q^+, which we denote by $\mathcal{B}_\infty$.

THEOREM 14.2.6 *Kashiwara's basis $\mathcal{B}_\infty$ is identical to Lusztig's basis $\mathcal{B}$.* ∎

This is proved in Lusztig (1990e).

B The topological construction To describe the second construction of Lusztig's canonical basis, we need to recall the discussion of quivers in Subsection 9.3D.

Fix an orientation of the Dynkin graph of g, and denote the resulting quiver by Γ. We must choose a reduced decomposition $w_0 = s_{i_1} \ldots s_{i_N}$ which is *adapted* to this orientation, in the following sense. Say that a vertex i is a *sink* of Γ if there is no oriented edge $\overset{i}{\bullet} \longrightarrow \overset{j}{\bullet}$ in Γ; in this case, let $s_i(\Gamma)$ be the orientation obtained by reversing every oriented edge of Γ that has vertex i as its head. Then, $w_0 = s_{i_1} \ldots s_{i_N}$ is adapted to the orientation Γ if i_1 is a sink of Γ and if, for each $r = 1, \ldots, N-1$, i_{r+1} is a sink of $\Gamma_r = s_{i_r}(\cdots (s_{i_2}(s_{i_1}(\Gamma))) \cdots)$.

It can be shown that there is at least one reduced decomposition of w_0 adapted to any given orientation of the Dynkin graph. It is easiest to describe such a decomposition when w_0 lies in the centre of W (this holds when g is of type A_1, D_n for n even, E_7 and E_8). Then, one can define a Coxeter element

$c = s_{i_1} \ldots s_{i_n}$ for which $\{i_1, \ldots, i_n\} = \{1, \ldots, n\}$ and such that ${}^{j}_{\bullet} \longrightarrow {}^{j'}_{\bullet}$ in Γ implies $j < j'$ (see Bernstein, Gel'fand & Ponomarev (1973)). In this case, the order h of c is even and

$$w_0 = (s_{i_1} \ldots s_{i_n})(s_{i_1} \ldots s_{i_n}) \ldots (s_{i_1} \ldots s_{i_n}) \qquad (h/2 \text{ times})$$

is a reduced decomposition adapted to Γ.

We reformulate slightly the notion of a representation of Γ, introduced in Subsection 9.3D; we work over an algebraic closure $\overline{\mathbb{F}}$ of a finite field $\mathbb{F}$. If $\boldsymbol{m} = (m_1, \ldots, m_n) \in \mathbb{N}^n$, let

$$\overline{\mathbb{F}}^{\boldsymbol{m}} = \bigoplus_{i=1}^{n} \overline{\mathbb{F}}^{m_i}.$$

A representation of Γ on $\overline{\mathbb{F}}^{\boldsymbol{m}}$ is a linear map $T : \overline{\mathbb{F}}^{\boldsymbol{m}} \to \overline{\mathbb{F}}^{\boldsymbol{m}}$ such that $T(\overline{\mathbb{F}}^{m_i}) \subseteq \overline{\mathbb{F}}^{m_j}$ whenever ${}^{i}_{\bullet} \longrightarrow {}^{j}_{\bullet}$ is an oriented edge of Γ; in other words, T is an element of the vector space

$$E_{\boldsymbol{m}} = \bigoplus_{{}^{i}_{\bullet} \longrightarrow {}^{j}_{\bullet}} \operatorname{Hom}_{\overline{\mathbb{F}}}(\overline{\mathbb{F}}^{m_i}, \overline{\mathbb{F}}^{m_j}).$$

Moreover, the group

$$GL_{\boldsymbol{m}}(\overline{\mathbb{F}}) = \prod_{i=1}^{n} GL_{m_i}(\overline{\mathbb{F}})$$

acts on $E_{\boldsymbol{m}}$ by $(g.T)_{ij} = g_j T_{ij} g_i^{-1}$, where $g = (g_1, \ldots, g_n) \in GL_{\boldsymbol{m}}(\overline{\mathbb{F}})$, and T_{ij} is the component of T in $\operatorname{Hom}_{\overline{\mathbb{F}}}(\overline{\mathbb{F}}^{m_i}, \overline{\mathbb{F}}^{m_j})$. The set of isomorphism classes of representations of Γ of dimension $\boldsymbol{m}$ is in one-to-one correspondence with the set of orbits of $GL_{\boldsymbol{m}}(\overline{\mathbb{F}})$ on $E_{\boldsymbol{m}}$.

Let $\beta_r = s_{i_1} s_{i_2} \ldots s_{i_{r-1}}(\alpha_{i_r})$, $r = 1, \ldots, N$, be the positive roots of $\mathfrak{g}$. Let e_{β_r} be the indecomposable representation of Γ corresponding to β_r, as in Gabriel's theorem 9.3.13, and, for any $\boldsymbol{t} = (t_1, \ldots, t_N) \in \mathbb{N}^N$, let

$$V^{\boldsymbol{t}} = \bigoplus_{r=1}^{N} e_{\beta_r}^{\oplus t_r}.$$

Let $\boldsymbol{m} = \dim(V^{\boldsymbol{t}})$, and let $\mathcal{O}_{\boldsymbol{t}}$ be the $GL_{\boldsymbol{m}}(\overline{\mathbb{F}})$-orbit in $E_{\boldsymbol{m}}$ corresponding to $V^{\boldsymbol{t}}$. Define a partial order on $\mathbb{N}^N$ by saying that $\boldsymbol{t}' \leq \boldsymbol{t}$ if and only if $\mathcal{O}_{\boldsymbol{t}'}$ is contained in the Zariski closure of $\mathcal{O}_{\boldsymbol{t}}$ and $\dim(V^{\boldsymbol{t}}) = \dim(V^{\boldsymbol{t}'})$.

Fix a prime $p \in \mathbb{N}$ which is non-zero in $\mathbb{F}$, and denote by $\mathbb{Q}_p$ the field of p-adic numbers. Let x be an $\mathbb{F}$-rational point of $\mathcal{O}_{\boldsymbol{t}}$, and let $IC_x^k(\overline{\mathcal{O}_{\boldsymbol{t}}})$ be the stalk at x of the kth cohomology sheaf with coefficients in $\mathbb{Q}_p$ of the intersection cohomology complex of the closure of $\mathcal{O}_{\boldsymbol{t}}$ (see Beilinson, Bernstein & Deligne (1982)).

THEOREM 14.2.7 *Let σ be a reduced decomposition of w_0 adapted to the chosen orientation of Γ, and let the other notation be as above. Then:*

(i) $IC_x^k(\overline{\mathcal{O}_t}) = 0$ *if k is odd.*

(ii) *If $t, t' \in \mathbf{N}^N$, define $\zeta_t^{t'} \in \mathbf{Z}[q, q^{-1}]$ by*

$$\sum_{k \text{ even}} \dim(IC_x^k(\overline{\mathcal{O}_t}))q^k = q^{\dim(\mathcal{O}_{t'}) - \dim(\mathcal{O}_t)} \zeta_t^{t'}$$

if $t' \leq t$, and $\zeta_t^{t'} = 0$ otherwise. Then, the canonical basis element $\tilde{\pi}(X_{\sigma,t}^+)$ is given by

$$\tilde{\pi}(X_{\sigma,t}^+) = \sum_{t' \leq t} \zeta_{t'}^{t} X_{\sigma,t'}^+.$$

(cf. 14.2.1). ∎

This result is a consequence of a more abstract construction of the canonical basis in terms of perverse sheaves. For each $t \in \mathbf{N}^N$, one defines a certain category of perverse sheaves on $\overline{\mathbf{F}}^t$, and forms the free $\mathcal{A}$-module $\mathcal{A}(\mathcal{P})$ with basis the set $\mathcal{P}$ of objects of these categories, for all $t \in \mathbf{N}^N$. One then defines, by cohomological methods, an associative algebra structure on $\mathcal{A}(\mathcal{P})$, and proves that it is isomorphic to $U_{\mathcal{A}}^{\mathrm{res}\,+}$. Since $\mathcal{A}(\mathcal{P})$ has a natural basis, namely $\mathcal{P}$, so does $U_{\mathcal{A}}^{\mathrm{res}\,+}$. The final step is to show that this basis is independent of the choice of orientation of the Dynkin graph.

For complete details, see Lusztig (1990d, 1991, 1993).

C Some combinatorial formulas We conclude this chapter by describing a combinatorial formula for the dimension of any finite-dimensional irreducible U_q-module which can be derived using Lusztig's canonical basis.

To state it, we first choose a partition of the set of vertices of the Dynkin diagram of $\mathfrak{g}$ into a disjoint union of two subsets, say $\mathcal{V}'$ and $\mathcal{V}''$, with the property that $a_{ij} = 0$ whenever $i \neq j$ are both in $\mathcal{V}'$, or both in $\mathcal{V}''$ (it is easy to see that there is a unique such partition for any connected Dynkin diagram). Let $|\mathcal{V}'| = k$, $|\mathcal{V}''| = n - k$. We may assume that the vertices are numbered so that $\mathcal{V}' = \{1, \ldots, k\}$ and $\mathcal{V}'' = \{k + 1, \ldots, n\}$. Then

$$w_0 = s_1 s_2 \ldots s_n s_1 s_2 \ldots s_n s_1 \ldots$$

is a reduced decomposition, say σ, and

$$w_0 = s_n s_{n-1} \ldots s_1 s_n s_{n-1} \ldots s_1 s_n \ldots$$

is another reduced decomposition, say σ'. These reduced decompositions are both adapted to the unique orientation of the Dynkin graph for which the vertices in $\mathcal{V}'$ are all sinks, and those of $\mathcal{V}''$ are all sources (i.e. sinks for the orientation obtained by reversing all the arrows).

Now recall the bijection $R_\sigma^{\sigma'}$ introduced in the remark following 14.2.1. For any $\boldsymbol{m} \in \mathbf{N}^n$, let $Z(\boldsymbol{m})$ be the set of pairs $(\boldsymbol{t}, \boldsymbol{t}') \in \mathbf{N}^N \times \mathbf{N}^N$ satisfying the following conditions:

(i) $R_\sigma^{\sigma'}(\boldsymbol{t}) = \boldsymbol{t}'$,
(ii) $t_r \leq m_r$ if $1 \leq r \leq k$,
(iii) $t_r' \leq m_{n-r+1}$ if $1 \leq r \leq n-k$.

Then, we have

PROPOSITION 14.2.8 *Let* $V_q(\lambda)$ *be the finite-dimensional irreducible* U_q-*module with highest weight* λ, *and define* $\boldsymbol{m} = (m_1, \ldots, m_n) \in \mathbf{N}^n$ *by* $m_i = (\lambda, \alpha_i)$. *Then,*
$$\dim(V_q(\lambda)) = |Z(\boldsymbol{m})|. \quad \blacksquare$$

Note that, by the results of Chapter 10, the dimension of $V_q(\lambda)$ is the same as that of the $\mathfrak{g}$-module $V(\lambda)$ with the same highest weight.

As it stands, 14.2.8 does not give a purely combinatorial way of computing $\dim(V_q(\lambda))$, since the bijection $R_\sigma^{\sigma'}$ is defined rather implicitly in terms of the relation between the canonical basis of U_q^+ and its Poincaré–Birkhoff–Witt bases. However, a combinatorial description of $R_\sigma^{\sigma'}$, for any two reduced decompositions σ and σ', can be given as follows.

We first define $R_\sigma^{\sigma'}$ in two special cases:

(a) σ' can be obtained from σ by replacing 3 consecutive reflections $s_i s_j s_i$ in σ, such that $a_{ij} = -1$, by $s_j s_i s_j$: then $R_\sigma^{\sigma'}(\boldsymbol{t}) = \boldsymbol{t}'$, where $\boldsymbol{t}'$ is the same as $\boldsymbol{t}$ except in the three positions where σ and σ' differ, while if (u, v, w) are these three positions in $\boldsymbol{t}$, then those in $\boldsymbol{t}'$ are

$$(v + w - \min(u, w),\ \min(u, w),\ u + v - \min(u, w)).$$

(b) σ' can be obtained from σ by replacing two consecutive reflections $s_i s_j$ in σ, such that $a_{ij} = 0$, by $s_j s_i$: then $R_\sigma^{\sigma'}(\boldsymbol{t}) = \boldsymbol{t}'$, where $\boldsymbol{t}'$ is the same as $\boldsymbol{t}$ except in the two positions where σ and σ' differ, while if (u, v) are these 2 positions in $\boldsymbol{t}$, then those in $\boldsymbol{t}'$ are (v, u).

It is classical that, given any two reduced decompositions σ and σ', there is a sequence $\sigma = \sigma_1,\ \sigma_2, \ldots,\ \sigma_\ell = \sigma'$ such that any pair of consective reduced decompositions in the sequence satisfy one of the conditions (a) or (b); then

$$R_\sigma^{\sigma'} = R_{\sigma_{\ell-1}}^{\sigma_\ell} R_{\sigma_{\ell-2}}^{\sigma_{\ell-1}} \cdots R_{\sigma_1}^{\sigma_2}.$$

Example 14.2.9 Let $\mathfrak{g}$ be of type A_2. We take $\mathcal{V}' = \{1\}$, $\mathcal{V}'' = \{2\}$. Then, the two reduced decompositions are

$$\sigma\ :\quad w_0 = s_1 s_2 s_1, \qquad \sigma'\ :\quad w_0 = s_2 s_1 s_2.$$

From the preceding description, the bijection $R_\sigma^{\sigma'}$ is given by

$$R_\sigma^{\sigma'}(t_1, t_2, t_3) = \begin{cases} (t_2 + t_3 - t_1, t_1, t_2) & \text{if } t_1 \leq t_3, \\ (t_2, t_3, t_1 + t_2 - t_3) & \text{if } t_1 \geq t_3. \end{cases}$$

We use 14.2.8 to compute the dimension of the adjoint representation of $\mathfrak{g}$. Since $\boldsymbol{m} = (1, 1)$, conditions (ii) and (iii) above become

$$t_1 \leq 1, \qquad t_1' \leq 1.$$

If $t_1 = 0$, then $t_1' = t_2 + t_3$, giving the three possibilities $\boldsymbol{t} = (0, 0, 0)$, $(0, 1, 0)$, $(0, 0, 1)$. If $t_1 = 1$ and $t_3 = 0$, then $t_1' = t_2$, giving $\boldsymbol{t} = (1, 0, 0)$ or $(1, 1, 0)$. Similarly, if $t_1 = t_3 = 1$, then $\boldsymbol{t} = (1, 0, 1)$ or $(1, 1, 1)$, and if $t_1 = 1$ and $t_3 = 2$, then $\boldsymbol{t} = (1, 0, 2)$. Thus, $|Z(\boldsymbol{m})| = 8$. $\Diamond$

Remark Lusztig (1990e) proves a formula similar to that in 14.2.8 for the multiplicities of the irreducible components in the tensor product of two finite-dimensional irreducible U_q-modules. This is of a different nature from Kashiwara's tensor product algorithm 14.1.18. $\triangle$

Bibliographical notes

14.1 The notion of a crystal basis was introduced in Kashiwara (1990b), where their existence was proved when $\mathfrak{g}$ is of type A, B, C or D. Misra & Miwa (1990) exhibited a crystal basis of the quantum analogue of the basic representation of affine sl_n, indicating the possibility of extension to infinite-dimensional $\mathfrak{g}$. Finally, Kashiwara (1990c) announced the proof of the existence of crystal bases for integrable representations of arbitrary symmetrizable quantum Kac–Moody algebras. Detailed proofs appeared in Kashiwara (1991). (A representation of $U_q(\mathfrak{g})$ is 'integrable' if it is the direct sum of its weight spaces, these weight spaces being finite dimensional, and if, when regarded as a representation of the $U_{q_i}(sl_2(\mathbb{C}))$-subalgebra of $U_q(\mathfrak{g})$ generated by $X_i^\pm$, $K_i^{\pm 1}$, for each $i = 1, \ldots, n$, it is a sum of finite-dimensional representations.)

Although we have presented Kashiwara's results linearly, he proves most of them simultaneously, by an inductive argument, in what he calls the 'grand loop'.

For an application of crystal bases to vertex models in statistical mechanics (see Section 7.5), see Kang *et al.* (1992a–c).

14.2 Most of the results of this section are taken from Lusztig (1990d,e), where further details and complete proofs can be found. In these papers (and in our treatment), it is assumed that $\mathfrak{g}$ is of type A, D or E, but Lusztig (1991) deals with the case where $\mathfrak{g}$ is any Kac–Moody algebra with a symmetric generalized Cartan matrix. The fact, proved in Lusztig (1990e), that his basis coincides with Kashiwara's basis in the A, D, E cases, together with the fact that the latter is defined for any symmetrizable quantum Kac–Moody algebra, strongly suggested that Lusztig's results should hold in this generality. This was shown to be the case in Lusztig (1993), where a very thorough treatment of the whole subject of canonical bases can be found. To pass from symmetric to merely symmetrizable generalized Cartan matrices, one has to take into account the action of a certain cyclic group.

See Lakshmibai & Reshetikhin (1991, 1992) for a quantum analogue of standard monomial theory.

See Mathieu (1992) for a simplified discussion of the topological construction in Subsection 14.2B, dealing directly with the specialization to $q = 1$.

15

Quantum group invariants
of knots and 3-manifolds

One of the most remarkable developments in the story of quantum groups was the discovery, due essentially to V. F. R. Jones, that they can be used to construct new isotopy invariants of knots (and, more generally, of links) in the 3-sphere S^3. In fact, the crucial ingredient in Jones's construction is a solution of the quantum Yang–Baxter equation, from which it is easy to obtain a representation of Artin's braid group $\mathcal{B}_m$ on m strands, for all m. Since every link can be obtained by joining the free ends of a braid (in a standard way), one has an operator associated to each link. This operator will depend on the braid used to represent the given link, but Jones observed that, by taking a suitable 'Markov trace' of the operator, one could obtain a scalar which depends only on the isotopy class of the link.

We have seen on several occasions throughout this book that quantum groups provide a machine for producing solutions of the QYBE. Although Jones actually obtained his R-matrix by another method, it turns out that it is the R-matrix associated to the two-dimensional irreducible representation of $U_q(sl_2(\mathbb{C}))$. One thus obtains an isotopy invariant of links which depends on the deformation parameter q. After suitable rescaling and reparametrization, this becomes the Jones polynomial. It is natural to try to generalize this construction to any finite-dimensional representation of $U_q(\mathfrak{g})$, for any simple Lie algebra $\mathfrak{g}$. The difficulty lies in the construction of a suitable Markov trace, and, although it has not yet been carried out in complete generality, V. Turaev succeeded in the case of the natural representations of $U_q(\mathfrak{g})$ when $\mathfrak{g}$ is of classical type. He showed that, when $\mathfrak{g}$ is of type A_n, the resulting sequence of polynomial invariants can be organized into a single two-variable invariant, the HOMFLY polynomial; when $\mathfrak{g}$ is of type B_n, C_n and D_n, he obtained the Kauffman polynomial.

Meanwhile, E. Witten gave, in the heuristic language of Feynman integrals, a construction of invariants of links in an arbitrary compact oriented 3-manifold. In the case when the ambient 3-manifold is the sphere, these invariants reduce to the HOMFLY polynomial; at the opposite extreme, when the 3-manifold is arbitrary but the link is empty, they give new invariants of 3-manifolds. N. Yu. Reshetikhin and V. Turaev showed how Witten's invariants could be obtained from the quantum group approach, by using the fact that any 3-manifold can be obtained by performing surgery along a 'framed link' in S^3. A framed link is the same thing as a ribbon tangle, to which invariants were attached in Section 5.3 using ribbon Hopf algebras, and Reshetikhin and Turaev showed that there is an essentially unique combination of these ribbon tangle invariants which depends only on the given 3-manifold, and not on the surgery used to construct it.

After summarizing, in Section 15.1, the background results on links and 3-manifolds we shall need, in Section 15.2 we describe Turaev's construction of the HOMFLY polynomial: using what we now know about quantum groups and their associated R-matrices, the construction reduces to a fairly easy exercise in linear algebra. We also describe another construction, due to Jones but based heavily on ideas of L. H. Kauffman, which works directly with a plane projection of a link, and does not depend on the relation with braids. This essentially involves formulating vertex models, which we studied in Section 7.5, on an arbitrary 4-valent graph instead of a rectangular lattice (such a formulation was given earlier by R. J. Baxter for other purposes).

3-Manifold invariants are discussed in Section 15.3. The ribbon tangle invariants, from which the 3-manifold invariants are constructed, depend on a choice of a 'colouring' of the ribbon tangle, i.e. a choice of a representation of some fixed ribbon Hopf algebra for each component of the link. One has to impose some restrictions on the allowable colourings, which lead to the notion of a *modular Hopf algebra*. We show that the 'simply-connected covering' of the finite-dimensional subalgebra $U_\epsilon^{\text{fin}}(sl_2(\mathbb{C}))$ of the restricted specialization of $U_q(sl_2(\mathbb{C}))$ at any root of unity ϵ of odd (and sufficiently large) order gives a modular Hopf algebra, and hence a 3-manifold invariant. The proof of this relies on the properties of tilting modules for $U_\epsilon^{\text{res}}(sl_2(\mathbb{C}))$, discussed in Section 11.3. The work of Witten strongly suggests that this result should generalize to arbitrary $\mathfrak{g}$.

Since the quantum group constructions we are going to describe are all ultimately two-dimensional in nature, it is perhaps worth recalling that part of the motivation for Witten's work on link invariants was the desire to give an intrinsically three-dimensional description of the Jones polynomial, independent of a choice of plane projection. Witten accomplished this by his quantum field theoretic approach, but it seems that this has not yet been completely translated into rigorous mathematics. On the other hand, J. S. Birman and X. S. Lin have show very recently that the quantum group invariants are closely related to invariants constructed by V. A. Vassiliev by a completely different method.

15.1 Knots, links and 3-manifolds: a quick review

The purpose of this section is to collect the definitions and notations we shall use in the two sections which follow. For proofs and further details, the reader should consult the references mentioned in the Bibliographical notes.

A From braids to links A *link* in $\mathbb{R}^3$ (or the 3-sphere S^3) is a one-dimensional smooth compact submanifold of $\mathbb{R}^3$ (or S^3), in other words, a disjoint union of embedded circles. A *knot* is a connected link. A link is *oriented* if each of its connected components is oriented. Two oriented links L and L' are *equivalent* (or ambient isotopic) if there exists an orientation-preserving diffeomorphism f of $\mathbb{R}^3$ (or S^3) such that $f(L) = L'$, and such that f takes the orientation of L into that of L'.

Links L can be pictured in two dimensions by projecting perpendicularly onto a plane, chosen so that the projection $\overline{L}$ has only transversal crossings, none of which have more than two branches (we shall always assume that our link projections have these properties). The equivalence class of the link can be recovered from a projection provided one includes certain extra information at each crossing in the projection (cf. the discussion of braids in 5.1.7). Namely, if we think of the direction perpendicular to the chosen plane as 'vertical', the branch of the link which lies 'below' the branch which it crosses in the projection is shown broken in the two-dimensional picture. See Fig. 23.

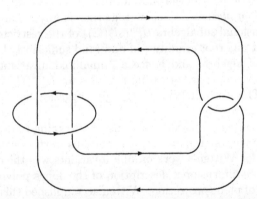

Fig. 23 Plane projection of the Whitehead link

Two oriented link projections define equivalent links if and only if they can be obtained from each other by applying an isotopy of the plane together with a sequence of *Reidemeister moves*:

Reidemeister I

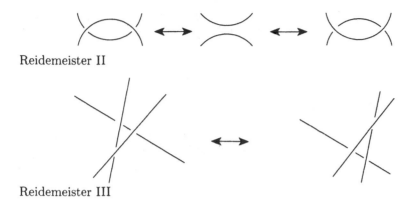

Reidemeister II

Reidemeister III

Here, the two link projections are supposed to be the same except for the portion shown. All possible orientations are allowed.

Links are distinguished from each other by means of invariants. An *(oriented) link invariant* is an assignment of an element of a set S to every (oriented) link in such a way that equivalent links are assigned the same element of S. In view of the result of the preceding paragraph, defining such an invariant is the same thing as associating an element of S to every (oriented) link projection in such a way that it is unchanged by plane isotopy and the Reidemeister moves.

Link invariants are usually easier to define than to calculate. However, there is a special class of invariants, the calculation of which can be effectively carried out by an inductive procedure. Three oriented link projections $\overline{L}_+$, $\overline{L}_-$ and $\overline{L}_0$ are said to be *skein related* if they are the same outside a neighbourhood of some crossing point, near which they are as shown below:

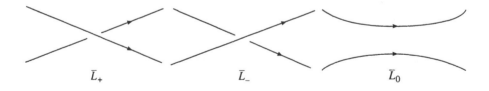

It is intuitively plausible (and also true!) that any link projection can be transformed into that of an unlink, i.e. a disjoint union of circles, by successively changing overcrossings to undercrossings, or vice versa. Thus, of the three link projections $\overline{L}_+$, $\overline{L}_-$ and $\overline{L}_0$, one of them (either $\overline{L}_+$ or $\overline{L}_-$) is 'more complicated' than the other two, either because it has more crossings, or because it is further from an unlink.

THEOREM 15.1.1 *There exists a unique invariant P of oriented links with* values in the ring $\mathbb{Z}[x, x^{-1}, y, y^{-1}]$ of Laurent polynomials in two variables x and y (with integer coefficients) such that $P = 1$ for a single unknotted circle and

$$(1) \qquad xP(\overline{L}_+) + x^{-1}P(\overline{L}_-) + yP(\overline{L}_0) = 0$$

whenever $\overline{L}_+$, $\overline{L}_-$ and $\overline{L}_0$ are skein related. ∎

The invariant P is called the HOMFLY polynomial, after its discoverers Freyd, Hoste, Lickorish, Millett, Ocneanu & Yetter (1985) and Przytycki & Traczyk (1987) (unfortunately, P had been named before the contribution of the last two authors was recognized).

By the remarks preceding this theorem, the value of P on a link L can be calculated inductively by successively simplifying the projection $\overline{L}$ using the skein relations. This proves uniqueness. The problem is to show that P exists, i.e. that two different ways of simplifying the link projection lead to the same value of $P(L)$. We shall establish this in the next section by using quantum groups.

Some special cases of the HOMFLY polynomial were discovered earlier. For example, the *Alexander polynomial* is

$$A(L)(t) = P(L)(\sqrt{-1}, -\sqrt{-1}(t^{1/2} - t^{-1/2})),$$

and the *Jones polynomial*, whose discovery by Jones (1985) motivated the construction of P, is

$$J(L)(t) = P(L)(\sqrt{-1}t^{-1}, -\sqrt{-1}(t^{1/2} - t^{-1/2})).$$

Example 15.1.2 Let us compute the value of $P(L)$ when L is the trefoil knot. The skein relation,

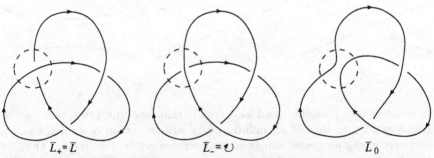

$$\overline{L}_+ = \overline{L} \qquad\qquad \overline{L}_- = \mho \qquad\qquad \overline{L}_0$$

applied to the crossing enclosed by the dashed circle, gives

$$xP(\overline{L}) = -x^{-1} - yP(\overline{L}_0),$$

while

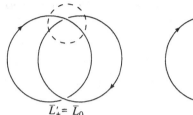 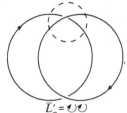 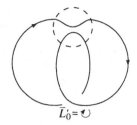

gives

$$xP(\overline{L}_0) = -x^{-1}P(\overline{L}'_-) - y.$$

Finally,

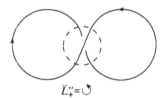 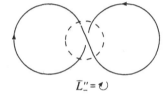 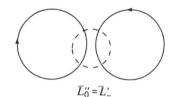

gives

$$P(\overline{L}'_-) = -(x + x^{-1})/y.$$

Combining these equations, we find that

$$P(\overline{L}) = y^2 x^{-2} - x^{-4} - 2x^{-2}. \quad \Diamond$$

There is another polynomial invariant of oriented links whose existence can be proved using quantum groups, namely the *Kauffman polynomial*. To define it, we need to recall the notion of the *writhe* (or *Tait number*) of an oriented link projection. We associate a sign to each crossing according to the scheme

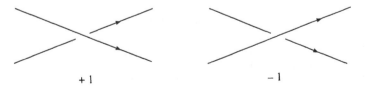

Then, the writhe $w(\overline{L})$ of a link projection $\overline{L}$ is obtained by adding up the signs of all the crossings in a projection of $\overline{L}$. Note that the writhe is not a link invariant, since it is obviously changed by Reidemeister I. In general, the

equivalence relation on links generated by plane isotopy and Reidemeister II and III is called *regular isotopy*. Thus, writhe is a regular isotopy invariant.

THEOREM 15.1.3 *There exists a unique invariant K of oriented links, with values in the ring $\mathbb{Z}[z, z^{-1}, a, a^{-1}]$ of Laurent polynomials in 2 variables z and a, such that, if $\bar{L}$ is any plane projection of a link L, then $\tilde{K}(\bar{L}) = a^{-w(\bar{L})} K(L)$ has the following properties:*

(i) $\tilde{K}(\circlearrowleft) = 1$;
(iia) if link projections $\bar{L}$ and $\bar{L}'$ are related by

$$\bar{L}' \qquad\qquad \bar{L}$$

then $\tilde{K}(\bar{L}') = a\tilde{K}(\bar{L})$;
(iib) if link projections $\bar{L}$ and $\bar{L}''$ are related by

$$\bar{L}'' \qquad\qquad \bar{L}$$

then $\tilde{K}(\bar{L}'') = a^{-1}\tilde{K}(\bar{L})$;
(iii) if link projections $\bar{L}_+$, $\bar{L}_-$, $\bar{L}_0$ and $\bar{L}_\infty$ are related by

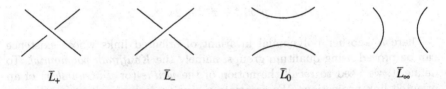

$$\bar{L}_+ \qquad\qquad \bar{L}_- \qquad\qquad \bar{L}_0 \qquad\qquad \bar{L}_\infty$$

then $\tilde{K}(\bar{L}_+) + \tilde{K}(\bar{L}_-) = z(\tilde{K}(\bar{L}_0) + \tilde{K}(\bar{L}_\infty))$. ∎

Remarks [1] Conditions (i)–(iii) do not involve the orientation of the link, which is used only to compute its writhe.

[2] The Alexander polynomial is not a specialization of the Kauffman polynomial, but the Jones polynomial is:

$$J(L)(t) = -K(L)(-t^{-1/4} - t^{1/4}, t^{-3/4}).$$

[3] An invariant closely related to the writhe is the *linking number* $\mathrm{lk}(L, L')$ of two disjoint oriented links L and L'. To define it, choose a plane projection

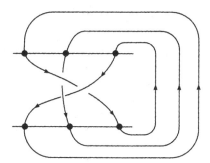

Fig. 24 The closure of a braid

of the link $L \cup L'$, and give a sign to each crossing as above. Then, $\operatorname{lk}(L, L')$ is the sum of the signs associated to each crossing where L passes under L'.

[4] Since the writhe is only an invariant of regular isotopy, the same is true of $\tilde{K}$, and the purpose of 'rescaling' $\tilde{K}$ by the factor $a^{w(L)}$ is to convert $\tilde{K}$ into a genuine isotopy invariant. There is another way to accomplish this, which is based on a result of Trace (1983).

Define the *rotation number* of an oriented link L to be $\int_L d\angle$, where $d\angle$ is the pull-back of the canonical 'angle' 1-form on the circle S^1 by the map $L \to S^1$ which associates to a point of L the unit tangent vector at the point. Then, Trace's result asserts that two oriented link projections are regular isotopic if and only if they are isotopic and corresponding components have the same writhe and rotation number. The writhe and rotation numbers of a link can be changed, without changing the isotopy class of the link, by introducing 'kinks' as indicated below:

Thus, one can convert a regular isotopy invariant into a full isotopy invariant by first adjusting the writhe and rotation numbers suitably. $\triangle$

Braids in $\mathbb{R}^3$ were introduced in 5.1.7; we use the notation defined there. Their relevance to this circle of ideas is that they provide a method of finding invariants which allows one to avoid the use of Reidemeister moves. If β is a braid, one can obtain a link by joining together its endpoints p_i^0 and p_i^1, for all i, as shown in Fig. 24 (in a plane projection). This link is called the *closure* of β. Note that any braid has a canonical orientation, reading a plane representation from top to bottom, so that its closure is also naturally oriented. It was proved by J. Alexander that, up to equivalence, every oriented link is the closure of a braid.

It is obvious that equivalent braids have equivalent closures, but the converse is false. The precise conditions under which two braids have equivalent closures were found by A. Markov. To state his result, note first that there are obvious embeddings $\mathcal{B}_m \hookrightarrow \mathcal{B}_{m+1}$ given by adding an $(m+1)$th strand to any braid $\beta \in \mathcal{B}_m$, sufficiently far away from the strands of β so that it is not linked with any of them:

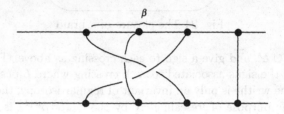

Let us say that two braids β, $\beta' \in \coprod_m \mathcal{B}_m$ are *Markov equivalent* if they can be obtained from each other by a sequence of *Markov moves*:

(2)
$$\text{Markov I} \quad \beta' \sim \beta\beta'\beta^{-1} \quad (\beta, \beta' \in \mathcal{B}_m),$$
$$\text{Markov II} \quad \beta \sim \beta T_m^{\pm 1} \quad (\beta \in \mathcal{B}_m).$$

Then, Markov's theorem asserts that two braids have equivalent closures if and only if they are Markov equivalent. Thus, defining a link invariant with values in a set S is the same thing as defining a map $\coprod_m \mathcal{B}_m \to S$ which is constant on Markov equivalence classes.

B From links to 3-manifolds The construction of 3-manifold invariants from quantum groups is based on the classical result that every compact, oriented 3-manifold (without boundary) can be obtained by performing surgery along a framed link in S^3. We shall now explain what this means, and how to tell when two such surgeries give rise to the same 3-manifold (up to orientation-preserving homeomorphism).

A *framing* of an oriented link $L \subset S^3$ is an assignment of an integer to each component of L. This is the same thing as choosing a ribbon tangle T whose core is L (cf. 5.1.9). In fact, let S be a connected component of T with core a component K of L. Then, if K' is one of the two components of the boundary ∂S of S, the integer associated to K is $\mathrm{lk}(K, K')$.

Let $K_1, \ldots, K_n$ be the components of a framed link L in S^3, and let f_i be the integer associated to K_i by the framing. Choose disjoint tubular neighbourhoods $N_1, \ldots, N_n$ of $K_1, \ldots, K_n$ in S^3. One then attaches n copies of $D^2 \times D^2$ to the 4-ball D^4 with boundary S^3 by gluing the part $\partial D^2 \times D^2 =$

$S^1 \times D^2$ of the ith copy onto N_i in such a way that, for each $x \in D^2$, the circle $S^1 \times x$ goes to a curve which runs 'parallel' to K_i and whose linking number with K_i is f_i. There results a compact 4-manifold D_L whose boundary is a compact 3-manifold S_L. Fixing an orientation of D^4 induces an orientation of D_L, and hence also an orientation of S_L. One says that S_L *is obtained from S^3 by surgery along the framed link L.*

Of course, surgeries along two different framed links may lead to the same 3-manifold. Two framed links L and L' are said to be related by a *Kirby ϵ-move*, where $\epsilon = \pm 1$, if they are the same except for pieces which look as in Fig. 25. There, K is an unknotted component of L with framing ϵ, and the

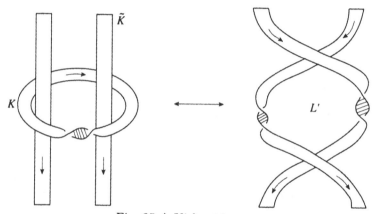

Fig. 25 A Kirby $+1$-move

vertical strands represent the components of L which have non-zero linking number with K, other than K itself. The framings have been represented by ribbons as described above. To obtain L' from L, we delete K and give each of the vertical strands $\tilde{K}$ a right-hand (resp. left-hand) twist if $\epsilon = -1$ (resp. if $\epsilon = 1$), at the same time adding $-\epsilon \, \mathrm{lk}(\tilde{K}, K)^2$ to its framing. In particular, if K is contained in a ball which does not intersect $L \backslash K$, then L' is obtained from L simply by deleting K: this is called a *special Kirby ϵ-move*. We emphasize that any Kirby move can be applied in either direction, to obtain L' from L or vice versa.

It was proved by Kirby (1978) and Fenn & Rourke (1979) that the 3-manifolds obtained by surgery along framed links L and L' are the same, up to orientation-preserving homeomorphism, if and only if L and L' are related by a sequence of Kirby ± 1-moves. In fact, it was pointed out by Reshetikhin & Turaev (1991) that only Kirby $+1$-moves and special Kirby -1-moves are needed. Thus, one can try to define a 3-manifold invariant by constructing an invariant of framed links which has the additional property

of being unchanged by these Kirby moves. We shall carry out this program
in Section 15.3.

15.2 Link invariants from quantum groups

In this section, we describe two ways of constructing link invariants, starting
from representations of quantum groups. The first is based on the relation
between links and braids discussed at the end of Subsection 15.1A, while the
second makes use of the vertex models of statistical mechanics, which we
discussed in Section 7.5.

A Link invariants from R-matrices Suppose first that V is a free module
of finite rank over a commutative ring k and that $R \in \text{End}(V \otimes V)$ is an
invertible solution of the quantum Yang–Baxter equation:

$$R_{12}R_{13}R_{23} = R_{23}R_{13}R_{12}.$$

Then, we obtain a representation ρ_m of the braid group $\mathcal{B}_m$ on $V^{\otimes m}$ by setting

$$\rho_m(T_i)(v_1 \otimes \cdots \otimes v_m) = v_1 \otimes \cdots \otimes v_{i-1} \otimes \sigma R(v_i \otimes v_{i+1}) \otimes v_{i+2} \otimes \cdots \otimes v_m,$$

where $\sigma \in \text{End}(V \otimes V)$ is the interchange of the two factors. Indeed, it is
easy to check that the QYBE is equivalent to the braid group relations,

$$\rho_m(T_i)\rho_m(T_j)\rho_m(T_i) = \rho_m(T_j)\rho_m(T_i)\rho_m(T_j), \quad \text{if } |i - j| = 1,$$

and the remaining relations

$$\rho_m(T_i)\rho_m(T_j) = \rho_m(T_j)\rho_m(T_i), \quad \text{if } |i - j| > 1,$$

are obvious because if $|i - j| > 1$ the operators $\rho_m(T_i)$ and $\rho_m(T_j)$ act on
disjoint pairs of factors in $V^{\otimes m}$. Note that the representations ρ_m are compatible, in the sense that they fit into commutative diagrams

$$
\begin{array}{ccc}
\mathcal{B}_m & \xrightarrow{\rho_m} & \text{End}(V^{\otimes m}) \\
\downarrow & & \downarrow{\cdot \otimes \text{id}_V} \\
\mathcal{B}_{m+1} & \xrightarrow{\rho_{m+1}} & \text{End}(V^{\otimes m+1})
\end{array}
$$

in which the left-hand vertical arrow is the natural inclusion $T_i \mapsto T_i$, $i =
1, \ldots, m - 1$.

Example 15.2.1 Let $V = \mathbb{C}^2$ with basis $\{v_0, v_1\}$, and let R be the endomorphism given by the matrix

$$\begin{pmatrix} 1 & 0 & 0 & 0 \\ 0 & \lambda & 0 & 0 \\ 0 & 1 - \lambda^2 & \lambda & 0 \\ 0 & 0 & 0 & 1 \end{pmatrix}$$

with respect to the basis $\{v_0 \otimes v_0, v_1 \otimes v_0, v_0 \otimes v_1, v_1 \otimes v_1, \}$ of $V \otimes V$. Up to a rescaling and a change of parameter $\lambda = e^{-h}$, this is the exactly the result of evaluating the universal R-matrix of $U_h(sl_2(\mathbb{C}))$ in its two-dimensional indecomposable representation (cf. 6.4.12). Thus, R satisfies the QYBE. The associated representation ρ_m of $\mathcal{B}_m$ resembles the well-known *Burau representation* of $\mathcal{B}_m$. In fact, note that

$$\rho_m(T_i) = \begin{pmatrix} 1 & & & & & & & \\ & 1 & & & & & & \\ & & \ddots & & & & & \\ & & & 1 - \lambda^2 & \lambda & & & \\ & & & \lambda & 0 & & & \\ & & & & & 1 & & \\ & & & & & & \ddots & \\ & & & & & & & 1 \end{pmatrix},$$

with the $1 - \lambda^2$ in the $(2^{i-1} + 2, 2^{i-1} + 2)$ position. By an obvious change of basis in $V^{\otimes m}$, we obtain an equivalent representation in which

$$(3) \qquad T_i \mapsto \begin{pmatrix} 1 & & & & & & & \\ & 1 & & & & & & \\ & & \ddots & & & & & \\ & & & 1 - \lambda^2 & \lambda^2 & & & \\ & & & 1 & 0 & & & \\ & & & & & 1 & & \\ & & & & & & \ddots & \\ & & & & & & & 1 \end{pmatrix}.$$

One can check that the matrices (3) still give a representation of $\mathcal{B}_m$ if we interpret them as being $m \times m$, with the $1 - \lambda^2$ being in the (i, i) position. The line spanned by the vector $(1, 1, \ldots, 1)$ is obviously preserved by all these operators, and the $(m - 1)$-dimensional quotient representation can be seen to be irreducible; this is the Burau representation. $\Diamond$

Suppose now that a link L is the closure of a braid β (see Fig. 24). In view of Markov's theorem, a first attempt to construct an invariant of L might be to take the trace of $\rho_m(\beta)$ (we shall not distinguish between β and its equivalence class in $\mathcal{B}_m$). This is independent of the choice of m because of the compatibility of the representations ρ_m, and is invariant under Markov I because conjugating an operator does not change its trace. But it will not, in general, be invariant under Markov II. Following Turaev (1988), we therefore make

DEFINITION 15.2.2 *Let V be a free module of finite rank over a commutative ring k. An enhanced quantum Yang–Baxter operator on $V \otimes V$ is a quadruple (R, f, λ, μ) consisting of*

(i) an invertible solution $R \in \mathrm{End}(V \otimes V)$ of the QYBE;
(ii) an invertible operator $f \in \mathrm{End}(V)$ such that $f \otimes f$ commutes with R;
(iii) invertible elements λ, $\mu \in k$ such that

$$\mathrm{trace}_2(I \circ (f \otimes f)) = \lambda\mu f, \quad \mathrm{trace}_2(I^{-1} \circ (f \otimes f)) = \lambda^{-1}\mu f,$$

where $I = \sigma \circ R$.

Remark It is no loss of generality to assume that $\lambda = \mu = 1$ in 15.2.2, for if (R, f, λ, μ) is an enhanced quantum Yang–Baxter operator, $(\lambda^{-1}R, \mu^{-1}f, 1, 1)$ is one too. However, it is not always convenient to make this normalization. $\triangle$

Here, and in the following discussion, if $0 \leq j \leq m$ and $F \in \mathrm{End}(V^{\otimes m}) \cong \mathrm{End}(V)^{\otimes j-1} \otimes \mathrm{End}(V) \otimes \mathrm{End}(V)^{\otimes m-j}$, then $\mathrm{trace}_j(F) \in \mathrm{End}(V^{\otimes m-1})$ is the result of applying the usual trace in the jth factor of the tensor product. We record two simple properties of these traces for later use. The proofs are easy exercises.

LEMMA 15.2.3 *Let $F \in \mathrm{End}(V^{\otimes m+1})$, $G \in \mathrm{End}(V^{\otimes m})$. Then, we have*

(4) $$\mathrm{trace}(\mathrm{trace}_{m+1}(F)) = \mathrm{trace}(F),$$
(5) $$\mathrm{trace}_{m+1}((G \otimes \mathrm{id}_V) \circ F) = G \circ \mathrm{trace}_{m+1}(F). \quad \blacksquare$$

There is an invariant of oriented links associated to any enhanced quantum Yang-Baxter operator (R, f, λ, μ). In fact, let $\alpha : \mathcal{B}_m \to \mathbb{Z}$ be the augmentation homomorphism, i.e. $\alpha(T_i^{\pm 1}) = \pm 1$ for $i = 1, \ldots, m - 1$; here, $\mathbb{Z}$ is thought of as an additive group. Set

$$P(\beta) = \lambda^{-\alpha(\beta)}\mu^{-m+1}\mathrm{trace}(\rho_m(\beta) \circ f^{\otimes m}),$$

where $\rho_m : \mathcal{B}_m \to \mathrm{End}(V^{\otimes m})$ is the representation associated to R.

PROPOSITION 15.2.4 *The map* $P : \coprod_m \mathcal{B}_m \to k$ *is invariant under the Markov moves of types I and II, and hence defines an invariant of oriented links.*

PROOF We give the proof to emphasize how straightforward Turaev's construction is.

Indeed, invariance under Markov I follows from the property 15.2.2(ii) of f. For that implies that $f^{\otimes m}$ commutes with $I_{i\,i+1}$ for all i, and hence with the whole of $\rho_m(\mathcal{B}_m)$. As for Markov II, note that, if $\beta \in \mathcal{B}_m$, then

$$\rho_m(\beta.T_m) = (\rho_m(\beta) \otimes \mathrm{id}_V) \circ I_{m\,m+1}.$$

Since $\alpha(\beta.T_m) = \alpha(\beta) + 1$,

$$
\begin{aligned}
P(\beta.T_m) &= \lambda^{-\alpha(\beta)-1}\mu^{-m}\mathrm{trace}((\rho_m(\beta) \otimes \mathrm{id}_V) \circ I_{m\,m+1} \circ (\mathrm{id}_V^{\otimes m-1} \otimes f \otimes f) \\
&\qquad \circ (f^{\otimes m-1} \otimes \mathrm{id}_V \otimes \mathrm{id}_V)) \\
&= \lambda^{-\alpha(\beta)-1}\mu^{-m}\mathrm{trace}(\mathrm{trace}_{m+1}((\rho_m(\beta) \otimes \mathrm{id}_V) \\
&\qquad \circ (\mathrm{id}_V^{\otimes m-1} \otimes (I \circ f \otimes f)) \circ (f^{\otimes m-1} \otimes \mathrm{id}_V \otimes \mathrm{id}_V))) \qquad \text{by (4)} \\
&= \lambda^{-\alpha(\beta)-1}\mu^{-m}\mathrm{trace}(\rho_m(\beta) \circ (\mathrm{id}_V^{\otimes m-1} \otimes \mathrm{trace}_2(I \circ f \otimes f)) \\
&\qquad \circ (f^{\otimes m-1} \otimes \mathrm{id}_V)) \quad \text{by (5)} \\
&= \lambda^{-\alpha(\beta)-1}\mu^{-m}\mathrm{trace}(\rho_m(\beta) \circ (\mathrm{id}_V^{\otimes m-1} \otimes \lambda\mu f) \circ (f^{\otimes m-1} \otimes \mathrm{id}_V)) \\
&\qquad \text{by 15.2.2(iii)} \\
&= \lambda^{-\alpha(\beta)}\mu^{-m+1}\mathrm{trace}(\rho_m(\beta) \circ f^{\otimes m}) \\
&= P(\beta).
\end{aligned}
$$

One deals with $\beta.T_m^{-1}$ in a similar way. ∎

The problem now is to find a supply of enhanced quantum Yang–Baxter operators. We are familiar by now with the fact that there is a solution of the QYBE associated to any finite-dimensional representation V of the standard quantization $U_h(\mathfrak{g})$ of a finite-dimensional complex simple Lie algebra $\mathfrak{g}$, which can be obtained, in principle at least, by evaluating the universal R-matrix of $U_h(\mathfrak{g})$ in the representation. We do not know if all R-matrices which arise in this way can be enhanced. However, this is so if $\mathfrak{g}$ is of type A_n, B_n, C_n or D_n and V is the natural representation; the R-matrices in these cases were written down in Section 7.3. Here, we concentrate on the A_n case.

PROPOSITION 15.2.5 *Let* $R \in (M_{n+1}(\mathbb{C}) \otimes M_{n+1}(\mathbb{C}))[[h]]$ *be the matrix*

$$R = e^h \sum_{i=0}^{n} E_{ii} \otimes E_{ii} + \sum_{i \neq j} E_{ii} \otimes E_{jj} + (e^h - e^{-h}) \sum_{i<j} E_{ij} \otimes E_{ji},$$

where E_{ij} is the matrix with a 1 in the (i,j)-position and zeros elsewhere, and let f be the diagonal matrix

$$f = \begin{pmatrix} e^{-nh} & & & \\ & e^{-(n-2)h} & & \\ & & \ddots & \\ & & & e^{nh} \end{pmatrix}.$$

Finally, let $\lambda = e^{-(n+1)h}$ and $\mu = -1$. Then, $(-R, f, \lambda, \mu)$ is an enhanced Yang–Baxter operator.

Moreover, the associated oriented link invariant P_n satisfies the equation

$$(6) \qquad e^{(n+1)h} P_n(\overline{L}_+) - e^{-(n+1)h} P_n(\overline{L}_-) - (e^h - e^{-h}) P_n(\overline{L}_0) = 0,$$

whenever $\overline{L}_+$, $\overline{L}_-$ and $\overline{L}_0$ are skein related (see Subsection 15.1A).

PROOF The proof of the first part is a straightforward calculation. For the second part, note that, if $\overline{L}_+$, $\overline{L}_-$ and $\overline{L}_0$ are skein related links, we may find a braid $\beta \in \mathcal{B}_m$ (say) such that $\overline{L}_+$, $\overline{L}_-$ and $\overline{L}_0$ are equivalent to the closures of $T_1^{-1}\beta$, $T_1\beta$ and β, respectively. Hence,

$$\begin{aligned}
&e^{(n+1)h} P_n(\overline{L}_+) - e^{-(n+1)h} P_n(\overline{L}_-) - (e^h - e^{-h}) P_n(\overline{L}_0) \\
&= (-1)^{-m+1} e^{(n+1)h} \lambda^{-\alpha(\beta)+1} \operatorname{trace}(\rho_m(T_1^{-1}\beta) \circ f^{\otimes m}) \\
&\qquad - e^{-(n+1)h} \lambda^{-\alpha(\beta)-1} \operatorname{trace}(\rho_m(T_1\beta) \circ f^{\otimes m}) \\
&\qquad - (e^h - e^{-h}) \lambda^{-\alpha(\beta)} \operatorname{trace}(\rho_m(\beta) \circ f^{\otimes m})) \\
&= (-1)^{-m+1} \lambda^{-\alpha(\beta)} \operatorname{trace}((I^{-1} - I - (e^h - e^{-h})\operatorname{id}^{\otimes 2}) \otimes \operatorname{id}^{\otimes m-2} \\
&\qquad\qquad \circ \rho_m(\beta) \circ f^{\otimes m})) \\
&= 0,
\end{aligned}$$

since one easily checks that $I = -\sigma \circ R$ satisfies $I - I^{-1} = -(e^h - e^{-h})\operatorname{id}^{\otimes 2}$. ∎

Before proceeding further, note that

$$P_n(\mathbb{O}) = \operatorname{trace}(f) = [n+1]_{e^h}.$$

Together with the skein relation (6), and the discussion in Subsection 15.1A, this implies that P_n takes values in the subring $\mathbb{Z}[q, q^{-1}]$ of $\mathbb{C}[[h]]$, where $q = e^h$.

The final step is to regard the sequence of (Laurent) polynomials P_n in the single variable q as a single polynomial in two variables. To make this precise, we need

LEMMA 15.2.6 *Let L be a link with r components. Choose a plane projection $\overline{L}$ of L, and suppose that it has c crossings. Then, if $n \geq 4c + 2r$, one has*

$$(q - q^{-1})^{c+r} P_n(L) = \sum_{a,b \in \mathbf{Z}} p_{a,b} q^{a+(n+1)b},$$

where the integers $p_{a,b}$ have the following properties:

(i) $p_{a,b} = 0$ for all but finitely many pairs (a, b);
(ii) $p_{a,b} = 0$ if $|a| > 2c + r$;
(iii) the $p_{a,b}$ do not depend on n.

PROOF Applying the skein relation (6) inductively as described in Subsection 15.1A, one sees that $P_n(\overline{L})$ is a linear combination, with integer coefficients, of terms of the form

$$P_n(\mho^d) q^{(n+1)e} (q - q^{-1})^f,$$

where $e, f \in \mathbf{Z}$, $0 \leq f \leq c$ and $d \leq c + r$. Next, one uses the skein relation to compute

$$P_n(\mho^d) = \left(\frac{q^{n+1} - q^{-n-1}}{q - q^{-1}} \right)^d$$

(see 15.1.2). Thus, $(q - q^{-1})^{c+r} P_n(\overline{L})$ is a linear combination of terms of the form $q^{(n+1)e}(q^{n+1} - q^{-n-1})^d(q - q^{-1})^{f+c+r-d}$. Since $f + c + r - d \leq f + c + r \leq 2c + r$, the lemma is proved. ∎

Now, with the notation in the lemma, let t be an indeterminate independent of q and set

$$\tilde{P}(\overline{L}) = (q - q^{-1})^{-c-r} \sum_{a,b \in \mathbf{Z}} p_{a,b} q^a t^b \in \mathbf{Z}[q, q^{-1}, t, t^{-1}].$$

It follows from (6) that $\tilde{P}$ satisfies the skein relation

$$t\tilde{P}(\overline{L}_+) - t^{-1}\tilde{P}(\overline{L}_-) - (q - q^{-1})\tilde{P}(\overline{L}_0) = 0.$$

Furthermore, it is clear that

$$\tilde{P}(\mho) = \frac{t - t^{-1}}{q - q^{-1}}.$$

Thus, if we set

$$P(\overline{L}) = \left(\frac{q - q^{-1}}{t - t^{-1}} \right) \tilde{P}(\overline{L}),$$

and write $x = \sqrt{-1}t$, $y = -\sqrt{-1}(q - q^{-1})$, then P satisfies the skein relation (1). We have therefore proved the existence of the HOMFLY polynomial (and 15.1.1).

Theorem 15.1.3 can be proved in a similar way. One shows that the R-matrices associated to the natural representations of $U_h(\mathfrak{g})$, when $\mathfrak{g}$ is of type B_n, C_n and D_n, can be enhanced, and hence define invariants. One thus obtains three sequences of invariants which are (Laurent) polynomials in a single variable q. By combining them in a suitable way, one obtains a two-variable invariant which satisfies the conditions in 15.1.3. See Turaev (1988) for the details.

B Link invariants from vertex models In Section 7.4, we discussed certain 2-dimensional models in statistical mechanics, called vertex models, in which 'atoms' located at the vertices of a rectangular lattice 'interact' with their nearest neighbours. The energy $\mathcal{E}_{ij}^{kl}$ of a particular atom is a certain function of the 'states' i, j, k, l of the edges of the lattice incident with the atom. The physical properties of the model can be deduced from its partition function

$$Z = \sum_{\text{states}} \prod_{\text{vertices}} R_{ij}^{kl},$$

where the Boltzmann weights

$$R_{ij}^{kl} = \exp(-\beta \mathcal{E}_{ij}^{kl}),$$

β being a constant, and the sum is over all possible assignments of states to the edges of the lattice.

There is an obvious generalization of these rectangular lattice vertex models, for the same functions $\mathcal{E}_{ij}^{kl}$ can be used to associate a vertex model to any oriented 4-valent graph, provided one decides on a way of numbering the edges which meet at a given vertex. Since a plane projection of any link is a 4-valent graph (forgetting the over/under information), the partition function associates a number to every oriented link. One may then try to choose the Boltzmann weights in such a way that the partition function depends only on the equivalence class of the link.

To carry out this program, we shall have to impose some conditions on the Boltzmann weights beyond the quantum Yang–Baxter equation required in Section 7.5. Unfortunately, although these conditions are necessary for the topological applications we have in mind, they do not seem to be satisfied in physically interesting models.

DEFINITION 15.2.7 *A vertex model consists of*

(a) *a finite-dimensional complex vector space V equipped with a distinguished basis $\mathcal{S}$;*

(b) *a family of endomorphisms $R_\pm(\theta) \in \mathrm{End}(V \otimes V)$ parametrized by $\theta \in \mathbb{C}$;*

(c) *a family of invertible endomorphisms $L(\theta) \in \mathrm{End}(V)$, also parametrized by $\theta \in \mathbb{C}$, of the form*

$$L(\theta)(s) = \exp(f(s)\theta)s \qquad \text{for all } s \in \mathcal{S},$$

where $f : S \to \mathbb{C}$.

These data are required to satisfy the following conditions:

(i) $(L(-\delta) \otimes \mathrm{id}_V) R_\pm(\theta) (L(-\delta) \otimes \mathrm{id}_V)^{-1} = (\mathrm{id}_V \otimes L(\delta)) R_\pm(\theta) (\mathrm{id}_V \otimes L(\delta))^{-1} = R_\pm(\theta + \delta)$, *for all $\theta, \delta \in \mathbb{C}$;*

(ii) $I_-(0) = I_+(0)^{-1}$ *(here, $I_\pm = \sigma \circ R_\pm$, σ being as usual the interchange of the factors in the tensor product);*

(iii) $(\sigma \circ R_+(\pi)^{\mathrm{tr}_1}).(\sigma \circ R_-(\pi)^{\mathrm{tr}_2}) = \mathrm{id}_V$ *(here, $\mathrm{tr}_i : \mathrm{End}(V^{\otimes 2}) \cong \mathrm{End}(V)^{\otimes 2} \to \mathrm{End}(V)^{\otimes 2}$ takes the transpose of the ith factor, $i = 1, 2$);*

(iv) $R_\pm(\theta)$ *satisfy the quantum Yang–Baxter equation with spectral parameters*

$$R_{12}(\theta) R_{13}(\theta + \varphi) R_{23}(\varphi) = R_{23}(\varphi) R_{13}(\theta + \varphi) R_{12}(\theta),$$

for all $\theta, \varphi \in \mathbb{C}$ ($R = R_+$ or R_-).

Remarks [1] Condition (i) implies that $R_\pm(\theta)$ are completely determined by $R_\pm(0)$ and $L(\theta)$, and that $L(\theta) \otimes L(\theta)$ commutes with $R_\pm(0)$. Conversely, if these latter conditions hold and we define

$$R_\pm(\theta) = (L(-\theta) \otimes \mathrm{id}_V) R_\pm(0) (L(-\theta) \otimes \mathrm{id}_V)^{-1}$$
$$= (\mathrm{id}_V \otimes L(\theta)) R_\pm(0) (\mathrm{id}_V \otimes L(\theta))^{-1},$$

then $R_\pm(\theta)$ satisfies (i) (one makes use of the group property $L(\theta + \delta) = L(\theta) L(\delta)$).

[2] By (i) again, the QYBE with spectral parameters in (iv) is equivalent to the QYBE

(7) $$R_{12}(0) R_{13}(0) R_{23}(0) = R_{23}(0) R_{13}(0) R_{12}(0)$$

($R = R_+$ or R_-). By (ii), $R_-(0)$ is determined by $R_+(0)$, and obviously satisfies the QYBE if $R_-(0)$ does. Thus, $R_\pm(\theta)$ are, in fact, determined by $L(\theta)$ and the single (constant) solution $R_+(0)$ of the QYBE.

[3] The condition that $L(\theta) \otimes L(\theta)$ commutes with $R_+(0)$ is reminiscent of condition (ii) in the definition 15.2.2 of an enhanced Yang–Baxter operator. We shall discuss the braid-theoretic interpretation of vertex models later in this subsection. $\triangle$

Forgetting the over/under information in a plane projection $\overline{L}$ of an oriented link L gives an oriented 4-valent graph. Given a vertex model as above, a *state* of this graph is a map Σ from the edges of the graph to S. The discussion in Section 7.5 suggests that we define the partition function of $\overline{L}$ by

$$\tilde{Z}(\overline{L}) = \sum_{\text{states}} \prod_{\text{crossings}} R_\pm(\theta)_{ij}^{kl},$$

where

$$R_\pm(\theta)(i \otimes j) = \sum_{k,l} R_\pm(\theta)_{ij}^{kl}(k \otimes l).$$

The choice of whether to use R_+ or R_- at a particular crossing, and the convention as to the way in which the labels i, j, k and l are attached to the edges, is indicated below:

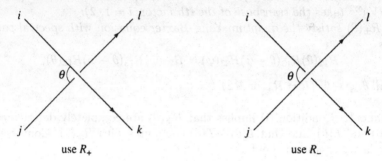

use R_+ use R_-

The way the angle θ is determined is also shown in these diagrams; it is chosen so that $0 \le \theta \le \pi$.

Unfortunately, $\tilde{Z}(\overline{L})$ is not an invariant in general. This is the reason for the extra datum in 15.2.7 consisting of the map $f : S \to \mathbb{R}$. If Σ is any state of $\overline{L}$, let f_Σ be the locally constant function on $\overline{L}$ which is equal to $f(\Sigma(e))$ along the edge e.

DEFINITION 15.2.8 *Given a vertex model as in 15.2.7, and a plane projection $\overline{L}$ of an oriented link L, the* partition function *of $\overline{L}$ is*

$$Z(\overline{L}) = \sum_{\substack{\text{states} \\ \Sigma}} \left(\prod_{\text{crossings}} R_\pm(\theta)_{ij}^{kl} \right) \exp\left(\int_{\overline{L}} f_\Sigma d\angle \right).$$

The angle 1-form $d\angle$ was defined in Remark [4] following 15.1.3.

PROPOSITION 15.2.9 *The partition function Z defined above is an invariant of regular isotopy.*

OUTLINE OF PROOF Two plane projections define oriented links which are equivalent under regular isotopy if and only if one can transform one into the other by an isotopy of the plane, together with Reidemeister moves II and III. Let us check invariance under the type II move:

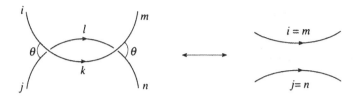

The contribution to the partition function from the portion of the diagram on the left is

$$R_+(\theta)^{kl}_{ij} R_-(\theta)^{nm}_{lk} \exp(\theta(f(k) - f(l))),$$

while that of the portion of the diagram on the right is

$$\exp(\theta(f(i) - f(j))).$$

Hence, we must prove that

$$\sum_{k,l} R_+(\theta)^{kl}_{ij} R_-(\theta)^{nm}_{lk} \exp(\theta(f(k) - f(l))) = \exp(\theta(f(i) - f(j))) \, \delta_{i,m} \delta_{j,n}.$$

Now,

$$\exp(\theta(f(j) - f(l))) R_+(\theta)^{kl}_{ij} = ((\mathrm{id}_V \otimes L(\theta))^{-1} R_+(\theta)(\mathrm{id}_V \otimes L(\theta)))^{kl}_{ij}$$
$$= R_+(0)^{kl}_{ij}$$

by condition (i) in 15.2.7. Similarly,

$$\exp(\theta(f(k) - f(i))) R_-(\theta)^{nm}_{lk} = R_-(0)^{nm}_{lk}.$$

Thus, the equation to be proved is

$$\sum_{k,l} R_+(0)^{kl}_{ij} R_-(0)^{nm}_{lk} = \delta_{i,m} \delta_{j,n}.$$

This is obviously equivalent to condition (ii) in 15.2.7.
    The other type II Reidemeister move

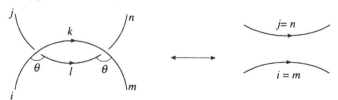

is dealt with in the same way. Invariance under the type III move

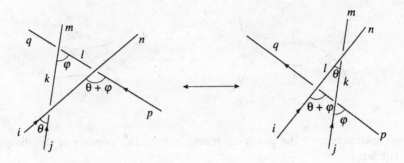

follows from conditions (i) and (iv), but for the other oriented type III moves one must work a little harder, making use of the other conditions. Finally, invariance of Z under plane isotopy follows from condition (i) (it essentially guarantees that Z does not depend on the angles at the crossings). ∎

The partition function Z is not invariant under type I Reidemeister moves in general, and so is not an isotopy invariant. Of course, one can obtain an isotopy invariant by introducing kinks as described in Remark [4] following 15.1.3, but there is actually a simple extra condition on the model which guarantees full isotopy invariance:

COROLLARY 15.2.10 *The partition function of a vertex model defines an oriented link invariant provided the following additional conditions hold:*
(8)
$$\text{trace}_1((L(2\pi) \otimes \text{id}_V)I_{\pm}(0)) = \text{id}_V, \quad \text{trace}_2(I_{\pm}(0)(\text{id}_V \otimes L(-2\pi))) = \text{id}_V. \quad \blacksquare$$

The proof is straightforward.

Before giving examples of vertex models, we indicate the relation between the last two results and the braid approach to constructing link invariants. The solution $R_+(0)$ of the QYBE defines, as usual, a representation of the braid group $\mathcal{B}_m$ on $V^{\otimes m}$ for all m by letting the generator T_i of $\mathcal{B}_m$ act as $I_+(0)$ in the ith and $(i+1)$th positions. We define a *Markov trace* on $\text{End}(V^{\otimes m})$ by

$$\text{mtr}(A) = \text{trace}(L(-2\pi)^{\otimes m} \circ A).$$

By Markov's theorem, these traces will define a link invariant provided they are unchanged by the Markov moves, i.e.

$$\text{mtr}(\beta\beta'\beta^{-1}) = \text{mtr}(\beta') \quad \text{if } \beta, \beta' \in \mathcal{B}_m,$$
$$\text{mtr}(\beta T_m^{\pm 1}) = \text{mtr}(\beta) \quad \text{if } \beta \in \mathcal{B}_m.$$

The first equation follows from the fact that $L(-2\pi) \otimes L(-2\pi)$ commutes with $I_+(0)$, which implies that $L(-2\pi)^{\otimes m}$ commutes with every element of $\mathcal{B}_m$:

$$\text{trace}(L(-2\pi)^{\otimes m}\beta\beta'\beta^{-1}) = \text{trace}(\beta L(-2\pi)^{\otimes m}\beta'\beta^{-1}) = \text{trace}(L(-2\pi)^{\otimes m}\beta').$$

The second equation follows from property (8): for example,

$$\text{trace}\,(L(-2\pi)^{\otimes m+1}\beta T_m)$$
$$= \text{trace}(\text{trace}_{m+1}(L(-2\pi)^{\otimes m}\beta(\text{id}^{\otimes m}\otimes L(-2\pi))(\text{id}^{\otimes m-1}\otimes I_+(0))))$$
$$= \text{trace}(L(-2\pi)^{\otimes m}\beta(\text{id}^{\otimes m-1}\otimes \text{trace}_2(I_+(0)(\text{id}\otimes L(-2\pi))))))$$
$$= \text{trace}(L(-2\pi)^{\otimes m}\beta)$$

by (8).

We see from this calculation that the reason the usual trace does not define a link invariant in general, because it is not invariant under Markov II, is the same as the reason for including the 'angular term' $\exp(\int_L f_\Sigma\, d\mathcal{L})$ in the definition of a vertex model.

The next result shows that there is a vertex model associated to any finite-dimensional representation of the standard quantization $U_h(\mathfrak{g})$ of any finite-dimensional complex simple Lie algebra $\mathfrak{g}$.

Let $\mathcal{R}_h$ be the universal R-matrix of $U_h(\mathfrak{g})$ (see Chapter 8). If $\sum_i r_i\alpha_i$ is the sum of the positive roots of $\mathfrak{g}$, set $H_{\rho^\bullet} = \sum_i d_i r_i H_i$ (see Subsection 11.3B). We recall that the square of the antipode S_h of $U_h(\mathfrak{g})$ is given by

$$S_h^2(x) = \exp(hH_{\rho^\bullet}).x.\exp(-hH_{\rho^\bullet}) \quad (x\in U_h(\mathfrak{g})).$$

We shall write the right-hand side of this equation as $\text{Ad}(\exp(hH_{\rho^\bullet}))(x)$.

PROPOSITION 15.2.11 *Let* $T : U_h(\mathfrak{g}) \to \text{End}_{\mathbb{C}[[h]]}(V)$ *be a finite-dimensional representation of* $U_h(\mathfrak{g})$, *and set*

$$R_+(0) = (T\otimes T)(\mathcal{R}_h), \quad R_-(0) = \sigma(R_+(0)^{-1}),$$
$$\mathcal{L}(\theta) = \exp(-h\theta H_{\rho^\bullet}/2\pi), \quad L(\theta) = T(\mathcal{L}(\theta)).$$

Then, $R_\pm(0)$ *and* $L(\theta)$ *define a vertex model in the sense of 15.2.7.*

It is understood that $R_\pm(\theta)$ are defined in terms of $R_\pm(0)$ as described in the remarks following 15.2.7.

PROOF Since $\mathcal{R}_h$ satisfies the QYBE, so does $R_+(0)$. Next, we have

$$\mathcal{R}_h(\mathcal{L}_h(\theta)\otimes \mathcal{L}_h(\theta))\mathcal{R}_h^{-1} = \mathcal{R}_h\Delta_h(\mathcal{L}_h(\theta))\mathcal{R}_h^{-1}$$
$$= \Delta_h^{\text{op}}(\mathcal{L}_h(\theta))$$
$$= \mathcal{L}_h(\theta)\otimes \mathcal{L}_h(\theta).$$

Applying T shows that $L_h(\theta)\otimes L_h(\theta)$ commutes with $R_+(0)$.

It remains to prove property (iii) in 15.2.7. For this, we shall interpret the transpose on the algebra level.

Let Φ_h be the anti-involution of $U_h(\mathfrak{g})$ given on generators by

$$\Phi_h(X_i^\pm) = X_i^\mp, \quad \Phi_h(H_i) = H_i.$$

As in the classical case, V admits a non-degenerate symmetric $\mathbb{C}[[h]]$-bilinear form $(\ ,\ )$ such that

$$(9) \qquad\qquad (x.v_1,\, v_2) = (v_1,\, \Phi_h(x).v_2)$$

for all v_1, $v_2 \in V$, $x \in U_h(\mathfrak{g})$. We recall the construction briefly. We may assume that V is indecomposable. Let v_λ be a highest weight vector in V of weight $\lambda \in \mathfrak{h}^*$, say, and extend λ to a homomorphism of $\mathbb{C}[[h]]$-algebras $U_h^0 \to \mathbb{C}[[h]]$ (see Subsection 8.1A for the notation). Now, if $v_1 = x_1.v_\lambda$ and $v_2 = x_2.v_\lambda$, with x_1, $x_2 \in U_h(\mathfrak{g})$, one defines (v_1, v_2) to be the image of $\Phi_h(x_1)x_2$ under the composite map

$$U_h(\mathfrak{g}) = U_h^- \otimes U_h^0 \otimes U_h^+ \xrightarrow{\ \epsilon_h \otimes \mathrm{id} \otimes \epsilon_h\ } U_h^0 \xrightarrow{\ \lambda\ } \mathbb{C}[[h]],$$

where we have used the Poincaré–Birkhoff–Witt factorization 8.1.8.

It follows from (9) that, if we work in an orthonormal basis of V with respect to $(\ ,\ )$, the matrix of $T(\Phi_h(x))$ is the transpose of that of $T(x)$, for all $x \in U_h(\mathfrak{g})$.

We can now proceed with the proof. Define

$$\mathcal{R}_h^+(\theta) = \mathrm{Ad}(\mathcal{L}(-\theta) \otimes 1)(\mathcal{R}_h), \quad \mathcal{R}_h^-(\theta) = \mathrm{Ad}(\mathcal{L}(-\theta) \otimes 1)((\mathcal{R}_h^{21})^{-1}).$$

Then, from the remarks following 15.2.7, $R_\pm(\theta) = (T \otimes T)(\mathcal{R}_h^\pm(\theta))$, and condition (iii) will follow if we prove that

$$((\Phi_h \otimes \mathrm{id})(\mathcal{R}_h^+(\pi)))^{21}.(\mathrm{id} \otimes \Phi_h)(\mathcal{R}_h^-(\pi)) = 1 \otimes 1;$$

in other words, $\mathcal{R}_h^+(\pi)\mathcal{R}_h^-(\pi)^{21} = 1 \otimes 1$. Using the definitions, and the group property of $\mathcal{L}(\theta)$, this is equivalent to

$$\mathrm{Ad}(\mathcal{L}(-2\pi) \otimes 1)(\mathcal{R}_h) = \mathcal{R}_h.$$

But, we observed above that $\mathrm{Ad}(\mathcal{L}(-2\pi)) = \mathrm{Ad}(e^{hH_{\rho^*}}) = S_h^2$, so our assertion follows from (15), Chapter 4. ∎

Remark The vertex model given by this proposition does not satisfy the additional property (8), even in the sl_2 case. However, the invariants of regular isotopy given by 15.2.9 and 15.2.11 can often be 'rescaled' to obtain genuine isotopy invariants. In particular, this can be done when $\mathfrak{g}$ is of type A_n, B_n, C_n or D_n, and V is the natural representation, and the resulting sequences of invariants can be combined to give the HOMFLY and Kauffman polynomials in a manner similar to that described in the previous subsection. $\triangle$

15.3 Modular Hopf algebras and 3-manifold invariants

Suppose that a 3-manifold M is obtained by surgery along a framed link L in S^3. The methods of Section 5.3 enable us to associate to this link, equipped with a suitable colouring, an isotopy invariant. In this section, we show that, by taking suitable combinations of such invariants, one can construct an invariant of M itself. We begin by describing the class of colourings we shall need.

A Modular Hopf algebras Recall from Subsection 4.2C that a ribbon Hopf algebra $(A, \mathcal{R}, v)$ over a field k (say) is a quasitriangular Hopf algebra $(A, \mathcal{R})$ equipped with an invertible central element v satisfying certain conditions (given in 4.2.8) which imply that the element $g = uv^{-1}$ is group-like, where $u = \mu(S \otimes \mathrm{id})(\mathcal{R}_{21})$. Recall further that, if $\rho_V : A \to \mathrm{End}(V)$ is a finite-dimensional A-module, the quantum trace of a k-linear map $f : V \to V$ is defined to be $\mathrm{qtr}(f) = \mathrm{trace}(\rho_V(g) \circ f)$; in particular, taking $f = \mathrm{id}_V$ gives the quantum dimension $\mathrm{qdim}(V)$. Recall finally that the left and right duals of V are canonically isomorphic or, equivalently, the second (left) dual V^{**} is canonically isomorphic to V: in fact, the isomorphism $V^{**} \to V$ is essentially given by the action of g.

With these definitions in mind, we can state

DEFINITION 15.3.1 *A* modular Hopf algebra *is a ribbon Hopf algebra* $(A, \mathcal{R}, v)$ *over a field* k *equipped with a distinguished family* $\{V(\lambda)\}_{\lambda \in \mathcal{C}}$ *of irreducible A-modules, indexed by a finite set* $\mathcal{C}$. *The set* $\mathcal{C}$ *is equipped with an involution* $\lambda \mapsto \lambda^*$ *and a distinguished element* 0 *such that* $0^* = 0$ *and* $V(0)$ *is the trivial module. Finally, for each* $\lambda \in \mathcal{C}$, *one is given an isomorphism of A-modules*

$$\omega_\lambda : V(\lambda)^* \to V(\lambda^*).$$

These data are required to satisfy the following conditions:

(i) $\mathrm{qdim}(V(\lambda)) \neq 0$ *for all* $\lambda \in \mathcal{C}$;
(ii) the composite map

$$V(\lambda) \xrightarrow{\omega_{\lambda^*}^{-1}} V(\lambda^*)^* \xrightarrow{\omega_\lambda^*} V(\lambda)^{**} \to V(\lambda),$$

the last map being the canonical isomorphism of vector spaces, is given by the action of the element $g \in A$;

(iii) for any $\lambda_1, \ldots, \lambda_r \in \mathcal{C}$, we have

$$V(\lambda_1) \otimes \cdots \otimes V(\lambda_r) \cong Z \oplus \bigoplus_{\lambda \in \mathcal{C}} V(\lambda)^{\oplus m_\lambda}$$

as A-modules, where $m_\lambda \in \mathbf{N}$ and the module Z has the property

$$\mathrm{qtr}(f) = 0$$

for all A-module endomorphisms f of Z;
(iv) if $s_{\lambda\mu}$ is the quantum trace of the map $V(\lambda) \otimes V(\mu) \to V(\lambda) \otimes V(\mu)$ given by the action of $\mathcal{R}_{21}\mathcal{R}$, then the matrix $(s_{\lambda\mu})_{\lambda,\mu \in \mathcal{C}}$ is invertible.

Remarks [1] In condition (iv), we are using the obvious fact that, if $(A, \mathcal{R}^A, v^A)$ and $(B, \mathcal{R}^B, v^B)$ are two ribbon Hopf algebras, then their tensor product $A \otimes B$, equipped with the usual Hopf algebra structure, is itself a ribbon Hopf algebra, the universal R-matrix being $\mathcal{R}_{13}^A \mathcal{R}_{24}^B$, and the distinguished central element $v^A \otimes v^B$.

[2] Reshetikhin & Turaev (1991) included an additional condition when they introduced the notion of a modular Hopf algebra. Suppose that the central element v acts on $V(\lambda)$ by the scalar v_λ; this will hold if k is algebraically closed or, more generally, if the $V(\lambda)$ remain irreducible over an algebraic closure of k. Since the matrix $(s_{\lambda\mu})$ is invertible, there are unique scalars $(d_\lambda)_{\lambda \in \mathcal{C}}$ such that

$$(10) \qquad \sum_{\lambda \in \mathcal{C}} v_\lambda s_{\lambda\mu} d_\lambda = v_\mu^{-1} \mathrm{qdim}(V(\mu)), \qquad (\mu \in \mathcal{C}).$$

Then, the additional condition was that the scalar

$$(11) \qquad c = \sum_{\lambda \in \mathcal{C}} v_\lambda^{-1} \mathrm{qdim}(V(\lambda)) d_\lambda$$

is non-zero. We have omitted it because it was proved by Turaev & Wenzl (1993) that it is actually a consequence of the other conditions in 15.3.1. △

In the next subsection, we show how to associate a 3-manifold invariant to any modular Hopf algebra, but first we describe an example.

Conditions (i)–(iii) are reminiscent of the properties of tilting modules obtained in Section 11.3, so natural candidates for modular Hopf algebras are the quantum groups $U_\epsilon^{\mathrm{res}}(\mathfrak{g})$ studied there, ϵ being a root of unity, together with the family of irreducible representations $V_\epsilon^{\mathrm{res}}(\lambda)$ with highest weights λ lying in the set

$$\mathcal{C}_\ell = \{\lambda \in P^+ \mid \langle \lambda + \rho, \check{\alpha} \rangle < \ell \text{ for all } \alpha \in \Delta^+\}.$$

These representations are irreducible under the finite-dimensional subalgebra $U_\epsilon^{fin}(\mathfrak{g})$ of $U_\epsilon^{res}(\mathfrak{g})$ by 11.2.10, and $U_\epsilon^{fin}(\mathfrak{g})$ is quasitriangular by 9.3.9. Unfortunately, however, these data do not define a modular Hopf algebra, since condition (iv) is not satisfied (even when $\mathfrak{g} = sl_2(\mathbb{C})$). The way to proceed was shown by Reshetikhin & Turaev (1991), at least in the sl_2 case.

We work with the 'simply-connected' quantum group (see Subsection 9.1A). Namely, we let $\hat{U}_\epsilon^{fin}$ be the associative algebra over $\mathbb{C}$ with generators $L^{\pm 1}$, $X^\pm$ and defining relations

$$LX^\pm L^{-1} = \epsilon^{\pm 2} X^\pm, \quad [X^+, X^-] = \frac{L^2 - L^{-2}}{\epsilon^2 - \epsilon^{-2}},$$
$$(X^\pm)^\ell = 0, \quad L^{4\ell} = 1.$$

Here, $\epsilon = e^{\sqrt{-1}\pi/2\ell}$, where ℓ is any integer greater than one (not necessarily odd). The Hopf algebra structure is given by

$$\Delta(L) = L \otimes L,$$
$$\Delta(X^+) = X^+ \otimes L^2 + 1 \otimes X^+, \quad \Delta(X^-) = X^- \otimes 1 + L^{-2} \otimes X^-,$$
$$S(L) = L^{-1}, \quad S(X^+) = -X^+ L^{-2}, \quad S(X^-) = -L^2 X^-,$$
$$\epsilon(L) = 1, \quad \epsilon(X^\pm) = 0.$$

The following formulas define a representation of $\hat{U}_\epsilon^{fin}$ on the vector space $\hat{V}_\epsilon^{fin}(\lambda)$ with basis $\{v_0, v_1, \ldots, v_\lambda\}$:

$$L.v_r = \epsilon^{\lambda - 2r} v_r, \quad X^+.v_r = [\lambda - r + 1]_{\epsilon^2} v_{r-1}, \quad X^-.v_r = [r+1]_{\epsilon^2} v_{r+1}.$$

It is easy to see that the representation $\hat{V}_\epsilon^{fin}(\lambda)$ is irreducible if and only if $0 \le \lambda \le \ell - 1$, and that all such representations are isomorphic to their (left) duals.

THEOREM 15.3.2 *The Hopf algebra $\hat{U}_\epsilon^{fin}$, together with the family of representations $\{\hat{V}_\epsilon^{fin}(\lambda)\}_{0 \le \lambda < \ell - 1}$, is a modular Hopf algebra.*

OUTLINE OF PROOF The first step is to show that $\hat{U}_\epsilon^{fin}$ is quasitriangular. For this, one verifies by a direct calculation, analogous to that used in the proof of 6.4.8, that

$$\mathcal{R} = \frac{1}{4\ell} \sum_{n=0}^{\ell-1} \sum_{a,b=0}^{4\ell-1} \frac{(\epsilon^2 - \epsilon^{-2})^n}{[n]_{\epsilon^2}!} \epsilon^{n(n-1)-ab} L^a (X^+)^n \otimes L^b (X^-)^n$$

satisfies the conditions in 4.2.1 and 4.2.6(ii) (see 9.3.9). That $\hat{U}_\epsilon^{fin}$ is actually a ribbon Hopf algebra now follows easily from the general results on quasitriangular Hopf algebras in Section 4.2 and the observation that the square of the antipode S of $\hat{U}_\epsilon^{\mathrm{fin}}$ is given by conjugation by the group-like element $g = L^2$.

Properties (i)–(iii) in 15.3.1 follow by arguments similar to those used in Section 11.3. To verify the crucial condition (iv), we must compute the matrix $(s_{\lambda\mu})$. Now, using the functor $\tilde{\mathcal{F}}$ constructed in 5.3.7, with $X = \{\hat{V}_\epsilon^{\mathrm{fin}}(\lambda)\}$, we have

$$s_{\lambda\mu} = \mathrm{qtr}_{\hat{V}_\epsilon^{\mathrm{fin}}(\lambda) \otimes \hat{V}_\epsilon^{\mathrm{fin}}(\mu)}(\mathcal{R}_{21}\mathcal{R}_{12})$$

$$= \tilde{\mathcal{F}}\left(\text{the Hopf link} \quad \right)$$

$$= \tilde{\mathcal{F}}\left(\text{closure of} \quad \right)$$

$$= \mathrm{qtr}\left(\tilde{\mathcal{F}}\left(\quad \right) \right)$$

$$(12) \qquad = [\lambda + 1]_{\epsilon^2} . \tilde{\mathcal{F}}\left(\quad \right)$$

(the second equality uses 5.3.6). For the last equation, we used the fact that the endomorphism in question is a scalar, since $\hat{V}_\epsilon^{\mathrm{fin}}(\lambda)$ is irreducible. It may therefore be computed using the diagram

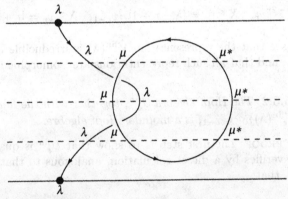

to be the composite map

$$\hat{V}_\epsilon^{\mathrm{fin}}(\lambda) \cong \hat{V}_\epsilon^{\mathrm{fin}}(\lambda) \otimes \mathbb{C} \quad \longrightarrow \quad \hat{V}_\epsilon^{\mathrm{fin}}(\lambda) \otimes \hat{V}_\epsilon^{\mathrm{fin}}(\mu) \otimes \hat{V}_\epsilon^{\mathrm{fin}}(\mu)^*$$

$$\downarrow$$

$$\hat{V}_\epsilon^{\mathrm{fin}}(\lambda) \otimes \hat{V}_\epsilon^{\mathrm{fin}}(\mu) \otimes \hat{V}_\epsilon^{\mathrm{fin}}(\mu)^* \quad \longleftarrow \quad \hat{V}_\epsilon^{\mathrm{fin}}(\mu) \otimes \hat{V}_\epsilon^{\mathrm{fin}}(\lambda) \otimes \hat{V}_\epsilon^{\mathrm{fin}}(\mu)^*$$

$$\downarrow$$

$$\hat{V}_\epsilon^{\mathrm{fin}}(\lambda) \otimes \hat{V}_\epsilon^{\mathrm{fin}}(\mu)^{**} \otimes \hat{V}_\epsilon^{\mathrm{fin}}(\mu)^* \quad \longrightarrow \quad \hat{V}_\epsilon^{\mathrm{fin}}(\lambda)$$

given by

$$(\mathrm{id}_{\hat{V}_\epsilon^{\mathrm{fin}}(\lambda)} \otimes \mathrm{ev}_{\hat{V}_\epsilon^{\mathrm{fin}}(\mu)^*}) \circ (\mathrm{id}_{\hat{V}_\epsilon^{\mathrm{fin}}(\lambda)} \otimes g \otimes \mathrm{id}_{\hat{V}_\epsilon^{\mathrm{fin}}(\mu)^*}) \circ (R_{\hat{V}_\epsilon^{\mathrm{fin}}(\mu)\hat{V}_\epsilon^{\mathrm{fin}}(\lambda)} \otimes \mathrm{id}_{\hat{V}_\epsilon^{\mathrm{fin}}(\mu)^*})$$

$$\circ (R_{\hat{V}_\epsilon^{\mathrm{fin}}(\lambda)\hat{V}_\epsilon^{\mathrm{fin}}(\mu)} \otimes \mathrm{id}_{\hat{V}_\epsilon^{\mathrm{fin}}(\mu)^*}) \circ (\mathrm{id}_{\hat{V}_\epsilon^{\mathrm{fin}}(\lambda)} \otimes \pi_{\hat{V}_\epsilon^{\mathrm{fin}}(\mu)})$$

$$= (\rho_{\hat{V}_\epsilon^{\mathrm{fin}}(\lambda)} \otimes \mathrm{qtr}\rho_{\hat{V}_\epsilon^{\mathrm{fin}}(\mu)})(\mathcal{R}_{21}\mathcal{R}_{12})$$

(the notation is that used in Subsection 5.3A). We can compute this scalar endomorphism by evaluating it on any non-zero vector. If we choose the highest weight vector v_λ of $\hat{V}_\epsilon^{\mathrm{fin}}(\lambda)$, then only the $n = 0$ terms in $\mathcal{R}_{21}$ do not annihilate v_λ, and then only the $n = 0$ terms in $\mathcal{R}_{12}$ need be considered, otherwise the quantum trace will vanish. Thus, we get

$$\tilde{\mathcal{F}}\left(\quad\overset{\lambda}{\underset{\mu}{\circlearrowleft}}\quad\right) = \frac{1}{16\ell^2} \sum_{a,b,a',b'=0}^{4\ell-1} \epsilon^{-ab-a'b'+\lambda(a'+b)} \mathrm{qtr}_{\hat{V}_\epsilon^{\mathrm{fin}}(\mu)}(L^{a+b'}).$$

Now,

$$\frac{1}{16\ell^2} \sum_{a',b=0}^{4\ell-1} \epsilon^{-ab-a'b'+\lambda(a'+b)} = \frac{1}{16\ell^2}\left(\sum_{a'=0}^{4\ell-1} \epsilon^{a'(\lambda-b')}\right)\left(\sum_{b=0}^{4\ell-1} \epsilon^{b(\lambda-a)}\right) = \delta_{a,\lambda}\delta_{b',\lambda}.$$

Thus, the desired scalar is equal to

$$\mathrm{qtr}_{\hat{V}_\epsilon^{\mathrm{fin}}(\mu)}(L^{2\lambda}) = \mathrm{tr}_{\hat{V}_\epsilon^{\mathrm{fin}}(\mu)}(L^{2\lambda+2}) = \frac{[(2\lambda+2)(\mu+1)]_\epsilon}{[2\lambda+2]_\epsilon} = \frac{[(\lambda+1)(\mu+1)]_{\epsilon^2}}{[\lambda+1]_{\epsilon^2}}.$$

Comparing with (12), we have finally

$$s_{\lambda\mu} = [(\lambda+1)(\mu+1)]_{\epsilon^2}.$$

The proof of condition (iv) is now straightforward. For the easily proved identity

$$\sum_{\nu=0}^{\ell-2} [(\lambda+1)(\nu+1)]_{\epsilon^2}[(\nu+1)(\mu+1)]_{\epsilon^2} = \frac{\ell}{2}\delta_{\lambda,\mu}$$

shows that the square of the matrix $(s_{\lambda\mu})$ is equal to $\ell/2$ times the identity matrix. ∎

Remark Turaev & Wenzl (1993) show that one can associate to $U_\epsilon^{\mathrm{fin}}(\mathfrak{g})$, for all $\mathfrak{g}$ of type A, B, C or D, a 'quasi-modular Hopf algebra'. This is a slightly weaker notion than that of a modular Hopf algebra, but it still leads to 3-manifold invariants. However, the methods of Turaev and Wenzl make use of the special properties of certain representations of $U_\epsilon^{\mathrm{fin}}(\mathfrak{g})$ when $\mathfrak{g}$ is of classical type, and it would be desirable to find a proof which works for all $\mathfrak{g}$ simultaneously. △

B The construction of 3-manifold invariants Let M be a compact, connected, oriented 3-manifold without boundary. Choose a framed link L in S^3, with components $K_1, \ldots, K_m$, such that the manifold S_L obtained by performing surgery along L is orientation-preserving homeomorphic to M; recall from Subsection 15.1B that S_L is the boundary of a compact 4-manifold D_L. Fix an orientation ω of L. As we explained in Subsection 15.1B, we may view L as a directed ribbon tangle (with annuli but no bands); by an isotopy of S^3, we can arrange that $L \subset \mathbb{R}^2 \times (0, 1) \subset \mathbb{R}^3 \subset S^3$. Fix a modular Hopf algebra as in 15.3.1, and let $\boldsymbol{\lambda} = (\lambda_1, \ldots, \lambda_m) \in \mathcal{C}^m$ be a colouring of L with the elements of $\mathcal{C}$. Theorem 5.3.7 associates to these data a homomorphism of A-modules $k \to k$, i.e. a scalar $\tilde{\mathcal{F}}(L, \boldsymbol{\lambda}) \in k$.

We need one further piece of topological data before we can state the main result of this section. Consider the intersection pairing on the homology group $H_2(D_L; \mathbb{R})$. This is a symmetric bilinear form; let $\sigma_{\leq 0}(L)$ be the number of non-positive eigenvalues in a diagonal representation.

THEOREM 15.3.3 *Fix a modular Hopf algebra A as in 15.3.1. Let S_L be the compact, connected, oriented 3-manifold without boundary which is obtained from S^3 by surgery along an oriented, framed link L with m components coloured with representations $V(\lambda_1), \ldots, V(\lambda_m)$ of A, where $\lambda_1, \ldots, \lambda_m \in \mathcal{C}$. Let*

$$\tau(S_L, \omega, L) = c^{-\sigma_{\leq 0}(L)} \sum_{\boldsymbol{\lambda}} \prod_{i=1}^{m} d_{\lambda_i} \tilde{\mathcal{F}}(L, \boldsymbol{\lambda}) \in k,$$

where $\sigma_{\leq 0}(L)$ and $\tilde{\mathcal{F}}(L, \boldsymbol{\lambda})$ are as defined above and the scalar c is defined in (11). *Then, $\tau(S_L, \omega, L)$ is a topological invariant of S_L.*

The last sentence means that, if L' is another framed link in S^3 with orientation ω', and if $S_{L'}$ is orientation-preserving homeomorphic to S_L, then $\tau(S_{L'}, \omega', L') = \tau(S_L, \omega, L)$.

OUTLINE OF PROOF By Subsection 15.1B, to show that $\tau(M, \omega, L)$ depends only on S_L, and not on L, it is enough to show that it is unchanged by special Kirby -1-moves and by Kirby $+1$-moves (we shall not discuss the proof that $\tau(M, \omega, L) = \tau(M, L)$ is independent of the orientation ω of L).

In the former case, we start with a ribbon tangle which is the union of a ribbon tangle (= framed link) L and an unknotted annulus K, not linked to L, with one full right-hand twist:

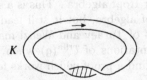 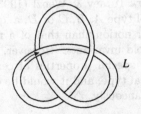

The special Kirby -1-move consists simply in deleting (or inserting) K. Let

λ_L be any colouring of L, λ_K any colouring of K, and denote the corresponding colouring of $L \cup K$ by (λ_L, λ_K). Since the functor $\widetilde{\mathcal{F}}$ constructed in 5.3.7 is a monoidal functor,

$$\widetilde{\mathcal{F}}(L \cup K, (\lambda_L, \lambda_K)) = \widetilde{\mathcal{F}}(L, \lambda_L)\widetilde{\mathcal{F}}(K, \lambda_K)$$

(we have omitted the tensor product since all three endomorphisms are scalars). Since $\sigma_{\leq 0}(L \cup K) = \sigma_{\leq 0}(L) + 1$, the definition of τ gives

$$\tau(S_{L \cup K}, L \cup K) = \tau(S_L, L)c^{-1} \sum_{\lambda_K \in \mathcal{C}} d_{\lambda_K}\widetilde{\mathcal{F}}(K, \lambda_K).$$

But, by 5.3.5,

$$\widetilde{\mathcal{F}}(K, \lambda_K) = v_{\lambda_K}^{-1}\mathrm{qdim}(V(\lambda_K)),$$

so $\widetilde{\mathcal{F}}(K, \lambda_K) = 1$ by the definition of c in (11).

In the case of a Kirby $+1$-move, we have the situation in Fig. 25, where K is now an unknotted annulus with one full left-hand twist, and the r (say) vertical strands represent the components of the complement L of K which are linked with K. We can assume that L is the closure of a tangle $T' \circ T$, where T is the tangle shown in the left-hand diagram in Fig. 25, and T' is some tangle with r free ends at the top and bottom. The effect of the Kirby $+1$-move is to delete K and replace T with a tangle $\tilde{T}$ obtained by performing a single left-hand twist of the r strands (cf. Fig. 25):

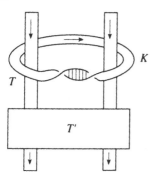
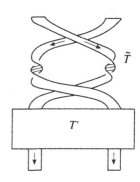

Let $\tilde{L}$ be the closure of $T' \circ \tilde{T}$. Denoting colourings of $L \cup K$ as before, we have

$$\tau(S_{L \cup K}, L \cup K) = \sum_{\lambda_K \in \mathcal{C}} d_{\lambda_K} \sum_{\lambda_L \in \mathcal{C}^r} \prod_{i=1}^{r} d_{\lambda_i}\mathrm{qtr}\big(\widetilde{\mathcal{F}}(T' \circ T, (\lambda_L, \lambda_K))\big),$$

by 5.3.4, and, similarly,

$$\tau(S_{\tilde{L}}, \tilde{L}) = \sum_{\lambda_L \in \mathcal{C}^r} \prod_{i=1}^{r} d_{\lambda_i}\mathrm{qtr}\big(\widetilde{\mathcal{F}}(T' \circ \tilde{T}, \lambda_L)\big).$$

So it is enough to prove that

$$(13) \qquad \sum_{\lambda_K \in \mathcal{C}} d_{\lambda_K} \mathrm{qtr}\big(\widetilde{\mathcal{F}}(T' \circ T, (\boldsymbol{\lambda}_L, \lambda_K))\big) = \mathrm{qtr}\big(\widetilde{\mathcal{F}}(T' \circ \tilde{T}, \boldsymbol{\lambda}_L)\big).$$

Suppose first that $r = 1$. In this case, we have the stronger result

$$(14) \qquad \sum_{\lambda_K \in \mathcal{C}} d_{\lambda_K} \widetilde{\mathcal{F}}(T, (\lambda_L, \lambda_K)) = \widetilde{\mathcal{F}}(\tilde{T}, \lambda_L),$$

from which (13) follows by using the functoriality of $\widetilde{\mathcal{F}}$ and the linearity of the quantum trace. To prove (14), let $\widetilde{\mathcal{F}}(T, (\lambda_L, \lambda_K)) = f_{\lambda_L \lambda_K}$; note that $f_{\lambda_L \lambda_K}$ is an A-module endomorphism of the irreducible module $V(\lambda_L)$, and so is a scalar. To compute it, note that, if $\overline{T}$ denotes the closure of T, then

$$\widetilde{\mathcal{F}}(\overline{T}, (\lambda_L, \lambda_K)) = f_{\lambda_L, \lambda_K} \mathrm{qdim}(V(\lambda_L))$$

by 5.3.4, while a computation similar to that in 5.3.6 shows, on the other hand, that

$$\widetilde{\mathcal{F}}(\overline{T}, (\lambda_L, \lambda_K)) = v_{\lambda_K} s_{\lambda_L \lambda_K}$$

in the notation of 15.3.1. Thus,

$$\widetilde{\mathcal{F}}(T, (\lambda_L, \lambda_K)) = v_{\lambda_K} s_{\lambda_L \lambda_K} (\mathrm{qdim}(V(\lambda_L)))^{-1}.$$

On the other hand, we found in 5.3.5 that $\widetilde{\mathcal{F}}(\tilde{T}, \lambda_L) = v_{\lambda_L}^{-1}$. Thus, (14) is a consequence of (10).

To prove (13) for arbitrary r, let $\lambda_1, \ldots, \lambda_r$ be the colours of the strands incident with the top of T', and $\epsilon_1, \ldots, \epsilon_r$ the signs indicating their directions. Note that $\widetilde{\mathcal{F}}(T' \circ T, (\boldsymbol{\lambda}_L, \lambda_K))$ and $\widetilde{\mathcal{F}}(T' \circ \tilde{T}, \boldsymbol{\lambda}_L)$ are A-module endomorphisms of $V_1^{\epsilon_1} \otimes \cdots \otimes V_r^{\epsilon_r}$ (recall that $V^\epsilon = V$ or V^* according as $\epsilon = +1$ or -1; we have written V_i for $V(\lambda_i)$). Write

$$(15) \qquad V_1^{\epsilon_1} \otimes \cdots \otimes V_r^{\epsilon_r} \cong Z \oplus \bigoplus_{\lambda \in \mathcal{C}} V(\lambda)^{\oplus m_\lambda}$$

as in 15.3.1(iii). Writing the endomorphisms in question in block matrix form with respect to the decomposition (15), we must compute the sum of the quantum traces of the diagonal blocks. Those of the Z–Z blocks vanish by 15.3.1(iii), since they are composites

$$Z \xrightarrow{\iota_Z} \otimes V(\lambda_i)^{\epsilon_i} \to \otimes V(\lambda_i)^{\epsilon_i} \xrightarrow{\pi_Z} Z,$$

where the first and last homomorphisms are the obvious inclusions and projections. As for the $V(\lambda)^{\otimes m_\lambda}$–$V(\lambda)^{\otimes m_\lambda}$ blocks, it is easy to reduce to the

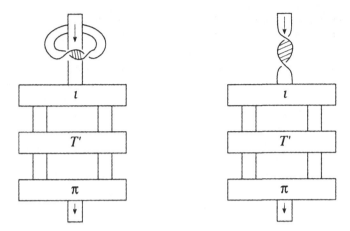

case $m_\lambda = 1$ by writing $V(\lambda)^{\otimes m_\lambda} = V(\lambda) \otimes M(\lambda)$ and choosing a basis of the multiplicity space $M(\lambda)$, and in that case it suffices to prove that

$$(16) \quad \sum_{\lambda_K \in \mathcal{C}} d_{\lambda_K}(\pi_{V(\lambda)} \circ \widetilde{\mathcal{F}}(T' \circ T, (\boldsymbol{\lambda}_L, \boldsymbol{\lambda}_K))) \circ \iota_{V(\lambda)}) = \pi_{V(\lambda)} \circ \widetilde{\mathcal{F}}(T' \circ \widetilde{T}, \boldsymbol{\lambda}_L) \circ \iota_{V(\lambda)},$$

(or in fact, just that both sides have the same quantum trace), where $\iota_{V(\lambda)}$ and $\pi_{V(\lambda)}$ are the obvious inclusions and projections. The pictures necessary to compute the two sides of (16) are shown above, in which the upper and lower boxes are coupons representing $\iota_{V(\lambda)}$ and $\pi_{V(\lambda)}$, respectively (and here we need the generalization 5.3.7 of 5.3.2). Equation (16) follows immediately from these diagrams, using the $r = 1$ case together with the functoriality of $\widetilde{\mathcal{F}}$. ∎

The construction of invariants of 3-manifolds we have outlined can be extended without much difficulty to the case of coloured ribbon tangles in arbitrary compact, oriented 3-manifolds. See Reshetikhin & Turaev (1991).

Bibliographical notes

15.1 This section merely summarizes the terminology and basic results used in the remainder of the chapter. For further information on braids and links, the reader should consult one of the many standard sources, such as Birman (1976) for braids and links, and Rolfsen (1976) for links and 3-manifolds.

15.2 The polynomial link invariant discovered by Jones (1985) arises from representations of the braid groups $\mathcal{B}_m$ which factor through the Hecke algebra $\mathcal{H}_m$ (see 10.2.1), and the Markov trace is provided by a canonical trace on the Hecke algebra itself, discovered by A. Ocneanu. As we saw in Section 10.2, the Birman–Murakami–Wenzl algebra is to Lie algebras of type B, C and D as $\mathcal{H}_m$ is to type A, and in fact Birman & Wenzl (1989) showed how to associate link invariants to representations of the BMW algebra.

Our discussion of the braid-theoretic approach to link invariants is based on Turaev (1988). 'Universal' approaches to this question have been given by Lawrence (1989) and Drinfel'd (1990b). The former depends on constructing an invariant in a certain enlargement of a quantized universal enveloping algebra which gives scalar invariants on applying suitable representations. The latter is based on quasi-Hopf algebras, discussed in the next chapter.

Date, Jimbo, Miki & Miwa (1990, 1991c) gave an analogous construction of link invariants for the non-restricted specialization $U_\epsilon(\mathfrak{g})$ of $U_q(\mathfrak{g})$ at a root of unity ϵ (for some $\mathfrak{g}$ of low rank). As we saw in Section 11.1, even though $U_\epsilon(\mathfrak{g})$ is not quasitriangular, certain families of its representations commute under tensor product, and the intertwiners may still give solutions of the QYBE. If one can construct a Markov trace, one obtains link invariants in the usual way. Rosso (1992c) gives an alternative approach by showing that one can construct a universal R-matrix for the 'part' of $U_\epsilon(\mathfrak{g})$ relevant to the representations one is interested in.

Our discussion of the vertex model approach to link invariants is based on Jones (1989) and Rosso (1988c). For further information, see Kauffman (1991b).

15.3 The work of Reshetikhin & Turaev (1991) summarized here was motivated, as we said in the introduction, by Witten (1989), although only later were the two constructions proved to be equivalent, by Walker (1991). The invariants arising from quantum sl_2 are studied in more detail by Kirby & Melvin (1991) and Kirby, Melvin & Zhang (1993). The properties of the invariants associated to Lie algebras of classical type by Turaev & Wenzl (1993) do not seem to have been investigated in detail yet.

As these words are being written, an understanding is beginning to emerge of the meaning of the invariants of Witten, and Reshetikhin and Turaev, in terms of 'classical' algebraic topology. This is based on the construction by Vassiliev (1990, 1993) of knot invariants by analysing the singularities of the 'space of all knots'. It appears that, if one replaces q (or ϵ) by e^h in a quantum group invariant, the coefficients of powers of h in its power series expansion are Vassiliev invariants. See Birman (1993) and Birman & Lin (1993).

16

Quasi-Hopf algebras
and the
Knizhnik–Zamolodchikov equation

We saw in Section 5.1 that the category of representations of a Hopf algebra is monoidal, which means essentially that it is equipped with a tensor product which is associative up to isomorphism. In fact, the coassociativity property of a Hopf algebra means that the associativity isomorphisms $U \otimes (V \otimes W) \to (U \otimes V) \otimes W$ can be taken to be the natural isomorphisms of vector spaces. This suggests that one might be able to relax the coassociativity condition and still obtain a monoidal category, perhaps with non-trivial associativity isomorphisms. The rigidity of the category of representations of a Hopf algebra, which depends on the properties of its antipode, might also be retained by replacing these properties by some suitable analogue. These suggestions are realized in the notion of a *quasi-Hopf algebra*, which we define in Section 16.1. Although the category of quasi-Hopf algebras is larger than the category of Hopf algebras, it turns out to be more manageable because it admits a kind of 'gauge group', which sometimes allows one to 'twist' a given quasi-Hopf algebra into an apparently simpler one, but one which has the 'same' representation theory as the original quasi-Hopf algebra.

Section 16.2 describes the relation, due to V. G. Drinfel'd but implicit in earlier work of T. Kohno, between quasi-Hopf algebras and a certain system of first order partial differential equations for a function of m (say) complex variables, called the Knizhnik–Zamolodchikov (KZ) equation. The KZ equation first appeared in conformal field theory, as the equation satisfied by the 'm-point functions' of the theory, but it has since proved to play an important role in several areas of mathematics. Of most importance to us is that, by considering the monodromy of its solutions along closed paths, it leads to representations of braid groups. There is actually one KZ equation for each solution of the classical Yang–Baxter equation with spectral parameters, but we shall need only the one corresponding to the r-matrix underlying the Yangians $Y(\mathfrak{g})$, where $\mathfrak{g}$ is a finite-dimensional complex simple Lie algebra (see Chapter 12).

The main result of Section 16.2 is that, if $U_h(\mathfrak{g})$ is the standard QUE algebra associated to $\mathfrak{g}$, and $\mathcal{R}_h$ is its universal R-matrix, the braid group representation defined by $\mathcal{R}_h$ (see Subsection 15.2A) is equivalent to that defined by the KZ equation. This can be viewed as a 'quantization' theorem, since the two braid group representations arise essentially from a solution of the QYBE and a solution of the CYBE, respectively, and indeed the proof proceeds by using the KZ equation to construct a quantization of a certain 'quasi-Lie bialgebra'.

The Kohno–Drinfel'd theorem shows that the category of finite-dimensional representations of $U_h(\mathfrak{g})$ is equivalent, as a quasitensor category, to the category of finite-dimensional representations of $\mathfrak{g}$ itself, equipped with a quasitensor structure defined using the KZ equation. D. Kazhdan and G. Lusztig proved an extension of this result, relating the category of finite-dimensional representations of $U_\epsilon^{\mathrm{res}}(\mathfrak{g})$, where $\epsilon \in \mathbb{C}^\times$, to a certain category of representations of level κ of the (classical) affine Lie algebra $\tilde{\mathfrak{g}}$, where $\epsilon = e^{-\sqrt{-1}\pi/\kappa}$; the Kohno–Drinfel'd theorem can be viewed essentially as the $\epsilon \to \infty$ limit of Kazhdan & Lusztig's result. It goes without saying that the quasitensor structure on the latter category is not given by the usual tensor product of representations of $\hat{\mathfrak{g}}$, since under that tensor product the levels add. The definition of the required tensor product, and the proof that it gives rise to a quasitensor category, is rather difficult, and again makes use of (a generalization of) the KZ equation. We outline this work of Kazhdan and Lusztig in Section 16.3.

It is a consequence of Kazhdan and Lusztig's theorem that the conjecture of Lusztig stated in Subsection 11.2C, which determines the characters of the finite-dimensional irreducible representations of the quantum groups $U_\epsilon^{\mathrm{res}}(\mathfrak{g})$ when ϵ is a root of unity, is true.

A connection between quantum groups and classical affine Lie algebras was predicted in the literature on conformal field theory, although physicists are mainly interested in unitary representations of $\tilde{\mathfrak{g}}$, which are excluded from the category considered by Kazhdan and Lusztig. However, there is a parallel result, very recently proved by M. Finkelberg, which establishes an equivalence between a category of representations of $\tilde{\mathfrak{g}}$ which includes these 'physical' representations (but not those considered by Kazhdan and Lusztig!), and the 'truncated' category of tilting modules $\overline{\mathbf{tilt}}_\epsilon$ discussed in Section 11.3.

Section 16.4 sketches a remarkable relation between quasi-Hopf algebras and algebraic number theory, specifically the structure of the 'absolute Galois group' $\mathrm{Gal}(\overline{\mathbb{Q}}/\mathbb{Q})$. This relation was discovered by Drinfel'd, but some aspects of it were certainly foreseen by A. Grothendieck in 1984, in his 'Esquisse d'un programme'.

16.1 Quasi-Hopf algebras

In this section, we define a generalization of the notion of a Hopf algebra, called a quasi-Hopf algebra, and study some of their basic properties.

A Definitions The relation between quasi-Hopf algebras and Hopf algebras is similar to that between almost cocommutative Hopf algebras and cocommutative Hopf algebras. In the latter case, one weakens cocommutativity to 'cocommutativity up to conjugacy', while in the former case one does something similar with associativity.

DEFINITION 16.1.1 *A quasi-bialgebra over a commutative ring k is an associative algebra A over k together with homomorphisms $\epsilon_A : A \to k$ (the counit) and $\Delta_A : A \to A \otimes A$ (the comultiplication), and an invertible element $\Phi_A \in A \otimes A \otimes A$ (the coassociator), satisfying the following conditions (we omit the subscripts):*

(1)
$$(\mathrm{id} \otimes \Delta)(\Delta(a)) = \Phi^{-1}.(\Delta \otimes \mathrm{id})(\Delta(a)).\Phi,$$

for all $a \in A$;

(2)
$$(\Delta \otimes \mathrm{id} \otimes \mathrm{id})(\Phi).(\mathrm{id} \otimes \mathrm{id} \otimes \Delta)(\Phi) = (\Phi \otimes 1).(\mathrm{id} \otimes \Delta \otimes \mathrm{id})(\Phi).(1 \otimes \Phi);$$

(3)
$$(\epsilon \otimes \mathrm{id}) \circ \Delta = \mathrm{id} = (\mathrm{id} \otimes \epsilon) \circ \Delta;$$

(4)
$$(\mathrm{id} \otimes \epsilon \otimes \mathrm{id})(\Phi) = 1.$$

A quasi-Hopf algebra is a quasi-bialgebra A equipped with an anti-homomorphism $S_A : A \to A$ (the antipode), and elements $\alpha_A, \beta_A \in A$, such that

(5)
$$\sum_r S(a_r)\alpha a'_r = \epsilon(a)\alpha, \quad \sum_r a_r \beta S(a'_r) = \epsilon(a)\beta,$$

for all $a \in A$,

(6)
$$\sum_s S(\varphi_s)\alpha\varphi'_s \beta S(\varphi''_s) = 1, \quad \sum_t \bar{\varphi}_t \beta S(\bar{\varphi}'_t)\alpha\bar{\varphi}''_t = 1,$$

where

$$\Delta(a) = \sum_r a_r \otimes a'_r, \quad \Phi = \sum_s \varphi_s \otimes \varphi'_s \otimes \varphi''_s, \quad \Phi^{-1} = \sum_t \bar{\varphi}_t \otimes \bar{\varphi}'_t \otimes \bar{\varphi}''_t.$$

If B is another quasi-bialgebra (resp. quasi-Hopf algebra), an algebra homomorphism $f : A \to B$ is a homomorphism of quasi-bialgebras (resp. of quasi-Hopf algebras) if

$$\Delta_B \circ f = (f \otimes f) \circ \Delta_A, \quad \epsilon_B \circ f = \epsilon_A, \quad \Phi_B = (f \otimes f \otimes f)(\Phi_A),$$

(resp. and $S_B \circ f = f \circ S_A$).

Remarks [1] A Hopf algebra is the same thing as a quasi-Hopf algebra for which $\alpha = \beta = 1$ and $\Phi = 1 \otimes 1 \otimes 1$.

[2] If u is an invertible element of A, and if S, α and β satisfy (5) and (6), then so do $\tilde{S}$, $\tilde{\alpha}$ and $\tilde{\beta}$, where

$$(7) \qquad \tilde{S}(a) = uS(a)u^{-1}, \quad \tilde{\alpha} = u\alpha, \quad \tilde{\beta} = \beta u^{-1}.$$

Conversely, one can show that if S, α, β and $\tilde{S}$, $\tilde{\alpha}$, $\tilde{\beta}$ both satisfy (5) and (6), there is a unique $u \in A$ for which (7) holds. Thus, S, α and β are essentially uniquely determined, if they exist. It can be shown further that the two equations in (6) are equivalent, in the presence of conditions (1)–(5). $\triangle$

PROPOSITION 16.1.2 *If A is a quasi-Hopf algebra over a commutative ring k, then the category* $\mathbf{rep}_A$ *of A-modules, which are finitely generated and projective as k-modules, is a rigid, k-linear, monoidal category.*

OUTLINE OF PROOF The associativity isomorphisms

$$\alpha_{U,V,W} : U \otimes (V \otimes W) \to (U \otimes V) \otimes W$$

are given by the action of Φ. That the $\alpha_{U,V,W}$ are A-module homomorphisms follows from (1), and the commutativity of the pentagon diagram in 5.1.1(i) follows from (2).

The trivial module k and the tensor product of A-modules are defined using the counit and comultiplication of A, as in the case of Hopf algebras. The fact that, for any A-module U, the natural maps $\rho_U : U \otimes k \to U$ and $\lambda_U : k \otimes U \to U$ are A-module isomorphisms follows from (3), and the commutativity of the diagram in 5.1.1(ii) follows from (4). The (left) dual U^* of an A-module U is also defined as in the case of Hopf algebras, by using the antipode. But the isomorphisms $\mathrm{ev}_U : U^* \otimes U \to k$ and $\pi_U : k \to U \otimes U^*$ are now given by

$$\mathrm{ev}_U(\xi \otimes u) = \langle \xi, \alpha.u \rangle, \quad \pi_U(1) = \sum_i \beta.u_i \otimes \xi^i,$$

if $\langle \, , \, \rangle$ is the natural pairing and $\{u_i\}$ and $\{\xi^i\}$ are dual bases of U and U^*. That these maps are A-module homomorphisms follows from (5), and the commutativity of the diagrams in Subsection 5.1C follows from (6). ∎

In view of this result, and of those in Section 5.1, it is natural to ask what further conditions on a quasi-Hopf algebra A will ensure that $\mathbf{rep}_A$ is a tensor or quasitensor category.

DEFINITION 16.1.3 *A quasitriangular quasi-Hopf algebra is a quasi-Hopf algebra A equipped with an invertible element $\mathcal{R} \in A \otimes A$, called its universal R-matrix, such that*

$$(8) \qquad \Delta^{\mathrm{op}}(a) = \mathcal{R}.\Delta(a).\mathcal{R}^{-1},$$

$$(9) \qquad (\Delta \otimes \mathrm{id})(\mathcal{R}) = \Phi_{231}^{-1} \mathcal{R}_{13} \Phi_{132} \mathcal{R}_{23} \Phi_{123}^{-1},$$

$$(10) \qquad (\mathrm{id} \otimes \Delta)(\mathcal{R}) = \Phi_{312} \mathcal{R}_{13} \Phi_{213}^{-1} \mathcal{R}_{12} \Phi_{123},$$

for all $a \in A$. Here, if $\Phi = \sum_r \varphi_r \otimes \varphi_r' \otimes \varphi_r''$, we define $\Phi_{231} = \sum_r \varphi_r' \otimes \varphi_r'' \otimes \varphi_r$, etc. If, in addition, $\mathcal{R}$ satisfies the relations

$$(11) \qquad \mathcal{R}_{21} = \mathcal{R}_{12}^{-1}, \quad (\epsilon \otimes \epsilon)(\mathcal{R}) = 1,$$

then A is said to be triangular.

Quasitriangular and triangular quasi-bialgebras are defined in the same way, dropping the requirement of the existence of an antipode.

Remark The universal R-matrix of a quasitriangular quasi-Hopf algebra is a solution of the 'quasi-quantum Yang–Baxter equation':

$$(12) \qquad \mathcal{R}_{12} \Phi_{231}^{-1} \mathcal{R}_{13} \Phi_{132} \mathcal{R}_{23} \Phi_{123}^{-1} = \Phi_{132}^{-1} \mathcal{R}_{23} \Phi_{312} \mathcal{R}_{13} \Phi_{213}^{-1} \mathcal{R}_{12}.$$

In fact, using (9) and (10), the left-hand side is equal to $\mathcal{R}_{12}(\Delta \otimes \mathrm{id})(\mathcal{R})$, and the right-hand is equal to $(\Delta^{\mathrm{op}} \otimes \mathrm{id})(\mathcal{R})\mathcal{R}_{12}$, so the result follows from (8). $\triangle$

From 16.1.2, we obtain

COROLLARY 16.1.4 *If A is a quasitriangular (resp. triangular) quasi-bialgebra, then rep_A is a quasitensor (resp. tensor) category.*

PROOF The commutativity maps $\sigma_{U,V} : U \otimes V \to V \otimes U$, for any objects U and V in rep_A, are defined exactly as in the case of Hopf algebras (cf. 5.2.2). The axioms in 5.2.1 follow immediately from (8)–(11). ∎

At first sight, the category of quasi-Hopf algebras appears to be more difficult to work with than the category of Hopf algebras, but in fact the opposite is true, because the former category admits a kind of 'gauge symmetry', which we have already encountered in a restricted form in Subsection 4.2E. Namely, let A be any quasi-bialgebra and let $\mathcal{F}$ be an invertible element of $A \otimes A$ such that

$$(13) \qquad (\epsilon \otimes \mathrm{id})(\mathcal{F}) = 1 = (\mathrm{id} \otimes \epsilon)(\mathcal{F}).$$

Set

$$(14) \qquad \Delta^{\mathcal{F}}(a) = \mathcal{F}.\Delta(a).\mathcal{F}^{-1},$$

$$(15) \qquad \Phi^{\mathcal{F}} = \mathcal{F}_{12}.(\Delta \otimes \mathrm{id})(\mathcal{F}).\Phi.(\mathrm{id} \otimes \Delta)(\mathcal{F})^{-1}.\mathcal{F}_{23}^{-1}.$$

PROPOSITION 16.1.5 *Let A be a quasi-bialgebra (resp. quasi-Hopf algebra) and $\mathcal{F} \in A \otimes A$ an invertible element satisfying (13), and let $A^{\mathcal{F}}$ be the algebra A, equipped with the same counit (resp. and antipode) as A, but with the comultiplication $\Delta^{\mathcal{F}}$ and coassociator $\Phi^{\mathcal{F}}$ defined in (14) and (15). Then, $A^{\mathcal{F}}$ is a quasi-bialgebra (resp. a quasi-Hopf algebra).*

If A is quasitriangular with universal R-matrix $\mathcal{R}$, then $A^{\mathcal{F}}$ is quasitriangular with universal R-matrix $\mathcal{R}^{\mathcal{F}} = \mathcal{F}_{21} \mathcal{R} \mathcal{F}_{12}^{-1}$.

PROOF A straightforward verification. The elements α and β in (5) and (6) appropriate to $A^{\mathcal{F}}$ are given by

$$\alpha^{\mathcal{F}} = \sum_s S(\bar{f}_s) \alpha \bar{f}'_s, \qquad \beta^{\mathcal{F}} = \sum_r f_r \beta S(f'_r),$$

where $\mathcal{F} = \sum_r f_r \otimes f'_r$, $\mathcal{F}^{-1} = \sum_s \bar{f}_s \otimes \bar{f}'_s$. ∎

We shall say that two quasi-bialgebras (or quasi-Hopf algebras) A and B are *twist equivalent* if $B = A^{\mathcal{F}}$ for some $\mathcal{F} \in A \otimes A$ as above. This is an equivalence relation, for it is easy to see that twisting with $\mathcal{F}_1$ and then with $\mathcal{F}_2$ is the same as twisting with $\mathcal{F}_2 \mathcal{F}_1$.

Twisting has a simple category-theoretic interpretation:

PROPOSITION 16.1.6 *Suppose that two quasi-bialgebras A and B are twist equivalent. Then, rep_A and rep_B are equivalent as monoidal categories. If A and B are quasitriangular, then rep_A and rep_B are equivalent as quasitensor categories.*

PROOF We have to construct an equivalence of categories $f : \mathrm{rep}_A \to \mathrm{rep}_B$, together with a natural isomorphism $\mathcal{I}$ between the bifunctors

$$(U, V) \mapsto f(U \otimes V) \quad \text{and} \quad (U, V) \mapsto f(U) \otimes f(V),$$

and an isomorphism $\iota : f(k) \to k$, such that the following diagrams commute:

$$
\begin{array}{ccc}
f((U \otimes V) \otimes W) & \xrightarrow{(\mathcal{I}_{U,V} \otimes \mathrm{id}_{f(W)}) \circ \mathcal{I}_{U \otimes V, W}} & (f(U) \otimes f(V)) \otimes f(W) \\
\Big\uparrow{\scriptstyle f(\alpha_{U,V,W})} & & \Big\uparrow{\scriptstyle \alpha_{f(U),f(V),f(W)}} \\
f(U \otimes (V \otimes W)) & \xrightarrow[(\mathrm{id}_{f(U)} \otimes \mathcal{I}_{V,W}) \circ \mathcal{I}_{U,V \otimes W}]{} & f(U) \otimes (f(V) \otimes f(W))
\end{array}
$$

$$
\begin{array}{ccccc}
f(k \otimes U) & \xrightarrow{\mathcal{I}_{k,U}} & f(k) \otimes f(U) & \quad\quad f(U \otimes k) & \xrightarrow{\mathcal{I}_{U,k}} & f(U) \otimes f(k) \\
\Big\downarrow{\scriptstyle \cong} & & \Big\downarrow{\scriptstyle \iota \otimes \mathrm{id}_{f(U)}} & \quad\quad \cong \Big\downarrow & & \Big\downarrow{\scriptstyle \mathrm{id}_{f(U)} \otimes \iota} \\
f(U) & \xrightarrow[\cong]{} & k \otimes f(U) & \quad\quad f(U) & \xrightarrow[\cong]{} & f(U) \otimes k
\end{array}
$$

If $B = A^{\mathcal{F}}$, we can take f and ι to be the identity, and $\mathcal{I}_{U,V}$ to be the image of $\mathcal{F}$ in $\mathrm{End}_k(U \otimes V)$.

We leave it to the reader to write down, and prove commutative, the diagrams necessary to establish the second part of the proposition. ∎

B An example from conformal field theory We illustrate the preceding definitions and results with a very interesting example due to Dijkgraaf, Pasquier & Roche (1992).

Example 16.1.7 Let G be a finite group, $\mathbb{C}[G]$ its group algebra over the complex numbers, and $\mathcal{F}(G)$ the algebra of complex-valued functions on G. Let $c : G \times G \times G \to \mathbb{T}$ be a 3-cocycle on G with values in the circle group $\mathbb{T}$. This means that

$$c(y, z, w)c(xy, z, w)^{-1}c(x, yz, w)c(x, y, zw)^{-1}c(x, y, z) = 1$$

for all x, y, z, $w \in G$. We assume that c is *normalized*, i.e. that $c(x, y, z) = 1$ if x, y or z is equal to the identity element $e \in G$. We define a quasi-Hopf algebra $\mathcal{D}^c(G)$ as follows.

As a vector space, $\mathcal{D}^c(G) = \mathcal{F}(G) \otimes \mathbb{C}[G]$. It has a basis consisting of the elements $\boldsymbol{\delta}_g \otimes x$, where g, $x \in G$, and $\boldsymbol{\delta}_g$ denotes the 'Dirac delta function,' i.e. $\boldsymbol{\delta}_g(h) = 1$ if $h = g$ and $\boldsymbol{\delta}_g(h) = 0$ otherwise. Let

$$\theta_g(x, y) = \frac{c(g, x, y)c(x, y, (xy)^{-1}gxy)}{c(x, x^{-1}gx, y)},$$

$$\gamma_x(g, h) = \frac{c(g, h, x)c(x, x^{-1}gx, x^{-1}hx)}{c(g, x, x^{-1}hx)}.$$

The multiplication of $\mathcal{D}^c(G)$ is defined by

$$(\boldsymbol{\delta}_g \otimes x).(\boldsymbol{\delta}_h \otimes y) = \begin{cases} \theta_g(x, y)(\boldsymbol{\delta}_g \otimes xy) & \text{if } gx = xh, \\ 0 & \text{otherwise,} \end{cases}$$

the unit element being $1 = \sum_g (\boldsymbol{\delta}_g \otimes e)$. The comultiplication, counit and antipode are defined by

$$\Delta(\boldsymbol{\delta}_g \otimes x) = \sum_{\{h,k \in G | hk=g\}} \gamma_x(h, k)(\boldsymbol{\delta}_h \otimes x) \otimes (\boldsymbol{\delta}_k \otimes x), \quad \epsilon(\boldsymbol{\delta}_g \otimes x) = \delta_{g,e} 1,$$

$$S(\boldsymbol{\delta}_g \otimes x) = \theta_{g^{-1}}(x, x^{-1})^{-1} \gamma_x(g, g^{-1})^{-1}(\boldsymbol{\delta}_{x^{-1}gx} \otimes x^{-1}).$$

Finally,

$$\Phi = \sum_{g,h,k \in G} c(g, h, k)(\boldsymbol{\delta}_g \otimes e) \otimes (\boldsymbol{\delta}_h \otimes e) \otimes (\boldsymbol{\delta}_k \otimes e).$$

One checks that condition (2) in 16.1.1 is equivalent to the cocyle property of c. Conditions (3) and (4) follow from the normalization of c, while the associativity of multiplication, the fact that Δ is an algebra homomorphism, and condition (1), respectively, are consequences of the following identities:

$$\theta_g(x,y)\theta_g(xy,z) = \theta_g(x,yz)\theta_{x^{-1}gx}(y,z),$$

$$\theta_g(x,y)\theta_h(x,y)\gamma_x(g,h)\gamma_y(x^{-1}gx, x^{-1}hx) = \theta_{gh}(x,y)\gamma_{xy}(g,h),$$

$$\gamma_x(g,h)\gamma_x(gh,k)c(x^{-1}gx, x^{-1}hx, x^{-1}kx) = \gamma_x(h,k)\gamma_x(g,hk)c(g,h,k).$$

These identities in turn follow by repeated use of the cocycle property of c.

It is easy to see that, if c is the trivial cocycle $c(g,h,k) \equiv 1$, then $\mathcal{D}^c(G)$ is exactly the quantum double of $\mathcal{F}(G)$ (see Subsection 4.2D). Since quantum doubles form the most important examples of quasitriangular Hopf algebras, it is natural to expect that, for any 3-cocycle c, $\mathcal{D}^c(G)$ is a quasitriangular quasi-Hopf algebra. This is indeed the case, the universal R-matrix being given by the same formula as for the quantum double:

$$\mathcal{R} = \sum_{g\in G} (\boldsymbol{\delta}_g \otimes e) \otimes (1 \otimes g).$$

It is also natural to ask when two of the quasi-Hopf algebras $\mathcal{D}^c(G)$ and $\mathcal{D}^{c'}(G)$ are twist equivalent. It is not hard to see that this happens precisely when the 3-cocycles c and c' are cohomologous. In fact, if $c' - c = db$ for some map $b : G \times G \to \mathbb{T}$, then $\mathcal{D}^{c'}(G) = \mathcal{D}^c(G)^{\mathcal{F}}$, where

$$\mathcal{F} = \sum_{g,h\in G} b(g,h)^{-1}(\boldsymbol{\delta}_g \otimes e) \otimes (\boldsymbol{\delta}_h \otimes e).$$

Thus, the quasi-Hopf algebras $\mathcal{D}^c(G)$ are classified, up to twist equivalence, by the cohomology group $H^3(G, \mathbb{T})$.

Dijkgraaf *et al.* (1992) show that the category of finite-dimensional representations of $\mathcal{D}^c(G)$ is semisimple, and that its fusion ring is isomorphic to that of an 'orbifold' conformal field theory with symmetry group G (see Subsection 5.2E). In view of the result of the preceding paragraph and 16.1.6, this suggests that such conformal field theories might themselves be classified by $H^3(G, \mathbb{T})$. Dijkgraaf & Witten (1990) proved that this is indeed the case. (We have actually reversed the historical order of the development of these ideas.) ◇

C Quasi-Hopf QUE algebras The quasi-Hopf algebras of most interest to us are deformations of universal enveloping algebras. By a *deformation* of a quasi-bialgebra A over a field k, we mean a topological quasi-bialgebra A_h over $k[[h]]$ such that $A_h/hA_h \cong A$ as quasi-Hopf algebras over k, and $A_h \cong A[[h]]$ as $k[[h]]$-modules. We assume also that the coassociator Φ_h of A_h is $\equiv 1 \otimes 1 \otimes 1 \pmod{h}$. As in the case of Hopf algebras, the adjective 'topological' means that all tensor products appearing in 16.1.1 are to be understood in the h-adically completed sense. Furthermore, again as in the Hopf case (see Subsection 6.1A), one can show that any deformation of a quasi-Hopf algebra as a quasi-bialgebra is, in fact, a quasi-Hopf algebra.

DEFINITION 16.1.8 *A quasi-Hopf QUE algebra is a quasi-Hopf deformation A_h of the universal enveloping algebra $U(\mathfrak{g})$ of a Lie algebra $\mathfrak{g}$ over a field k, such that the coassociator Φ_h of A_h satisfies the following condition:*

$$(16) \qquad \sum_{\pi \in \Sigma_3} (\Phi_h)_{\pi(1)\pi(2)\pi(3)} \equiv 1 \otimes 1 \otimes 1 \quad (mod\ h^2).$$

A quasi-Hopf QUE algebra A_h is topologically quasitriangular *if it is quasitriangular as a quasi-Hopf algebra and its universal R-matrix $\mathcal{R}_h \in A_h \otimes A_h$ (completed tensor product) satisfies $\mathcal{R}_h \equiv 1 \otimes 1$ (mod h).*

Remarks [1] If A_h is a quasi-Hopf deformation of $U(\mathfrak{g})$ which has a universal R-matrix $\mathcal{R}_h \equiv 1 \otimes 1$ (mod h), then condition (16) follows automatically from (9) and (10).

[2] When we consider twists of quasi-Hopf QUE algebras A_h by elements $\mathcal{F}_h \in A_h \otimes A_h$ (completed tensor product), we shall always assume that $\mathcal{F}_h \equiv 1 \otimes 1$ (mod h). Twisting preserves condition (16).

[3] By twisting, one can always ensure that $\Phi_h \equiv 1 \otimes 1 \otimes 1$ (mod h^2). This result is proved by cohomological methods – see Drinfel'd (1990b).

[4] Twisting also allows one to assume that $\mathcal{R}_h^{21} = \mathcal{R}_h$ (we write $\mathcal{R}_h^{21}$ instead of $(\mathcal{R}_h)_{21}$ here and below for typographical reasons). In fact, $\tilde{\mathcal{R}}_h = \mathcal{F}_h^{21} \mathcal{R}_h \mathcal{F}_h^{-1}$ is symmetric if and only if $\mathcal{F}_h = \mathcal{F}'_h.(\mathcal{R}_h(\mathcal{R}_h^{21}\mathcal{R}_h)^{-1/2})^{1/2}$, where $\mathcal{F}'_h$ is symmetric (for, the equation $(\mathcal{R}_h^{\mathcal{F}_h})^{21} = \mathcal{R}_h^{\mathcal{F}_h}$ is equivalent to $\mathcal{R}_h^{21}\mathcal{R}_h = (\mathcal{F}_h^{-1}\mathcal{F}_h^{21}\mathcal{R}_h)^2)$. Note that the square roots make sense because $\mathcal{R}_h \equiv 1 \otimes 1$ (mod h). △

If A_h is a quasi-Hopf QUE algebra, we define a map $\delta : U(\mathfrak{g}) \to U(\mathfrak{g}) \otimes U(\mathfrak{g})$ as in the case of QUE algebras:

$$\delta(x) = \frac{\Delta_h(a) - \Delta_h^{\mathrm{op}}(a)}{h} \quad (mod\ h),$$

where $a \in A_h$ is equal to $x \in U(\mathfrak{g})$ (mod h). Arguing as in the Hopf case (cf. the proof of 6.2.3), one shows that δ is determined by its restriction to $\mathfrak{g}$ by the equation in 6.2.2(iii), and that $\delta : \mathfrak{g} \to \wedge^2(\mathfrak{g})$ is a 1-cocycle. Moreover, if we have arranged that $\Phi_h \equiv 1 \otimes 1 \otimes 1$ (mod h^2), and we set $\varphi = -h^{-2}\mathrm{Alt}(\Phi_h)$ (mod h), then similar arguments show that $\varphi \in \wedge^3(\mathfrak{g})$ and satisfies

$$(17) \qquad \frac{1}{2}\mathrm{Alt}(\delta \otimes \mathrm{id})\delta(x) = [x \otimes 1 \otimes 1 + 1 \otimes x \otimes 1 + 1 \otimes 1 \otimes x\,,\,\varphi],$$

$$(18) \qquad \mathrm{Alt}(\delta \otimes \mathrm{id} \otimes \mathrm{id})(\varphi) = 0.$$

Here,

$$\mathrm{Alt}(x_1 \otimes x_2 \otimes x_3) = \sum_{\pi \in \Sigma_3} \mathrm{sign}(\pi)\, x_{\pi(1)} \otimes x_{\pi(2)} \otimes x_{\pi(3)}.$$

DEFINITION–PROPOSITION 16.1.9 *A quasi-Lie bialgebra is a Lie algebra* $\mathfrak{g}$ *equipped with a 1-cocycle* $\delta : \mathfrak{g} \to \wedge^2(\mathfrak{g})$ *and an element* $\varphi \in \wedge^3(\mathfrak{g})$ *satisfying (17) and (18).*

If $r \in \wedge^2(\mathfrak{g})$ *and we set*

$$\delta^r(x) = \delta(x) + [x \otimes 1 + 1 \otimes x, r],$$

$$\varphi^r = \varphi + \frac{1}{2}\mathrm{Alt}(\delta \otimes \mathrm{id})(r) - [[r, r]],$$

where $[[r, r]] = [r_{12}, r_{13}] + [r_{12}, r_{23}] + [r_{13}, r_{23}]$ *is the left-hand side of the classical Yang–Baxter equation, then* $\mathfrak{g}$ *equipped with* δ^r *and* φ^r *is also a quasi-Lie bialgebra, called the twist of* $\mathfrak{g}$ *by* r, *and denoted by* $\mathfrak{g}^r$.

If A_h *is a quasi-Hopf QUE algebra with* $A_h/hA_h \cong U(\mathfrak{g})$ *and* $\Phi_h \equiv 1 \otimes 1 \otimes 1$ *(mod* h^2), *and, if we define* δ *and* φ *as above, then* $\mathfrak{g}$ *becomes a quasi-Lie bialgebra, called the classical limit of* A_h. *If* $A_h^{\mathcal{F}_h}$ *is the quasi-Hopf QUE algebra obtained by twisting* A_h *by an element* $\mathcal{F}_h$, *and if* $\Phi_h^{\mathcal{F}_h} \equiv 1 \otimes 1 \otimes 1$ *(mod* h^2), *then the classical limit of* $A_h^{\mathcal{F}_h}$ *is* $\mathfrak{g}^r$, *where* $r = -h^{-1}\mathrm{Alt}(\mathcal{F}_h)$ *(mod* h). ∎

The only non-trivial point is to show that the element $r = -h^{-1}\mathrm{Alt}(\mathcal{F}_h)$ (mod h) in the last part, which *a priori* lies in $U(\mathfrak{g}) \otimes U(\mathfrak{g})$, actually belongs to $\wedge^2(\mathfrak{g})$. This is proved by using the same cohomological argument as was used to establish Remark [3] following 16.1.8.

We shall be particularly interested in quasitriangular quasi-Hopf QUE algebras. In that case, it turns out that the classical limit is determined by a $\mathfrak{g}$-invariant symmetric element of $\mathfrak{g} \otimes \mathfrak{g}$.

PROPOSITION 16.1.10 *Let* A_h *be a quasitriangular quasi-Hopf QUE algebra with universal R-matrix* $\mathcal{R}_h$ *and* $A_h/hA_h \cong U(\mathfrak{g})$. *Set*

$$t = h^{-1}(\mathcal{R}_h^{21}\mathcal{R}_h^{12} - 1 \otimes 1) \quad (\mathrm{mod}\ h).$$

Then, t *is a symmetric* $\mathfrak{g}$-*invariant element of* $\mathfrak{g} \otimes \mathfrak{g}$, *and is unchanged by twisting* A_h.

Moreover, by twisting if necessary, we can assume that $h^{-1}(\mathcal{R}_h - 1 \otimes 1) \equiv t/2$ *(mod* h) *and* $\Phi_h \equiv 1 \otimes 1 \otimes 1$ *(mod* h^2), *and then the classical limit of* A_h *is given by* $\delta = 0$, $\varphi = -\frac{1}{4}[t_{12}, t_{23}]$.

OUTLINE OF PROOF From (10) and the fact that $\Phi_h \equiv 1 \otimes 1 \otimes 1$ (mod h), it follows that

$$(\Delta \otimes \mathrm{id})(t) = t_{13} + t_{23}.$$

As in the proof of 6.2.3, this implies that $t \in \mathfrak{g} \otimes U(\mathfrak{g})$. Since t is obviously symmetric, $t \in \mathfrak{g} \otimes \mathfrak{g}$. Finally, $\mathfrak{g}$-invariance follows from the fact that $\mathcal{R}_h^{21}\mathcal{R}_h^{12}$ commutes with $\Delta_h(A_h)$.

For the second part, we may assume by the third remark following 16.1.8 that $\Phi_h \equiv 1 \otimes 1 \otimes 1 \pmod{h^2}$. The stated conditions can then be achieved by a further twist by an element $\mathcal{F}_h$ such that $\mathcal{F}_h \equiv 1 \otimes 1 + \frac{h}{2}(r_{12} - r_{21})$ $\pmod{h^2}$. The formula for φ now follows from 16.1.3, and the vanishing of δ follows from the $\mathfrak{g}$-invariance of t. ∎

By abuse of language, in the situation of the preceding proposition, we shall call the pair $(\mathfrak{g}, t)$ the *classical limit* of the quasitriangular quasi-Hopf algebra A_h. Note that, if A_h happens to be a quasitriangular (Hopf) QUE algebra, its classical limit as a QUE algebra is the Lie bialgebra $(\mathfrak{g}, \delta)$, where $\delta(x) = [x \otimes 1 + 1 \otimes x \,,\, r]$ and $r = h^{-1}(\mathcal{R}_h - 1 \otimes 1) \pmod{h}$. On the other hand, its classical limit as a quasitriangular quasi-Hopf QUE algebra is $(\mathfrak{g}, r_{12} + r_{21})$ (with $\delta = 0$).

Given a Lie algebra $\mathfrak{g}$ and a symmetric $\mathfrak{g}$-invariant element $t \in \mathfrak{g} \otimes \mathfrak{g}$, it is natural to ask whether a *quantization* of $(\mathfrak{g}, t)$ exists, i.e. whether there is a quasitriangular quasi-Hopf QUE algebra with classical limit $(\mathfrak{g}, t)$. This question is completely answered by

THEOREM 16.1.11 *Let $\mathfrak{g}_h$ be a Lie algebra over the ring of formal power series $k[[h]]$, where k is a field of characteristic zero, such that $\mathfrak{g}_h \cong \mathfrak{g}[[h]]$ as $k[[h]]$-modules, for some Lie algebra $\mathfrak{g}$ over k. Then, to any $\mathfrak{g}_h$-invariant symmetric element t_h of $\mathfrak{g}_h \otimes \mathfrak{g}_h$ is associated a quasitriangular quasi-Hopf algebra $A_{\mathfrak{g}_h, t_h}$ over $k[[h]]$, which is isomorphic to $U(\mathfrak{g}_h)$ (completed in the h-adic topology) as a $k[[h]]$-algebra. In particular, if $\mathfrak{g}_h = \mathfrak{g}[[h]]$ and $t_h = t \in \mathfrak{g} \otimes \mathfrak{g}$, then the classical limit of $A_{\mathfrak{g}_h, t_h} = A_{\mathfrak{g}, t}$ is $(\mathfrak{g}, t)$.*

Conversely, if A_h is a quasitriangular quasi-Hopf QUE algebra such that $A_h / h A_h \cong U(\mathfrak{g})$ as quasi-Hopf algebras, then A_h is isomorphic to a twisting of $A_{\mathfrak{g}_h, t_h}$, for some $(\mathfrak{g}_h, t_h)$ satisfying the above conditions, and with $t_h \equiv t$ $\pmod{h}$. Moreover, the isomorphism class of $(\mathfrak{g}_h, t_h)$ is uniquely determined by A_h.

We shall outline the proof of the existence part of this theorem in Subsection 16.2D. Uniqueness was proved first in Drinfel'd (1991), and then by a simpler method in Drinfel'd (1992b).

16.2 The Kohno–Drinfel'd monodromy theorem

The main topic of this section is the Knizhnik–Zamolodchikov equation and its application to the 'quantization problem' for quasi-Hopf algebras. But we must begin by giving yet another interpretation of braid groups, which will also be needed in Section 16.4.

A Braid groups and configuration spaces If M is any connected smooth manifold, let $\mathcal{D}_m(M)$ be the space of all ordered m-tuples of distinct points of

M. Note that the symmetric group Σ_m acts freely on $\mathcal{D}_m(M)$ by permuting the points; the *configuration space* $\mathcal{C}_m(M)$ is the orbit space $\mathcal{D}_m(M)/\Sigma_m$. We define the '$M$-braid group' $\mathcal{B}_m(M)$ to be the fundamental group of $\mathcal{C}_m(M)$.

It is clear that $\mathcal{B}_m(M)$ is of most interest when M is two-dimensional. For, if $\dim(M) > 2$, then $\mathcal{D}_m(M)$ differs from M^m by a submanifold of codimension ≥ 3, so that $\pi_1(\mathcal{D}_m(M)) \cong \pi_1(M)^m$; it follows that $\pi_1(\mathcal{C}_m(M))$ is the semidirect product of Σ_m with $\pi_1(M)^m$ (with the obvious diagonal action).

It is fairly clear that $\mathcal{B}_m(\mathbb{C})$ is Artin's braid group $\mathcal{B}_m$ on m strands (the fundamental group of $\mathcal{D}_m(\mathbb{C})$ is called the *pure braid group*, and denoted by $\mathcal{P}_m$): we recall that $\mathcal{B}_m$ has generators $T_1, \ldots, T_{m-1}$ and defining relations

$$T_i T_j = T_j T_i \quad \text{if } |i - j| > 1,$$
$$T_i T_j T_i = T_j T_i T_j \quad \text{if } |i - j| = 1.$$

In fact, in the notation of 5.1.7, if we take the intersection of such a braid with the plane P^t and then identify P^t with P^0 by vertical translation, we obtain m curves in $\mathbb{C}$, no two of which intersect at the same value of t, and whose endpoints are a permutation of their initial points. This is the same thing as a closed curve in $\mathcal{C}_m(\mathbb{C})$. By a classical result of R. H. Fox, the universal covering space of $\mathcal{C}_m(\mathbb{C})$ is contractible, so that $\mathcal{C}_m(\mathbb{C})$ is actually the classifying space of $\mathcal{B}_m$.

The other case which will be of interest to us is the *sphere braid group* $\mathcal{B}_m(\mathbb{CP}^1)$, which we shall denote by $\mathcal{SB}_m$ (the fundamental group of $\mathcal{D}_m(\mathbb{CP}^1)$ is the *pure sphere braid group* $\mathcal{SP}_m$). It is clear that $\mathcal{SB}_m$ is a quotient of $\mathcal{B}_m$ (and similarly for the pure groups), since any loop in $\mathbb{CP}^1$ can be deformed so as to lie in $\mathbb{C} = \mathbb{CP}^1 \backslash \{\infty\}$. In fact, $\mathcal{SB}_m$ is obtained from $\mathcal{B}_m$ by imposing the additional relation

$$(T_1 T_2 \ldots T_{m-2} T_{m-1})(T_{m-1} T_{m-2} \ldots T_2 T_1) = 1.$$

Using this topological interpretation, it is easy to describe the structure of $\mathcal{P}_m$ as an abstract group. Consider the projection $\mathcal{D}_{m+1}(\mathbb{C}) \to \mathcal{D}_m(\mathbb{C})$ which forgets the $(m + 1)$th point. This is a fibration whose fibre over a point $(z_1^0, \ldots, z_m^0) \in \mathcal{D}_m(\mathbb{C})$ is $\mathbb{C} \backslash \{z_1^0, \ldots, z_m^0\}$. Now, in any fibration, lifting closed paths in the base to paths in the total space defines an action of the fundamental group of the base on that of the fibre. In our case, the fundamental group of the fibre is obviously the free group F_m on m generators, say $f_1, \ldots, f_m$, and that of the base is, by definition, the pure braid group $\mathcal{P}_m$. It is not difficult to see that $\mathcal{P}_m$ consists exactly of the automorphisms of F_m which fix the product $f_1 \ldots f_m$ and take each f_i into a conjugate of f_i and, by considering the exact sequence of fundamental groups associated to the fibration, that $\mathcal{P}_{m+1}$ is isomorphic to the semidirect product of $\mathcal{P}_m$ and F_m. Thus, $\mathcal{P}_m$ is an iterated semidirect product of free groups.

There is a similar description of $\mathcal{B}_m$, namely as the group of automorphisms of F_m which fix $f_1 \ldots f_m$ and, for some $\sigma \in \Sigma_m$, take each f_i into a conjugate of $f_{\sigma(i)}$. It was known to Artin (1947) that $\mathcal{P}_m$ is generated by the elements $P_{ij} = (T_{j-1} \ldots T_{i+1}) T_i^2 (T_{j-1} \ldots T_{i+1})^{-1}$ for $1 \leq i, j \leq m$ (the relations satisfied by the P_{ij} are also known, but we shall not need them). In particular, $\mathcal{P}_2$ is generated by T^2, where T is the generator of the infinite cyclic group $\mathcal{B}_2$, and $\mathcal{P}_3$ is generated by T_1^2, T_2^2 and $(T_1 T_2)^3$; moreover, $(T_1 T_2)^3 = (T_2 T_1)^3$ generates the centre of $\mathcal{B}_3$ and of $\mathcal{P}_3$.

B The Knizhnik–Zamolodchikov equation Let $r(z)$ be a holomorphic function defined for z lying in some open set $M \subset \mathbb{C}$ and with values in the $\mathfrak{g} \otimes \mathfrak{g}$, for some Lie algebra $\mathfrak{g}$. Let $\rho_k : \mathfrak{g} \to \text{End}(V_k)$, $k = 1, \ldots, m$, be finite-dimensional representations. Then, r defines a connection in the trivial bundle $\mathbf{V}^{(m)}$ over $\mathcal{D}_m(M)$ with fibre $V_1 \otimes \cdots \otimes V_m$, i.e. a 1-form θ^r on $\mathcal{D}_m(M)$ with values in $\text{End}(V_1 \otimes \cdots \otimes V_m)$, by

$$(19) \qquad \theta^r = \sum_{j<k} r_{jk}(z_j - z_k)(dz_j - dz_k),$$

where $r_{jk}(z_j - z_k)$ means, as usual, the endomorphism of $V_1 \otimes \cdots \otimes V_m$ which acts as $(\rho_j \otimes \rho_k)(r(z_j - z_k))$ on the jth and kth factors and as the identity on the others.

PROPOSITION 16.2.1 *The connection defined by the matrix of 1-forms θ^r defined in (19) is flat if and only if r satisfies the classical Yang–Baxter equation with spectral parameters:*

$$[r_{12}(z_1 - z_2), r_{13}(z_1 - z_3)] + [r_{12}(z_1 - z_2), r_{23}(z_2 - z_3)]$$
$$+ [r_{13}(z_1 - z_3), r_{23}(z_2 - z_3)] = 0$$

for all $(z_1, z_2, z_3) \in \mathcal{D}_3(M)$.

PROOF We have to show that

$$d\theta^r + \theta^r \wedge \theta^r = 0.$$

It is obvious that $d\theta^r = 0$. On the other hand, if $s < t$, the coefficient of $dz_s \wedge dz_t$ in $\theta^r \wedge \theta^r$ is easily found to be

$$(20) \quad \sum_{j>s,k>t} [r_{sj}, r_{tk}] + \sum_{j>s,k<t} [r_{kt}, r_{sj}] + \sum_{j<s,k>t} [r_{tk}, r_{js}] + \sum_{j<s,k<t} [r_{js}, r_{kt}],$$

where we have written r_{sj} for $r_{sj}(z_s - z_j)$, etc. Now, the operators r_{jk} and $r_{j'k'}$ commute if j, k, j' and k' are distinct, since in that case they act on disjoint pairs of factors in the tensor product $V_1 \otimes \cdots \otimes V_m$. The only non-zero terms in (20) are, therefore,

$$(21) \quad \begin{aligned} &\sum_{k>t}[r_{st}, r_{tk}] + \sum_{k>t}[r_{sk}, r_{tk}] + \sum_{j>s}[r_{st}, r_{sj}] + \sum_{s<k<t}[r_{kt}, r_{sk}] \\ &\quad + \sum_{k<t}[r_{kt}, r_{st}] + \sum_{j<s}[r_{js}, r_{jt}] + \sum_{j<s}[r_{js}, r_{st}]. \end{aligned}$$

We can split up the terms in this sum according to which factors in the tensor product they act on. There are three types: those which act on V_j, V_s and V_t for some $j < s$; those for which $s < j < t$; and those for which $t < j$. The terms of the first type come from the last three terms in (21), namely

$$[r_{jt}, r_{st}] + [r_{js}, r_{jt}] + [r_{js}, r_{st}],$$

the vanishing of which is equivalent to (19). Similarly, one finds that (19) is also equivalent to the vanishing of the other two types of terms. ∎

DEFINITION 16.2.2 *The* Knizhnik–Zamolodchikov (KZ) *equation associated to a classical r-matrix r as above is the system of differential equations for a function f on $\mathcal{D}_m(M)$ with values in $V_1 \otimes \cdots \otimes V_m$ given by*

$$(22) \quad \frac{\partial f}{\partial z_j} = \sum_{\substack{k=1 \\ k \neq j}}^{m} r_{jk}(z_j - z_k)(f), \qquad j = 1, \ldots, m.$$

In the special case when the V_j are all equal to some representation V of $\mathfrak{g}$, the connection (19) is invariant under the obvious action of the symmetric group Σ_m on $\mathcal{D}_m(M) \times V^{\otimes m}$ given by permuting the coordinates of a point in $\mathcal{D}_m(M)$ and the factors on the tensor product, and so defines a flat connection on a bundle over $\mathcal{D}_m(M)/\Sigma_m = \mathcal{C}_m(M)$ with fibre $V^{\otimes m}$. Equation (22) is the condition that f is a covariant constant section of this bundle. In general, such sections will only exist locally; global sections will be well defined only on the universal covering space of $\mathcal{C}_m(\mathbb{C})$. More precisely, if $\gamma : [0, 1] \to \mathcal{C}_m(M)$ is a smooth curve, then, by analytically continuing a solution f_1 of (22) defined near $\gamma(0)$ along the curve, we obtain a solution f_2 defined near $\gamma(1)$. If γ is closed, then, since (22) is homogeneous, f_1 and f_2 differ by an element $M(\gamma) \in GL(V^{\otimes m})$, independent of f_1, called the *monodromy* of the KZ equation along γ. Since the connection θ^r is flat, $M(\gamma)$ depends only on the homotopy class of γ. Thus, associating to a closed curve γ in $\mathcal{C}_m(M)$ the monodromy $M(\gamma)$ defines a representation of the braid group $\pi_1(\mathcal{C}_m(M)) = \mathcal{B}_m(M)$ on $V^{\otimes m}$.

The case of interest to us is the KZ equation associated to the r-matrix

$$(23) \quad r(z_1 - z_2) = \frac{t}{z_1 - z_2},$$

where t is a $\mathfrak{g}$-invariant symmetric element of $\mathfrak{g} \otimes \mathfrak{g}$. We saw in 2.1.9 that r is indeed a solution of the CYBE. In this case, $M = \mathbb{C}$, so, by the results of Subsection 11.2A, the monodromy of the KZ equation gives us a representation of Artin's braid group $\mathcal{B}_m$ on $V^{\otimes m}$, for any representation V of $\mathfrak{g}$.

We have seen in Subsection 5.2B that, in any quasitensor category $\mathbf{C}$, the braid group $\mathcal{B}_3$ acts on triple tensor products (the three objects involved should be the same, otherwise it is only the pure braid group which acts). If $\mathbf{C}$ is the category of representations of a quasitriangular Hopf (or quasi-Hopf) algebra A, this action is given by that of the universal R-matrix $\mathcal{R}$ of A (see Subsection 5.2B). If A is a quantization of a universal enveloping algebra $U(\mathfrak{g})$, it is natural to expect that this action is closely related to that given by the monodromy of the KZ equation associated to the classical r-matrix r which is the classical limit of $\mathcal{R}$. That this is indeed the case is the main result of this section, and to prove it we shall actually use the KZ equation to carry out the quantization (see Subsection 16.2D).

C The KZ equation and affine Lie algebras We allow ourselves a small digression at this point to explain the relation between the KZ equation and the representation theory of (classical) affine Lie algebras, which is one of the main reasons for its importance in conformal field theory. This is not strictly necessary for our purposes, but it provides motivation for the developments in Section 16.3.

For convenience, we work over the field $\mathbb{C}((z))$ of formal power series, and take the affine Lie algebra $\hat{\mathfrak{g}}$ associated to a finite-dimensional complex simple Lie algebra $\mathfrak{g}$ to be $\mathfrak{g}((z)) \oplus \mathbb{C}.c$ as a vector space, where $\mathfrak{g}((z)) = \mathfrak{g} \otimes \mathbb{C}((z))$ and c is a central element. The bracket on $\hat{\mathfrak{g}}$ is given by

$$[f, g](z) = [f(z), g(z)] + \mathrm{res}_0\left(f(z), \frac{dg}{dz}\right)c,$$

where $f, g \in \mathfrak{g}((z))$, $(\,,\,)$ is the standard invariant bilinear form on $\mathfrak{g}$, and res_0 means the z^{-1} coefficient. Note that, if we view $\mathfrak{g} \subset \mathfrak{g}((z))$ as the constant functions of z, the restriction of the cocycle

$$(f, g) \mapsto \mathrm{res}_0\left(f(z), \frac{dg}{dz}\right)$$

of $\hat{\mathfrak{g}}$ to $\mathfrak{g} \times \mathfrak{g}$ vanishes; thus, $\mathfrak{g}$ is canonically embedded in $\hat{\mathfrak{g}}$.

As usual, we denote by D the derivation of $\hat{\mathfrak{g}}$ such that

$$[D, c] = 0, \quad [D, f](z) = z\frac{df}{dz}.$$

There are two types of representations of $\hat{\mathfrak{g}}$ which are compatible with the derivation D, in the sense that they extend to representations of the semidirect product of $\hat{\mathfrak{g}}$ with the one-dimensional Lie algebra $\mathbb{C}.D$. These are the highest (and lowest) weight representations, together with the *loop representations*.

We recall (see the Appendix) that a highest weight representation W of $\hat{\mathfrak{g}}$ is generated by a vector $w \in W$ which is annihilated by the subalgebra $z\mathfrak{g}[[z]]$, is a highest weight vector in the usual sense for the action of the subalgebra $\mathfrak{g}$ of $\hat{\mathfrak{g}}$, and is an eigenvector of the central element c. Any irreducible highest weight representation of $\hat{\mathfrak{g}}$ is characterized, up to isomorphism, by the eigenvalue κ of c (called the level) and the weight λ of w for the action of $\mathfrak{g}$. The *conformal weight* of w is defined to be

$$h_\kappa(\lambda) = \frac{C_\lambda}{2(\kappa + g)},$$

where C_λ is the eigenvalue of the Casimir element of $U(\mathfrak{g})$ on w and g is the dual Coxeter number of $\mathfrak{g}$ (we assume that $\kappa + g \neq 0$).

A loop representation of $\hat{\mathfrak{g}}$, on the other hand, is of the form $V((z))$, where V is a finite-dimensional irreducible representation of $\mathfrak{g}$, with the obvious pointwise action of $\mathfrak{g}((z))$ and with c acting as zero.

In conformal field theory, a crucial role is played by the intertwining operators

$$\mathcal{I}_V^{W',W}(z) : W \to W' \otimes V((z)),$$

where W and W' are highest weight representations and $V((z))$ is a loop representation. Of course, for the existence of a (non-zero) intertwiner of this kind, it is necessary that W and W' have the same level. It will be convenient to work with the rescaled operators

$$\mathcal{J}_V^{W',W}(z) = \mathcal{I}_V^{W',W}(z) z^{h_\kappa(\lambda') - h_\kappa(\lambda) + h_\kappa(\mu)},$$

where κ is the level of W and W', λ and λ' are the $\mathfrak{g}$-weights of their highest weight vectors, and μ is the highest weight of V.

PROPOSITION 16.2.3 *Let $\mathfrak{g}$ be a finite-dimensional complex simple Lie algebra, and let $\hat{\mathfrak{g}}$ be the associated affine Lie algebra. Let $\rho_j : \mathfrak{g} \to \mathrm{End}(V_j)$, $j = 1, \ldots, m$, be finite-dimensional irreducible representations of $\mathfrak{g}$, and let $W_0, \ldots, W_m$ be highest weight representations of $\hat{\mathfrak{g}}$ of level κ. Define an operator*

$$F : W_m \to W_0 \otimes V_1((z_1)) \otimes \cdots \otimes V_m((z_m))$$

by

$$F = (\mathcal{J}_{V_1}^{W_0,W_1}(z_1) \otimes \mathrm{id}^{\otimes m-1}) \circ \cdots \circ (\mathcal{J}_{V_{m-1}}^{W_{m-2},W_{m-1}}(z_{m-1}) \otimes \mathrm{id}) \circ \mathcal{J}_{V_m}^{W_{m-1},W_m}(z_m),$$

where the rescaled intertwining operators $\mathcal{J}_V^{W',W}(z)$ are as defined above. Regard F as a linear map $W_m \to W_0$ with values in $(V_1 \otimes \cdots \otimes V_m)((z_1, \ldots, z_m))$, and let f be the matrix element of F between the highest weight vectors of W_0

and W_m (with respect to bases of W_0 and W_m consisting of weight vectors). Then, f satisfies the KZ equation

$$\frac{\partial f}{\partial z_j} = \frac{1}{\kappa + g} {\sum_k}' \frac{(\rho_j \otimes \rho_k)(t)}{z_j - z_k}(f),$$

where $t \in \mathfrak{g} \otimes \mathfrak{g}$ is the canonical element associated to the standard bilinear form on $\mathfrak{g}$, and the $\sum'$ indicates that the term with $k = j$ is omitted from the sum. ∎

The function f in the statement of this result is an example of an *m-point function*. For a proof of this result and further discussion of the role played by the KZ equation in conformal field theory, see, for example, Tsuchiya & Kanie (1988) and Frenkel & Reshetikhin (1992a).

D Quantization and the KZ equation In this subsection, we shall outline the proof of the existence part of 16.1.11. We denote by $\mathfrak{g}_h$ and t_h the objects defined there, and take A_h to be the completed universal enveloping algebra $U(\mathfrak{g}_h)$ as an algebra over $k[[h]]$, and with universal R-matrix $\mathcal{R}_{KZ} = e^{ht_h/2}$; note that the comultiplication Δ_h of $U(\mathfrak{g}_h)$ satisfies $\Delta_h^{\mathrm{op}} = \mathcal{R}_{KZ}.\Delta_h.\mathcal{R}_{KZ}^{-1}$ by the $\mathfrak{g}_h$-invariance of t_h. In view of Remark [4] following 16.1.8, to prove the existence part of 16.1.11 it is sufficient to construct an element $\Phi_h \in U(\mathfrak{g}_h)^{\otimes 3}$ satisfying (1), (2), (4), (9) and (10).

We assume initially that $k = \mathbb{C}$, and consider the 'universal' KZ equation

$$(\mathrm{KZ}_m) \qquad \frac{\partial f}{\partial z_j} = \hbar \sum_{\substack{k=1 \\ k \neq j}}^{m} \frac{t_h^{jk}}{z_j - z_k} f, \qquad j = 1, \dots, m,$$

where $\hbar = h/2\pi\sqrt{-1}$ and f is a function of $z_1, \dots, z_m$ with values in $U(\mathfrak{g}_h)^{\otimes m}$. Of course, the usual KZ equation can be obtained from this by applying a representation of $U(\mathfrak{g}_h)^{\otimes m}$ to both sides.

Observe first that the R-matrix $\mathcal{R}_{KZ} = e^{ht_h/2}$ is the monodromy of (KZ_2). More precisely, since $\partial f/\partial z_1 + \partial f/\partial z_2 = 0$, any solution of (KZ_2) is of the form $f(z_1, z_2) = g(z_1 - z_2)$, where $g(z)$ satisfies the ordinary differential equation

$$g'(z) = \frac{\hbar t_h}{z} g(z).$$

The monodromy of the solution $g(z) = z^{\hbar t_h} = e^{\hbar t_h \ln z}$ from the region $z \gg 0$ to the region $z \ll 0$ is $e^{\sqrt{-1}\pi\hbar t} = e^{ht/2}$. Note that this is also the monodromy of (KZ_2), in its original form, from the asymptotic region $z_1 - z_2 \gg 0$ to the region $z_1 - z_2 \ll 0$. (Of course, g should really be considered as a function on the universal covering of $\mathbb{C}\backslash\{0\}$, and the solution f of (KZ_2) as a function on the universal covering of $\mathcal{C}_2(\mathbb{C})$.)

The coassociator is constructed in a similar way using (KZ$_3$). We assume initially that z_1, z_2 and z_3 are real. Since (KZ$_3$) is unchanged if we subject z_1, z_2 and z_3 simultaneously to a transformation of the form $z_i \mapsto az_i + b$, it follows that any solution of (KZ$_3$) in the region $\{z_1 < z_2 < z_3\}$ is of the form

$$f(z_1, z_2, z_3) = (z_3 - z_1)^{h(t_h^{12} + t_h^{13} + t_h^{23})} g\left(\frac{z_2 - z_1}{z_3 - z_1}\right),$$

where g satisfies the differential equation

$$(24) \qquad g'(x) = \hbar\left(\frac{t_h^{12}}{x} + \frac{t_h^{23}}{x - 1}\right) g(x)$$

for $0 < x < 1$. Similar reductions are, of course, possible in each of the other five connected components of the complement in $\mathbb{R}^3$ of the hyperplanes $z_i = z_j$, $1 \le i < j \le 3$.

According to the classical theory of fuchsian differential equations, there are unique analytic solutions g_1 and g_2 of (24), defined for $x \in \mathbb{C}\backslash\{0, 1\}$, and with asymptotic behaviour

$$g_1(x) \sim x^{ht_h^{12}} \qquad \text{as } x \to 0,$$
$$g_2(x) \sim (1 - x)^{ht_h^{23}} \qquad \text{as } x \to 1.$$

We should explain the meaning of 'analytic' and of the symbol $\sim$. First, to say that g is analytic means that, for all $n \in \mathbb{N}$, $g(x)$ is equal (mod h^n) to a finite sum of products of analytic functions of x with elements of $U(\mathfrak{g}_h)^{\otimes 3}$. We shall be most interested in the case where $\mathfrak{g}_h$ is the null deformation $\mathfrak{g}_h = \mathfrak{g}[[h]]$ of a Lie algebra $\mathfrak{g}$ over k, and then 'analytic' simply means that

$$g(x) = \sum_{n=0}^{\infty} g^{(n)}(x)h^n,$$

where each $g^{(n)}$ is an analytic function with values in a finite-dimensional subspace of $U(\mathfrak{g})^{\otimes 3}$. Saying that $g(x) \sim x^{ht_h^{12}}$, for example, means that $g(x)x^{-ht_h^{12}}$ extends to an analytic function, in the sense we have just described, which is defined in a neighbourhood of $x = 0$ and equal to 1 at that point.

Since (24) is homogeneous in g, it follows that $\Phi_{\text{KZ}} = g_1^{-1}g_2$ is a constant element of $U(\mathfrak{g}_h)^{\otimes 3}$. This is the desired coassociator of A_h. Note that $f_1 \Phi_{\text{KZ}} = f_2$, where f_1 and f_2 are the solutions of (KZ$_3$) in $\{z_1 < z_2 < z_3\}$ with asymptotic behaviour

$$f_1(z_1, z_2, z_3) \sim (z_2 - z_1)^{ht_h^{12}}(z_3 - z_1)^{h(t_h^{13} + t_h^{23})} \qquad \text{for } z_2 - z_1 \ll z_3 - z_1,$$
$$f_2(z_1, z_2, z_3) \sim (z_3 - z_2)^{ht_h^{23}}(z_3 - z_1)^{h(t_h^{12} + t_h^{13})} \qquad \text{for } z_3 - z_2 \ll z_3 - z_1.$$

Thus, Φ_{KZ} is the monodromy of (KZ$_3$) from the asymptotic region $z_2 - z_1 \ll z_3 - z_1$ to the region $z_3 - z_2 \ll z_3 - z_1$.

To verify that Φ_{KZ} has the desired properties, note first that (1) is the statement that Φ_{KZ} is $\mathfrak{g}_h$-invariant, and that this follows from the uniqueness of f_1 and f_2 and the $\mathfrak{g}_h$-invariance of t_h. Similarly, (4) follows from the fact that $(\epsilon_h \otimes \mathrm{id})(t_h) = (\mathrm{id} \otimes \epsilon_h)(t_h) = 0$. Further, by replacing x by $1 - x$ in (24), we see that $\Phi_{KZ}^{321} = \Phi_{KZ}^{-1}$, so that (10) follows from (9). Thus, we have to prove equations (2) and (9).

We recall that (2) guarantees the commutativity of the pentagon diagram in 5.1.1(i). We consider the system (KZ$_4$) in the region $\{z_1 < z_2 < z_3 < z_4\} \subset \mathbb{R}^4$, and distinguish five zones in this region corresponding to the vertices of the pentagon diagram, according to the following rule:

Zone I	$z_2 - z_1 \ll z_3 - z_1 \ll z_4 - z_1$	$((U \otimes V) \otimes W) \otimes Z,$
Zone II	$z_3 - z_2 \ll z_3 - z_1 \ll z_4 - z_1$	$(U \otimes (V \otimes W)) \otimes Z,$
Zone III	$z_3 - z_2 \ll z_4 - z_2 \ll z_4 - z_1$	$U \otimes ((V \otimes W) \otimes Z),$
Zone IV	$z_4 - z_3 \ll z_4 - z_2 \ll z_4 - z_1$	$U \otimes (V \otimes (W \otimes Z)),$
Zone V	$z_2 - z_1 \ll z_4 - z_1, \quad z_4 - z_3 \ll z_4 - z_1$	$(U \otimes V) \otimes (W \otimes Z).$

In these zones, we look for solutions f of (KZ$_4$) with the following asymptotic behaviour:

$$
\begin{aligned}
f_{\mathrm{I}} &\sim (z_2 - z_1)^{\hbar t_h^{12}}(z_3 - z_1)^{\hbar(t_h^{13}+t_h^{23})}(z_4 - z_1)^{\hbar(t_h^{14}+t_h^{24}+t_h^{34})}, \\
f_{\mathrm{II}} &\sim (z_3 - z_2)^{\hbar t_h^{23}}(z_3 - z_1)^{\hbar(t_h^{12}+t_h^{13})}(z_4 - z_1)^{\hbar(t_h^{14}+t_h^{24}+t_h^{34})}, \\
f_{\mathrm{III}} &\sim (z_3 - z_2)^{\hbar t_h^{23}}(z_4 - z_2)^{\hbar(t_h^{24}+t_h^{34})}(z_4 - z_1)^{\hbar(t_h^{12}+t_h^{13}+t_h^{14})}, \\
f_{\mathrm{IV}} &\sim (z_4 - z_3)^{\hbar t_h^{34}}(z_4 - z_2)^{\hbar(t_h^{23}+t_h^{24})}(z_4 - z_1)^{\hbar(t_h^{12}+t_h^{13}+t_h^{14})}, \\
f_{\mathrm{V}} &\sim (z_2 - z_1)^{\hbar t_h^{12}}(z_4 - z_3)^{\hbar t_h^{34}}(z_4 - z_1)^{\hbar(t_h^{13}+t_h^{14}+t_h^{23}+t_h^{24})}.
\end{aligned}
$$

(25)

Replacing the representations at the vertices of the pentagon diagram 5.1.1(i) by these solutions according to the above rule, and the associativity maps between them by the appropriate elements of $U(\mathfrak{g}_h)^{\otimes 4}$ built out of Φ_{KZ}, we obtain the following diagram:

$$
\begin{array}{ccccc}
f_{\mathrm{IV}} & \xrightarrow{(\mathrm{id}\otimes\mathrm{id}\otimes\Delta_h)(\Phi_{KZ})} & f_{\mathrm{V}} & \xrightarrow{(\Delta_h\otimes\mathrm{id}\otimes\mathrm{id})(\Phi_{KZ})} & f_{\mathrm{I}} \\
\Big\downarrow{\scriptstyle 1\otimes\Phi_{KZ}} & & & & \Big\uparrow{\scriptstyle \Phi_{KZ}\otimes 1} \\
f_{\mathrm{III}} & & \xrightarrow[(\mathrm{id}\otimes\Delta_h\otimes\mathrm{id})(\Phi_{KZ})]{} & & f_{\mathrm{II}}
\end{array}
$$

Equation (2) is a consequence of the following relations:

$$
\begin{aligned}
&f_{\mathrm{IV}} = f_{\mathrm{V}}.(\mathrm{id} \otimes \mathrm{id} \otimes \Delta_h)(\Phi_{KZ}), \qquad f_{\mathrm{V}} = f_{\mathrm{I}}.(\Delta_h \otimes \mathrm{id} \otimes \mathrm{id})(\Phi_{KZ}), \\
&f_{\mathrm{IV}} = f_{\mathrm{III}}.(1 \otimes \Phi_{KZ}), \qquad\qquad\qquad f_{\mathrm{II}} = f_{\mathrm{I}}.(\Phi_{KZ} \otimes 1), \\
&\qquad\qquad f_{\mathrm{III}} = f_{\mathrm{II}}.(\mathrm{id} \otimes \Delta_h \otimes \mathrm{id})(\Phi_{KZ}).
\end{aligned}
$$

(26)

The existence and uniqueness of solutions of (KZ_4) with the asymptotics in (25) follow by standard methods in the theory of differential equations. For example,

$$f_V = \varphi_V \left(\frac{z_2 - z_1}{z_4 - z_1}, \frac{z_4 - z_3}{z_4 - z_1} \right),$$

where the $U(\mathfrak{g}_h)^{\otimes 3}$-valued function $\varphi_V(u, v)$ is analytic in a neighbourhood of $(0,0)$ and equal to $1 \otimes 1 \otimes 1$ at that point, and satisfies a system of differential equations of the form

$$\frac{\partial \varphi_V}{\partial u} = A(u, v)\varphi_V, \qquad \frac{\partial \varphi_V}{\partial v} = B(u, v)\varphi_V,$$

where A and B are analytic in a neighbourhood of $(0,0)$. Thus, proving the existence and uniqueness of a function f_V with the desired properties reduces to solving an initial value problem for a first order system of partial differential equations with real-analytic coefficients.

Each of the equations (26) is proved by comparing two solutions to the same initial value problem. We first observe that $f_I, \ldots, f_V$ can be analytically continued to the whole region $\{z_1 < z_2 < z_3 < z_4\}$. To prove, for example, that $f_I.(\Phi_{KZ} \otimes 1) = f_{II}$, let

$$F_I = f_I.(z_4 - z_1)^{-\hbar(t_h^{14} + t_h^{24} + t_h^{34})},$$
$$F_{II} = f_{II}.(z_4 - z_1)^{-\hbar(t_h^{14} + t_h^{24} + t_h^{34})}.(\Phi_{KZ} \otimes 1)^{-1}.$$

It is enough to prove that $F_I = F_{II}$, for $[\Phi_{KZ}, t_h^{14} + t_h^{24} + t_h^{34}] = 0$ by the $\mathfrak{g}_h$-invariance of Φ_{KZ}, so that

$$F_{II} = f_{II}.(\Phi_{KZ} \otimes 1)^{-1}.(z_4 - z_1)^{-\hbar(t_h^{14} + t_h^{24} + t_h^{34})}.$$

From the asymptotics of F_I and F_{II}, it is clear that these functions extend analytically to $z_4 = \infty$. One checks that they both satisfy the differential system

$$\frac{\partial F}{\partial z_1} = \hbar \left(\frac{t_h^{12}}{z_1 - z_2} F + \frac{t_h^{13}}{z_1 - z_3} F - F \frac{t_h^{14} + t_h^{24} + t_h^{34}}{z_1 - z_4} \right),$$
$$\frac{\partial F}{\partial z_2} = \hbar \left(\frac{t_h^{12}}{z_2 - z_1} + \frac{t_h^{23}}{z_2 - z_3} \right) F, \qquad \frac{\partial F}{\partial z_3} = \hbar \left(\frac{t_h^{13}}{z_3 - z_1} + \frac{t_h^{23}}{z_3 - z_2} \right) F.$$

Thus, at $z_4 = \infty$, both F_I and F_{II} satisfy (KZ_3); in fact, in view of their asymptotics, we have, in the notation introduced earlier in this subsection,

$$F_I = g_1(x) \otimes 1, \qquad F_{II} = (g_2(x) \otimes 1).(\Phi_{KZ} \otimes 1)^{-1},$$

where $x = (z_2 - z_1)/(z_3 - z_1)$. Recalling the definition of Φ_{KZ}, we see that $F_I = F_{II}$ throughout the region $\{z_1 < z_2 < z_3, z_4 = \infty\}$. But, it is easy to check that F_I and F_{II} also satisfy

$$\frac{\partial F}{\partial z_4} = \hbar \sum_{j=1}^{3} \frac{[t_h^{j4}, F]}{z_4 - z_j},$$

so, using the uniqueness of solutions of initial value problems once again, we have that $F_I = F_{II}$ throughout $\{z_1 < z_2 < z_3 < z_4\}$.

The proof of (9) proceeds along similar lines, but this time one must allow the z_j to take complex values. In fact, we consider (KZ$_3$) in the region

$$\{(z_1, z_2, z_3) \in \mathbb{C}^3 \mid \Im(z_1) \leq \Im(z_2) \leq \Im(z_3),\ z_1,\ z_2,\ z_3 \text{ distinct}\},$$

where $\Im(z)$ denotes the imaginary part of a complex number z. Let $\tilde{f}_1, \ldots, \tilde{f}_{VI}$ be the solutions of (KZ$_3$) in this region for which

$$\tilde{f}_I \sim (z_2 - z_1)^{\hbar t_h^{12}}(z_3 - z_1)^{\hbar(t_h^{13}+t_h^{23})} \quad \text{for } |z_2 - z_1| \ll |z_3 - z_1|,$$

$$\tilde{f}_{II} \sim (z_3 - z_2)^{\hbar t_h^{23}}(z_3 - z_1)^{\hbar(t_h^{12}+t_h^{13})} \quad \text{for } |z_3 - z_2| \ll |z_3 - z_1|,$$

$$\tilde{f}_{III} \sim (z_2 - z_3)^{\hbar t_h^{23}}(z_2 - z_1)^{\hbar(t_h^{12}+t_h^{13})} \quad \text{for } |z_3 - z_2| \ll |z_2 - z_1|,$$

$$\tilde{f}_{IV} \sim (z_3 - z_1)^{\hbar t_h^{13}}(z_2 - z_1)^{\hbar(t_h^{12}+t_h^{23})} \quad \text{for } |z_3 - z_1| \ll |z_2 - z_1|,$$

$$\tilde{f}_V \sim (z_1 - z_3)^{\hbar t_h^{13}}(z_2 - z_3)^{\hbar(t_h^{12}+t_h^{23})} \quad \text{for } |z_1 - z_3| \ll |z_2 - z_3|,$$

$$\tilde{f}_{VI} \sim (z_2 - z_1)^{\hbar t_h^{12}}(z_2 - z_3)^{\hbar(t_h^{13}+t_h^{23})} \quad \text{for } |z_2 - z_1| \ll |z_3 - z_1|.$$

If the z_j are real, one shows, using the definition of Φ_{KZ} as above, that $\tilde{f}_I.\Phi_{KZ} = \tilde{f}_{II}$ in the region $\{z_1 < z_2 < z_3\}$, and hence everywhere by analytic continuation. Similarly, one shows that $\tilde{f}_{IV}.\Phi_{KZ}^{132} = \tilde{f}_{III}$ and $\tilde{f}_V.\Phi_{KZ}^{231} = \tilde{f}_{VI}$ by considering the two regions $\{z_1 < z_3 < z_2\}$ and $\{z_3 < z_1 < z_2\}$, respectively.

By considering complex values of the z_j, on the other hand, one shows that $\tilde{f}_{II} = \tilde{f}_{III}.e^{\hbar t_h^{23}/2}$, $\tilde{f}_{IV} = \tilde{f}_V.e^{\hbar t_h^{13}/2}$ and $\tilde{f}_I = \tilde{f}_{VI}.e^{\hbar(t_h^{13}+t_h^{23})/2}$. Combining these formulas with those in the preceding paragraph immediately gives (9) (one needs to recall that $\mathcal{R}_h = e^{\hbar t_h/2}$ and to note that $\Delta_h(t_h) = t_h^{13} + t_h^{23}$). To prove the first formula, for example, choose the branch of the logarithm which is real on the positive real axis and holomorphic on the complement of the negative imaginary axis (see the next diagram). We compare $\tilde{f}_{II}$ and $\tilde{f}_{III}$ when $z_2 - z_3 > 0$ (and hence $\Im(z_2) = \Im(z_3)$). Thus,

$$(z_3 - z_2)^{\hbar t_h^{23}} = \exp((\ln(z_2 - z_3) + \sqrt{-1}\pi)\hbar t_h^{23}) = (z_2 - z_3)^{\hbar t_h^{23}} e^{\hbar t_h^{23}/2}.$$

Since the other factors in $\tilde{f}_{II}$ and $\tilde{f}_{III}$ are asymptotically equal in the region being considered, and since $[t_h^{23}, t_h^{12} + t_h^{13}] = 0$ by the $\mathfrak{g}_h$-invariance of t_h, it

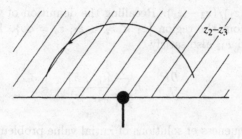

follows that $\tilde{f}_{II}$ and $\tilde{f}_{III} \cdot e^{ht_h^{23}/2}$ have the same asymptotic behaviour in their common region of definition. Since they are both solutions of (KZ$_3$), they coincide.

This completes our outline of the proof of the existence part of 16.1.11 in the case $k = \mathbb{C}$. The construction of Φ_{KZ} we have described is transcendental, and obviously does not make sense unless the ground field $k \supseteq \mathbb{C}$. Of course, it is possible *a priori* that the resulting Φ_{KZ} is actually defined over $\mathbb{Q}$ in some sense, but unfortunately this is not the case. Drinfel'd (1991) computes a series expansion of Φ_{KZ}, among the coefficients of which feature the numbers $\zeta(n)/n(2\pi\sqrt{-1})^n$, $n \in \mathbb{N}$, half of which are imaginary and all of which are probably transcendental. Thus, a different method of proof must be used if $k \not\supseteq \mathbb{C}$. Drinfel'd (1991) uses a cohomological argument which produces a 'universal' solution defined over $\mathbb{Q}$, but we shall not describe it here.

Remark Since the representations of the quasi-Hopf algebra $A_{\mathfrak{g}h,t_h}$, whose existence we have established, form a quasitensor category, there is an action of the braid group $\mathcal{B}_m$ on $V_h^{\otimes m}$, for any representation $\rho_h : A_{\mathfrak{g}h,t_h} \to \text{End}(V_h)$ and any $m \geq 2$. This action is the same as that given by the monodromy of (KZ$_m$) (with t_h replaced by $(\rho_h \otimes \rho_h)(t_h)$). To see this, it is enough to consider the case $m = 3$. We fix a basepoint $(0, 1, \infty) \in \mathcal{C}_3(\mathbb{C})$, where '$\infty$' means a very large positive number. We associate the representations $(V_h \otimes V_h) \otimes V_h$ and $V_h \otimes (V_h \otimes V_h)$ with the following asymptotic zones in $\mathcal{C}_3(\mathbb{C})$:

Zone I	$z_2 - z_1 \ll z_3 - z_1$	$(V_h \otimes V_h) \otimes V_h,$
Zone II	$z_3 - z_2 \ll z_3 - z_1$	$V_h \otimes (V_h \otimes V_h).$

The generator $T_1 \in \mathcal{B}_3$ is represented by a path from $(0, 1, \infty)$ to $(1, 0, \infty)$ lying entirely in zone I, and since (KZ$_3$) degenerates to (KZ$_2$) when $z_3 = \infty$, it follows that the monodromy along this path is $\mathcal{R}_{KZ} = e^{ht_h/2}$. This is the same as the action of T_1 on $(V_h \otimes V_h) \otimes V_h$. On the other hand, the action of the generator T_2 is given by the composite

$$(V_h \otimes V_h) \otimes V_h \xrightarrow{\Phi_{KZ}^{-1}} V_h \otimes (V_h \otimes V_h) \xrightarrow{\mathcal{R}_{KZ}^{23}} V_h \otimes (V_h \otimes V_h) \xrightarrow{\Phi_{KZ}} (V_h \otimes V_h) \otimes V_h.$$

Now T_2 can be represented by the conjunction of three paths (see Fig. 26):

(a) a path from $(0, 1, \infty)$ to $(-\infty, 0, 1)$, which, in terms of $z = \frac{z_2 - z_1}{z_3 - z_1}$, is a straight line path from a point close to 0 to a point close to 1, followed by
(b) a small path in zone II from $(-\infty, 0, 1)$ to $(-\infty, 1, 0)$, followed by
(c) a path from $(-\infty, 1, 0)$ to $(0, 1, \infty)$, which, in terms of z, is a straight line path from a point close to 1 to a point close to 0.

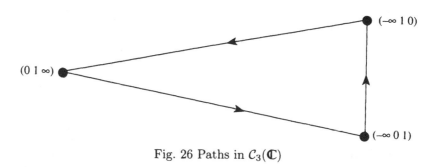

Fig. 26 Paths in $\mathcal{C}_3(\mathbb{C})$

The monodromy of (KZ$_2$) along these paths is Φ_{KZ}^{-1}, $\mathcal{R}_{\mathrm{KZ}}^{23}$ and Φ_{KZ}, respectively, by the definition of $\mathcal{R}_{\mathrm{KZ}}$ and Φ_{KZ}. Hence, the action of T_2 is also given by the monodromy of the KZ equation. $\triangle$

E The monodromy theorem In this subsection, $\mathfrak{g}$ is a finite-dimensional complex simple Lie algebra, and $U_h(\mathfrak{g})$ is the associated standard (Hopf) QUE algebra defined in Section 6.5. We can associate to any finite-dimensional representation $\rho_h : U_h(\mathfrak{g}) \to \mathrm{End}(V_h)$ two representations of the braid group $\mathcal{B}_m$ on $V_h^{\otimes m}$, for any $m \geq 2$, as follows.

First, let $\mathcal{R}_h \in U_h(\mathfrak{g}) \otimes U_h(\mathfrak{g})$ be the universal R-matrix of $U_h(\mathfrak{g})$ (see Section 8.3). Then, we described in Subsection 15.2A how to use the matrix $(\rho_h \otimes \rho_h)(\mathcal{R}_h) \in \mathrm{End}(V_h \otimes V_h)$ to obtain a representation ρ^{QYBE} of $\mathcal{B}_m$ on $V_h^{\otimes m}$.

On the other hand, let t be the symmetric $\mathfrak{g}$-invariant element of $\mathfrak{g} \otimes \mathfrak{g}$ associated to the standard inner product on $\mathfrak{g}$. Recall that V_h is a deformation of a representation $\rho : \mathfrak{g} \to \mathrm{End}(V \otimes V)$; in particular, we may assume that $V_h = V[[h]]$ as a $\mathbb{C}[[h]]$-module. We consider the KZ equation

$$\frac{\partial f}{\partial z_j} = h \sum_{\substack{k=1 \\ k \neq j}}^{m} \frac{(\rho_j \otimes \rho_k)(t)}{z_j - z_k}(f).$$

We saw in Subsection 16.2B that the monodromy of this equation defines a representation ρ^{KZ} of $\mathcal{B}_m$ on $V^{\otimes m}$. We extend this representation $\mathbb{C}[[h]]$-linearly to obtain a representation, also denoted by ρ^{KZ}, of $\mathcal{B}_m$ on $V_h^{\otimes m}$.

Kohno and Drinfel'd independently proved the following beautiful result:

THEOREM 16.2.4 *The representations ρ^{QYBE} and ρ^{KZ} of $\mathcal{B}_m$ defined above are equivalent.*

OUTLINE OF PROOF We can consider $U_h(\mathfrak{g})$ as a quasitriangular quasi-Hopf QUE algebra over $\mathbb{C}[[h]]$ with comultiplication Δ_h, universal R-matrix $\mathcal{R}_h$ and trivial coassociator. By 16.1.6 and the remark at the end of the previous subsection, it is enough to show that $U_h(\mathfrak{g})$ is isomorphic as a quasi-Hopf QUE algebra to a twisting of $A_{\mathfrak{g}_h,t}$, where $\mathfrak{g}_h = \mathfrak{g}[[h]]$ is the null deformation of $\mathfrak{g}$ and t is the invariant symmetric element of $\mathfrak{g} \otimes \mathfrak{g} \subset \mathfrak{g}_h \otimes \mathfrak{g}_h$ associated to the standard inner product on $\mathfrak{g}$.

By 6.5.4, $U_h(\mathfrak{g}) \cong U(\mathfrak{g})[[h]]$ as an algebra over $\mathbb{C}[[h]]$. Let $\tilde{\Delta}_h$ be the composite homomorphism

$$U(\mathfrak{g})[[h]] \xrightarrow{\cong} U_h(\mathfrak{g}) \xrightarrow{\Delta_h} U_h(\mathfrak{g}) \otimes U_h(\mathfrak{g}) \xrightarrow{\cong} (U(\mathfrak{g}) \otimes U(\mathfrak{g}))[[h]].$$

Since $\tilde{\Delta}_h \equiv \Delta \pmod{h}$, where Δ is the usual comultiplication of $U(\mathfrak{g})$, and since $H^1(\mathfrak{g}, U(\mathfrak{g}) \otimes U(\mathfrak{g})) = 0$ (see the first remark in Subsection 6.5B), there exists an element $\mathcal{F}_h \in (U(\mathfrak{g}) \otimes U(\mathfrak{g}))[[h]]$ such that $\mathcal{F}_h \equiv 1 \otimes 1 \pmod{h}$ and $\tilde{\Delta}_h(x) = \mathcal{F}_h^{-1} \Delta(x) \mathcal{F}_h$ for all $x \in U(\mathfrak{g})$. Let $\tilde{\mathcal{R}}_h$ be the element of $(U(\mathfrak{g}) \otimes U(\mathfrak{g}))[[h]]$ corresponding to $\mathcal{R}_h \in U_h(\mathfrak{g}) \otimes U_h(\mathfrak{g})$. Then, twisting the quasitriangular Hopf QUE algebra $(U(\mathfrak{g})[[h]], \tilde{\Delta}_h, \tilde{\mathcal{R}}_h)$ by $\mathcal{F}_h$, we obtain a quasitriangular quasi-Hopf QUE algebra $(U(\mathfrak{g})[[h]], \Delta, \overline{\Phi}_h, \overline{\mathcal{R}}_h)$, say. By Remark [4] following 16.1.8, we can assume that $\overline{\mathcal{R}}_h^{21} = \overline{\mathcal{R}}_h$ by further twisting if necessary. But, by 4.2.8 and the results of Subsection 8.3F, we have

$$\mathcal{R}_h^{21} \mathcal{R}_h = \Delta_h(e^{hC_h/2})(e^{-hC_h/2} \otimes e^{-hC_h/2}),$$

where C_h is the element of $U_h(\mathfrak{g})$ which corresponds to the Casimir element C of $U(\mathfrak{g})$. Hence, $(\overline{\mathcal{R}}_h)^2 = \overline{\mathcal{R}}_h^{21} \overline{\mathcal{R}}_h = \Delta_h(e^{hC/2})(e^{-hC/2} \otimes e^{-hC/2})$. Finally, it is clear that $\Delta(C) = C \otimes 1 + 1 \otimes C + t$, so that $\overline{\mathcal{R}}_h = e^{ht/2}$. Since Φ_h is unique up to twisting by a $\mathfrak{g}$-invariant element of $(U(\mathfrak{g}) \otimes U(\mathfrak{g}))[[h]]$ by 16.1.11, this completes the proof. ∎

16.3 Affine Lie algebras and quantum groups

The results of the previous section already indicate a connection between quantum groups and affine Lie algebras, for we saw in Subsection 16.2C that the KZ equation is intimately related to representations of affine Lie algebras, and in the Kohno–Drinfel'd monodromy theorem that it is also related to QUE algebras. In this section, we describe Kazhdan and Lusztig's extension of this latter result, which gives an explicit equivalence between certain categories of representations of quantum groups and affine Lie algebras. It is beyond the scope of this book to discuss this deep result in detail, and even to state it will require considerable preparation.

Throughout this section, $\mathfrak{g}$ denotes a finite-dimensional complex simple Lie algebra of type A, D or E.

A The category $\mathcal{O}_\kappa$ Let $\hat{\mathfrak{g}}$ be the affine Lie algebra defined in Subsection 16.2A: thus, $\hat{\mathfrak{g}}$ is a central extension

$$0 \to \mathbb{C}.c \to \hat{\mathfrak{g}} \to \mathfrak{g}((z)) \to 0.$$

Let $\tilde{\mathfrak{g}}$ be the subalgebra of $\hat{\mathfrak{g}}$ given by the pre-image of $\mathfrak{g}[z, z^{-1}]$ in $\hat{\mathfrak{g}}$. Our aim is to define a quasitensor structure on a suitable category of representations of $\tilde{\mathfrak{g}}$. One difficulty which must be overcome is that the usual tensor product of two 'randomly selected' representations of $\tilde{\mathfrak{g}}$ will be 'too large'. Thus, our first step will be to select a category of 'small' (but still infinite-dimensional) representations of $\tilde{\mathfrak{g}}$.

DEFINITION 16.3.1 *For any $\kappa \in \mathbb{C}$, let $\mathcal{O}_\kappa$ be the full subcategory of the category of representations of $\tilde{\mathfrak{g}}$ whose objects V satisfy the following conditions:*

(i) the central element c acts as the scalar $\kappa - h$ on V, where h is the Coxeter number of $\mathfrak{g}$;
(ii) $\mathfrak{g}$ acts locally finitely on V (i.e. for any $v \in V$, $\dim(U(\mathfrak{g}).v) < \infty$);
(iii) $z\mathfrak{g}[[z]]$ acts locally nilpotently on V (i.e. if $v \in V$, $x \in z\mathfrak{g}[[z]]$, then $x^N.v = 0$ for sufficiently large N);
(iv) V is finitely generated as a $\tilde{\mathfrak{g}}$-module.

Example 16.3.2 Let $V(\lambda)$ be the irreducible finite-dimensional $\mathfrak{g}$-module with highest weight $\lambda \in P^+$. Since $\mathfrak{g} \cong \mathfrak{g}[[z]]/z\mathfrak{g}[[z]]$ as Lie algebras, we can regard $V(\lambda)$ as a $\mathfrak{g}[[z]]$-module on which $z\mathfrak{g}[[z]]$ acts trivially. Extend to a representation of $\mathfrak{g}[[z]] \oplus \mathbb{C}.c$ (direct sum of Lie algebras) by making c act as $(\kappa - h).\text{id}$, and define

$$W^\kappa(\lambda) = U(\tilde{\mathfrak{g}}) \underset{U(\mathfrak{g}[[z]] \oplus \mathbb{C}.c)}{\otimes} V(\lambda).$$

It is not difficult to see that $W^\kappa(\lambda) \in \mathcal{O}_\kappa$ for all $\lambda \in P^+$. In fact, the only condition in 16.3.1 which is not completely obvious is (iii), which follows from the fact that every element of $W^\kappa(\lambda)$ can be written as $y \otimes v_\lambda$, where $y \in U(\tilde{\mathfrak{g}})$ and v_λ is a highest weight vector of $V(\lambda)$, together with the observation that, for any $x \in z\mathfrak{g}[[z]]$, the N-fold iterated bracket

$$[x, [x, [x, \ldots, [x, y] \cdots]]] \in z\mathfrak{g}[[z]]$$

for sufficiently large N. It is not difficult to show further that $W^\kappa(\lambda)$ has a unique irreducible quotient module $V^\kappa(\lambda)$, and that every irreducible object of $\mathcal{O}_\kappa$ is isomorphic to some $V^\kappa(\lambda)$. $\Diamond$

The properties of $\mathcal{O}_\kappa$ depend strongly on whether κ is a (positive) rational number:

PROPOSITION 16.3.3 *(i) If $\kappa \notin \mathbb{Q}$, then $\mathcal{O}_\kappa$ is semisimple (i.e. every object of $\mathcal{O}_\kappa$ is completely reducible);*
(ii) if $\kappa \notin \mathbb{Q}_{\geq 0}$, then every object of $\mathcal{O}_\kappa$ has a composition series of finite length. ∎

For the existence of the kind of tensor product on $\mathcal{O}_\kappa$ we are looking for, it turns out to be crucial that modules in $\mathcal{O}_\kappa$ have finite length. Thus, *for the remainder of this section, we assume that $\kappa \notin \mathbb{Q}_{\geq 0}$.* On the other hand, the case $\kappa \in \mathbb{Q}_{<0}$ will be of particular interest to us.

If V is any $\tilde{\mathfrak{g}}$-module and $N \in \mathbb{N}$, let $V(N)$ be the subspace of V consisting of those vectors which are annihilated by all products of N elements of $z\mathfrak{g} \subset \tilde{\mathfrak{g}}$. Obviously, $V(1) \subset V(2) \subset \cdots$; we define the set of *smooth vectors in V* to be

$$V(\infty) = \bigcup_{N=1}^{\infty} V(N),$$

and say that V is *smooth* if $V = V(\infty)$.

PROPOSITION 16.3.4 *A $\tilde{\mathfrak{g}}$-module V is in $\mathcal{O}_\kappa$ if and only if V is smooth and $V(1)$ is finite dimensional.* ∎

One of the reasons for the importance of smoothness is that any smooth $\tilde{\mathfrak{g}}$-module V is naturally a $\hat{\mathfrak{g}}$-module. In fact, let $f \in \mathbb{C}((z))$ and let $v \in V(N)$. Choose $g \in \mathbb{C}[z, z^{-1}]$ such that $f - g \in z^N \mathbb{C}[[z]]$ and set $(fx).v = (gx).v$ for all $x \in \mathfrak{g}$. This is well defined, for, if $h \in z^N \mathbb{C}[[z]]$, then hx is a linear combination of monomials, each of which is a product of N elements of $z\mathfrak{g}$. (This is proved by induction on N: for the inductive step, one has to show that, for any $x \in \mathfrak{g}$, $zx \in [z\mathfrak{g}, \mathfrak{g}]$, and this is an immediate consequence of the simplicity of $\mathfrak{g}$.)

B The tensor product We shall now define duality and tensor product operations on $\mathcal{O}_\kappa$, beginning with the duality, which is the simpler of the two.

The usual dual of an object V in $\mathcal{O}_\kappa$ is not, of course, in $\mathcal{O}_\kappa$, since it has the wrong level, namely $-(\kappa - h)$, and because it is not smooth, in general. To overcome the first difficulty, let V^ω be the result of twisting V with the automorphism ω of $\tilde{\mathfrak{g}}$ given by

$$\omega(z^k x) = (-z)^{-k} x, \quad \omega(c) = -c,$$

for $x \in \mathfrak{g}$, $k \in \mathbb{Z}$. Then, V^ω has level $-(\kappa - h)$, and we define

$$V^* = \mathrm{Hom}_{\mathbb{C}}(V^\omega, \mathbb{C})(\infty)$$

to be the smooth part of the vector space dual of V^ω.

Example 16.3.5 If $\lambda \in P^+$ and $V^\kappa(\lambda)$ is the irreducible object of $\mathcal{O}_\kappa$ defined in 16.3.2, then

$$V^\kappa(\lambda)^* \cong V^\kappa(-w_0(\lambda)). \quad \diamondsuit$$

PROPOSITION 16.3.6 *(i) If V is in $\mathcal{O}_\kappa$, so is V^*;*
(ii) $V \mapsto V^*$ is an exact contravariant functor $\mathcal{O}_\kappa \to \mathcal{O}_\kappa$;
(iii) the natural map $V \to V^{**}$, which associates to a vector $v \in V$ the linear functional on V^* given by evaluation at v, is an isomorphism of $\tilde{\mathfrak{g}}$-modules. ∎

We now give an abstract characterization of the duality operation we have defined.

Let $\hat{\mathfrak{g}}^{(2)}$ be the quotient of the direct sum of Lie algebras $\hat{\mathfrak{g}} \oplus \hat{\mathfrak{g}}$ by the one-dimensional central ideal spanned by $(c, -c)$. Let $\Gamma^{(2)}$ be the image of the embedding $\mathfrak{g}[z, z^{-1}] \hookrightarrow \mathfrak{g}((z)) \oplus \mathfrak{g}((z))$ given by

$$z^k x \mapsto (z^k x, z^{-k} x), \qquad (x \in \mathfrak{g},\ k \in \mathbb{Z}).$$

If $f, g \in \mathbb{C}[z, z^{-1}]$, the differential 1-form $f\,dg$ can have poles only at 0 and ∞, so by the residue theorem,

$$\mathrm{res}_0(f\,dg) + \mathrm{res}_\infty(f\,dg) = 0.$$

Thus, the cocycle defining the central extension $\hat{\mathfrak{g}}^{(2)}$ vanishes on $\Gamma^{(2)} \times \Gamma^{(2)}$, so that $\Gamma^{(2)}$ can be regarded naturally as a subalgebra of $\hat{\mathfrak{g}}^{(2)}$.

If V and W are in $\mathcal{O}_\kappa$, the usual tensor product $W \otimes V$ is naturally a module for $\hat{\mathfrak{g}} \oplus \hat{\mathfrak{g}}$, and, since the difference of the levels of V and W is zero, this action factors through an action of $\hat{\mathfrak{g}}^{(2)}$. Hence, $W \otimes V$ may be regarded as a $\Gamma^{(2)}$-module. Fix V in $\mathcal{O}_\kappa$ and consider the functor $\mathcal{D}_V : \mathcal{O}_\kappa \to \mathbf{vec}$ from $\mathcal{O}_\kappa$ to the category of vector spaces over $\mathbb{C}$ defined by

$$\mathcal{D}_V(W) = (W \otimes V)^{\Gamma^{(2)}},$$

the right-hand side denoting the largest subspace of $W \otimes V$ on which $\Gamma^{(2)}$ acts trivially.

PROPOSITION 16.3.7 *The functor $\mathcal{D}_V$ is represented by V^*, i.e. there are natural isomorphisms*

$$(W \otimes V)^{\Gamma^{(2)}} \cong \mathrm{Hom}_{\mathcal{O}_\kappa}(W, V^*)$$

for all objects W of $\mathcal{O}_\kappa$. ∎

Remark It follows from this result that the space of invariants $(W \otimes V)^{\Gamma^{(2)}}$ is finite dimensional, for one can show that, for any objects X and Y in $\mathcal{O}_\kappa$, $\mathrm{Hom}_{\mathcal{O}_\kappa}(X, Y)$ is finite-dimensional. $\triangle$

We now turn to the definition of the tensor product on $\mathcal{O}_\kappa$. The usual tensor product fails to preserve $\mathcal{O}_\kappa$ for the same reasons as the usual dual: for example, the level of the usual tensor product of two objects of $\mathcal{O}_\kappa$ is 2κ. We begin by giving an abstract characterization of the desired tensor product analogous to that given for the dual in 16.3.7.

Let $\hat{\mathfrak{g}}^{(3)}$ be the quotient of the direct sum of Lie algebras $\hat{\mathfrak{g}}^{\oplus 3}$ by the two-dimensional central ideal $\{(\alpha_1 c, \alpha_2 c, \alpha_3 c) \mid \alpha_1 + \alpha_2 + \alpha_3 = 0\}$. If R is the algebra of regular functions on $\mathbb{P}^1 \backslash \{0, 1, \infty\}$, there are three obvious embeddings $R \hookrightarrow \mathbb{C}((z))$ given by associating to a function $f \in R$ its power series expansions in the local parameters z at 0, $z - 1$ at 1 and $1/z$ at ∞. Let $\Gamma^{(3)}$ be the image of the resulting embedding $\mathfrak{g} \otimes R \hookrightarrow \mathfrak{g}((z))^{\otimes 3}$. If $f, g \in R$, the sum of the residues of $f \, dg$ at 0, 1 and ∞ is zero, so that $\Gamma^{(3)}$ can be regarded naturally as a subalgebra of $\hat{\mathfrak{g}}^{(3)}$.

If U, V and W are any three objects in $\mathcal{O}_\kappa$, then $\hat{\mathfrak{g}}^{(3)}$ acts naturally on $U \otimes V \otimes W$, so this tensor product can be regarded as a $\Gamma^{(3)}$-module. For fixed U, V in $\mathcal{O}_\kappa$, define a functor $\mathcal{T}_{U,V} : \mathcal{O}_\kappa \to \mathbf{vec}$ by

$$\mathcal{T}_{U,V}(W) = (U \otimes V \otimes W)/\Gamma^{(3)}.(U \otimes V \otimes W).$$

DEFINITION–PROPOSITION 16.3.8 *For every pair of objects U and V in $\mathcal{O}_\kappa$, there exists an object $U \overset{\bullet}{\otimes} V$ in $\mathcal{O}_\kappa$, unique up to isomorphism, such that $\mathrm{Hom}_\mathbb{C}(\mathcal{T}_{U,V}(W), \mathbb{C})$ is naturally isomorphic to $\mathrm{Hom}_{\mathcal{O}_\kappa}(U \overset{\bullet}{\otimes} V, W^*)$ for all objects W in $\mathcal{O}_\kappa$.* ∎

To describe $U \overset{\bullet}{\otimes} V$ concretely, let $\hat{\mathfrak{g}}^R$ be the pull-back of the extension $\hat{\mathfrak{g}}$ by the homomorphism $\mathfrak{g} \otimes R \to \mathfrak{g}((z))$ induced by the embedding $R \hookrightarrow \mathbb{C}((z))$ which associates to a regular function its expansion at 1. Associating to such a function its expansions at 0 and ∞, and mapping $c \mapsto -c$, defines a homomorphism of Lie algebras $\hat{\mathfrak{g}}^R \to \hat{\mathfrak{g}}^{(2)}$ (that this is a homomorphism follows once again from the fact that the sum of the residues of a differential 1-form regular on $\mathbb{P}^1 \backslash \{0, 1, \infty\}$ is zero). The usual tensor product $W = U \otimes V$ is naturally a $\hat{\mathfrak{g}}^{(2)}$-module, and hence also a $\hat{\mathfrak{g}}^R$-module; note that the level of the latter module is $-(\kappa - h)$. Let R_1 be the subspace of R consisting of the functions which vanish at 1 (and, in particular, are regular there). For any $k \in \mathbb{N}$, let W_k be the subspace of W spanned by the elements $(f_1 x_1) \dots (f_k x_k).w$ for $f_1, \dots, f_k \in R_1$, $x_1, \dots, x_k \in \mathfrak{g}$ and $w \in W$. Then,

$$W \supseteq W_1 \supseteq W_2 \supseteq \cdots,$$

and we define $\hat{W}$ to be the projective limit

$$\hat{W} = \varprojlim (W/W_k).$$

We now show that $\hat{W}$ has a natural $\hat{\mathfrak{g}}$-module structure. Note that an element of $\hat{W}$ is represented by a sequence $(w_1, w_2, \dots)$ of elements of W such that $w_{k+1} - w_k \in W_k$ for $k = 1, 2, \ldots$. Let $f \in \mathbb{C}((z))$, fix $r \geq 0$ such that $f \in z^{-r}\mathbb{C}[[z]]$, and choose elements $g_1, g_2, \ldots \in R$ such that, for $k = 1, 2, \ldots$, the expansion of g_k at 1 differs from f by an element of $(z-1)^k\mathbb{C}[[z]]$. It is not difficult to show that such elements exist, and that setting

$$(fx).(w_1, w_2, \ldots) = ((g_1 x).w_{r+1}, (g_2 x).w_{r+2}, \ldots),$$

gives a well-defined action of $\hat{\mathfrak{g}}$ on $\hat{W}$. We may therefore form the $\tilde{\mathfrak{g}}$-module $\hat{W}^\omega(\infty)$, which clearly has level $\kappa - h$.

PROPOSITION 16.3.9 *With the above notation, the $\tilde{\mathfrak{g}}$-module $\hat{W}^\omega(\infty)$ is in $\boldsymbol{\mathcal{O}}_\kappa$ and represents the functor $T_{U,V} : \boldsymbol{\mathcal{O}}_\kappa \to$ vec in the sense of 16.3.8. Thus,* $U \overset{\bullet}{\otimes} V \cong \hat{W}^\omega(\infty)$. ∎

Remark The most difficult point of the proof that $\hat{W}^\omega(\infty)$ is in $\boldsymbol{\mathcal{O}}_\kappa$ is to show that it is finitely generated. It is here that the assumption that $\kappa \notin \mathbb{Q}_{\geq 0}$ is crucial. △

We can now state

THEOREM 16.3.10 *With the duality operation $*$ and the tensor product $\overset{\bullet}{\otimes}$, $\boldsymbol{\mathcal{O}}_\kappa$ becomes a rigid, $\mathbb{C}$-linear, quasitensor category.* ∎

We cannot discuss the proof of this result here, but the reader who has come this far will not be surprised to learn that it involves (a generalization of) the Knizhnik–Zamolodchikov equation.

C The equivalence theorem We turn finally to the relation between quantum groups and the quasitensor category $\boldsymbol{\mathcal{O}}_\kappa$ defined in the last two subsections. Let $\epsilon(\kappa) = e^{-\pi\sqrt{-1}/\kappa}$, and recall the Hopf algebra $U^{\mathrm{res}}_{\epsilon(\kappa)}(\mathfrak{g})$ studied in Sections 9.3, 10.1 and 11.2. Let $\tilde{\boldsymbol{\mathcal{O}}}_\kappa$ be the category of finite-dimensional $U^{\mathrm{res}}_{\epsilon(\kappa)}(\mathfrak{g})$-modules V of type $\mathbf{1}$ which are the direct sums of their weight spaces. We have shown in 10.1.20 and 11.2.12 that the tensor product and duality operations induced by the Hopf structure of $U^{\mathrm{res}}_{\epsilon(\kappa)}(\mathfrak{g})$ make $\tilde{\boldsymbol{\mathcal{O}}}_\kappa$ into a quasitensor category.

THEOREM 16.3.11 *There is an equivalence of quasitensor categories $\boldsymbol{\mathcal{O}}_\kappa \cong \tilde{\boldsymbol{\mathcal{O}}}_\kappa$.* ∎

The construction of the equivalence is outlined in Kazhdan & Lusztig (1991).

This theorem is the confirmation of the connection, conjectured by Lusztig (1989), between the representation theories of affine Lie algebras and quantum groups. In fact, taking $\kappa \in \mathbb{Q}_{<0}$ in 16.3.11, one deduces

THEOREM 16.3.12 *Conjecture 11.2.14, on the characters of the irreducible representations $V_\epsilon^{res}(\lambda)$ of $U_\epsilon^{res}(\mathfrak{g})$ when $\lambda \in P^+$ and ϵ is a root of unity, is true.* ∎

A relationship between quantum groups and classical affine algebras has been widely predicted in the literature on conformal field theory – see especially Moore & Seiberg (1988). Physicists are interested mainly in unitary representations of affine Lie algebras, for which the level must be a non-negative integer. Unfortunately, such representations are excluded from our discussion of $\mathcal{O}_\kappa$, since we have always assumed that $\kappa \notin \mathbb{Q}_{\geq 0}$. However, there is an analogue of 16.3.11 which applies to such representations.

Let $\mathbf{int}_\kappa$ be the category of *integrable* $\tilde{\mathfrak{g}}$-modules of level $\kappa - h$ and of finite length: this forces $\kappa - h \in \mathbb{N}$. (A representation V of $\tilde{\mathfrak{g}}$ is integrable if it is the direct sum of its weight spaces and if the Chevalley generators $X_i^{\pm}$ of $\tilde{\mathfrak{g}}$ act locally nilpotently on V.) One can define a tensor product $\overset{\bullet}{\otimes}$ on $\mathbf{int}_\kappa$ in a manner analogous to that used for $\mathcal{O}_\kappa$. At first sight, this seems paradoxical, since, to define the tensor product on $\mathcal{O}_\kappa$, we had to assume that $\kappa \notin \mathbb{Q}_{\geq 0}$. But the finiteness property, which the assumption $\kappa \notin \mathbb{Q}_{\geq 0}$ guaranteed in the case of $\mathcal{O}_\kappa$, holds automatically in $\mathbf{int}_\kappa$. Georgiev & Mathieu (1993) made the following

CONJECTURE 16.3.13 *Provided that $\kappa \in \mathbb{N}$ is sufficiently large, $\mathbf{int}_\kappa$ and $\overline{\mathbf{tilt}}_\epsilon$ are equivalent as $\mathbb{C}$-linear quasitensor categories, where $\epsilon = e^{2\pi\sqrt{-1}/\kappa}$.* ∎

We understand that this has been proved by Finkelberg (1993).

Remarks [1] We saw in Subsection 11.3D that the category $\overline{\mathbf{tilt}}_\ell$ is semisimple. This is consistent with 16.3.13, for it follows from results of Deodhar, Gabber and Kac that $\mathbf{int}_\kappa$ is semisimple. On the other hand, neither of the categories $\mathcal{O}_\kappa$ and $\tilde{\mathcal{O}}_\kappa$ in 16.3.11 is semisimple when $-\kappa \in \mathbb{N}$.

[2] Also implicit in the literature of conformal field theory is a connection between quantum groups and the Virasoro algebra. Such a result has not been precisely formulated at the time of writing, but it is expected that there is an equivalence of quasitensor categories, analogous to that in 16.3.11, between a certain category of representations of the Virasoro algebra and the category of finite-dimensional representations of the tensor product of two copies of quantum $sl_2(\mathbb{C})$ at a root of unity. △

16.4 Quasi-Hopf algebras and Grothendieck's Esquisse

We conclude this chapter, and the book, by sketching a remarkable connection between quasi-Hopf algebras and algebraic number theory. The first serious results in this area were obtained in Drinfel'd (1991), although to some extent the connection was foreseen by A. Grothendieck in 1984 in his 'Esquisse d'un programme'.

A $\mathrm{Gal}(\overline{\mathbb{Q}}/\mathbb{Q})$ and pro-finite fundamental groups Let X be an algebraic variety defined over $\mathbb{Q}$, and, for any field k with $\mathbb{Q} \subseteq k \subseteq \mathbb{C}$, denote by $X(k)$ the set of k-rational points of X. We assume that $X(\mathbb{C})$ is connected in the complex topology. By the generalized Riemann existence theorem, all finite coverings of $X(\mathbb{C})$ are algebraic and defined over $\overline{\mathbb{Q}}$, the field of all algebraic numbers in $\mathbb{C}$ (see Hartshorne (1977), page 442, for example). Hence, if $\tau \in \mathrm{Gal}(\overline{\mathbb{Q}}/\mathbb{Q})$ is an automorphism of $\overline{\mathbb{Q}}$ which fixes $\mathbb{Q}$ pointwise, any finite Galois covering of $X(\mathbb{C})$ can be 'twisted' by τ to give another finite Galois covering of $X(\mathbb{C})$. This implies that, provided the basepoint $x_0 \in X(\mathbb{Q})$, $\mathrm{Gal}(\overline{\mathbb{Q}}/\mathbb{Q})$ acts as a group of automorphisms of the pro-finite completion

$$\hat{\pi}_1(X(\mathbb{C}), x_0) = \varprojlim \pi_1(X(\mathbb{C}), x_0)/S$$

of the fundamental group of $X(\mathbb{C})$ (in the projective limit, S runs through all the normal subgroups of $\pi_1(X(\mathbb{C}), x_0)$ of finite index).

To explain this in a little more detail, we must give another description of the pro-finite completion $\hat{\pi}_1(X(\mathbb{C}), x_0)$. Note that normal subgroups of $\pi_1(X(\mathbb{C}), x_0)$ of finite index correspond to finite Galois coverings $p : X(\mathbb{C})_p \to X(\mathbb{C})$, and that lifting a closed path γ in $X(\mathbb{C})$ based at x_0 to $X(\mathbb{C})_p$ gives a bijection $\gamma_p : p^{-1}(x_0) \to p^{-1}(x_0)$. These bijections are *compatible*, in the sense that, if p' is another finite Galois covering of $X(\mathbb{C})$ and $f : X(\mathbb{C})_p \to X(\mathbb{C})_{p'}$ is a covering map such that the induced map $X(\mathbb{C}) \to X(\mathbb{C})$ fixes x_0, then the following diagram commutes:

$$
\begin{array}{ccc}
p^{-1}(x_0) & \xrightarrow{\ f\ } & {p'}^{-1}(x_0) \\
\gamma_p \downarrow & & \downarrow \gamma_{p'} \\
p^{-1}(x_0) & \xrightarrow{\ f\ } & {p'}^{-1}(x_0)
\end{array}
$$

From these remarks, it follows that giving an element of $\hat{\pi}_1(X(\mathbb{C}), x_0)$ is the same thing as giving an assignment, to each covering p of $X(\mathbb{C})$, of a bijection $\gamma_p : p^{-1}(x_0) \to p^{-1}(x_0)$, these bijections being compatible in the above sense. The group operation on $\hat{\pi}_1(X(\mathbb{C}), x_0)$ corresponds to the obvious way of composing such bijections. An element $\tau \in \mathrm{Gal}(\overline{\mathbb{Q}}/\mathbb{Q})$ acts on a family of bijections $\{\gamma_p\}$ by

$$\tau.\{\gamma_p\} = \{\tau \circ \gamma_{\tau^{-1}p} \circ \tau^{-1}\}.$$

It is clear that this defines an action of $\mathrm{Gal}(\overline{\mathbb{Q}}/\mathbb{Q})$ as a group of automorphisms of $\hat{\pi}_1(X(\mathbb{C}), x_0)$.

Since $\hat{\pi}_1(X(\mathbb{C}), x_0)$ is independent of x_0 up to inner automorphisms, the homomorphisms $\varphi_{X,x_0} : \mathrm{Gal}(\overline{\mathbb{Q}}/\mathbb{Q}) \to \mathrm{Aut}(\hat{\pi}_1(X(\mathbb{C}), x_0))$ we have constructed induce a homomorphism

$$\varphi_X : \mathrm{Gal}(\overline{\mathbb{Q}}/\mathbb{Q}) \to \mathrm{Out}(\hat{\pi}_1(X(\mathbb{C})))$$

independent of a choice of basepoint ($\mathrm{Out}(\hat{\pi}_1)$ is the group of outer automorphisms of $\hat{\pi}_1$, i.e. the quotient of the group of automorphisms by the group of inner automorphisms).

According to Grothendieck (1984), the most important varieties X to consider are the configuration spaces

$$X_m = \mathcal{D}_m(\mathbb{P}^1)/PGL_2(\mathbb{C}),$$

where $PGL_2(\mathbb{C})$ acts diagonally on $(\mathbb{P}^1)^m$. The fundamental group $\pi_1(X_m(\mathbb{C}))$ is obviously isomorphic to the quotient of the pure sphere braid group $\mathcal{SP}_m$ by its centre ($= \pi_1(PGL_2(\mathbb{C})) \cong \mathbb{Z}_2$). In particular, since $PGL_2(\mathbb{C})$ acts triply transitively on $\mathbb{P}^1$, $X_4 \cong \mathbb{P}^1 \backslash \{0, 1, \infty\}$; thus, $\pi_1(X_4(\mathbb{C}))$ is the free group F_2 on two generators X and Y, representing loops which go once round 0 and 1, respectively, but do not enclose any other point of $\{0, 1, \infty\}$.

Belyi (1987) proved that the homomorphism

$$\varphi_{X_4} : \mathrm{Gal}(\overline{\mathbb{Q}}/\mathbb{Q}) \to \mathrm{Out}(\hat{\pi}_1(X_4(\mathbb{C})))$$

is *injective*; hence, the homomorphisms φ_{X_4, x_0} are also injective. Although the precise image of φ_{X_4, x_0} is not known, it was proved by Ihara (1988) that it is contained in a certain remarkable subgroup of $\mathrm{Aut}(\hat{F}_2)$, which we shall now describe.

Following Drinfel'd (1991), we define the *Grothendieck–Teichmüller group* $\widehat{GT}$ to be the set of automorphisms of $\hat{F}_2$ of the form

$$X \mapsto X^m, \quad Y \mapsto f^{-1}.Y^m.f,$$

where f lies in the commutator subgroup $(\hat{F}_2, \hat{F}_2)$ of $\hat{F}_2$, and m lies in the pro-finite completion

$$\hat{\mathbb{Z}} = \varprojlim \mathbb{Z}/N\mathbb{Z}$$

of the integers, and where m and f satisfy conditions (i), (ii) and (iii) below:

(i) $f(X, Y)f(Y, X) = 1$;

(ii) $f(X_3, X_1)X_3^m f(X_2, X_3)X_2^m f(X_1, X_2)X_1^m = 1$ whenever $X_1 X_2 X_3 = 1$ in $\hat{F}_2$;

(iii) the relation

$$f(P_{12}, P_{23}P_{24})f(P_{13}P_{23}, P_{34}) = f(P_{23}, P_{34})f(P_{12}P_{13}, P_{24}P_{34})f(P_{12}, P_{23})$$

holds in the pro-finite completion of the pure braid group $\mathcal{P}_4$, where the P_{ij} are the generators of $\mathcal{P}_4$ defined in Subsection 16.2A.

A little surprisingly, $\widehat{GT}$ is actually a subgroup of $\mathrm{Aut}(\hat{F}_2)$.

Since $\mathrm{Gal}(\overline{\mathbb{Q}}/\mathbb{Q})$ is a subgroup of $\widehat{GT}$, one would like to understand more about $\widehat{GT}$ itself. In the next subsection, we describe Drinfel'd's insight, that $\widehat{GT}$ is essentially the 'universal symmetry group' of quasitriangular quasi-Hopf algebras.

B The Grothendieck–Teichmüller group and quasitriangular quasi-Hopf algebras Suppose that **C** is a quasitensor category. We saw in Subsection 5.1D that, if V_1, V_2 and V_3 are objects of **C**, then every element $\beta \in \mathcal{B}_3$ defines an isomorphism $(V_1 \otimes V_2) \otimes V_3 \to (V_{\sigma(1)} \otimes V_{\sigma(2)}) \otimes V_{\sigma(3)}$, where $\sigma \in \Sigma_3$ is the permutation associated to β^{-1}. In particular, $\mathcal{P}_3$ acts on $(V_1 \otimes V_2) \otimes V_3$. Composing the given associativity isomorphisms with the action of an element of $\mathcal{P}_3$ therefore gives rise to new candidates for the associativity isomorphisms. Similarly, composing the given commutativity isomorphisms with the action of elements of $\mathcal{P}_2$ on tensor products $V_1 \otimes V_2$ gives new candidates for the commutativity isomorphisms. Of course, in order that these new commutativity and associativity isomorphisms should define a quasitensor category, some restrictions must be satisfied.

From Subsection 16.2A, we know that every element of $\mathcal{P}_2$ is of the form T^{2m}, where T is the generator of $\mathcal{B}_2$, and that every element of $\mathcal{P}_3$ is of the form $f(T_1^2, T_2^2)(T_1 T_2)^{3n}$, where $f(X, Y)$ is an element of the free group F_2 with generators X and Y. The commutativity of the hexagon diagrams in 5.2.1 forces $n = 0$ and the relations (i) and (ii) in Subsection 16.4A (but now the X_i are in $\mathcal{P}_3$), while the commutativity of the pentagon diagram in 5.1.1 forces condition (iii) (in $\mathcal{P}_4$ rather than its pro-finite completion). Twisting a quasitensor category by a pair (λ, f), where $\lambda = 1 + 2m$, and then by (λ', f') is the same as twisting by (λ'', f''), where

$$(27) \qquad \lambda'' = \lambda\lambda', \quad f''(X, Y) = f'(f(X, Y)X^\lambda f(X, Y)^{-1}, Y^\lambda).f(X, Y),$$

$(x, y) = xyx^{-1}y^{-1}$ denoting the commutator of two group elements x and y. This makes the set of pairs (λ, f) satisfying the above conditions into a semigroup, which we denote by $\underline{GT}$.

If A is a quasitriangular quasi-bialgebra over a field k of characteristic zero, then, by 16.1.4, the category $\mathbf{rep}_A$ is quasitensor, and we may 'twist' it with pairs $(\lambda, f) \in \underline{GT}$ by using the preceding construction. This is equivalent to changing the universal R-matrix $\mathcal{R}$ and the coassociator Φ of A to $(\lambda, f).\mathcal{R}$

and $(\lambda, f).\Phi$, where

$$(\lambda, f).\mathcal{R} = \mathcal{R}.(\mathcal{R}_{21}\mathcal{R})^m = (\mathcal{R}.\mathcal{R}_{21})^m.\mathcal{R}, \qquad m = (\lambda - 1)/2,$$

$$(28) \qquad (\lambda, f).\Phi = f(\mathcal{R}_{21}\mathcal{R}_{12}, \Phi.\mathcal{R}_{32}\mathcal{R}_{23}\Phi^{-1})^{-1}.\Phi$$

$$= \Phi.f(\Phi^{-1}\mathcal{R}_{21}\mathcal{R}_{12}\Phi, \mathcal{R}_{32}\mathcal{R}_{23})^{-1}.$$

It is time to admit that, unfortunately, conditions (i)–(iii) (in P_3 and P_4) have only two solutions, namely $m = \pm 1$, $f = 1$ (this is not difficult to see)! However, suppose that we replace A by a deformation A_h of A as a quasi-bialgebra; we assume that the universal R-matrix $\mathcal{R}$ of A is $\equiv 1 \otimes 1 \pmod{h}$. By the remarks at the beginning of Subsection 16.1C, A_h is then a quasitriangular quasi-Hopf algebra. Then, the right-hand sides of the formulas in (28) make sense for any $m \in k$ and any $f(X, Y)$ in the pro-nilpotent completion F_2^{nil} of F_2. In fact, since $\mathcal{R} \equiv 1 \otimes 1 \pmod{h}$, $\ln(\mathcal{R}_{21}.\mathcal{R})$ is a well-defined element of the completed tensor product $A_h \otimes A_h$, as is $(\mathcal{R}_{21}.\mathcal{R})^m = \exp(m\ln(\mathcal{R}_{21}.\mathcal{R}))$. Moreover, F_2^{nil} consists of expressions of the form $\exp(\mathcal{F}(\ln X, \ln Y))$, where $\mathcal{F}$ is a formal Lie series over k (see Bourbaki (1972), Chap. II, § 6, no. 3). Since $\Phi_h \equiv 1 \otimes 1 \otimes 1 \pmod{h}$, the right-hand sides of the second equation in (28) also make sense. Finally, conditions (i)–(iii) are also meaningful if we work in the appropriate nilpotent completions.

We therefore have an action, on the category of quasitriangular quasi-Hopf algebras over $k[[h]]$ with the above properties, of the semigroup $\underline{GT}(k)$ of pairs (λ, f) satisfying conditions (i)–(iii), where $\lambda \in k$ and $f \in F_2^{\mathrm{nil}}$. The invertible elements in $\underline{GT}(k)$ are precisely those for which $\lambda \neq 0$, and form what Drinfel'd (1991) calls the k-pro-unipotent version $GT(k)$ of the Grothendieck-Teichmüller group. This group is highly non-trivial: for example, it is known that the homomorphism $GT(k) \to k^\times$ given by $(\lambda, f) \mapsto \lambda$ is surjective.

Remark The twisting of quasi-Hopf algebras over $k[[h]]$ by elements of $GT(k)$ we have defined should not be confused with the twists introduced in Subsection 16.1A. The two are quite separate: in fact, they commute with each other. $\triangle$

Bibliographical notes

16.1 The results of this section are taken from Drinfel'd (1990b).

16.2 The main sources for this section are Drinfel'd (1989, 1990b, 1991) and the papers of T. Kohno in the references. The KZ equation was introduced in Knizhnik & Zamolodchikov (1984); see also Belavin, Polyakov & Zamolodchikov (1984) and Tsuchiya & Kanie (1988). For more information on the relation between the KZ equation and the topology of complements of hyperplanes, see Schechtman & Varchenko (1991a,b) and Schechtman (1992a,b). The solutions of the KZ equation can be expressed in terms of (generalized) hypergeometric functions – see Schechtman & Varchenko (1990). For further information on monodromy representations, see Cherednik (1991a). Frenkel & Reshetikhin (1992a) define a q-analogue of the KZ equation, show that its solutions are given in terms of q-hypergeometric functions (see Chapter 13), and study its relation to quantum affine algebras (see Section 12.3).

16.3 Theorems 16.3.11 and 16.3.12 were announced in Kazhdan & Lusztig (1991), and the details of this work have appeared in Kazhdan & Lusztig (1993a,b, 1994a,b) and Lusztig (1994). Conjecture 16.3.13 is the subject of the thesis of Finkelberg (1993).

16.4 The literature on Grothendieck's Esquisse is extensive. As well as Grothendieck (1984), one should consult Deligne (1989) and Ihara (1991), and the references there. The relation to quasi-Hopf algebras is given in Drinfel'd (1991).

Appendix
Kac–Moody algebras

In this Appendix, we summarize the results and notation relating to Kac-Moody algebras used throughout the book.

A 1 Generalized Cartan matrices A generalized Cartan matrix is a square matrix $A = (a_{ij})_{i,j=1,\ldots,n}$ of integers such that, for all i, j,

$$a_{ii} = 2, \quad a_{ij} \leq 0 \text{ if } i \neq j, \quad a_{ij} = 0 \text{ if and only if } a_{ji} = 0.$$

A is symmetrizable if there exist coprime positive integers $d_1,\ldots,d_n$ such that the matrix $(d_i a_{ij})$ is symmetric. The d_i are then uniquely determined.

A is decomposable if, by applying a permutation to the rows of A and the same permutation to the columns, A can be put in the form

$$\begin{pmatrix} B & 0 \\ 0 & C \end{pmatrix},$$

where B and C are non-zero square matrices.

The Dynkin diagram of A is the graph with vertices (or nodes) labelled by $\{1,\ldots,n\}$, nodes i and j being joined by $a_{ij}a_{ji}$ edges (or bonds) if $i \neq j$, and carrying an arrow pointing towards node i if $|a_{ij}| > |a_{ji}|$. A is determined by its Dynkin diagram if $a_{ij}a_{ji} \leq 4$ for all i, j.

A symmetrizable generalized Cartan matrix A is of finite (resp. affine) type if and only if the matrix $(d_i a_{ij})$ is positive-definite (resp. positive semi-definite of rank $n-1$) ; in the former case, A is called a Cartan matrix.

Note In the remainder of the Appendix, A denotes an indecomposable symmetrizable generalized Cartan matrix.

A 2 Kac–Moody algebras Denote by $\mathfrak{g}'(A)$ the Lie algebra over $\mathbb{C}$ (or any field of characteristic zero) with generators H_i, $X_i^{\pm}$, $i = 1,\ldots,n$, and defining relations

$$[H_i, H_j] = 0, \quad [H_i, X_j^{\pm}] = a_{ij}X_j^{\pm}, \quad [X_i^{+}, X_j^{-}] = \delta_{i,j}H_i,$$
$$(\mathrm{ad}_{X_i^{\pm}})^{1-a_{ij}}(X_j^{\pm}) = 0, \quad i \neq j.$$

The $X_i^{\pm}$ are called the Chevalley generators of $\mathfrak{g}'(A)$. The subspace $\mathfrak{g}^{(i)}$ of $\mathfrak{g}'(A)$ spanned by H_i, X_i^{+} and X_i^{-} is a subalgebra isomorphic to $sl_2(\mathbb{C})$.

Let $\mathfrak{h}'$ be the linear span of the H_i. Choose a vector space $\mathfrak{h}''$ of dimension $n - r$, $r = \text{rank}(A)$, with basis $\{D_{r+1}, \ldots, D_n\}$. Assume, by applying a permutation to the rows and the same permutation to the columns of A, if necessary, that the first r rows of A are linearly independent. Let $\mathfrak{g}(A)$ be the Lie algebra with generators H_i, $X_i^{\pm}$, $i = 1, \ldots, n$, and D_i, $i = r + 1, \ldots, n$, and with defining relations those of $\mathfrak{g}'(A)$ together with

$$[D_i, D_j] = 0, \quad [D_i, H_j] = 0, \quad [D_i, X_j^{\pm}] = \pm \delta_{i,j} X_j^{\pm}.$$

The direct sum $\mathfrak{h} = \mathfrak{h}' \oplus \mathfrak{h}''$ is called the Cartan subalgebra of $\mathfrak{g}(A)$.

For any A, $\mathfrak{g}'(A) = [\mathfrak{g}(A), \mathfrak{g}(A)]$. We have $\mathfrak{g}(A) = \mathfrak{g}'(A)$ if and only if $\det(A) \neq 0$ if and only if $\mathfrak{g}(A)$ is simple; in particular, if A is of finite type, then $\mathfrak{g}'(A) = \mathfrak{g}(A)$ is simple. Further, $\dim(\mathfrak{g}'(A)) < \infty$ if and only if $\dim(\mathfrak{g}(A)) < \infty$ if and only if A is of finite type.

Note We denote $\mathfrak{g}(A)$ (resp. $\mathfrak{g}'(A)$) simply by $\mathfrak{g}$ (resp. $\mathfrak{g}'$) from now on.

A 3 The invariant bilinear form There exists a unique non-degenerate invariant symmetric bilinear form $(\, , \,) : \mathfrak{g} \times \mathfrak{g} \to \mathbb{C}$ such that, for all i, j,

$$(H_i, H_j) = d_j^{-1} a_{ij}, \quad (D_i, D_j) = 0, \quad (H_i, D_j) = d_i^{-1} \delta_{i,j}, \quad (D_i, X_j^{\pm}) = 0,$$
$$(H_i, X_j^{\pm}) = 0, \quad (X_j^{\pm}, X_j^{\pm}) = 0, \quad (X_i^+, X_j^-) = d_i^{-1} \delta_{i,j}.$$

The restriction of $(\, , \,)$ to $\mathfrak{g}'$ is non-degenerate if and only if $\det(A) \neq 0$. Its restriction to $\mathfrak{h}$ is always non-degenerate.

A 4 Roots The simple roots of $\mathfrak{g}$ are the linear functionals $\alpha_i : \mathfrak{h} \to \mathbb{C}$, $i = 1, \ldots, n$, given by

$$\alpha_i(H_j) = a_{ji}, \quad \alpha_i(D_j) = \delta_{i,j}.$$

They are linearly independent.

The bilinear form $(\, , \,)$ induces an isomorphism of vector spaces $\mathfrak{h} \cong \mathfrak{h}^*$, under which $\alpha_i \in \mathfrak{h}^*$ corresponds to $d_i H_i \in \mathfrak{h}$. The induced bilinear form on $\mathfrak{h}^*$ is given by $(\alpha_i, \alpha_j) = d_i a_{ij}$.

Set $\Pi = \{\alpha_1, \ldots, \alpha_n\}$, and

$$Q = \bigoplus_{i=1}^{n} \mathbb{Z}.\alpha_i, \quad Q^+ = \bigoplus_{i=1}^{n} \mathbb{N}.\alpha_i.$$

If $\alpha = \sum_i k_i \alpha_i \in Q$, its height is $\text{height}(\alpha) = \sum_i k_i$. Let

$$P = \{\lambda \in \mathfrak{h}^* \mid \lambda(H_i) \in \mathbb{Z} \text{ for all } i\}, \quad P^+ = \{\lambda \in \mathfrak{h}^* \mid \lambda(H_i) \in \mathbb{N} \text{ for all } i\}.$$

We have $Q \subset P$. There is a natural partial order on P given by $\lambda \geq \mu$ if and only if $\lambda - \mu \in Q^+$.

If $\alpha \in Q$, define the root space

$$\mathfrak{g}_\alpha = \{x \in \mathfrak{g} \mid [h, x] = \alpha(h)x \text{ for all } h \in \mathfrak{h}\}.$$

Set $\Delta = \{\alpha \in Q \mid \alpha \neq 0,\ \mathfrak{g}_\alpha \neq 0\}$. Then, Δ is the set of roots of $\mathfrak{g}$, $\Delta^+ = \Delta \cap Q^+$ the set of positive roots, $\Delta^- = -\Delta^+$ the set of negative roots. We have $\Delta = \Delta^+ \coprod \Delta^-$ (disjoint union).

We have the following orthogonality relations between the root spaces:

$$(\mathfrak{g}_\alpha, \mathfrak{g}_\beta) = 0 \quad \text{if } \alpha + \beta = 0, \qquad (\mathfrak{g}_\alpha, \mathfrak{h}) = 0 \quad \text{if } \alpha \neq 0.$$

Let $\mathfrak{n}_\pm$ be the subalgebras of $\mathfrak{g}$ generated by the $X_i^\pm$. Then, we have, as direct sums of vector spaces,

$$\mathfrak{n}_\pm = \bigoplus_{\alpha \in \Delta^\pm} \mathfrak{g}_\alpha, \quad \mathfrak{h} = \mathfrak{g}_0, \quad \mathfrak{g} = \mathfrak{n}_- \oplus \mathfrak{h} \oplus \mathfrak{n}_+.$$

The subalgebras $\mathfrak{b}_\pm = \mathfrak{n}_\pm \oplus \mathfrak{h}$ are the (positive and negative) Borel subalgebras of $\mathfrak{g}$.

If A is of finite type, define $\lambda_i \in \mathfrak{h}^*$, $i = 1, \ldots, n$, by $\lambda_i(H_j) = \delta_{i,j}$. Then, the λ_i form a basis of P and

$$\alpha_j = \sum_{i=1}^n a_{ij}\lambda_i.$$

If ρ is half the sum of the positive roots of $\mathfrak{g}$, then $\rho = \sum_i \lambda_i$ and $(\rho, \alpha_i) = d_i$ for all i. There is a unique maximal root $\theta \in \Delta^+$ with respect to the partial order on Q. The dual Coxeter number of $\mathfrak{g}$ is $g = 2\frac{(\rho,\theta)}{(\theta,\theta)} + 1$.

A 5 The Weyl group Define linear maps $s_i : \mathfrak{h} \to \mathfrak{h}$, called the fundamental (or simple) reflections, by

$$s_i(h) = h - \alpha_i(h)H_i, \quad (h \in \mathfrak{h}).$$

The Weyl group W of $\mathfrak{g}$ is the subgroup of $GL(\mathfrak{h})$ generated by $s_1, \ldots, s_n$. The action of W preserves the bilinear form $(\ ,\ )$ on $\mathfrak{h}$.

As an abstract group, W is a Coxeter group with generators $s_1, \ldots, s_n$ and defining relations

$$s_i^2 = 1, \quad (s_i s_j)^{m_{ij}} = 1, \quad \text{if } i \neq j,$$

where the integers m_{ij} are given by

$a_{ij}a_{ji}$	0	1	2	3	≥ 4
m_{ij}	2	3	4	6	∞

(if $m_{ij} = \infty$, the relation $(s_i s_i)^{m_{ij}} = 1$ is omitted). The element $c = s_1 s_2 \ldots .s_n$ is called the Coxeter element: its order is the Coxeter number h of $\mathfrak{g}$.

An expression $w = s_{i_1} \ldots .s_{i_k}$ of an element $w \in W$ as a product of simple reflections is called a reduced expression (or a reduced decomposition) if k is the minimal number of simple reflections which appear in any such expression of w, and then k is called the length of w, and is denoted by $\ell(w)$. If w' is another element of W, we say that $w' \leq w$ if $w' = s_{j_1} s_{j_2} \ldots s_{j_\ell}$, where the sequence $j_1, j_2, \ldots , j_\ell$ is obtained from the sequence $i_1, i_2, \ldots , i_k$ by deleting some (possibly none) of its terms. Then, $\leq$ is a partial ordering on W, called the Bruhat ordering.

Identifying $\mathfrak{h}$ with $\mathfrak{h}^*$ using the bilinear form, we get an action of W on $\mathfrak{h}^*$ given by

$$s_i(\xi) = \xi - \xi(H_i)\alpha_i, \qquad (\xi \in \mathfrak{h}^*).$$

We have $W(\Delta) = \Delta$ and $\dim(\mathfrak{g}_\alpha) = \dim(\mathfrak{g}_{w(\alpha)})$ for all $w \in W$, $\alpha \in \Delta$.

A 6 Root vectors In this subsection we assume that A is of finite type. There is a unique element $w_0 \in W$ of maximal length; this length is $N = |\Delta^+|$; and we have $w_0^2 = 1$. If $w_0 = s_{i_1} \ldots .s_{i_N}$ is a reduced expression, then

$$\Delta^+ = \{\alpha_{i_1}, s_{i_1}(\alpha_{i_2}), \ldots, s_{i_1} \ldots s_{i_{N-1}}(\alpha_{i_N})\},$$

each positive root occurring exactly once on the right-hand side.

There are automorphisms $T_1, \ldots, T_n$ of $\mathfrak{g}$ such that

$$T_i(X_i^\pm) = -X_i^\mp, \quad T_i(H_j) = H_j - a_{ji}H_i,$$
$$T_i(X_j^+) = (-a_{ij})!^{-1}(\text{ad}_{X_i^+})^{-a_{ij}}(X_j^+), \quad \text{if } i \neq j,$$
$$T_i(X_j^+) = (-1)^{a_{ij}}(-a_{ij})!^{-1}(\text{ad}_{X_i^-})^{-a_{ij}}(X_j^-), \quad \text{if } i \neq j.$$

They satisfy the defining relations of the braid group $\mathcal{B}_\mathfrak{g}$:

$$T_i T_j = T_j T_i, \quad T_i T_j T_i = T_j T_i T_j, \quad (T_i T_j)^2 = (T_j T_i)^2, \quad (T_i T_j)^3 = (T_j T_i)^3,$$

according as $a_{ij} a_{ji} = 0$, 1, 2 or 3, respectively. Each T_i preserves $\mathfrak{h}$ and induces on it the action of the reflection s_i.

If $\beta = s_{i_1} s_{i_2} \ldots s_{i_{k-1}}(\alpha_{i_k}) \in \Delta^+$, where $w_0 = s_{i_1} \ldots s_{i_N}$ is a reduced expression, define

$$X_\beta^\pm = T_{i_1} T_{i_2} \ldots T_{i_{k-1}}(X_{i_k}^\pm).$$

Then, $X_\beta^\pm$ is a non-zero element of $\mathfrak{g}_{\pm\beta}$ for all $\beta \in \Delta^+$. The $X_\beta^\pm$ are independent of the choice of reduced expression of w_0, up to sign.

A 7 Affine Lie algebras Let $\mathfrak{g} = \mathfrak{g}(A)$, where A is of finite type. The extended Cartan matrix of $\mathfrak{g}$ is $\tilde{A} = (a_{ij})_{i,j=0,1,\ldots,n}$, where $A = (a_{ij})_{i,j=1,\ldots,n}$, $a_{00} = 2$ and

$$a_{0i} = -2\frac{(\theta, \alpha_i)}{(\alpha_i, \alpha_i)}, \qquad a_{i0} = -2\frac{(\theta, \alpha_i)}{(\theta, \theta)}$$

for $i = 1, \ldots, n$. Then, $\tilde{A}$ is a generalized Cartan matrix of affine type.

The Lie algebra $\mathfrak{g}'(\tilde{A})$ is the affine Lie algebra $\tilde{\mathfrak{g}}$, which is isomorphic to $\mathfrak{g}[z, z^{-1}] \oplus \mathbb{C}.c$ as a vector space with bracket

$$[Xz^k, Yz^l] = [X, Y]z^{k+l} + \delta_{k,-l}(X, Y)c, \quad (X, Y \in \mathfrak{g}, \ k, l \in \mathbb{Z}),$$

the element c being central. The Lie algebra $\mathfrak{g}(\tilde{A})$ is the semidirect product of $\tilde{\mathfrak{g}}$ with the one-dimensional algebra spanned by the derivation D given by $D(Xz^k) = kXz^k$, $D(c) = 0$.

If c acts by a scalar ℓ on a $\tilde{\mathfrak{g}}$-module V, one says that V is of level ℓ.

A 8 Highest weight modules If V is a $\mathfrak{g}$-module and $\lambda \in \mathfrak{h}^*$, the weight space

$$V_\lambda = \{v \in V \mid h.v = \lambda(h)v \text{ for all } h \in \mathfrak{h}\}.$$

Define the set of weights of V to be $P(V) = \{\lambda \in \mathfrak{h}^* \mid V_\lambda \neq 0\}$.

A $\mathfrak{g}$-module V has highest weight $\lambda \in \mathfrak{h}^*$ if there exists a vector $v_\lambda \in V_\lambda$ such that $V = U(\mathfrak{g}).v_\lambda$ and $\mathfrak{n}_+.v_\lambda = 0$. The Verma module $M(\lambda)$ is the quotient of $U(\mathfrak{g})$ by the left ideal generated by $\mathfrak{n}_+$ and the elements $h - \lambda(h).1$ for $h \in \mathfrak{h}$; it has highest weight λ and is free as a $U(\mathfrak{n}_-)$-module. Moreover, $M(\lambda)$ has a unique irreducible quotient module $V(\lambda)$; up to isomorphism, $V(\lambda)$ is the unique irreducible $\mathfrak{g}$-module of highest weight λ. If $\mu \in P(M(\lambda))$, then $\mu \leq \lambda$.

A $\mathfrak{g}$-module V is integrable if

$$V = \bigoplus_{\mu \in P(V)} V_\mu$$

and V is a sum of finite-dimensional $\mathfrak{g}^{(i)}$-modules for each i. If V is integrable, then $W(P(V)) = P(V)$. The module $V(\lambda)$ is integrable iff $\lambda \in P^+$. If A is of finite type, $V(\lambda)$ is integrable if and only if it is finite dimensional.

References

Abe, E. (1980) *Hopf Algebras*, Cambridge Tracts in Mathematics 74, Cambridge University Press, Cambridge

Abe, E. & Takeuchi, M. (1992) Groups associated with some types of infinite dimensional Lie algebras, *J. Algebra* **146**, 385-404

Abraham, R. & Marsden, J. E. (1978) *Foundations of Mechanics*, Benjamin, Reading, MA

Accardi, L., Schürmann, M. & von Waldenfels, W. (1988) Quantum independent increment processes on superalgebras, *Math. Z.* **198**, 451-77

Adler, M. (1979) On a trace functional for formal pseudo-differential operators and the symplectic structure of the KdV-type equations, *Invent. Math.* **50**, 219-48

Adler, M. & van Moerbecke, P. (1980) Completely integrable systems, euclidean Lie algebras, and curves, *Adv. Math.* **38**, 267-317

Agarwal, G. S. & Chaturvedi, S. (1992) Atomic transitions in q-deformed fields, *Modern Phys. Lett. A* **7**, 2407-14

Aghamohammadi, A., Karimipour, V. & Rouhari, S. (1993) The multiparametric nonstandard deformation of A_{n-1}, *J. Phys. A* **26**, L75-82

Aizawa, N. (1993a) $q \leftrightarrow q^{-1}$ invariance of q-oscillators and new realizations of quantum algebras, *J. Phys. A* **26**, 1115-22

Aizawa, N. (1993b) Tensor operators and Clebsch-Gordan coefficients for the quantum algebra $su_q(1,1)$, *J. Math. Phys.* **34**, 1937-63

Aizawa, N. & Sato, H. (1991) q-deformation of the Virasoro algebra with central extension, *Phys. Lett. B* **256**, 185-90

Akutsu, Y., Deguchi, T. & Wadati, M. (1989) The Yang-Baxter equation: a new tool for knot theory, in *Braid Group, Knot Theory and Statistical Mechanics*, C. N. Yang & M. L. Ge (eds), pp. 151-200, World Scientific, Teaneck, NJ

Alcalde, C. & Cadavid, A. C. (1992) Quantization, characters and the star-exponential, in *Topological and Geometrical Methods in Field Theory*, J. Mickelsson & O. Pekonen (eds), pp. 1-12, World Scientific, Singapore

Alcaraz, F. C., Köberle, R. & Lima-Santos, A. (1992) All exactly solvable $U(1)$-invariant quantum spin 1 chains from Hecke algebra, *Internat. J. Modern Phys. A* **7**, 7615—28

Aldaya, V., de Azcarraga, J. A., Bisquert, J. & Cervero, J. M. (1990) Dynamics on $SL(2,\mathbb{R})\tilde{\times}U(1)$, *J. Phys. A* **23**, 707-20

Aldaya, V., Loll, R. & Navarro-Salas, J. (1989) BRST supergroups and quantization, *Phys. Lett. B* **225**, 340-6

Aldaya, V. & Navarro-Salas, J. (1990) Quantization of the Virasoro group, *Comm. Math. Phys.* **126**, 575-95

Alekseev, A. Yu., Faddeev, L. D. & Semenov-Tian-Shansky, M. A. (1992a) Hidden quantum groups inside Kac-Moody algebras, in *Quantum Groups, Proceedings of Workshops held in the Euler International Mathematical Institute 1990*, P. P. Kulish (ed.), Lecture Notes in Mathematics 1510, pp. 148-58, Springer, Berlin

Alekseev, A. Yu., Faddeev, L. D. & Semenov-Tian-Shansky, M. A. (1992b) Hidden quantum groups inside Kac-Moody algebras, *Comm. Math. Phys.* **149**, 335-46

Alekseev, A. & Shatashvili, S. (1989) From geometric quantization to conformal field theory, in *Problems of Modern Quantum Field Theory, Alushta. 1989*, A. A. Belavin, A. U. Klimyk & A. B. Zamolodchikov (eds), Research Reports in Physics, pp. 22-42, Springer, Berlin

Alekseev, A. & Shatashvili, S. (1990) Quantum groups and WZNW models. *Comm. Math. Phys.* **133**, 353-68

Alishauskes, S. I. & Kulish, P. P. (1985) Spectral expansion of $SU(3)$ invariant solutions of the Yang-Baxter equation, *Zap. Nauchn. Sem. Leningrad Otdel. Mat. Inst. Steklov* **145** (Russian)

Altschuler, D. & Coste, A. (1992) Quasi-quantum groups, knots, three-manifolds and topological field theory, *Comm. Math. Phys.* **150**, 83–108

Alvarez-Gaumé, L. (1989) Quantum groups and conformal field theories, *Phil. Trans. Roy. Soc. London Ser. A* **329**, 343–7

Alvarez-Gaumé, L. (1990a) Quantum groups and conformal field theories, in *The Interface of Mathematics and Particle Physics, Oxford, 1988*, D. Quillen, G. B. Segal & S. T. Tsou (eds), Institute of Mathematics and its Applications Conference Series New Series 24, pp. 1–7, Oxford University Press, New York

Alvarez-Gaumé, L. (1990b) Quantum group approach to conformal field theory, *Nucl. Phys. B Proc. Suppl.* **16**, 571–3

Alvarez-Gaumé, L. & Gomez, C. (1989) Hidden quantum symmetries in rational conformal field theories, *Nucl. Phys. B* **319**, 155–86

Alvarez-Gaumé, L., Gomez, C. & Sierra, G. (1989) Quantum group interpretation of some conformal field theories, *Phys. Lett. B* **220**, 142–52

Alvarez-Gaumé, L., Gomez, C. & Sierra, G. (1990) Duality and quantum groups, *Nucl. Phys. B* **330**, 347–98

Aminou, R. (1987) Bigèbres de Lie, structures de Poisson et variantes de l'équation de Yang-Baxter, *Publ. IRMA Lille* **9**, no. 3

Aminou, R. & Kosmann-Schwarzbach, Y. (1988) Bigèbres de Lie, doubles et carrés, *Ann. Inst. Henri Poincaré Phys. Théor.* **49**, 461–78

Andersen, H. H. (1992a) Tensor products of quantized tilting modules, *Comm. Math. Phys.* **149**, 149–59

Andersen, H. H. (1992b) Finite dimensional representations of quantum groups, preprint no. 14, Matematisk Institut Aarhus Universitet

Andersen, H. H. (1992c) Quantum groups, invariants of 3-manifolds and semisimple tensor categories, preprint no. 15, Matematisk Institut Aarhus Universitet

Andersen, H. H., Jantzen, J. C. & Soergel, W. (1994) Representations of quantum groups at a p-th root of unity and of semisimple groups in characteristic p: independence of p, *Astérisque* **220**, Société Mathématique de France, Paris

Andersen, H. H., Polo, P. & Wen Kexin (1991) Representations of quantum algebras, *Invent. Math.* **104**, 1–59

Andersen, H. H., Polo, P. & Wen Kexin (1992) Injective modules for quantum algebras, *Amer. J. Math.* **114**, 571–604

Andersen, H. H. & Wen Kexin (1992) Representations of quantum algebras, the mixed case, *J. reine angew. Math.* **427**, 35–50

Andrews, G. E. (1976) *The Theory of Partitions*, Addison-Wesley, London

Andrews, G. E. (1986) *q-Series: Their Development and Application in Analysis, Number Theory, Physics and Computer Algebra*, CBMS Regional Conference Series in Mathematics 66, American Mathematical Society, Providence, RI

Andrews, G. E., Baxter, R. J. & Forrester, P. J. (1984) Eight vertex SOS model and generalized Rogers-Ramanujan-type identities, *J. Stat. Phys.* **35**, 193–266

Andruskiewitsch, A. (1992) Some exceptional compact matrix pseudogroups, *Bull. Soc. Math. France* **120**, 297–326

Andruskiewitsch, A. & Enriquez, B. (1992) Examples of compact matrix pseudogroups arising from the twisting operation, *Comm. Math. Phys.* **149**, 195–208

Aneva, B. (1991) A deformed quantum $SU(2)$ superalgebra, *J. Phys. A* **24**, L455–8

Aratyn, H., Nissimov, E. & Pacheva, S. (1992) Classical r-matrices and Poisson bracket structures on infinite-dimensional groups, *Phys. Lett. B* **284**, 273–82

Archer, F. J. (1992) The $U_q sl(3)$ 6-j symbols and state sum invariants, *Phys. Lett. B* **295**, 199–208

Aref'eva, I. Ya. & Volovich, I. V. (1991a) Quantum group chiral fields and differential Yang-Baxter equations, *Phys. Lett. B* **264**, 62–8

Aref'eva, I. Ya. & Volovich, I. V. (1991b) Quantum group gauge fields, *Modern Phys. Lett. A* **6**, 893–907

Aref'eva, I. Ya. & Volovich, I. V. (1991c) Quantum group particles and non-archimedean geometry, *Phys. Lett. B* **268**, 179–87

Arik, M. (1991) The q-difference operator, the quantum hyperplane, Hilbert spaces of analytic functions and q-oscillators, *Z. Phys. C* **51**, 627–32

Arik, M. & Coon, D. D. (1976) Hilbert spaces of analytic functions and generalized coherent states, *J. Math. Phys.* **17**, 524–7

Arnaudon, D. (1990) Periodic and flat irreducible representations of $SU(3)_q$, *Comm. Math. Phys.* **134**, 523–37

Arnaudon, D. (1991) Fusion rules and R-matrix for the composition of regular spins with semi-periodic representations of $SL(2)_q$, *Phys. Lett. B* **268**, 217–21

Arnaudon, D. (1992a) On periodic representations of quantum groups, in *Topological and Geometrical Methods in Field Theory*, J. Mickelsson & O. Pekonen (eds), pp. 13–21, World Scientific, Singapore

Arnaudon, D. (1992b) New fusion rules and R-matrices for $SL(N)_q$ at roots of unity, *Phys. Lett. B* **280**, 31–8

Arnaudon, D. & Chakrabarti, A. (1991a) Flat periodic representations of $U_q(\mathcal{G})$, *Comm. Math. Phys.* **139**, 605–17

Arnaudon, D. & Chakrabarti, A. (1991b) Periodic and partially periodic representations of $SU(N)_q$, *Comm. Math. Phys.* **139**, 461–78

Arnaudon, D. & Chakrabarti, A. (1991c) Periodic representations of $SO(5)_q$, *Phys. Lett. B* **262**, 68–70

Arnaudon, D. & Chakrabarti, A. (1991d) q-analogue of $IU(n)$ for q a root of unity, *Phys. Lett. B* **255**, 242–8

Arnaudon, D. & Rittenberg, V. (1993) Quantum chains with $U_q(sl(2))$ symmetry and unrestricted representations, *Phys. Lett. B* **306**, 86–90

Arnold, V. I. (1978) *Mathematical Methods of Classical Mechanics*, Graduate Texts in Mathematics 60, Springer, Berlin

Artin, E. (1947) Theory of braids, *Ann. of Math.* **48**, 101–26

Artin, M., Schelter, W. & Tate, J. (1991) Quantum deformations of GL_n, *Comm. Pure Appl. Math.* **44**, 879–95

Aschieri, P. & Castellani, L. (1992) Bicovariant differential geometry of the quantum group $GL_q(3)$, *Phys. Lett. B* **293**, 299–308

Aschieri, P. & Castellani, L. (1993) An introduction to noncommutative differential geometry on quantum groups, *Internat. J. Modern Phys. A* **8**, 1667–706

Atiyah, M. F. (1988) Topological quantum field theories, *Publ. Math. IHES* **68**, 175–86

Atiyah, M. F. (1989) The Jones-Witten invariants of knots, Séminaire Bourbaki exp. no. 715, *Astérisque* **189–190**, 7–16, Société Mathématique de France, Paris

Atiyah, M. F. (1991) Magnetic monopoles and the Yang-Baxter equation, *Internat. J. Modern Phys. A* **6**, 2761–74

Au-Yang, H., McCoy, B. M., Perk, J., Tang, S. & Yan, M. L. (1987) Commuting transfer matrices in the chiral Potts models: solutions of star-triangle equations for genus > 1, *Phys. Lett. A* **123**, 219–23

Au-Yang, H. & Perk, J. (1989) Onsager's star-triangle equation: master key to integrability, in *Integrable Systems in Quantum Field Theory and Statistical Mechanics*, M. Jimbo, T. Miwa & A. Tsuchiya (eds), Advanced Studies in Pure Mathematics 19, pp. 57–94, Academic Press, Boston

Au-Yang, H. & Perk, J. (1991) Star-triangle equations and multicomponent chiral Potts models, in *Current Problems in Statistical Mechanics, Washington, DC, 1991*, E. Domany & D. Jasnow (eds), pp. 139–45

Avan, J. (1990a) Current algebra realization of R-matrices associated to $\mathbb{Z}_2$-graded Lie algebras, *Phys. Lett. B* **252**, 230–6

Avan, J. (1990b) Graded Lie algebras in the Yang-Baxter equation, *Phys. Lett. B* **245**, 491–6

Avan, J. (1990c) Rational and trigonometric constant non-antisymmetric R-matrices, *Phys. Lett. B* **241**, 77–82

Avan, J. (1991) From rational to trigonometric R-matrices, *Phys. Lett. A* **156**, 61–8

Avan, J. & Tolan, M. (1991) Graded R-matrices for integrable systems, *Nucl. Phys. B* **352**, 215–49

Avancini, S. S. & Brunelli, J. C. (1993) q-deformed variational study of the Lipkin-Meshkov-Glick model via coherent states, *Phys. Lett. A* **174**, 358–62

Avancini, S. S. & Menezes, D. P. (1993) Generating function for Clebsch-Gordan coefficients of the $su_q(2)$ quantum algebra, *J. Phys. A* **26**, 1139–48

Baaj, S. (1992) Representation reguliere du groupe quantique $E_\mu(2)$ de Woronowicz, *C. R. Acad. Sci. Paris Sér. I* **314**, 1021–6

Babelon, O. (1984) Representations of the Yang-Baxter algebra associated to Toda field theory, *Nucl. Phys. B* **230**, 241–9

Babelon, O. (1988a) Jimbo's q-analogues and current algebras, *Lett. Math. Phys.* **15**, 111–7

Babelon, O. (1988b) Extended conformal algebra and the Yang-Baxter equation, *Phys. Lett. B* **215**, 523–9

Babelon, O. (1989) The role of the Yang-Baxter equation in conformal field theory, in *Knots, Topology and Quantum Field Theories, Florence, 1989*, L. Lusanna (ed.), pp. 447–68, World Scientific, River Edge, NJ

Babelon, O. (1990) From integrable to conformal field theory, *Nucl. Phys. B Proc. Suppl.* **18 A**, 1–22

Babelon, O. (1991) Universal exchange algebra for Bloch waves and Liouville theory, *Comm. Math. Phys.* **139**, 619–43

Babelon, O. (1992) Liouville theory on the lattice and universal exchange algebra for Bloch waves, in *Quantum Groups, Proceedings of Workshops held in the Euler International Mathematical Institute 1990*, P. P. Kulish (ed.), Lecture Notes in Mathematics 1510, pp. 159–75, Springer, Berlin

Babelon, O. & Bernard, D. (1991) Dressing transformations and the origin of quantum group symmetries, *Phys. Lett. B* **356**, 387–438

Babelon, O. & Bernard, D. (1992) Dressing symmetries, *Comm. Math. Phys.* **149**, 279–306

Babelon, O. & Bonora, L. (1991) Quantum Toda theory, *Phys. Lett. B* **253**, 365–72

Babelon, O. & Viallet, C. M. (1990) Hamiltonian structures and Lax equations, *Phys. Lett. B* **237**, 411–6

Bacry, H. (1993a) The problem of mass in the LNR quantum Poincaré algebra, *Phys. Lett. B* **306**, 41–3

Bacry, H. (1993b) Classical electrodynamics on a quantum Poincaré group, *Phys. Lett. B* **306**, 44–8

Baez, J. C. (1991) Differential calculi on quantum vector spaces with Hecke type relations, *Lett. Math. Phys.* **23**, 133–41

Baez, J. C. (1992) R-commutative geometry and quantization of Poisson algebras, *Adv. Math.* **95**, 61–91

Ballesteros, A., Gadella, M. & Del Olmo, M. A. (1992) Moyal quantization of $2+1$ dimensional Galilean systems, *J. Math. Phys.* **33**, 3379–86

Ballesteros, A. & Negro, J. (1992) A characterization of functional realizations of three-dimensional quantum groups, *J. Phys. A* **25**, 5945–62

Balog, J., Dabrowski, L. & Feher, L. (1990) Classical r-matrix and exchange algebra in WZNW and Toda theories, *Phys. Lett. B* **244**, 227–34

Balog, J., Dabrowski, L. & Feher, L. (1991) A new quantum deformation of $SL(3)$, *Phys. Lett. B* **257**, 74–8

Bantay, P. (1990) Orbifolds and Hopf algebras, *Phys. Lett. B* **245**, 477–9

Bantay, P. (1991) Orbifolds, Hopf algebras and moonshine, *Lett. Math. Phys.* **22**, 187–94

Barouch, E. (1984) Lax pair for the free-fermion eight-vertex model, *Stud. Appl. Math.* **70**, 151–62

Basu-Mallick, B. & Kundu, A. (1991) Single q-oscillator mode realization of the quantum group through canonical bosonisations of $SU_q(2)$ and q-oscillators, *Modern Phys. Lett. A* **6**, 701–5

Basu-Mallick, B. & Kundu, A. (1992a) Hidden quantum group structure in relativistic quantum integrable model, *Phys. Lett. B* **287**, 149–53

Basu-Mallick, B. & Kundu, A. (1992b) Spectral parameter-dependent approach to quantized algebra, its multiparameter deformations and their q-oscillator realizations, *J. Phys. A* **25**, 4147–56

Batchelor, M. (1991) Measuring coalgebras, quantum group-like objects and noncommutative geometry, in *Differential Geometric Methods in Theoretical Physics, Rapallo, 1990*, C. Bartocci, U. Bruzzo & R. Cianci (eds), Lecture Notes in Physics 375, pp. 47–60, Springer, Berlin

Batchelor, M. & Kuniba, A. (1991/2) Temperley-Lieb lattice models from quantum groups, *J. Phys. A* **24**, 2599–614; corrigendum, *ibid.* **25**, 1019

Batchelor, M., Mezincescu, L., Nepomechie, R. & Rittenberg, V. (1990) q-deformations of the $O(3)$ symmetric spin-1 Heisenberg chain, *J. Phys. A* **23**, L141–4

Baulieu, L. & Floratos, G. (1991) Path integral on the quantum plane, *Phys. Lett. B* **258**, 171–8

Baxter, R. J. (1972) Partition function of the eight-vertex lattice model, *Ann. Phys.* **70**, 193–228

Baxter, R. J. (1973) Eight vertex model in lattice statistics and one-dimensional anisotropic Heisenberg chain, *Ann. Phys.* **76**, 1–24; 25–47; 48–71

Baxter, R. J. (1982) *Exactly Solved Models in Statistical Mechanics*, Academic Press, New York

Baxter, R. J., Perk, J. H. H. & Au-Yang, H. (1988) New solutions of the star-triangle relations for the chiral Potts model, *Phys. Lett. A* **128**, 138–42

Bayen, F., Flato, M., Fronsdal, C., Lichnerowicz, A. & Sternheimer, D. (1978) Deformation theory and quantization. I. Deformations of symplectic structures, *Ann. Phys.* **111**, 61–110; II. Physical applications, *ibid.* **111**, 111–151

Bazhanov, V. V. (1985a) Trigonometric solutions of triangle equations and classical Lie algebras, *Phys. Lett. B* **159**, 321–4

Bazhanov, V. V. (1985b) Hidden symmetry of the free fermion model, I, *Theoret. Math. Phys.* **62**, 253–60; II, *ibid.* **63**, 519–27; III, *ibid.* **63**, 604–11

Bazhanov, V. V. (1987) Integrable quantum systems and classical Lie algebras, *Comm. Math. Phys.* **113**, 471–503

Bazhanov, V. V. & Kashaev, R. M. (1991) Cyclic L operators related with 3-state R-matrix, *Comm. Math. Phys.* **136**, 607–23

Bazhanov, V. V., Kashaev, R. M., Mangazeev, V. V. & Stroganov, Yu. G. (1991) $(\mathbb{Z}_N)^{\times n-1}$-generalization of the chiral Potts model, *Comm. Math. Phys.* **138**, 393–408

Bazhanov, V. V. & Reshetikhin, Yu. N. (1989) Critical RSOS models and conformal field theory, *Internat. J. Modern Phys. A* **4**, 115–42

Bazhanov, V. V. & Reshetikhin, Yu. N. (1990) Restricted solid-on-solid models connected with simply-laced algebras and conformal field theory, *J. Phys. A* **23**, 1477–92

Bazhanov, V. V. & Reshetikhin, Yu. N. (1991) Thermodynamic Bethe ansatz for RSOS scattering theories, *Nucl. Phys.* B **358**, 497–523

Bazhanov, V. V. & Stroganov, Yu. G. (1982) Trigonometric and S_n symmetric solutions of triangle equations with variables on the faces, *Nucl. Phys.* B **205**, 505–26

Bazhanov, V. V. & Stroganov, Yu. G. (1990) Chiral Potts models as a descendant of the six-vertex models, *J. Stat. Phys.* **51**, 799–817

Beck, J. (1993) Braid group action and quantum affine algebras, preprint, MIT

Beckers, J. & Debergh, N. (1992) From $N = 2$ supersymmetry to quantum deformations, *Phys. Lett.* B **286**, 290–2

Bednar, M., Burdik, C., Couture, M. & Hlavaty, L. (1992) On the quantum symmetries associated with the two-parameter free fermion model, *J. Phys.* A **25**, L341–6

Beggs, E. & Majid, S. (1990) Matched pairs of topological Lie algebras corresponding to Lie bialgebra structures on $diff(S^1)$ and $diff(\mathbb{R})$, *Ann. Inst. Henri Poincaré Phys. Théor.* **53**, 15–34

Beilinson, A., Bernstein, J. & Deligne, P. (1982) *Faisceaux Pervers*, Astérisque **100**, Société Mathématique de France, Paris

Beilinson, A. A., Lusztig, G. & MacPherson, R. (1990) A geometric setting for the quantum deformation of GL_n, *Duke Math. J.* **61**, 655–77

Belavin, A. A. (1981) Dynamical symmetry of integrable quantum systems, *Nucl. Phys.* B **180**, 189–200

Belavin, A. A. & Drinfel'd, V. G. (1982) Solutions of the classical Yang-Baxter equation for simple Lie algebras, *Funct. Anal. Appl.* **16**, 159–80

Belavin, A. A. & Drinfel'd, V. G. (1984a) Triangle equations and simple Lie algebras, *Soviet Scientific Reviews Sect.* C **4**, 93–165, Harwood Academic Publishers, Chur, Switzerland

Belavin, A. A. & Drinfel'd, V. G. (1984b) Classical Young-Baxter (*sic*) equation for simple Lie algebras, *Funct. Anal. Appl.* **17**, 220–1

Belavin, A. A., Polyakov, A. M. & Zamolodchikov, A. B. (1984) Infinite conformal symmetry in two-dimensional quantum field theory, *Nucl. Phys.* B **241**, 333–80

Bellon, M. P., Maillard, J. M. & Viallet, C. (1991) Infinite discrete symmetry group for the Yang-Baxter equations. Vertex models, *Phys. Lett.* B **260**, 87–100

Belov, A. A. & Chaltikian, K. D. (1993) Q-deformation of Virasoro algebra and lattice conformal theories, *Modern Phys. Lett.* A **8**, 1233–42

Belyi, A. (1987) On the commutator of the absolute Galois group, in *Proceedings of the International Congress of Mathematicians, Berkeley, 1986*, A. M. Gleason (ed.), pp. 346–9, American Mathematical Society, Providence, RI (Russian)

Berger, R. (1991) Quantification de l'identité de Jacobi, *C. R. Acad. Sci. Paris Sér. I* **312**, 721–4

Berger, R. (1992) The quantum Poincaré-Birkhoff-Witt theorem, *Comm. Math. Phys.* **143**, 215–34

Bergman, G. M. (1978) The diamond lemma for ring theory, *Adv. Math.* **29**, 178–218

Bergman, G. M. (1985) Everybody knows what a Hopf algebra is, in *Group Actions on Rings*, S. Montgomery (ed.), Contemporary Mathematics 43, pp. 25–48, American Mathematical Society, Providence, RI

Berkovich, A., Gomez, C. & Sierra, G. (1993) q-magnetism at roots of unity, *J. Phys.* A **26**, L45–52

Bernard, D. (1989) Vertex operator representations of the quantum affine algebra $U_q(B_r^{(1)})$, *Lett. Math. Phys.* **17**, 239–45

Bernard, D. (1990) Quantum Lie algebras and differential calculus on quantum groups, *Progr. Theor. Phys. Suppl.* **102**, 49–66

Bernard, D. (1991a) A remark on quasi-triangular quantum Lie algebras, *Phys. Lett.* B **260**, 389–93

Bernard, D. (1991b) Hidden Yangians in 2D massive current algebras, *Comm. Math. Phys.* **137**, 191–208

Bernard, D. (1992) A propos du calcul differentiel sur les groupes quantiques, *Ann. Inst. Henri Poincaré Phys. Théor.* **56**, 443–8

Bernard, D. & Felder, G. (1991) Quantum group symmetries in two-dimensional lattice quantum field theory, *Nucl. Phys. B* **365**, 98–120

Bernard, D. & LeClair, A. (1989) q-deformation of $SU(1, 1)$, conformal Ward identities and q-strings, *Phys. Lett. B* **227**, 417–23

Bernard, D. & LeClair, A. (1990) Residual quantum symmetries of the restricted sine-Gordan theories, *Nucl. Phys. B* **340**, 721–51

Bernard, D. & LeClair, A. (1991) Quantum group symmetries and non-local currents in 2D QFT, *Comm. Math. Phys.* **142**, 99–138

Bernard, D. & LeClair, A. (1992) Non-local currents in 2D QFT: an alternative to the quantum scattering method, in *Quantum Groups, Proceedings of Workshops held in the Euler International Mathematical Institute 1990*, P. P. Kulish (ed.), Lecture Notes in Mathematics 1510, pp. 176–96, Springer, Berlin

Bernard, D. & Pasquier, V. (1990) Exchange algebra and exotic supersymmetry in the chiral Potts model, *Internat. J. Modern Phys. B* **4**, 913–27

Bernstein, I. N., Gel'fand, I. M. & Ponomarev, V. A. (1973) Coxeter functors and Gabriel's theorem, *Russian Math. Surveys* **28** (2), 17–32

Bernstein, I. N. & Zelevinsky, A. V. (1977) Induced representations of p-adic groups I, *Ann. Sci. Ecole Norm. Sup. 4ᵉ Sér.* **10**, 441–72

Biedenharn, L. C. (1989) The quantum group $SU_q(2)$ and a q-analogue of the boson operators, *J. Phys. A* **22**, L873–8

Biedenharn, L. C. (1990) A q-boson realization of the quantum group $SU_q(2)$ and the theory of q-tensor operators, in *Quantum Groups, Clausthal, Germany, 1989*, H.-D. Doebner & J.-D. Hennig (eds), Lecture Notes in Physics 370, pp. 67–88, Springer, Berlin

Biedenharn, L. C. (1991) An overview of quantum groups, in *Group Theoretical Methods in Physics, Moscow, 1990*, V. V. Dodonov & V. I. Manko (eds), Lecture Notes in Physics 382, pp. 147–63, Springer, Berlin

Biedenharn, L. C. & Lohe, M. A. (1990) Some remarks on quantum groups - a new variation on the theme of symmetry, in *From Symmetries to Strings, Rochester NY, 1990*, A. Das (ed.), pp. 189–206, World Scientific, Teaneck, NJ

Biedenharn, L. C. & Lohe, M. A. (1991) Quantum groups and basic hypergeometric functions, in *Quantum Groups*, T. Curtright, D. Fairlie & C. Zachos (eds), 123–32, World Scientific, Singapore

Biedenharn, L. C. & Lohe, M. A. (1992a) Induced representations and tensor operators for quantum groups, in *Quantum Groups, Proceedings of Workshops held in the Euler International Mathematical Institute 1990*, P. P. Kulish (ed.), Lecture Notes in Mathematics 1510, pp. 197–209, Springer, Berlin

Biedenharn, L. C. & Lohe, M. A. (1992b) An extension of the Borel-Weil construction to the quantum group $U_q(n)$, *Comm. Math. Phys.* **146**, 483–504

Biedenharn, L. C. & Tarlini, M. (1990) On q-tensor operators for quantum groups, *Lett. Math. Phys.* **20**, 271–8

Bilal, A. (1991) Toda theories, W-algebras and their branching and fusion properties, in *Quantum Groups*, T. Curtright, D. Fairlie & C. Zachos (eds), 258–74, World Scientific, Singapore

Bincer, A. M. (1991) Casimir operators for $su_q(n)$, *J. Phys. A* **24**, L1133–8

Birman, J. S. (1976) *Braids, Links and Mapping Class Groups*, Annals of Mathematics Studies 82, Princeton University Press, Princeton

Birman, J. S. (1993) New points of view in knot theory, *Bull. Amer. Math. Soc.* **28**, 253–87

Birman, J. S. & Lin, X. S. (1993) Knot polynomials and Vassiliev's invariants, *Invent. Math.* **111**, 225–70

Birman, J. S. & Wenzl, H. (1989) Braids, links and a new algebra, *Trans. Amer. Math. Soc.* **313**, 239–45

Bogoliubov, N. M. & Bullough, R. K. (1992) Completely integrable model of interacting q-bosons, *Phys. Lett. A* **168**, 264–9

Bogolyubov, N. N., Mikityuk, I. V. & Prikarpatskii, A. K. (1991) Verma modules over the quantum Lie algebra of currents on the circle, *Soviet Math. Dokl.* **42**, 424–8

Bogoyavlenskii, O. I. (1976) On perturbations of the periodic Toda lattice, *Comm. Math. Phys.* **51**, 201–9

Bonatsos, D., Argyres, E. N., Drenska, S. B., Raychev, P. P., Roussev, R. P. & Smirnov, Yu. F. (1990) $SU_q(2)$ description of rotational spectra and its relation to the variable moment of the inertia model, *Phys. Lett. B* **251**, 477–82

Bonatsos, D., Argyres, E. N. & Raychev, P. P. (1991) $SU_q(1,1)$ description of vibrational molecular spectra, *J. Phys. A* **24**, L403–8

Bonatsos, D., Brito, L. & Menezes, D. (1993) The q-deformed Moszkowski model: RPA modes, *J. Phys. A* **26**, 895–904

Bonatsos, D. & Daskaloyannis, C. (1993) Equivalence of deformed fermionic algebras, *J. Phys. A* **26**, 1589–600

Bonatsos, D., Daskaloyannis, C. & Kokkotas, K. (1992) WKB equivalent potentials for q-deformed harmonic and anharmonic oscillators, *J. Math. Phys.* **33**, 2958–65

Bonatsos, D., Faessler, A., Raychev, P. P., Roussev, R. P. & Smirnov, Yu. F. (1992a) An exactly soluble nuclear model with $SU_q(3) \supset SU_q(2) \supset SO_q(2)$ symmetry, *J. Phys. A* **25**, L267–74

Bonatsos, D., Faessler, A., Raychev, P. P., Roussev, R. P. & Smirnov, Yu. F. (1992b) $B(E2)$ transition probabilities in the q-rotator model with $SU_q(2)$ symmetry, *J. Phys. A* **25**, 3275–86

Bonechi, F., Celeghini, E., Giachetti, R., Sorace, E. & Tarlini, M. (1992) Heisenberg XXZ model and quantum Galilei group, *J. Phys. A* **25**, L939–44

Bonneau, P. (1992) Cohomology and associated deformations for not necessarily co-associative bialgebras, *Lett. Math. Phys.* **26**, 277–84

Bonneau, P., Flato, M. & Pinczon, G. (1992) A natural and rigid model of quantum groups, *Lett. Math. Phys.* **25**, 75–84

Bordemann, M. (1990) Generalized Lax pairs, the modified classical Yang-Baxter equation and affine geometry of Lie groups, *Comm. Math. Phys.* **135**, 201–16

Borisov, N. V., Ilinski, K. N. & Uzdin, V. M. (1992) Quantum group particles and parastatistical excitations, *Phys. Lett. A* **169**, 427–32

Bourbaki, N. (1968) *Groupes et Algèbres de Lie, Chapitres 4, 5 et 6*, Hermann, Paris

Bourbaki, N. (1970), *Algèbre, Chapitres 1 à 3*, Hermann, Paris

Bourbaki, N. (1972) *Groupes et Algèbres de Lie, Chapitres 2 et 3*, Hermann, Paris

Bourbaki, N. (1985) *Algèbre Commutative, Chapitres 5 à 7*, Masson, Paris

Bouwknegt, P., McCarthy, J., Nemeschansky, D. & Pilch, K. (1991) Vertex operators and fusion rules in the free field realizations of WZNW theories, *Phys. Lett. B* **258**, 127–33

Bouwknegt, P., McCarthy, J. & Pilch, K. (1990a) Quantum group structure in the Fock space resolutions of $\hat{sl}(n)$ representations, *Comm. Math. Phys.* **131**, 125–55

Bouwknegt, P., McCarthy, J. & Pilch, K. (1990b) Free field realizations of WZNW models. The BRST complex and its quantum group structure, *Phys. Lett. B* **234**, 297–303

Bowcock, P. & Watts, G. M. T. (1992) On the classification of quantum W-algebras, *Nucl. Phys. B* **379**, 63–95

Bozejko, M. & Speicher, R. (1991) An example of a generalized Brownian motion, *Comm. Math. Phys.* **137**, 519–31

Bracken, A. J., Gould, M. D. & Tsohantjis, I. (1993) Boson-fermion models for $osp(1,2)$ and $U_q(osp(1,2))$, *J. Math. Phys.* **34**, 1654–64

Bracken, A. J., Gould, M. D. & Zhang, R. B. (1990) Quantum supergroups and solutions of the Yang-Baxter equation, *Modern Phys. Lett. A* **5**, 831–40

Bracken, A. J., McAnally, D. S., Zhang, R. B. & Gould, M. D. (1991) A q-analogue of Bargmann space and its scalar product, *J. Phys. A* **24**, 1379–91

Braden, H. W., Corrigan, E., Dorey, P. E. & Sasaki, R. (1992) Affine Toda field theory: S-matrix vs. perturbation, in *Quantum Groups, Proceedings of Workshops held in the Euler International Mathematical Institute 1990*, P. P. Kulish (ed.), Lecture Notes in Mathematics 1510, pp. 210–20, Springer, Berlin

Braden, H. W., Corrigan, E., Dorey, P. E. & Sasalu, R. (1991) Aspects of affine Toda field theory, in *Quantum Groups*, T. Curtright, D. Fairlie & C. Zachos (eds), pp. 275–305, World Scientific, Singapore

Bragiel, K. (1989) The twisted $SU(3)$ group. Irreducible *-representations of the C*-algebra $C(S_\mu U(3))$, *Lett. Math. Phys.* **17**, 37–44

Bragiel, K. (1990) The twisted $SU(N)$ group. On the C*-algebra $C(S_\mu U(N))$, *Lett. Math. Phys.* **20**, 251–7

Bragiel, K. (1991a) On the Wigner-Eckart theorem for tensor operators connected with compact matrix quantum groups, *Lett. Math. Phys.* **21**, 181–91

Bragiel, K. (1991b) On the spherical and zonal functions on the compact quantum groups, *Lett. Math. Phys.* **22**, 195–202

Bratelli, O., Elliot, G. A., Evans, D. E. & Kishimoto, A. (1991) Non-commutative spheres I, *Internat. J. Math.* **2**, 139–66

Brauer, R. (1937) On algebras which are connected with the semisimple continuous groups, *Ann. of Math.* **38**, 857–72

Brzezinski, T. (1993) Quantum group related to the space of differential operators on the quantum hyperplane, *J. Phys. A* **26**, 921–6

Brzezinski, T., Dabrowski, H. & Rembielinski, J. (1992) On the quantum differential calculus and the quantum holomorphicity, *J. Math. Phys.* **33**, 19–24

Brzezinski, T. & Majid, S. (1993) Quantum group gauge theory on classical spaces, *Phys. Lett. B* **298**, 339–43

Brzezinski, T. & Rembielinski, J. (1992) q-integrals on the quantum complex plane, *J. Phys. A* **25**, 1945–52

Brzezinski, T., Rembielinski, J. & Smolinski, K. A. (1993) Quantum particle on a quantum circle, *Modern Phys. Lett. A* **8**, 409–16

Bullough, R. K., Pilling, D. J. & Timonen, J. (1988) Soliton statistical mechanics, in *Solitons, Tiruchirapalli, 1987*, M. Lakshmanan (ed.), Springer Series in Nonlinear Dynamics, pp. 250–81, Springer, Berlin

Bullough, R. K. & Timonen, J. (1991) Quantum groups and quantum complete integrability: theory and experiment, in *Differential Geometric Methods in Theoretical Physics, Rapallo, 1990*, C. Bartocci, U. Bruzzo & R. Cianci (eds), Lecture Notes in Physics 375, pp. 71–90, Springer, Berlin

Burdik, C., Cerny, L. & Navratil, O. (1993) The q-boson realization of the quantum group $U_q(sl(n+1,C))$, *J. Phys. A* **26**, L83–6

Burdik, C., Havlicek, M. & Vancura, A. (1992) Irreducible highest weight representations of quantum groups $U_q(gl(n,\mathbb{C}))$, *Comm. Math. Phys.* **148**, 417–23

Burdik, C. & Hellinger, P. (1992a) Universal R-matrix for a two-parametric quantization of $gl(2)$, *J. Phys. A* **25**, L629–32

Burdik, C. & Hellinger, P. (1992b) The universal R-matrix and the Yang-Baxter equation with parameters, *J. Phys. A* **25**, L1023–8

Burdik, C. & Hlavaty, L. (1991) A two-parametric quantization of $sl(2)$, *J. Phys. A* **24**, L165–8

Burdik, C. & Navratil, O. (1990) The boson realizations of the quantum group $U_q(sl(2))$, *J. Phys. A* **23**, L2105–8

Burroughs, N. (1990a) The universal R-matrix for $U_q sl(3)$ and beyond, *Comm. Math. Phys.* **127**, 109–28

Burroughs, N. (1990b) Relating the approaches to quantized algebras and quantum groups, *Comm. Math. Phys.* **133**, 91–117

Burroughs, N. (1991) The quantum group methods of quantizing the special linear group $SL_2(\mathbb{C})$, in *Differential Geometric Methods in Theoretical Physics, Davis, 1988*, L. L. Chau & W. Nahm (eds), NATO Advanced Science Institutes Series B: Physics 245, pp. 513–39, Plenum, New York

Buzek, V. (1991) Dynamics of a q-analogue of the quantum harmonic oscillator, *J. Modern Optics* **38**, 801–12

Buzek, V. (1992) The Jaynes-Cummings model with a q-analogue of a coherent state, *J. Modern Optics* **39**, 949–59

Cahen, M., Gutt, S., Ohn, C. & Parker, M. (1990) Lie-Poisson groups: remarks and examples, *Lett. Math. Phys.* **19**, 343–53

Cahen, M., Gutt, S. & Rawnsley, J. (1992) Nonlinearizability of the Iwasawa Poisson-Lie structure, *Lett. Math. Phys.* **24**, 79–83

Caldero, P. (1993) Eléments ad-finis de certains groupes quantiques, *C. R. Acad. Sci. Paris Sér. I* **316**, 327–9

Caldi, D. G. (1991) $SU_q(2)$ and $SU(2)$ as hamiltonian symmetries, in *Quantum Groups*, T. Curtright, D. Fairlie & C. Zachos (eds), pp. 236–46, World Scientific, Singapore

Caldi, D. G., Chodos, A., Zhu, Z. & Barth, A. (1991) The classical $su(2)$ invariance of the $su(2)_q$-invariant XXZ chain, *Lett. Math. Phys.* **22**, 163–6

Carow-Watamura, U., Schlieker, M., Scholl, M. & Watamura, S. (1990) Tensor representation of the quantum group $SL_q(2, \mathbb{C})$ and quantum Minkowski space, *Z. Phys. C* **48**, 159–65

Carow-Watamura, U., Schlieker, M., Scholl, M. & Watamura, S. (1991) A quantum Lorentz group, *Internat. J. Modern Phys. A* **6**, 3081–108

Carow-Watamura, U., Schlieker, M. & Watamura, S. (1991) $SO_q(N)$ covariant differential calculus on quantum space and quantum deformation of Schrodinger equation, *Z. Phys. C* **49**, 439–46

Carow-Watamura, U., Schlieker, M., Watamura, S. & Weich, W. (1991) Bicovariant differential calculus on quantum groups $SU_q(N)$ and $SO_q(N)$, *Comm. Math. Phys.* **142**, 605–41

Carow-Watamura, U. & Watamura, S. (1993) Complex quantum group, dual algebra and bicovariant differential calculus, *Comm. Math. Phys.* **151**, 487–514

Cartier, P. (1979) Representations of p-adic groups: a survey, in *Automorphic Forms, Representations and L-Functions*, Proceedings of Symposia in Pure Mathematics 33, Part I, pp. 111–55, American Mathematical Society, Providence, RI

Cartier, P. (1990) Développements récents sur les groupes de tresses. Applications à la topologie et à l'algèbre, Séminaire Bourbaki exp. no. 716, *Astérisque* **189–190**, 17–67, Société Mathématique de France, Paris

Casian, L. (1990) Kazhdan-Lusztig multiplicity formulas for Kac-Moody algebras, *C. R. Acad. Sci. Paris Sér. I* **310**, 333–7

Casian, L. (1993) Kazhdan-Lusztig conjecture in the negative level case (Kac–Moody algebras of affine type), preprint

Castellani, L. (1992a) Bicovariant differential calculus on the quantum $D = 2$ Poincaré group, *Phys. Lett. B* **279**, 291–8

Castellani, L. (1992b) Gauge theories of quantum groups, *Phys. Lett. B* **292**, 93–8

Castellani, L. (1993) R-matrix and bicovariant calculus for the inhomogeneous quantum groups $IGL_q(n)$, *Phys. Lett. B* **298**, 335–8

Cateau, H. & Saito, S. (1990) Braids of strings, *Phys. Rev. Lett.* **65**, 2487–90

Celeghini, E., Giachetti, R., Kulish, P., Sorace, E. & Tarlini, M. (1991) Hopf superalgebra contractions and R-matrix for fermions, *J. Math. Phys. A* **24**, 5675–82

Celeghini, E., Giachetti, R., Reyman, A., Sorace, E. & Tarlini, M. (1991) $SO_q(n+1, n-1)$ as a real form of $SO_q(2n, \mathbb{C})$, *Lett. Math. Phys.* **23**, 45–9

Celeghini, E., Giachetti, R., Sorace, E. & Tarlini, M. (1990) Three dimensional quantum groups from contractions of $SU(2)_q$, *J. Math. Phys.* **31**, 2548–51

Celeghini, E., Giachetti, R., Sorace, E. & Tarlini, M. (1991a) The quantum Heisenberg group $H(1)_q$, *J. Math. Phys.* **32**, 1155–8

Celeghini, E., Giachetti, R., Sorace, E. & Tarlini, M. (1991b) The three dimensional euclidean quantum group $E(3)_q$ and its R-matrix, *J. Math. Phys.* **32**, 1159–65

Celeghini, E., Giachetti, R., Sorace, E. & Tarlini, M. (1992) Contractions of quantum groups, in *Quantum Groups, Proceedings of Workshops held in the Euler International Mathematical Institute 1990*, P. P. Kulish (ed.), Lecture Notes in Mathematics 1510, pp. 221–44, Springer, Berlin

Celeghini, E., Palev, T. D. & Tarlini, M. (1991) The quantum superalgebra $B_q(0|1)$ and q-deformed creation and annihilation operators, *Modern Phys. Lett. B* **5**, 187–93

Celeghini, E., Rasetti, M. & Vitiello, G. (1991) Squeezing and quantum groups, *Phys. Rev. Lett.* **66**, 2056–9

Chaichian, M., De Azcarraga, J. A., Presnajder, P. & Rodenas, F. (1992) Oscillator realization of the q-deformed anti-de Sitter algebra, *Phys. Lett. B* **291**, 411–7

Chaichian, M. & Demichev, A. P. (1993) Quantum Poincaré group, *Phys. Lett. B* **304**, 220–4

Chaichian, M. & Ellinas, D. (1990) On the polar decomposition of the quantum group $SU(2)$ algebra, *J. Phys. A* **23**, L291–6

Chaichian, M., Ellinas, D. & Kulish, P. (1990) Quantum algebra as the dynamical symmetry of the deformed Jaynes-Cummings model, *Phys. Rev. Lett.* **65**, 980–3

Chaichian, M., Ellinas, D. & Popovich, Z. (1990) Quantum conformal algebra with central extension, *Phys. Lett. B* **248**, 95–9

Chaichian, M., Isaev, A. P., Lukierski, J., Popovich, Z. & Presnajder, P. (1992) q-deformations of Virasoro algebra and conformal dimension, *Phys. Lett. B* **262**, 32–8

Chaichian, M. & Kulish, P. (1990) Quantum Lie superalgebras and q-oscillators, *Phys. Lett. B* **234**, 72–80

Chaichian, M., Kulish, P. & Lukierski, J. (1990) q-deformed Jacobi identity, q-oscillators and q-deformed infinite-dimensional algebras, *Phys. Lett. B* **237**, 401–6

Chaichian, M., Kulish, P. & Lukierski, J. (1991a) Supercovariant q-oscillators, in *Nonlinear Fields: Classical, Random, Semiclassical, Karpacz, 1991*, P. Garbaczewski & Z. Popowicz (eds), pp. 336–45, World Scientific, Singapore

Chaichian, M., Kulish, P. & Lukierski, J. (1991b) Supercovariant systems of q-oscillators and q-supercovariant Hamiltonians, *Phys. Lett. B* **262**, 43–8

Chaichian, M., Popovich, Z. & Presnajder, P. (1990) q-Virasoro algebra and its relation to the q-deformed KdV system, *Phys. Lett. B* **249**, 63–5

Chaichian, M. & Presnajder, P. (1992a) Sugawara construction and the q-deformation of Virasoro algebra, in *Quantum Groups and Related Topics*, R. Gierlak *et al.* (eds), pp. 3–12, Kluwer, Dordrecht

Chaichian, M. & Presnajder, P. (1992b) Sugawara construction and the q-deformation of Virasoro (super) algebra, *Phys. Lett. B* **277**, 109–18

Chair, N. & Zhu, C. J. (1991) Tetrahedra and polynomial equations in topological field theory, *Internat. J. Modern Phys. A* **6**, 3571–98

Chakrabarti, A. (1991) q-analogs of $IU(n)$ and $U(n, 1)$, *J. Math. Phys.* **32**, 1227–34

Chakrabarti, A. (1993) Canonical structure in $SO(4)_q$ and relations to $E(3)_q$ and $SO(3, 1)_q$, *J. Math. Phys.* **34**, 1964–85

Chakrabarti, A. & Jagannathan, R. (1991a) On the representations of $GL_{p,q}(2)$, $GL_{p,q}(1|1)$ and noncommutative spaces, *J. Phys. A* **24**, 5683–701

Chakrabarti, A. & Jagannathan, R. (1991b) On the representations of $GL_q(n)$ using the Heisenberg-Weyl relations, *J. Phys. A* **24**, 1709–20

Chakrabarti, A. & Jagannathan, R. (1991c) A (p, q)-oscillator realization of two-parameter quantum algebras, *J. Phys. A* **24**, L711–8

Chakrabarti, A. & Jagannathan, R. (1992a) A (p, q)-deformed Virasoro algebra, *J. Phys. A* **25**, 2607–14

Chakrabarti, A. & Jagannathan, R. (1992b) On the number operators of single mode q-oscillators, *J. Phys. A* **25**, 6393–8

Chakrabarti, A. & Jagannathan, R. (1992c) A two-parameter deformation of the Jaynes-Cummings model: path integral representation, *J. Phys. A* **25**, 6399–408

Chang, D., Phillips, I. & Rozansky, L. (1992) R-matrix approach to quantum superalgebras $su_q(m|n)$, *J. Math. Phys.* **33**, 3710–5

Chang, Z. (1992a) The quantum multiboson algebra and generalized coherent states of $SU_q(1, 1)$, *J. Phys. A* **25**, L707–12

Chang, Z. (1992b) The quantum q-deformed symmetric top: an exactly solved model, *J. Phys. A* **25**, L781–8

Chang, Z. (1992c) Quantum group realized in a symmetric top system, *Phys. Rev. A* **45**, 4303–11

Chang, Z., Chen, W. & Guo, H. Y. (1990) $SU_{q\to0}(2)$ and $SU_q(2)$, the classical and quantum q-deformations of the $SU(2)$ algebra, *J. Phys. A* **23**, L4185–90

Chang, Z., Chen, W., Guo, H. Y. & Yan, H. (1990a) $SU_{q\to0}(2)$ and $SU_q(2)$, the classical and quantum q-deformations of the $SU(2)$ algebra II, *J. Phys. A* **23**, 5371–82

Chang, Z., Chen, W., Guo, H. Y. & Yan, H. (1990b) $SU_{q\to0}(2)$ and $SU_q(2)$, the classical and quantum q-deformations of the $SU(2)$ algebra III, *J. Phys. A* **24**, 1427–34

Chang, Z., Fei, S. M., Guo, H. Y. & Yan, H. (1991) $SU_{q\to0}(2)$ and $SU_q(2)$, the classical and quantum q-deformations of the $SU(2)$ algebra IV, *J. Phys. A* **24**, 5435–44

Chang, Z., Guo, H. Y. & Yan, H. (1991) The q-deformed oscillator model and the vibrational spectra of diatomic molecules, *Phys. Lett. A* **156**, 192–6

Chang, Z., Guo, H. Y. & Yan, H. (1992) The q-Hermite polynomial and the representations of Heisenberg and quantum Heisenberg algebras, *J. Phys. A* **25**, 1517–25

Chang, Z., Wang, J. X. & Yan, H. (1991) The realization of quantum groups of A_{n-1} and C_n types in q-deformed oscillator systems at classical and quantum levels, *J. Math. Phys.* **32**, 3241–5

Chang, Z. & Yan, H. (1991a) The $SU_q(2)$ quantum group symmetry and diatomic molecules, *Phys. Lett. A* **154**, 254–8

Chang, Z. & Yan, H. (1991b) Quantum group-theoretic approach to vibrating and rotating diatomic molecules, *Phys. Lett. A* **158**, 242–6

Chang, Z. & Yan, H. (1991c) $H_q(4)$ and $SU_q(2)$ symmetries in diatomic molecules, *Phys. Rev. A (3)* **43**, 6043–52

Chari, V. (1985) Integrable representations of affine Lie algebras, *Invent. Math.* **81**, 317–37

Chari, V. & Ilangovan, S. (1984) On the Harish-Chandra homomorphism for infinite-dimensional Lie algebras, *J. Algebra* **90**, 476–94

Chari, V. & Premet, A. (1993) Indecomposable restricted representations of quantum sl_2, *Publ. Res. Inst. Math. Sci.* (to appear)

Chari, V. & Pressley, A. N. (1986) New unitary representations of loop groups, *Math. Ann.* **275**, 87–104

Chari, V. & Pressley, A. N. (1990a) Notes on quantum groups, *Nucl. Phys. B Proc. Suppl.* **18 A**, 207–28

Chari, V. & Pressley, A. N. (1990b) Yangians and R-matrices, *L'Enseignement Math.* **36**, 267–302

Chari, V. & Pressley, A. N. (1991a) Fundamental representations of Yangians and rational R-matrices, *J. reine angew. Math.* **417**, 87–128

Chari, V. & Pressley, A. N. (1991b) Quantum affine algebras, *Comm. Math. Phys.* **142**, 261–83

Chari, V. & Pressley, A. N. (1991c) Minimal cyclic representations of quantum groups at roots of 1, *C. R. Acad. Sci. Paris Sér. I* **313**, 429–34

Chari, V. & Pressley, A. N. (1991d) Introduction to quantum groups, in *Proceedings of International Conference on Algebraic Groups, Hyderabad, India, December, 1989*, S. Ramanan, C. Musili & N. Mohan Kumar (eds), pp. 81–122, Manoj Prakashan, Madras

Chari, V. & Pressley, A. N. (1992) Fundamental representations of quantum groups, *Lett. Math. Phys.* **26**, 133–46

Chari, V. & Pressley, A. N. (1993a) Representations of modular Lie algebras through quantum groups, *C. R. Acad. Sci. Paris Sér. I* **317**, 728–9

Chari, V. & Pressley, A. N. (1993b) Quantum affine algebras and affine Hecke algebras, *Pacific J. Math.* (to appear)

Chari, V. & Pressley, A. N. (1994a) Representations of quantum $SO(8)$ and related quantum algebras, *Comm. Math. Phys. Comm. Math. Phys.* **159** 29–49

Chari, V. & Pressley, A. N. (1994b) Small representations of quantum affine algebras, *Lett. Math. Phys.* **30**, 131–45

Chari, V. & Pressley, A. N. (1994c) Quantum affine algebras and their representations, in *Proceedings of the Canadian Mathematical Society Annual Seminar - Representations of Groups, Banff, 1994* (to appear)

Chen, Y. X. & Ni, G. J. (1991) General theory of quantum statistics with internal degrees of freedom in two dimensions, *Phys. Rev. D* **43**, 4133–41

Cheng, Y., Ge, M. L. & Xue, K. (1991) Yang-Baxterization of braid group representations, *Comm. Math. Phys.* **136**, 195–208

Cherednik, I. V. (1980) On the method of constructing factorized S-matrices in elementary functions, *Theoret. Math. Phys.* **43**, 117–9 (Russian)

Cherednik, I. V. (1982) On the properties of factorized S-matrices in elliptic functions, *Soviet J. Nucl. Phys.* **36**, 320–4

Cherednik, I. V. (1983) Bäcklund-Darboux transformations for classical Yang-Baxter bundles, *Funct. Anal. Appl.* **17**, 155–7

Cherednik, I. V. (1984) Factorizing particles on a half line and root systems, *Theoret. Math. Phys.* **61**, 977–83

Cherednik, I. V. (1985) Some finite-dimensional representations of generalized Sklyanin algebras, *Funct. Anal. Appl.* **19**, 77–9

Cherednik, I. V. (1986a) On R-matrix quantization of formal loop groups, in *Group Theoretical Methods in Physics, Vol. II, Yurmala, USSR, 1985)*, V. V. Dodonov, V. I. Manko & M. A. Markov (eds), pp. 161–80, VNU Science Press, Utrecht

Cherednik, I. V. (1986b) On the quantum deformations of irreducible finite-dimensional representations of gl_N, *Soviet Math. Dokl.* **33**, 507–10

Cherednik, I. V. (1987a) A new interpretation of Gel'fand-Tzetlin bases, *Duke Math. J.* **54**, 563–77

Cherednik, I. V. (1987b) On irreducible R-algebras, *Soviet Math. Dokl.* **34**, 446–50

Cherednik, I. V. (1988a) q-analogues of Gel'fand-Tzetlin bases, *Funct. Anal. Appl.* **22**, 78–9

Cherednik, I. V. (1988b) On special bases of irreducible finite-dimensional representations of the degenerate affine Hecke algebra, *Funct. Anal. Appl.* **20**, 87–8

Cherednik, I. V. (1989) Quantum groups as hidden symmetries of classic representation theory, in *Differential Geometric Methods in Mathematical Physics, Proceedings of the 17th International Conference, Chester, 1988*, A. I. Solomon (ed.), pp. 47–55, World Scientific, Teaneck, NJ

Cherednik, I. V. (1990a) Generalised braid groups and local r-matrix systems, *Soviet Math. Dokl.* **40**, 43–8

Cherednik, I. V. (1990b) Calculation of the monodromy of some W-invariant local systems of type B, C and D, *Funct. Anal. Appl.* **24**, 78–9

Cherednik, I. V. (1991a) Monodromy representations for generalized Knizhnik-Zamolodchikov equations and Hecke algebras, *Publ. Res. Inst. Math. Sci.* **27**, 711–26

Cherednik, I. V. (1991b) Integral solutions of trigonometric Knizhnik-Zamolodchikov equations and Kac-Moody algebras, *Publ. Res. Inst. Math. Sci.* **27**, 727–44

Cherednik, I. V. (1991c) A unification of Knizhnik-Zamolodchikov and Dunkl operators via affine Hecke algebras, *Invent. Math.* **106**, 411–31

Cherednik, I. V. (1991d) Affine extensions of Knizhnik-Zamolodchikov equations and Lusztig's isomorphisms, in *Special Functions, Okayama, 1990*, M. Kashiwara & T. Miwa (eds), pp. 63–77, Springer, Tokyo

Cherednik, I. V. (1992a) Double affine Hecke algebras, Knizhnik-Zamolodchikov equations, and Macdonald's operators, *Duke Math. J. Internat. Math. Res. Notices* **2**, 171–80

Cherednik, I. V. (1992b) Degenerate affine Hecke algebras and two-dimensional particles, in *Infinite Analysis*, A. Tsuchiya, T. Eguchi & M. Jimbo (eds), Advanced Series in Mathematical Physics 16, pp. 109–40, World Scientific, Singapore

Cherednik, I. V. (1992c) Quantum Knizhnik-Zamolodchikov equations and affine root systems, *Comm. Math. Phys.* **150**, 109–36

Chernyak, V. Ya., Grigorishin, K. I. & Ogievetsky, E. I. (1991) Triangle equation and quantum algebra representations, *Phys. Lett. A* **158**, 291–4

Chernyak, V. Ya., Grigorishin, K. I. & Ogievetsky, E. I. (1992a) Universal R-matrix and representation theory for quantum groups, *Phys. Lett. A* **164**, 389–97

Chernyak, V. Ya., Grigorishin, K. I. & Ogievetsky, E. I. (1992b) Quantum group bootstrap approach and infinite-dimensional factorized scattering matrices, *Phys. Lett. A* **164**, 398–407

Chernyak, V. Ya., Grigorishin, K. I. & Ogievetsky, E. I. (1992c) Quantum group bootstrap approach. Application to extended Heisenberg group, *Phys. Lett. A* **164**, 408–15

Chernyak, V. Ya., Kozhekin, A. E. & Ogievetsky, E., (1993) $SU(1,1)$-invariant solutions of the quantum Yang-Baxter equation, *J. Phys. A* **26**, 1313–6

Chiu, S. H., Gray, R. W. & Nelson, C. A. (1992) The q-analogue quantized radiation field and its uncertainty relations, *Phys. Lett. A* **164**, 237–42

Chryssomalokos, C. (1992) Coproduct of vector fields on the quantum Lorentz group, *Modern Phys. Lett. A* **7**, 2851–6

Chryssomalokos, C., Drabant, B., Schlieker, M., Weich, W. & Zumino, B. (1992) Vector fields on complex quantum groups, *Comm. Math. Phys.* **147**, 635–46

Chung, H. J. & Koh, I. G. (1991) Solutions to the quantum Yang-Baxter equation for the exceptional Lie algebra with a spectral parameter, *J. Math. Phys.* **32**, 2406–8

Chung, Z., Guo, H. Y. & Yan, H. (1991) The q-deformed oscillator model and the vibrational spectra of diatomic molecules, *Phys. Lett. A* **156**, 192–6

Codrianski, S. (1991) Comments on q-algebras, *Internat. J. Theoret. Phys.* **30**, 59–76

Codrianski, S. (1992) $SU(2)_q$ in a Hilbert space of analytic functions, *Internat. J. Theoret. Phys.* **31**, 907–24

Cohen, M. (1992) Hopf algebra actions - revisited, in *Deformation Theory and Quantum Groups with Applications to Mathematical Physics*, M. Gerstenhaber & J. Stasheff (eds), Contemporary Mathematics 134, pp. 1–18, American Mathematical Society, Providence, RI

Connes, A., Flato, M. & Sternheimer, D. (1992) Closed star products and cyclic cohomology, *Lett. Math. Phys.* **24**, 1–12

Cornwell, J. F. (1992) Multiparameter deformations of the universal enveloping algebras of the simple Lie algebras A_l for all $l \geq 2$ and the Yang-Baxter equation, *J. Math. Phys.*

33, 3963-77

Corrigan, E., Fairlie, D. B., Fletcher, P. & Sasaki, R. (1990) Some aspects of quantum groups and supergroups, *J. Math. Phys.* **31**, 776-80

Coste, A., Dazord, P. & Weinstein, A. (1987) Groupoides symplectiques, Publications du Département de Mathématiques, Université Claude Bernard-Lyon I 2A, pp. 1-62

Cotta-Ramasino, P. (1992) Link diagrams, Yang-Baxter equations and quantum holonomy, in *Deformation Theory and Quantum Groups with Applications to Mathematical Physics*, M. Gerstenhaber & J. Stasheff (eds), Contemporary Mathematics 134, pp. 19-44, American Mathematical Society, Providence, RI

Cotta-Ramasino, P. & Rinaldi, M. (1991) Multi-parameter quantum groups related to link diagrams, *Comm. Math. Phys.* **142**, 589-604

Couture, M. (1991) On some quantum R matrices associated with representations of $U_q(sl(2,\mathbb{C}))$ when q is a root of unity, *J. Phys. A* **24**, L103-7

Couture, M., Cheng, Y., Ge, M. L. & Xue, K. (1991) New solutions of the Yang-Baxter equation and their Yang-Baxterization, *Internat. J. Modern Phys. A* **6**, 559-76

Crane, L. (1991) $2 - d$ physics and $3 - d$ topology, *Comm. Math. Phys.* **135**, 615-40

Cremmer, E. & Gervais, J.-L. (1990) The quantum group structure associated with non-linearly extended Virasoro algebras, *Comm. Math. Phys.* **134**, 619-32

Crivelli, M., Felder, G. & Wieczerkowski, C. (1993) Generalized hypergeometric functions on the torus and the adjoint representation of $U_q(sl_2)$, *Comm. Math. Phys.* **154**, 1-24

Cuerno, R. (1991) Quantum symmetries in the free field realization of W_n algebras, *Phys. Lett. B* **271**, 314-20

Cummins, C. J. & King, R. C. (1992) Characters of A_{n-1} Hecke algebras at roots of unity, *J. Phys. A* **25**, L789-98

Curtright, T. I. (1991) Deformations, coproducts and U, in *Quantum Groups*, T. Curtright, D. Fairlie & C. Zachos (eds), pp. 72-96, World Scientific, Singapore

Curtright, T. I., Ghandour, G. I. & Zachos, C. K. (1991) Quantum algebra deforming maps, Clebsch-Gordan coefficients, coproducts, R and U matrices, *J. Math. Phys.* **32**, 676-88

Curtright, T. I. & Zachos. C. K. (1990) Deforming maps for quantum algebras, *Phys. Lett. B* **243**, 237-44

Dabrowski, L. & Dobrev, V. K. (1993) Positive energy representations of the conformal quantum algebra, *Phys. Lett. B* **302**, 215-22

Dabrowski, L. & Wang, L. Y. (1991) Two-parameter quantum deformation of $GL(1|1)$, *Phys. Lett. B* **266**, 51-4

Dai, J. H., Guo, H. Y. & Yan, H. (1991) The q-deformed differential operator algebra, a new solution to the Yang-Baxter equation and quantum plane, *J. Phys. A* **24**, L409-14

Damaskinskii, E. V. & Kulish, P. P. (1991) Deformed oscillators and their applications, *Zap. Nauchn. Sem. Leningrad Otdel. Mat. Inst. Steklov* **189** (Russian)

Damaskinskii, E. V. & Kulish, P. P. (1992) Hermite q-polynomials and q-oscillators, *Zap. Nauchn. Sem. Leningrad Otdel. Mat. Inst. Steklov* **199** (Russian)

Daskaloyannis, C. (1992) Generalized deformed Virasoro algebras, *Modern Phys. Lett. A* **7**, 809-16

Date, E., Jimbo, M., Kuniba, A., Miwa, T. & Okado. M. (1987) Exactly solvable SOS models: Local height probabilities and theta function identities, *Nucl. Phys. B* **290**, 231-73

Date, E., Jimbo, M., Kuniba, A., Miwa, T. & Okado, M. (1988) Exactly solvable SOS models: II. Proof of the star-triangle relation and combinatorial identities, in *Conformal Field Theory and Solvable Lattice Models, Kyoto, 1986*, A. Tsuchiya, T. Eguchi & M. Jimbo (eds), Advanced Studies in Pure Mathematics 16, pp. 17-122, Academic Press, Boston

Date, M., Jimbo, M. & Miki, K. (1990) A note on the branching rule for cyclic representations of $U_q(gl_n)$, preprint, RIMS, Kyoto

Date, E., Jimbo, M., Miki, K. & Miwa, T. (1990) R-matrix for cyclic representations of $U_q(\hat{sl}(3, \mathbb{C}))$ at $q^3 = 1$, *Phys. Lett. A* **148**, 45–9

Date, M., Jimbo, M., Miki, K. & Miwa, T. (1991a) Generalized chiral Potts models and minimal cyclic representations of $U_q\hat{gl}(n, \mathbb{C})$, *Comm. Math. Phys.* **137**, 133–47

Date, E., Jimbo, M., Miki, K. & Miwa, T. (1991b) Cyclic representations of $U_q(sl(n + 1))$ at $q^N = 1$, *Publ. Res. Inst. Math. Sci.* **27**, 347–66

Date, E., Jimbo, M., Miki, K. & Miwa, T. (1991c) New R-matrices associated with cyclic representations of $U_q(A_2^{(2)})$, *Publ. Res. Inst. Math. Sci.* **27**, 639–55

Date, M., Jimbo, M., Miki, K. & Miwa, T. (1992) Braid group representations arising from the generalized chiral Potts models, *Pacific J. Math.* **154**, 37–66

Date, E., Jimbo, M. & Miwa, T. (1990) Representations of $U_q(gl(n, \mathbb{C}))$ at $q = 0$ and the Robinson-Schensted correspondence, in *Physics and Mathematics of Strings*, L. Brink, D. Friedan & A. M. Polyakov (eds), pp. 185–211, World Scientific, Teaneck, NJ

De Concini, C. & Kac, V. G. (1990) Representations of quantum groups at roots of 1, in *Operator Algebras, Unitary Representations, Enveloping Algebras and Invariant Theory*, A. Connes, M. Duflo, A. Joseph & R. Rentschler (eds), pp. 471–506, Birkhäuser, Boston

De Concini, C., Kac, V. G. & Procesi, C. (1992a) Quantum coadjoint action, *J. Amer. Math. Soc.* **5**, 151–89

De Concini, C., Kac, V. G. & Procesi, C. (1992b) Representations of quantum groups at roots of 1: reduction to the exceptional case, in *Infinite Analysis*, A. Tsuchiya, T. Eguchi & M. Jimbo (eds), Advanced Series in Mathematical Physics 16, pp. 141–50, World Scientific, Singapore

De Concini, C., Kac, V. G. & Procesi, C. (1993a) Some remarkable degenerations of quantum groups, *Comm. Math. Phys.* **157**, 405–27

De Concini, C., Kac, V. G. & Procesi, C. (1993b) Some quantum analogues of solvable groups, preprint, RIMS, Kyoto

De Concini, C. & Lyubashenko, V. (1993) Quantum function algebra at roots of 1, preprint

Deguchi, T. (1989) Braid group representations and link polynomials derived from generalized $SU(n)$ vertex models, *J. Phys. Soc. Japan* **58**, 3441–4

Deguchi, T. (1990) Braids, link polynomials and transformations of solvable models, *Internat. J. Modern Phys. A* **5**, 2195–239

Deguchi, T. & Akutsu, Y. (1992) Colored braid matrices from infinite-dimensional representations of $U_q(g)$, *Modern Phys. Lett. A* **7**, 767–79

Deguchi, T., Fujii, A. & Ito, K. (1992) Quantum superalgebra $U_q(osp(2, 2))$, *Phys. Lett. B* **238**, 242–6

De la Harpe, P. (1988) Introduction to knot and link polynomials, in *Fractals, Quasicrystals, Chaos, Knots and Algebraic Quantum Mechanics, Maratea, 1987*, A. Amann, L. Cederbaum & W. Gans (eds), NATO Advanced Science Institutes Series C: Mathematical and Physical Sciences 235, pp. 233–63, Kluwer, Dordrecht

Delbecq, C. & Quesne, C. (1993a) Nonlinear deformations of $su(2)$ and $su(1, 1)$ generalizing Witten's algebra, *J. Phys. A* **26**, L127–34

Delbecq, C. & Quesne, C. (1993b) Representation theory and q-boson realizations of Witten's $su(2)$ and $su(1, 1)$ deformations, *Phys. Lett. B* **300**, 227–33

Delbecq, C. & Quesne, C. (1993c) A cubic deformation of $su(2)$, *Modern Phys. Lett. A* **8**, 961–6

Deligne, P. (1989) Le groupe fondamental de la droite projective moins trois points, in *Galois groups over* $\mathbb{Q}$, P. Cartier *et al.* (eds), Publications of the Mathematical Sciences Research Institute 16, pp. 79–298, Springer, Berlin

Deligne, P. (1991) Catégories Tannakiennes, in *Grothendieck Festschrift, Vol. 2*, P. Cartier *et al.* (eds), pp. 111–95, Birkhäuser, Boston

Deligne, P. & Milne, J. (1982) Tannakian categories, in *Hodge Cycles, Motives and Shimura Varieties*, P. Deligne, J. S. Milne, A. Ogus & K. Shih (eds), Lecture Notes in Mathematics

900, pp. 101–228, Springer, Berlin

Del Sol Mesa, A., Loyola, G., Moshinsky, M. & Velasquez, V. (1993) Quantum groups and the recovery of $U(3)$ symmetry in the Hamiltonian of the nuclear shell model, *J. Phys. A* **26**, 1147–60

Demidov, E. E. (1991a) Function algebras on quantum matrix supergroups, *Funct. Anal. Appl.* **24**, 238–40

Demidov, E. E. (1991b) Multiparameter quantum deformations of the group $GL(n)$, *Russian Math. Surveys* **46** (4), 169–71

Demidov, E., Manin, Yu. I., Mukhin, E. E. & Zhdanovich, D. V. (1990) Non-standard quantum deformations of $GL(n)$ and constant solutions of the Yang-Baxter equation, *Progr. Theor. Phys. Suppl.* **102**, 203–18

Deodhar, V. V., Gabber, O. & Kac, V. G. (1982) Structure of some categories of representations of infinite-dimensional Lie algebras, *Adv. Math.* **45**, 92–116

Destri, C. & de Vega, H. J. (1992a) Bethe ansatz and quantum groups: the light cone lattice approach, I. Six vertex and SOS models, *Nucl. Phys. B* **374**, 692–719

Destri, C. & de Vega, H. J. (1992b) Bethe ansatz and quantum groups: the light cone lattice approach, II. From RSOS $(p + 1)$-models to p-restricted sine Gordon field theories, *Nucl. Phys. B* **385**, 361–91

Devchand, Ch. & Saveliev, M. V. (1991) Comultiplication for quantum deformations of the centreless Virasoro algebra in the continuum formulation, *Phys. Lett. B* **258**, 364–8

De Vega, H. J. (1988) Quantum groups (YBZF algebras), integrable field theories and statistical mechanics, in *Two-dimensional Models and String Theories*, E. Abdalla & M. C. B. Abdalla (eds), pp. 86–160, World Scientific, Singapore

De Vega, H. J. (1989a) Yang-Baxter algebras, integrable theories and quantum groups, *Internat. J. Modern Phys. A* **4**, 2371–463

De Vega, H. J. (1989b) Yang-Baxter algebras, integrable quantum field theories and conformal models, in *Differential Geometric Methods in Theoretical Physics, Chester, 1988*, A. I. Solomon (ed.), pp. 3–11, World Scientific, Teaneck, NJ

De Vega, H. J. (1990a) Yang-Baxter algebras, integrable theories and Bethe ansatz, in *Quantum Groups, Clausthal, Germany, 1989*, H.-D. Doebner & J.-D. Hennig (eds), pp. 129–82, Lecture Notes in Physics 370, Springer, Berlin

De Vega, H. J. (1990b) Yang-Baxter algebras, integrable theories and Bethe ansatz, *Internat. J. Modern Phys.* **4**, 735–801

De Vega, H. J. & Fateev, V. A. (1991) Factorizable S-matrices for perturbed W-invariant theories, *Internat. J. Modern Phys. A* **6**, 3221–34

De Vega, H. J. & Nicolai, H. (1990) The octionic S-matrix, *Phys. Lett. B* **244**, 295–8

De Vega, H. J. & Sanchez, N. (1989) Quantum group generalization of string theory, *Phys. Lett. B* **216**, 97–102

De Wilde, M. & Lecomte, P. (1988) Formal deformations of the Poisson Lie algebra of a symplectic manifold and star-products. Existence, equivalence, derivations, in *Deformation Theory of Algebras and Structures and Applications*, M. Hazewinkel (ed.), pp. 897–960, Kluwer, Dordrecht

D'Hoker, E., Floreanini, R. & Vinet, L. (1991) q-oscillator realizations of the metaplectic representation of quantum $osp(3, 2)$, *J. Math. Phys.* **32**, 1427–9

Dijkgraaf, R. (1989) *A Geometrical Approach to Two-dimensional Conformal Field Theory*, Doctoral Thesis, University of Utrecht

Dijkgraaf, R., Pasquier, V. & Roche, P. (1992) Quasi-quantum groups related to orbifold models, in *Integrable Systems and Quantum Groups*, M. Carfora, M. Martinelli & A. Marzuoli (eds), pp. 75–98, World Scientific, Singapore

Dijkgraaf, R. & Witten, E. (1990) Topological gauge theories and group cohomology, *Comm. Math. Phys.* **129**, 393–429

Dimakis, A. & Müller-Hoissen, F. (1992) Quantum mechanics on a lattice and q-deformations, *Phys. Lett.* B **295**, 242–7

Dimakis, A., Müller-Hoissen, F. & Striker, T. (1993) Non-commutative differential calculus and lattice gauge theory, *J. Phys.* A **26**, 1927–50

Dipper, R. (1991) Polynomial representations of finite general linear groups in nondescribing characteristic, in *Representation Theory of Finite Groups and Finite-dimensional Algebras, Bielefeld, 1991*, G. O. Michler & C. M. Ringel (eds), Progress in Mathematics 95, pp. 343–70, Birkhäuser, Basel

Dipper, R. & Donkin, S. (1991) Quantum GL_n, *Proc. London Math. Soc.* **63**, 165–211

Dipper, R. & James, G. D. (1989) The q-Schur algebra, *Proc. London Math. Soc.* **59**, 23–50

Dipper, R. & James, G. D. (1991) q-tensor space and q-Weyl modules, *Trans. Amer. Math. Soc.* **327**, 251–82

Dixmier, J. (1977) *Enveloping Algebras*, North-Holland, Amsterdam

Djemai, A. E. F. (1992) Quantum mechanical Galilei group and Q-planes, *Modern Phys. Lett.* A **7**, 3169–78

Dobrev, V. K. (1989) Character formulae for $U_q(sl(3, \mathbb{C}))$ representations, in *Knots, Topology and Quantum Field Theories, Florence, 1989*, L. Lusanna (ed.), pp. 539–47, World Scientific, Singapore

Dobrev, V. K. (1990) Classification and characters of $U_q(sl(3, \mathbb{C}))$ representations, in *Quantum Groups, Clausthal, Germany, 1989*, H.-D. Doebner & J.-D. Hennig (eds), Lecture Notes in Physics 370, pp. 107–17, Springer, Berlin

Dobrev, V. K. (1991) Singular vectors of quantum group representations for straight Lie algebra roots, *Lett. Math. Phys.* **22**, 251–66

Dobrev, V. K. (1992a) Singular vectors of representations of quantum groups, *J. Phys.* A **25**, 149–60

Dobrev, V. K. (1992b) Duality for the matrix quantum group $GL_{p,q}(2)$, *J. Math. Phys.* **33**, 3419–30

Dobrev, V. K. (1993) Canonical q-deformations of non-compact Lie (super-) algebras, *J. Phys.* A **26**, 1317–34

Doebner, H.-D., Hennig, J.-D. & Lücke, W. (1990) Mathematical guide to quantum groups, in *Quantum Groups, Clausthal, Germany, 1989*, H.-D. Doebner & J.-D. Hennig (eds), Lecture Notes in Physics 370, pp. 29–63, Springer, Berlin

Donin, J. & Gurevich, D. (1993) Quasi-Hopf algebras and R-matrix structures in line bundles over flag manifolds, *Selecta Mathematica* **12**, 37–48

Donkin, S. (1993) On tilting modules for algebraic groups, *Math. Z.* **212**, 39–60

Doplicher, S. & Roberts, J. E. (1990) Why there is a field algebra with a compact gauge group describing the superselection structure in particle physics, *Comm. Math. Phys.* **131**, 51–107

Dorey, P. (1991) Root systems and purely elastic S-matrices, *Nucl. Phys.* B **358**, 654–76

Dorey, P. (1992) Root systems and purely elastic S-matrices II, *Nucl. Phys.* B **374**, 746–61

Dorey, P. (1993) Partition functions, intertwiners and the Coxeter element, *Internat. J. Modern Phys.* A **8**, 193–208

Dorfel, B. D. (1988) Quantum integrable systems and the inverse scattering method, *Fortschr. Phys.* **36**, 281–333

Drabant, B., Schlieker, M., Weich, W. & Zumino, B. (1992a) Complex quantum groups and their quantum universal enveloping algebras, *Comm. Math. Phys.* **147**, 625–33

Drabant, B., Schlieker, M., Weich, W. & Zumino, B. (1992b) Complex quantum groups and their dual Hopf algebras, in *Quantum Groups and Related Topics*, R. Gierlak *et al.* (eds), pp. 13–22, Kluwer, Dordrecht

Drinfel'd, V. G. (1983a) Hamiltonian structures on Lie groups, Lie bialgebras and the geometric meaning of the classical Yang-Baxter equations, *Soviet Math. Dokl.* **27**, 68–71

Drinfel'd, V. G. (1983b) On constant quasiclassical solutions of the Yang-Baxter quantum equation, *Soviet Math. Dokl.* **28**, 667–71

Drinfel'd, V. G. (1985) Hopf algebras and the quantum Yang-Baxter equation, *Soviet Math. Dokl.* **32**, 254–8

Drinfel'd, V. G. (1986a) On quadratic commutation relations in the quasiclassical case, in *Mathematical Physics, Functional Analysis*, V. A. Marchenko (ed.), pp. 25–34, Naukova Dumka, Kiev (Russian)

Drinfel'd, V. G. (1986b) Degenerate affine Hecke algebras and Yangians, *Funct. Anal. Appl.* **20**, 62–4

Drinfel'd, V. G. (1987) Quantum groups, in *Proceedings of the International Congress of Mathematicians, Berkeley, 1986*, A. M. Gleason (ed.), pp. 798–820, American Mathematical Society, Providence, RI

Drinfel'd, V. G. (1988a) Quantum groups, *J. Soviet Math.* **41**, 18–49

Drinfel'd, V. G. (1988b) A new realization of Yangians and quantized affine algebras, *Soviet Math. Dokl.* **36**, 212–6

Drinfel'd, V. G. (1989) Quasi-Hopf algebras and the Knizhnik-Zamolodchikov equations, in *Problems of Modern Quantum Field Theory*, A. A. Belavin, A. U. Klimyk & A. B. Zamolodchikov (eds), pp. 1–13, Springer, Berlin

Drinfel'd, V. G. (1990a) On almost cocommutative Hopf algebras, *Leningrad Math. J.* **1**, 321–42

Drinfel'd, V. G. (1990b) Quasi-Hopf algebras, *Leningrad Math. J.* **1**, 1419–57

Drinfel'd, V. G. (1991) On quasitriangular quasi-Hopf algebras and a certain group closely connected with $Gal(\bar{\mathbb{Q}}, \mathbb{Q})$, *Leningrad Math. J.* **2**, 829–60

Drinfel'd, V. G. (1992a) On some unsolved problems in quantum group theory, in *Quantum Groups, Proceedings of Workshops held in the Euler International Mathematical Institute 1990*, P. P. Kulish (ed.), Lecture Notes in Mathematics 1510, pp. 1–8, Springer, Berlin

Drinfel'd, V. G. (1992b) On the structure of the quasitriangular quasi-Hopf algebras, *Funct. Anal. Appl.* **26**, 63–5

Du, J. (1991) The modular representation theory of q-Schur algebras II, *Math. Z.* **208**, 503–36

Du, J. (1992a) Canonical bases for irreducible representations of quantum GL_n, *Bull. London Math. Soc.* **24**, 325–34

Du, J. (1992b) The modular representation theory of q-Schur algebras, *Trans. Amer. Math. Soc.* **329**, 253–71

Du, J., Parshall, B. & Wang, J. P. (1991) Two-parameter quantum linear groups and the hyperbolic invariance of q-Schur algebras, *J. London Math. Soc.* **44**, 420–36

Dubois-Violette, M. (1988) Dérivations et calcul differentiel non commutatif, *C. R. Acad. Sci. Paris Sér. I* **307**, 403–8

Dubois-Violette, M. (1990) On the theory of quantum groups, *Lett. Math. Phys.* **19**, 121–6

Dubois-Violette, M., Kerner, R. & Madore, J. (1990) Noncommutative differential geometry of matrix algebras, *J. Math. Phys.* **31**, 323–30

Dubois-Violette, M. & Launer, G. (1990) The quantum group of a non-degenerate bilinear form, *Phys. Lett. B* **245**, 175–7

Durhuus, B., Jakobsen, P. & Nest, R. (1992) Topological quantum field theories from generalized $6j$-symbols, in *Topological and Geometrical Methods in Field Theory*, J. Mickelsson & O. Pekonen (eds), pp. 121–34, World Scientific, Singapore

Dutta-Roy, B. & Ghosh, G. (1993a) Geometric phase in q-deformed 'quantum mechanics', *Phys. Lett. A* **173**, 439–41

Dutta-Roy, B. & Ghosh, G. (1993b) Curvature of Hilbert space and q-deformed quantum mechanics, *Modern Phys. Lett. A* **8**, 1427–32

Efthimou, C. J. (1993) Quantum group symmetry for the Φ_{12}-perturbed and Φ_{21}-perturbed minimal models of conformal field theory, *Nucl. Phys. B* **398**, 697–740

Egusquiza, I. L. (1992) Quantum group invariance in quantum sphere valued statistical models, *Phys. Lett. B* **276**, 465–71

Elashvili, A. (1983) Frobenius Lie algebras, *Funct. Anal. Appl.* **16**, 94–5

Ellinas, D. (1990) Studies in Lie and quantum algebras, *Soc. Sci. Fenn. Comment. Phys.-Math.* **120**, 11pp.

Enock, M. & Schwartz, J.-M. (1992) *Kac-Algebras and Duality of Locally Compact Groups*, Springer, Berlin

Enriquez, B. (1992) Rational forms for twistings of enveloping algebras of simple Lie algebras, *Lett. Math. Phys.* **25**, 111–20

Ewen, H. & Ogievetsky, O. (1992) Jordanian solution of simplex equations, *Lett. Math. Phys.* **26**, 307–14

Ewen, H., Ogievetsky, O. & Wess, J. (1991) Quantum matrices in two dimensions, *Lett. Math. Phys.* **22**, 297–305

Exton, H. (1983) *q-Hypergeometric Functions and Applications*, Ellis Horwood, Chichester, UK

Faddeev, L. (1980) Quantum completely integrable models in field theory, *Soviet Scientific Reviews Sect. C* **1**, 107–55, Harwood Academic Publishers, Chur, Switzerland

Faddeev, L. D. (1984) Integrable models in $(1 + 1)$-dimensional quantum field theory, in *Recent Advances in Field Theory and Statistical Mechanics, Les Houches, Session XXXIX, 1982,* J.-B. Zuber & R. Stora (eds), pp. 561–608, North-Holland, Amsterdam

Faddeev, L. D. (1989) Quantum groups, *Bol. Soc. Brasil. Mat.* (New Series) **20**, 47–54

Faddeev, L. D. (1990) On the exchange matrix for WZNW model, *Comm. Math. Phys.* **132**, 131–8

Faddeev, L., Reshetikhin, N. & Takhtajan, L. (1988) Quantization of Lie groups and Lie algebras, in *Algebraic analysis, Vol. 1*, M. Kashiwara & T. Kawai (eds), pp. 129–39, Academic Press, New York

Faddeev, L., Reshetikhin, N. & Takhtajan, L. (1989) Quantum groups, in *Braid Group, Knot Theory and Statistical Mechanics*, C. N. Yang & M. L. Ge (eds), pp. 97–110, Advanced Series in Mathematical Physics 9, World Scientific, Singapore

Faddeev, L., Sklyanin, E. K. & Takhtajan, L.A. (1979) The quantum inverse problem I, *Theoret. Math. Phys.* **40**, 194–220 (Russian)

Faddeev, L. D. & Takhtajan, L. (1979) The quantum inverse scattering method of the inverse problem and the Heisenberg XYZ model, *Russian Math. Surveys* **34** (5), 11–68

Faddeev, L. & Takhtajan, L. (1987) *Hamiltonian Methods in the Theory of Solitons*, Springer, Berlin

Fairlie, D. B. (1990a) Quantum deformations of $SU(2)$, *J. Phys. A* **23**, L183–7

Fairlie, D. B. (1990b) Quantum groups or the uses of algebraic computational programs in the suggestion and verification of algebraic conjectures, in *Strings '89*, R. Arnowitt *et al.* (eds), pp. 334–40, World Scientific, River Edge, NJ

Fairlie, D. B. (1991a) Polynomial algebras with q-Heisenberg operators, in *Quantum Groups*, T. Curtright, D. Fairlie & C. Zachos (eds), pp. 133–42, World Scientific, Singapore

Fairlie, D. B. (1991b) q-analysis and quantum groups, in *Symmetries in Science V*, B. Gruber & H.-D. Doebner (eds), pp. 147–57, Plenum, New York

Fairlie, D. B. & Zachos, C. (1991) Multiparameter associative generalizations of canonical commutation relations and quantized planes, *Phys. Lett. B* **256**, 43–9

Fateev, V. A. & Lukyanov, S. L. (1992a) Poisson-Lie groups and classical W-algebras, *Internat. J. Modern Phys. A* **7**, 853–76

Fateev, V. A. & Lukyanov, S. L. (1992b) Vertex operators and representations of quantum universal enveloping algebras, *Internat. J. Modern Phys. A* **7**, 1325–59

Fateev, V. A. & Zamolodchikov, A. V. (1982) Self-dual solutions of the star-triangle relations in $\mathbb{Z}_N$-models, *Phys. Lett. A* **92**, 37–9

Fei, S. M. (1991) Hopf algebraic structures of $SU_{q\to 0}(2)$ and $SU_q(2)$ algebras, monopoles and symplectic geometry on 2D manifolds, *J. Phys. A* **24**, 5195–214

Fei, S. M. & Guo, H. Y. (1991) Symplectic geometry of $SU_{q\to 0}(2)$ and $SU_q(2)$ algebras, *J. Phys. A* **24**, 1–10

Fei, S. M., Guo, H. Y. & Shi, H. (1992) Multiparameter solutions of the Yang-Baxter equation, *J. Phys. A* **25**, 2711–20

Fei, S. M. & Yue, R. H. (1993) New algebra related to non-standard R-matrix, *J. Phys. A* **26**, L217–20

Feigin, B. & Frenkel, E. (1990) Quantization of Drinfeld-Sokolov reduction, *Phys. Lett. B* **246**, 75–81

Feinsilver, P. (1987) Discrete analogues of the Heisenberg-Weyl algebra, *Monatsh. Math.* **104**, 89–108

Feinsilver, P. (1989) Elements of q-harmonic analysis, *J. Math. Anal. Appl.* **141**, 509–26

Feinsilver, P. (1990) Lie algebras and recurrence relations III: q-analogs and quantized algebras, *Acta Appl. Math.* **19**, 207–51

Felder, G., Fröhlich, J. & Keller, G. (1989) Braid matrices and structure constants for minimal conformal models, *Comm. Math. Phys.* **124**, 647–64

Felder, G. & Leclair, A. (1992) Restricted quantum affine symmetry of perturbed minimal conformal models, in *Infinite Analysis*, A. Tsuchiya, T. Eguchi & M. Jimbo (eds), Advanced Series in Mathematical Physics 16, pp. 239–78, World Scientific, Singapore

Felder, G. & Wieczerkowski, C. (1991) Topological representations of the quantum group $U_q(sl_2)$, *Comm. Math. Phys.* **138**, 583–605

Felder, G. & Wieczerkowski, C. (1992) Fock space representations of $A_1^{(1)}$ and topological representations of $U_q(sl_2)$, in *New Symmetry Principles in Quantum Field Theory*, J. Fröhlich *et al.* (eds), NATO Advanced Science Institutes Series B: Physics 295, pp. 513–22, Plenum, New York

Feng, P. & Tsygan, B. (1991) Hochschild and cyclic cohomology of quantum groups, *Comm. Math. Phys.* **140**, 481–521

Fenn, R. & Rourke, C. (1979) On Kirby's calculus of links, *Topology* **18**, 1–15

Filippov, A. T., Gangopadhyay, D. & Isaev, A. P. (1991) Harmonic oscillator realization of the canonical q-transformation, *J. Phys. A* **24**, L63–8

Filippov, A. T. & Isaev, A. P. (1992) Para-grassmann analysis and quantum groups, *Modern Phys. Lett. A* **7**, 2129–41

Finkelberg, M. (1993) Ph. D. Thesis, Harvard University, Cambridge, MA

Finkelstein, R. J. (1992) q-deformations of the oscillator, *J. Math. Phys.* **33**, 4259–66

Fiore, G. (1992) $SO_q(N,\mathbb{R})$-symmetric harmonic oscillator on the N dim real quantum euclidean space, *Internat. J. Modern Phys. A* **7**, 7597–614

Fivel, D. (1990) Interpolation between Fermi and Bose statistics using generalized commutators, *Phys. Rev. Lett.* **65**, 3361–4

Fivel, D. (1991) Quasi-coherent states and the spectral resolution of the q-Bose field operator, *J. Phys. A* **24**, 3575–86

Flato, M., Lichnerowicz, A. & Sternheimer, D. (1975) Déformations 1-différentiables des algèbre de Lie attachées à une varieté symplectique ou de contact, *Compositio Math.* **31**, 47–82

Flato, M. & Lu, Z. (1991) Remarks on quantum groups, *Lett. Math. Phys.* **21**, 85–8

Flato, M. & Sternheimer, D. (1991) On a possible origin of quantum groups, *Lett. Math. Phys.* **22**, 155–60

Fletcher, P. (1990) The uniqueness of the Moyal algebra, *Phys. Lett. B* **248**, 323–8

Fletcher, P. (1991) The Moyal bracket, in *Quantum Groups*, T. Curtright, D. Fairlie & C. Zachos (eds), pp. 143–57, World Scientific, Singapore

Floratos, E. G. (1989) Representations of the quantum group $GL_q(2)$ for values of q on the unit circle, *Phys. Lett. B* **233**, 395–9

Floratos, E. G. (1990) Manin's quantum spaces and standard quantum mechanics, *Phys. Lett. B* **252**, 97–100

Floratos, E. G. (1991) The two anyon system and the q-oscillator, in *Quantum Groups*, T. Curtright, D. Fairlie & C. Zachos (eds), pp. 158–65, World Scientific, Singapore

Floreanini, R., Leites, D. A. & Vinet, L. (1991) On the defining relations of quantum superalgebras, *Lett. Math. Phys.* **23**, 127–31

Floreanini, R., Spiridonov, V. P. & Vinet, L. (1990) Bosonic realization of the quantum superalgebra $osp_q(1, 2n)$, *Phys. Lett. B* **242**, 383–6

Floreanini, R., Spiridonov, V. P. & Vinet, L. (1991a) q-oscillator realizations of the quantum superalgebras $sl_q(m, n)$ and $osp_q(m, 2n)$, *Comm. Math. Phys.* **137**, 149–60

Floreanini, R., Spiridonov, V. P. & Vinet, L. (1991b) q-oscillator realizations of quantum superalgebras, in *Group Theoretical Methods in Physics, Moscow, 1990*, V. V. Dodonov & V. I. Manko (eds), Lecture Notes in Physics 382, pp. 208–14, Springer, Berlin

Floreanini, R. & Vinet, L. (1990) q-analogues of the para-Bose and para-Fermi oscillators and representations of quantum algebras, *J. Phys. A* **23**, L1019–23

Floreanini, R. & Vinet, L. (1991a) q-orthogonal polynomials and the oscillator quantum group, *Lett. Math. Phys.* **22**, 45–54

Floreanini, R. & Vinet, L. (1991b) Braid group action on the q-Weyl algebra, *Lett. Math. Phys.* **23**, 151–8

Floreanini, A. & Vinet, L. (1992a) q-oscillators and quantum superalgebras, in *Topological and Geometrical Methods in Field Theory*, J. Mickelsson & O. Pekonen (eds), pp. 147–52, World Scientific, Singapore

Floreanini, A. & Vinet, L. (1992b) q-conformal quantum mechanics and q-special functions, *Phys. Lett. B* **277**, 442–6

Floreanini, A. & Vinet, L. (1992c) Addition formulas for q-Bessel functions, *J. Math. Phys.* **33**, 2984–8

Floreanini, R. & Vinet, L. (1992d) The metaplectic representation of $sl_q(2)$ and the q-Gegenbauer polynomials, *J. Math. Phys.* **33**, 1358–62

Floreanini, R. & Vinet, L. (1992e) Using quantum algebras in q-special function theory, *Phys. Lett. A* **170**, 21–8

Foda, O. & Miwa, T. (1992) Corner transfer matrices and quantum affine algebras, in *Infinite Analysis*, A. Tsuchiya, T. Eguchi & M. Jimbo (eds), Advanced Series in Mathematical Physics 16, pp. 279–302, World Scientific, Singapore

Fordy, A. P., Reyman, A. G. & Semenov-Tian-Shansky, M. A. (1989) Classical r-matrices and compatible Poisson brackets for coupled KdV systems, *Lett. Math. Phys.* **17**, 25–9

Forger, M. (1989) Solutions of the Yang-Baxter equation from field theory, in *Differential Geometric Methods in Theoretical Physics, Chester, 1988*, A. I. Solomon (ed.), pp. 36–46, World Scientific, Teaneck, NJ

Fradkin, E. S. & Netsaev, R. R. (1990) Quantum R-matrix in the relativistic string model in a space of constant curvature, *Modern Phys. Lett. A* **5**, 1329–38

Frahm, H. (1991) On the construction of integrable XXZ Heisenberg models with arbitrary spin, *Contemp. Math.* **122**, 41–5

Frappat, L. & Sciarrino, A. (1992) A quasi-parafermionic realization of G_2 and $U_q(G_2)$, *J. Phys. A* **25**, L383–6

Frappat, L., Sorba, P. & Sciarrino, A. (1991) q-fermionic operators and quantum exceptional algebras, *J. Phys. A* **24**, L179–83

Freidel, L. & Maillet, J.-M. (1991a) Quantum algebras and integrable systems, *Phys. Lett. B* **262**, 278–84

Freidel, L. & Maillet, J.-M. (1991b) On classical and quantum integrable field theories associated to Kac-Moody current algebras, *Phys. Lett. B* **263**, 403–10

Freidel, L. & Maillet, J.-M. (1992) The universal R-matrix and its associated quantum algebra as functionals of the classical r-matrix. The sl_2 case, *Phys. Lett. B* **296**, 353–60

Frenkel, E., Kac, V. & Wakimoto, M. (1992) Characters and fusion rules for W-algebras via quantized Drinfeld-Sokolov reduction, *Comm. Math. Phys.* **147**, 295–328

Frenkel, I. B. & Jing, N. (1988) Vertex representations of quantum affine algebras, *Proc. Natl Acad. Sci. USA* **85**, 9373–7

Frenkel, I. B. & Moore, G. (1991) Simplex equations and their solutions, *Comm. Math. Phys.* **138**, 259–71

Frenkel, I. B. & Reshetikhin, N. Yu. (1992a) Quantum affine algebras and holonomic difference equations, *Comm. Math. Phys.* **146**, 1–60

Frenkel, I. B. & Reshetikhin, N. Yu. (1992b) Quantum affine algebras, commutative systems of difference equations and elliptic solutions to the Yang-Baxter equation, in *Proceedings of the XX International Conference on Differential Geometric Methods in Theoretical Physics*, S. Catto & A. Rocha (eds), World Scientific, Singapore

Freund, P. G. O. & Zabrodin, A. V. (1992a) Macdonald polynomials for Sklyanin algebras: a conceptual basis for the p-adics – quantum group connection, *Comm. Math. Phys.* **147**, 277–94

Freund, P. G. O. & Zabrodin, A. V. (1992b) $\mathbb{Z}_n$-Baxter models and quantum symmetric spaces, *Phys. Lett. B* **284**, 283–8

Freyd, P. J. & Yetter, D. N. (1989) Braided compact closed categories with applications to low dimensional topology, *Adv. Math.* **77**, 156–82

Freyd, P. J., Hoste, J., Lickorish, W. B. R., Millet, K. C., Ocneanu, A. & Yetter, D. N. (1985) A new polynomial invariant of knots and links, *Bull. Amer. Math. Soc.* **12**, 239–46

Friedlander, E. & Parshall, B. (1986) Cohomology of Lie algebras and algebraic groups, *Amer. J. Math.* **108**, 235–53

Frishman, Y., Lukierski, J. & Zakrzewski, W. J. (1993) Quantum group σ models, *J. Phys. A* **26**, 301–12

Fröhlich, J. (1988) Statistics of fields, the Yang-Baxter equation and the theory of knots and links, in *Non-perturbative Quantum Field Theory, Cargese, 1987*, G. 't Hooft, A. Jaffe, G. Mack, P. K. Mitter & R. Stora (eds), NATO Advanced Science Institutes Series B: Physics 185, pp. 71–100, Plenum, New York

Fröhlich, J. & Gabbiani, F. (1990) Braid statistics in local quantum theory, *Rev. Math. Phys.* **2**, 251–353

Fröhlich, J. & Gabbiani, F. (1992) Operator algebras and conformal field theory, preprint ETH-TH/92-30, Zürich

Fröhlich, J. & Kerler, T. (1993) *Quantum Groups, Quantum Categories and Quantum Field Theory*, Lecture Notes in Mathematics 1542, Springer, Berlin

Fröhlich, J. & King, C. (1989) Two-dimensional conformal field theory and three-dimensional topology, *Internat. J. Modern Phys. A* **4**, 5321–99

Fronsdal, C. (1991) Normal ordering and quantum groups, *Lett. Math. Phys.* **22**, 225–8

Fronsdal, C. (1992) $gl_q(n)$ and quantum monodromy, *Lett. Math. Phys.* **24**, 73–8

Fronsdal, C. & Galindo, A. (1993) The dual of a quantum group, *Lett. Math. Phys.* **27**, 59–72

Fu, H. C. & Ge, M. L. (1992a) Verma modules of the quantum group $GL(n)_q$ and its q-boson and Heisenberg-Weyl realizations, *J. Phys. A* **25**, L389–96

Fu, H. C. & Ge, M. L. (1992b) The q-boson realization of parametrized cyclic representations of quantum algebras at $q^p = 1$, *J. Math. Phys.* **33**, 427–35

Fu, H. C. & Ge, M. L. (1992c) q-boson realization of quadratic algebra A_1 and its representations, *J. Phys. A* **25**, L1233–8

Fu, H. C. & Ge, M. L. (1993) Reflection quadratic algebra associated with Z_2 model, *J. Phys. A* **26**, L233–8

Fuchs, J. & van Driel, P. (1990) WZW fusion rules, quantum groups and the modular matrix S, *Nucl. Phys. B* **346**, 632–48

Furlan, P., Ganchev, A. Ch. & Petkova, V. B. (1990) Quantum groups and fusion rules multiplicities, *Nucl. Phys. B* **343**, 205–27

Furlan, P., Ganchev, A. Ch. & Petkova, V. B. (1991) Remarks on the quantum group structure of the rational $c < 1$ conformal field theories, *Internat. J. Modern Phys. A* **6**, 4859–84

Furlan, P., Hadjiivanov, L. K. & Todorov, I. T. (1992) Quantum deformation of the ladder representations of $U(1, 1)$ for $|q| = 1$, *J. Math. Phys.* **33**, 4255–8

Furlan, P., Stanev, Y. S. & Todorov, I. T. (1991) Coherent state operators and n-point invariants for $U_q(sl(2))$, *Lett. Math. Phys.* **22**, 307–19

Gabriel, P. (1972) Unzerlegbare Darstellungen I, *Manuscripta Math.* **6**, 71–103

Ganchev, A. Ch. & Petkova, V. B. (1989) $U_q(sl(2))$ invariant operators and minimal theories fusion matrices, *Phys. Lett. B* **233**, 374–82

Ganchev, A. Ch. & Petkova, V. B. (1990) $U_q(sl(2))$-invariant operators and reduced polynomial identities, in *Quantum Groups, Clausthal, Germany, 1989*, H.-D. Doebner & J.-D. Hennig (eds), Lecture Notes in Physics 370, pp. 96–106, Springer, Berlin

Gangopadhyay, D. (1991) On canonical q-transformations with two q-oscillators, *Modern Phys. Lett. A* **6**, 2909–16

Gasper, G. & Rahman, M. (1990) *Basic Hypergeometric Series*, Cambridge University Press, Cambridge

Gavrilik, A. M. & Klimyk, A. U. (1991) q-deformed orthogonal and pseudo-orthogonal algebras and their representations, *Lett. Math. Phys.* **21**, 215–20

Gawedzki, K. (1991) Classical origin of quantum group symmetries in Wess-Zumino-Witten conformal field theory, *Comm. Math. Phys.* **139**, 201–13

Ge, M. L. (1992) New solutions of Yang-Baxter equations and quantum group structures, in *Quantum Groups, Proceedings of Workshops held in the Euler International Mathematical Institute 1990*, P. P. Kulish (ed.), Lecture Notes in Mathematics 1510, pp. 245–58, Springer, Berlin

Ge, M. L., Gwa, L. H., Piao, F. & Xue, K. (1990) Braid group representation associated with the 10-dimensional representation of $SU(5)$ and its Yang-Baxterization, *J. Phys. A* **23**, 2273–86

Ge, M, L., Jing, N. H. & Liu, G. G. (1992) On quantum groups for $\mathbf{Z}_N$-models, *J. Phys. A* **25**, L799–806

Ge, M. L., Li, Y. Q., Wang, Y. Y. & Xue, K. (1990) The braid group representations associated with some non-fundamental representations of Lie algebras, *J. Phys. A* **23**, 605–18

Ge, M. L. & Liu, G. G. (1992a) A new kind of R-matrix associated with the Hopf algebra $U_q(A_2(\mu))$, *Phys. Lett. A* **167**. 161–4

Ge, M. L. & Liu, G. G. (1992b) Construction of a new quantum double and new solutions to the Yang-Baxter equation, *Lett. Math. Phys.* **24**, 197–203

Ge, M. L., Liu, G. G. & Sun, C. P. (1991a) Nonstandard R-matrices for the Yang-Baxter equation and boson representations of the quantum algebra $sl_q(2)$ with $q^p = 1$, *Phys. Lett. A* **155**. 137–9

Ge, M. L., Liu, G. G. & Sun, C. P. (1991b) The q-boson realization of new solutions of the Yang-Baxter equation associated with the quantum algebra $U_q(sl(3))$ at $q^p = 1$, *Phys. Lett. A* **160**, 433–6

Ge, M. L., Liu, G. G. & Sun, C. P. (1991c) New boson representations of the $sl_q(2)$ with multiplicity two and new solutions to the Yang-Baxter equation at $q^p = 1$, *Lett. Math. Phys.* **23**, 169–78

Ge, M. L., Liu, G. G. & Sun, C. P. (1992) Representations of quantum matrix algebra $M_q(2)$ and its q-boson realization, *J. Math. Phys.* **33**, 2541–5

Ge, M. L., Liu, G. G., Sun, C. P. & Xue, K. (1991) General solutions of $R^{j\frac{1}{2}}(x)$ and quantum group structure, *J. Phys. A* **24**, 4955–63

Ge, M. L., Liu, G. G. & Xue, K. (1991) New solutions of the Yang-Baxter equations: Birman-Wenzl algebra and quantum group structures, *J. Phys. A* **24**, 2679–90

Ge, M. L., Liu, X. F. & Sun, C. P. (1992) The cyclic representation of the quantum algebra $U_q(osp(2,1))$ in terms of the Z^n-algebra, *J. Phys. A* **25**, 2907–10

Ge, M. L. & Su, G. (1991) The statistical distribution function of the q-deformed harmonic oscillator, *J. Phys. A* **24**, L721–3

Ge, M. L., Sun, C. P. & Wang, L. Y. (1990) Weight conservation and quantum group construction of the braid group representation, *J. Phys. A* **23**, L645–8

Ge, M. L., Sun, C. P. & Xue, K. (1992a) New R-matrices for the Yang-Baxter equation associated with the representations of the quantum superalgebra $U_q osp(1,2)$ with q a root of unity, *Phys. Lett. A* **163**, 176–80

Ge, M. L., Sun, C. P. & Xue, K. (1992b) Construction of general colored R matrices for the Yang-Baxter equation and q-boson realization of quantum algebra $sl_q(2)$ when q is a root of unity, *Internat. J. Modern Phys. A* **7**, 6609–22

Ge, M. L. & Wang, Y. W. (1993) More about the non-standard R-matrix associated with $SU_q(2)$, *J. Phys. A* **26**, 443–7

Ge, M. L., Wang, L. Y. & Kong, X. P. (1991) Yang-Baxterization of braid group representation associated with the seven-dimensional representation of G_2, *J. Phys. A* **24**, 569–79

Ge, M. L., Wang, L. Y., Xue, K. & Wu, Y. S. (1989) Akutsu-Wadati link polynomials from Feynman-Kauffman diagrams, *Internat. J. Modern Phys. A* **4**, 3351–73

Ge, M. L. & Wu, A. C. T. (1992/3) Quantum groups constructed from the non-standard braid group representations in the Faddeev-Reshetikhin-Takhtajan approach: I, *J. Phys. A* **24**, L725–32; II, *J. Phys. A* **25**, L807–16

Ge, M. L., Wu, Y. S. & Xue, K. (1991) Explicit trigonometric Yang-Baxterization, *Internat. J. Modern Phys. A* **6**, 3735–79

Ge, M. L. & Xue, K. (1990) Braid group representations related to the generalized Toda system, *Phys. Lett. A* **146**, 245–51

Ge, M. L. & Xue, K. (1991a) Exotic solutions of the Yang-Baxter equation and the quantum group structure, in *Quantum Groups*, T. Curtright, D. Fairlie & C. Zachos (eds), pp. 97–112, World Scientific, Singapore

Ge, M. L. & Xue, K. (1991b) New solutions of braid group representations associated with Yang-Baxter equation, *J. Math. Phys.* **32**, 1301–9

Ge, M. L. & Xue, K. (1991c) Rational Yang-Baxterization and factorized S matrix, *Phys. Lett. A* **152**, 266–72

Ge, M. L. & Xue, K. (1993a) Trigonometric Yang-Baxterization of coloured $\breve{R}$-matrix, *J. Phys. A* **26**, 281–92

Ge, M. L. & Xue, K. (1993b) Extended Birman-Wenzl algebra and Yang-Baxterization, *J. Phys. A* **26**, 1865–74

Ge, M. L., Zhao, Q. & Zhang, Y. J. (1993) Yang-Baxter equations associated with scattering from kaleidoscopes, *Phys. Lett. A* **175**, 199–202

Gel'fand, I. M. & Cherednik, I. V. (1983) Abstract hamiltonian formalism for classical Yang-Baxter bundles, *Russian Math. Surveys* **38** (3), 1–22

Gel'fand, I. M. & Dickii, L. A. (1976) Fractional powers of operators and hamiltonian systems, *Funct. Anal. Appl.* **10**, 259–73

Gel'fand, I. M. & Dorfman, I. Ya. (1979) Hamiltonian operators and algebraic structures connected with them, *Funct. Anal. Appl.* **15**, 173–87

Gel'fand, I. M. & Dorfman, I. Ya. (1982) Hamiltonian operators and the classical Yang-Baxter equation, *Funct. Anal. Appl.* **16**, 241–8

Gel'fand, I. M. & Fairlie, D. B. (1991) The algebra of Weyl symmetrized polynomials and its quantum extension, *Comm. Math. Phys.* **136**, 487–99

Gel'fand, I. M. & Tsetlin, M. L. (1959a) Finite dimensional representations of the group of unimodular matrices, *Dokl. Akad. Nauk. USSR* **71**, 825–8 (Russian); English translation in *I. M. Gel'fand. Collected Papers, Vol. II*, pp. 653–6, Springer, Berlin, 1988

Gel'fand, I. M. & Tsetlin, M. L. (1959b) Finite dimensional representations of the group of orthogonal matrices, *Dokl. Akad. Nauk. USSR* **71**, 1017–20 (Russian); English translation in *I. M. Gel'fand. Collected Papers, Vol. II*, pp. 657–61, Springer, Berlin, 1988

Gel'fand, S. I. & Kazhdan, D. (1992) Examples of tensor categories, *Invent. Math.* **109**, 595–617

Georgiev, G. & Mathieu, O. (1992) Catégorie de fusion pour les groupes de Chevalley, *C. R. Acad. Sci. Paris Sér. I* **315**, 659–62

Gepner, D. (1991) Fusion rings and geometry, *Comm. Math. Phys.* **141**, 381–411

Gerdjikov, V. S. (1987) The Zakharov-Shabat dressing method and the representation theory of semisimple Lie algebras, *Phys. Lett. A* **121**, 184–8

Gerstenhaber, M., Giaquinto, A. & Schack, S. D. (1992) Quantum symmetry, in *Quantum Groups, Proceedings of Workshops held in the Euler International Mathematical Institute 1990*, P. P. Kulish (ed.), Lecture Notes in Mathematics 1510, pp. 9–46, Springer, Berlin

Gerstenhaber, M. & Schack, S. D. (1988) Algebraic cohomology and deformation theory, in *Deformation Theory of Algebras and Structures and Applications*, M. Hazewinkel (ed.), pp. 11–264, Kluwer, Dordrecht

Gerstenhaber, M. & Schack, S. D. (1990) Bialgebra cohomology, deformations and quantum groups, *Proc. Natl Acad. Sci. USA* **87**, 478–81

Gerstenhaber, M. & Schack, S. D. (1992) Algebras, bialgebras, quantum groups and algebraic deformations, in *Deformation Theory and Quantum Groups with Applications to Mathematical Physics*, M. Gerstenhaber & J. Stasheff (eds), Contemporary Mathematics 134, pp. 51–92, American Mathematical Society, Providence, RI

Gervais, J.-L. (1985) Transport matrices associated with the Virasoro algebra, *Phys. Lett. B* **160**, 279–82

Gervais, J.-L. (1990a) The quantum group structure of 2D gravity and minimal models I, *Comm. Math. Phys.* **130**, 257–83

Gervais, J.-L. (1990b) The quantum group of Virasoro conformal theories, in *Recent Developments in Conformal Field Theories, Trieste, 1989*, S. Randjbar-Daemi, E. Sezgin & J.-B. Zuber (eds), pp. 143–59, World Scientific, Singapore

Gervais, J.-L. (1990c) Lie and quantum group structures common to integrable and conformal theories, in *Mathematical Physics, Islamabad, 1989*, F. Hussain & A. Qadir (eds), pp. 138–53, World Scientific, Singapore

Gervais, J.-L. (1991a) Solving the strongly coupled 2D gravity: I. Unitary truncation and quantum group structure, *Comm. Math. Phys.* **138**, 301–38

Gervais, J.-L. (1991b) On the algebraic structure of quantum gravity in two dimensions, *Internat. J. Modern Phys. A* **6**, 2805–27

Gervais, J.-L. (1992a) Quantum group symmetry of 2D gravity, in *Quantum Groups, Proceedings of Workshops held in the Euler International Mathematical Institute 1990*, P. P. Kulish (ed.), Lecture Notes in Mathematics 1510, pp. 259–76, Springer, Berlin

Gervais, J.-L. (1992b) Two-dimensional gravity: quantum group structure of the continuum theory, *Classical and Quantum Gravity* **9**, S97–116

Gervais, J.-L. & Rostand, B. (1991) On two-dimensional supergravity and the quantum super-Mobius group, *Comm. Math. Phys.* **143**, 175–99

Giaquinto, A. (1992) Quantization of tensor representations and deformations of matrix bialgebras, *J. Pure Appl. Algebra* **79**, 169–90

Giler, S., Kosinski, P., Majewski, M., Maslanka, P. & Kunz, J. (1992) More about the q-deformed Poincaré algebra, *Phys. Lett. B* **286**, 57–62

Giler, S., Kosinski, P. & Maslanka, P. (1991) Remarks on differential calculi on Manin's plane, *Modern Phys. Lett. A* **6**, 3251–4

Ginzburg, V. L. & Kumar, S. (1993) Cohomology of quantum groups at roots of unity, *Duke Math. J.* **69**, 179–98

Ginzburg, V. L. & Weinstein, A. (1992) Lie-Poisson structure on some Poisson-Lie groups, *J. Amer. Math. Soc.* **5**, 445–53

Glockner, P. & von Waldenfels, W. (1989) The relations of the non-commutative coefficient algebra of the unitary group, in *Quantum Probability and Applications IV*, L. Accardi & W. von Waldenfels (eds), Lecture Notes in Mathematics 1396, pp. 182–220, Springer, Berlin

Golenishcheva-Kutuzova, M. & Lebedev, D. (1992) Vertex operator representations of some quantum tori Lie algebras, *Comm. Math. Phys.* **148**, 403–16

Gomez, C., Ruiz-Altaba, M. & Sierra, G. (1991) New R-matrices associated with finite-dimensional representations of $U_q(sl(2))$ at roots of unity, *Phys. Lett. B* **265**, 95–8

Gomez, C. & Sierra, G. (1990a) Quantum group meaning of the Coulomb gas, *Phys. Lett. B* **240**, 149–57

Gomez, C. & Sierra, G. (1990b) Comments on rational conformal field theory, quantum groups and towers of algebras, in *Quantum Groups, Clausthal, Germany, 1989*, H.-D. Doebner & J.-D. Hennig (eds), Lecture Notes in Physics 370, pp. 278–306, Springer, Berlin

Gomez, C. & Sierra, G. (1991a) The quantum group symmetry of rational conformal field theories, *Nucl. Phys. B* **352**, 791–828

Gomez, C. & Sierra, G. (1991b) Quantum groups, Riemann surfaces and conformal field theory, in *Differential Geometric Methods in Theoretical Physics, Rapallo, 1990*, C. Bartocci, U. Bruzzo & R. Cianci (eds), Lecture Notes in Physics 375, pp. 120–30, Springer, Berlin

Gomez, C. & Sierra, G. (1991c) Integrability and uniformization in Liouville theory: the geometrical origin of quantized symmetries, *Phys. Lett. B* **255**, 51–60

Gomez, C. & Sierra, G. (1992) Integrability and quantum symmetries, in *Integrable Systems and Quantum Groups*, M. Carfora, M. Martinelli & A. Marzuoli (eds), pp. 37–50, World Scientific, Singapore

Gomez, C. & Sierra, G. (1993) Quantum harmonic oscillator algebra and link invariants, *J. Math. Phys.* **34**, 2119–31

Gong, R. (1992) Path integral formalism for $SU_q(2)$ coherent states, *J. Phys. A* **25**, L1145–50

Gonzalez-Ruiz, A. & Ibort, L. A. (1992) Induction of quantum group representations, *Phys. Lett. B* **296**, 104–8

Goodman, F. M., de la Harpe, P. & Jones, V. F. R. (1989) *Coxeter Graphs and Towers of Algebras*, Mathematical Sciences Research Institute Publications 14, Springer, New York

Goodman, F. M. & Wenzl, H. (1990) Littlewood-Richardson coefficients for Hecke algebras at roots of unity, *Adv. Math.* **82**, 244–65

Gora, J. (1992) Two models of a q-deformed hydrogen atom, *J. Phys. A* **25**, L1281–6

Gould, M. D. (1992a) Quantum groups and diagonalization of the braid generator, *Lett. Math. Phys.* **24**, 183–96

Gould, M. D. (1992b) Reduced Wigner coefficients for $U_q(gl(n))$, *J. Math. Phys.* **33**, 1023–31

Gould, M. D. & Biedenharn, L. C. (1992) The pattern calculus for tensor operators on quantum groups, *J. Math. Phys.* **33**, 3613–35

Gould, M. D., Links, J. & Bracken, A. J. (1992) Matrix elements and Wigner coefficients for $U_q(sl(n))$, *J. Math. Phys.* **33**, 1008–22

Gould, M. D., Zhang, R. B. & Bracken, A. J. (1991a) Generalized Gel'fand invariants and characteristic identities for quantum groups, *J. Math. Phys.* **32**, 2298–303

Gould, M. D., Zhang, R. B. & Bracken, A. J. (1991b) Lie bi-superalgebras and the graded classical Yang-Baxter equation, *Rev. Math. Phys.* **3**, 223–40

Grabowski, J. (1990) Quantum $SU(2)$ group of Woronowicz and Poisson structures, in *Differential Geometry and its Applications, Brno, 1989*, J. Janyska & D. Krupka (eds), pp. 313–22, World Scientific, Teaneck, NJ

Gracia-Bondia, J. M. (1992) Generalized Moyal quantization and homogeneous symplectic spaces, in *Deformation Theory and Quantum Groups with Applications to Mathematical Physics*, M. Gerstenhaber & J. Stasheff (eds), Contemporary Mathematics 134, pp. 93–114, American Mathematical Society, Providence, RI

Granovskii, Ya. I. & Zhedanov, A. S. (1992) 'Twisted' Clebsch-Gordan coefficients for $SU_q(2)$, *J. Phys. A* **25**, L1029–30

Granovskii, Ya. I. & Zhedanov, A. S. (1993a) Linear covariance algebra for $SL_q(2)$, *J. Phys. A* **26**, L357–60

Granovskii, Ya. I. & Zhedanov, A. S. (1993b) Production of Q-bosons by a classical current: an exactly solvable model, *Modern Phys. Lett. A* **8**, 1029–36

Gray, R. W. & Nelson, C. A. (1990) A completeness relation for the q-analogue coherent states by q-integration, *J. Phys. A* **23**, L945–50

Greenberg, O. W. (1991) q-mutators and violations of statistics, in *Quantum Groups*, T. Curtright, D. Fairlie & C. Zachos (eds), pp. 166–80, World Scientific, Singapore

Gromov, N. A. (1993) The matrix quantum unitary Cayley-Klein groups, *J. Phys. A* **26**, L5–8

Gromov, N. A. & Manko, V. I. (1991) Contractions and analytic continuations of the irreducible representations of the quantum algebra $SU_q(2)$, in *Group Theoretical Methods in Physics, Moscow, 1990*, V. V. Dodonov & V. I. Manko (eds), Lecture Notes in Physics 382, pp. 225–8, Springer, Berlin

Gromov, N. A. & Manko, V. I. (1992) Contractions of the irreducible representations of the quantum algebras $su_q(2)$ and $so_q(3)$, *J. Math. Phys.* **33**, 1374–8

Grossman, R. & Radford, D. (1992) A simple construction of bialgebra deformations, in *Deformation Theory and Quantum Groups with Applications to Mathematical Physics*, M. Gerstenhaber & J. Stasheff (eds), Contemporary Mathematics 134, pp. 115–7, American Mathematical Society, Providence, RI

Grothendieck, A. (1984) Esquisse d'un programme, mimeographed notes, Université de Montpelier

Groza, V. A. & Kachurik, I. I. (1990) Addition and multiplication theorems for Krawtchouk, Hahn and Racah q-polynomials, *Dokl. Akad. Nauk. Ukrain. SSR Ser. A* **89**, 3–6 (Russian)

Groza, V. A., Kachurik, I. I. & Klimyk, A. U. (1990) On Clebsch-Gordan coefficients and matrix elements of representations of the quantum algebra $U_q(su_2)$, *J. Math. Phys.* **31**, 2769–80

Gruber, B. & Smirnov, Yu. F. (1991) On quantized Verma modules, in *Symmetries in Science V, Lochau, 1990*, B. Gruber, L. C. Biedenharn & H.-D. Doebner (eds), pp. 293–304, Plenum, New York

Guadagnini, E. (1992) The universal link polynomial, *Internat. J. Modern Phys. A* **7**, 877–946

Guadagnini, E., Martellini, M. & Mintchev, M. (1990a) Chern-Simons holonomies and the appearance of quantum groups, *Phys. Lett. B* **235**, 275–81

Guadagnini, E., Martellini, M. & Mintchev, M. (1990b) Braids and quantum group symmetry in Chern-Simons theory, *Nucl. Phys. B* **336**, 581–609

Guadagnini, E., Martellini, M. & Mintchev, M. (1990c) Chern-Simons field theory and quantum groups, in *Quantum Groups, Clausthal, Germany, 1989*, H.-D. Doebner & J.-D. Hennig (eds), pp. 307–17, Lecture Notes in Physics 370, Springer, Berlin

Guan, X. W., Xiong Zhuang & Zhou, H. Q. (1992) Classical Yang-Baxter equations and quantum $gl(p, q)$-Gaudin model, *Modern Phys. Lett.* **7**, 1647–50

Guil, F. (1990) Commuting differential operators over integrable hierarchies, in *Integrable and Superintegrable Systems*, B. Kuperschmidt (ed.), pp. 307–20, World Scientific, Teaneck, NJ

Gurevich, D. I. (1986) The Yang-Baxter equation and a generalization of formal Lie theory, *Soviet Math. Dokl.* **33**, 758–62

Gurevich, D. I. (1989a) Hecke symmetries and quantum determinants, *Soviet Math. Dokl.* **38**, 555–9

Gurevich, D. I. (1989b) On Poisson brackets associated with the classical Yang-Baxter equation, *Soviet J. Funct. Anal.* **23**, 68–9

Gurevich, D. I. (1990) Equation de Yang-Baxter et quantification de cocycles, *C. R. Acad. Sci. Paris Sér. I* **310**, 845–8

Gurevich, D. I. (1991) Algebraic aspects of the quantum Yang-Baxter equation, *Leningrad Math. J.* **2**, 801–28

Gurevich, D. I., Radul, A. & Rubtsov, V. (1992) Noncommutative differential geometry related to the Yang-Baxter equation, *Zap. Nauchn. Sem. Leningrad Otdel. Mat. Inst. Steklov* **199** (Russian)

Gurevich, D. I. & Rubtsov, V. (1992) Yang-Baxter equation and deformation of associative and Lie algebras, in *Quantum Groups, Proceedings of Workshops held in the Euler International Mathematical Institute 1990*, P. P. Kulish (ed.), Lecture Notes in Mathematics 1510, pp. 47–55, Springer, Berlin

Gurevich, D. I., Rubtsov, V. & Zobin, N. (1992) Quantization of Poisson pairs: the R-matrix approach, *J. Geom. Phys.* **9**, 25–44

Gyoja, A. (1986) A q-analogue of Young symmetrizer, *Osaka J. Math.* **23**, 841–52

Haag, R. (1992) *Local Quantum Physics*, Springer, Berlin

Hadjiivanov, L. K., Paunov, R. R. & Todorov, I. T. (1991) Quantum group extended chiral p-models, *Nucl. Phys. B* **356**, 387–438

Hadjiivanov, L. K., Paunov, R. R. & Todorov, I. T. (1992a) Extended chiral conformal field theories with a quantum symmetry, in *Quantum Groups, Proceedings of Workshops held in the Euler International Mathematical Institute 1990*, P. P. Kulish (ed.), Lecture Notes in Mathematics 1510, pp. 277–302, Springer, Berlin

Hadjiivanov, L. K., Paunov, R. R. & Todorov, I. T. (1992b) U_q covariant oscillators and vertex operators, *J. Math. Phys.* **33**, 1379–94

Hadjiivanov, L. K. & Stoyanov, D. T. (1991) On a class of infinite-dimensional representations of $U_q(sl(2))$, *J. Phys. A* **24**, L907–12

Hakobyan, T. S. & Sedrakyan, A. G. (1993a) R-matrices for highest weight representations of $\widehat{sl_q(2, \mathbb{C})}$ at roots of unity, *Phys. Lett. B* **303**, 27–32

Hakobyan, T. S. & Sedrakyan, A. G. (1993b) Some new spinor representations of quantum groups $B_q(n)$, $C_q(n)$, $G_q(2)$, *J. Math. Phys.* **34**, 2554–60

Hartshorne, R. (1977) *Algebraic Geometry*, Graduate Texts in Mathematics 72, Springer, Berlin

Hayashi, T. (1990) q-analogues of Clifford and Weyl algebras - spinor and oscillator representations of quantum enveloping algebras, *Comm. Math. Phys.* **127**, 129–44

Hayashi, T. (1991) An algebra related to the fusion rules of Wess-Zumino–Witten models, *Lett. Math. Phys.* **22**, 291–6

Hayashi, T. (1992a) Quantum groups and quantum determinants, *J. Algebra* **152**, 146–65

Hayashi, T. (1992b) Quantum deformations of classical groups, *Publ. Res. Inst. Math. Sci.* **28**, 57–81

Hazewinkel, M. (1978) *Formal Groups and Applications*, Academic Press, New York

Hazewinkel, M. (1991) Introductory recommendations for the study of Hopf algebras in mathematics and physics, *CWI Quart.* **4**, 3–26

Helgason, S. (1978) *Differential Geometry, Lie Groups and Symmetric Spaces*, Academic Press, New York

Hennings, M. A. (1993) On solutions to the braid equation identified by Woronowicz, *Lett. Math. Phys.* **27**, 13–6

Hibi, T. & Wakayama, M. (1991) A q-analogue of Capelli's identity for $GL(2)$, preprint, Hokkaido University

Hietarinta, J. (1992) All solutions to the constant quantum Yang-Baxter equation in two dimensions, *Phys. Lett. A* **165**, 245–51

Hietarinta, J. (1993a) Some constant solutions to Zamolodchikov's tetrahedra equations, *J. Phys. A* **26**, L9–16

Hietarinta, J. (1993b) Solving the 2-dimensional constant quantum Yang-Baxter equation, *J. Math. Phys.* **34**, 1725–56

Hinrichsen, H. & Rittenberg, V. (1992) A two-parameter deformation of the $SU(1|1)$ superalgebra and the XY quantum chain in a magnetic field, *Phys. Lett. B* **275**, 350–4

Hinrichsen, H. & Rittenberg, V. (1993) Quantum groups, correlation functions and infrared divergences, *Phys. Lett. B* **304**, 115–20

Hlavaty, L. (1985) Solution to the Yang-Baxter equation corresponding to the XXZ models in an external magnetic field, Communications of the Joint Institute for Nuclear Research, 5 pp., Dubna

Hlavaty, L. (1987) Unusual solutions to the Yang-Baxter equations, *J. Phys. A* **20**, 1661–7

Hlavaty, L. (1991) Two-dimensional quantum spaces corresponding to solutions of the Yang-Baxter equations, *J. Phys. A* **24**, 2903–12

Hlavaty, L. (1992a) Yang-Baxter matrices and differential calculi on quantum hyperplanes, *J. Phys. A* **25**, 485–94

Hlavaty, L. (1992b) New constant and trigonometric 4×4 solutions to the Yang-Baxter equations, *J. Phys. A* **25**, L63–8

Hlavaty, L. (1992c) On the solutions of the Yang-Baxter equations, in *Quantum Groups and Related Topics*, R. Gierlak *et al.* (eds), pp. 179–88, Kluwer, Dordrecht

Hlavaty, L. (1992d) On solutions of the Yang-Baxter equation without additivity, *J. Phys. A* **25**, 1395–7

Hlavaty, L. (1993) A remark on quantum supergroups, *Modern Phys. Lett. A* **7**, 3365–72

Hochschild, G. P. (1965) *The Structure of Lie Groups*, Holden–Day, San Francisco

Hodges, T. J. (1990) Ring theoretical aspects of the Bernstein-Beilinson theorem, in *Noncommutative Ring Theory, Athens, Ohio, 1989*, S. K. Jain & S. R. Lopez-Permouth (eds), Lecture Notes in Mathematics 1448, pp. 155–63, Springer, New York

Hodges, T. J. & Levasseur, T. (1993) Primitive ideals of $C_q[SL(3)]$, *Comm. Math. Phys.* **156**, 581–605

Hojman, S. A. (1991) Quantum algebras in classical mechanics, *J. Phys. A* **24**, L249–54

Hollowood, T. (1993) Quantizing $SL(N)$ solitons and the Hecke algebra, *Internat. J. Modern Phys. A* **8**, 947–81

Hollowood, T. & Mansfield, P. (1990) Quantum group structure of quantum Toda conformal field theories (I), *Nucl. Phys. B* **330**, 720–40

Hopf, H. (1941) Über die Topologie der Gruppen-mannigfaltigkeiten und ihre Verallgemeinerungen, *Ann. of Math.* **42**, 22–52

Hou, B. Y., Hou, B. Y. & Ma, Z. Q. (1991a) Solutions to the Yang-Baxter equation for the spinor representations of q-B_1, *J. Phys. A* **24**, 1363–77

Hou, B. Y., Hou, B. Y. & Ma, Z. Q. (1991b) The XXZ model with Beraha values, *J. Phys. A* **24**, 2847–61

Hou, B. Y., Hou, B. Y. & Ma, Z. Q. (1992) Quantum Clebsch-Gordan coefficients for non-generic q-values, *J. Phys. A* **25**, 1211–22

Hou, B. Y., Hou, B. Y., Ma, Z. Q. & Yin, Y. D. (1991) Solutions of Yang-Baxter equation in the vertex model and the face model for octet representation, *J. Math. Phys.* **32**, 2210–8

Hou, B. Y., Li, K. & Wang, P. (1990) General solutions to fundamental problems of $SU(2)_k$ WZW models, *J. Phys. A* **23**, 3431–46

Hou, B. Y., Lie, D. P. & Yue, R. H. (1989) Quantum group structure in the unitary minimal model, *Phys. Lett. B* **229**, 45–50

Hou, B. Y. & Ma, Z. Q. (1991) Solutions to the Yang-Baxter equation for the spinor representations of q-B_l, *J. Phys. A* **24**, 1363–77

Hou, B. Y., Shi, K. J., Wang, P. & Yue, R. H. (1990) The crossing matrices of WZW $SU(2)$ model and minimal models with the quantum $6j$ symbols, *Nucl. Phys. B* **345**, 659–84

Hou, B. Y., Shi, K. J. & Yang, Z. Z. (1992) $sl_q(n)$ quantum algebra as a limit of the elliptic case, in *Infinite Analysis*, A. Tsuchiya, T. Eguchi & M. Jimbo (eds), Advanced Series in Mathematical Physics 16, pp. 391–404, World Scientific, Singapore

Hou, B. Y., Shi, K. J., Yang, Z. Z. & Yue, R. H. (1991) Integrable quantum chain and the representation theory of the quantum group $SU_q(2)$, *J. Phys. A* **24**, 3825–36

Hou, B. Y., Shi, K. J. & Yue, R. H. (1990) Explicit expressions for fusion and braid matrices in the unitary minimal model, *High Energy Phys. Nucl. Phys.* **14**, 802–9

Hou, B. Y., Wang, P. & Yue, R. H. (1990) The crossing matrices of WZW $SU(2)$ model and minimal models with the quantum $6j$ symbols, *Nucl. Phys. B* **345**, 659–84

Hou, B. Y. & Wei, H. (1989) Algebras connected with the Z_n elliptic solution of the Yang-Baxter equation, *J. Math. Phys.* **30**, 2750–5

Hou, B. Y. & Yang, Z. X. (1991) The Bethe ansatz solutions to the XXZ model and the highest weight states of quantum groups, *J. Northwest Univ.* **21**, 7–11 (Chinese)

Hou, B. Y. & Zhou, Y. K. (1990) Fusion procedure and Sklyanin algebra, *J. Phys. A* **23**, 1147–54

Howe, R. (1988) *The Classical Groups* and invariants of binary forms, in *The Mathematical Heritage of Hermann Weyl*, R. O. Wells Jr. (ed.), Proceedings of Symposia in Pure Mathematics 48, pp. 133–66, American Mathematical Society, Providence, RI

Huang, Y. Z. & Lepowsky, J. (1992) Toward a theory of tensor products for representations of a vertex operator algebra, in *Proceedings of the 20th International Conference on Differential Geometric Methods in Theoretical Physics*, S. Catto & A. Rocha-Caridi (eds), pp. 344–54, World Scientific, Singapore

Huebschmann, J. (1990) Poisson cohomology and quantization, *J. reine angew. Math.* **408**, 57–113

Humphreys, J. E. (1972) *Introduction to Lie Algebras and Representation Theory*, Graduate Texts in Mathematics 9, Springer, New York

Humphreys, J. E. (1990) *Reflection Groups and Coxeter Groups*, Cambridge University Press, Cambridge

Idzumi, M., Tokohiro, T., Iohara, K., Jimbo, M., Miwa, T. & Nakashima, T. (1993) Quantum affine symmetry in vertex models, *Internat. J. Modern Phys. A* **8**, 1479–511

Ihara, Y. (1988) Galois groups over $\mathbb{Q}$ and monodromy, in *Prospects of Algebraic Analysis*, RIMS Report 675, pp. 23–34 (Japanese)

Ihara, Y. (1991) Braids, Galois groups and some arithmetic functions, in *Proceedings of the International Congress of Mathematicians, Berkeley, 1986*, A. M. Gleason (ed.), pp. 99–120, American Mathematical Society, Providence, RI

Ikeda, K. (1991) The Hamiltonian systems on the Poisson structure of the quasi-classical limit of $GL_q(\infty)$, *Lett. Math. Phys.* **23**, 121–6

Iosifescu, M. & Scutaru, H. (1988a) Degenerate representations from quantum kinematical constraints, in *Group Theoretical Methods in Physics, Varna, 1987*, H.-D. Doebner, J.-D. Hennig & T. D. Palev (eds), Lecture Notes in Physics 313, pp. 230–7, Springer, Berlin

Iosifescu, M. & Scutaru, H. (1988b) Second degree kinematical constraints associated with dynamical symmetries, *J. Math. Phys.* **29**, 742–57

Iotov, M. S. & Todorov, I. T. (1991) On the universal R-matrix for $U_q(sl_{r+1})$, *Lett. Math. Phys.* **23**, 241–50

Isaev, A. P. & Popowicz, Z. (1992) q-trace for quantum groups and q-deformed Yang-Mills theory, *Phys. Lett. B* **281**, 271—8

Isaev, A. P. & Malik, R. P. (1992) Deformed traces and covariant quantum algebras for quantum groups $GL_{qp}(2)$ and $GL_{qp}(1|1)$, *Phys. Lett. B* **280**, 219–26

Itoh, T. & Yamada, Y. (1991a) Explicit exchange relations in the $SL(2)$ Wess-Zumino-Witten model, *Internat. J. Modern Phys. A* **6**, 3283–91

Itoh, T. & Yamada, Y. (1991b) Exchange relations in Wess-Zumino-Novikov-Witten model and quantum groups, *Progr. Theoret. Phys.* **85**, 751–8

Itoyama, H. (1989) Symmetries of the generalized Toda system, *Phys. Lett. A* **140**, 391–4

Itoyama, H. & Moxhay, P. (1991) Thermodynamic Bethe ansatz, sine Gordan theory and minimal models, in *Quantum Groups*, T. Curtright, D. Fairlie & C. Zachos (eds), pp. 306–20, World Scientific, Singapore

Itoyama, H. & Sevrin, A. (1990) Braiding matrices of conformal blocks and coset models, *Internat. J. Modern Phys. A* **5**, 211–22

Ivanov, I. T. & Uglov, D. B. (1992) R-matrices for the semicyclic representations of $U_q(\widehat{sl(2)})$, *Phys. Lett. A* **167**, 459–64

Iwahori, N. (1964) On the structure of a Hecke ring of a Chevalley group over a finite field, *J. Fac. Sci. Univ. Tokyo Sect. I* **10**, 215–36

Iwahori, N. & Matsumoto, H. (1965) On some Bruhat decompositions and the structure of the Hecke ring of p-adic Chevalley groups, *Publ. Math. IHES* **25**, 5–48

Iwao, S. (1990) Spectroscopy in Witten's quantum universal enveloping algebra of SU_2 and strength of deformation of its representation space, *Progr. Theor. Phys.* **83**, 363–72

Izergin, A. G. & Korepin, V. E. (1981) The inverse scattering method approach to the quantum Shabat-Mikhailov model, *Comm. Math. Phys.* **79**, 303–16

Izergin, A. G. & Korepin, V. E. (1982) Quantum inverse scattering method, *Soviet J. Particles & Nuclei* **13**, 207–23

Izergin, A. G. & Korepin, V. E. (1984a) The quantum inverse scattering approach to correlation functions, *Comm. Math. Phys.* **94**, 67–92

Izergin, A. G. & Korepin, V. E. (1984b) The most general L-operator for the R-matrix of the XXX model, *Lett. Math. Phys.* **8**, 259–65

Izergin, A. G. & Korepin, V. E. (1985) Correlation functions for the Heisenberg XXZ ferromagnet, *Comm. Math. Phys.* **99**, 271–302

Jacobson, N. (1962) *Lie Algebras*, Wiley (Interscience), New York

Jagannathan, R., Sridhar, R., Vasudevan, R., Chaturvedi, S., Krishnakamuri, M., Shanta, P. & Srinivasan, V. (1992) On the number operators of multimode systems of deformed oscillators covariant under quantum groups, *J. Phys. A* **25**, 6429–54

Jain, V. & Ogievetsky, O. (1992) Classical isomorphisms for quantum groups, *Modern Phys. Lett. A* **7**, 2199–209

Jannussis, A., Brodimas, G. & Mignani, R. (1991) Quantum groups and Lie-admissible time evolution, *J. Phys. A* **24**, L775–8

Jannussis, A., Brodimas, G., Sourlas, D., Papaloucas, L., Poulopoulos, P. & Siafarikas, P. (1982) Generalized q Bose operators, *Lett. Nuovo Cim.* **34**, 375–9

Jannussis, A., Brodimas, G., Sourlas, D., Vlachos, K., Siafarikas, P. & Papaloucas, L. (1983) Some properties of q-analysis and application to noncanonical quantum mechanics, *Hadronic J.* **6**, 1653–86

Jannussis, A., Brodimas, G., Sourlas, D. & Zisis, V. (1981) Remarks on the q-quantization, *Lett. Nuovo Cim.* **30**, 123–7

Jantzen, J. C. (1987) *Representations of Algebraic Groups*, Academic Press, Boston

Jarvis, P. D. & Baker, T. (1993) q-deformation of radial problems: the simple harmonic oscillator in two dimensions, *J. Phys. A* **26**, 883–94

Jarvis, P. D., Warner, R. C., Yang, C. M. & Zhou, R. B. (1992) BRST cohomology for $U_q(sl(2))$ representations, *J. Phys. A* **25**, L895–900

Jimbo, M. (1985) A q-difference analogue of $U(\mathfrak{g})$ and the Yang-Baxter equation, *Lett. Math. Phys.* **10**, 63–9

Jimbo, M. (1986a) A q-analogue of $U(gl(N+1))$, Hecke algebra and the Yang-Baxter equation, *Lett. Math. Phys.* **11**, 247–52

Jimbo, M. (1986b) Quantum R-matrix related to the generalized Toda system: an algebraic approach, in *Field Theory, Quantum Gravity and Strings*, H. J. de Vega & N. Sanchez (eds), Lecture Notes in Physics 246, pp. 335–61, Springer, Berlin

Jimbo, M. (1986c) Quantum R-matrix for the generalized Toda system, *Comm. Math. Phys.* **102**, 537–47

Jimbo, M. (ed.) (1989a) *Yang-Baxter Equation in Integrable Systems*, World Scientific, Singapore

Jimbo, M. (1989b) Introduction to the Yang-Baxter equation, in *Braid Group, Knot Theory and Statistical Mechanics*, C. N. Yang & M. L. Ge (eds), Advanced Series in Physics 9, pp. 111–34, World Scientific, Singapore

Jimbo, M. (1989c) Introduction to the Yang-Baxter equation, *Internat. J. Modern Phys. A* **4**, 3759–77

Jimbo, M., Kuniba, A., Miwa, T. & Okado, M. (1988) An $A^{(1)}_{n-1}$ family of solvable lattice models, *Comm. Math. Phys.* **119**, 543–65

Jimbo, M., Misra, K. C. & Miwa, T. (1990) Crystal base for the basic representation of $U_q(\hat{sl}_n)$, *Comm. Math. Phys.* **134**, 79–88

Jimbo, M., Misra, K. C., Miwa, T. & Okado, M. (1991) Combinatorics of representations of $U_q(\hat{sl}(n))$ at $q = 0$, *Comm. Math. Phys.* **136**, 543–66

Jimbo, M. & Miwa, T. (1985) Classification of solutions to the star triangle relation for a class of 3 and 4 state IRF models, *Nucl. Phys. B* **257**, 1–18

Jimbo, M., Miwa, T. & Okado, M. (1988a) Local state probabilities of solvable lattice models: an $A^{(1)}_{n-1}$ family, *Nucl. Phys. B* **300**, 74–108

Jimbo, M., Miwa, T. & Okado, M. (1988b) Solvable lattice models related to the vector representation of classical simple Lie algebras, *Comm. Math. Phys.* **116**, 507–25

Jimenez, F. (1990) Quantum group symmetry of $N = 1$ superconformal field theories, *Phys. Lett. B* **252**, 577–85

Jimenez, F. (1991) The quantized symmetries of $N = 2$ superconformal field theories, *Phys. Lett. B* **273**, 399–408

Jing, N. H. (1990) Twisted vertex representations of quantum affine algebras, *Invent. Math.* **102**, 663–90

Jing, N. H. (1991) On a trace of q-analog vertex operators, in *Quantum Groups*, T. Curtright, D. Fairlie & C. Zachos (eds), pp. 113–22, World Scientific, Singapore

Jing, N. H. (1992) Quantum groups with two parameters, in *Deformation Theory and Quantum Groups with Applications to Mathematical Physics*, M. Gerstenhaber & J. Stasheff (eds), Contemporary Mathematics 134, pp. 129–38, American Mathematical Society, Providence, RI

Jing, N. H., Ge, M. L. & Wu, Y. S. (1991) A new quantum group associated with a 'nonstandard' braid group representation, *Lett. Math. Phys.* **21**, 193–203

Jing, N. H. & Xu, J. J. (1991) Comment on the q-deformed fermionic oscillator, *J. Phys. A* **24**, L891–4

Jing, S. (1993) The Jordan-Schwinger realization of two-parametric quantum group $sl_{q,s}(2)$, *Modern Phys. Lett. A* **8**, 543–8

Jing, S. & Fan, H. (1993) A new completeness relation in the q-deformed two-mode Fock space, *J. Phys. A* **26**, L69–74

Jones, V. F. R. (1985) A polynomial invariant for knots via von Neumann algebras, *Bull. Amer. Math. Soc.* **12**, 103–11

Jones, V. F. R. (1987) Hecke algebra representations of braid groups and link polynomials, *Ann. of Math.* **126**, 335–88

Jones, V. F. R. (1988) Subfactors and related topics, in *Operator Algebras and Applications, Vol. 2*, D. E. Evans & M. Takesaki (eds), London Mathematical Society Lecture Note Series 136, pp. 103–18, Cambridge University Press, Cambridge

Jones, V. F. R. (1989) On knot invariants related to some statistical mechanical models, *Pacific J. Math.* **137**, 311–34

Jones, V. F. R. (1990) Baxterization, in *Yang-Baxter Equations, Conformal Invariance and Integrability in Statistical Mechanics and Field Theory*, M. N. Barber & P. A. Pearce (eds), pp. 1–13, World Scientific, Singapore

Jones, V. F. R. (1991) *Subfactors and Knots*, CBMS Regional Conference Series in Mathematics 80, American Mathematical Society, Providence, RI

Jordan, D. A. (1993) Iterated skew polynomial rings and quantum groups, *J. Algebra* **156**, 194–218

Joseph, A. & Letzter, G. (1992) Local finiteness of the adjoint action for quantized enveloping algebras, *J. Algebra* **153**, 289–318

Joyal, A. & Street, R. (1986) Braided monoidal categories, Macquarie Mathematics Reports no. 860081

Joyal, A. & Street, R. (1991a) Tortile Yang-Baxter operators in tensor categories, *J. Pure Appl. Algebra* **71**, 43–51

Joyal, A. & Street, R. (1991b) The geometry of tensor calculus I, *Adv. Math.* **88**, 55–112

Joyal, A. & Street, R. (1992) An introduction to Tannaka duality and quantum groups, in *Category Theory, Como, 1990*, A. Carboni, M. C. Pedicchio & G. Rosolini (eds), Lecture Notes in Mathematics 1488, pp. 413–92, Springer, Berlin

Jurčo, B. (1990) Classical Yang-Baxter equations and quantum integrable systems (Gaudin models), in *Quantum Groups, Clausthal, Germany, 1989*, H.-D. Doebner & J.-D. Hennig (eds), pp. 219–27, Lecture Notes in Physics 370, Springer, Berlin

Jurčo, B. (1991a) On coherent states for the simplest quantum groups, *Lett. Math. Phys.* **21**, 51–8

Jurčo, B. (1991b) Differential calculus on quantized simple Lie groups, *Lett. Math. Phys.* **22**, 177–86

Jurčo, B. & Stovicek, P. (1993) Quantum dressing orbits on compact groups, *Comm. Math. Phys.* **152**, 97–126

Kac, V. G. (1983) *Infinite Dimensional Lie Algebras*, Birkhaüser, Boston

Kac, V. G. & Wakimoto, M. (1988) Modular and conformal invariance constraints in representation theory of affine algebras, *Adv. Math.* **70**, 156–236

Kac, V. G. & Weisfeiler, B. (1976) Coadjoint action of a semi-simple algebraic group and the center of the enveloping algebra in characteristic p, *Indag. Math.* **38**, 136–51

Kachurik, I. I. & Klimyk, A. U. (1990) On Racah coefficients of the quantum algebra $U_q(su_2)$, *J. Phys. A* **23**, 2717–28

Kachurik, I. I. & Klimyk, A. U. (1991) General recurrence relations for Clebsch-Gordan coefficients of the quantum algebra $U_q(su_2)$, *J. Phys. A* **24**, 4009–15

Kalnins, E. G., Manocha, H. L. & Miller, W. (1992) Models of q-algebra representations: I. Tensor products of special unitary and oscillator algebras, *J. Math. Phys.* **33**, 2365–83

Kalnins, E. G., Miller, W. & Mukherjee, S. (1994) Models of q-algebra representations: the group of plane motions, *SIAM J. Math. Anal.* **25**, 513–27

Kang, S. J., Kashiwara, M., Misra, K. C., Miwa, T., Nakashima, T. & Nakayashiki, A. (1992a) Vertex models and crystals, *C. R. Acad. Sci. Paris Sér. I* **315**, 375–80

Kang, S. J., Kashiwara, M., Misra, K. C., Miwa, T., Nakashima, T. & Nakayashiki, A. (1992b) Affine crystals and vertex models, in *Infinite Analysis*, A. Tsuchiya, T. Eguchi & M. Jimbo (eds), Advanced Series in Mathematical Physics 16, pp. 449–84, World Scientific, Singapore

Kang, S. J., Kashiwara, M., Misra, K. C., Miwa, T., Nakashima, T. & Nakayashiki, A. (1992c) Perfect crystals of quantum affine Lie algebras, *Duke Math. J.* **68**, 499–607

Kang, S. J., Misra, K. C. & Miwa, T. (1993) Fock space representations of the quantized universal enveloping algebras $U_q(C_l^{(1)})$, $U_q(A_{2l}^{(2)})$ and $U_q(D_{l+1}^{(2)})$, *J. Algebra* **155**, 238–51

Karasev, M. V. (1987) Analogues of objects of Lie group theory for non-linear Poisson brackets, *Math. USSR Izvestiya* **28**, 497–527

Karasev, M. V. & Maslov, V. P. (1993) *Nonlinear Poisson Brackets: Geometry and Quantization*, Translations of Mathematical Monographs 119, American Mathematical Society, Providence, RI

Karimipour, V. (1993) The quantum double and the universal R-matrix for non-standard deformation of $A_{(n-1)}$, *J. Phys. A* **26**, L239–44

Karimipour, V. & Aghamohammadi, A. (1993) Multiparametric quantization of the special linear superalgebra, *J. Math. Phys.* **34**, 2561–71

Karowski, M. (1990) Yang-Baxter algebra - Bethe ansatz - conformal quantum field theories - quantum groups, in *Quantum Groups, Clausthal, Germany, 1989*, H.-D. Doebner & J.-D. Hennig (eds), pp. 183–218, Lecture Notes in Physics 370, Springer, Berlin

Karowski, M., Müller, W. & Schrader, R. (1992) State sum invariants of compact 3-manifolds with boundary and 6j-symbols, *J. Phys. A* **25**, 4847–60

Karowski, M. & Schrader, R. (1992) A quantum group version of quantum gauge theories in two dimensions, *J. Phys. A* **25**, L1151–4

Karowski, M. & Schrader, R. (1993) A combinatorial approach to topological quantum field theories and invariants of graphs, *Comm. Math. Phys.* **151**, 355–402

Kashaev, R. M. & Mangazeev, V. V. (1990) The four state solution of the Yang-Baxter equation, *Phys. Lett. A* **150**, 375–9

Kashaev, R. M. & Mangazeev, V. V. (1992) $N^{n(n-1)/2}$-state intertwiner related to $U_q(sl(n))$ algebra at $q^{2N} = 1$, *Modern Phys. Lett. A* **7**, 2827–36

Kashaev, R. M., Mangazeev, V. V. & Stroganov, Yu. G. (1991) Cyclic eight-state R-matrix related to $U_q(sl(3))$ algebra at $q^2 = -1$, *Modern Phys. Lett. A* **6**, 3437–43

Kashaev, R. M., Mangazeev, V. V. & Stroganov, Yu. G. (1992) N^3-state R-matrix related with $U_q(sl(3))$ algebra at $q^{2N} = 1$, in *Infinite Analysis*, A. Tsuchiya, T. Eguchi & M. Jimbo (eds), Advanced Series in Mathematical Physics 16, pp. 485–92, World Scientific, Singapore

Kashiwara, M. (1990a) Kazhdan-Lusztig conjecture for a symmetrizable Kac-Moody algebra, in *Grothendieck Festschrift, Vol. II*, P. Cartier *et al.* (eds), Progress in Mathematics 87, pp. 407–33, Birkhäuser, Boston

Kashiwara, M. (1990b) Bases cristallines, *C. R. Acad. Sci. Paris. Sér. I* **311**, 277–80

Kashiwara, M. (1990c) Crystalizing the q-analogue of universal enveloping algebras, *Comm. Math. Phys.* **133**, 249–60

Kashiwara, M. (1991) On crystal bases of the q-analogue of universal enveloping algebras, *Duke Math. J.* **63**, 465–516

Kashiwara, M. & Miwa, T. (1986) A class of elliptic solutions to the star-triangle equation, *Nucl. Phys. B* **275**, 121–34

Kashiwara, M. & Tanisaki, T. (1990) Kazhdan-Lusztig conjecture for a symmetrizable Kac-Moody algebra II, in *Operator Algebras, Unitary Representations, Enveloping Algebras and Invariant Theory*, A. Connes, M. Duflo, A. Joseph & R. Rentschler (eds), Progress in Mathematics 92, pp. 159–95, Birkhäuser, Boston

Kashiwara, M. & Tanisaki, T. (1994) Characters of negative level highest weight modules for affine Lie algebras, preprint

Kassel, Ch. (1992) Cyclic homology of differential operators, the Virasoro algebra and a q-analogue, *Comm. Math. Phys.* **146**, 343–56

Kato, M. (1990) Flat connection, quantum field theory and quantum group, in *Differential Geometric Methods in Theoretical Physics, Davis, 1988*, L. L. Chau & W. Nahm (eds), NATO Advanced Science Institutes Series B: Physics 245, pp. 251–6, Plenum, New York

Katriel, J. & Solomon, A. I. (1991) A q-analogue of the Campbell-Baker-Hausdorff expansion, *J. Phys. A* **24**, L1139–42

Kauffman, L. H. (1991a) From knots to quantum groups and back, in *Quantum Groups*, T. Curtright, D. Fairlie & C. Zachos (eds), pp. 1–32, World Scientific, Singapore

Kauffman, L. H. (1991b) *Knots and Physics*, World Scientific, Singapore

Kauffman, L. H & Saleur, H. (1992) Fermions and link invariants, in *Infinite Analysis*, A. Tsuchiya, T. Eguchi & M. Jimbo (eds), Advanced Series in Mathematical Physics 16, pp. 493–532, World Scientific, Singapore

Kazhdan, D. & Lusztig, G. (1979) Representations of Coxeter groups and Hecke algebras, *Invent. Math.* **53**, 165–84

Kazhdan, D. & Lusztig, G. (1991) Affine Lie algebras and quantum groups, *Duke Math. J. Internat. Math. Res. Notices* **2**, 21–9

Kazhdan, D. & Lusztig, G. (1993a) Tensor structures arising from affine Lie algebras I, *J. Amer. Math. Soc.* **6**, 905–47

Kazhdan, D. & Lusztig, G. (1993b) Tensor structures arising from affine Lie algebras II, *J. Amer. Math. Soc.* **6**, 949–1011

Kazhdan, D. & Lusztig, G. (1994a) Tensor structures arising from affine Lie algebras III, *J. Amer. Math. Soc.* **7**, 335–81

Kazhdan, D. & Lusztig, G. (1994b) Tensor structures arising from affine Lie algebras IV, *J. Amer. Math. Soc.* **7**, 383–453

Kazhdan, D. & Soibelman, Ya. S. (1992) Representations of the quantized function algebras, 2-categories and Zamolodchikov tetrahedra equation, pre-print, Harvard University

Kazhdan, D. & Wenzl, H. (1993) Reconstructing monoidal categories, pre-print, Harvard University

Kekäläinen, P. (1992) Step algebras of quantum $sl(n)$, *J. Algebra* **150**, 245–53

Keller, G. (1991) Fusion rules of $U_q(sl(2,\mathbb{C}))$, $q^m = 1$, *Lett. Math. Phys.* **21**, 273–86

Kempf, A. (1993) Quantum group symmetric Bargmann-Fock space: integral kernels, Green functions, driving forces, *J. Math. Phys.* **34**, 969–87

Kennedy, T. (1992) Solutions of the Yang-Baxter equation for isotropic quantum spin chains, *J. Phys. A* **25**, 2809–18

Kerler, T. (1992) Non-Tannakian categories in quantum field theory, in *New Symmetry Principles in Quantum Field Theory*, J. Fröhlich *et al.* (eds), NATO Advanced Science Institutes Series B: Physics 295, pp. 449–85, Plenum, New York

Kerov, S. V. (1992a) Characters of Hecke and Birman-Wenzl algebras, in *Quantum Groups, Proceedings of Workshops held in the Euler International Mathematical Institute 1990*, P. P. Kulish (ed.), Lecture Notes in Mathematics 1510, pp. 335–40, Springer, Berlin

Kerov, S. V. (1992b) A q-analog of the Hook walk algorithm and random Young tableaux, *Funct. Anal. Appl.* **26**, 179–87

Kerov, S. V. & Vershik, A. M. (1989) Characters and realizations of representations of an infinite dimensional Hecke algebra and knot invariants, *Soviet Math. Dokl.* **38**, 134–7

Khesin, B. & Zakharevich, I. (1992) Lie-Poisson group of pseudodifferential symbols and fractional KP-KdV hierarchies, preprint IHES/M/92/71

Khoroshkin, S. M., Radul, A. & Rubtsov, V. (1993) A family of Poisson structures on hermitian symmetric spaces, *Comm. Math. Phys.* **152**, 299–315

Khoroshkin, S. M. & Tolstoy, V. N. (1991a) Universal R-matrix for quantized (super) algebras, *Comm. Math. Phys.* **141**, 599–617

Khoroshkin, S. M. & Tolstoy, V. N. (1991b) Universal R-matrix for quantum supergroups, in *Group Theoretical Methods in Physics, Moscow, 1990*, V. V. Dodonov & V. I. Manko (eds), Lecture Notes in Physics 382, pp. 229–32, Springer, Berlin

Khoroshkin, S. M. & Tolstoy, V. N. (1992a) Extremal projector and universal R-matrix for quantized contragredient Lie (super) algebras, in *Quantum Groups and Related Topics*, R. Gierlak *et al.* (eds), pp. 23–32, Kluwer, Dordrecht

Khoroshkin, S. M. & Tolstoy, V. N. (1992b) Uniqueness theorem for universal R-matrix, *Lett. Math. Phys.* **24**, 231–44

Kim, J. D., Koh, I. G. & Ma, Z. Q. (1991) Quantum Ř matrix for E_7 and F_4 groups, *J. Math. Phys.* **32**, 845–56

Kirby, A. (1978) A calculus for framed links, *Invent. Math.* **45**, 35–56

Kirby, R. & Melvin, P. (1991) The 3-manifold invariants of Witten and Reshetikhin-Turaev for $sl(2, \mathbb{C})$, *Invent. Math.* **105**, 473–505

Kirby, R., Melvin, P. & Zhang, X. (1993) Quantum invariants at the sixth root of unity, *Comm. Math. Phys.* **151**, 607–18

Kirillov, A. A. (1976) Local Lie algebras, *Russian Math. Surveys* **31** (4), 55–75

Kirillov, A. N. (1988) Clebsch-Gordan quantum coefficients, *Zap. Nauchn. Sem. Leningrad Otdel. Mat. Inst. Steklov* **9**, 67–84 (Russian)

Kirillov, A. N. & Reshetikhin, N. Yu. (1986) The Yangians, Bethe ansatz and combinatorics, *Lett. Math. Phys.* **12**, 199–208

Kirillov, A. N. & Reshetikhin, N. Yu. (1987) Exact solution of the integrable XXZ Heisenberg model with arbitrary spin. II. Thermodynamics of the system, *J. Phys. A* **20**, 1587–97

Kirillov, A. N. & Reshetikhin, N. Yu. (1989) Representations of the algebra $U_q(sl(2))$, q-orthogonal polynomials and invariants of links, in *Infinite-dimensional Lie Algebras and Groups*, V. G. Kac (ed.), pp. 285–339, World Scientific, Singapore

Kirillov, A. N. & Reshetikhin, N. Yu. (1990a) q-Weyl group and a multiplicative formula for universal R-matrices, *Comm. Math. Phys.* **134**, 421–31

Kirillov, A. N. & Reshetikhin, N. Yu. (1990b) Representations of Yangians and multiplicities of occurrence of the irreducible components of the tensor product of representations of simple Lie algebras, *J. Soviet Math.* **52**, 3156–64

Klimcik, C. & Klimcik, E. (1993) On macroscopic energy gap for q-quantum mechanical systems, *J. Phys. A* **26**, L289–92

Klimek, M. (1993) Extension of q-deformed analysis and q-deformed models of classical mechanics, *J. Phys. A* **26**, 955–68

Klimek, S. & Lesniewski, A. (1992a) Quantum Riemann surfaces I. The unit disc, *Comm. Math. Phys.* **146**, 103–22

Klimek, S. & Lesniewski, A. (1992b) Quantum Riemann surfaces II. The discrete series, *Lett. Math. Phys.* **24**, 125–40

Klimyk, A. U. (1992) Wigner-Eckart theorem for tensor operators of the quantum group $U_q(n)$, *J. Phys. A* **25**, 2919–27

Klimyk, A. U. & Pakuliak, S. Z. (1992) Representations of the quantum algebras $U_q(u_{r,s})$ and $U_q(u_{r+s})$ related to the quantum hyperboloid and sphere, *J. Math. Phys.* **33**, 1937–47

Klimyk, A. U., Smirnov, Yu. F. & Gruber, B. (1991) Representations of the quantum algebras $U_q(su(2))$ and $U_q(su(1,1))$, in *Symmetries in Science V*, B. Gruber & H.-D. Doebner (eds), pp. 341–68, Plenum, New York

Knizhnik, V. & Zamolodchikov, A. (1984) Current algebra and Wess-Zumino model in two dimensions, *Nucl. Phys. B* **247**, 83–103

Kobayashi, T. & Uematsu, T. (1993) q-deformed superconformal algebra on quantum superspace, *Phys. Lett. B* **306**, 27–33

Koelink, H. T. (1991a) On ∗-representations of the Hopf ∗-algebra associated with the quantum group $U_q(n)$, *Compositio Math.* **77**, 199–231

Koelink, H. T. (1991b) On quantum groups and q-special functions, Thesis, University of Leiden

Koelink, H. T. (1992) Quantum group theoretical proof of the addition formula for continuous q-Legendre polynomials, in *Deformation Theory and Quantum Groups with Applications to Mathematical Physics*, M. Gerstenhaber & J. Stasheff (eds), Contemporary Mathematics 134, pp. 139–40, American Mathematical Society, Providence, RI

Koelink, H. T. (1994a) The quantum group of plane motions and the Hahn–Exton q-Bessel function, *Duke Math. J.* **76**, 483–508

Koelink, H. T. (1994b) The quantum group of plane motions and basic Bessel functions, *Indagationes Math.* (to appear)

Koelink, H. T. & Koornwinder, T. H. (1989) The Clebsch-Gordon coefficients for the quantum group $S_\mu U(2)$ and q-Hahn polynomials, *Nederl. Akad. Wetensch. Proc. Ser. A* **92**, 443–56

Koelink, H. T. & Koornwinder, T. H. (1992) q-special functions: a tutorial, in *Deformation Theory and Quantum Groups with Applications to Mathematical Physics*, M. Gerstenhaber & J. Stasheff (eds), Contemporary Mathematics 134, pp. 141–2, American Mathematical Society, Providence, RI

Koelink, H. T. & Swarttouw, R. F. (1994) A q-analogue of Graf's addition formula for the Hahn–Exton q-Bessel function, *J. Approx. Theory* (to appear)

Koh, I. G. & Ma, Z. Q. (1990) Exceptional quantum groups, *Phys. Lett. B* **234**, 480–6

Kohno, T. (1985) Série de Poincaré-Koszul associée aux groups de tresses pures, *Invent. Math.* **82**, 57–75

Kohno, T. (1986) Homology of a local system on a complement of hyperplanes, *Proc. Japan Acad. Sci. Ser. A* **62**, 144–7

Kohno, T. (1987a) Monodromy representations of braid groups and Yang-Baxter equations, *Ann. Inst. Fourier (Grenoble)* **37**, 139–60

Kohno, T. (1987b) One parameter family of linear representations of Artin's braid groups, *Adv. Stud. in Pure Math.* **12**, 189–200

Kohno, T. (1988a) Linear representations of braid groups and classical Yang-Baxter equations, in *Braids, Santa Cruz, 1986*, J. S. Birman & S. Libgober (eds), Contemporary Mathematics 78, pp. 339–63, American Mathematical Society, Providence, RI

Kohno, T. (1988b) Hecke algebra representations of braid groups and classical Yang-Baxter equations, in *Conformal Field Theories and Solvable Lattice Models, Kyoto, 1986*, M. Jimbo, T. Miwa & A. Tsuchiya (eds), Advanced Studies in Pure Mathematics 16, pp. 255–69, Academic Press, Boston

Kohno, T. (1989a) Integrable systems related to braid groups and Yang-Baxter equation, in *Braid Group, Knot Theory and Statistical Mechanics*, C. N. Yang & M. L. Ge (eds), Advanced Series in Mathematical Physics 9, pp. 139–45, World Scientific, Singapore

Kohno, T. (1989b) Monodromy representations of braid groups, *Sugaku* **41**, 305–19

Komata, S., Mohri, K. & Nohara, H. (1991) Classical and quantum extended superconformal algebra, *Nucl. Phys. B* **359**, 168–200

Konishi, Y. (1992) A note on actions of compact matrix quantum groups on von Neumann algebras, *Nihonkai Math. J.* **3**, 23–9

Konishi, Y., Nagisa, M. & Watatani, Y. (1992) Some remarks on actions of compact matrix quantum groups on C^*-algebras, *Pacific J. Math.* **153**, 119–27

Koornwinder, T. H. (1989a) Representations of the twisted $SU(2)$ quantum group and some q-hypergeometric orthogonal polynomials, *Nederl. Akad. Wetensch. Proc. Ser. A* **92**, 97–117

Koornwinder, T. H. (1989b) Continuous q-Legendre polynomials are spherical matrix elements of irreducible representations of the quantum $SU(2)$ group, *CWI Quart.* **2**, 171–3

Koornwinder, T. H. (1990) Orthogonal polynomials in connection with quantum groups, in *Orthogonal Polynomials: Theory and Practice*, P. Nevai (ed.), NATO Advanced Science Institutes Series C: Mathematical and Physical Sciences 294, pp. 257–92, Kluwer, Dordrecht

Koornwinder, T. H. (1991a) The addition formula for little q-Legendre polynomials and the $SU(2)$ quantum group, *SIAM J. Math. Anal.* **22**, 295–301

Koornwinder, T. H. (1991b) Positive convolution structures associated with quantum groups, in *Probability Measures on Groups X, Oberwolfach, 1990*, H. Heyer (ed.), pp. 249–68, Plenum, New York

Koornwinder, T. H. (1992) q-special functions and their occurrence in quantum groups, in *Deformation Theory and Quantum Groups with Applications to Mathematical Physics*, M.

Gerstenhaber & J. Stasheff (eds), Contemporary Mathematics 134, pp. 143–4, American Mathematical Society, Providence, RI

Koornwinder, T. H. & Swarttouw, R. F. (1992) On q-analogues of the Fourier and Hankel transforms, *Trans. Amer. Math. Soc.* **333**, 445–61

Korenskii, S. V. (1991) Representations of the quantum group $SL_J(2)$, *Russian Math. Surveys* **46** (6), 221–2

Korepin, V. E. (1983) The analysis of the bilinear relation of the six-vertex model, *Soviet Phys. Dokl.* **27**, 1050–1

Korogodskii, L. I. & Vaksman, L. L. (1992) Quantum G-spaces and Heisenberg algebra, in *Quantum Groups, Proceedings of Workshops held in the Euler International Mathematical Institute 1990*, P. P. Kulish (ed.), Lecture Notes in Mathematics 1510, pp. 56–66, Springer, Berlin

Kosmann-Schwarzbach, V. (1990a) Quantum and classical Yang-Baxter equations, *Modern Phys. Lett. A* **5**, 981–90; errata, *ibid.* **6** (1991), 3373

Kosmann-Schwarzbach, V. (1990b) Groupes de Lie-Poisson quasitriangulaires, in *Géométrie Symplectique et Mécanique, Colloque International La Grande Motte, France, 23-28 Mai, 1988*, C. Albert (ed.), Lecture Notes in Mathematics 1416, pp. 161–7, Springer, Berlin

Kosmann-Schwarzbach, V. (1991a) From 'quantum groups' to 'quasi-quantum groups', in *Symmetry in Science V*, B. Gruber & H.-D. Doebner (eds), pp. 369–93, Plenum, New York

Kosmann-Schwarzbach, V. (1991b) Quasi-bigèbres de Lie et groupes de Lie quasi-Poisson, *C. R. Acad. Sci. Paris Sér. I* **312**, 391–4

Kosmann-Schwarzbach, V. & Magri, F. (1988) Poisson-Lie groups and complete integrability, I. Drinfeld algebras, dual extensions and their canonical representations, *Ann. Inst. Henri Poincaré Phys. Théor.* **49**, 433–60

Kosmann-Schwarzbach, V. & Magri, F. (1990) Poisson-Nijenhuis structures, *Ann. Sci. Henri Poincaré Phys. Théor.* **53**, 35–81

Kostant, B. (1979) The solution to the generalized Toda lattice and representation theory, *Adv. Math.* **34**, 195–338

Kosuda, M. & Murakami, J. (1992) The centralizer algebras of mixed tensor representations of $U_q(gl_n)$ and the HOMFLY polynomial, *Proc. Japan Acad. Ser. A Math. Sci.* **68**, 148–51

Kowalski, K. & Rembielinski, J. (1993) Coherent states for the quantum complex plane, *J. Math. Phys.* **34**, 2153–65

Kruglyak, S. A. (1984) Representations of a quantum algebra connected with the Yang-Baxter equation, in *Spectral Theory of Operators and Infinite-dimensional Analysis*, Yu. M. Berezanskii (ed.), pp. 111–20, Akad. Nauk Ukrain. SSR, Inst. Mat., Kiev

Kuang, L. M. & Wang, F. B. (1993) The $su_q(1,1)$ q-coherent states and their nonclassical properties, *Phys. Lett. A* **173**, 221–7

Kulish, P. P. (1985) Integrable graded magnets, *Zap. Nauchn. Sem. Leningrad Otdel. Mat. Inst. Steklov* **145** (Russian)

Kulish, P. P. (1989) Quantum Lie superalgebras and supergroups, in *Problems of Modern Quantum Field Theory*, A. A. Belavin, A. U. Klimyk & A. B. Zamolodchikov (eds), Research Reports in Physics, pp. 14–21, Springer, Berlin

Kulish, P. P. (1990a) A two-parameter quantum group and a gauge transformation, *Zap. Nauchn. Sem. Leningrad Otdel. Mat. Inst. Steklov* **23**, 89–93 (Russian)

Kulish, P. P. (1990b) Clebsch-Gordan coefficients for a quantum superalgebra of rank one I, *Zap. Nauchn. Sem. Leningrad Otdel. Mat. Inst. Steklov* **180** (Russian)

Kulish, P. P. (1991a) The quantum superalgebra $osp(2|1)$, *J. Soviet Math.* **54**, 923–30

Kulish, P. P. (1991b) Finite-dimensional Fateev-Zamolodchikov algebra and q-oscillators, *Phys. Lett. A* **161**, 50–2

Kulish, P. P. (1991c) Contraction of quantum algebras and q-oscillators, *Theoret. Math. Phys.* **86**, 108–10

Kulish, P. P. (1991d) Quantum algebras and symmetries of dynamical systems, in *Group Theoretical Methods in Physics, Moscow, 1990*, V. V. Dodonov & V. I. Manko (eds), Lecture Notes in Physics 382, pp. 195–8, Springer, Berlin

Kulish, P. P. & Damashinsky (1990) On the q oscillator and the quantum algebra $su_q(1,1)$, *J. Phys. A* **23**, L415–9

Kulish, P. P. & Reshetikhin, N. Yu. (1982) On GL_3-invariant solutions of the Yang-Baxter equation and associated quantum systems, *Zap. Nauchn. Sem. Leningrad Otdel. Mat. Inst. Steklov* **120** (Russian)

Kulish, P. P. & Reshetikhin, N. Yu. (1983) Quantum linear problem for the sine-Gordon equation and higher representations, *J. Soviet Math.* **23**, 2435–41

Kulish, P. P. & Reshetikhin, N. Yu. (1989) Universal R-matrix of the quantum superalgebra $osp(2|1)$, *Lett. Math. Phys.* **18**, 143–9

Kulish, P. P., Reshetikhin, N. Yu. & Sklyanin, E. K. (1981) Yang-Baxter equation and representation theory I, *Lett. Math. Phys.* **5**, 393–403

Kulish, P. P., Sasaki, R. & Schwiebert, C. (1993) Constant solutions of reflection equations and quantum groups, *J. Math. Phys.* **34**, 286–304

Kulish, P. & Sklyanin, E. K. (1982a) Solutions of the Yang-Baxter equation, *J. Soviet Math.* **19**, 1596–620

Kulish, P. & Sklyanin, E. K. (1982b) Quantum spectral transform method. Recent developments, in *Integrable Quantum Field Theories, Tvarminne, 1981*, J. Hietarinta & C. Montonen (eds), Lecture Notes in Physics 151, pp. 61–119, Springer, Berlin

Kulish, P. & Sklyanin, E. K. (1991) The general $U_q(sl(2))$ invariant XXZ integrable quantum spin chain, *J. Phys. A* **24**, L435–9

Kulish, P. & Sklyanin, E. K. (1992) Algebraic structure related to the reflection equations, *J. Phys. A* **25**, 5963–76

Kumari, M. K., Shanta, P., Chaturvedi, S. & Srinivasan, V. (1992) On q-deformed para-oscillators and para-q-oscillators, *Modern Phys. Lett. A* **7**, 2593–600

Kundu, A. & Basu-Mallick, B. (1991) q-deformation of the Holstein-Primakoff transformation and other bosonisations of the quantum group, *Phys. Lett. A* **156**, 175–8

Kundu, A. & Basu-Mallick, B. (1992a) Solutions of the non-additive YBE by spectral parameter factorization and re-Yang-Baxterization, *J. Phys. A* **25**, 6307–16

Kundu, A. & Basu-Mallick, B. (1992b) Construction of integrable quantum lattice models through Sklyanin-like algebras, *Modern Phys. Lett. A* **7**, 61–9

Kundu, A. & Basu-Mallick, B. (1993) Classical and quantum integrability of a derivative non-linear Schrödinger model related to quantum groups, *J. Math. Phys.* **34**, 1052–62

Kuniba, A. (1990) Quantum R-matrix for G_2 and a solvable 175 vertex model, *J. Phys. A* **23**, 1349–62

Kuniba, A. (1993) Thermodynamics of the $U_q(X_r^{(1)})$ Bethe ansatz system with q a root of unity, *Nucl. Phys. B* **389**, 209–46

Kuniba, A. & Nakanishi, T. (1992) Fusion RSOS models and rational coset models, in *Quantum Groups, Proceedings of Workshops held in the Euler International Mathematical Institute 1990*, P. P. Kulish (ed.), Lecture Notes in Mathematics 1510, pp. 303–11, Springer, Berlin

Kuniba, A., Nakanishi, T. & Suzuki, J. (1991) Ferro- and antiferro-magnetizations in RSOS models, *Nucl. Phys. B* **356**, 750–74

Kunz, J., Maslanka, P. & Giler, S. (1991) Quantum BRST cohomology, *J. Phys. A* **24**, 4235–9

Kuperschmidt, B. A. (1991) Quasiclassical limit of quantum matrix groups, in *Mechanics, Analysis and Geometry: 200 Years after Lagrange*, M. Francaviglia (ed.), pp. 171–99, North-Holland, Amsterdam

Kuperschmidt, B. A. (1992a) The quantum group $T_hGL(n)$, *J. Phys. A* **25**, L911–4

Kuperschmidt, B. A. (1992b) Trace formulae for the quantum group $GL_q(2)$, *J. Phys. A* **25**, L915-20

Kuperschmidt, B. A. (1992c) The quantum group $GL_h(2)$, *J. Phys. A* **25**, L1239-44

Kuperschmidt, B. A. (1993a) A q-analogue of the dual space to the Lie algebras $gl(2)$ and $sl(2)$, *J. Phys. A* **26**, L1-4

Kuperschmidt, B. A. (1993b) All quantum group structures on the supergroup $GL(1|1)$, *J. Phys. A* **26**, L251-6

Kuperschmidt, B. A. (1993c) A quantum analogue of the group of motions on the plane and its coadjoint representation, *Phys. Lett. A* **174**, 81-4

Lacki, J. (1989) Algèbres de dimension infinie, groupes de tresses et groupes quantiques dans les théories conformes à deux dimensions, *Riv. Nuovo Cim.* **12** (3), 52pp.

Lakshmibai, V. & Reshetikhin, N. Yu. (1991) Quantum deformations of flag and Schubert schemes, *C. R. Acad. Sci. Paris Sér. I* **313**, 121-6

Lakshmibai, V. & Reshetikhin, N. Yu. (1992) Quantum flag and Schubert schemes, in *Deformation Theory and Quantum Groups with Applications to Mathematical Physics*, M. Gerstenhaber & J. Stasheff (eds), Contemporary Mathematics 134, pp. 145-81, American Mathematical Society, Providence, RI

Lakshmibai, V. & Seshadri, C. S. (1991) Standard monomial theory, in *Proceedings of International Conference on Algebraic groups, Hyderabad, India, December, 1989*, S. Ramanan, C. Musili & N. Mohan Kumar (eds), pp. 279-322, Manoj Prakashan, Madras

Lambe, L. A. (1992) Homological perturbation theory, Hochschild homology and formal groups, in *Deformation Theory and Quantum Groups with Applications to Mathematical Physics*, M. Gerstenhaber & J. Stasheff (eds), Contemporary Mathematics 134, pp. 183-218, American Mathematical Society, Providence, RI

Lambe, L. A. & Radford, D. E. (1993) Algebraic aspects of the quantum Yang-Baxter equation, *J. Algebra* **154**, 228-88

Larson, R. G. (1971) Characters of Hopf algebras, *J. Algebra* **17**, 352-68

Larson, R. G. & Towber, J. (1991) Two dual classes of bialgebras related to the concepts of 'quantum group' and 'quantum Lie algebra', *Comm. Algebra* **19**, 3295-345

Larsson, T. A. (1992) Universal solutions to the simplex equations, *Nucl. Phys. B* **380**, 575-87

Lawrence, R. J. (1989) A universal link invariant, in *The Interface of Mathematics and Particle Physics, Oxford, 1988*, D. Quillen, G. B. Segal & S. T. Tsou (eds), Institute of Mathematics and its Applications Conference Series New Series 24, pp. 151-6, Oxford University Press, New York

Lawrence, R. J. (1990) Homological representations of Hecke algebras, *Comm. Math. Phys.* **135**, 141-91

Lawrence, R. J. (1992) On algebras and triangle relations, in *Topological and Geometrical Methods in Field Theory*, J. Mickelsson & O. Pekonen (eds), pp. 429-48, World Scientific, Singapore

Lawrence, R. J. (1993) A functorial approach to the one-variable Jones polynomial, *J. Diff. Geom.* **37**, 689-710

Lazarev, A. Yu. & Movshev, M. V. (1991a) Deformations of Hopf algebras, *Russian Math. Surveys* **46** (1), 253-4

Lazarev, A. Yu. & Movshev, M. V. (1991b) On the quantization of certain Lie groups and Lie algebras, *Russian Math. Surveys* **46** (6), 225-6

Leblanc, Y. & Wallet, J.-C. (1993) R-matrix and q-covariant oscillators for $U_q(sl(n|m))$, *Phys. Lett. B* **304**, 89-97

LeClair, A. (1989) Restricted sine-Gordon theory and the minimal conformal series, *Phys. Lett. B* **230**, 103-7

LeClair, A. (1991) Taming the integrable zoo, in *Quantum Groups*, T. Curtright, D. Fairlie & C. Zachos (eds), pp. 247-57, World Scientific, Singapore

LeClair, A. & Smirnov, F. A. (1992) Infinite quantum group symmetry in massive 2D quantum field theory, *Internat. J. Modern Phys. A* **7**, 2997–3022

Lee, H. C. (1992a) On Seifert circles and functors for tangles, in *Infinite Analysis*, A. Tsuchiya, T. Eguchi & M. Jimbo (eds), Advanced Series in Mathematical Physics 16, pp. 581–610, World Scientific, Singapore

Lee, H. C. (1992b) Tangle invariants and the centre of the quantum group, in *Knots 90, Osaka, 1990*, A. Kawauchi (ed.), pp. 341–61, De Gruyter, Berlin

Lee, H.-C., Ge, M.-L., Couture, M. J. & Wu, Y. S. (1989) Strange statistics, braid group representations and multipoint functions in the N-component model, *Internat. J. Modern Phys. A* **9**, 2333–70

Leites, D. A. & Serganova, V. V. (1983) On classical Yang-Baxter equation for simple Lie superalgebras, *Soviet J. Theor. Math. Phys.* **17**, 69–70

Le Man Kuang (1992) The q-supercoherent states of the q-deformed $SU(2)$ superalgebra, *J. Phys. A* **25**, 4827–34

Lenczewski, R. (1993a) A q-analog of the quantum central limit theorem for $SU_q(2)$, q complex, *J. Math. Phys.* **34**, 480–9

Lenczewski, R. (1993b) On sums of q-independent $SU_q(2)$ quantum variables, *Comm. Math. Phys.* **154**, 127–34

Lenczewski, R. & Podgorski, K. (1992) A q-analog of the quantum central limit theorem for $SU_q(2)$, *J. Math. Phys.* **33**, 2768–78

Lesniewski, A. & Rinaldi, M. (1993) Tensor products of representations of $C(SU_q(2))$, *J. Math. Phys.* **34**, 305–14

Levendorskii, S. Z. (1990) Twisted function algebras on a compact quantum group and their representation, *Funct. Anal. Appl.* **24**, 330–2

Levendorskii, S. Z. (1992) Twisted function algebras on a compact quantum group and their representation, *St. Petersburg Math. J.* **3**, 405–23

Levendorskii, S. Z. (1993) On PBW bases for Yangians, *Lett. Math. Phys.* **27**, 37–42

Levendorskii, S. Z. & Soibelman, Ya. S. (1990) Some applications of quantum Weyl groups, *J. Geom. Phys.* **7**, 241–54

Levendorskii, S. Z. & Soibelman, Ya. S. (1991a) Algebras of functions on compact quantum groups, Schubert cells and quantum tori, *Comm. Math. Phys.* **139**, 141–70

Levendorskii, S. Z. & Soibelman, Ya. S. (1991b) Quantum group A_∞, *Comm. Math. Phys.* **140**, 399–414

Levendorskii, S. Z. & Soibelman, Ya. S. (1991c) The quantum Weyl group and a multiplicative formula for the R-matrix of a simple Lie algebra, *Funct. Anal. Appl.* **25**, 143–5

Levendorskii, S. Z., Soibelman, Ya. S. & Stukopin, V. (1993) The quantum Weyl group and the universal quantum R-matrix for affine Lie algebra $A_1^{(1)}$, *Lett. Math. Phys.* **27**, 253–64

Leznov, A. N. & Mukhtarov, M. A. (1987) The internal symmetry algebra of completely integrable dynamical systems in the quantum regime, *Theoret. Math. Phys.* **71**, 370–5

Li, Y. Q. (1992a) Representations of a braid group with transpose symmetry and the related link invariants, *J. Phys. A* **25**, 6713–22

Li, Y. Q. (1992b) General form of BGR associated with A_n and its Yang-Baxterization, *Phys. Lett. A* **161**, 411–5

Li, Y. Q. (1993) Yang-Baxterization, *J. Math. Phys.* **34**, 864–74

Li, Y. Q. & Ge, M. L. (1991) Polynomials from non-standard braid group representations, *Phys. Lett. A* **152**, 273–5

Li, Y. Q., Ge, M. L., Xue, K. & Wang, L. Y. (1991) Weight conservation condition and structure of the braid group representation, *J. Phys. A* **24**, 3443–53

Li, Y. Q. & Sheng, Z. (1992) A deformation of quantum mechanics, *J. Phys. A* **25**, 6779–88

Li, Y. Q., Wang, L. Y. & Zhang, J. (1992) Representations of braid group and Birman-Wenzl algebra, *Phys. Lett. A* **162**, 449–52

Liao, L. & Song, X. C. (1991a) Quantum Lie superalgebras and 'nonstandard' braid group representations, *Modern Phys. Lett. A* **6**, 959–68

Liao, L. & Song, X. C. (1991b) q-deformation of Lie superalgebras $B(m,n)$, $B(0,n)$, $C(1+n)$ and $D(m,n)$ in their boson-fermion representations, *J. Phys. A* **24**, 5451–63

Lichnerowicz, A. (1981) Existence and invariance of twisted products on a symplectic manifold, *Lett. Math. Phys.* **5**, 117–26

Lichnerowicz, A. (1982) Déformations d'algèbres associées à une variété symplectique (les $*_\nu$-produits), *Ann. Inst. Fourier* **32**, 157–209

Lickorish, W. B. R. (1991) Three-manifold invariants and the Temperley-Lieb algebra, *Math. Ann.* **290**, 657–70

Lickorish, W. B. R. (1993) Distinct 3-manifolds with all $SU(2)_q$ invariants the same, *Proc. Amer. Math. Soc.* **117**, 285–92

Lie, S. (1888-93) *Theorie der Transformationsgruppen*, Bd. 1-3, Teubner, Leipzig

Lienert, C. R. & Butler, P. H. (1992a) Racah-Wigner algebra for q-deformed algebras, *J. Phys. A* **25**, 1223–36

Lienert, C. R. & Butler, P. H. (1992b) Recursive calculation of the R-matrices of q-deformed algebras, *J. Phys. A* **25**, 5577–86

Liguori, A. & Mintchev, M. (1992) Spectral parameters of the quantum Yang-Baxter equation, *Phys. Lett. B* **275**, 371–4

Lin, Z. (1993) Induced representations of Hopf algebras: applications to quantum groups at roots of 1, *J. Algebra* **154**, 152–87

Links, J. R. & Gould, M. D. (1993) Raising and lowering operators for $U_q(sl(n))$, *J. Math. Phys.* **34**, 1577–86

Links, J. R., Gould, M. D. & Tsohantjis, I. (1993) Baxterization of the R-matrix for the adjoint representation of $U_q[D(2,1;\alpha)]$, *Lett. Math. Phys.* **27**, 51–8

Liskova, N. A. & Kirillov, A. N. (1992) Clebsch-Gordan and Racah-Wigner coefficients for $U_q(SU(1,1))$, in *Infinite Analysis*, A. Tsuchiya, T. Eguchi & M. Jimbo (eds), Advanced Series in Mathematical Physics 16, pp. 611–22, World Scientific, Singapore

Liu, K. (1991) Quantum central extensions, *C. R. Math. Rep. Acad. Sci. Canada* **13**, 135–40

Liu, K. (1992) Characterizations of the quantum Witt algebra, *Lett. Math. Phys.* **24**, 257–66

Liu, Z. J. & Qian, M. (1990) Classical R-matrix and semi-simple Lie algebras, in *Nonlinear Physics, Shanghai, 1989*, Chao Hao Gu, Yi Shen Li & Gui Zhang Tu (eds), Research Reports in Physics, pp. 146–51, Springer, Berlin

Liu, Z. J. & Qian, M. (1992) Generalized Yang-Baxter equations, Koszul operators and Poisson-Lie groups, *J. Diff. Geom.* **35**, 399–414

Lu, J. F. & Ge, M. L. (1987) The Yang-Baxter algebra in a nonlinear chiral model, *Kexue-Tongbao* **32**, 1853–6 (Chinese)

Lu, J. H. & Weinstein, A. (1989) Groupoides symplectiques doubles des groupes de Lie-Poisson, *C. R. Acad. Sci. Sér. I* **309**, 951–4

Lu, J. H. & Weinstein, A. (1990) Poisson-Lie groups, dressing transformations and Bruhat decompositions, *J. Diff. Geom.* **31**, 501–26

Lu, J. H. & Weinstein, A. (1991) *Comm. Math. Phys.* **135**, 229–31; appendix to Sheu (1991)

Lu, Z. C. (1992a) Star products and phase space realizations of quantum groups, *Lett. Math. Phys.* **25**, 51–60

Lu, Z. C. (1992b) Quantum algebras on phase space, *J. Math. Phys.* **33**, 446–53

Lukierski, J. & Nowicki, A. (1992a) Real forms of $U_q(OSp(1|2))$ and quantum $D = 2$ supersymmetry algebras, *J. Phys. A* **25**, L161–70

Lukierski, J. & Nowicki, A. (1992b) Quantum deformations of $D = 4$ Poincaré algebra, in *Quantum Groups and Related Topics*, R. Gierlak *et al.* (eds), pp. 33–44, Kluwer, Dordrecht

Lukierski, J. & Nowicki, A. (1992c) Quantum deformations of $D = 4$ Poincaré and Weyl algebra from q-deformed $D = 4$ conformal algebra, *Phys. Lett.* B **279**, 299–307

Lukierski, J., Nowicki, A. & Ruegg, H. (1991) Real forms of complex quantum anti-de Sitter algebra $U_q(Sp(4, \mathbb{C}))$ and their contraction schemes, *Phys. Lett.* B **271**, 321–8

Lukierski, J., Nowicki, A. & Ruegg, H. (1992a) Quantum deformations of Poincaré algebra and the supersymmetric extensions, in *Topological and Geometrical Methods in Field Theory*, J. Mickelsson & O. Pekonen (eds), pp. 202–26, World Scientific, Singapore

Lukierski, J., Nowicki, A. & Ruegg, H. (1992b) New quantum Poincaré algebra and κ-deformed field theory, *Phys. Lett.* B **293**, 344–52

Lukierski, J., Ruegg, H., Nowicki, A. & Tolstoy, V. N. (1991) q-deformation of the Poincaré algebra, *Phys. Lett.* B **264**, 331–8

Lukyanov, S. & Shatashvili, S. L. (1993) Free field representation of the classical limit of the quantum affine algebra, *Phys. Lett.* B **298**, 111–5

Lusztig, G. (1980) Some problems in the representation theory of finite Chevalley groups, in *Finite Groups, Santa Cruz, 1979*, J. S. Birman & S. Libgober (eds), Proceedings of Symposia in Pure Mathematics 37, pp. 313–7, American Mathematical Society, Providence, RI

Lusztig, G. (1988) Quantum deformations of certain simple modules over enveloping algebras, *Adv. Math.* **70**, 237–49

Lusztig, G. (1989) Modular representations and quantum groups, in *Classical Groups and Related Topics, Beijing, 1987*, A. J. Hahn, D. G. James & Zhe Xian Wan (eds), Contemporary Mathematics 82, pp. 59–77, American Mathematical Society, Providence, RI

Lusztig, G. (1990a) Finite-dimensional Hopf algebras arising from quantized universal enveloping algebras, *J. Amer. Math. Soc.* **3**, 257–96

Lusztig, G. (1990b) Quantum groups at roots of 1, *Geom. Dedicata* **35**, 89–113

Lusztig, G. (1990c) On quantum groups, *J. Algebra* **131**, 466–75

Lusztig, G. (1990d) Canonical bases arising from quantized enveloping algebras, *J. Amer. Math. Soc.* **3**, 447–98

Lusztig, G. (1990e) Canonical bases arising from quantized enveloping algebras II, *Progr. Theor. Phys. Suppl.* **102**, 175–201

Lusztig, G. (1991) Quivers, perverse sheaves and quantized enveloping algebras, *J. Amer. Math. Soc.* **4**, 365–421

Lusztig, G. (1992a) Introduction to quantized enveloping algebras, in *New Developments in Lie Theory and their Applications*, J. Tirao & N. Wallach (eds), Progress in Mathematics 105, pp. 49–66, Birkhäuser, Boston

Lusztig, G. (1992b) Problems on canonical bases, preprint, MIT

Lusztig, G. (1992c) Affine quivers and canonical bases, *Publ. Math. IHES* **76**, 111–63

Lusztig, G. (1993) *Introduction to Quantum Groups*, Progress in Mathematics 110, Birkhäuser, Boston

Lusztig, G. (1994) Monodromic systems on flag manifolds, preprint, MIT

Lyakhovsky, V. D. & Mudrov, A. (1992) Generalized quantization scheme for Lie algebras, *J. Phys. A* **25**, L1139–54

Lyubashenko, V. V. (1986) Hopf algebras and vector symmetries, *Russian Math. Surveys* **41** (5), 153–4

Lyubashenko, V. V. (1992) Real and imaginary forms of quantum groups, in *Quantum Groups, Proceedings of Workshops held in the Euler International Mathematical Institute 1990*, P. P. Kulish (ed.), Lecture Notes in Mathematics 1510, pp. 67–78, Springer, Berlin

Lyubashenko, V. V. & Majid, S. (1992) Fourier transform identities in quantum mechanics and the quantum line, *Phys. Lett.* B **284**, 66–70

Ma, Z. Q. (1990a) Representations of the braid group obtained from quantum $sl(3)$ enveloping algebra, *J. Math. Phys.* **31**, 550–6

Ma, Z. Q. (1990b) New link polynomials obtained from octet representation of quantum $sl(3)$ enveloping algebra, *J. Math. Phys.* **31**, 3079–84

Ma, Z. Q. (1990c) Rational solution for the minimal representation of G_2, *J. Phys. A* **23**, 4415–6

Ma, Z. Q. (1990d) The spectrum-dependent solutions to the Yang-Baxter equation for quantum E_6 and E_7, *J. Phys. A* **23**, 5513–22

Ma, Z. Q. (1991) The embedding e_0 and the spectrum dependent R-matrix for q-F_4, *J. Phys. A* **24**, 433–49

McConnell, J. C. (1990) Quantum groups, filtered rings and Gel'fand-Kirillov dimension, in *Non-commutative Ring Theory, Athens, Ohio, 1989*, S. K. Jain & S. R. Lopez-Permouth (eds), Lecture Notes in Mathematics 1448, pp. 139–47, Springer, Berlin

McConnell, J. C. & Robson, J. C. (1987) *Noncommutative Noetherian Rings*, Wiley, Chichester, UK

MacFarlane, A. (1989) On q-analogues of the quantum harmonic oscillator and the quantum group $SU(2)_q$, *J. Phys. A* **22**, 4581–8

MacFarlane, A. (1992) Spectrum generating quantum group of the harmonic oscillator, *Internat. J. Modern Phys. A* **7**, 4377–93

MacFarlane, A. & Majid, S. (1991) Quantum group structure in a fermionic extension of the quantum harmonic oscillator, *Phys. Lett. B* **268**, 71–4

MacFarlane, A. & Majid, S. (1992a) The superalgebra $osp(1/2)$ and a related quantum group structure for harmonic oscillator systems, in *Topological and Geometrical Methods in Field Theory*, J. Mickelsson & O. Pekonen (eds), pp. 227–36, World Scientific, Singapore

MacFarlane, A. & Majid, S. (1992b) Spectrum generating quantum group of the harmonic oscillator, *Internat. J. Modern Phys. A* **7**, 4377–94

Mack, G. & Schomerus, V. (1990) Conformal field algebras with quantum symmetry from the theory of superselection sectors, *Comm. Math. Phys.* **134**, 139–96

Mack, G. & Schomerus, V. (1991) Quasi quantum group symmetry and local braid relations in the conformal Ising model, *Phys. Lett. B* **267**, 207–13

Mack, G. & Schomerus, V. (1992a) Quasi Hopf quantum symmetry in quantum theory, *Nucl. Phys. B* **370**, 185–230

Mack, G. & Schomerus, V. (1992b) Action of truncated quantum group algebras on quasi-quantum planes and a quasi-associative differential calculus, *Comm. Math. Phys.* **149**, 513–48

MacKay, N. H. (1991a) Rational R-matrices in irreducible representations, *J. Phys. A* **24**, 4017–26

MacKay, N. H. (1991b) New factorized S-matrices associated with $SO(N)$, *Nucl. Phys. B* **356**, 729–49

MacKay, N. H. (1992a) On the classical origins of Yangian symmetry in integrable field theory, *Phys. Lett. B* **281**, 90–7

MacKay, N. H. (1992b) The full set of C_n-invariant factorized S-matrices, *J. Phys. A* **25**, L1343–9

MacKay, N. H. (1992c) The fusion of R-matrices using the Birman-Wenzl-Murakami algebra, *J. Math. Phys.* **33**, 1529–37

MacLane, S. (1971) *Categories for the Working Mathematician*, Graduate Texts in Mathematics 5, Springer, Berlin

Madivanane, S. & Venkata-Satyanarayana, M. (1984) A note on the relation between q-algebras without interaction and Weyl commutation relations, *Lett. Nuovo Cim.* **40**, 19–22

Maekawa, T. (1991) On the Wigner and Racah coefficients of $su_q(2)$ and $su_q(1,1)$, *J. Math. Phys.* **32**, 2598–604

Maillard, J. M. (1986) Automorphisms of algebraic varieties and Yang-Baxter equations, *J. Math. Phys.* **27**, 2776–81; erratum, *ibid.* **28** (1987), 1209

Maillet, J. M. (1985) Kac-Moody algebra and extended Yang-Baxter relations in the $O(N)$ nonlinear σ-model, *Phys. Lett. B* **162**, 137–42

Maillet, J. M. (1986) New integrable canonical structures in two-dimensional models, *Nucl. Phys. B* **269**, 54–76

Maillet, J. M. (1990) Lax equations and quantum groups, *Phys. Lett. B* **245**, 480–6

Maillet, J. M. & Nijhoff, F. (1989) Gauging the quantum groups, *Phys. Lett. B* **229**, 71–8

Majid, S. (1988) Hopf algebras for physics at the Planck scale, *J. Classical and Quantum Gravity* **5**, 1587–606

Majid, S. (1989) Matched pairs of Lie groups and Hopf algebra bicrossproducts, *Nucl. Phys. B Suppl.* **6**, 422–4

Majid, S. (1990a) Quasi-quantum groups as internal symmetries of topological quantum field theories, *Lett. Math. Phys.* **21**, 330–2

Majid, S. (1990b) Matched pairs of Lie groups associated to solutions of the Yang-Baxter equation, *Pacific J. Math.* **141**, 311–32

Majid, S. (1990c) Representation-theoretic rank and double Hopf algebras, *Comm. Algebra* **18**, 3705–12

Majid, S. (1990d) On q-regularization, *Internat. J. Modern Phys. A* **5**, 4689–96

Majid, S. (1990e) Physics for algebraists: non-commutative and non-cocommutative Hopf algebras by a bicrossproduct construction, *J. Algebra* **130**, 17–64

Majid, S. (1990f) More examples of bicrossproduct and double crossproduct Hopf algebras, *Israel J. Math.* **72**, 133–48

Majid, S. (1990g) Quasitriangular Hopf algebras and Yang-Baxter equations, *Internat. J. Modern Phys. A* **5**, 1–91

Majid, S. (1990h) Fourier transforms on $A(G)$ and knot invariants, *J. Math. Phys.* **31**, 924–7

Majid, S. (1990i) Quantum group duality in vertex models and other results in the theory of quasitriangular Hopf algebras, in *Differential Geometric Methods in Theoretical Physics, Davis, 1988*, L. L. Chau & W. Nahm (eds), NATO Advanced Science Institutes Series B: Physics 245, pp. 373–85, Plenum, New York

Majid, S. (1991a) Some Physical applications of category theory, in *Differential Geometric Methods in Theoretical Physics, Rapallo, 1990*, C. Bartocci, U. Bruzzo & R. Cianci (eds), Lecture Notes in Physics 375, pp. 131–42, Springer, Berlin

Majid, S. (1991b) Braided groups and algebraic quantum field theories, *Lett. Math. Phys.* **22**, 167–76

Majid, S. (1991c) Hopf-von Neumann algebra bicrossproducts, Kac algebra bicrossproducts and the classical Yang-Baxter equations, *J. Func. Anal.* **95**, 291–319

Majid, S. (1991d) Reconstruction theorems and rational conformal field theories, *Internat. J. Modern Phys. A* **6**, 4359–74

Majid, S. (1991e) Representations, duals and quantum doubles of monoidal categories, *Suppl. Rend. Circ. Mat. Palermo Ser. II* **26**, 197–206

Majid, S. (1991f) Examples of braided groups and braided matrices, *J. Math. Phys.* **32**, 3246–53

Majid, S. (1991g) Quantum groups and quantum probability, in *Quantum Probability and Related Topics*, L. Accardi (ed.), pp. 333–58, World Scientific, Singapore

Majid, S. (1991h) Quasi-quantum groups and internal symmetries of topological quantum field theories, *Lett. Math. Phys.* **22**, 83–90

Majid, S. (1991i) Doubles of quasitriangular Hopf algebras, *Comm. Algebra* **19**, 3061–73

Majid, S. (1992a) Tannaka-Krein theorem for quasi-Hopf algebras and other results, in *Deformation Theory and Quantum Groups with Applications to Mathematical Physics*, M. Gerstenhaber & J. Stasheff (eds), Contemporary Mathematics 134, pp. 219–32, American Mathematical Society, Providence, RI

Majid, S. (1992b) Braided groups and duals of monoidal categories, *Canadian Math. Soc. Conf. Proc.* **13**, 329–43

Majid, S. (1992c) Rank of quantum groups and braided groups in dual form, in *Quantum Groups, Proceedings of Workshops held in the Euler International Mathematical Institute 1990*, P. P. Kulish (ed.), Lecture Notes in Mathematics 1510, pp. 79–89, Springer, Berlin

Majid, S. (1992d) An open problem in quantum groups, in *Quantum Groups, Proceedings of Workshops held in the Euler International Mathematical Institute 1990*, P. P. Kulish (ed.), Lecture Notes in Mathematics 1510, pp. 391–2, Springer, Berlin

Majid, S. (1992e) C-statistical quantum groups and Weyl algebras, *J. Math. Phys.* **33**, 3431–44

Majid, S. (1993a) Braided momentum in the q-Poincaré group, *J. Math. Phys.* **34**, 2045–58

Majid, S. (1993b) Anyonic quantum groups, in *Proceedings of the 2nd Max Born Syposium, Wroclaw, Poland, 1992*, A. Borowiec et al. (eds), Kluwer, Dordrecht

Majid, S. (1993c) Beyond supersymmetry and quantum symmetry (An introduction to braided groups and braided matrices), in *Proceedings of the 5th Nankai Workshop, Tianjin, China, June, 1992*, M. L. Ge (ed.), World Scientific, Singapore

Majid, S. & Soibelman, Ya. S. (1991a) Chern-Simons theory, modular functions and quantum mechanics in an alcove, *Internat. J. Modern Phys. A* **6**, 1815–27

Majid, S. & Soibelman, Ya. S. (1991b) Rank of quantized universal enveloping algebras and modular functions, *Comm. Math. Phys.* **137**, 249–62

Malikov, F. (1992) Quantum groups: singular vectors and BGG resolution, in *Infinite Analysis*, A. Tsuchiya, T. Eguchi & M. Jimbo (eds), Advanced Series in Mathematical Physics 16, pp. 623–44, World Scientific, Singapore

Maltsiniotis, G. (1990) Groupes quantiques et structures différentielles, *C. R. Acad. Sci. Paris Sér I* **311**, 831–44

Maltsiniotis, G. (1992) Groupoides quantiques, *C. R. Acad. Sci. Paris Sér I* **314**, 249–52

Maltsiniotis, G. (1993) Le langage des espaces et des groupes quantiques, *Comm. Math. Phys.* **151**, 275–302

Manin, Yu. I. (1987) Some remarks on Koszul algebras and quantum groups, *Ann. Inst. Fourier (Grenoble)* **37**, 191–205

Manin, Yu. I. (1988) *Quantum Groups and Non-commutative Geometry*, Centre de Recherches Mathématiques, Montréal

Manin, Yu. I. (1989) Multiparameter quantum deformation of the general linear supergroup, *Comm. Math. Phys.* **123**, 163–75

Manin, Yu. I. (1990) Quantized theta functions, *Progr. Theor. Phys. Suppl.* **102**, 219–28

Manin, Yu. I. (1991a) Quantum groups, *Nederl. Akad. Wetensch. Versla Afd. Natuurk.* **100**, 55–68

Manin, Yu. I. (1991b) *Topics in Non-commutative Geometry*, Princeton University Press, Princeton

Manin, Yu. I. (1991c) Quantum groups and non-commutative de Rham complexes, preprint MPI/91-47, Bonn

Manin, Yu. I. (1991d) Notes on quantum groups and quantum de Rham complexes, preprint, MPI/91-60, Bonn

Manin, Yu. I. & Schechtman, V. V. (1986) Higher Bruhat orderings connected with the symmetric group, *Funkts. Anal. i Prilozhen* **20**, 74–5 (Russian)

Manko, V. I., Marmo, G., Solimeno, S. & Zaccaria, F. (1993) Correlation functions of quantum q-oscillators, *Phys. Lett. A* **176**, 173–5

Manocha, H. L. (1990) On models of irreducible q-representations of $sl(2, \mathbb{C})$, *Appl. Anal.* **37**, 19–47

Marchiafava, S. & Rembielinski, J. (1992) Quantum quaternions, *J. Math. Phys.* **33**, 171–3

Marcinek, W. (1992) Graded algebras and geometry based on Yang-Baxter operators, *J. Math. Phys.* **33**, 1631–5

Markov, A. (1941) Über die freie Äquivalenz geschlossener Zöpfe, *Recueil Math. Moscow* **1**, 73–8

Martin, P. P. (1992) On Schur-Weyl duality, A_n Hecke algebras and quantum $sl(N)$ on $\otimes^{n+1}\mathbb{C}^N$, in *Infinite Analysis*, A. Tsuchiya, T. Eguchi & M. Jimbo (eds), Advanced Series in Mathematical Physics 16, pp. 645–74, World Scientific, Singapore

Martin-Delgado, M. A. (1991) A physical interpretation of the quantum group $U_q(SU(2))$, *J. Phys. A* **24**, L807–13

Masuda, T., Mimachi, K., Nakagami, Y., Noumi, M. & Ueno, K. (1988) Representations of quantum groups and a q-analogue of orthogonal polynomials, *C. R. Acad. Sci. Paris Sér. I* **307**, 559–64

Masuda, T., Mimachi, K., Nakagami, Y., Noumi, M. & Ueno, K. (1991) Representations of the quantum group $SU_q(2)$ and the little q-Jacobi polynomials, *J. Func. Anal.* **99**, 357–86

Masuda, T., Mimachi, K., Nakagami, Y., Noumi, M., Saburi, Y. &. & Ueno, K. (1990a) Unitary representations of the quantum group $SU_q(1,1)$: structure of the dual space of $U_q(sl(2))$, *Lett. Math. Phys.* **19**, 187–94

Masuda, T., Mimachi, K., Nakagami, Y., Noumi, M., Saburi, Y. & Ueno, K. (1990b) Unitary representations of the quantum group $SU_q(1,1)$: II - matrix elements of unitary representations and the basic hypergeometric functions, *Lett. Math. Phys.* **19**, 195–204

Masuda, T., Nakagami, Y. & Watanabe, J. (1990) Non-commutative differential geometry on the quantum $SU(2)$, I. An algebraic viewpoint, *K-theory* **4**, 157–80

Masuda, T., Nakagami, Y. & Watanabe, J. (1991) Non-commutative differential geometry on the quantum 2-spheres of Podleś, I. An algebraic viewpoint, *K-theory* **5**, 151–75

Masuda, T. & Watanabe, J. (1991) Sur les espaces vectoriels topologiques associés aux groupes quantiques $SU_q(2)$ et $SU_q(1,1)$, *C. R. Acad. Sci. Paris Sér. I* **312**, 827–30

Matheus-Valle, J. L. & Monteiro, M. A. R. (1993a) Quantum group generalization of the heterotic QFT, *Phys. Lett. B* **300**, 66–72

Matheus-Valle, J. L. & Monteiro, M. A. R. (1993b) Quantum group generalization of the classical supersymmetric point particle, *Modern Phys. Lett. A* **7**, 3023–8

Mathieu, O. (1992) Bases des représentations des groupes simples complexes, Séminaire Bourbaki exp. no. 743, *Astérisque* **201–203**, 421–42, Société Mathématique de France, Paris

Mathur, S. D. (1992) Quantum Kac-Moody symmetry in integrable field theories, *Nucl. Phys. B* **369**, 433–60

Matsuo, A. (1993) Jackson integrals of Jordan-Pockhammer type and quantum Knizhnik-Zamolodchikov equation, *Comm. Math. Phys.* **151**, 263–74

Matsuzaki, T. & Suzuki, T. (1992) A representation of $U_q(su(1,1))$ on the space of quasi-primary fields and correlation functions, *Phys. Lett. B* **296**, 33–9

Matthes, R. (1991) An example of a differential calculus on the quantum complex n-space, *Seminar Sophus Lie* **1**, 23–30

Matthes, R. (1992) 'Quantum group' structure and 'covariant' differential calculus on symmetric algebras corresponding to commutation factors on $\mathbb{Z}^n$, in *Quantum Groups and Related Topics*, R. Gierlak *et al.* (eds), pp. 45–54, Kluwer, Dordrecht

Mayer, M. E. (1988) Groupoids and Lie bigebras in gauge and string theories, in *Differential Geometrical Methods in Theoretical Physics*, K. Bleuler & M. Werner (eds), NATO Advanced Science Institutes Series C: Mathematical and Physical Sciences 250, pp. 149–64, Kluwer, Dordrecht

Mayer, M. E. (1991) From Poisson groupoids to quantum groupoids and back, in *Differential Geometric Methods in Theoretical Physics, Rapallo, 1990*, C. Bartocci, U. Bruzzo & R. Cianci (eds), Lecture Notes in Physics 375, pp. 143–54, Springer, Berlin

Meister, A. (1992) The C^*-algebra obtained by completing the quantum group $Fun_q(SU(2))$, *J. Math. Phys.* **33**, 4177–89

Menezes, D. P., Avancini, S. S. & Providenca, C. (1992) Quantum algebraic description of the Moszkowski model, *J. Phys. A* **25**, 6317–22

Merkulov, S. A. (1991a) Quantum $m \times n$ matrices and the q-deformed Binet-Cauchy formula, *J. Phys. A* **24**, L1243–7

Merkulov, S. A. (1991b) Quantum deformation of compactified Minkowski space, *Z. Phys. C* **52**, 583–8

Mezincescu, L. & Nepomechie, R. I. (1990) Unitarity and irrationality for the quantum algebra $U_q[SU(2)]$, *Phys. Lett. B* **246**, 412–6

Mezincescu, L. & Nepomechie, R. I. (1991a) Integrability of open spin chains with quantum algebra symmetry, *Internat. J. Modern Phys. A* **6**, 5231–48; addendum, *ibid.* **7**, 5657–9

Mezincescu, L. & Nepomechie, R. I. (1991b) Integrable higher-spin chains with $SU_q(2)$ symmetry, in *Quantum Groups*, T. Curtright, D. Fairlie & C. Zachos (eds), pp. 206–35, World Scientific, Singapore

Mezincescu, L. & Nepomechie, R. I. (1991c) Quantum algebra structure of exactly soluble quantum spin chains, *Modern Phys. Lett. A* **6**, 2497–508

Mezincescu, L. & Nepomechie, R. I. (1991d) Integrable open spin chains with non-symmetric R-matrices, *J. Phys. A* **24**, L17–23

Michaelis, W. (1980) Lie coalgebras, *Adv. Math.* **38**, 1–54

Mikami, K. (1991) Symplectic double groupoids over Poisson $(ax+b)$ groups, *Trans. Amer. Math. Soc.* **324**, 447–63

Milnor, J. W. & Moore, J. C. (1965) On the structure of Hopf algebras, *Ann. of Math.* **81**, 211–64

Mir-Kasimov, R. M. (1991a) $SU_q(1, 1)$ and the relativistic oscillator, *J. Phys. A* **24**, 4283–302

Mir-Kasimov, R. M. (1991b) Relativistic oscillator = q-oscillator, in *Group Theoretical Methods in Physics, Moscow, 1990*, V. V. Dodonov & V. I. Manko (eds), Lecture Notes in Physics 382, pp. 215–20, Springer, Berlin

Misra, K. C. & Miwa, T. (1990) Crystal base for the basic representation of $U_q(\hat{sl}(n))$, *Comm. Math. Phys.* **134**, 79–88

Molev, A. I. (1992) Representations of twisted Yangians, *Lett. Math. Phys.* **26**, 211–8

Moore, G. & Reshetikhin, N. Yu. (1989) A comment on quantum group symmetry in conformal field theory, *Nucl. Phys. B* **328**, 557–74

Moore, G. & Seiberg, N. (1988) Classical and quantum conformal field theory, *Comm. Math. Phys.* **123**, 177–254

Montgomery, S. & Smith, S. P. (1990) Skew derivations and $U_q(sl(2))$, *Israel J. Math.* **72**, 158–66

Moreno, C. (1990) Produits star sur certain G/K kähleriens. Equation de Yang-Baxter et produits star sur G, in *Géométrie Symplectique et Mécanique, Colloque International La Grande Motte, France, 23-28 Mai, 1988*, C. Albert (ed.), Lecture Notes in Mathematics 1416, pp. 210–34, Springer, Berlin

Morosi, C. (1992) The R-matrix theory and the reduction of Poisson manifolds, *J. Math. Phys.* **33**, 941–52

Morton, H. R. & Strickland, P. (1991) Jones polynomial invariants for knots and satellites, *Math. Proc. Camb. Phil. Soc.* **109**, 83–103

Moyal, J. E. (1949) Quantum mechanics as a statistical theory, *Proc. Camb. Phil. Soc.* **45**, 99–124

Mukhin, E. E. (1991) Yang-Baxter operators and the non-commutative de Rham complex, *Russian Math. Surveys* **46** (4), 192–3

Müller, A. (1992) Classifying spaces for quantum principal bundles, *Comm. Math. Phys.* **149**, 495–512

Muller-Hoissen, F. (1992) Differential calculi on the quantum group $GL_{p,q}(2)$, *J. Phys. A* **25**, 1703–4

Murakami, J. (1987) The Kauffman polynomial of links and representation theory, *Osaka J. Math.* **24**, 745–58

Murakami, J. (1989) Solvable lattice models and algebras of face operators, in *Integrable Systems in Quantum Field Theory and Statistical Mechanics*, M. Jimbo, T. Miwa & A. Tsuchiya (eds), Advanced Studies in Pure Mathematics 19, pp. 399–415, Academic Press, Boston

Murakami, J. (1990) The representations of the q-analogue of Brauer's centralizer algebras and the Kauffman polynomial of links, *Publ. Res. Inst. Math. Sci.* **26**, 935–45

Murakami, J. (1992a) The multi-variable Alexander polynomial and a one-parameter family of representations of $U_q(sl(2,C))$ at $q^2 = -1$, in *Quantum Groups, Proceedings of Workshops held in the Euler International Mathematical Institute 1990*, P. P. Kulish (ed.), Lecture Notes in Mathematics 1510, pp. 350–3, Springer, Berlin

Murakami, J. (1992b) The free-fermion model in presence of field related to the quantum group $U_q(\hat{sl}_2)$ of affine type and the multi-variable Alexander polynomial of links, in *Infinite Analysis*, A. Tsuchiya, T. Eguchi & M. Jimbo (eds), Advanced Series in Mathematical Physics 16, pp. 765–72, World Scientific, Singapore

Myrberg, P. J. (1922) Über Systeme analytischer Funktionen welche ein Additions-theorem besitzen; Preisschrift gekrönt und herausgegeben von der Fürstlich Jablonowskischen Gesellschaft zu Leipzig

Nagy, G. (1993) On the Haar measure of the quantum $SU(N)$ group, *Comm. Math. Phys.* **153**, 217–28

Nakabo, S. (1993) An invariant of links in a handlebody associated with the spin j representation of $U_q(sl(2,\mathbb{C}))$, *Proc. Amer. Math. Soc.* **118**, 645–56

Nakashima, T. (1990) A basis of symmetric tensor representations for the quantum analogue of the Lie algebras B_n, C_n and D_n, *Publ. Res. Inst. Math. Sci.* **26**, 723–33

Nakashima, T. (1993) Crystal base and a generalization of the Littlewood-Richardson rule for classical Lie algebras, *Comm. Math. Phys.* **154**, 215–43

Nakatsu, T. (1991) Quantum group approach to affine Toda field theory, *Nucl. Phys. B* **356**, 499–529

Nanhua, X. (1993) Special bases of irreducible modules of the quantized universal enveloping algebra $U_v(gl(n))$, *J. Algebra* **154**, 377–86

Narganes-Quijano, F. J. (1991a) The quantum deformation of $SU(1|1)$ and the dynamical symmetry of the anharmonic oscillator, *J. Phys. A* **24**, 1699–707

Narganes-Quijano, F. J. (1991b) Cyclic representations of a q-deformation of the Virasoro algebra, *J. Phys. A* **24**, 593–601

Nazarov, M. L. (1991) Quantum berezinian and the classical Capelli identity, *Lett. Math. Phys.* **21**, 123–31

Nazarov, M. L. (1992) Yangians of the 'strange' Lie superalgebras, in *Quantum Groups, Proceedings of Workshops held in the Euler International Mathematical Institute 1990*, P. P. Kulish (ed.), Lecture Notes in Mathematics 1510, pp. 47–55, Springer, Berlin

Nazarov, M. L. (1993a) Yangians of the queer Lie superalgebra, in *Quantum Groups, Proceedings of the 2nd Wigner Symposium, Goslar, 1991*, Springer, Berlin

Nazarov, M. L. (1993b) Young's symmetrizers for projective representations of the symmetric group, preprint, RIMS, Kyoto

Nazarov, M. L. (1993c) Yangians and Gel'fand-Zetlin bases, preprint, RIMS, Kyoto

Ndimubandi, J. (1993) A note on q-deformations and supersymmetric quantum mechanics, *Modern Phys. Lett. A* **8**, 429–34

Nelson, J. E. & Regge, T. (1991) Quantum gravity and quantum groups, in *Mechanics, Analysis and Geometry: 200 Years after Lagrange*, M. Francaviglia (ed.), pp. 503–12, North-Holland, Amsterdam

Nelson, J. E. & Regge, T. (1992) $SU(2)_q$ and $2+1$ quantum gravity, in *Integrable Systems and Quantum Groups*, M. Carfora, M. Martinelli & A. Marzuoli (eds), pp. 173–81, World

Scientific, Singapore

Neskovic, P. V. & Urosevic, B. V. (1992) Quantum oscillators: applications in statistical mechanics, *Internat. J. Modern Phys. A* **7**, 3379–88

Ng, Y. J. (1990) Comment on the q-analogues of the harmonic oscillator, *J. Phys. A* **23**, 1023–7

Nijhoff, F. W. & Capel, H. W. (1992) Integrable quantum mappings and non-ultralocal Yang-Baxter structures, *Phys. Lett. A* **163**, 49–56

Nomura, M. (1988) Relations among n-j symbols in forms of the star-triangle relation, *J. Phys. Soc. Japan* **57**, 3653–6

Nomura, M. (1989a) Relations for Clebsch-Gordan coefficients in $su_q(2)$ and Yang-Baxter equations, *J. Math. Phys.* **30**, 2397–405

Nomura, M. (1989b) Yang-Baxter relations in terms of n-j symbols of $su_q(2)$ algebra, *J. Phys. Soc. Japan* **58**, 2694–704

Nomura, M. (1990a) An alternative description of the quantum group $SU_q(2)$ and the q-analog Racah-Wigner algebra, *J. Phys. Soc. Japan* **59**, 439–48

Nomura, M. (1990b) Recursion relations for the Clebsch-Gordon coefficient of quantum group $SU_q(2)$, *J. Phys. Soc. Japan* **59**, 1954–61; addendum, *ibid.* **60**, 726

Nomura, M. (1990c) Various kinds of relations for $3n$-j symbols of quantum group $SU_q(2)$, *J. Phys. Soc. Japan* **59**, 3851–60

Nomura, M. (1990d) Representation functions d^j_{mk} of $U[sl_q(2)]$ as wave functions of 'quantum symmetric tops' and relationship to braiding matrices, *J. Phys. Soc. Japan* **59**, 4260–71

Nomura, M. (1990e) A Jordan-Schwinger representation of quadratic relations for $SU_q(2)$ operators and of the q-analog Wigner-Eckart theorem, *J. Phys. Soc. Japan* **59**, 2345–54; erratum, *ibid.* **59**, 3805

Nomura, M. (1991a) Concepts of tensors in $U_q(sl(2))$ and a van der Waerden method for quantum Clebsch-Gordan coefficients, *J. Phys. Soc. Japan* **60**, 789–97

Nomura, M. (1991b) Co- and contra-variant tensors of a general rank in quantum matrix algebras $su_q(2)$ expressed by creation and annihilation operators I, *J. Phys. Soc. Japan* **60**, 3260–70; II, *ibid.* **60**, 4060–70

Nomura, M. (1992) Covariant exchange algebras and quantum groups I, *J. Phys. Soc. Japan* **61**, 1485–94

Nomura, M. & Biedenharn, L. C. (1992) On the q-symplecton realization of the quantum group $SU_q(2)$, *J. Math. Phys.* **33**, 3636–48

Northcott, D. G. (1968) *Lessons on Rings, Modules and Multiplicities*, Cambridge University Press, Cambridge

Noumi, M. (1991) Quantum groups and q-orthogonal polynomials - towards a realization of Askey-Wilson polynomials on $SU_q(2)$, in *Special Functions, Okayama, 1990*, M. Kashiwara & T. Miwa (eds), pp. 260–88, Springer, Berlin

Noumi, M. & Mimachi, K. (1990a) Big q-Jacobi polynomials, q-Hahn polynomials and a family of quantum 3-spheres, *Lett. Math. Phys.* **19**, 299–305

Noumi, M. & Mimachi, K. (1990b) Askey-Wilson polynomials and the quantum group $SU_q(2)$, *Proc. Japan Acad. Ser. A Math. Sci.* **66**, 146–9

Noumi, M. & Mimachi, K. (1990c) Quantum 2-spheres and big q-Jacobi polynomials, *Comm. Math. Phys.* **128**, 521–31

Noumi, M. & Mimachi, K. (1991) Rogers' q-ultraspherical polynomials on a quantum 2-sphere, *Duke Math. J.* **63**, 65–80

Noumi, M. & Mimachi, K. (1992a) Askey-Wilson polynomials as spherical functions on the quantum group $SU_q(2)$, in *Quantum Groups, Proceedings of Workshops held in the Euler International Mathematical Institute 1990*, P. P. Kulish (ed.), Lecture Notes in Mathematics 1510, pp. 98–103, Springer, Berlin

Noumi, M. & Mimachi, K. (1992b) Spherical functions on a family of quantum 3-spheres, *Compositio Math.* **83**, 19–42

Noumi, M., Yamada, H. & Mimachi, K. (1989) Zonal spherical functions on the quantum homogeneous space $SU_q(n+1)/SU_q(n)$, *Proc. Japan Acad. Ser. A Math. Sci.* **65**, 169–71

Nowicki, A., Sorace, E. & Tarlini, M. (1993) The quantum deformed Dirac equation from the κ-Poincaré algebra, *Phys. Lett. B* **302**, 419–22

Ocneanu, A. (1988) Quantized groups, string algebras and Galois theory for algebras, in *Operator Theory and Applications, Vol. 2*, D. E. Evans & M. Takesaki (eds), London Mathematical Society Lecture Note Series 136, pp. 119–72, Cambridge University Press, Cambridge

Odaka, K. (1992) Fermionic q-oscillators and associated canonical q-transformations, *J. Phys. A* **25**, L39–44

Odesskii, A. V. (1986) An analogue of the Sklyanin algebra, *Funct. Anal. Appl.* **20**, 152–4

Odesskii, A. V. & Feigin, B. L. (1989) Sklyanin's elliptic algebras, *Funct. Anal. Appl.* **23**, 207–14

Oevel, W. (1988) R-matrices and related involution theorems, in *Finite-dimensional Integrable Nonlinear Dynamical Systems, Johannesburg, 1988*, P. G. L. Leach & W. H. Steeb (eds), pp. 60–73, World Scientific, Teaneck, NJ

Oevel, W. (1989) R-matrices and higher Poisson brackets for integrable systems, in *Nonlinear Physics, Shanghai, 1989*, Chao Hao Gu, Yi Shen Li & Gui Zhang Tu (eds), pp. 136–45, Springer, Berlin

Oevel, W. & Ragnisco, O. (1989) R-matrices and higher Poisson brackets for integrable systems, *J. Phys. A* **161**, 181–220

Ogievetsky, E. (1992) Differential operators on quantum spaces for $GL_q(n)$ and $SO_q(n)$, *Lett. Math. Phys.* **24**, 245–55

Ogievetsky, E., Reshetikhin, N. Yu. & Wiegmann, P. (1987) The principal chiral field in two dimensions and classical Lie algebras, *Nucl. Phys. B* **280**, 45–96

Ogievetsky, E., Schmidke, W. B. & Wess, J. (1991) Six generator q-deformed Lorentz algebra, *Lett. Math. Phys.* **23**, 233–40

Ogievetsky, E., Schmidke, W. B., Wess, J. & Zumino, B. (1992) q-deformed Poincaré algebra. *Comm. Math. Phys.* **150**, 495–518

Ogievetsky, E. & Wess, J. (1991) Relations between $GL_{p,q}$'s, *Z. Phys. C* **50**, 123–31

Ogievetsky, E. & Wiegmann, P. (1986) Factorized S-matrix and the Bethe ansatz for simple Lie groups, *Phys. Lett. B* **168**, 360–6

Ogievetsky, E. & Zumino, B. (1992) Reality in the differential claculus on q-Euclidean spaces, *Lett. Math. Phys.* **25**, 121–30

Oh, C. H. & Singh, K. (1992) Conformal properties of primary fields in a q-deformed theory, *J. Phys. A* **25**, L149–56

Ohn, C. (1992) A ∗-product on $SL(2)$ and the corresponding nonstandard quantum $U(sl(2))$, *Lett. Math. Phys.* **25**, 85–8

Okado, M. (1990) Quantum R-matrices related to the spin representations of B_n and D_n, *Comm. Math. Phys.* **134**, 467–86

Okado, M. & Yamane, H. (1991) R-matrices with gauge parameters and multi-parameter quantized enveloping algebras, *Special Functions, Okayama, 1990*, M. Kashiwara & T. Miwa (eds), pp. 289–93, Springer, Tokyo

Olive, D. & Turok, N. (1985) Local conserved densities and zero-curvature conditions for Toda lattice field theories, *Nucl. Phys. B* **257**, 277–301

Olshanskii, G. I. (1988) Extension of the algebra $U(\mathfrak{g})$ for the infinite-dimensional classical Lie algebras $\mathfrak{g}$ and the Yangians $Y(gl(m))$, *Soviet Math. Dokl.* **36**, 569–73

Olshanskii, G. I. (1989) Yangians and universal enveloping algebras, *J. Soviet Math.* **47**, 2466–73

Olshanskii, G. I. (1991) Representations of infinite-dimensional classical groups, limits of enveloping algebras and Yangians, in *Topics in Representation Theory*, A. A. Kirillov (ed.),

Advances in Soviet Mathematics 2, pp. 1–66, American Mathematical Society, Providence, RI

Olshanskii, G. I. (1992a) Quantized enveloping superalgebra of type Q and a super extension of the Hecke algebra, *Lett. Math. Phys.* **24**, 93–102

Olshanskii, G. I. (1992b) Twisted Yangians and infinite-dimensional classical Lie algebras, in *Quantum Groups, Proceedings of Workshops held in the Euler International Mathematical Institute 1990*, P. P. Kulish (ed.), Lecture Notes in Mathematics 1510, pp. 104–19, Springer, Berlin

Omori, H., Maeda, Y. & Yoshioka, A. (1992a) Existence of a closed star product, *Lett. Math. Phys.* **26**, 284–94

Omori, H., Maeda, Y. & Yoshioka, A. (1992b) Deformation quantization of Poisson algebras, *Proc. Japan Acad. Ser. A Math. Sci.* **68**, 97–100

Ottinger, H. C. & Honerkamp, J. (1982) Note on the Yang-Baxter equations for generalized Baxter models, *Phys. Lett. A* **88**, 339–43

Ozawa, T. (1990) Quantization of a Poisson algebra and polynomials associated to links, *Nagoya Math. J.* **120**, 113–27

Palev, T. D. & Tolstoy, V. N. (1991) Finite-dimensional irreducible representations of the quantum superalgebra $U_q(gl(n/1))$, *Comm. Math. Phys.* **141**, 549–58

Pan, F. (1991) Irreducible tensor operators for the quantum algebra $su(2)_q$, *J. Phys. A* **24**, L803–6

Pan, F. (1993) Quantum deformation of $SU(3)$ and subalgebras, *J. Phys. A* **26**, L257–62

Pan, F. & Chen, J. Q. (1992) Symmetric irreducible representations in the $SU(2)_q \times U(1)$ basis and some extensions to $SU(N)_q \times U(1)$, *J. Phys. A* **25**, 4017–24

Paradowski, J. (1992) Filtrations of modules over quantum algebras, preprint

Parashar, P., Bhasin, V. S. & Soni, S. K. (1993) Covariant differential calculi on quantum symplectic and orthogonal planes, *Modern Phys. Lett. A* **8**, 389–98

Parashar, P. & Soni, S. K. (1992) Covariant differential calculus on the quantum exterior vector space, *Z. Phys. C* **53**, 609–11

Pareigis, B. (1981) A noncommutative, noncocommutative Hopf algebra in 'nature', *J. Algebra* **70**, 356–74

Parshall, B. & Wang, J. P. (1990) On bialgebra cohomology, *Bull. Soc. Math. Belg. Sér. A* **42**, 607–42

Parshall, B. & Wang, J. P. (1991) *Quantum Linear Groups*, Memoirs of the American Mathematical Society 439, American Mathematical Society, Providence, RI

Parshall, B. & Wang, J. P. (1992) Cohomology of infinitesimal quantum groups I, *Tohoku Math. J.* **44**, 395–423

Parthasarathy, R. & Viswanathan, K. S. (1991) A q-analogue of the supersymmetric oscillator and its q-superalgebra, *J. Phys. A* **24**, 613–7

Pasquier, V. (1988a) Etiology of IRF models, *Comm. Math. Phys.* **118**, 355–64

Pasquier, V. (1988b) Continuum limit of lattice models built on quantum groups, *Nucl. Phys. B* **295**, 491–510

Pasquier, V. (1992) Quantum groups in lattice models, in *New Symmetry Principles in Quantum Field Theory*, J. Fröhlich *et al.* (eds), NATO Advanced Science Institutes Series B: Physics 295, pp. 355–81, Plenum, New York

Pasquier, V. & Saleur, H. (1990) Common structures between finite systems and conformal field theories through quantum groups, *Nucl. Phys. B* **330**, 523–56

Perk, J. H. H. (1989) Star-triangle equations, quantum Lax pairs, and higher genus curves, in *Theta Functions, Bowdoin, ME, 1987*, L. Ehrenpreis & R. C. Gunning (eds), Proceedings of Symposia in Pure Mathematics 49, Part I, pp. 341–54, American Mathematical Society, Providence, RI

Pierce, R. S. (1982) *Associative Algebras*, Graduate Texts in Mathematics 88, Springer, Berlin

Podleś, P. (1987) Quantum spheres, *Lett. Math. Phys.* **14**, 193–202

Podleś, P. (1989) Differential calculus on quantum spheres, *Lett. Math. Phys.* **18**, 107–19

Podleś, P. (1991) Complex quantum groups and their real representations, preprint RIMS-754, Kyoto

Podleś, P. (1992a) Quantization reinforces interaction. Quantum mechanics of two particles on a quantum sphere, in *Infinite Analysis*, A. Tsuchiya, T. Eguchi & M. Jimbo (eds), Advanced Series in Mathematical Physics 16, pp. 805–12, World Scientific, Singapore

Podleś, P. (1992b) The classification of differential structures on quantum 2-spheres, *Comm. Math. Phys.* **150**, 167–80

Podleś, P. (1992c) Complex quantum groups and their real representations, *Publ. Res. Inst. Math. Sci.* **28**, 709–46

Podleś, P. & Woronowicz, S. L. (1990) Quantum deformation of Lorentz group, *Comm. Math. Phys.* **130**, 381–431

Polychronakos, A. P. (1990) A classical realization of quantum algebras, *Modern Phys. Lett. A* **5**, 2325–33

Polychronakos, A. P. (1991a) Aspects of q-Virasoro algebra, in *Quantum Groups*, T. Curtright, D. Fairlie & C. Zachos (eds), pp. 194–205, World Scientific, Singapore

Polychronakos, A. P. (1991b) Consistence conditions and representations of a q-deformed Virasoro algebra, *Phys. Lett. B* **256**, 35–40

Pressley, A. N. (1992) Quantum groups, Hamburger Beiträge zur Mathematik, Heft 16

Przytycki, J. (1992) Quantum group of links in a handlebody, in *Deformation Theory and Quantum Groups with Applications to Mathematical Physics*, M. Gerstenhaber & J. Stasheff (eds), Contemporary Mathematics 134, pp. 235–45, American Mathematical Society, Providence, RI

Przytycki, J. & Traczyk, P. (1987) Conway algebras and skein equivalence of links, *Proc. Amer. Math. Soc.* **100**, 744–8

Pusz, W. (1989) Twisted canonical anticommutation relations, *Rep. Math. Phys.* **27**, 349–60

Pusz, W. (1991) On the implementation of $S_\mu U(2)$ action in the irreducible representations of twisted canonical commutation relations, *Lett. Math. Phys.* **21**, 59–67

Pusz, W. (1993) Irreducible unitary representations of quantum Lorentz group, *Comm. Math. Phys.* **152**, 591–626

Pusz, W. & Woronowicz, S. L. (1989) Twisted second quantization, *Rep. Math. Phys.* **27**, 231–57

Qian, Z. H., Qian, M. & Guo, M. (1992) A new type of Hopf algebra which is neither commutative nor cocommutative, *J. Phys. A* **25**, 1237–46

Quan, X. C. (1991) Compact quantum groups and group duality, *Acta Appl. Math.* **25**, 277–99

Quano, Y. H. & Fujii, A. (1991) Generalized Sklyanin algebra, *Modern Phys. Lett. A* **6**, 3635–40

Quesne, C. (1991) Coherent states, K-matrix theory and q-boson realizations of the quantum algebra $su_q(2)$, *Phys. Lett. A* **153**, 303–7

Quesne, C. (1992) Complementarity of $su_q(3)$ and $u_q(2)$ and q-boson realization of the $su_q(3)$ irreducible representations, *J. Phys. A* **25**, 5977–98

Quesne, C. (1993a) Raising and lowering operators for $u_q(n)$, *J. Phys. A* **26**, 357–72

Quesne, C. (1993b) Sets of covariant and contravariant spinors for $SU_q(2)$ and alternative quantizations, *J. Phys. A* **26**, L299–306

Quesne, C. (1993c) Two-parameter versus one-parameter quantum deformation of $su(2)$, *Phys. Lett. A* **174**, 19–24

Quesne, C. (1993d) q-bosonic vector operators for $so_q(3)$ and a q-deformed $u(3)$ algebra, *Phys. Lett. B* **304**, 81–8

Quesne, C. (1993e) q-bosonic operators as double irreducible tensors for $u_q(n) + u_q(m)$, *Phys. Lett. B* **298**, 344–50

Quillen, D. (1969) Rational homotopy theory, *Ann. of Math.* **90** (2), 205–95

Quispel, G. R. W. & Nijhoff, F. W. (1992) Integrable two-dimensional quantum mappings, *Phys. Lett. A* **161**, 419–22

Radford, D. E. (1976) The order of the antipode of a finite dimensional Hopf algebra is finite, *Amer. J. Math.* **98**, 333–5

Radford, D. E. (1992) On the antipode of a quasitriangular Hopf algebra, *J. Algebra* **151**, 1–11

Radford, D. E. (1993) Minimal quasitriangular Hopf algebras, *J. Algebra* **157**, 285–315

Rajagopal, A. K. & Gupta, V. (1992) Uncertainty principle, squeezing and quantum groups, *Modern Phys. Lett. A* **7**, 3759–64

Rajeswari, V. & Srinivasa Rao, K. (1991a) Generalized basic hypergeometric functions and the q-analogues of $3 - j$ and $6 - j$ coefficients, *J. Phys. A* **24**, 3761–80

Rajeswari, V. & Srinivasa Rao, K. (1991b) Note on the explicit forms of the Clebsch-Gordan coefficients of the quantum group $SU_q(2)$, *J. Phys. Soc. Japan* **60**, 3583–4

Ralchenko, Yu. V. (1992) q-deformed paracommutation relations, *J. Phys. A* **25**, L1155–8

Ram, A. (1991) A Frobenius formula for the characters of the Hecke algebras, *Invent. Math.* **106**, 461–88

Ramirez, C., Ruegg, H. & Ruiz-Altaba, M. (1990) Explicit quantum symmetries of WZNW theories, *Phys. Lett. B* **247**, 499–508

Ramirez, C., Ruegg, H. & Ruiz-Altaba, M. (1991a) The contour picture of quantum groups: conformal field theories, *Nucl. Phys. B* **364**, 195–233

Ramirez, C., Ruegg, H. & Ruiz-Altaba, M. (1991b) Coulomb gas realization of simple quantum groups, in *Group Theoretical Methods in Physics, Moscow, 1990*, V. V. Dodonov & V. I. Manko (eds), Lecture Notes in Physics 382, pp. 199–203, Springer, Berlin

Recknagel, A. (1993) Fusion rules from algebraic K-theory, *Internat. J. Modern Phys. A* **8**, 1345–57

Reidemeister, K. (1932) *Knottentheorie*, Springer, Berlin

Rembielinskii, J. (1992a) Differential and integral calculus on the quantum C-plane, in *Quantum Groups and Related Topics*, R. Gierlak *et al.* (eds), pp. 129–40, Kluwer, Dordrecht

Rembielinskii, J. (1992b) Quantum inhomogeneous groups related to Manin's plane, *Phys. Lett. B* **296**, 335–40

Rembielinskii, J. & Tybor, W. (1992) Polar decomposition of the twisted de Rham complex for $\mathbb{C}_q$, *J. Phys. A* **25**, L1209–11

Reshetikhin, N. Yu. (1983) The functional equation method in the theory of exactly solvable quantum systems, *Soviet Phys. JETP* **57**, 691–6

Reshetikhin, N. Yu. (1985a) Integrable models of quantum 1-dimensional magnetics with $O(n)$ and $Sp(2k)$-symmetry, *Theoret. Math. Phys.* **63**, 555–69

Reshetikhin, N. Yu. (1985b) $O(N)$-invariant quantum field theoretical models: exact solution, *Nucl. Phys. B* **251**, 565–80

Reshetikhin, N. Yu. (1987) The spectrum of the transfer matrices connected with Kac-Moody algebras, *Lett. Math. Phys.* **14**, 235–46

Reshetikhin, N. Yu. (1990a) Multiparameter quantum groups and twisted quasitriangular Hopf algebras, *Lett. Math. Phys.* **20**, 331–5

Reshetikhin, N. Yu. (1990b) Quasitriangular Hopf algebras and invariants of tangles, *Leningrad Math. J.* **1**, 491–513

Reshetikhin, N. Yu. (1992a) Quantization of Lie bialgebras, *Duke Math. J. Internat. Math. Res. Notices* **7**, 143–51

Reshetikhin, N. Yu. (1992b) The Knizhnik-Zamolodchikov system as a deformation of the isomonodromy problem, *Lett. Math. Phys* **26**, 153–65

Reshetikhin, N. Yu. (1992c) Jackson-type integrals, Bethe vectors, and solutions to a difference analog of the Knizhnik-Zamolodchikov system, *Lett. Math. Phys.* **26**, 167–77

Reshetikhin, N. Yu. & Faddeev, L. D. (1983) Hamiltonian structures for integrable field theory models, *Teoret. Mat. Fiz.* **56**, 323–43 (Russian)

Reshetikhin, N. Yu. & Semenov-Tian-Shansky, M. A. (1988) Quantum R-matrices and factorization problems, *J. Geom. Phys.* **5**, 533–50

Reshetikhin, N. Yu. & Semenov-Tian-Shansky, M. A. (1990) Central extensions of quantum current groups, *Lett. Math. Phys.* **19**, 133–42

Reshetikhin, N. Yu. & Smirnov, F. (1990) Hidden quantum group symmetry and integrable perturbations of conformal field theories, *Comm. Math. Phys.* **131**, 157–77

Reshetikhin, N. Yu., Takhtajan, L. A. & Faddeev, L. D. (1990) Quantization of Lie groups and Lie algebras, *Leningrad Math. J.* **1**, 193–225

Reshetikhin, N. Yu. & Turaev, V. G. (1990) Ribbon graphs and their invariants derived from quantum groups, *Comm. Math. Phys.* **127**, 1–26

Reshetikhin, N. Yu. & Turaev, V. G. (1991) Invariants of 3-manifolds via link polynomials and quantum groups, *Invent. Math.* **103**, 547–97

Reshetikhin, N. Yu. & Wiegmann, P. B. (1987) Towards the classification of completely integrable quantum field theories (the Bethe ansatz associated with Dynkin diagrams and their automorphisms), *Phys. Lett. B* **189**, 125–31

Reyman, A. G. & Semenov-Tian-Shansky, M. A. (1979) Reduction of hamiltonian systems, affine Lie algebras, and Lax equations I, *Invent. Math.* **54**, 81–100

Reyman, A. G. & Semenov-Tian-Shansky, M. A. (1981) Reduction of hamiltonian systems, affine Lie algebras, and Lax equations II, *Invent. Math.* **63**, 423–32

Reyman, A. G. & Semenov-Tian-Shansky, M. A. (1988) Compatible Poisson structures for Lax equations: an r-matrix approach, *Phys. Lett. A* **130**, 456–60

Rideau, G. (1992) On the representations of the quantum oscillator algebra, *Lett. Math. Phys.* **24**, 147–54

Rieffel, M. A. (1989) Deformation quantization of Heisenberg manifolds, *Comm. Math. Phys.* **122**, 531–62

Rieffel, M. A. (1990a) Non-commutative tori: a case study of non-commutative differentiable manifolds, in *Geometrical and Topological Invariants of Elliptic Operators, Brunswick, ME, 1988*, J. Kaminker (ed.), Contemporary Mathematics 105, pp. 191–211, American Mathematical Society, Providence, RI

Rieffel, M. A. (1990b) Deformation quantization and operator algebras, in *Operator Theory: Operator Algebras and Applications, Durham, NH, 1988*, W. V. Arveson & R. G. Douglas (eds), Proceedings of Symposia in Pure Mathematics 51, Part I, pp. 411–23, American Mathematical Society, Providence, RI

Rieffel, M. A. (1992) Category theory and quantum field theory, in *Noncommutative Rings*, S. Montgomery & L. W. Small (eds), Mathematical Sciences Research Institute Publications 24, pp. 115–29, Springer, Berlin

Ringel, C. M. (1990) Hall algebras and quantum groups, *Invent. Math.* **101**, 583–92

Ringel, C. M. (1991) The category of modules with good filtrations over a quasi-hereditary algebra has almost split sequences, *Math. Zeit.* **208**, 209–25

Ringel, C. M. (1992) From representations of quivers via Hall and Loewy algebras to quantum groups, in *Proceedings of the International Conference on Algebra, Part 2, Novosibirsk, 1989*, L. A. Bokut, Yu. L. Ershov & A. I. Kostrikin (eds), Contemporary Mathematics 141, pp. 381–401, American Mathematical Society, Providence, RI

Ringel, C. M. (1993) The composition algebra of a cyclic quiver. Towards an explicit description of the quantum group of type $\tilde{A}_n$, *Proc. London Math. Soc.* **66**, 507–38

Rittenberg, V. & Scheunert, M. (1992) Tensor operators for quantum groups and applications, *J. Math. Phys.* **33**, 436–45

Rittenberg, V. & Schutz, G. (1991) Patterns of $U_q(SU(2))$ symmetry breaking in Heisenberg quantum chains, *Helv. Phys. Acta* **64**, 871–6

Rocek, M. (1991) Representation theory of the nonlinear $SU(2)$ algebra, *Phys. Lett.* B **255**, 554–7

Roche, P. & Arnaudon, D. (1989) Irreducible representations of the quantum analogue of $SU(2)$, *Lett. Math. Phys.* **17**, 295–300

Rodriguez-Plaza, M. J. (1991) Casimir operators of $U_q sl(3)$, *J. Math. Phys.* **32**, 2020–7

Rodriguez-Plaza, M. J. (1992) Universal $\mathcal{R}$ matrix for quantum A_2, B_2 and G_2, in *Topological and Geometrical Methods in Field Theory*, J. Mickelsson & O. Pekonen (eds), pp. 337–46, World Scientific, Singapore

Rogawski, J. D. (1985) On modules over the Hecke algebra of a p-adic group, *Invent. Math.* **79**, 443–65

Roger, C. (1990) Déformations universelles des crochets de Poisson, in *Géométrie Symplectique et Mécanique, Colloque International La Grande Motte, France, 23-28 Mai, 1988*, C. Albert (ed.), Lecture Notes in Mathematics 1416, pp. 242–54, Springer, Berlin

Rolfsen, D. (1976) *Knots and Links*, Publish or Perish, Berkeley

Rosenberg, A. L. (1992) The unitary irreducible representations of the quantum Heisenberg group, *Comm. Math. Phys.* **144**, 41–52

Rosso, M. (1987a) Représentations irréductibles de dimension finie du q-analogue de l'algèbre enveloppante d'une algèbre de Lie simple, *C. R. Acad. Sci. Paris Sér. I* **305**, 587–90

Rosso, M. (1987b) Comparaison des groupes de Drinfel'd et de Woronowicz, *C. R. Acad. Sci. Paris Sér. I* **304**, 323–6

Rosso, M. (1988a) Finite dimensional representations of the quantum analog of the enveloping algebra of a complex simple Lie algebra, *Comm. Math. Phys.* **117**, 581–93

Rosso, M. (1988b) Groupes quantiques de Drinfel'd et Woronowicz, *Publ. Res. Inst. Math. Univ. Strasbourg* **38**, 67–82

Rosso, M. (1988c) Groupes quantiques et modèles à vertex de V. Jones en théorie des noeuds, *C. R. Acad. Sci. Paris Sér. I* **307**, 207–10

Rosso, M. (1989) An analogue of P.B.W. theorem and the universal R-matrix for $U_h sl(N+1)$, *Comm. Math. Phys.* **124**, 307–18

Rosso, M. (1990a) Algèbres enveloppantes quantifiées, groupes quantiques compacts de matrices et calcul différentiel non commutatif, *Duke Math. J.* **61**, 11–40

Rosso, M. (1990b) Analogues de la forme de Killing et du théorème d'Harish-Chandra pour les groupes quantiques, *Ann. Sci. Ecole Norm. Sup.* **23** (4), 445–67

Rosso, M. (1990c) Koszul resolutions and quantum groups, *Nucl. Phys.* B Proc. Suppl. **18** **B**, 269–76

Rosso, M. (1991) An analogue of the B. G. G. resolution for the quantum $SL(N)$ group, in *Symplectic Geometry and Mathematical Physics, Aix-en-Provence, 1990*, P. Donato, C. Duval, J. Elhadad & G. M. Tuynman (eds), Progress in Mathematics 99, pp. 422–32, Birkhäuser, Boston

Rosso, M. (1992a) Représentations des groupes quantiques, Séminaire Bourbaki exp. no. 744, *Astérisque* **201–203**, 443–83, Société Mathématique de France, Paris

Rosso, M. (1992b) Certaines formes bilinéaires sur les groupes quantiques et une conjecture de Schechtman et Varchenko, *C. R. Acad. Sci. Paris Sér. I* **314**, 5–8

Rosso, M. (1992c) Quantum groups at a root of 1 and tangle invariants, in *Topological and Geometrical Methods in Field Theory*, J. Mickelsson & O. Pekonen (eds), pp. 347–58, World Scientific, Singapore

Ruegg, H. (1990a) A simple derivation of the quantum Clebsch-Gordan coefficients for $SU_q(2)$, *J. Math. Phys.* **31**, 1085–7

Ruegg, H. (1990b) Polynomial basis for $SU_q(2)$ and Clebsch-Gordan coefficients, in *Quantum Groups, Clausthal, Germany, 1989*, H.-D. Doebner & J.-D. Hennig (eds), Lecture

Notes in Physics 370, pp. 89–95, Springer, Berlin

Ryang, S. (1992) Vertex operators for $U_q(su(2))$ correlation functions, *Phys. Rev. D* **45** (3), 3873–6

Saavedra Rivano, N. (1972) *Catégories Tannakiennes*, Lecture Notes in Mathematics 265, Springer, Berlin

Salam. M. A. & Wybourne, B. G. (1991) The q-deformation of symmetric functions and the symmetric group, *J. Phys. A* **24**, L317–21

Saleur, H. (1990a) Zeroes of chromatic polynomials. A new approach to Beraha conjecture using quantum groups, *Comm. Math. Phys.* **132**, 657–79

Saleur, H. (1990b) Representations of $U_q(sl(2))$ for q a root of unity, in *Number theory and Physics, Les Houches, 1989*, pp. 68–76, Springer, Berlin

Saleur, H. (1990c) Quantum $osp(1,2)$ and solutions of the graded Yang-Baxter equation, *Nucl. Phys. B* **336**, 363–76

Saleur, H. (1991) Level-rank duality in quantum groups, *Nucl. Phys. B* **354**, 579–613

Saleur, H. & Zuber, J. B. (1991) Integrable lattice models and quantum groups, in *String Theory and Quantum Gravity, Trieste, 1990*, M. Green, R. Iengo, S. Randjbar-Daemi, E. Sezgin & H. Verlinde (eds), pp. 1–53, World Scientific, River Edge, NJ

Sanchez, N. (1991) String theory, quantum gravity and quantum groups, in *Group Theoretical Methods in Physics, Moscow, 1990*, V. V. Dodonov & V. I. Manko (eds), Lecture Notes in Physics 382, pp. 204–7, Springer, Berlin

Schechtman, V. (1992a) Vanishing cycles and quantum groups I, *Duke Math. J. Internat. Math. Res. Notices* **2**, 39–49

Schechtman, V. (1992b) Vanishing cycles and quantum groups II, *Duke Math. J. Internat. Math. Res. Notices* **2**, 207–15

Schechtman, V. & Varchenko, A. (1990) Hypergeometric solutions of the Knizhnik-Zamolodchikov equations, *Lett. Math. Phys.* **20**, 279–83

Schechtman, V. & Varchenko, A. (1991a) Quantum groups and homology of local systems, in *Algebraic Geometry and Analytic Geometry, Proceedings of the International Congress of Mathematicians Satellite Conference, Tokyo, 1990*, pp. 182–91, Springer, Berlin

Schechtman, V. & Varchenko, A. (1991b) Arrangements of hyperplanes and Lie algebra homology, *Invent. Math.* **106**, 134–94

Schirrmacher, A. (1991a) The multiparametric deformation of $GL(n)$ and the covariant differential calculus on the quantum vector space, *Z. Phys. C* **50**, 321–7

Schirrmacher, A. (1991b) Multiparameter R-matrices and their quantum groups, *J. Phys. A* **24**, L1249–58

Schirrmacher, A. (1992) Remarks on the use of R-matrices, in *Quantum Groups and Related Topics*, R. Gierlak *et al.* (eds), pp. 55–66, Kluwer, Dordrecht

Schirrmacher, A., Wess, J. & Zumino, B. (1991) The two parameter deformation of $GL(2)$, its differential calculus and Lie algebra, *Z. Phys. C* **49**, 317–24

Schlieker, M. & Scholl, M. (1990) Spinor calculus for quantum groups, *Z. Phys. C* **47**, 625–8

Schlieker, M., Weich, W. & Weixler, R. (1992) Inhomogeneous quantum groups, *Z. Phys. C* **53**, 79–82

Schmeing, N. C. (1991) A Hopf algebra based on the Patera-Zassenhaus grading of A_n, *J. Phys. A* **24**, 1971–7

Schmidke, W. B., Vokos, S. P. & Zumino, B. (1990) Differential geometry of the quantum supergroup $GL_q(1|1)$, *Z. Phys. C* **48**, 249–55

Schmidke, W. B., Wess, J. & Zumino, B. (1991) A q-deformed Lorentz algebra, *Z. Phys. C* **52**, 471–6

Schmüdgen, K. (1991a) Eine Einführung in die Theorie der Quantenmatrixgruppen, *Seminar Sophus Lie* **1**, 3–21

Schmüdgen, K. (1991b) Noncommutative covariant differential calculi on quantum spaces and quantum groups, *Seminar Sophus Lie* **1**, 125–32

Schneider, H.-J. (1990) Principal homogeneous spaces for arbitrary Hopf algebras, *Israel Math. J.* **72**, 167–95

Schnizer, A. W. (1992) Roots of unity: representations for symplectic and orthogonal quantum groups, preprint RIMS-864, Kyoto

Schouten, J. A. (1940) Über differentialkomitanten zweier kontravarianter Grössen, *Nederl. Akad. Wetensch. Proc. Ser. A* **43**, 449–52

Schürmann, M. (1991a) White noise on involutive bialgebras, in *Quantum Probability and Related Topics*, L. Accardi (ed.), pp. 401–19, World Scientific, River Edge, NJ

Schürmann, M. (1991b) Quantum q-white noise and a q-central limit theorem, *Comm. Math. Phys.* **140**, 589–615

Schupp, P., Watts, P. & Zumino, B. (1992a) The two-dimensional quantum Euclidean algebra, *Lett. Math. Phys.* **24**, 141–6

Schupp, P., Watts, P. & Zumino, B. (1992b) Differential geometry on quantum linear groups, *Lett. Math. Phys.* **25**, 139–47

Schwenk, J. (1991) Differential calculus for the N-dimensional quantum plane, in *Quantum Groups*, T. Curtright, D. Fairlie & C. Zachos (eds), pp. 53–61, World Scientific, Singapore

Schwenk, J., Schmidke, W. B. & Vokos, S. P. (1990) Properties of 2×2 quantum matrices in $\mathbf{Z}_2$-graded spaces, *Z. Phys. C* **46**, 643–6

Schwenk, J. & Wess, J. (1992) A q-deformed quantum mechanical toy model, *Phys. Lett. B* **291**, 273–7

Sciarrino, A. (1992) q-oscillators realization of $F_q(4)$ and $G_q(3)$, *J. Phys. A* **25**, L219–26

Sie-Qwon Oh (1993) Finite dimensional simple modules over the coordinate ring of quantum matrices, *Bull. London Math. Soc.* **25**, 427–30

Semenov-Tian-Shansky, M. A. (1983) What is a classical R-matrix ?, *Funct. Anal. Appl.* **17**, 259–72

Semenov-Tian-Shansky, M. A. (1984) Classical r-matrices and quantization, *J. Soviet Math.* **31**, 3411–6

Semenov-Tian-Shansky, M. (1985) Dressing transformations and Poisson-Lie group actions, *Publ. Res. Inst. Math. Sci.* **21**, 1237–60

Semenov-Tian-Shansky, M. (1986) Poisson groups and dressing transformations, *Zap. Nauch. Sem. Leningrad Otdel. Mat. Inst. Steklov.* **150**, 119–42, 221–2 (Russian)

Sen, S. & Chowdhury, A. R. (1991) Yang-Baxter equation and non-ultralocal integrable system, *Phys. Scripta* **44**, 409–11

Sergeev, S. M. (1991) Spectral decomposition of R-matrices for exceptional Lie algebras, *Modern Phys. Lett. A* **10**, 923–7

Serre, J.-P. (1965) *Lie algebras and Lie Groups*, Benjamin, Reading, MA

Shabanov, S. V. (1992a) The Poisson bracket for q-deformed systems, *J. Phys. A* **25**, L1245–50

Shabanov, S. V. (1992b) Path integral for the q-deformed oscillator and its interpretation, *Phys. Lett. B* **293**, 117–22

Sheu, A. J. L. (1991) Quantization of the Poisson $SU(2)$ and its Poisson homogeneous space the 2-sphere, *Comm. Math. Phys.* **135**, 217–32

Sheu, A. J. L. (1992) Quantum Poisson $SU(2)$ and quantum Poisson spheres, in *Deformation Theory and Quantum Groups with Applications to Mathematical Physics*, M. Gerstenhaber & J. Stasheff (eds), Contemporary Mathematics 134, pp. 247–58, American Mathematical Society, Providence, RI

Shibukawa, Y. (1992) Clebsch-Gordan coefficients for $U_q(su(1,1))$ and $U_q(sl(2))$ and linearization formula of matrix elements, *Publ. Res. Inst. Math. Sci.* **28**, 775–808

Shibukawa, Y. & Ueno, K. (1992) Completely $\mathbf{Z}$-symmetric R-matrix, *Lett. Math. Phys.* **25**, 239–48

Shigano, H. (1977) A correspondence between observable Hopf ideals and left coideal sub-algebras, *Tsukuba Math. J.* **1**, 149–56

Shigano, H. (1982) On observable and strongly observable Hopf ideals, *Tsukuba Math. J.* **6**, 127–50

Shiraishi, J. (1993) Free boson representation of $U_q(\hat{sl}_2)$, *Phys. Lett. A* **171**, 243–8

Shnider, S. (1991) Bialgebra deformations, *C. R. Acad. Sci. Paris Sér. I* **312**, 7–12

Shnider, S. (1992) Deformation cohomology for bialgebras and quasi-bialgebras, in *Deformation Theory and Groups with Applications in Mathematical Physics*, M. Gerstenhaber & J. Stasheff (eds), Contemporary Mathematics 134, pp. 259–97, American Mathematical Society, Providence, RI

Siafarikas, P., Jannussis, A., Brodimas, G. & Papaloucas, L. (1983) q-algebras without interaction, *Lett. Nuovo Cim.* **37**, 119–23

Sierra, G. (1989) Duality and quantum groups, in *Knots, Topology and Quantum Field Theories, Florence, 1989*, L. Lusanna (ed.), pp. 417–45, World Scientific, River Edge, NJ

Siopsis, G. (1991a) Quantum groups in WZW and Chern-Simons theories, in *Quantum Groups*, T. Curtright, D. Fairlie & C. Zachos (eds), pp. 181–93, World Scientific, Singapore

Siopsis, G. (1991b) On the quantum group structure in WZW models and Chern-Simons theories, *Modern Phys. Lett. A* **6**, 1515–23

Sklyanin, E. K. (1982a) Quantum variant of the method of inverse scattering problem, *J. Soviet Math.* **19**, 1546–96

Sklyanin, E. K. (1982b) Some algebraic structures connected with the Yang-Baxter equation, *Funct. Anal. Appl.* **16**, 263–70

Sklyanin, E. K. (1983) Some algebraic structures connected with the Yang-Baxter equation. Representations of quantum algebras, *Funct. Anal. Appl.* **17**, 273–84

Sklyanin, E. K. (1984) The Goryachev-Chaplygin top and the method of the inverse scattering problem, *Zap. Nauchn. Sem. Leningrad Otdel. Mat. Inst. Steklov* **133**, 236–57 (Russian)

Sklyanin, E. K. (1985) Classical limits of $SU(2)$ invariant solutions of the Yang-Baxter equation, *Zap. Nauchn. Sem. Leningrad Otdel. Mat. Inst. Steklov* **146** (Russian)

Smirnov, F. A. (1990) Quantum groups and generalized statistics in integrable models, *Comm. Math. Phys.* **132**, 415–39

Smirnov, F. A. (1992) Dynamical symmetries of massive integrable models, in *Infinite Analysis*, A. Tsuchiya, T. Eguchi & M. Jimbo (eds), Advanced Series in Mathematical Physics 16, pp. 813–38 & 839–58, World Scientific, Singapore

Smirnov, Yu. F. & Tolstoy, V. N. (1990) Extremal projectors for usual, super and quantum algebras and their use for solving Yang-Baxter problem, in *Selected Topics in Quantum Field Theory and Mathematical Physics, Liblice, 1989*, J. Niederle & J. Fischer (eds), pp. 347–59, World Scientific, Singapore

Smirnov, Yu. F. & Tolstoy, V. N. (1991a) Projection-operator method and the q-analog of the quantum theory of angular momentum. Racah coefficients, $3j$ and $6j$ symbols, and their symmetry properties, *Soviet J. Nuclear Phys.* **53**, 1068–86

Smirnov, Yu. F. & Tolstoy, V. N. (1991b) Projection operator method and Q-analog of angular momentum theory, in *Group Theoretical Methods in Physics, Moscow, 1990*, V. V. Dodonov & V. I. Manko (eds), Lecture Notes in Physics 382, pp. 183–94, Springer, Berlin

Smirnov, Yu. F. & Tolstoy, V. N. (1992) The quantum algebra $U_q(3)$, *Soviet J. Nuclear Phys.* **54**, 437–45

Smirnov, Yu. F., Tolstoy, V. N. & Kharitonov, Yu. I. (1991) Method of projection operators and the q-analog of the quantum theory of angular momentum. Clebsch-Gordan coefficients and irreducible tensor operators, *Soviet J. Nuclear Phys.* **53**, 593–605

Smirnov, Yu. F. & Wehrhahn, R. F. (1992) The Clebsch-Gordan coefficients for the two-parameter quantum algebra $SU_{p,q}(2)$ in the Löwdin-Shapiro approach, *J. Phys. A* **25**, 5563–76

Smit, D.-J. (1990) A quantum group structure in integrable conformal field theories, *Comm. Math. Phys.* **128**, 1–37

Smith, S. P. (1992) Quantum groups: an introduction and survey for ring theorists, in *Noncommutative Rings*, S. Montgomery & L. W. Small (eds), Mathematical Sciences Research Institute Publications 24, pp. 131–78, Springer, Berlin

Sochen, N. (1991) Integrable models through representations of the Hecke algebra, *Nucl. Phys. B* **360**, 613–40

Soibelman, Ya. S. (1990a) Irreducible representations of the algebra of functions on the quantum group $SU(n)$ and Schubert cells, *Soviet Math. Dokl.* **40**, 34–8

Soibelman, Ya. S. (1990b) Gel'fand-Naimark-Segal states and Weyl group for the quantum group $SU(n)$, *Funct. Anal. Appl.* **24**, 253–5

Soibelman, Ya. S. (1991a) The algebra of functions on a compact quantum group, and its representations, *Leningrad Math. J.* **2**, 161–78; correction, *ibid.* **2**, 256

Soibelman, Ya. S. (1991b) Quantum Weyl group and some of its applications, *Rend. Circ. Mat. Palermo Suppl.* **26** (2), 233–5

Soibelman, Ya. S. (1992a) Two problems in quantized algebras of functions, in *Quantum Groups, Proceedings of Workshops held in the Euler International Mathematical Institute 1990*, P. P. Kulish (ed.), Lecture Notes in Mathematics 1510, p. 393, Springer, Berlin

Soibelman, Ya. S. (1992b) Selected topics in quantum groups, in *Infinite Analysis*, A. Tsuchiya, T. Eguchi & M. Jimbo (eds), Advanced Series in Mathematical Physics 16, pp. 859–88, World Scientific, Singapore

Soibelman, Ya. S. (1992c) On quantum flag manifolds, *Funct. Anal. Appl.* **26**, 225–7

Soibelman, Ya. S. & Vaksman, L. L. (1992) On some problems in the theory of quantum groups, in *Representation Theory and Dynamical Systems*, A. M. Vershik (ed.), Advances in Soviet Mathematics 9, pp. 3–55, American Mathematical Society, Providence, RI

Solomon, A. I. & Katriel, J. (1990), On q-squeezed states, *J. Phys. A* **23**, L1209–12

Song, X. C. (1992) Spinor analysis for the quantum group $SU_q(2)$, *J. Phys. A* **25**, 2929–44

Song, X. C. & Li Liao (1992) The quantum Schrödinger equation and the q-deformation of the hydrogen atom, *J. Phys. A* **25**, 623–34

Soni, S. K. (1990) q-analogues of some prototype Berry phase calculations, *J. Phys. A* **23**, L951–5

Soni, S. K. (1991a) Differential calculus on the quantum superplane, *J. Phys. A* **24**, 619–24

Soni, S. K. (1991b) Comment on the differential calculus on quantum planes, *J. Phys. A* **24**, L169–74

Soni, S. K. (1991c) Covariant differential calculus on the quantum superplane and generalization of the Wess-Zumino formalism, *J. Phys. A* **24**, L459–62

Soni, S. K. (1992) Classical spin dynamics and quantum algebras, *J. Phys. A* **25**, L837–42

Sorace, E. (1992) Construction of some Hopf algebras, in *Quantum Groups and Related Topics*, R. Gierlak *et al.* (eds), pp. 67–82, Kluwer, Dordrecht

Spiridonov, V. (1992a) Exactly solvable potentials and quantum algebras, *Phys. Rev. Lett.* **69**, 398–401

Spiridonov, V. (1992b) Deformed conformal and supersymmetric quantum mechanics, *Modern Phys. Lett. A* **7**, 1241–52

Srinivasa Rao, K. & Rajeswari, V. (1991) Some aspects of angular momentum coefficients in $SU_q(2)$, in *Group Theoretical Methods in Physics, Moscow, 1990*, V. V. Dodonov & V. I. Manko (eds), Lecture Notes in Physics 382, pp. 221–4, Springer, Berlin

Stachura, P. (1992) Bicovariant differential calculi on $S_\mu(2)$, *Lett. Math. Phys.* **25**, 175–88

Stanev, Ya. S., Todorov, I. T. & Hadjiivanov, L. K. (1992) Braid invariant rational conformal models with a quantum group symmetry, *Phys. Lett. B* **276**, 87–94

Stasheff, J. (1992a) Drinfel'd's quasi-Hopf algebras and beyond, in *Deformation Theory and Quantum Groups with Applications to Mathematical Physics*, M. Gerstenhaber &

J. Stasheff (eds), Contemporary Mathematics 134, pp. 297–307, American Mathematical Society, Providence, RI

Stasheff, J. (1992b) Differential graded Lie algebras, quasi-Hopf algebras and higher homotopy algebras, in *Quantum Groups, Proceedings of Workshops held in the Euler International Mathematical Institute 1990*, P. P. Kulish (ed.), Lecture Notes in Mathematics 1510, pp. 120–37, Springer, Berlin

Steinberg, R. (1965) Regular elements of semi-simple algebraic groups, *Publ. Math. IHES* **25**, 49–80

Stolin, A. (1991a) On rational solutions of the classical Yang-Baxter equation, Thesis, Matematiska Institutionen, Stockholms Universitet

Stolin, A. (1991b) On rational solutions of the Yang-Baxter equations. Maximal orders in loop algebra, *Comm. Math. Phys.* **141**, 533–48

Stolin, A. (1991c) Constant solutions of the Yang-Baxter equation for $sl(2)$ and $sl(3)$, *Math. Scand.* **69**, 81–8

Stolin, A. (1991d) On rational solutions of Yang-Baxter equation for $sl(n)$, *Math. Scand.* **69**, 57–80

Stovicek, P. (1993) Quantum line bundles on S^2 and the method of orbits for $SU_q(2)$, *J. Math. Phys.* **34**, 1606–13

Su, G. & Ge, M. L. (1993) Thermodynamic characteristics of the q-deformed ideal boson gas, *Phys. Lett. A* **173**, 17–20

Sudbery, A. (1990) Consistent multiparameter quantisation of $GL(n)$, *J. Phys. A* **23**, L697–704

Sudbery, A. (1991) Non-commuting coordinates and differential operators, in *Quantum Groups*, T. Curtright, D. Fairlie & C. Zachos (eds), pp. 33–52, World Scientific, Singapore

Sudbery, A. (1992) Canonical differential calculus on quantum general linear groups and supergroups, *Phys. Lett. B* **284**, 61–5

Sun. C. P. & Fu, H. C. (1989) The q-deformed boson realization of the quantum group $SU(n)_q$ and its representations, *J. Phys. A* **22**, L983–6

Sun, C. P., Fu, H. C. & Ge, M. L. (1991) The cyclic representations of the quantum superalgebra $U_q osp(2,1)$ with q a root of unity, *Lett. Math. Phys.* **23**, 19–27

Sun, C. P. & Ge, M. L. (1991a) The q-analog of the boson algebra, its representation on the Fock space, and applications to the quantum group, *J. Math. Phys.* **32**, 595–8

Sun, C. P. & Ge, M. L. (1991b) The cyclic boson operators and new representations of the quantum algebra $sl_q(2)$ for q a root of unity, *J. Phys. A* **24**, L969–73

Sun, C. P. & Ge, M. L. (1991c) The q-deformed boson realization of representations of quantized universal enveloping algebras for q a root of unity: I. The case of $U_q(SL(l))$, *J. Phys. A* **24**, 3265–80

Sun, C. P. & Ge, M. L. (1992) The q-deformed boson realization of representations of quantized universal enveloping algebras for q a root of unity: II. The subalgebra chain of $U_q(C_l)$, *J. Phys. A* **25**, 401–10

Sun, C. P., Liu, X. F. & Ge, M. L. (1991) A new q-deformed boson realization of quantum algebra $sl_q(2)$ and nongeneric $sl_q(2)$ R-matrices, *J. Math. Phys.* **32**, 2409–12

Sun, C. P., Liu, X. F. & Ge, M. L. (1992) Cyclic boson algebra and q-boson realization of cyclic representations of the quantum algebra $sl_q(3)$, *J. Phys. A* **25**, 161–8

Sun, C. P., Liu, X. F. & Ge, M. L. (1993) A new quasitriangular Hopf algebra as the nontrivial quantum double of a simplest $\mathbb{C}$ algebra and its universal R-matrix for Yang-Baxter equation, *J. Math. Phys.* **34**, 1218–22

Sun, C. P., Liu, X. F., Lu, J. F. & Ge, M. L. (1992) A q-deformed oscillator system with quantum group $SL_q(l)$ symmetry, *J. Phys. A* **25**, L35–8

Sun, C. P., Lu, J. F. & Ge, M. L. (1990) Structure of regular representation of quantum universal enveloping algebra $SL_q(2)$ with $q^p = 1$, *J. Phys. A* **23**, L1199–204

Sun, C. P., Lu, J. F. & Ge, M. L. (1991a) q-analogues of Verma representations of quantum algebras for q a root of unity: the case of $U_q(sl(3))$, *J. Phys. A* **24**, 3731–40

Sun, C. P., Lu, J. F. & Ge, M. L. (1991b) q-deformed Verma representations of quantum algebra $sl_q(3)$ and new type Yang-Baxter matrices, *Sci. China Ser. A* **34**, 1474–83

Sun, X. D. & Wang, S. K. (1992) Bicovariant differential calculus on the two-parameter quantum group $GL_{p,q}(2)$, *J. Math. Phys.* **33**, 3313–30

Sun, X. D. & Wang, S. K. (1993) Yang-Baxter matrix and *-calculi on quantum groups of A-series, *J. Phys. A* **26**, L293–8

Sun, C. P., Xue, K., Liu, X. F. & Ge, M. L. (1991) Boson realization of non-generic $sl_q(2)$. R-matrices for the Yang-Baxter equation, *J. Phys. A* **24**, L545–8

Sweedler, M. E. (1969) *Hopf Algebras*, Benjamin, New York

Symes, W. (1980a) Hamiltonian group actions and integrable systems, *Phys. D* **1**, 339–74

Symes, W. (1980b) Systems of Toda type, inverse spectral problems and representation theory, *Invent. Math.* **59**, 13–51; addendum, *ibid.* **63** (1981), 519

Szymczak, I. & Zakrzewski, S. (1990) Quantum deformations of the Heisenberg group obtained by geometric quantization, *J. Geom. Phys.* **7**, 553–69

Taft, E. J. (1992a) Quantum deformation of the flag variety, in *Quantum Groups, Proceedings of Workshops held in the Euler International Mathematical Institute 1990*, P. P. Kulish (ed.), Lecture Notes in Mathematics 1510, pp. 138–41, Springer, Berlin

Taft, E. J. (1992b) Unsolved problems, in *Quantum Groups, Proceedings of Workshops held in the Euler International Mathematical Institute 1990*, P. P. Kulish (ed.), Lecture Notes in Mathematics 1510, p. 394, Springer, Berlin

Taft, E. J. & Towber, J. (1991) Quantum deformation of flag schemes and Grassmann schemes I. A q-deformation of the shape algebra for $GL(n)$, *J. Algebra* **142**, 1–36

Taft, E. J. & Wilson, R. L. (1980) There exist finite dimensional Hopf algebras with antipodes of arbitrary even order, *J. Algebra* **62**, 283–91

Takata, T. (1992) Invariants of 3-manifolds associated with quantum groups and Verlinde's formula, *Publ. Res. Inst. Math. Sci.* **28**, 139–67

Takeuchi, M. (1981) Matched pairs of groups and bismash products of Hopf algebras, *Comm. Algebra* **9**, 841–82

Takeuchi, M. (1985) Topological coalgebras, *J. Algebra* **97**, 505–39

Takeuchi, M. (1989) Quantum orthogonal and symplectic groups and their embedding into quantum GL, *Proc. Japan Acad. Ser. A Math. Sci.* **65**, 55–8

Takeuchi, M. (1990a) A two-parameter quantization of GL_n (Summary), *Proc. Japan Acad. Ser. A Math. Sci.* **66**, 112–4

Takeuchi, M. (1990b) Matric bialgebras and quantum groups, *Israel J. Math.* **72**, 232–51

Takeuchi, M. (1990c) The q-bracket product and quantum enveloping algebras of classical types, *J. Math. Soc. Japan* **42**, 605–29

Takeuchi, M. (1992a) Some topics on $GL_q(n)$, *J. Algebra* **147**, 379–410

Takeuchi, M. (1992b) Finite dimensional representations of the quantum Lorentz group, *Comm. Math. Phys.* **144**, 557–80

Takeuchi, M. (1992c) Hopf algebra techniques applied to the quantum group $U_q(sl(2))$, in *Deformation Theory and Quantum Groups with Applications to Mathematical Physics*, M. Gerstenhaber & J. Stasheff (eds), Contemporary Mathematics 134, pp. 309–23, American Mathematical Society, Providence, RI

Takeuchi, M. & Tambara, D. (1991) A new one-parameter family of 2×2 quantum matrices, *Proc. Japan Acad. Ser. A Math. Sci.* **67**, 267–9

Takhtajan, L. A. (1984) Solutions of the triangle equations with $\mathbb{Z}_N \times \mathbb{Z}_N$ symmetry as matrix analogues of the Weierstrass zeta and sigma functions, *Zap. Nauchn. Sem. Leningrad Otdel. Mat. Inst. Steklov* **133**, 258–76 (Russian)

Takhtajan, L. (1985) Introduction to algebraic Bethe ansatz, in *Exactly Solvable Problems in Condensed Matter and Relativistic Field Theory, Panchgani, 1985*, B. S. Shastry. S. S.

Jha & V. Singh (eds), Lecture Notes in Physics 242, pp. 175–219, Springer, Berlin

Takhtajan, L. A. (1989a) Quantum groups and integrable models, *Adv. Stud. Pure Math.* **19**, 435-57

Takhtajan, L. A. (1989b) Noncommutative homology of quantum tori, *Funct. Anal. Appl.* **23**, 147-9

Takhtajan, L. A. (1990) Introduction to quantum groups, in *Quantum Groups, Clausthal, Germany, 1989*, H.-D. Doebner & J.-D. Hennig (eds), Lecture Notes in Physics 370, pp. 3--28, Springer, Berlin

Takhtajan, L. A. (1992) Lectures on quantum groups, in *Introduction to Quantum Group and Integrable Massive Models of Quantum Field Theory*, M. L. Ge & B. H. Zhao (eds), pp. 69--197, World Scientific, Singapore

Tanisaki, T. (1990) Harish-Chandra isomorphisms for quantum algebras, *Comm. Math. Phys.* **127**, 555-71

Tanisaki, T. (1991) Finite-dimensional representations of quantum groups, *Osaka J. Math.* **28**, 37-53

Tanisaki, T. (1992) Killing forms, Harish-Chandra homomorphisms and universal R-matrices for quantum algebras, in *Infinite Analysis*, A. Tsuchiya, T. Eguchi & M. Jimbo (eds), Advanced Series in Mathematical Physics 16, pp. 941-62, World Scientific, Singapore

Tarasov, V. O. (1985a) Irreducible monodromy matrices for the R-matrix of the XXZ model and local lattice quantum hamiltonians, *Theoret. Math. Phys.* **63**, 440-54

Tarasov, V. O. (1985b) Structure of the quantum L operators for the R-matrix of the XXZ model, *Theoret. Math. Phys.* **63**, 1065--72

Tarasov, V. O. (1988) An algebraic Bethe ansatz for the Izergin-Korepin R-matrix, *Theoret. Math. Phys.* **76**, 793–804

Tarasov, V. O. (1992) Cyclic monodromy matrices for the R-matrix of the six-vertex model and the chiral Potts model with fixed spin boundary conditions, in *Infinite Analysis*, A. Tsuchiya, T. Eguchi & M. Jimbo (eds), Advanced Series in Mathematical Physics 16, pp. 963--76, World Scientific, Singapore

Temperley, N. & Lieb, E. (1971) Relations between the 'percolation' and 'colouring' problem and other graph-theoretical problems associated with regular planar lattices: some exact results for the 'percolation' problem, *Proc. Roy. Soc. Ser. A* **322**, 251–80

Thams, L. (1993) Two classical results in the quantum mixed case, *J. reine angew. Math.* **436**, 129-53

Tiraboschi, A. (1991) Compact quantum groups G_2, F_4 and E_8, *C. R. Acad. Sci. Paris Sér. I* **313**, 913-8

Tjin, T. (1992) Introduction to quantized Lie groups and Lie algebras, *Internat. J. Modern Phys. A* **7**, 6175-214

Todorov, I. T. (1990) Quantum groups as symmetries of chiral conformal algebras, in *Quantum Groups, Clausthal, Germany, 1989*, H.-D. Doebner & J.-D. Hennig (eds), Lecture Notes in Physics 370, pp. 231--77, Springer, Berlin

Todorov, I. T. (1991) Extended chiral conformal models with a quantum group symmetry, in *Group Theoretical Methods in Physics, Moscow, 1990*, V. V. Dodonov & V. I. Manko (eds), Lecture Notes in Physics 382, pp. 299--313, Springer, Berlin

Tolstoy, V. N. (1990) Extremal projectors for quantized Kac-Moody superalgebras and some of their applications, in *Quantum Groups, Clausthal, Germany, 1989*, H.-D. Doebner & J.-D. Hennig (eds), Lecture Notes in Physics 370, pp. 118--25, Springer, Berlin

Tolstoy, V. N. & Khoroshkin, S. M. (1992) The universal R-matrix for quantum nontwisted affine Lie algebras, *Funct. Anal. Appl.* **26**, 69--71

Trace, B. (1983) Reidemeister moves of a classical knot, *Proc. Amer. Math. Soc.* **89**, 722--4

Tracy, C. A. (1985a) Complete integrability in statistical mechanics and the Yang-Baxter equations, *Phys. D* **14**, 253--64

Tracy, C. A. (1985b) Embedded elliptic curves and the Yang-Baxter equations, *Phys. D* **16**, 203-20

Truini, P. & Varadarajan, V. S. (1991) The concept of a quantum semisimple group, *Lett. Math. Phys.* **21**, 287-92

Truini, P. & Varadarajan, V. S. (1992) Universal deformations of simple Lie algebras, *Lett. Math. Phys.* **24**, 63-72

Tsuchiya, A. & Kanie, Y. (1988) Vertex operators in conformal field theory on P^1 and monodromy representations of braid groups, in *Conformal Field Theory and Solvable Lattice Models, Kyoto, 1986*, A. Tsuchiya, T. Eguchi & M. Jimbo (eds), Advanced Studies in Pure Mathematics 16, pp. 297-372, Academic Press, Boston

Tsukuda, H. (1992) Functional realization of quantum affine algebras, *J. Math. Phys.* **33**, 3848-64

Tsygan, B. (1993) Notes on differential forms on quantum groups, *Selecta Mathematica* **12**, 75-103

Turaev, V. G. (1988) The Yang-Baxter equation and invariants of links, *Invent. Math.* **92**, 527-53

Turaev, V. G. (1989a) Algebras of loops on surfaces, algebras of knots and quantization, in *Braid Group, Knot Theory and Statistical Mechanics*, C. N. Yang & M. L. Ge (eds), Advanced Series in Physics 9, pp. 59-95, World Scientific, Singapore

Turaev, V. G. (1989b) The category of oriented tangles and its representations, *Funct. Anal. Appl.* **23**, 254-5

Turaev, V. G. (1990a) The Yang-Baxter equation and invariants of links, in *New Developments in the Theory of Knots*, T. Kohno (ed.), pp. 175-201, World Scientific, Teaneck, NJ

Turaev, V. G. (1990b) Operator invariants of tangles and R-matrices, *Math. USSR Izv.* **35**, 411-44

Turaev, V. G. (1991) Quantum invariants of 3-manifolds and a glimpse of shadow topology, *C. R. Acad. Sci. Paris Sér. I* **313**, 396-8

Turaev, V. G. (1992) Shadow links and face models of statistical mechanics, *J. Diff. Geom.* **36**, 35-74

Turaev, V. G. & Viro, O. Y. (1992) State sum invariants of 3-manifolds and quantum $6j$-symbols, *Topology* **31**, 865-902

Turaev, V. G. & Wenzl, H. (1993) Quantum invariants of 3-manifolds associated with classical simple Lie algebras, preprint

Tuszynski, J. A., Rubin, J. L., Meyer, J. & Kibler, M. (1993) Statistical mechanics of a q-deformed boson gas, *Phys. Lett. A* **175**, 173-7

Twietmeyer, E. (1992) Real forms of $\mathcal{U}_q(\mathfrak{g})$, *Lett. Math. Phys.* **24**, 49-58

Ubriaco, M. R. (1992) Non-commutative differential calculus and q-analysis, *J. Phys. A* **25**, 169-74

Ubriaco, M. R. (1993) Quantum deformation of quantum mechanics, *Modern Phys. Lett. A* **8**, 89-96

Ueno, K. (1990a) Representation theory of quantum groups, *Sugaku* **42**, 68-73

Ueno, K. (1990b) Spectral analysis for the Casimir operator on the quantum group $SU_q(1,1)$, *Proc. Japan Acad. Ser. A Math. Sci.* **66**, 42-4

Ueno, K., Masuda, T., Mimachi, K., Nakagami, Y., Noumi, M. & Saburi, Y. (1990) Matrix elements of unitary representations of the quantum group $SU_q(1,1)$ and the basic hypergeometric functions, in *Differential Geometric Methods in Theoretical Physics, Davis, 1988*, L. L. Chau & W. Nahm (eds), NATO Advanced Science Institutes Series B: Physics 245, pp. 331-43, Plenum, New York

Ueno, K. & Shibukawa, Y. (1992) Character table of Hecke algebra of type A_{N-1} and representations of the quantum group $U_q(gl_{n+1})$, in *Infinite Analysis*, A. Tsuchiya, T.

Eguchi & M. Jimbo (eds), Advanced Series in Mathematical Physics 16, pp. 977–84, World Scientific, Singapore

Ueno, K. & Takebayashi, T. (1992) Zonal spherical functions on quantum symmetric spaces and Macdonald's polynomials, in *Quantum Groups, Proceedings of Workshops held in the Euler International Mathematical Institute 1990*, P. P. Kulish (ed.), Lecture Notes in Mathematics 1510, pp. 142–7, Springer, Berlin

Ueno, K., Takebayashi, T. & Shibukawa, Y. (1989) Gel'fand-Zetlin basis for $U_q(gl(N+1))$ modules, *Lett. Math. Phys.* **18**, 215–21

Ueno, K., Takebayashi, T. & Shibukawa, Y. (1990) Construction of Gel'fand-Zetlin basis for $U_q(gl(N+1))$ modules, *Publ. Res. Inst. Math. Sci.* **26**, 667–79

Ui, H. & Aizawa, N. (1990) The q-analogue of boson commutator and the quantum groups $SU_q(2)$ and $SU_q(1,1)$, *Modern Phys. Lett. A* **5**, 237–42

Ulbrich, K.-H. (1990) On Hopf algebras and rigid monoidal categories, *Israel J. Math.* **72**, 252–6

Vainerman, L. I. (1992) Relations between compact quantum groups and Kac algebras, in *Representation Theory and Dynamical Systems*, A. M. Vershik (ed.), Advances in Soviet Mathematics 9, pp. 151–60, American Mathematical Society, Providence, RI

Vaksman, L. L. (1989) q-analogues of Clebsch-Gordan coefficients and the algebra of functions on the quantum group $SU(2)$, *Soviet Math. Dokl.* **39**, 467–70

Vaksman, L. L. & Korogodskii, L. I. (1989) The algebra of bounded functions on the quantum group of the motions of the plane and q-analogues of Bessel functions, *Soviet Math. Dokl.* **39**, 173–7

Vaksman, L. L. & Korogodskii, L. I. (1991) Spherical functions on the quantum group $SU(1,1)$ and a q-analogue of the Mehler-Fock formula, *Funct. Anal. Appl.* **25**, 48–9

Vaksman, L. L. & Soibelman, Ya. S. (1988) Algebra of functions on the quantum group $SU(2)$, *Funct. Anal. Appl.* **22**, 170–81

Vaksman, L. L. & Soibelman, Ya. S. (1991) Algebra of functions on the quantum group $SU(N+1)$ and odd-dimensional quantum spheres, *Leningrad Math. J.* **2**, 1023–42

Van Assche, W. & Koornwinder, T. H. (1991) Asymptotic behaviour for Wall polynomials and the addition formula for little q-Legendre polynomials, *SIAM J. Math. Analysis* **22**, 302–11

Van Daele, A. (1991a) The operator $a \otimes b + b \otimes a^{-1}$ when $ab = \lambda ba$, preprint, K. U. Leuven

Van Daele, A. (1991b) Quantum deformation of the Heisenberg group, in *Current Topics in Operator Algebras, Nara, 1990*, pp. 314–25, World Scientific, Singapore

Van Daele, A. (1993) Dual pairs of Hopf *-algebras, *Bull. London Math. Soc.* **25**, 209–30

Van Daele, A. & Van Keer, S. (1991) The Yang-Baxter and pentagon equation, preprint, K. U. Leuven

Van der Jeugt, J. (1992a) The q-boson operator algebra and q-Hermite polynomials, *Lett. Math. Phys.* **24**, 267–74

Van der Jeugt, J. (1992b) On the principal subalgebras of quantum enveloping algebras $gl_q(l+1)$, *J. Phys. A* **25**, L213–8

Van der Jeugt, J. (1992c) Knot invariants and universal R-matrices from perturbative Chern-Simons theory in the almost axial gauge, *Nucl. Phys. B* **379**, 172–98

Van der Jeugt, J. (1993a) Dynamical algebra of the q-deformed three-dimensional algebra, *J. Math. Phys.* **34**, 1799–806

Van der Jeugt, J. (1993b) R-matrix formulation of deformed boson algebra, *J. Phys. A* **25**, L405–13

Varadarajan, V. S. (1974) *Lie Groups, Lie Algebras and their Representations*, Prentice-Hall, Englewood Cliffs, NJ

Vassiliev, V. A. (1990) Cohomology of knot spaces, in *Theory of Singularities and its Applications*, V. I. Arnold (ed.), pp. 23–69, American Mathematical Society, Providence, RI

Vassiliev, V. A. (1993) *Complements of Discriminants of Smooth Maps: Topology and Applications*, Translations of Mathematical Monographs 98, American Mathematical Society, Providence, RI

Vaysleb, E. (1992) Representations of quantum *-algebras $sl_t(N+1, \mathbb{R})$, in *Representation Theory and Dynamical Systems*, A. M. Vershik (ed.), Advances in Soviet Mathematics 9, pp. 57–65, American Mathematical Society, Providence, RI

Vaysleb, E. (1993) Infinite-dimensional *-representations of Sklyanin algebra and of the quantum algebra $U_q(sl(2))$, *Selecta Mathematica* **12**, 57–73

Verdier, J.-L. (1987) Groupes quantiques, Séminaire Bourbaki exp. no. 685, *Astérisque* **152–153**, 305–19, Société Mathématique de France, Paris

Verlinde, E. (1988) Fusion rules and modular transformations in 2D conformal field theory, *Nucl. Phys. B* **300**, 360–75

Vey, J. (1975) Déformations du crochet de Poisson d'une variété symplectique, *Comm. Math. Helv.* **50**, 421–54

Vinteler, E. (1991) Extended braid group in higher dimensional integrable theories, *Phys. Lett. B* **254**, 103–5

Viswanathan, K. S., Parthasarathy, R. & Jagannathan, R. (1992) Generalized q-fermion oscillators and q-coherent states, *J. Phys. A* **25**, L335–40

Vladimirov, A. A. (1993) A method for obtaining quantum doubles from the Yang-Baxter R-matrices, *Modern Phys. Lett. A* **8**, 1315–22

Vokos, S. P. (1991) Multiparameter deformations of $GL(N)$ and Jordan-Schwinger constructions, *J. Math. Phys.* **32**, 2979–85

Vokos, S. P., Zumino, B. & Wess, J. (1990) Analysis of the basic matrix representation of $GL_q(2, \mathbb{C})$, *Z. Phys. C* **48**, 65–74

Vorobev, Yu. M. & Karasev, M. V. (1988a) Corrections to classical dynamics and quantization conditions arising in the deformation of Poisson brackets, *Soviet Math. Dokl.* **36**, 594–8

Vorobev, Yu. M. & Karasev, M. V. (1988b) Poisson manifolds and the Schouten bracket, *Funct. Anal. Appl.* **22**, 1–9

Vorobev, Yu. M. & Karasev, M. V. (1990) Deformation and cohomologies of Poisson brackets, in *Global Analysis – Studies and Applications, 4*, Yu. G. Borisovich & Yu. E. Gliklikh (eds), Lecture Notes in Mathematics 1453, pp. 75–89, Springer, Berlin

Wadati, M. & Akutsu, Y. (1988) From solitons to knots and links, *Progr. Theoret. Phys. Suppl.* **94**, 1–41

Wadati, M., Deguchi, T. & Akutsu, Y. (1992) Yang-Baxter relation, exactly solvable models and link polynomials, in *Quantum Groups, Proceedings of Workshops held in the Euler International Mathematical Institute 1990*, P. P. Kulish (ed.), Lecture Notes in Mathematics 1510, pp. 373–88, Springer, Berlin

Walker, L. (1991) On Witten's three-manifold invariants, preprint

Wallet, J.-C. (1992) R-matrix and covariant q-superoscillators for $U_q(gl(1|1))$, *J. Phys. A* **25**, L1159–66

Wambst, M. (1992) Complexes de Koszul quantiques et homologie cyclique, *C. R. Acad. Sci. Paris Sér. I* **314**, 977–82

Wang, F. B. & Kuang, L. M. (1993) Even and odd q-coherent states and their optical staistics properties, *J. Phys. A* **26**, 293–300

Wang, J. P. (1991) A survey of the theory of quantum groups, *Adv. Math. (China)* **20**, 424–54

Wang, K. W. (1991) q-determinants and permutations, *Fibonacci Quart.* **29**, 160–3

Ward, R. S. (1985) Generalized Nahm equations and the quantum Yang-Baxter equation, *Phys. Lett. A* **112**, 3–5

Wasserman, A. (1988) Coactions and Yang-Baxter equations for ergodic actions and subfactors, in *Operator Algebras and Applications, Vol. 2*, D. E. Evans & M. Takesaki (eds),

London Mathematical Society Lecture Note Series 136, pp. 203–36, Cambridge University Press, Cambridge

Waterhouse, W. C. (1975) Antipodes and group-likes in finite Hopf algebras, *J. Algebra* **37**, 290–5

Wei, H. (1990) Structure constants of the quantum algebra and the Yang-Baxter relation for the Z_n symmetric statistical model, *Acta Phys. Sinica* **39**, 1357–62

Weinstein, A. (1983) The local structure of Poisson manifolds, *J. Diff. Geom.* **18**, 523–57

Weinstein, A. & Xu, P. (1991) Extensions of symplectic groupoids and quantization, *J. reine angew. Math.* **417**, 159–89

Weinstein, A. & Xu, P. (1992) Classical solutions of the quantum Yang-Baxter equation, *Comm. Math. Phys.* **148**, 309–43

Weisfeiler, B. & Kac, V. G. (1971) On irreducible representations of Lie p-algebras, *Funct. Anal. Appl.* **5**, 28–36

Wenzl, H. (1988a) Hecke algebras of type A_n and subfactors, *Invent. Math.* **92**, 349–83

Wenzl, H. (1988b) On the structure of Brauer's centralizer algebras, *Ann. of Math.* **128**, 173–93

Wenzl, H. (1988c) Derived link invariants and subfactors, in *Operator Algebras and Applications, Vol. 2*, D. E. Evans & M. Takesaki (eds), London Mathematical Society Lecture Note Series 136, pp. 237–40, Cambridge University Press, Cambridge

Wenzl, H. (1989) Unitarization of solutions of the quantum Yang-Baxter equation and subfactors, in *IXth International Congress on Mathematical Physics, Swansea, 1988*, B. Simon, A. Truman & I. M. Davies (eds), pp. 434–43, Adam Hilger, Bristol

Wenzl, H. (1990a) Quantum groups and subfactors of type B, C and D, *Comm. Math. Phys.* **133**, 383–432

Wenzl, H. (1990b) Representations of braid groups and the quantum Yang-Baxter equation, *Pacific J. Math.* **145**, 153–80

Wenzl, H. (1992) Unitary braid representations, in *Infinite Analysis*, A. Tsuchiya, T. Eguchi & M. Jimbo (eds), Advanced Series in Mathematical Physics 16, pp. 985–1006, World Scientific, Singapore

Wess, J. & Zumino, B. (1991) Covariant differential calculus on the quantum hyperplane, *Nucl. Phys. B Proc. Suppl.* **18 B**, 302–12

Weyers, J. (1990) The quantum group $GL_q(N)$ and Weyl-Heisenberg operators, *Phys. Lett. B* **240**, 396–400

Weyl, H. (1927) Quantenmechanik und Gruppentheorie, *Z. Phys.* **46**, 1–46

Weyl, H. (1939) *The Classical Groups*, Princeton University Press, Princeton

Witten, E. (1988a) Coadjoint orbits of the Virasoro group, *Comm. Math. Phys.* **114**, 1–53

Witten, E. (1988b) Topological quantum field theory, *Comm. Math. Phys.* **117**, 353–86

Witten, E. (1989) Quantum field theory and the Jones polynomial, *Comm. Math. Phys.* **121**, 351–99

Witten, E. (1990) Gauge theories, vector models and quantum groups, *Nucl. Phys. B* **330**, 285–346

Woronowicz, S. L. (1980) Pseudospaces, pseudogroups and Pontriagin duality, in *Proceedings of the International Conference on Mathematics and Physics, Lausanne, 1979*, K. Osterwalder (ed.), Lecture Notes in Physics 116, pp. 407–12, Springer, Berlin

Woronowicz, S. L. (1986) Group structure on noncommutative spaces, in *Fields and Geometry, Karpacz, 1986*, A. Jadczyk (ed.), pp. 478–96, World Scientific, Teaneck, NJ

Woronowicz, S. L. (1987a) Compact matrix pseudogroups, *Comm. Math. Phys.* **111**, 613–65

Woronowicz, S. L. (1987b) Twisted $SU(2)$ group. An example of a non-commutative differential calculus, *Publ. Res. Inst. Math. Sci.* **23**, 117–81

Woronowicz, S. L. (1988) Tannaka-Krein duality for compact matrix pseudogroups. Twisted $SU(N)$ group, *Invent. Math.* **93**, 35–76

Woronowicz, S. L. (1989) Differential calculus on compact matrix pseudogroups (quantum groups), *Comm. Math. Phys.* **122**, 125–70

Woronowicz, S. L. (1991a) A remark on compact matrix quantum groups, *Lett. Math. Phys.* **21**, 35–9

Woronowicz, S. L. (1991b) Unbounded elements affiliated with C^*-algebras and non-compact quantum groups, *Comm. Math. Phys.* **136**, 399–432

Woronowicz, S. L. (1991c) Quantum $E(2)$ group and its Pontryagin dual, *Lett. Math. Phys.* **23**, 251–63

Woronowicz, S. L. (1991d) New quantum deformation of $SL(2, \mathbb{C})$. Hopf algebra level, *Rep. Math. Phys.* **30**, 259–69

Woronowicz, S. L. (1991e) Solutions of the braid equation related to a Hopf algebra, *Lett. Math. Phys.* **23**, 143–5

Woronowicz, S. L. (1992a) Operator equalities related to the quantum $E(2)$ group, *Comm. Math. Phys.* **144**, 417–28

Woronowicz, S. L. (1992b) Quantum $SU(2)$ and $E(2)$ groups. Contraction procedure, *Comm. Math. Phys.* **149**, 637–52

Woronowicz, S. L. & Zakrzewski, S. (1992) Quantum Lorentz group having Gauss decomposition property, *Publ. Res. Inst. Math. Sci.* **28**, 809–24

Xi, N. H. (1990) Finite dimensional modules of some quantum groups over $F_p(v)$, *J. reine angew. Math.* **410**, 109–15

Yamane, H. (1988) A Poincaré-Birkhoff-Witt theorem for the quantum group of type A_N, *Proc. Japan Acad. Ser. A Math. Sci.* **64**, 385–6

Yamane, H. (1989) A Poincaré-Birkhoff-Witt theorem for quantized universal enveloping algebras of type A_N, *Publ. Res. Inst. Math. Sci.* **25**, 503–20

Yamane, H. (1991) Universal R-matrices for quantum groups associated to simple Lie superalgebras, *Proc. Japan Acad. Ser. A Math. Sci.* **67**, 108–12

Yan, H. (1990) q-deformed oscillator algebra as a quantum group, *J. Phys. A* **23**, L1155–60

Yan, H. (1991) On q-deformed oscillator systems and the q-oscillator algebra $H_q(4)$, *Phys. Lett. B* **262**, 459–62

Yan, H., Zhou, Y. & Zhu, T. H. (1993) Two-parametric solutions to graded Yang-Baxter equation and two-parametric deformed algebra $U_{uv}gl(1|1)$, *J. Phys. A* **26**, 935–42

Yang, C. M. (1993) The modified classical Yang-Baxter equation and supersymmetric Gel'fand-Dickii brackets, *Modern Phys. Lett. A* **8**, 129–38

Yang, C. N. (1967) Some exact results for the many-body problem in one dimension with repulsive delta-function interaction, *Phys. Rev. Lett.* **19**, 1312–4

Yang, Q. G. & Xu, B. W. (1993) Energy spectrum of a q-analogue of the hydrogen atom, *J. Phys. A* **26**, L365–8

Yeh, E. H. Y. (1989) Quantization of the semisimple Lie group, *Nucl. Phys. B Proc. Suppl.* **6**, 436–9

Yetter, D. N. (1988) Markov algebras, in *Braids, Santa Cruz, 1986*, J. S. Birman & A. Libgober (eds), Contemporary Mathematics 78, pp. 705–30, American Mathematical Society, Providence, RI

Yetter, D. N. (1990) Quantum groups and representations of monoidal categories, *Math. Proc. Camb. Phil. Soc.* **108**, 261–90

Yetter, D. N. (1992) Framed tangles and a theorem of Deligne on braided deformations of monoidal categories, in *Deformation Theory and Quantum Groups with Applications to Mathematical Physics*, M. Gerstenhaber & J. Stasheff (eds), Contemporary Mathematics 134, pp. 325–49, American Mathematical Society, Providence, RI

Yoshinaga, S. (1991) An invariant of spatial graphs associated with $U_q(sl(2, \mathbb{C}))$, *Kobe Math. J.* **8**, 25–40

You, H. Q. & Zhou, S. Y. (1990) On the quantum R-matrix on a $U_q(sl(2,\mathbb{C}))$-module, *Modern Phys. Lett. A* **5**, 2465–72

Yu, Y., Zhu, Z. & Lee, H. C. (1993) Flat connections on quantum bundles and fractional statistics in geometric quantization, *J. Math. Phys.* **34**, 988–96

Yu, Z. R. (1991a) Some realizations of the quantum algebra $U_q(SU(2))$, *J. Phys. A* **24**, L1321–5

Yu, Z. R. (1991b) A new method of determining an irreducible representation of quantum $SL_q(3)$ algebra, *J. Phys. A* **24**, L399–402

Yu, Z. R. (1992) q-Usui operator of some quantum algebras, *Phys. Lett. A* **162**, 5–6

Zabrodin, A. (1992) Integrable models of field theory and scattering on quantum hyperboloids, *Modern Phys. Lett. A* **7**, 441–6

Zachos, C. (1991) Quantum deformations, in *Quantum Groups*, T. Curtright, D. Fairlie & C. Zachos (eds), pp. 62–71, World Scientific, Singapore

Zachos, C. (1992a) Elementary paradigms of quantum algebras, in *Deformation Theory and Quantum Groups with Applications to Mathematical Physics*, M. Gerstenhaber & J. Stasheff (eds), Contemporary Mathematics 134, pp. 351–77, American Mathematical Society, Providence, RI

Zachos, C. (1992b) Altering the symmetry of wave functions in quantum algebras and supersymmetry, *Modern Phys. Lett. A* **7**, 1595–600

Zakrzewski, S. (1985) A differential structure for quantum mechanics, *J. Geom. Phys.* **2**, 135–45

Zakrzewski, S. (1990a) Quantum and classical pseudogroups, part I. Union pseudogroups and their quantization, *Comm. Math. Phys.* **134**, 347–70

Zakrzewski, S. (1990b) Quantum and classical pseudogroups, part II. Differential and symplectic pseudogroups, *Comm. Math. Phys.* **134**, 371–95

Zakrzewski, S. (1991a) A Hopf *-algebra of polynomials on the quantum $SL(2,\mathbb{R})$ for a "unitary" R-matrix, *Lett. Math. Phys.* **22**, 287–9

Zakrzewski, S. (1991b) Matrix pseudogroups associated with anti-commutative plane, *Lett. Math. Phys.* **21**, 309–21

Zakrzewski, S. (1992) Poisson-Lie groups and pentagonal transformations, *Lett. Math. Phys.* **24**, 13–20

Zelevinsky, A. V. (1980) Induced representations of reductive p-adic groups II. On irreducible representations of $GL(n)$, *Ann. Sci. Ecole Norm. Sup.* 4e Sér. **13**, 165–210

Zhang, M. Q. (1991) How to find the Lax pair from the quantum Yang-Baxter equation, *Comm. Math. Phys.* **141**, 523–31

Zhang, R. B. (1991a) Graded representations of the Temperley-Lieb algebra, quantum supergroups and the Jones polynomial, *J. Math. Phys.* **32**, 2605–13

Zhang, R. B. (1991b) Invariants of the quantum supergroup $U_q(gl(m/1))$, *J. Phys. A* **24**, L1327–32

Zhang, R. B. (1991c) Multiparameter dependent solutions of the Yang-Baxter equation, *J. Phys. A* **24**, L535–43

Zhang, R. B. (1992a) Universal L operator and invariants of the quantum supergroup $U_q(gl(m/n))$, *J. Math. Phys.* **33**, 1970–9

Zhang, R. B. (1992b) A two-parameter quantization of $osp(4/2)$, *J. Phys. A* **25**, L991–6

Zhang, R. B. (1992c) Braid group representations arising from quantum supergroups with arbitrary q and link polynomials, *J. Math. Phys.* **33**, 3918–30

Zhang, R. B. (1993) Finite-dimensional irreducible representations of the quantum supergroup $U_q(gl(m/n))$, *J. Math. Phys.* **34**, 1236–54

Zhang, R. B. & Gould, M. D. (1991) Universal R-matrices and invariants of quantum supergroups, *J. Math. Phys.* **32**, 3261–7

Zhang, R. B., Gould, M. D. & Bracken, A. J. (1991a) Quantum group invariants and link polynomials, *Comm. Math. Phys.* **137**, 13–27

Zhang, R. B., Gould, M. D. & Bracken, A. J. (1991b) Solutions of the graded classical Yang-Baxter equation and integrable models, *J. Phys. A* **24**, 1185–97

Zhang, R. B., Gould, M. D. & Bracken, A. J. (1991c) From representations of the braid group to solutions of the Yang-Baxter equation, *Nucl. Phys. B* **354**, 625–52

Zhang, R. B., Gould, M. D. & Bracken, A. J. (1991d) Solution of the graded Yang-Baxter equation associated with the vector representation of $U_q(osp(m|2n))$, *Phys. Lett. B* **257**, 133–9

Zhang, R. B., Gould, M. D. & Bracken, A. J. (1991e) Generalized Gel'fand invariants of quantum groups, *J. Phys. A* **24**, 937–43

Zhang, S. & Duan, Y. (1992) The quantum group $SU_q(2)$ and the q-analogue of angular momenta in classical mechanics, *J. Phys. A* **25**, 5135–40

Zhang Y. J. (1992) A remark on the representation theory of the algebra $U_q(sl(n))$ when q is a root of unity, *J. Phys. A* **25**, 851–60

Zhedanov, A. S. (1992a) 'Non-classical' q-oscillator realization of the quantum $SU(2)$ algebra, *J. Phys. A* **25**, L713–8

Zhedanov, A. S. (1992b) The 'Higgs algebra' as a 'quantum' deformation of $SU(2)$, *Modern Phys. Lett. A* **7**, 507–12

Zhedanov, A. S. (1992c) Quantum $SU_q(2)$ algebra: 'Cartesian' version and overlaps, *Modern Phys. Lett. A* **7**, 1589–93

Zhedanov, A. S. (1992d) Bogoliubov q-transformation and Clebsch-Gordan coefficients for a q-oscillator, *Phys. Lett. A* **165**, 53–7

Zhedanov, A. S. (1993a) On the realization of the Weyl commutation relation $HR = qRH$, *Phys. Lett. A* **176**, 300–2

Zhedanov, A. S. (1993b) Q rotations and other Q transformations as unitary nonlinear automorphisms of quantum algebras, *J. Math. Phys.* **34**, 2631–47

Zheng, Q. R. & Zhang, D. H. (1993) Solutions of the quantum three-simplex equation without spectral parameters, *Phys. Lett. A* **174**, 75–80

Zhong, Z. Z. (1992a) The hyperbolic complexification of quantum groups and the isomorphic relations, *J. Phys. A* **25**, L397–402

Zhong, Z. Z. (1992b) A remark on the classical realization of the double quantum group $SU_q(\eta, J)$, *J. Phys. A* **25**, L867–70

Zhong, Z. Z. (1993) Non-commutative analysis, quantum group gauge transformations and gauge fields, *J. Phys. A* **26**, L391–400

Zuber, J. B. (1990) Conformal field theories, Coulomb gas picture and integrable models, in *Champs, Cordes et Phénomenes Critiques, Les Houches, 1988*, E. Brezin & J. Zinn-Justin (eds), pp. 247–79, North-Holland, Amsterdam

Zumino, B. (1991) Deformation of the quantum mechanical phase space with bosonic or fermionic cooordinates, *Modern Phys. Lett. A* **6**, 1225–35

Index of notation

Miscellaneous symbols

General index